MECHANICAL VARIABLES MEASUREMENT

Solid, Fluid, and Thermal

MECHANICAL VARIABLES MEASUREMENT

Solid, Fluid, and Thermal

edited by

John G. Webster

CRC Press
Boca Raton London New York Washington, D.C.

Library of Congress Cataloging-in-Publication Data

Mechanical variables measurement : solid, fluid and thermal / John G. Webster, editor.
p. cm.
Includes bibliographical references.
ISBN 0-8493-0047-9 (alk. paper)
1. Engineering instruments. 2. Measuring instruments. I. Webster, John G., 1932–
TA165.M43 1999
681′.221—dc21 99-043783
CIP

Direct all inquiries to CRC Press LLC, 2000 N.W. Corporate Blvd., Boca Raton, Florida 33431.

International Standard Book Number 0-8493-0047-9
Library of Congress Card Number 99-043783
Printed in the United States of America 2 3 4 5 6 7 8 9 0
Printed on acid-free paper

Preface

Introduction

The purpose of *Mechanical Variables Measurement — Solid, Fluid, and Thermal* is to provide a reference that is both concise and useful for engineers in industry, scientists, designers, managers, research personnel and students, as well as many others who have measurement problems.

The book describes the use of instruments and techniques for practical measurements required in engineering, physics, and chemistry, and the life sciences. It includes sensors, techniques, hardware, and software. It also includes information processing systems, automatic data acquisition, reduction and analysis and their incorporation for control purposes.

Articles include descriptive information for professionals, students, and workers interested in measurement. Articles include equations to assist engineers and scientists who seek to discover applications and solve problems that arise in fields not in their specialty. They include specialized information needed by informed specialists who seek to learn advanced applications of the subject, evaluative opinions, and possible areas for future study. Thus, this book serves the reference needs of the broadcast group of users — from the advanced high school science student to industrial and university professionals.

Organization

The book is organized according to the *measurement problem.* Section I covers solid mechanical variables such as mass and strain. Section II comprises fluid mechanical variables such as pressure, flow, and velocity. Section III covers thermal mechanical variables such as temperature and heat flux.

Locating Your Topic

To find out how to measure a given variable, skim the Table of Contents, turn to that section and find the chapters that describe different methods of making the measurement. Consider the alternative methods of making the measurement and each of their advantages and disadvantages. Select a method, sensor, and signal processing method. Many articles list a number of vendors to contact for more information. You can also visit the http://www.sensorsmag.com site under Buyer's Guide to obtain a list of vendors.

For more detailed information, consult the index, since certain principles of measurement may appear in more than one chapter.

John G. Webster
Editor-in-Chief

Contributors

Marc J. Assael
Faculty of Chemical Engineering
Aristotle University of Thessaloniki
Thessalonika, Greece

William H. Bayles, Jr.
The Fredericks Company
Huntington Valley, Pennsylvania

B. Benhabib
Department of Mechanical and Industrial Engineering
University of Toronto
Toronto, Ontario, Canada

A. Bonen
University of Toronto
Toronto, Ontario, Canada

Howard M. Brady
The Fredericks Company
Huntington Valley, Pennsylvania

Christophe Bruttin
Rittmeyer, Ltd.
Zug, Switzerland

Jim Burns
Burns Engineering Inc.
Minnetonka, Minnesota

Kevin H. L. Chau
Micromachined Products Division
Analog Devices
Cambridge, Massachusetts

Brian Culshaw
Department of Electronic and Electrical Engineering
University of Strathclyde
Royal College Building
Glasgow, U.K.

Thomas E. Diller
Virginia Polytechnic Institute
Blacksburg, Virginia

Emil Drubetsky
The Fredericks Company
Huntington Valley, Pennsylvania

M. A. Elbestawi
Mechanical Engineering
McMaster University
Hamilton, Ontario, Canada

Halit Eren
Curtin University of Technology
Perth, WA, Australia

Jacob Fraden
Advanced Monitors Corporation
San Diego, California

Randy Frank
Semiconductor Products Sector
Transportation Systems Group
Motorola, Inc.
Phoenix, Arizona

Mark Fritz
Denver Instrument Company
Arvada, Colorado

Ivan J. Garshelis
Magnova, Inc.
Pittsfield, Massachusetts

Paolo Giordano
Rittmeyer, Ltd.
Zug, Switzerland

Ron Goehner
The Fredericks Company
Huntington Valley, Pennsylvania

Reinhard Haak
Universitaet Erlangen-Nuernberg
Erlangen, Germany

Emil Hazarian
Denver Instrument Company
Arvada, Colorado

Thomas Hossle
Rittmeyer, Ltd.
Zug, Switzerland

Zaki D. Husain
Daniel Flow Products, Inc.
Bellaire, Texas

John A. Kleppe
Electrical Engineering Department
University of Nevada
Reno, Nevada

Herbert Köchner
Universitaet Erlangen-Nuernberg
Erlangen, Germany

M. Kostic
Northern Illinois University
DeKalb, Illinois

G. E. Leblanc
School of Geography and Geology
McMaster University
Hamilton, Ontario, Canada

Christopher S. Lynch
Mechanical Engineering Department
The Georgia Institute of Technology
Atlanta, Georgia

Tolestyn Madaj
Technical University of Gdansk
Gdansk, Poland

Kin F. Man
Jet Propulsion Lab
California Institute of Technology
Pasadena, California

Wade M. Mattar
The Foxboro Company
Foxboro, Massachusetts

Adrian Melling
Universitaet Erlangen-Nuernberg
Erlangen, Germany

Rajan K. Menon
Laser Velocimetry Products
TSI Inc.
St. Paul, Minnesota

Jaroslaw Mikielewicz
Institute of Fluid Flow Machinery
Gdansk, Poland

Harold M. Miller
Data Industrial Corporation
Mattapoisett, Massachusetts

Roger Morgan
School of Engineering
Liverpool John Moores University
Liverpool, England

Armelle M. Moulin
University of Cambridge
Cambridge, England

J. V. Nicholas
The New Zealand Institute for Industrial Research and Development
Lower Hutt, New Zealand

Nam-Trung Nguyen
Berkeley Sensor and Actuator Center
University of California at Berkeley
Berkeley, California

John G. Olin
Sierra Instruments, Inc.
Monterey, California

Franco Pavese
CNR
Instituto di Metrologia
"G. Colonnetti"
Torino, Italy

Peder C. Pedersen
Electrical and Computer Engineering
Worcester Polytechnic Institute
Worcester, Massachusetts

Rekha Philip-Chandy
School of Engineering
Liverpool John Moores University
Liverpool, England

Per Rasmussen
GRAS Sound and Vibration
Vedback, Denmark

R. P. Reed
Proteun Services
Albuquerque, New Mexico

Herbert M. Runciman
Pilkington Optronics
Glasgow, Scotland, U.K.

Ricardo Saad
University of Toronto
Toronto, Ontario, Canada

Meyer Sapoff
MS Consultants
Princeton, New Jersey

Patricia J. Scully
School of Engineering
Liverpool John Moores University
Liverpool, England

R. A. Secco
The University of Western Ontario
Ontario, Canada

K. C. Smith
University of Toronto
Toronto, Ontario, Canada

Jan Stasiek
Mechanical Engineering Department
Technical University of Gdansk
Gdansk, Poland

Robert J. Stephenson
University of Cambridge
Cambridge, England

David B. Thiessen
California Institute of Technology
Pasedena, California

Richard Thorn
School of Engineering
University of Derby
Derby, U.K.

Sander van Herwaarden
Xensor Integration
Delft, The Netherlands

Hans-Peter Vaterlaus
Instrument Department
Rittmeyer, Ltd.
Zug, Switzerland

James H. Vignos
The Foxboro Company
Foxboro, Massachusetts

David Wadlow
Sensors Research Consulting, Inc.
Basking Ridge, New Jersey

William A. Wakeham
Imperial College
London, England

Donald J. Wass
Daniel Flow Products, Inc.
Houston, Texas

Mark E. Welland
University of Cambridge
Cambridge, England

Jesse Yoder
Automation Research Corporation
Dedham, Massachusetts

Contents

SECTION I Mechanical Variables Measurement — Solid

SECTION II Mechanical Variables Measurement — Fluid

I

Mechanical Variables Measurement — Solid

1

Mass and Weight Measurement

Mark Fritz
Denver Instrument Company

Emil Hazarian
Denver Instrument Company

Mass and weight are often used interchangeably; however, they are different. Mass is a quantitative measure of inertia of a body at rest. As a physical quantity, mass is the product of density and volume.

Weight or weight force is the force with which a body is attracted toward the Earth. Weight force is determined by the product of the mass and the acceleration of gravity.

$$M = V \times D \tag{1.1}$$

where M = mass
V = volume
D = density

Note: In most books, the symbol for density is the Greek letter ρ.

$$W = M \times G \tag{1.2}$$

where W = weight
G = gravity

The embodiment of units of mass are called weights; this increases the confusion between mass and weight. In the International System of Units (SI), the modernized metric measurement system, the unit for mass is called the kilogram and the unit for force is called the newton. In the United States, the customary system the unit for mass is called the slug and the unit for force is called the pound. When using the U.S. customary units of measure, people are using the unit pound to designate the mass of an object because, in the United States, the pound has been defined in terms of the kilogram since 1893.

1.1 Weighing Instruments

Weighing is one of the oldest known measurements, dating back to before written history. The equal arm balance was probably the first instrument used for weighing. It is a simple device in which two pans are suspended from a beam equal distance from a central pivot point. The standard is placed in one pan and the object to be measured is placed in the second pan; when the beam is level, the unknown is equal to the standard. This method of measurement is still in wide use throughout the world today. Figure 1.1 shows an Ainsworth equal arm balance.

0-8493-0047-9/00/$0.00+$.50

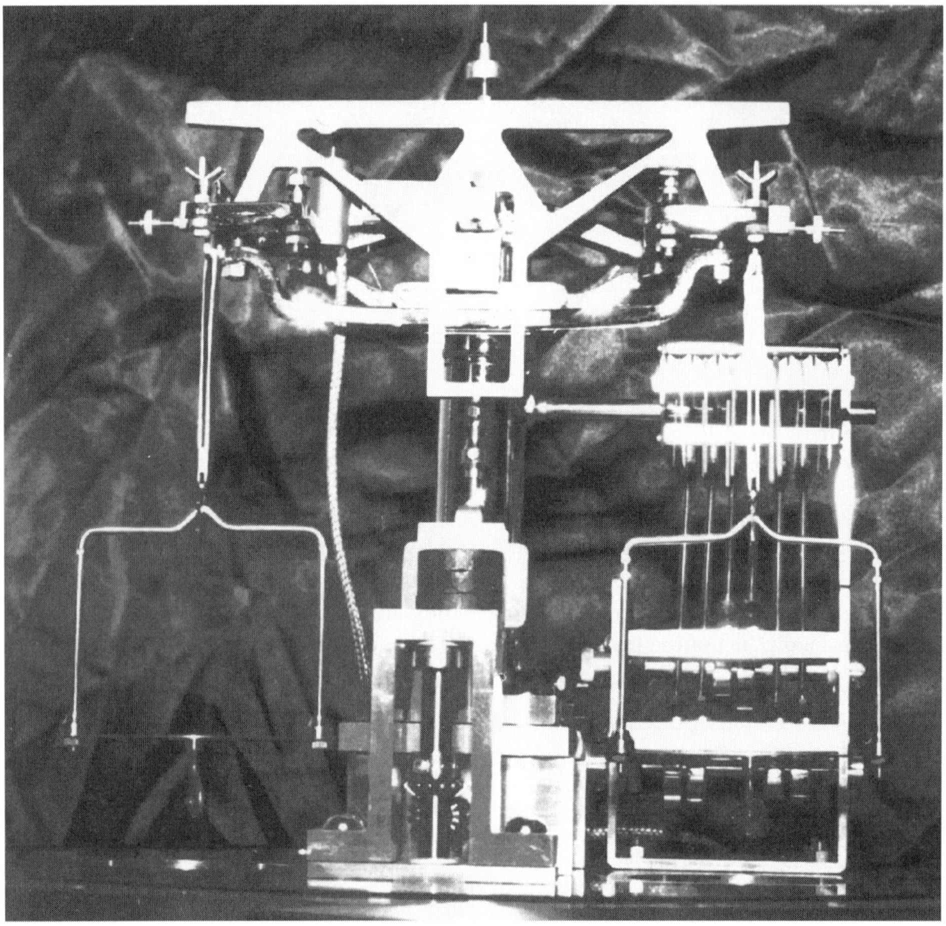

FIGURE 1.1 Ainsworth FV series equal arm balance. (Courtesy Denver Instrument Company.)

A balance is a measuring instrument used to determine the mass of an object by measuring the force exerted by the object on its support within the gravitational field of the Earth. One places a standard of known value on one pan of the balance. One then adds the unknown material to the second pan, until the gravitational force on the unknown material equals the gravitational force on the standard. This can be expressed mathematically as:

$$S \times G = U \times G \tag{1.3}$$

where S = mass of the standard
G = gravity
U = mass of the unknown

Given the short distance between pans, one assumes that the gravitational forces acting on them are equal. Another assumption is that the two arms of the balance are equal.

Since the gravitational force is equal, it can be removed from the equation and the standard and the unknown are said to be equal. This leads to one of the characteristics of the equal arm balance as well as of other weighing devices, the requirement to have a set of standards that allows for every possible measured value. The balance can only indicate if objects are equal and has a limited capability to determine how much difference there is between two objects.

Probably, the first attempt to produce direct reading balances was the introduction of the single pan substitution balance. A substitution balance is, in principle, similar to an equal arm balance. The object to be measured is placed in the weighing pan; inside the balance are a series of calibrated weights that can be added to the standard side of the balance, through the use of dials and levers. The standard weights can be added and subtracted through the use of the balance's mechanical system, to equal a large variety of weighing loads. Very small differences between the standard weights and the load are read out on an optical scale.

The spring scale is probably the least expensive device for making mass measurements. The force of gravity is once again used as the reference. The scale is placed so that the unknown object is suspended by the spring and the force of gravity can freely act on the object. The elasticity of the spring is assumed to be linear and the force required to stretch it is marked on the scale. When the force of gravity and the elastic force of the spring balance, the force is read from the scale, which has been calibrated in units of mass. Capacity can be increased by increasing the strength of the spring. However, as the strength of the spring increases, the resolution of the scale decreases. This decrease in resolution limits spring scales to relatively small loads of no more than a few kilograms. There are two kinds of springs used: spiral and cantilevered springs.

The torsion balance is a precise adaptation of the spring concept used to determine the mass indirectly. The vertical force produced by the load produces a torque on a wire or beam. This torque produces an angular deflection. As long as the balance is operated in the linear range, the angular deflection is proportional to the torque. Therefore, the angular deflection is also proportional to the applied load. When the torsion spring constant is calibrated, the angular deflection can be read as a mass unit. Unlike the crude spring scales, it is possible to make torsion balances capable of measuring in the microgram region. The torsion element could be a band, a wire, or a string.

The beam balance is probably the next step in accuracy and cost. The beam balance uses the same principle of operation as the equal arm balance. However, it normally has only one pan and the beam is offset. A set of sliding weights are mounted on the beam. As the weights slide out the beam, they gain a mechanical advantage due to the inequality of the distance from the pivot point of the balance. The weights move out along the beam until the balance is in equilibrium. Along the beam, there are notched positions that are marked to correspond to the force applied by the sliding weights. By adding up the forces indicated by the position of each weight, the mass of the unknown material is determined. Beam balances and scales are available in a wide range of accuracy's load capacities. Beam balances are available to measure in the milligram range and large beam scales are made with capacities to weigh trucks and trains. Once again, the disadvantage is that as load increases the resolution decreases. Figure 1.2 shows an example of a two pan beam balance.

The next progression in cost and accuracy is the strain gage load cell. A strain gage is an electrically resistive wire element that changes resistance when the length of the wire element changes. The gage is bonded to a steel cylinder that will shorten when compressed or lengthen when stretched. Because the gage is bonded to the cylinder, the length of the wire will lengthen or contract with the cylinder. The electrical resistance is proportional to the length of the wire element of the gage. By measuring the resistance of the strain gage, it is possible to determine the load on the load cell. The electric resistance is converted into a mass unit readout by the electric circuitry in the readout device.

The force restorative load cell is the heart of an electronic balance, shown in Figure 1.3. The force restorative load cell uses the principle of the equal arm balance. However, in most cases, the fulcrum is offset so it is no longer an equal arm balance, as one side is designed to have a mechanical advantage. This side of the balance is attached to an electric coil. The coil is suspended in a magnetic field. The other side is still connected to a weighing pan. Attached to the beam is a null indicating device, consisting of a photodiode and a light-emitting diode (LED) that are used to determine when the balance is in equilibrium. When a load is placed on the weighing pan, the balance goes out of equilibrium. The LED photodiode circuit detects that the balance is no longer in equilibrium, and the electric current in the coil is increased to bring the balance back to equilibrium. The electric current is then measured across a precision sense resistor and converted into a mass unit reading and displayed on the digital readout.

FIGURE 1.2 Beam balance. (Courtesy Denver Instrument Company.)

FIGURE 1.3 Force restorative load cell. (Courtesy Denver Instrument Company.)

A variation of the latter is the new generation of industrial scales, laboratory balances, and mass comparators. Mass comparators are no longer called balances because they always perform a comparison between known masses (standards) and unknown masses. These weighing devices are employing the electromagnetic force compensation principle in conjunction with joint flexures elements replacing the traditional knife-edge joints. Some of the advantages include a computer interfacing capability and a maintenance-free feature because there are no moving parts.

Another measuring method used in the weighing technology is the vibrating cord. A wire or cord of known length, which vibrates transversely, is tensioned by the force *F.* The vibration frequency changes in direct proportion to the load *F.* The piezoelectric effect is also used in weighing technology. Such weighing devices consist of the presence of an electric voltage at the surface of a crystal when the crystal is under load. Balances employing the gyroscopic effect are also used. This measuring device uses the output signal of a gyrodynamic cell similar to the frequency. Balances wherein the weight force of the load changes the reference distance of the capacitive or inductive converters are also known. As well, balances using the radioactivity changes of a body as a function of its mass under certain conditions exist.

1.2 Weighing Techniques

When relatively low orders of accuracy are required, reading mass or weight values directly from the weighing instrument are adequate. Except for the equal arm balance and some torsion balances, most modern weighing instruments have direct readout capability. For most commercial transactions and for simple scientific experiments, this direct readout will provide acceptable results.

In the case of equal arm balances, the balance will have a pointer and a scale. When relatively low accuracy is needed, the pointer and scale are used to indicate when the balance is close to equilibrium. The same is true when using a torsion balance. However, the equal arm balances of smaller (e.g., 30 g) or larger (e.g., 900 kg) capacity are also used for high-accuracy applications. Only the new generation of electronic balances are equal or better in terms of accuracy and benefit from other features.

Weighing is a deceptively simple process. Most people have been making and using weighing measurements for most of their lives. We have all gone to the market and purchased food that is priced by weight. We have weighed ourselves many times, and most of us have made weight or mass measurements in school. What could be simpler? One places an object on the weighing pan or platform and reads the result.

Unfortunately, the weighing process is very susceptible to error. There are errors caused by imperfections in the weighing instrument; errors caused by biases in the standards used; errors caused by the weighing method; errors caused by the operator; and errors caused by environmental factors. In the case of the equal arm balance, any difference between the lengths of the arms will result in a bias in the measurement result. Nearly all weighing devices will have some degree of error caused by small amounts of nonlinearity in the device. All standards have some amount of bias and uncertainty. Mass is the only base quantity in the International System of Units (SI) defined in relation with a physical artifact. The international prototype of the kilogram is kept at Sevres in France, under the custody of the International Bureau of Weights and Measures. All weighing measurements originate from this international standard. The international prototype of the kilogram is, by international agreement, exact; however, over the last century, it has changed in value. What one does not know is the exact magnitude or direction of the change. Finally, environmental factors such as temperature, barometric pressure, and humidity can affect the weighing process.

There are many weighing techniques used to reduce the errors in the weighing process. The simplest technique is *substitution weighing.* The substitution technique is used to eliminate some of the errors introduced by the weighing device. The single-substitution technique is one where a known standard and an unknown object are both weighed on the same device. The weighing device is only used to determine the difference between the standard and the unknown. First, the standard is weighed and the weighing device's indication is noted. (In the case of an equal arm balance, tare weights are added to the second pan to bring the balance to equilibrium.) The standard is then removed from the weighing device and the unknown is placed in the same position. Again, the weighing device's indication is noted. The first noted indication is subtracted from the second indication. This gives the difference between the standard and the unknown. The difference is then added to the known value of the standard to calculate the value of the unknown object. A variation of this technique is to use a small weight of known value to offset the weighing device by a small amount. The amount of offset is then divided by the known value of the small weight to calibrate the readout of the weighing device. The weighing results of this measurement is calculated as follows:

$$U = S + (O_2 - O_1)(SW)/(O_3 - O_2) \tag{1.4}$$

where U = value of the unknown
S = known value of the standard
SW = small sensitivity weight used to calibrate the scale divisions
O_1 = first observation (standard)
O_2 = second observation (unknown)
O_3 = third observation (unknown + SW)

These techniques remove most of the errors introduced by the weighing device, and are adequate when results to a few tenths of a gram are considered acceptable.

If results better than a few tenths of a gram are required, environmental factors begin to cause significant errors in the weighing process. Differences in density between the standard and the unknown object and air density combine together to cause significant errors in the weighing process.

It is the buoyant force that generates the confusion in weighing. What is called the "true mass" of an object is the mass determined in vacuum. The terms "true mass" and "mass in vacuum" are referring to the same notion of inertial mass or mass in the Newtonian sense. In practical life, the measurements are performed in the surrounding air environment. Therefore, the objects participating in the measurement process adhere to the Archimedean principle being lifted with a force equal to the weight of the displaced volume of air. Applying the buoyancy correction to the measurement requires the introduction of the term "apparent mass." The "apparent mass" of an object is defined in terms of "normal temperature" and "normal air density," conventionally chosen as 20°C and 1.2 mg cm^{-3}, respectively. Because of these conventional values, the "apparent mass" is also called the "conventional mass." The reference material is either brass (8.4 g cm^{-3}) or stainless steel (8.0 g cm^{-3}), for which one obtains an "apparent mass versus brass" and an "apparent mass versus stainless steel," respectively. The latter is preferred for reporting the "apparent mass" of an object.

Calibration reports from the National Institute of Standards and Technology will report mass in three ways: True Mass, Apparent Mass versus Brass, and Apparent Mass versus Stainless Steel. Conventional mass is defined as the mass of an object with a density of 8.0 g cm^{-3}, at 20°C, in air with a density of 1.2 mg cm^{-3}. However, most scientific weighings are of materials with densities that are different from 8.0 g cm^{-3}. This results in significant measurement errors.

As an example, use the case of a chemist weighing 1 liter of water. The chemist will first weigh a mass standard, a 1 kg weight made of stainless steel; then the chemist will weigh the water. The 1 kg mass standard made of 8.0 g cm^{-3} stainless steel will have a volume of 125 cm^3. The same mass of water will have a volume approximately equal to 1000 cm^3 (Volume = Mass/Density). The mass standard will displace 125 cm^3 of air, which will exert a buoyant force of 150 mg (125 cm^3 × 1.2 mg cm^{-3}). However, the water will displace 1000 cm^3 air, which will exert a buoyant force of 1200 mg (1000 cm^3 × 1.2 mg cm^{-3}). Thus, the chemist has introduced a significant error into the measurement by not taking the differing densities and air buoyancy into consideration.

Using 1.2 mg cm^{-3} for the density of air is adequate for measurements made close to sea level; it must be noted that air density decreases with altitude. For example, the air density in Denver, CO, is approximately 0.98 mg cm^{-3}. Therefore, to make accurate mass measurements, one must measure the air density at the time of the measurement if environmental errors in the measurement are to be reduced.

Air density can be calculated to an acceptable value using the following equations:

$$\rho_A \cong 0.0034848/(t + 273.15)(P - 0.0037960 \times U \times e_s) \tag{1.5}$$

where ρ_A = air density in mg cm^{-3}
t = temperature in °C
P = barometric pressure in pascals
U = relative humidity in percent
e_s = saturation vapor pressure

$$e_s \cong \left(1.7526 \times 10^{11}\right) \times e^{\left(-5315.56/\left(273.15+t\right)\right)} \tag{1.6}$$

where $e \cong 2.7182818$
t = temperature in °C

To apply an air buoyancy correction to the single substitution technique, use the following formulae:

$$M_u = \left(M_s\left(1-\rho A/\rho_s\right) + \left(O_2 - O_1\right)\left(M_{SW}\left(1-\rho_A/\rho_{SW}\right)/\left(O_3-O_2\right)\right)\right)/\left(1-\rho_A/\rho_u\right) \tag{1.7}$$

where M_u = mass of the unknown (in a vacuum)
M_s = mass of the standard (in a vacuum)
M_{sw} = mass of the sensitivity weight
ρ_A = air density
ρ_s = density of the standard
ρ_u = density of the unknown
ρ_{sw} = density of the sensitivity weight
O_1 = first observation (standard)
O_2 = second observation (unknown)
O_3 = third observation (unknown + SW)

$$\mathrm{CM} = M_u\left(1-0.0012/\rho_u\right)/0.99985 \tag{1.8}$$

where CM = conventional mass
M_u = mass of the unknown in a vacuum
ρ_u = density of the unknown

When very precise measurements are needed, the double-substitution technique coupled with an air buoyancy correction will provide acceptable results for nearly all scientific applications. The double-substitution technique is similar to the single-substitution technique using the sensitivity weight. In the double-substitution technique, the sensitivity weight is weighed with both the mass standard and the unknown. The main advantage of this technique over single substitution is that any drift in the weighing device is accounted for in the technique. Because of the precision of this weighing technique, it is only appropriate to use it on precision balances or mass comparators. As in the case of single substitution, one places the standard on the balance pan and takes a reading. The standard is then removed and the unknown object is placed on the balance pan and a second reading is taken. The third step is to add the small sensitivity weight to the pan with the unknown object and take a third reading. Then remove the unknown object and return the standard to the pan with the sensitivity weight and take a fourth reading. The mass is calculated using the following formulae:

$$M_u = \frac{\left(M_S\left(1-\rho_A/\rho_S\right) + \left(O_2 - O_1 + O_3 - O_4\right)\right)/2\left(M_{SW}\left(1-\rho_A/\rho_{SW}\right)/\left(O_3-O_2\right)\right)}{\left(1-\rho_A/\rho_u\right)} \tag{1.9}$$

where M_u = mass of the unknown (in a vacuum)
M_s = mass of the standard (in a vacuum)
M_{sw} = mass of the sensitivity weight
ρ_A = air density
ρ_s = density of the standard
ρ_u = density of the unknown
ρ_{sw} = density of the sensitivity weight
O_1 = first observation (standard)
O_2 = second observation (unknown)
O_3 = third observation (unknown + sensitivity weight)
O_4 = fourth observation (standard + sensitivity weight)

$$\mathrm{CM} = M_u\left(1-0.0012/\rho_u\right)/0.99985 \tag{1.10}$$

where CM = conventional mass
M_u = mass of the unknown in a vacuum
ρ_u = density of the unknown

To achieve the highest levels of accuracy, advanced weighing designs have been developed. These advanced weighing designs incorporate redundant weighing, drift compensation, statistical checks, and multiple standards. The simplest of these designs is the three-in-one design. It uses two standards to calibrate one unknown weight. In its simplest form, one would perform three double substitutions. The first compares the first standard and the unknown weight; the second double substitution compares the first standard against the second standard, which is called the check standard; and the third and final comparison compares the second (or check standard) against the unknown weight. These comparisons would then result in the following:

O_1 = reading with standard on the balance
O_2 = reading with unknown on the balance
O_3 = reading with unknown and sensitivity weight on the balance
O_4 = reading with standard and sensitivity weight on the balance
O_5 = reading with standard on the balance
O_6 = reading with check standard on the balance
O_7 = reading with check standard and sensitivity weight on the balance
O_8 = reading with standard and sensitivity weight on the balance
O_9 = reading with check standard on the balance
O_{10} = reading with unknown on the balance
O_{11} = reading with unknown and sensitivity weight on the balance
O_{12} = reading with check standard and sensitivity weight on the balance

The measured differences are calculated using the following formulae:

$$a=\left[\left(O_1-O_2+O_4-O_3\right)/2\right]\times\left[M_{SW}\left(1-\rho_A/\rho_{SW}\right)/O_3-O_2\right] \tag{1.11}$$

$$b=\left[\left(O_5-O_6+O_8-O_7\right)/2\right]\times\left[M_{SW}\left(1-\rho_A/\rho_{SW}\right)/O_7-O_6\right] \tag{1.12}$$

$$c=\left[\left(O_9-O_{10}+O_{12}-O_{11}\right)/2\right]\times\left[M_{SW}\left(1-\rho_A/\rho_{SW}\right)/O_{11}-O_{10}\right] \tag{1.13}$$

where a = difference between standard and unknown
b = difference between standard and check standard
c = difference between check standard and unknown
M_{sw} = mass of sensitivity weight
ρ_A = air density calculated using Equations 1.5 and 1.6
ρ_{sw} = density of sensitivity weight

The least-squares measured difference is computed for the unknown from:

$$d_u = \left(-2a - b - c\right)/3 \qquad (1.14)$$

Using the least-squares measured difference, the mass of the unknown is computed as:

$$U = \left(S\left(1 - \rho_A/\rho_S\right) + d_u\right)/\left(1 - \rho_A/\rho_U\right) \qquad (1.15)$$

where U = mass of unknown
S = mass of the standard
d_u = least-squares measured difference of the unknown
ρ_A = air density calculated using Equations 1.5 and 1.6
ρ_S = density of the standard
ρ_U = density of the unknown

The conventional mass of the unknown is now calculated as:

$$\text{CU} = U\left(1 - 0.0012/\rho_U\right)/0.99985 \qquad (1.16)$$

where CU = conventional mass
U = mass of unknown
ρ_U = density of unknown

The least-squares measured difference is now computed for the check standard as:

$$d_{CS} = \left(-a - 2b - c\right)/3 \qquad (1.17)$$

Using the least-squares measured difference, the mass of the check standard is computed from:

$$\text{CS} = \left(S\left(1 - \rho_A/\rho_S\right) + d_{CS}\right)/\left(1 - \rho_A/\rho\text{CS}\right) \qquad (1.18)$$

where CS = mass of check standard
s = mass of the standard
d_{CS} = least-squares measured difference of the check standard
ρ_A = air density calculated using Equations 1.5 and 1.6
ρ_S = density of the standard
ρCS = density of unknown

The mass of the check standard must lie within the control limits for the check standard. If it is out of the control limits, the measurement must be repeated.

The short-term standard deviation of the process is now computed:

$$\text{Short-term standard deviation} = 0.577\left(a - b + c\right) \qquad (1.19)$$

The short-term standard deviation is divided by the historical pooled short-time standard deviation to calculate the F-statistic:

$$F\text{-statistic} = \text{short-term standard deviation/pooled short-time standard deviation}$$

The F-statistic must be less than the value obtained from the student t-variant at the 99% confidence level for the number of degrees of freedom of the historical pooled standard deviation. If this test fails, the measurement is considered to be out of statistical control and must be repeated.

By measuring a check standard and by computing the short-term standard deviation of the process and comparing them to historical results, one obtains a high level of confidence in the computed value of the unknown.

There are many different weighing designs that are valid; the three-in-one (three equal weights) and four equal weights are the ones that can be easily calculated without the use of a computer. Primary calibration laboratories — private and government — are using these multiple intercomparisons, state-of-the-art mass calibration methods under the Mass Measurement Assurance Program using the Mass Code computer program provided by the National Institute of Standards and Technology. A full discussion of these designs can be found in the *National Bureau of Standards Technical Note 952.*

References

1. J. K. Taylor and H. V. Oppermann, *Handbook for the Quality Assurance of Metrological Measurements,* NBS Handbook 145, Washington, D.C.: U.S. Department of Commerce, National Bureau of Standards, 1986.
2. L. V. Judson, *Weights and Measures Standards of the United States, A Brief History,* NBS Special Publication 447, Washington, D.C.: Department of Commerce, National Bureau of Standards, 1976.
3. P. E. Pontius, *Mass and Mass Values,* NBS Monograph 133, Washington, D.C.: Department of Commerce, National Bureau of Standards, 1974.
4. K.B. Jaeger and R. S. Davis, *A Primer for Mass Metrology,* NBS Special Publication 700-1, Washington, D.C.: Department of Commerce, National Bureau of Standards, 1984.
5. J. M. Cameron, M. C. Croarkin, and R. C. Raybold, *Designs for the Calibration of Standards of Mass,* NBS Technical Note 952, Washington, D.C.: Department of Commerce, National Bureau of Standards, 1977.
6. G. L. Harris (Ed.), *Selective Publications for the Advanced Mass Measurements Workshop,* NISTIR 4941, Washington, D.C.: Department of Commerce, National Institute of Standards and Technology, 1992.
7. Metron Corporation, *Physical Measurements,* NAVAIR 17-35QAL-2, California: U.S. Navy, 1976.
8. R. S. Cohen, *Physical Science,* New York: Holt, Rinehart and Winston, 1976.
9. D. B. Prowse, The Calibration of Balances, CSIRO Division of Applied Physics, Australia, 1995.
10. E. Hazarian, Techniques of mass measurement, *Southern California Edison Mass Seminar Notebook,* Los Angeles, CA, 1994.
11. E. Hazarian, Analysis of mechanical convertors of electronic balances, *Measurement Sci. Conf.,* Anaheim, CA, 1993.
12. B. N. Taylor, *Guide for the use of the International System of Units (SI),* NIST SP811, 1995.

2

Density Measurement

Halit Eren
Curtin University of Technology

Density is a significant part of measurement and instrumentation. Density measurements are made for at least two important reasons: (1) for the determination of mass and volume of products, and (2) the quality of the product. In many industrial applications, density measurement ascertains the value of the product.

Density is defined as the mass of a given volume of a substance under fixed conditions. However, ultimate care must be exercised in measurements because density varies with pressure and temperature. The variation is much greater in gases.

In many modern applications, the densities of products are obtained by sampling techniques. In measurements, there are two basic concepts: *static density measurements* and *dynamic (on-line) density measurements*. Within each concept, there are many different methods employed. These methods are based on different physical principles. In many cases, the application itself and the characteristics of the process determine the best suitable method to be used. Generally, static methods are well developed, lower in cost, and more accurate. Dynamic samplers are expensive, highly automated, and use microprocessor-based signal processing devices. Nevertheless, nowadays, many static methods are also computerized, offering easy to use, flexible, and self-calibrating features.

There is no single universally applicable density measurement technique. Different methods must be employed for different products and materials. In many cases, density is normalized under reference conditions.

The density of a substance is determined by dividing the density of that substance by the density of a standard substance obtained under the same conditions. This dimensionless ratio is called the *specific gravity* (SG), also termed the *relative density*. The specific gravities of liquid and gases under reference conditions are given by:

$$\text{Liquid SG} = \text{density of liquid} \big/ \text{density of water} \tag{2.1}$$

$$\text{Gas SG} = \text{density of gas} \big/ \text{density of air} \tag{2.2}$$

0-8493-0047-9/00/$0.00+$.50

Commonly accepted sets of conditions are *normal temperature and pressure* (NTP) and *standard temperature and pressure* (STP). NTP is usually taken as the temperature of 0.00°C and a pressure of 760 mm Hg. The NTP is accepted as 15.00 or 15.56°C and 101.325 kPa.

2.1 Solid Density

If the uniformity is maintained, the determination of density of solids is a simple task. Once the volume of the solid and its mass are known, the density can be found using the basic ratio: density = mass/volume (kg m^{-3}).

However, in many applications, solids have different constituents and are made up from different materials having different ratios. Their volumes can also change often. In these cases, dynamic methods are employed, such as radioactive absorption types, ultrasonic, and other techniques. Some of these methods are described below.

2.2 Fluid Density

The measurement of densities of fluids is much more complex than for solids. For fluid densities, many different techniques are available. This is mainly due to complexities in processes, variations of fluid densities during the processes, and diverse characteristics of the process and the fluids themselves. Some of these methods are custom designed and applicable to special cases only. Others are very similar in principles and technology, and applicable to many different type of fluids. At present, apart from conventional methods, there are many novel and unusual techniques undergoing extensive development and research. For example, densitometers based on electromagnetic principles [1] can be given as part of an intelligent instrumentation system.

Fluids can be divided to liquids and gases. Extra care and further considerations are necessary in gas density measurements. For example, perfect gases contain an equal number of molecules under the same conditions and volumes. Therefore, molecular weights can be used in density measurements.

Depending on the application, fluid densities can be measured both in *static* and *dynamic* forms. In general, static density measurements of fluids are well developed, precise, and have greater resolution than most dynamic techniques. Pycnometers and buoyancy are examples of static techniques that can be adapted to cover small density ranges with a high resolution and precision. Nowadays, many manufacturers offer dynamic instruments previously known to be static. Also, many static density measurement devices are computerized and come with appropriate hardware and software. In general, static-type measurements are employed in laboratory conditions, and dynamic methods are employed for real-time measurements where the properties of a fluid vary from time to time.

In this chapter, the discussion will concentrate on the commonly applied, modern density measuring devices. These devices include:

1. Pycnometric densitometers
2. Buoyancy-type densitometers
3. Hydrometers
4. Hydrostatic weighing densitometers
5. Balance-type densitometers
6. Column-type densitometers
7. Vibrating element densitometers
8. Radioactive densitometers
9. Refractometer and index of reflection densitometers
10. Coriolis densitometers
11. Absorption-type densitometers

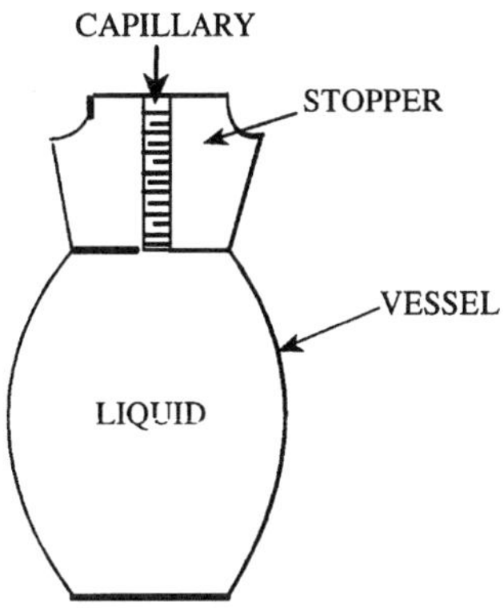

FIGURE 2.1 A pycnometer. A fixed volume container is filled with liquid and weighed accurately; capillary is used to determine the exact volume of the liquid.

Pycnometric Densitometers

Pycnometers are static devices. They are manufactured as fixed volume vessels that can be filled with the sample liquid. The density of the fluid is measured by weighing the sample. The simplest version consists of a vessel in the shape of a bottle with a long stopper containing a capillary hole, as shown in Figure 2.1. The capillary is used to determine the exact volume of the liquid, thus giving high resolution when filling the pycnometer. The bottle is first weighed empty, and then with distilled-aerated water to determine the volume of the bottle. The bottle is then filled with the process fluid and weighed again. The density is determined by dividing the mass by the volume. The specific gravity of the liquid is found by the ratio of the fluid mass to water mass. When pycnometers are used, for good precision, ultimate care must be exercised during the measurements; that is, the bottle must be cleaned after each measurement, the temperature must be kept constant, and precision balances must be used. In some cases, to ensure filling of the pycnometer, twin capillary tubes are used. The two capillaries, made of glass, are positioned such that the fluid can be driven into the vessel under vacuum conditions. Accurate filling to graduation marks on the capillary is then made.

The pycnometers have to be lightweight, strong enough to contain samples, and they need to be nonmagnetic for accurate weighing to eliminate possible ambient magnetic effects. Very high-resolution balances must be used to detect small differences in weights of gases and liquids. Although many pycnometers are made of glass, they are also made of metals to give enough strength for the density measurements of gases and liquids at extremely high pressures. In many cases, metal pycnometers are necessary for taking samples from the line of some rugged processes.

Pycnometers have advantages and disadvantages. Advantages are that if used correctly, they are accurate; and they can be used for both density and specific gravity measurements. The disadvantages include:

1. Great care must be exercised for accurate results.
2. The sample must be taken off-line, with consequent time lag in results. This creates problems of relating samples to the materials that exist in the actual process.
3. High-precision pycnometers are expensive. They require precision weighing scales and controlled laboratory conditions. Specialized techniques must be employed to take samples in high-pressure processes and hostile conditions, such as offshore installations.
4. Their good performances might depend on the skill of operator.

Buoyancy-Type Densitometers

The buoyancy method basically uses Archimedes principle. A suspended sinker, with a known mass and volume attached to a fine wire, is totally immersed in the sample liquid. A precision force balance is used to measure the force to support the sinker. Once the mass, volume, and supporting weight of the sinker

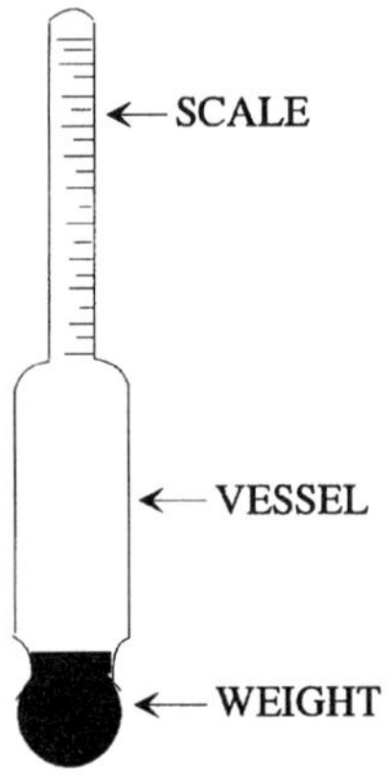

FIGURE 2.2 Hydrometer. A fixed weight and volume bulb is placed into the liquid. The bulb sinks in the liquid, depending on its density. The density is read directly from the scale. Temperature correction is necessary.

are known, the density of the liquid can be calculated. However, some corrections need to be made for surface tension on the suspension wire, the cubicle expansion coefficient of the sinker, and the temperature of process. Buoyancy-type densitometers give accurate results and are used for the calibration of the other liquid density transducers.

One advanced version of the buoyancy technique is the magnetic suspension system. The sinker is fully enclosed in a pressure vessel, thus eliminating surface tension errors. Their uses can also be extended to applications such as the specific gravity measurements under low vapor pressures and density measurements of hazardous fluids.

Hydrometers

Hydrometers are the most commonly used devices for measurement of the density of liquids. They are so commonly used that their specifications and procedure of use are described by national and international standards, such as ISO 387. The buoyancy principle is used as the main technique of operation. The volume of fixed mass is converted to a linear distance by a sealed bulb-shaped glass tube containing a long stem measurement scale, shown in Figure 2.2. The bulb is ballasted with a lead shot and pitch, the mass of which depends on the density range of the liquid to be measured. The bulb is simply placed into the liquid and the density is read from the scale. The scale is graduated in density units such as kg m^{-3}. However, many alternative scales are offered by manufacturers, such as specific gravity, API gravity, Brix, Brine, etc. Hydrometers can be calibrated for different ranges for surface tensions and temperatures. Temperature corrections can be made for set temperature such as 15°C, 20°C, or 25°C. ISO 387 covers a density range of 600 kg m^{-3} to 2000 kg m^{-3}. Hydrometers have a number of advantages and disadvantages. The advantages include:

1. Relatively low cost and easy to use
2. Good resolution for small range
3. Traceable to national and international standards

The disadvantages include:

1. They have small span; therefore, a number of meters are required to cover a significant range.
2. They are made from glass and fragile. Metal and plastic versions are not as accurate.
3. The fluid needs to be an off-line sample, not representing the exact conditions of the process. There are pressure hydrometers for low vapor pressure hydrocarbons, but this adds a need for accurately determining the pressure too.
4. If good precision is required, they are difficult to use, needing surface tension and temperature corrections. Further corrections could be required for opaque fluids.

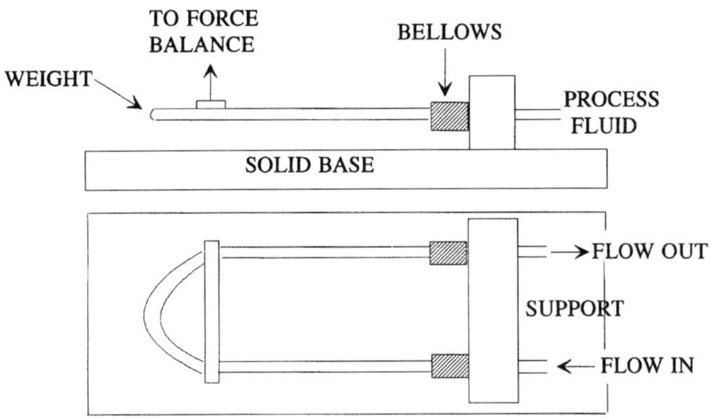

FIGURE 2.3 Hydrostatic weighing. The total weight of a fixed-volume tube containing liquid is determined accurately. The density is calculated using mass: volume ratio.

Hydrostatic Weighing Densitometers

The most common device using a hydrostatic weighing method consists of a U-tube that is pivoted on flexible end couplings. A typical example is shown in Figure 2.3. The total weight of the tube changes, depending on the density of fluid flowing through it. The change in the weight needs to be measured accurately, and there are a number of methods employed to do this. The most common commercial meters use a force balance system. The connectors are stainless steel bellows. In some cases, rubber or PTFE are used, depending on the process fluid characteristics and the accuracy required. There are temperature and pressure limitations due to bellows, and the structure of the system may lead to a reading offset. The meter must be securely mounted on a horizontal plane for optimal accuracy.

The advantages of this method include:

1. They give continuous reading and can be calibrated accurately.
2. They are rugged and can be used for two-phase liquids such as slurries, sugar solutions, powders, etc.

The disadvantages of these meters include:

1. They must be installed horizontally on a solid base. These meters are not flexible enough to adapt to any process; thus, the process must be designed around it.
2. They are bulky and cumbersome to use.
3. They are unsuitable for gas density measurements.

Balance-Type Densitometers

Balance-type densitometers are suitable for liquid and gas density measurements. Manufacturers offer many different types; four of the most commonly used ones are discussed below.

Balanced-Flow Vessel

A fixed volume vessel as shown in Figure 2.4 is employed for the measurements. While the liquid is flowing continuously through the vessel, it is weighed automatically by a sensitive scale — a spring balance system or a pneumatic force balance transmitter. Because the volume and the weight of the liquid are known, the density or specific gravity can easily be calculated and scaled in respective units. In the design process, extra care must be exercised for the flexible end connections.

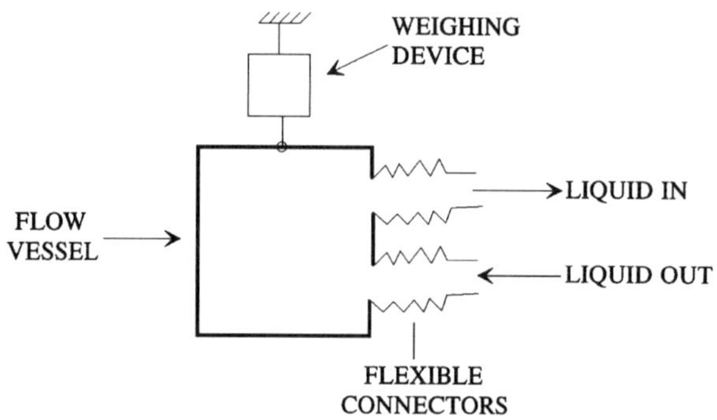

FIGURE 2.4 Balanced flow vessel. An accurate spring balance or force balance system is used to weigh the vessel as the liquid flows through it.

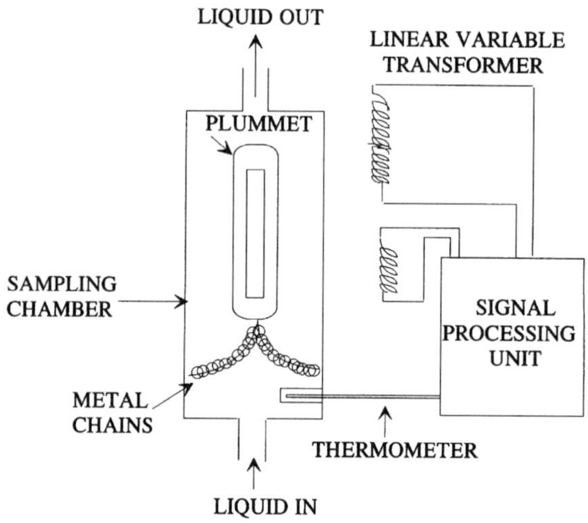

FIGURE 2.5 Chain balance float. The fixed volume and weight plummet totally suspended in the liquid assumes equilibrium position, depending on the density. The force exerted by the chains on the plummet is a function of plummet position; hence, the measured force is proportional to the density of the liquid.

Chain Balanced Float

In this system, a self-centering, fixed-volume, submerged plummet is used for density measurements, as illustrated in Figure 2.5. The plummet is located entirely under the liquid surface. At balance, the plummet operates without friction and is not affected by surface contamination. Under steady-state conditions, the plummet assumes a stable position. The effective weight of the chain on the plummet varies, depending on the position of the plummet, which in turn is a function of the density of the liquid. The plummet contains a metallic transformer core that transmits changes in the position to be measured by a pickup coil. The voltage differential, a function of plummet displacement, is calibrated as a measure of variations in specific gravity. A resistance thermometer bridge is used for the compensation of temperature effects on density.

Gas Specific Gravity Balance

A tall column of gas is weighed by the floating bottom of the vessel. This weight is translated into the motion of an indicating pointer, which moves over a scale graduated in units of density or specific gravity. This method can be employed for any gas density measurement.

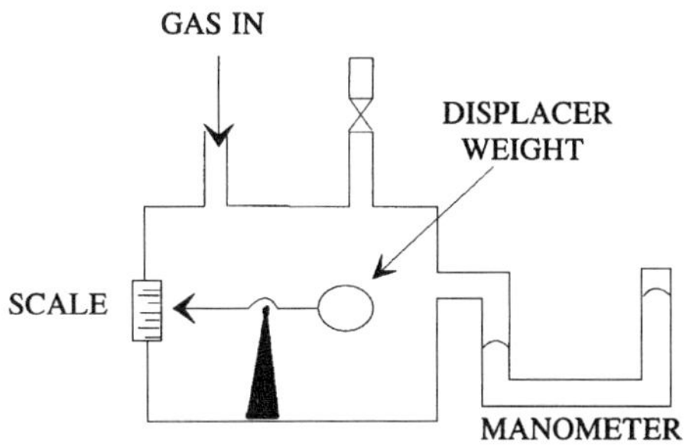

FIGURE 2.6 Buoyancy gas balance. The position of the balance beam is adjusted by a set pressure air, air is then displaced by gas of the same pressure. The difference in the reading of the balance beam gives the SG of the gas. The pressures are read on the manometer.

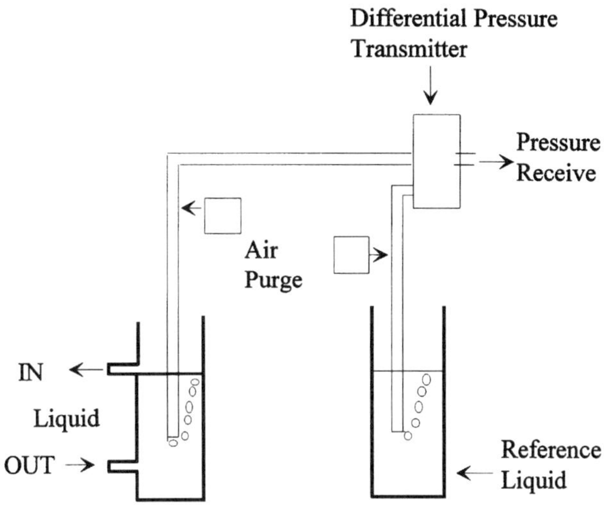

FIGURE 2.7 Reference column densitometer. Two identical tubes, having the same distance from the surface, are placed in water and liquid. Water with known density characteristics is used as the reference. The pressures necessary to displace the fluids inside the tubes are proportional to the densities of the fluids. The pressure difference at the differential pressure transmitter is translated into relative densities.

Buoyancy Gas Balance

In this instrument, a displacer is mounted on a balance beam in a vessel, as shown in Figure 2.6. The displacer is balanced for air, and the manometer reading is noted at the exact balance pressure. The air is then displaced by gas, and the pressure is adjusted until the same balance is restored. The ratio of the pressure of air to the pressure of gas is then the density of the gas relative to air. This method is commonly applied under laboratory conditions and is not suitable for continuous measurements.

Column-Type Densitometers

There are number of different versions of column methods. As a typical example, a reference column method is illustrated in Figure 2.7. A known head of sample liquid and water from the respective bubbler pipes are used. A differential pressure measuring device compares the pressure differences, proportional to relative densities of the liquid and the water. By varying the depth of immersion of the pipes, a wide

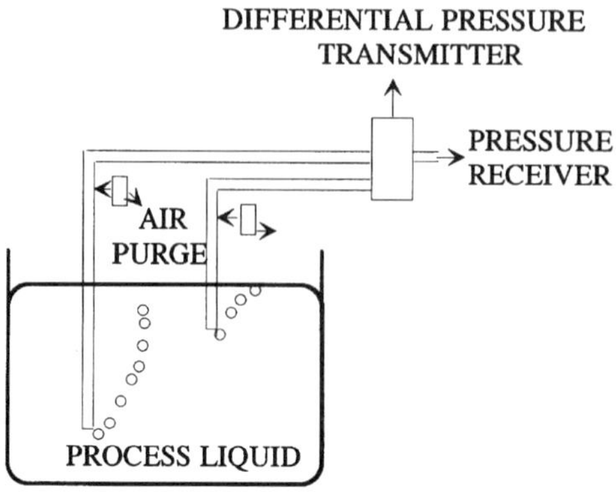

FIGURE 2.8 Two-tube column densitometer. The pressure difference at the differential pressure transmitter depends on the relative positions of the openings of the pipes and the density of liquid. Once the relative positions are fixed, the pressure difference can be related to the equivalent weight of the liquid column at the openings of the pipes, hence the density of the liquid.

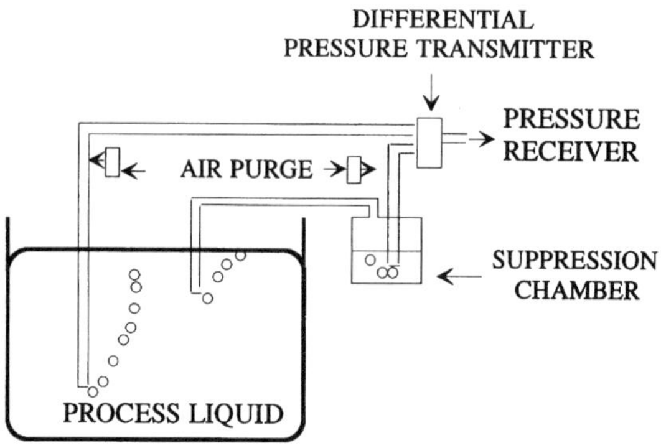

FIGURE 2.9 Suppression-type, two-tube column densitometer. Operation principle is the same as in Figures 2.7 and 2.8. In this case, the suppression chamber affords greater accuracy in readings.

range of measurements can be obtained. Both columns must be maintained at the same temperature to avoid the necessity for corrections of temperature effects.

A simpler and more widely used method of density measurement is achieved by the installation of two bubbler tubes as illustrated in Figure 2.8. The tubes are located in the sample fluid such that the end of one tube is higher than that of the other. The pressure required to bubble air into the fluid from both tubes is equal to the pressure of the fluid at the end of the bubbler tubes. The outlet of one tube higher than the other and the distances of the openings of the tubes are fixed; hence, the difference in the pressure is the same as the weight of a column of liquid between the ends. Therefore, the differential pressure measurement is equivalent to the weight of the constant volume of the liquid, and calibrations can be made that have a direct relationship to the density of the liquid. This method is accurate to within 0.1% to 1% specific gravity. It must be used with liquids that do not crystallize or settle in the measuring chamber during measurements.

Another version is the range suppression type, which has an additional constant pressure drop chamber as shown in Figure 2.9. This chamber is in series with the low-pressure side to give advantages in scaling and accurate readings of densities.

Vibrating Element Densitometers

If a body containing or surrounded by a fluid is set to resonance at its natural frequency, then the frequency of oscillation of the body will vary as the fluid properties and conditions change. The natural frequency is directly proportional to the stiffness of the body and inversely proportional to the combined mass of the body and the fluid. It is also dependent on the shape, size, and elasticity of the material, induced stress, mass, and mass distribution of the body. Basically, the vibration of the body can be equated to motion of a mass attached to a mechanical spring. Hence, an expression for the frequency can be written as:

$$\text{Resonant frequency} = \text{SQRT}\left(K/\left(M + k\rho\right)\right) \tag{2.3}$$

where K is the system stiffness, M is the transducer mass, k is the system constant, and ρ is the fluid density.

A factor common to all types of vibrating element densitometers is the problem of setting the element in vibration and maintaining its natural resonance. There are two drives for the purpose.

Magnetic Drives

Magnetic drives of the vibrating element and the pickup sensors of vibrations are usually achieved using small coil assemblies. Signals picked up by the sensors are amplified and fed back as a drive to maintain the disturbing forces on the vibrating body of the meter.

In order to achieve steady drives, the vibrating element sensor can be made from nonmagnetic materials. In this case, small magnetic armatures are attached.

The main advantage of magnetic drive and pickup systems is they are noncontact methods. They use conventional copper windings and they are reliable within the temperature range of –200 to +200°C.

Piezoelectric Drives

A wide range of piezoelectric materials are available to meet the requirements of driving vibrating elements. These materials demonstrate good temperature characteristics as do magnetic drive types. They also have the advantage of being low in cost. They have high impedance, making the signal conditioning circuitry relatively easy. They do not require magnetic sensors.

The piezoelectric drives are mechanically fixed on the vibrating element by adhesives. Therefore, attention must be paid to the careful placement of the mount in order to reduce the strain experienced by the piezo element due to thermal and pressure stresses while the instrument is in service.

A number of different types of densitometers have been developed that utilize this phenomenon. The three main commercial types are introduced here.

Vibrating Tube Densitometers

These devices are suitable for highly viscous liquids or slurry applications. The mode of operation of vibration tube meters is based on the transverse vibration of tubes as shown in Figure 2.10. The tube and the driving mechanisms are constrained to vibrate on a single plane. As the liquid moves inside the tube, the density of the entire mass of the liquid is measured. The tube length is approximately 20 times greater than the tube diameter.

A major design problem with the vibrating tube method is the conflict to limit the vibrating element to a finite length and accurately fix the nodes. Special attention must be paid to avoid any exchange of vibrational energy outside the sensory tube. The single tube has the disadvantage of presenting obstruction to the flow, thus experiencing some pressure losses. The twin tube, on the other hand, offers very small blockage (Figure 2.11) and can easily be inspected and cleaned. Its compact size is another distinct advantage. In some densitometers, the twin tube is designed to achieve a good dynamic balance, with the two tubes vibrating in antiphase. Their nodes are fixed at the ends, demonstrating maximum sensitivity to installation defects, clamping, and mass loading.

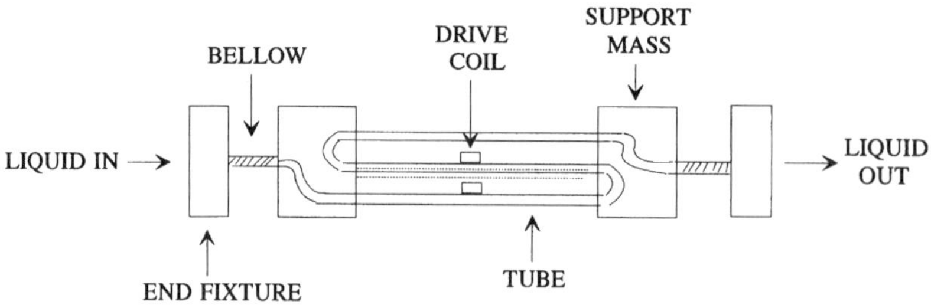

FIGURE 2.10 Vibrating tube densitometer. Tube containing fluid is vibrated at resonant frequency by electromagnetic vibrators. The resonant frequency, which is a function of the density of the fluid, is measured accurately. The tube is isolated from the fixtures by carefully designed bellows.

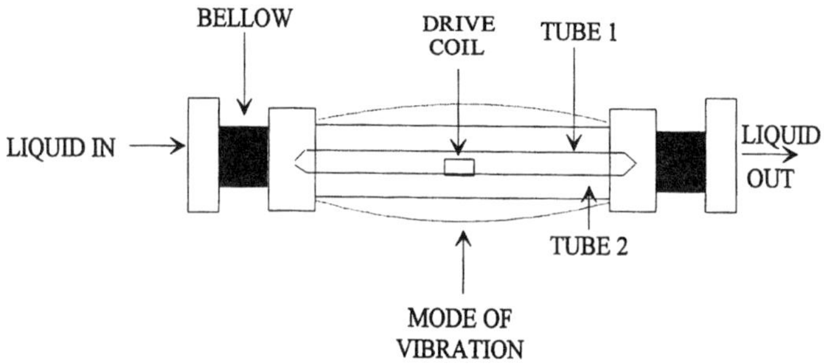

FIGURE 2.11 Two-tube vibrating densitometer. Two tubes are vibrated in antiphase for greater accuracy. Twin-tube densitometers are compact in size and easy to use.

The main design problems of the vibrating tube sensors are in minimizing the influence of end padding and overcoming the effects of pressure and temperature. Bellows are used at both ends of the sensor tubes to isolate the sensors from external vibrations. Bellows also minimize the end loadings due to differential expansions and installation stresses.

The fluid runs through the tubes; therefore, no pressure balance is required. Nevertheless, in some applications, the pressure stresses the tubes, resulting in stiffness changes. Some manufacturers modify the tubes to minimize the pressure effects. In these cases, corrections are necessary only when high accuracy is mandatory. The changes in the Young's modulus with temperature can be reduced to near-zero using Ni-span-C materials whenever corrosive properties of fluids permit. Usually, manufacturers provide pressure and temperature correction coefficients for their products.

It is customary to calibrate each vibration element densitometer against others as a transfer of standards. Often, the buoyancy method is used for calibration purposes. The temperature and pressure coefficients are normally found by exercising the transducer over a range of temperatures and pressures on a liquid with well-known properties. Prior to calibration, the vibration tube densitometers are subjected to a programmed burn-in cycle to stabilize them against temperatures and pressures.

Vibrating Cylinder Densitometers

A thin-walled cylinder, with a 3:1 length:diameter ratio, is fixed with stiff ends. The thickness of the cylinder wall varies from 25 μm to 300 μm, depending on the density range and type of fluid used. The cylinder can be excited to vibrate in a hoop mode by magnetic drives mounted either in or outside the cylinder.

For good magnetic properties, the cylinder is made of corrosion-resistant magnetic materials. Steel such as FV520 is often used for this purpose. Such materials have good corrosion-resistance characteristics; unfortunately, due to their poor thermoelastic properties, they need extensive temperature corrections.

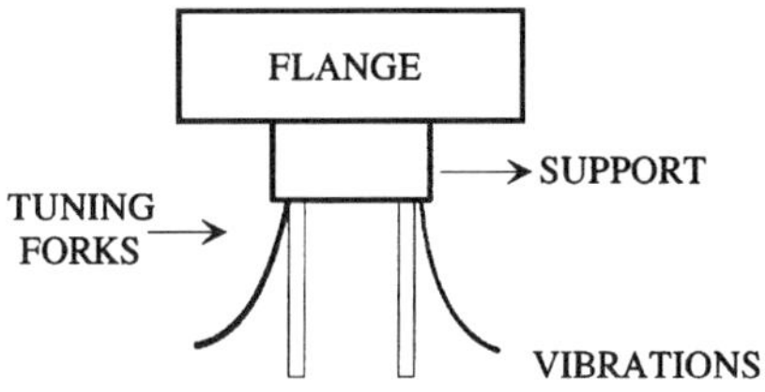

FIGURE 2.12 Tuning fork densitometer. Twin forks are inserted into the fluid and the natural frequencies are measured accurately. The natural frequency of the forks is a function of the density of the fluid.

Nickel-iron alloys such as Ni-span-C are often used to avoid temperature effects. Once correctly treated, the Ni-span-C alloy has near-zero Young's modulus properties. Because the cylinder is completely immersed in the fluid, there are no pressure coefficients.

The change in the resonant frequency is determined by the local mass loading of the fluid in contact with the cylinder. The curve of frequency against density is nonlinear and has a parabolic shape, thus requiring linearization to obtain practical outputs. The resonant frequency range varies from 2 kHz to 5 kHz, depending on the density range of the instrument. The cylinders need precision manufacturing and thus are very expensive to construct. Each meter needs to be calibrated individually for different temperatures and densities to suit specific applications. In the case of gas density applications, gases with well-known properties (e.g., pure argon or nitrogen) are used for calibrations. In this case, the meters are subjected to a gas environment with controlled temperature and pressure. The calibration curves are achieved by repetitions to suit the requirements of individual customers for particular applications. In the case of liquids, the meters are calibrated with liquids of known density, or they are calibrated against another standard (e.g., pycnometer or buoyancy type densitometers).

Vibration cylinder-type densitometers have zero pressure coefficients and they are ideal for liquefied gas products or refined liquids. Due to relatively small clearances between cylinder and housing, they require regular cleaning. They are not suitable for liquids or slurries with high viscous properties.

Tuning Fork Densitometers

These densitometers make use of the natural frequency of low-mass tuning forks, shown in Figure 2.12. In some cases, the fluid is taken into a small chamber in which the electromechanically driven forks are situated. In other cases, the fork is inserted directly into the liquid. Calibration is necessary in each application.

The advantages of vibrating element meters include:

1. They are suitable for both liquids and gases with reasonable accuracy.
2. They can be designed for real-time measurements.
3. They can easily be interfaced because they operate on frequencies and are inherently digital.
4. They are relatively robust and easy to install.
5. Programmable and computerized versions are available. Programmed versions make all the corrections automatically. They provide the output of live density, density at reference conditions, relative density, specific gravity, concentration, solid contents, etc.

The disadvantages include:

1. They do not relate directly to primary measurements; therefore, they must be calibrated.
2. They all have problems in measuring multiphase liquids.

Radioactive Densitometers

As radioactive isotopes decay, they emit radiation in the form of particles or waves. This physical phenomenon can be used for the purposes of density measurement. For example, γ rays are passed through the samples and their rate of arrivals are measured using ion- or scintillation-based detection [2]. Generally, γ-ray mass absorption rate is independent of material composition; hence they can be programmed for a wide range materials. Densitometers based on radiation methods can provide accuracy up to +0.0001 g mL^{-1}. Many of these devices have self-diagnostic capabilities and are able to compensate for drift caused by source decay, thus pinpointing any signaling problems.

If γ rays of intensity J_0 penetrate a material of a density ρ and thickness d then the intensity of the radiation after passing through the material can be expressed by:

$$J = J_0 \exp\left(n \rho d\right) \tag{2.4}$$

where n is the mass absorption coefficient.

The accuracy of the density measurement depends on the accuracy of the measurement of the intensity of the radiation and the path length d. A longer path length through the material gives a stronger detection signal.

For accurate operations, there are many arrangements for relative locations of transmitters and detectors, some of which are illustrated in Figures 2.13 and 2.14. Generally, the source is mounted in a lead container clamped onto the pipe or the container wall. In many applications, the detector is also clamped onto the wall.

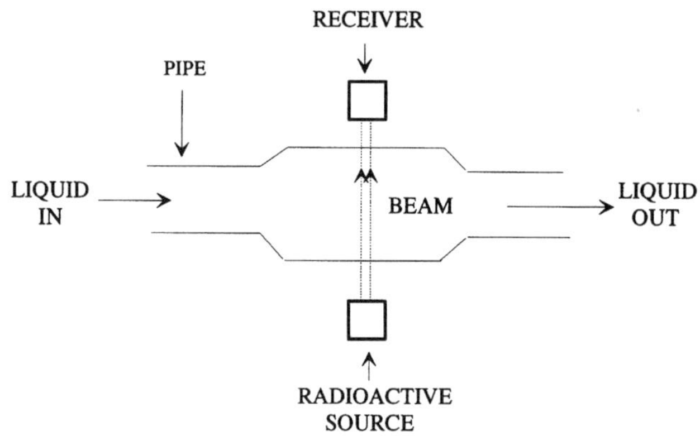

FIGURE 2.13 Fixing radioactive densitometer on an enlarged pipe. The pipe is enlarged to give longer beam length through the liquid, and hence better attenuation of the radioactive energy.

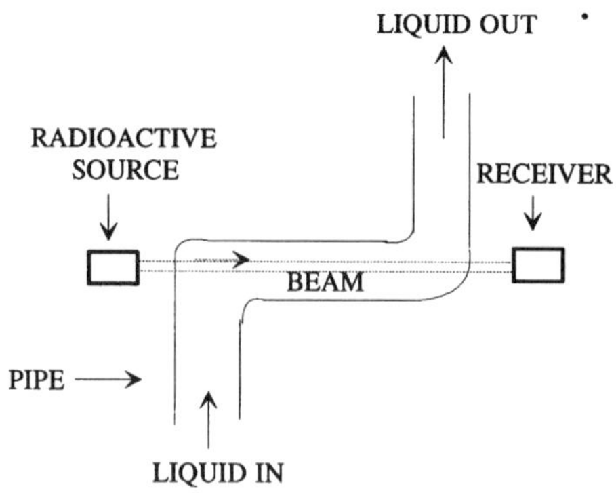

FIGURE 2.14 Fixing radioactive densitometer on an elongated pipe. Elongated path yields a longer path length of the radioactive energy through the liquid; hence, a stronger attenuation.

The advantages of using radioactive methods include:

1. The sensor does not touch the sample; hence, there is no blockage to the path of the liquid.
2. Multiphase liquids can be measured.
3. They come in programmable forms and are easy to interface.
4. They are most suitable in difficult applications, such as mining and heavy process industries.

The disadvantages include:

1. A radioactive source is needed; hence, there is difficulty in handling.
2. For reasonable accuracy, a minimum path length is required.
3. There could be long time constants, making them unsuitable in some applications.
4. They are suitable only for solid and liquid density measurements.

Refractometer and Index of Refraction Densitometers

Refractometers are essentially optical instruments operating on the principles of refraction of light traveling in liquid media. Depending on the characteristics of the samples, measurement of refractive index can be made in a variety of ways (e.g., critical angle, collimation, and displacement techniques). Usually, an in-line sensing head is employed, whereby a sensing window (commonly known as a prism) is wetted by the product to be measured. In some versions, the sensing probes must be installed inside the pipelines or in tanks and vessels. They are most effective in reaction-type process applications where blending and mixing of liquids take place. For example, refractometers can measure dissolved soluble solids accurately.

Infrared diodes, lasers, and other lights may be used as sources. However, this measurement technique is not recommended in applications in processes containing suspended solids, high turbidity, entrained air, heavy colors, poor transparency and opacity, or extremely high flow rates. The readings are automatically corrected for variations in process temperature. The processing circuitry can include signal outputs adjustable in both frequency and duration.

Another version of a refractometer is the index of refraction type densitometer. For example, in the case of position-sensitive detectors, the index of refraction of liquid under test is determined by measuring the lateral displacement of a laser beam. When the laser beam impinges on the cell at an angle of incidence, as in Figure 2.15, the axis of the emerging beam is displaced by the cell wall and by the inner liquid. The lateral displacement can accurately be determined by position-sensitive detectors. For maximum sensitivity, the devices need to be calibrated with the help of interferometers.

Refractometers are often used for the control of adulteration of liquids of common use (e.g., edible oils, wines, and gasoline). They also find application in pulp and paper, food and beverage, sugar, dairy, and other chemical industries.

Coriolis Densitometers

The Coriolis density metering systems are similar to vibrating tube methods, but with slight variations in the design. They are comprised of a sensor and a signal-processing transmitter. Each sensor consists of one or two flow tubes enclosed in a sensor housing. They are manufactured in various sizes and shapes [3]. The sensor tubes are securely anchored at the fluid inlet and outlet points and force is vibrated at the free end, as shown in Figure 2.16. The sensor operates by applying Newton's second law of motion ($F = ma$).

Inside the housing, the tubes are vibrated in their natural frequencies using drive coils and a feedback circuit. This resonant frequency of the assembly is a function of the geometry of the element, material of construction, and mass of the tube assembly. The tube mass comprises two parts: the mass of the tube itself and the mass of the fluid inside the tube. The mass of the tube is fixed for a given sensor. The mass of fluid in the tube is equal to the fluid density multiplied by volume. Because the tube volume is constant, the frequency of oscillation can be related directly to the fluid density. Therefore, for a given geometry

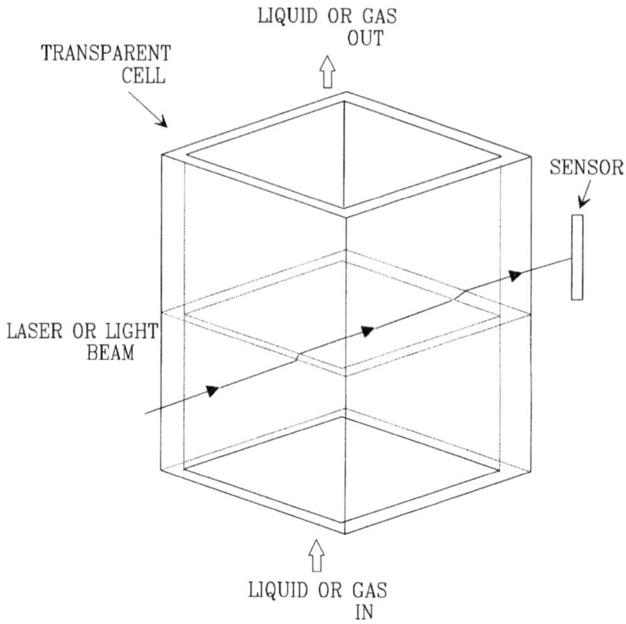

FIGURE 2.15 Index of refraction-type densitometer. The angle of refraction of the beam depends on the shape, size, and thickness of the container, and the density of fluid in the container. Because the container has the fixed characteristics, the position of the beam can be related to density of the fluid. Accurate measurement of the position of the beam is necessary.

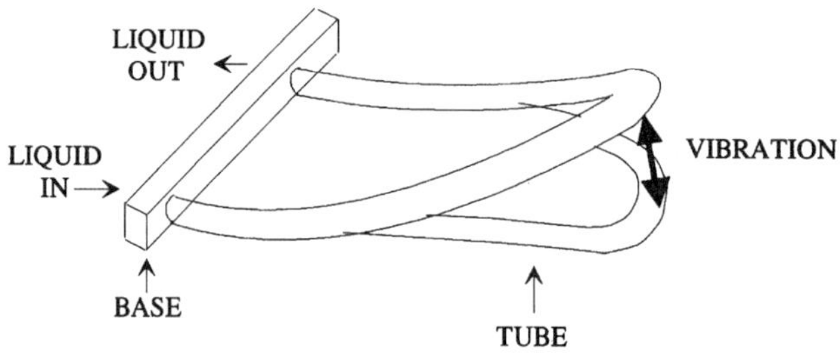

FIGURE 2.16 Coriolis densitometer. Vibration of the tube is detected and related to the mass and flow rate of the fluid. Further calibrations and calculations must be made to determine the densities.

of tube and the material of the construction, the density of the fluid can be determined by measuring the resonant frequency of vibration. Temperature sensors are used for overcoming the effects of changes in modulus of elasticity of the tube. The fluid density is calculated using a linear relationship between the density and the tube period and calibration constants.

Special peripherals, based on microprocessors, are offered by various manufacturers for a variety of measurements. However, all density peripherals employ the natural frequency of the sensor coupled with the sensor temperature to calculate on-line density of process fluid. Optional communication, interfacing facilities, and appropriate software are also offered.

Absorption-Type Densitometers

Absorption techniques are also used for density measurements in specific applications. X-rays, visible light, UV light, and sonic absorptions are typical examples of this method. Essentially, attenuation and phase shift of a generated beam going through the sample is sensed and related to the density of the sample. Most absorption-type densitometers are custom designed for applications having particular characteristics. Two typical examples are: (1) UV absorption or X-ray absorptions are used for determining the local densities of mercury deposits in arc discharge lamps, and (2) ultrasonic density sensors are used in connection with difficult density measurements (e.g., density measurement of slurries). The lime slurry, for example, is a very difficult material to handle. It has a strong tendency to settle out and coat all equipment with which it comes in contact. An ultrasonic density control sensor can fully be emerged into an agitated slurry, thus avoiding the problems of coating and clogging. Inasmuch as the attenuation of the ultrasonic beam is proportional to the suspended solids, the resultant electronic signal is proportional to the specific gravity of the slurry. Such devices can give accuracy up to 0.01%. The ultrasonic device measures the percentage of the suspended solids in the slurry by providing a close approximation of the specific gravity.

References

1. H. Eren, Particle concentration characteristics and density measurements of slurries using electromagnetic flowmeters, *IEEE Trans. Instr. Meas.*, 44, 783-786, 1995.
2. Micro Motion Product Catalogue, Mount Prospect, IL: Fisher-Rosemount, 1995.
3. Kay-Ray, Solution for Process Measurement, Mount Prospect, IL: Fisher-Rosemount, 1995.

Appendix

List of Manufacturers

ABB K-Flow Inc.
Drawer M Box 849
Millville, NJ 08332
Tel: (800) 825-3569

American Density Materials Inc.
Rd. 2, Box 38E
Belvidere, J 07823
Tel: (908) 475-2373

Anton Paar U.S.A.
13, Maple Leaf Ct.
Ashland, VA 23005
Tel: (800) 221-0174

Arco Instrument Company, Inc.
1745 Production Circle
Riverside, CA 92509
Tel: (909) 788-2823
Fax: (909) 788-2409

Cambridge Applied Systems, Inc.
196 Boston Avenue
Medford, MA 02155
Tel: (617) 393-6500

Dynatron
Automation Products, Inc.
3032 Max Roy Street
Houston, TX 77008
Tel: (800) 231-2062
Fax: (713) 869-7332

Kay-Ray/Sensall, Fisher-Rosemount
1400 Business Center Dr.
Mount Prospect, IL 60056
Tel: (708) 803-5100
Fax: (708) 803-5466

McGee Engineering Co., Inc.
Tujunga Canyon Blvd.
Tujunga, CA 91042
Tel: (800) 353-6675

Porous Materials, Inc.
Cornell Business & Technology Park
Ithaca, NY 14850
Tel: (800) 825-5764

Princo Instruments Inc
1020 Industrial Hwy., Dept L
Southampton, PA 18966-4095
Tel: (800) 496-5343

Quantachrome Corp.
1900-T Corporate Drive
Boynton Beach, FL 33426
Tel: (800) 966-1238

Tricor Systems, Inc.
400-T River Ridge Rd.
Elgin, IL 60123
Tel: (800) 575-0161

X-rite, Inc.
3100-T 44th St. S.W
Grandville, MI 49418
Tel: (800) 545-0694

3

Strain Measurement

Christopher S. Lynch
The Georgia Institute of Technology

This chapter begins with a review of the fundamental definitions of strain and ways it can be measured. This is followed by a review of the many types of strain sensors and their application, and sources for strain sensors and signal conditioners. Next, a more detailed look is taken at operating principles of various strain measurement techniques and the associated signal conditioning.

3.1 Fundamental Definitions of Strain

Stress and strain are defined in many elementary textbooks about the mechanics of deformable bodies [1, 2]. The terms *stress* and *strain* are used to describe loads on and deformations of solid materials. The simplest types of solids to describe are homogeneous and isotropic. *Homogeneous* means the material properties are the same at different locations and *isotropic* means the material properties are independent of direction in the material. An annealed steel bar is homogeneous and isotropic, whereas a human femur is not homogeneous because the marrow has very different properties from the bone, and it is not isotropic because its properties are different along the length and along the cross-section.

The concepts of stress and strain are introduced in the context of a long homogeneous isotropic bar subjected to a tensile load (Figure 3.1). The stress σ, is the applied force F, divided by the cross-sectional area A. The resulting strain ε, is the length change ΔL, divided by the initial length L. The bar elongates in the direction the force is pulling (longitudinal strain ε_L) and contracts in the direction perpendicular to the force (transverse strain ε_t).

When the strain is not too large, many solid materials behave like linear springs; that is, the displacement is proportional to the applied force. If the same force is applied to a thicker piece of material, the spring is stiffer and the displacement is smaller. This leads to a relation between force and displacement that depends on the dimensions of the material. Material properties, such as the density and specific heat, must be defined in a manner that is independent of the shape and size of the specimen. Elastic material properties are defined in terms of stress and strain. In the linear range of material response, the stress is proportional to the strain (Figure 3.2). The ratio of stress to strain for the bar under tension is an elastic constant called the Young's modulus E. The negative ratio of the transverse strain to longitudinal strain is the Poisson's ratio ν.

Forces can be applied to a material in a manner that will cause distortion rather than elongation (Figure 3.3). A force applied tangent to a surface divided by the cross-sectional area is described as a shear stress τ. The distortion can be measured by the angle change produced. This is the shear strain γ

0-8493-0047-9/00/$0.00+$.50

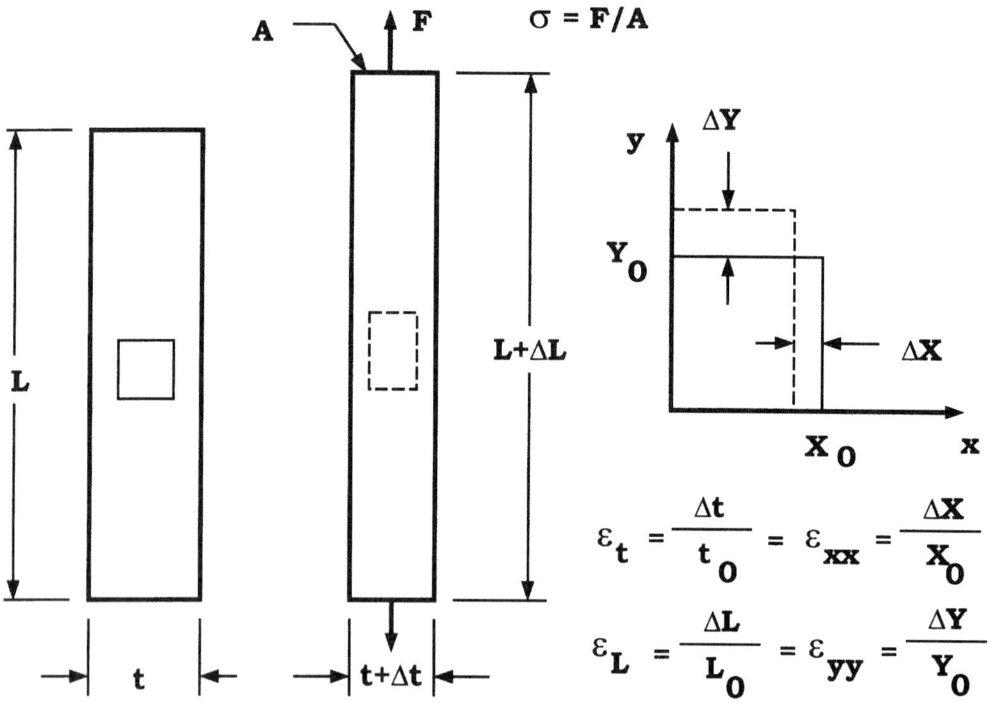

FIGURE 3.1 When a homogeneous isotropic bar is stretched by a uniaxial force, it elongates in the direction of the force and contracts perpendicular to the force. The relative elongation and contraction are defined as the longitudinal and transverse strains, respectively.

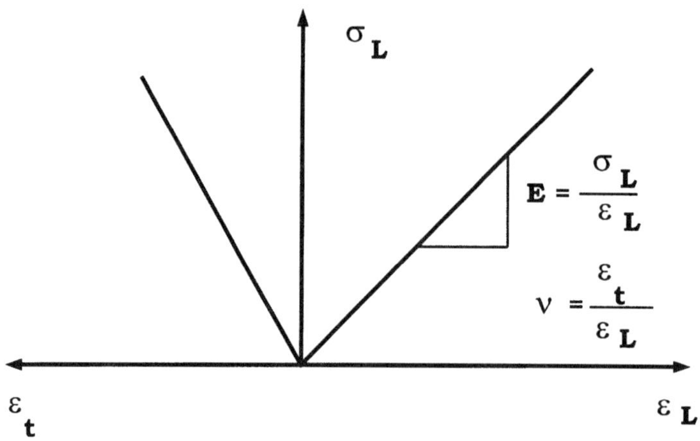

FIGURE 3.2 The uniaxial force shown in Figure 3.1 produces uniaxial stress in the bar. When the material response is linear, the slope of the stress vs. strain curve is the Young's modulus. The negative ratio of the transverse to longitudinal strain is the Poisson's ratio.

when the angle change is small. When the relation between shear stress and shear strain is linear, the ratio of the shear stress to shear strain is the shear modulus *G*.

Temperature change also induces strain. This is thermal expansion. In most materials, thermal strain increases with temperature. Over a limited temperature range, the relationship between thermal strain

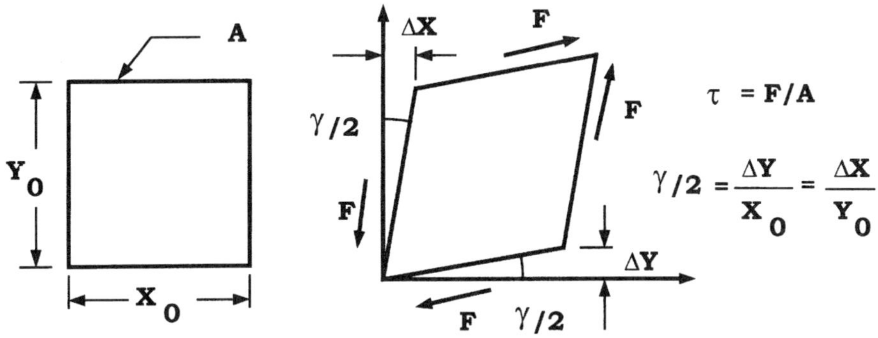

FIGURE 3.3 When a block of material is subjected to forces parallel to the sides as shown, it distorts. The force per unit area is the shear stress τ, and the angle change is the shear strain γ.

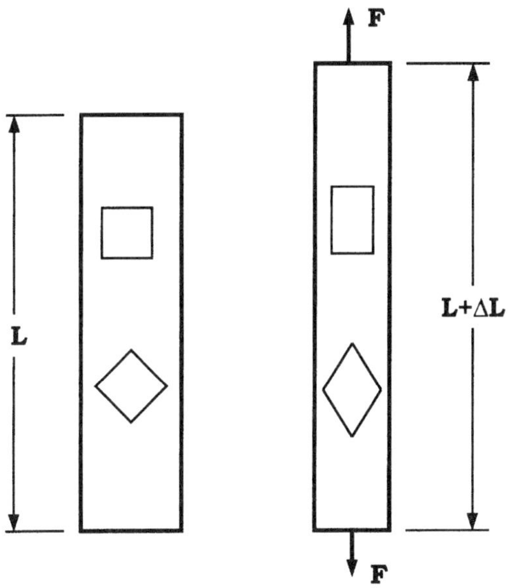

FIGURE 3.4 Some elements of a bar under uniaxial tension undergo elongation and contraction. These elements lie in principal directions. Other elements undergo distortion as well.

and temperature is linear. In this case, the strain divided by the temperature change is the thermal expansion coefficient α. In isotropic materials, thermal expansion only produces elongation strain, no shear strain.

Principal directions in a material are directions that undergo elongation but no shear. On any particular surface of a solid, there are always at least two principal directions in which the strain is purely elongation. This is seen if two squares are drawn on the bar under uniform tension (Figure 3.4). When the bar is stretched, the square aligned with the load is elongated, whereas the square at 45° is distorted (angles have changed) and elongated. If the principal directions are known, as with the bar under tension, then strain gages can be applied in these directions. If the principal directions are not known, such as near a hole or notch, in an anisotropic specimen, or in a structure with complicated geometry, then additional strain gages are needed to fully characterize the strain state.

The elastic and thermal properties can be combined to give Hooke's law, Equations 3.1 to 3.6.

$$\varepsilon_{xx} = \frac{\sigma_{xx}}{E} - \frac{\nu}{E}\left(\sigma_{yy} + \sigma_{zz}\right) \tag{3.1}$$

$$\varepsilon_{yy} = \frac{\sigma_{yy}}{E} - \frac{\nu}{E}\left(\sigma_{xx} + \sigma_{zz}\right) \tag{3.2}$$

$$\varepsilon_{xx} = \frac{\sigma_{zz}}{E} - \frac{\nu}{E}\left(\sigma_{xx} + \sigma_{yy}\right) \tag{3.3}$$

$$\gamma_{xy} = \frac{\tau_{xy}}{G} \tag{3.4}$$

$$\gamma_{xz} = \frac{\tau_{xz}}{G} \tag{3.5}$$

$$\gamma_{yz} = \frac{\tau_{yz}}{G} \tag{3.6}$$

Several types of sensors are used to measure strain. These include piezoresistive gages (foil or wire strain gages and semiconductor strain gages), piezoelectric gages (polyvinylidene fluoride (PVDF) film and quartz), fiber optic gages, birefringent films and materials, and Moiré grids. Each type of sensor requires its own specialized signal conditioning. Selection of the best strain sensor for a given measurement is based on many factors, including specimen geometry, temperature, strain rate, frequency, magnitude, as well as cost, complexity, accuracy, spatial resolution, time resolution, sensitivity to transverse strain, sensitivity to temperature, and complexity of signal conditioning. Table 3.1 describes typical characteristics of several sensors. Table 3.2 lists some manufacturers and the approximate cost of the sensors and associated signal conditioning electronics.

The data in Table 3.1 are to be taken as illustrative and by no means complete. The sensor description section describes only the type of sensor, not the many sizes and shapes. The longitudinal strain sensitivity is given as sensor output per unit longitudinal strain in the sensor direction. If the signal conditioning is included, the sensitivities can all be given in volts out per unit strain [3, 4], but this is a function of amplification and the quality of the signal conditioner. The temperature sensitivity is given as output change due to a temperature change. In many cases, higher strain resolution can be achieved, but resolving smaller strain is more difficult and may require vibration and thermal isolation. For the Moiré technique, the strain resolution is a function of the length of the viewing area. This technique can resolve a displacement of 100 nm (1/4 fringe order). This is divided by the viewing length to obtain the strain resolution. The spatial resolution corresponds to the gage length for most of the sensor types. The measurable strain range listed is the upper limit for the various sensors. Accuracy and reliability are usually reduced when sensors are used at the upper limit of their capability.

Manufacturers of the various sensors provide technical information that includes details of using the sensors, complete calibration or characterization data, and details of signal conditioning. The extensive technical notes and technical tips provided by Measurements Group, Inc. address such issues as thermal effects [5], transverse sensitivity corrections [6], soldering techniques [7], Rosettes [8], and gage fatigue [9]. Strain gage catalogs include information about gage materials, sizes, and selection. Manufacturers of other sensors provide similar information.

TABLE 3.1 Comparison of Strain Sensors

Description	Longitudinal strain sensitivity	Transverse strain sensitivity	Temperature sensitivity	Strain resolution	Spatial resolution	Time resolution	Measurable strain range
Piezoresistive constantan foil	$\Delta R/R/\Delta\varepsilon_L = 2.1$	$\Delta R/R/\Delta\varepsilon_t = <0.02$	$\Delta R/R/\Delta T = 2 \times 10^{-6}/°C$	<1 μstrain[a]	5–100 mm[b]	<1 μs[c]	0–3%
Annealed constantan foil[d]	$\Delta R/R/\Delta\varepsilon_L = 2.1$	$\Delta R/R/\Delta\varepsilon_t = <0.02$	$\Delta R/R/\Delta T = 2 \times 10^{-6}/°C$	<11 μstrain	5–100 mm	<1 μs	0–10%
Piezoresistive semiconductor	$\Delta R/R/\Delta\varepsilon_L = 150$	$\Delta R/R/\Delta\varepsilon_t = ???$	$\Delta R/R/\Delta T = 1.7 \times 10^{-3}/°C$	<0.1 μstrain	1–15 mm	<1 μs	0–0.1%
Piezoelectric PVDF	$\Delta Q/A/\Delta\varepsilon_L = 120\ nC/m^2/\mu\varepsilon$	$\Delta Q/A/\Delta\varepsilon_t = 60\ nC/m^2/\mu\varepsilon$	$\Delta Q/A/\Delta T = -27\ \mu C/m^2/°C$	1–10 μstrain	Gage size	<1 μs	0–30%
Piezoelectric quartz	$\Delta Q/A/\Delta\varepsilon_L = 150\ nC/m^2/\mu\varepsilon$ bonded to steel		$\Delta Q/A/\Delta T = 0$	<0.01 μstrain 20 mm gage	Gage size	<10 μs	0–0.1%
Fiber optic Fabry–Perot	2 to 1000 μstrain/V	Near zero		<1 μstrain	2–10 mm	<20 μs	
Birefringent Film	$K^e = 0.15$–0.002				0.5 mm[f]	<5 μs	0.05–5%
Moiré	1 fringe order/417 nm displ.	1 fringe order/417 nm displ.	Not defined	41.7 με over 10 mm	full field[g]	Limited by signal conditioning	0.005–5%

[a] With good signal conditioning.
[b] Equal to grid area.
[c] Gage response is within 100 ns. Most signal conditioning limits response time to far less than this.
[d] Annealed foil has a low yield stress and a large strain to failure. It also has hysteresis in the unload and a zero shift under cyclic load.
[e] This technique measures a difference in principal strains. $\varepsilon_2 - \varepsilon_1 = N\lambda/2tK$
[f] Approximately the film thickness.
[g] The spatial strain resolution depends on the strain level. This is a displacement measurement technique.

TABLE 3.2 Sources and Prices of Strain Sensors

Supplier	Address	Sensor Types	Sensor Cost	Signal Conditioning	Cost
Micro Measurements	P.O. Box 27777 Raleigh, NC 27611	Piezoresistive foil Birefringent film	From $5.00 From $10.00	Wheatstone bridge Polariscope	From $500 $5000 to 10,000
Texas Measurements	P.O. Box 2618 College Station, TX 77841	Piezoresistive foil and wire Load cells	From $5.00		
Omega Engineering	P.O. Box 4047 Stamford, CT 06907-0047	Piezoresistive foil	From $5.00	Strain meter Wheatstone bridge	From $550 From $2700
Dynasen, Inc.	20 Arnold Pl. Goleta, CA 93117	Piezoresistive foil Specialty gages for shock wave measurements Piezoelectric PVDF (calibrated)	From $55.00 From $55.00	2-Channel pulsed Wheatstone bridge Passive charge integrator	$5000 $250.00
Entran Sensors and Electronics	Entran Devices, Inc. 10 Washington Ave. Fairfield, NJ 07004-3877	Piezoresistive semiconductor	From $15.00		
Amp Inc.	Piezo Film Sensors P.O. Box 799 Valley Forge, PA 19482	Piezoelectric PVDF (not calibrated)	From $5.00		
Kistler Instrument Corp.	Amherst, NY 14228-2171	Piezoelectric quartz		Electrometer Charge amplifier	
F&S Inc.	Fiber and Sensor Technologies P.O. Box 11704, Blacksburg, VA 24062-1704	Fabry–Perot strain sensors	From $75	Electronics	From $3500.00
Photomechanics, Inc.	512 Princeton Dr. Vestal, NY 13850-2912	Moiré interferometer			$60,000

3.2 Principles of Operation of Strain Sensors

Piezoresistive Foil Gages

Piezoresistive foil and wire gages comprise a thin insulating substrate (usually polyimide film), a foil or wire grid (usually constantan) bonded to the substrate, lead wires to connect the grid to a resistance measuring circuit, and often an insulating encapsulation (another sheet of polyimide film) (Figure 3.5). The grid is laid out in a single direction so that strain will stretch the legs of the grid in the length direction. The gages are designed so that strain in the width or transverse direction separates the legs of the grid without straining them. This makes the gage sensitive to strain only along its length. There is always some sensitivity to transverse strain, and almost no sensitivity to shear strain. In most cases, the transverse sensitivity can be neglected.

When piezoresistive foil or wire strain gages are bonded to a specimen and the specimen is strained, the gage strains as well. The resistance change is related to the strain by a gage factor, Equation 3.7.

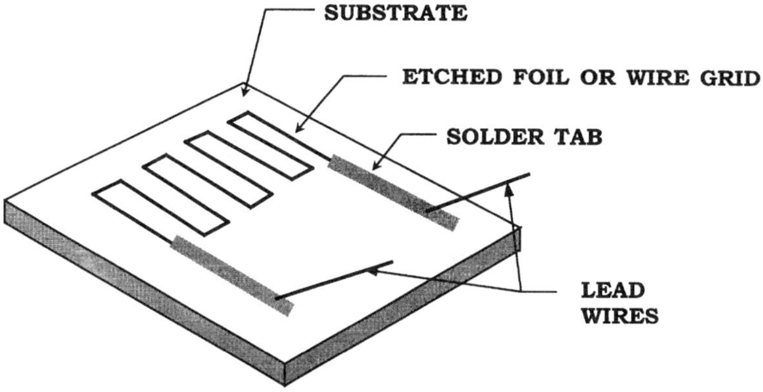

FIGURE 3.5 Gage construction of a foil or wire piezoresistive gage.

$$\frac{\Delta R}{R} = G_L \varepsilon_L \tag{3.7}$$

where $\Delta R/R$ = Relative resistance change
G = Gage factor
ε = Strain

These gages respond to the average strain over the area covered by the grid [10]. The resistance change is also sensitive to temperature. If the temperature changes during the measurement period, a correction must be made to distinguish the strain response from the thermal response. The gage response to longitudinal strain, transverse strain, and temperature change is given by Equation 3.8.

$$\frac{\Delta R}{R} = G_L \varepsilon_L + G_t \varepsilon_t + G_T \Delta T \tag{3.8}$$

where G_L, G_t, and G_T are the longitudinal, transverse, and temperature sensitivity, respectively. Micromeasurements, Inc. uses a different notation. Their gage data is provided as $G_L = F_G$, $G_t = K_t F_G$, $G_T = \beta_g$. When a strain gage is bonded to a specimen and the temperature changes, the strain used in Equation 3.8 is the total strain, thermal plus stress induced, as given by Equation 3.7.

The temperature contribution to gage output must be removed if the gages are used in tests where the temperature changes. A scheme referred to as self-temperature compensation (STC) can be used. This is accomplished by selecting a piezoresistive material whose thermal output can be canceled by the strain induced by thermal expansion of the test specimen. Gage manufacturers specify STC numbers that match the thermal expansion coefficients of common specimen materials.

Strain of piezoresistive materials produces a relative resistance change. The resistance change is the result of changes in resistivity and dimensional changes. Consider a single leg of the grid of a strain gage with a rectangular cross-section (Figure 3.6). The resistance is given by Equation 3.9.

$$R = \rho \frac{L}{A} \tag{3.9}$$

where R = Resistance
ρ = Resistivity
L = Length
A = Area of the cross-section

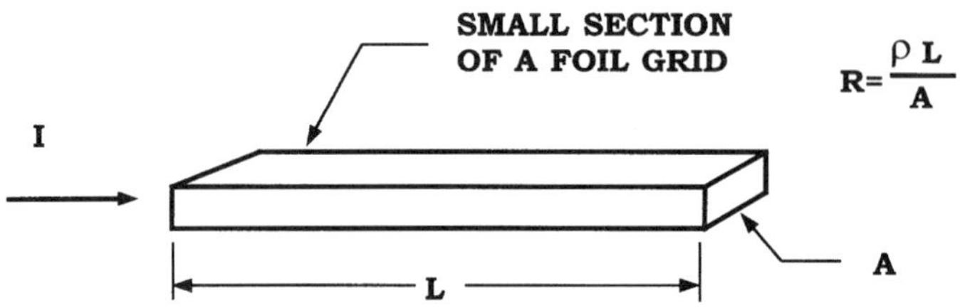

FIGURE 3.6 A single leg of a piezoresistive gage is used to explain the source of the relative resistance change that occurs in response to strain.

A small change in resistance is given by the first-order terms of a Taylor's series expansion, Equation 3.10.

$$\Delta R = \frac{\partial R}{\partial \rho}\Delta\rho + \frac{\partial R}{\partial L}\Delta L + \frac{\partial R}{\partial A}\Delta A \tag{3.10}$$

Differentiating Equation 3.9 to obtain each term of Equation 3.10 and then dividing by the initial resistance leads to Equation 3.11.

$$\frac{\Delta R}{R_0} = \frac{\Delta\rho}{\rho_0} + \frac{\Delta L}{L_0} - \frac{\Delta A}{A_0} \tag{3.11}$$

The relative resistance change is due to a change in resistivity, a change in length strain, and a change in area strain.

The strain gage is a composite material. The metal in a strain gage is like a metal fiber in a polymer matrix. When the polymer matrix is deformed, the metal is dragged along in the length direction; but in the width and thickness directions, the strain is not passed to the metal. This results in a stress state called uniaxial stress. This state was discussed in the examples above. The mathematical details involve an inclusion problem [11, 12]. Accepting that the stress state is uniaxial, the relationship between the area change and the length change in Equation 3.11 is found from the Poisson's ratio. The area strain is the sum of the width and thickness strain, Equation 3.12.

$$\frac{\Delta A}{A_0} = \frac{\Delta w}{w_0} + \frac{\Delta t}{t_0} \tag{3.12}$$

The definition of the Poisson's ratio gives Equation 3.13.

$$\frac{\Delta A}{A_0} = -2v\frac{\Delta L}{L_0} \tag{3.13}$$

where $\Delta w/w$ = width strain and $\Delta t/t$ = thickness strain. Substitution of Equation 3.13 into Equation 3.11 gives Equation 3.14 for the relative resistance change.

$$\frac{\Delta R}{R_0} = \frac{\Delta\rho}{\rho_0} + \frac{\Delta L}{L_0}(1+2v) \tag{3.14}$$

The relative resistivity changes in response to stress. The resistivity is a second-order tensor [13], and the contribution to the overall resistance change can be found in terms of strain using the elastic constitutive law [14]. The results lead to an elastic gage factor just over 2 for constantan gages. If the strain is large, the foil or wire in the gage will experience plastic deformation. When the deformation is plastic, the resistivity change is negligible and the dimensional change dominates. In this case, Poisson's ratio is 0.5 and the gage factor is 2. This effect is utilized in manufacturing gages for measuring strains larger than 1.5%. In this case, annealed foil is used. The annealed foil undergoes plastic deformation without failure. These gages are capable of measuring strain in excess of 10%. When metals undergo plastic deformation, they do not unload to the initial strain. This shows up as hysteresis in the gage response, that is, on unload, the resistance does not return to its initial value.

Foil and wire strain gages can be obtained in several configurations. They can be constructed with different backing materials, and left open faced or fully encapsulated. Backing materials include polyimide and glass fiber-reinforced phenolic resin. Gages can be obtained with solder tabs for attaching lead wires, or with lead wires attached. They come in many sizes, and in multiple gage configurations called rosettes.

Strain gages are mounted to test specimens with adhesives using a procedure that is suitable for bonding most types of strain sensors. This is accomplished in a step-by-step procedure [15] that starts with surface preparation. An overview of the procedure is briefly described. To successfully mount strain gages, the surface is first degreased. The surface is abraded with a fine emery cloth or 400 grit paper to remove any loose paint, rust, or deposits. Gage layout lines are drawn (not scribed) on the surface in a cross pattern with pen or pencil, one line in the grid direction and one in the transverse direction. The surface is then cleaned with isopropyl alcohol. This can be done with an ultrasonic cleaner or with wipes. If wiped, the paper or gauze wipe should be folded and a small amount of alcohol applied. The surface should be wiped with one pass and the wipe discarded. This should be repeated, wiping in the other direction. The final step is to neutralize the surface, bringing the alkalinity to a pH of 7 to 7.5. A surface neutralizer is available from most adhesive suppliers. The final step is to apply the gage.

Gage application is accomplished with cellophane tape, quick-set glue, and a catalyst. The gage is placed on a clean glass or plastic surface with bonding side down, using tweezers. (Never touch the gage. Oils from skin prevent proper adhesion.) The gage is then taped down with a 100 mm piece of cellophane tape. The tape is then peeled up with the gage attached. The gage can now be taped onto its desired location on the test specimen. Once the gage has been properly aligned, the tape is peeled back from one side, lifting the gage from the surface. The tape should remain adhered to the surface about 1 cm from the gage. Note that one side of the tape is still attached to the specimen so that the gage can be easily returned to its desired position. A thin coating of catalyst is applied to the exposed gage surface. A drop of glue is placed at the joint of the tape and the specimen. Holding the tape at about a 30° angle from the surface, the tape can be slowly wiped down onto the surface. This moves the glue line forward. After the glue line has passed the gage, the gage should be pressed in place and held for approximately 1 min. The tape can now be peeled back to expose the gage, and lead wires can be attached.

The relative resistance change of piezoresistive gages is usually measured using a Wheatstone bridge [16]. This allows a small change of resistance to be measured relative to an initial zero value, rather than relative to a large resistance value, with a corresponding increase in sensitivity and resolution. The Wheatstone bridge is a combination of four resistors and a voltage source (Figure 3.7). One to four of the resistors in the bridge can be strain gages. The output of the bridge is the difference between the voltage at points B and D. Paths ABC and ADC are voltage dividers so that V_B and V_D are given by Equations 3.15a and b.

$$V_B = V_{in} \frac{R_2}{R_1 + R_2} \tag{3.15a}$$

$$V_D = V_{in} \frac{R_3}{R_3 + R_4} \tag{3.15b}$$

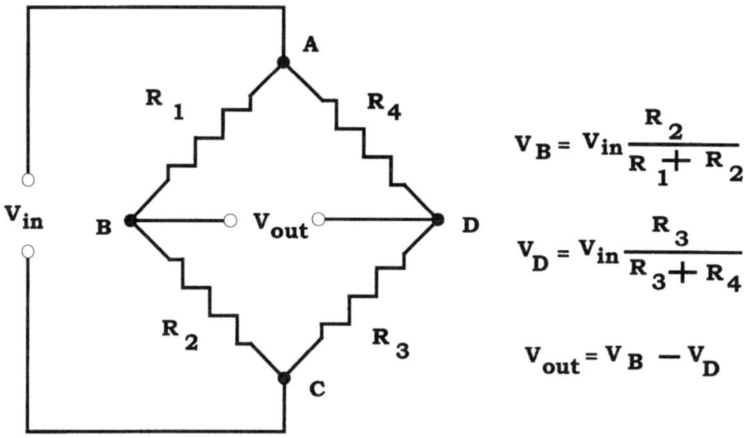

FIGURE 3.7 The Wheatstone bridge is used to measure the relative resistance change of piezoresistive strain gages.

The bridge output, Equation 3.16, is zero when the balance condition, Equation 3.17, is met.

$$V_0 = V_B - V_D \tag{3.16}$$

$$R_1 R_3 = R_2 R_4 \tag{3.17}$$

Wheatstone bridge signal conditioners are constructed with a way to "balance" the bridge by adjusting the ratio of the resistances so that the bridge output is initially zero.

The balance condition is no longer met if the resistance values undergo small changes ΔR_1, ΔR_2, ΔR_3, ΔR_4. If the values $R_1 + \Delta R_1$, etc. are substituted into Equation 3.15, the results substituted into Equation 3.16, condition (3.17) used, and the higher order terms neglected, the result is Equation 3.18 for the bridge output.

$$V_{out} = V_{in} \frac{R_1 R_3}{(R_1 + R_2)(R_3 + R_4)} \left(-\frac{\Delta R_1}{R_1} + \frac{\Delta R_2}{R_2} - \frac{\Delta R_3}{R_3} + \frac{\Delta R_4}{R_4} \right) \tag{3.18}$$

The Wheatstone bridge can be used to directly cancel the effect of thermal drift. If R_1 is a strain gage bonded to a specimen and R_2 is a strain gage held onto a specimen with heat sink compound (a thermally conductive grease available at any electronics store), then R_1 will respond to strain plus temperature, and R_2 will only respond to temperature. Since the bridge subtracts the output of R_1 from that of R_2, the temperature effect cancels.

The sensitivity of a measuring system is the output per unit change in the quantity to be measured. If the resistance change is from a strain gage, the sensitivity of the Wheatstone bridge system is proportional to the input voltage. Increasing the voltage increases the sensitivity. There is a practical limitation to increasing the voltage to large values. The power dissipated (heat) in the gage is $P = I^2R$, where I, the current through the gage, can be found from the input voltage and the bridge resistances. This heat must go somewhere or the temperature of the gage will continuously rise and the resistance will change due to heating. If the gage is mounted on a good thermal conductor, more power can be conducted away than if the gage is mounted on an insulator. The specimen must act as a heat sink.

Heat sinking ability is proportional to the thermal conductivity of the specimen material. A reasonable temperature gradient to allow the gage to induce in a material is 40°C per meter (about 1°C per 25 mm).

For thick specimens (thickness several times the largest gage dimension), this can be conducted away to the grips or convected to the surrounding atmosphere. If the four bridge resistances are approximately equal, the power to the gage in terms of the bridge voltage is given by Equation 3.19.

$$P_g = \frac{V_{in}^2}{4R} \tag{3.19}$$

The power per unit grid area, A_g, or power density to the gage can be equated to the thermal conductivity of the specimen and the allowable temperature gradient in the specimen by Equation 3.20.

$$\frac{P_g}{A_g} = \frac{V_{in}^2}{4RA_g} = K\nabla T \tag{3.20}$$

Thermal conductivities of most materials can be found in tables or on the Web. Some typical values are Al: $K = 204$ W m^{-1} °C^{-1}, steel: $K = 20$ to 70 W m^{-1} °C^{-1}, glass: $K = 0.78$ W m^{-1} °C^{-1} [17].

The acceptable bridge voltage can be calculated from Equation 3.21.

$$V_{in} = \sqrt{K\nabla T 4RA_g} \tag{3.21}$$

A sample calculation shows that for a 0.010 m × 0.010 m 120 Ω grid bonded to a thick piece of aluminum with a thermal conductivity of 204 W m^{-1} °C^{-1} and an acceptable temperature gradient of 40°C per meter, the maximum bridge voltage is 19 V. If thin specimens are used, the allowable temperature gradient will be smaller. If smaller gages are used for better spatial resolution, the bridge excitation voltage must be reduced with a corresponding reduction in sensitivity.

A considerably higher bridge voltage can be used if the bridge voltage is pulsed for a short duration. This dissipates substantially less energy in the gage and thus increases the sensitivity by a factor of 10 to 100. Wheatstone bridge pulse power supplies with variable pulse width from 10 μs and excitation of 350 V are commercially available [18].

The strain measurement required is often in a complex loading situation where the directions of principal strain are not known. In this case, three strain gages must be bonded to the test specimen at three angles. This is called a strain rosette. The angle between the rosette and the principal directions, as well as the magnitude of the principal strains, can be determined from the output of the rosette gages. This is most easily accomplished using the construct of a Mohr's circle (Figure 3.8).

A common rosette is the 0–45–90° pattern. The rosette is bonded to the specimen with relative rotations of 0°, 45°, and 90°. These will be referred to as the x, x', and y directions. The principal directions are labeled the 1 and 2 directions. The unknown angle between the x direction and the 1 direction is labeled θ. The Mohr's circle is drawn with the elongational strain on the horizontal axis and the shear strain on the vertical axis. The center of the circle is labeled C and the radius R. The principal directions correspond to zero shear strain. The principal values are given by Equations 3.22 and 3.23.

$$\varepsilon_1 = C + R \tag{3.22}$$

$$\varepsilon_2 = C - R \tag{3.23}$$

A rotation through an angle 2θ on the Mohr's circle corresponds to a rotation of the rosette of θ relative to the principal directions. The center of the circle is given by Equation 3.24 and the output of the strain gages is given by Equations 3.25 to 3.27.

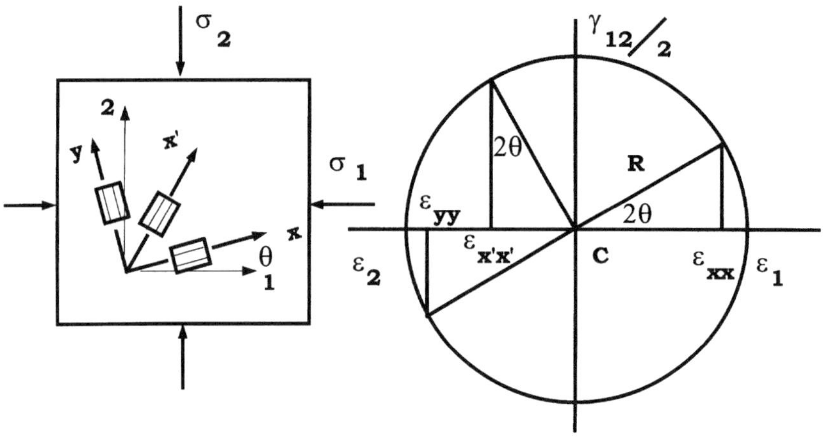

FIGURE 3.8 A three-element rosette is used to measure strain when the principal directions are not known. The Mohr's circle is used to find the principal directions and the principal strain values.

$$C = \frac{\varepsilon_{xx} + \varepsilon_{yy}}{2} \tag{3.24}$$

$$\varepsilon_{xx} - C = R\cos 2\theta \tag{3.25}$$

$$\varepsilon_{x'x'} - C = -R\sin 2\theta \tag{3.26}$$

$$\varepsilon_{yy} - C = -R\cos 2\theta \tag{3.27}$$

Dividing Equation 3.25 by Equation 3.26 leads to θ and then to R, Equations 3.28 and 3.29.

$$R^2 = \left(\varepsilon_{xx} - C\right)^2 + \left(\varepsilon_{x'x'} - C\right)^2 \tag{3.28}$$

$$\tan 2\theta = \frac{C - \varepsilon_{x'x'}}{\varepsilon_{xx} - C} \tag{3.29}$$

The principal directions and principal strain values have been found from the output of the three rosette gages.

Piezoresistive Semiconducting Gages

Piezoresistive semiconductor strain gages, like piezoresistive foil and wire gages, undergo a resistance change in response to strain, but with nearly an order of magnitude larger gage factor [19]. The coupling between resistance change and temperature is very large, so these gages have to be temperature compensated. The change of resistance with a small temperature change can be an order of magnitude larger than that induced by strain. Semiconductor strain gages are typically used to manufacture transducers such as load cells. They are fragile and require great care in their application.

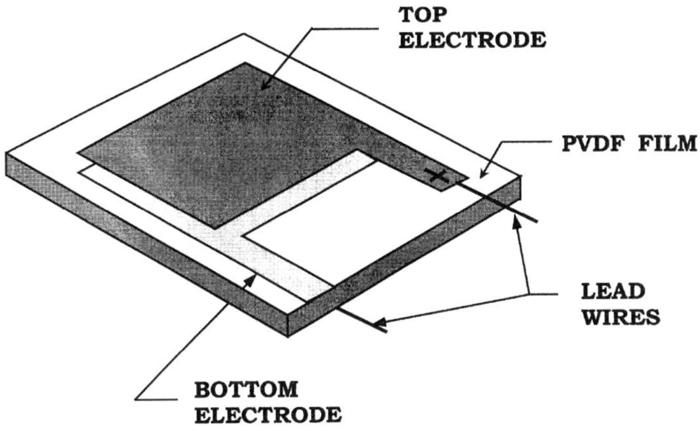

FIGURE 3.9 Typical gage construction of a piezoelectric gage.

Piezoelectric Gages

Piezoelectric strain gages are, effectively, parallel plate capacitors whose dielectric changes polarization in response to strain [14]. When the polarization changes, a charge proportional to the strain is produced on the electrodes. PVDF film strain gages are inexpensive, but not very accurate and subject to depoling by moderate temperature. They make good sensors for dynamic measurements such as frequency and logarithm decrement, but not for quantitative measurements of strain. When used for quasistatic measurements, the charge tends to drain through the measuring instrument. This causes the signal to decay with a time constant dependent on the input impedance of the measuring instrument. Quartz gages are very accurate, but also lose charge through the measuring instrument. Time constants can be relatively long (seconds to hours) with electrometers or charge amplifiers.

The PVDF gage consists of a thin piezoelectric film with metal electrodes (Figure 3.9). Lead wires connect the electrodes to a charge measuring circuit. Gages can be obtained with the electrodes encapsulated between insulating layers of polyimide.

The gage output can be described in terms of a net dipole moment per unit volume. If the net dipole moment is the total charge, Q, on the electrodes multiplied by spacing, d, between the electrodes, then the polarization is given by Equation 3.30.

$$P = \frac{Qd}{V} \tag{3.30}$$

From Equation 3.30, it is seen that the polarization P (approximately equal to the electric displacement D) is the charge per unit electrode area (Figure 3.10).

A Taylor's series expansion of Equation 3.30 gives Equation 3.31.

$$\Delta P = \frac{\partial P}{\partial V}\Delta V + \frac{\partial P}{\partial\left(Qd\right)}\Delta Qd \tag{3.31}$$

Which, after differentiating Equation 3.30 and substituting becomes Equation 3.31.

$$\Delta P = \frac{\Delta Qd}{V_0} - \frac{\Delta V}{V_0}P_0 \tag{3.32}$$

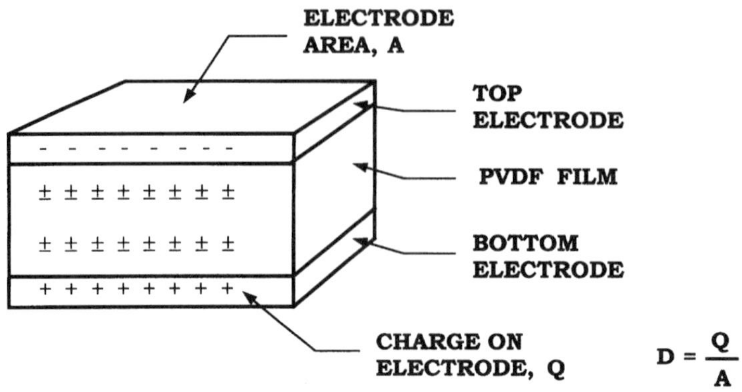

FIGURE 3.10 A representative cross-section of a piezoelectric material formed into a parallel plate capacitor. The piezoelectric material is polarized. This results in charge on the electrodes. When the material is strained, the polarization changes and charge flows.

For PVDF film, the second term in Equation 3.32 dominates. The output is proportional to the remanent polarization P_0. The remanent polarization slowly decays with time, has a strong dependence on temperature, and decays rapidly at temperatures around 50°C. This makes accuracy a problem. If the sensors are kept at low temperature, accuracy can be maintained within ±3%.

Strain sensors can also be constructed from piezoelectric ceramics like lead zirconate titanate (PZT) or barium titanate. Ceramics are brittle and can be depoled by strain so should only be used at strains less than 200 microstrain. PZT loses some of its polarization with time and thus has accuracy problems, but remains polar to temperatures of 150°C or higher. "Hard" PZT (usually iron doped) is the best composition for polarization stability and low hysteresis. Quartz has the best accuracy. It is not polar, but polarization is induced by strain. Quartz has excellent resolution and accuracy over a broad temperature range but is limited to low strain levels. It is also brittle, so is limited to small strain.

Two circuits are commonly used for piezoelectric signal conditioning: the electrometer and the charge amplifier (Figure 3.11). In the electrometer circuit, the piezoelectric sensor is connected to a capacitor with a capacitance value *C*, at least 1000 times that of the sensor C_g. There is always some resistance in

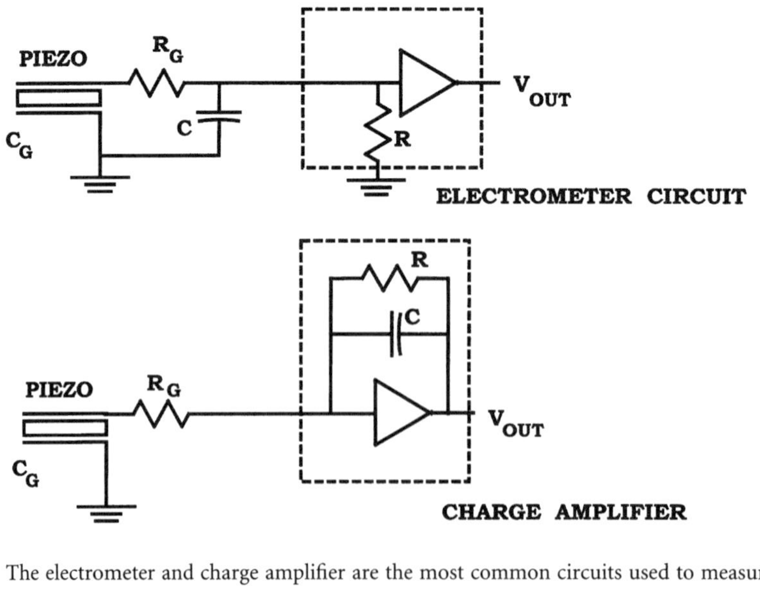

FIGURE 3.11 The electrometer and charge amplifier are the most common circuits used to measure charge from piezoelectric transducers.

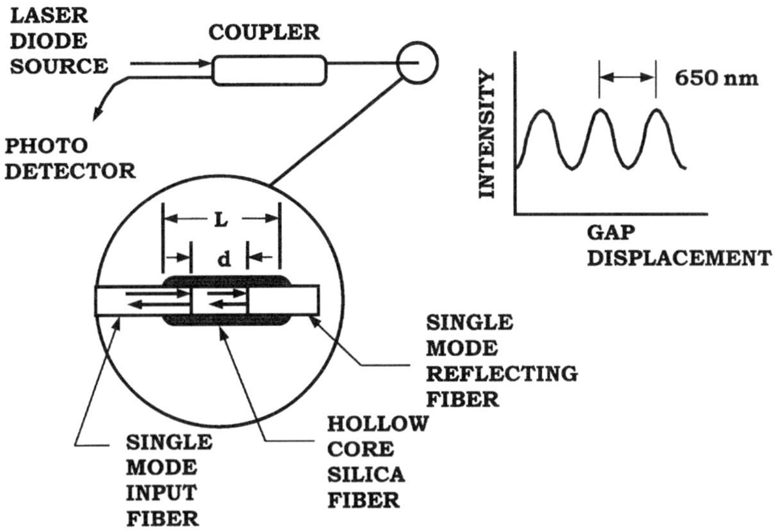

FIGURE 3.12 A schematic of the Fabry–Perot fiber optic strain gage. When the cavity elongates, alternating constructive and destructive interference occur.

the cable that connects the sensor to the capacitor. The circuit is simply two capacitors in parallel connected by a resistance. The charge equilibrates with a time constant given by $R_g C_g$. This time constant limits the fastest risetime that can be resolved to about 50 ns, effectively instantaneous for most applications. The charge is measured by measuring the voltage on the capacitor, then using Equation 3.33.

$$Q = CV \tag{3.33}$$

The difficulty is that measuring devices drain the charge, causing a time decay with a time constant *RC*. This causes the signal to be lost rapidly if conventional op amps are used. FET input op amps have a very high input impedance and can extend this time constant to many hours. The charge amplifier is another circuit used to measure charge. This is usually an FET input op amp with a capacitor feedback. This does not really amplify charge, but produces a voltage proportional to the input charge. Again, the time constant can be many hours, allowing use of piezoelectric sensors for near static measurements.

High input impedance electrometer and charge amplifier signal conditioners for near static measurements are commercially available [20] as well as low-cost capacitive terminators for high-frequency (high kilohertz to megahertz) measurements [18]. An advantage of piezoelectric sensors is that they are active sensors that do not require any external energy source.

Fiber Optic Strain Gages

Fiber optic strain gages are miniature interferometers [21, 22]. Many commercially available sensors are based on the Fabry–Perot interferometer. The Fabry–Perot interferometer measures the change in the size of a very small cavity.

Fabry–Perot strain sensors (Figure 3.12) comprise a laser light source, single-mode optical fibers, a coupler (the fiber optic equivalent of a beam splitter), a cavity that senses strain, and a photodetector. Light leaves the laser diode. It passes down the fiber, through the coupler, and to the cavity. The end of the fiber is the equivalent of a partially silvered mirror. Some of the light is reflected back up the fiber and some is transmitted. The transmitted light crosses the cavity and then is reflected from the opposite end back into the fiber where it recombines with the first reflected beam. The two beams have a phase difference related to twice the cavity length. The recombined beam passes through the coupler to the photodetector. If the two reflected beams are in phase, there will be constructive interference. If the two beams are out of phase, there will be destructive interference. The cavity is bonded to a specimen. When the specimen is strained, the cavity stretches. This results in a phase change of the cavity beam, causing a cycling between constructive and destructive interference. For a 1.3 μm light source, each peak in output corresponds to a 650 nm gap displacement. The gap displacement divided by the gap length gives the strain. The output is continuous between peaks so that a 3 mm gage can resolve 1 μstrain.

Birefringent Film Strain Sensing

Birefringent film strain sensors give a full field measurement of strain. A nice demonstration of this effect can be achieved with two sheets of inexpensive Polaroid film, a 6 mm thick, 25 mm × 200 mm bar of Plexiglas (polymethylmethacrylate or PMMA), and an overhead projector. Place the two Polaroid sheets at 90° to one another so that the light is blocked. Place the PMMA between the Polaroid sheets. Apply a bending moment to the bar and color fringes will appear. Birefringent materials have a different speed of light in different directions. This means that if light is polarized in a particular direction and passed through a birefringent specimen, if the fast direction is aligned with the electric field vector, the light passes through faster than if the slow direction is aligned with the electric field vector. This effect can be used to produce optical interference. In some materials, birefringence is induced by strain. The fast and slow directions correspond to the directions of principal strain, and the amount of birefringence corresponds to the magnitude of the strain. One component of the electric field vector travels through the specimen faster than the other. They emerge with a phase difference. This changes the relative amplitude and thus rotates the polarization of the light. If there is no birefringence, no light passes through the second polarizer. As the birefringence increases with strain, light passes through. As it further increases, the polarization rotation will be a full 180° and again no light will pass through. This produces a fringe that corresponds to a constant difference in principal strains. The difference in principal strains is given by Equation 3.34.

$$\varepsilon_2 - \varepsilon_1 = \frac{N\lambda}{tK} \tag{3.34}$$

where ε_1, ε_2 = Principal strains
N = Fringe order
λ = Wavelength
t = Specimen thickness
K = Strain-optical coefficient of the photoelastic material

A similar technique can be used with a birefringent plastic film with a silvered backing laminated to the surface of a specimen. Polarized light is passed through the film; it reflects from the backing, passes back through the film, and through the second polarizer. In this case, because light passes twice through the film, the equation governing the difference in principal strains is Equation 3.35.

$$\varepsilon_2 - \varepsilon_1 = \frac{N\lambda}{2tK} \tag{3.35}$$

If the polarizers align with principal strain directions, no birefringence is observed. Rotation of both polarizers allows the principal directions to be found at various locations on the test specimen. If a full view of the fringes is desired, quarter wave plates are used (Figure 3.13). In this arrangement, light is passed through the first polarizer, resulting in plane polarization; through the quarter wave plate, resulting in circular polarization; through the test specimen, resulting in phase changes; through the second quarter wave plate to return to plane polarization; and then through the final polarizer.

The optical systems for viewing birefringence are commercially available as "Polariscopes" [23]. Optical components to construct custom systems are available from many optical components suppliers.

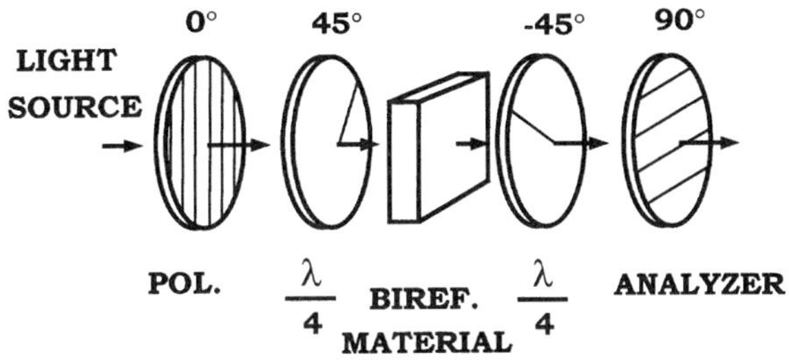

FIGURE 3.13 A schematic of the polariscope, a system for measuring birefringence. This technique gives a full field measure of the difference in principal strains.

Moiré Strain Sensing

Moiré interference is another technique that gives a full field measurement, but it measures displacement rather than strain. The strain field must be computed from the displacement field. This technique is based on the interference obtained when two transparent plates are covered with equally spaced stripes. If the plates are held over one another, they can be aligned so that no light will pass through or so that all light will pass through. If one of the plates is stretched, the spacing of the lines is wider on the stretched plate. Now, if one plate is placed over the other, in some regions light will pass through and in some regions it will not (Figure 3.14). The dark and light bands produced give information about the displacement field.

Moiré is defined as a series of broad dark and light patterns formed by the superposition of two regular gratings [24]. The dark or light regions are called fringes. Examples of pure extension and pure rotation are shown. In both cases, some of the light that would emerge from the first grating is obstructed by the superimposed grating. At the centers of the dark fringes, the bar of one grating covers the space of the other and no light comes through. The emergent intensity, I, is zero. Proceeding from there toward the next dark fringe, the amount of obstruction diminishes linearly and the amount of light increases linearly until the bar of one grating falls above the bar of the other. There, the maximum amount of light passes through the gratings.

Both geometric interference and optical interference are used. This discussion is restricted to geometric interference. Geometric moiré takes advantage of the interference of two gratings to determine displacements and rotations in the plane of view. In-plane moiré is typically conducted with two gratings, one applied to the specimen (specimen grating) and the other put in contact with the specimen grating (reference grating). When the specimen is strained, interference patterns or fringes occur. N is the moiré fringe order. Each fringe corresponds to an increase or decrease of specimen displacement by one grating pitch. The relationship between displacement and fringes is $\delta = gN$, where δ is component of the displacement perpendicular to the reference grating lines, g is reference grating pitch, and N is the fringe order.

For convenience, a zero-order fringe is designated assuming the displacement there is zero. With the reference grating at 0° and 90°, the fringe orders N_x and N_y are obtained. The displacements in x, y directions are then obtained from Equations 3.36 and 3.37.

$$u_x(x, y) = gN_x(x, y) \tag{3.36}$$

$$u_y(x, y) = gN_y(x, y) \tag{3.37}$$

Differentiation of Equations 3.36 and 3.37 gives the strains, Equations 3.38 through 3.40.

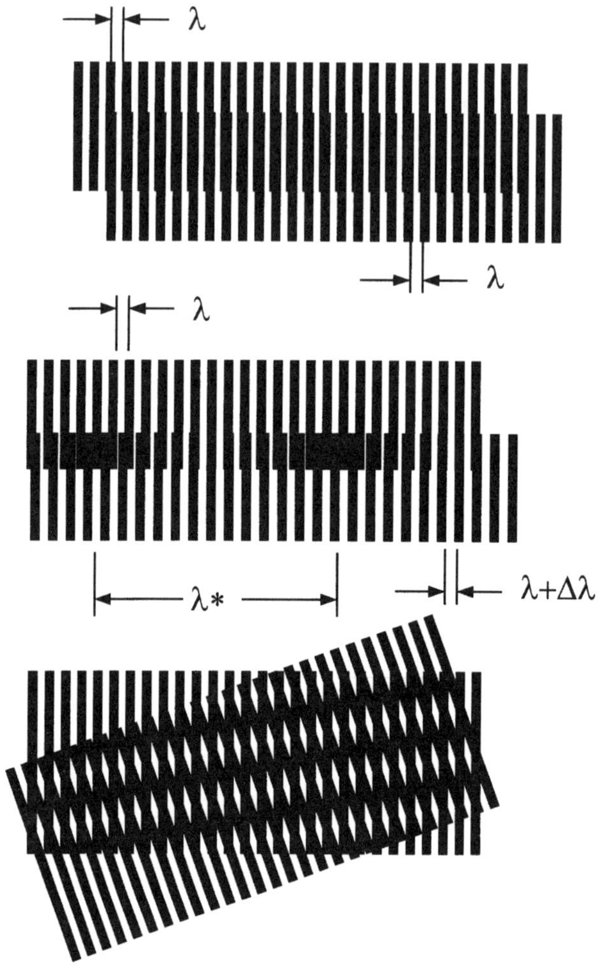

FIGURE 3.14 A demonstration of moiré fringes formed by overlapping gratings. The fringes are the result of stretching and relative rotation of the gratings. The fringe patterns are used to determine displacement fields.

$$\varepsilon_x = \frac{\partial u_x}{\partial x} = g\frac{\partial N_x}{\partial x} \tag{3.38}$$

$$\varepsilon_{xy} = \frac{1}{2}\left(\frac{\partial u_y}{\partial x} + \frac{\partial u_x}{\partial y}\right) = \frac{1}{2}\left(g\frac{\partial N_y}{\partial x} + g\frac{\partial N_x}{\partial y}\right) \tag{3.39}$$

$$\varepsilon_y = \frac{\partial u_y}{\partial y} = g\frac{\partial N_y}{\partial y} \tag{3.40}$$

In most cases, the sensitivity of geometric moiré is not adequate for determination of strain distributions. Strain analysis should be conducted with high-sensitivity measurement of displacement using moiré interferometry [24, 25]. Moiré interferometers are commercially available [26]. Out-of-plane measurement can be conducted with one grating (the reference grating). The reference grating is made to interfere with either its reflection or its shadow [27, 28].

References

1. N. E. Dowling, *Mechanical Behavior of Materials,* Englewood Cliffs, NJ: Prentice-Hall, 1993, 99-108.
2. R. C. Craig, *Mechanics of Materials,* New York: John Wiley & Sons, 1996.
3. A. Vengsarkar, Fiber optic sensors: a comparative evaluation, *The Photonics Design and Applications Handbook,* 1991, 114-116.
4. H. U. Eisenhut, Force measurement on presses with piezoelectric strain transducers and their static calibration up to 5 MN, *New Industrial Applications of the Piezoelectric Measurement Principle,* July 1992, 1-16.
5. TN-501-4, Strain Gauge Temperature Effects, Measurements Group, Inc., Raleigh, NC 27611.
6. TN-509, Transverse Sensitivity Errors, Measurements Group, Inc., Raleigh, NC 27611.
7. TT-609, Soldering Techniques, Measurements Group, Inc., Raleigh, NC 27611.
8. TN-515, Strain Gage Rosettes, Measurements Group, Inc., Raleigh, NC 27611.
9. TN-08-1, Fatigue of Strain Gages, Measurements Group, Inc., Raleigh, NC 27611.
10. C. C. Perry and H. R. Lissner, *The Strain Gage Primer,* New York: McGraw-Hill, 1962.
11. J. B. Aidun and Y. M. Gupta, Analysis of Lugrangian gauge measurements of simple and nonsimple plain waves, *J. Appl. Phys.,* 69, 6998-7014, 1991.
12. Y. M. Gupta, Stress measurement using piezoresistance gauges: modeling the gauge as an elastic-plastic inclusion, *J. Appl. Phys.,* 54, 6256-6266, 1983.
13. D. Y. Chen, Y. M. Gupta, and M. H. Miles, Quasistatic experiments to determine material constants for the piezoresistance foils used in shock wave experiments, *J. Appl. Phys.,* 55, 3984, 1984.
14. C. S. Lynch, Strain compensated thin film stress gauges for stress wave measurements in the presence of lateral strain, *Rev. Sci. Instrum.,* 66(11), 1-8, 1995.
15. B-129-7 M-Line Accessories Instruction Bulletin, Measurements Group, Inc., Raleigh, NC 27611.
16. J. W. Dally, W. F. Riley, and K. G. McConnell, *Instrumentation for Engineering Measurements,* 2nd ed., New York: John Wiley & Sons, 1993.
17. J. P. Holman, *Heat Transfer,* 7th ed., New York: McGraw Hill, 1990.
18. Dynasen, Inc. 20 Arnold Pl., Goleta, CA 93117.
19. M. Dean (ed.) and R. D. Douglas (assoc. ed.), *Semiconductor and Conventional Strain Gages,* New York: Academic Press, 1962.
20. Kistler, Instruments Corp., Amhurst, NY, 14228-2171.
21. J. S. Sirkis, Unified approach to phase strain temperature models for smart structure interferometric optical fiber sensors. 1. Development, *Opt. Eng.,* 32(4), 752-761, 1993.
22. J. S. Sirkis, Unified approach to phase strain temperature models for smart structure interferometric optical fiber sensors. 2. Applications, *Optical Engineering,* 32(4), 762-773, 1993.
23. Photoelastic Division, Measurements Group, Inc., P.O. Box 27777, Raleigh, NC 27611.
24. T. Valis, D. Hogg, and R. M. Measures, Composite material embedded fiber-optic Fabry-Perot strain rosette, *SPIE,* 1370, 154-161, 1990.
25. D. Post, B. Han, and P. Lfju, *High Sensitivity Moiré,* New York: Springer-Verlag, 1994.
26. V. J. Parks, Geometric Moiré, *Handbook on Experimental Mechanics,* A. S. Kobayashi, Ed., VCH Publisher, Inc., 1993.
27. Photomechanics, Inc. 512 Princeton Dr. Vestal, NY, 13850-2912.
28. T. Y. Kao and F. P. Chiang, Family of grating techniques of slope and curvature measurements for static and dynamic flexure of plates, *Opt. Eng.,* 21, 721-742, 1982.
29. D. R. Andrews, Shadow moiré contouring of impact craters, *Opt. Eng.,* 21, 650-654, 1982.

4

Force Measurement

M. A. Elbestawi
McMaster University

Force, which is a vector quantity, can be defined as an action that will cause an acceleration or a certain reaction of a body. This chapter will outline the methods that can be employed to determine the magnitude of these forces.

4.1 General Considerations

The determination or measurement of forces must yield to the following considerations: if the forces acting on a body do not produce any acceleration, they must form a *system of forces in equilibrium*. The system is then considered to be in static equilibrium. The forces experienced by a body can be classified into two categories: internal, where the individual particles of a body act on each other, and external otherwise. If a body is supported by other bodies while subject to the action of forces, deformations and/or displacements will be produced at the points of support or contact. The internal forces will be distributed throughout the body until equilibrium is established, and then the body is said to be in a state of tension, compression, or shear. In considering a body at a definite section, it is evident that all the internal forces act in pairs, the two forces being equal and opposite, whereas the external forces act singly.

4.2 Hooke's Law

The basis for force measurement results from the physical behavior of a body under external forces. Therefore, it is useful to review briefly the mechanical behavior of materials. When a metal is loaded in uniaxial tension, uniaxial compression, or simple shear (Figure 4.1), it will behave elastically until a critical value of normal stress (S) or shear stress (τ) is reached, and then it will deform plastically [1]. In the elastic region, the atoms are temporarily displaced but return to their equilibrium positions when the load is removed. Stress (S or τ) and strain (e or γ) in the elastic region are defined as indicated in Figure 4.2.

$$\nu = -\frac{e_2}{e_1} \tag{4.1}$$

0-8493-0047-9/00/$0.00+$.50

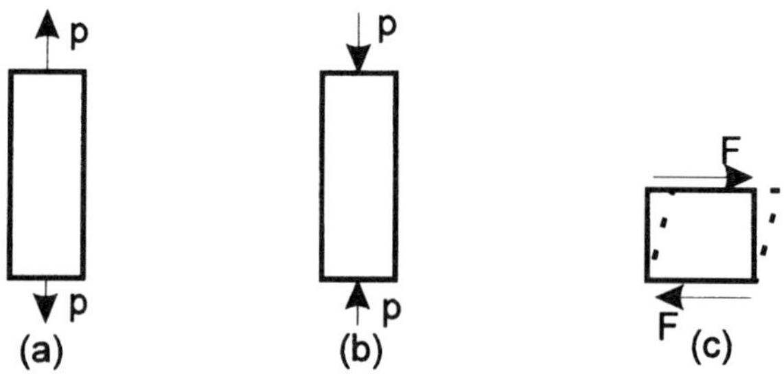

FIGURE 4.1 When a metal is loaded in uniaxial tension (*a*) uniaxial compression (*b*), or simple shear (*c*), it will behave elastically until a critical value of normal stress or shear stress is reached.

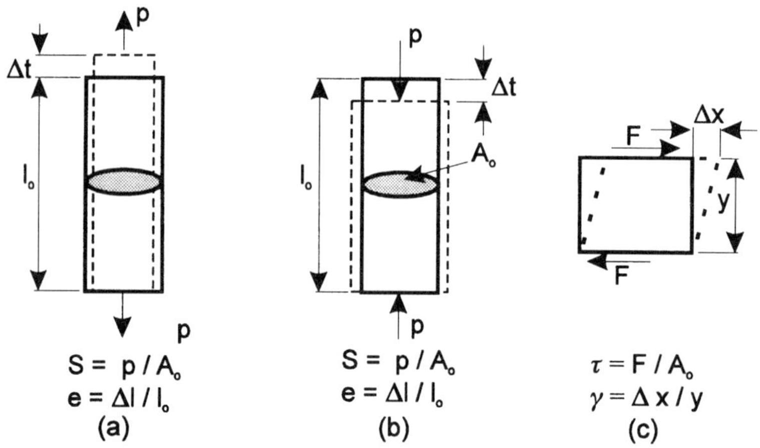

FIGURE 4.2 Elastic stress and strain for: (*a*) uniaxial tension; (*b*) uniaxial compression; (*c*) simple shear [1].

Poisson's ratio (v) is the ratio of transverse (e_2) to direct (e_1) strain in tension or compression. In the elastic region, v is between 1/4 and 1/3 for metals. The relation between stress and strain in the elastic region is given by Hooke's law:

$$S = E\,e\left(\text{tension or compression}\right) \tag{4.2}$$

$$\tau = G\gamma\left(\text{simple shear}\right) \tag{4.3}$$

where E and G are the Young's and shear modulus of elasticity, respectively. A small change in specific volume ($\Delta Vol/Vol$) can be related to the elastic deformation, which is shown to be as follows for an isotropic material (same properties in all directions).

$$\frac{\Delta Vol}{Vol} = e_1\left(1-2v\right) \tag{4.4}$$

The bulk modulus (K = reciprocal of compressibility) is defined as follows:

$$K = \Delta p \Big/ \left(\frac{\Delta Vol}{Vol}\right) \tag{4.5}$$

where Δp is the pressure acting at a particular point. For an elastic solid loaded in uniaxial compression (S):

$$K = S \Big/ \left(\frac{\Delta Vol}{Vol} \right) = \frac{S}{e_1 \left(1 - 2v \right)} = \frac{E}{1 - 2v} \tag{4.6}$$

Thus, an elastic solid is compressible as long as v is less than 1/2, which is normally the case for metals. Hooke's law (Equation 4.2) for uniaxial tension can be generalized for a three-dimensional elastic condition.

The theory of elasticity is well established and is used as a basis for force measuring techniques. Note that the measurement of forces in separate engineering applications is very application specific, and care must be taken in the selection of the measuring techniques outlined below.

Basic Methods of Force Measurement

An unknown force may be measured by the following means:

1. Balancing the unknown force against a standard mass through a system of levers.
2. Measuring the acceleration of a known mass.
3. Equalizing it to a magnetic force generated by the interaction of a current-carrying coil and a magnet.
4. Distributing the force on a specific area to generate pressure, and then measuring the pressure.
5. Converting the applied force into the deformation of an elastic element.

The aforementioned methods used for measuring forces yield a variety of designs of measuring equipment. The challenge involved with the task of measuring force resides primarily in sensor design. The basics of sensor design can be resolved into two problems:

1. Primary geometric, or physical constraints, governed by the application of the force sensor device.
2. The means by which the force can be converted into a workable signal form (such as electronic signals or graduated displacements).

The remaining sections will discuss the types of devices used for force to signal conversion and finally illustrate some examples of applications of these devices for measuring forces.

4.3 Force Sensors

Force sensors are required for a basic understanding of the response of a system. For example, cutting forces generated by a machining process can be monitored to detect a tool failure or to diagnose the causes of this failure in controlling the process parameters, and in evaluating the quality of the surface produced. Force sensors are used to monitor impact forces in the automotive industry. Robotic handling and assembly tasks are controlled by detecting the forces generated at the end effector. Direct measurement of forces is useful in controlling many mechanical systems.

Some types of force sensors are based on measuring a deflection caused by the force. Relatively high deflections (typically, several micrometers) would be necessary for this technique to be feasible. The excellent elastic properties of helical springs make it possible to apply them successfully as force sensors that transform the load to be measured into a deflection. The relation between force and deflection in the elastic region is demonstrated by Hooke's law. Force sensors that employ strain gage elements or piezoelectric (quartz) crystals with built-in microelectronics are common. Both impulsive forces and slowly varying forces can be monitored using these sensors.

Of the available force measuring techniques, a general subgroup can be defined as that of load cells. Load cells are comprised generally of a rigid outer structure, some medium that is used for measuring

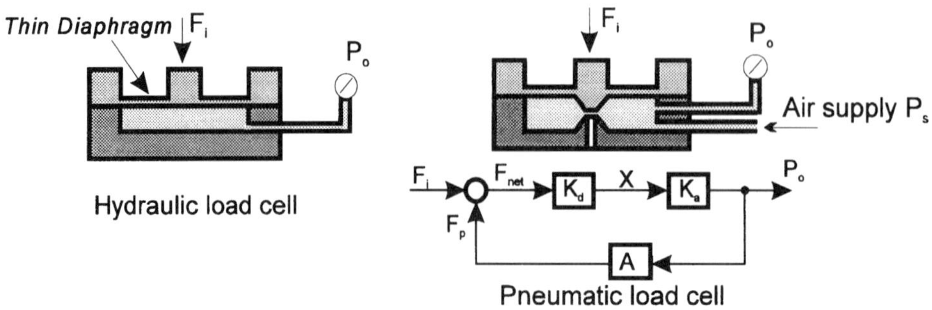

FIGURE 4.3 Different types of load cells [2].

the applied force, and the measuring gage. Load cells are used for sensing large, static or slowly varying forces with little deflection and are a relatively accurate means of sensing forces. Typical accuracies are of the order of 0.1% of the full-scale readings. Various strategies can be employed for measuring forces that are strongly dependent on the design of the load cell. For example, Figure 4.3 illustrates different types of load cells that can be employed in sensing large forces for relatively little cost. The hydraulic load cell employs a very stiff outer structure with an internal cavity filled with a fluid. Application of a load increases the oil pressure, which can be read off an accurate gage.

Other sensing techniques can be utilized to monitor forces, such as piezoelectric transducers for quicker response of varying loads, pneumatic methods, strain gages, etc. The proper sensing technique needs special consideration based on the conditions required for monitoring.

Strain Gage Load Cell

The strain gage load cell consists of a structure that elastically deforms when subjected to a force and a strain gage network that produces an electrical signal proportional to this deformation. Examples of this are beam and ring types of load cells.

Strain Gages

Strain gages use a length of gage wire to produce the desired resistance (which is usually about 120 Ω) in the form of a flat coil. This coil is then cemented (bonded) between two thin insulating sheets of paper or plastic. Such a gage cannot be used directly to measure deflection. It has to be first fixed properly to a member to be strained. After bonding the gage to the member, they are baked at about 195°F (90°C) to remove moisture. Coating the unit with wax or resin will provide some mechanical protection. The resistance between the member under test and the gage itself must be at least 50 MΩ. The total area of all conductors must remain small so that the cement can easily transmit the force necessary to deform the wire. As the member is stressed, the resulting strain deforms the strain gage and the cross-sectional area diminishes. This causes an increase in resistivity of the gage that is easily determined. In order to measure very small strains, it is necessary to measure small changes of the resistance per unit resistance ($\Delta R/R$). The change in the resistance of a bonded strain gage is usually less than 0.5%. A wide variety of gage sizes and grid shapes are available, and typical examples are shown in Figure 4.4.

The use of strain gages to measure force requires careful consideration with respect to rigidity and environment. By virtue of their design, strain gages of shorter length generally possess higher response frequencies (examples: 660 kHz for a gage of 0.2 mm and 20 kHz for a gage of 60 mm in length). The environmental considerations focus mainly on the temperature of the gage. It is well known that resistance is a function of temperature and, thus, strain gages are susceptible to variations in temperature. Thus, if it is known that the temperature of the gage will vary due to any influence, temperature compensation is required in order to ensure that the force measurement is accurate.

A Wheatstone bridge (Figure 4.5) is usually used to measure this small order of magnitude. In Figure 4.5, no current will flow through the galvanometer (G) if the four resistances satisfy a certain

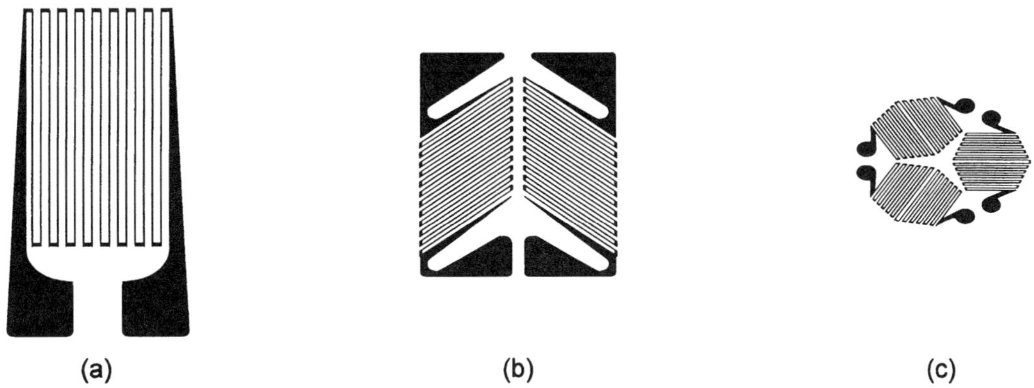

FIGURE 4.4 Configuration of metal-foil resistance strain gages: (*a*) single element; (*b*) two element; and (*c*) three element.

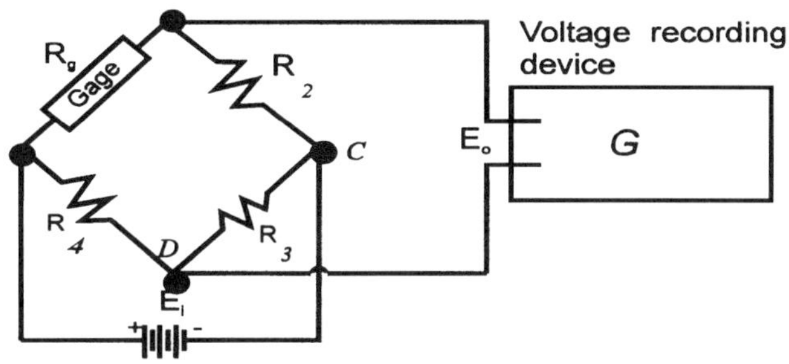

FIGURE 4.5 The Wheatstone bridge.

condition. In order to demonstrate how a Wheatstone bridge operates [3], a voltage scale has been drawn at points C and D of Figure 4.5. Assume that R_1 is a bonded gage and that initially Equation 4.7 is satisfied. If R_1 is now stretched so that its resistance increases by one unit ($+\Delta R$), the voltage at point D will be increased from zero to plus one unit of voltage ($+\Delta V$), and there will be a voltage difference of one unit between C and D that will give rise to a current through C. If R_4 is also a bonded gage, and at the same time that R_1 changes by $+\Delta R$, R_4 changes by $-\Delta R$, the voltage at D will move to $+2\Delta V$. Also, if at the same time, R_2 changes by $-\Delta R$, and R_3 changes by $+\Delta R$, then the voltage of point C will move to $-2\Delta V$, and the voltage difference between C and D will now be $4\Delta V$. It is then apparent that although a single gage can be used, the sensitivity can be increased fourfold if two gages are used in tension while two others are used in compression.

$$\frac{R_1}{R_4} = \frac{R_2}{R_3} \tag{4.7}$$

The grid configuration of the metal-foil resistance strain gages is formed by a photo-etching process. The shortest gage available is 0.20 mm; the longest is 102 mm. Standard gage resistance are 120 Ω and 350 Ω. A strain gage exhibits a resistance change $\Delta R/R$ that is related to the strain in the direction of the grid lines by the expression in Equation 4.8 (where S_g is the gage factor or calibration constant for the gage).

$$\frac{\Delta R}{R} = S_g \varepsilon \tag{4.8}$$

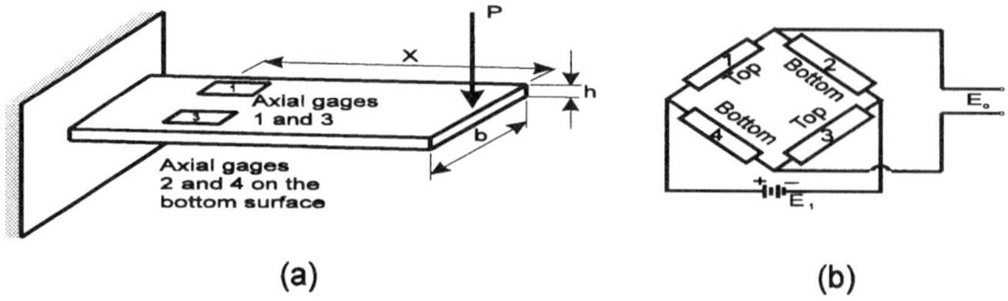

FIGURE 4.6 Beam-type load cells: (*a*) a selection of beam-type load cells (elastic element with strain gages); and (*b*) gage positions in the Wheatstone bridge [3].

Beam-Type Load Cell

Beam-type load cells are commonly employed for measuring low-level loads [3]. A simple cantilever beam (see Figure 4.6(*a*)) with four strain gages, two on the top surface and two on the bottom surface (all oriented along the axis of the beam) is used as the elastic member (sensor) for the load cell. The gages are wired into a Wheatstone bridge as shown in Figure 4.6(*b*). The load P produces a moment $M = Px$ at the gage location (x) that results in the following strains:

$$\varepsilon_1 = -\varepsilon_2 = \varepsilon_3 = -\varepsilon_4 = \frac{6M}{Ebh^2} = \frac{6Px}{Ebh^2} \tag{4.9}$$

where b is the width of the cross-section of the beam and h is the height of the cross-section of the beam. Thus, the response of the strain gages is obtained from Equation 4.10.

$$\frac{\Delta R_1}{R_1} = -\frac{\Delta R_2}{R_2} = \frac{\Delta R_3}{R_3} = -\frac{\Delta R_4}{R_4} = \frac{6S_g Px}{Ebh^2} \tag{4.10}$$

The output voltage E_o from the Wheatstone bridge, resulting from application of the load P, is obtained from Equation 4.11. If the four strain gages on the beam are assumed to be identical, then Equation 4.11 holds.

$$E_o = \frac{6S_g PxE_1}{Ebh^2} \tag{4.11}$$

The range and sensitivity of a beam-type load cell depends on the shape of the cross-section of the beam, the location of the point of application of the load, and the fatigue strength of the material from which the beam is fabricated.

Ring-Type Load Cell

Ring-type load cells incorporate a proving ring (see Figure 4.7) as the elastic element. The ring element can be designed to cover a very wide range of loads by varying the diameter D, the thickness t, or the depth w of the ring. Either strain gages or a linear variable-differential transformer (LVDT) can be used as the sensor.

The load P is linearly proportional to the output voltage E_o. The sensitivity of the ring-type load cell with an LVDT sensor depends on the geometry of the ring (R, t, and w), the material from which the ring is fabricated (E), and the characteristics of the LVDT (S and E_i). The range of a ring-type load cell is controlled by the strength of the material used in fabricating the ring.

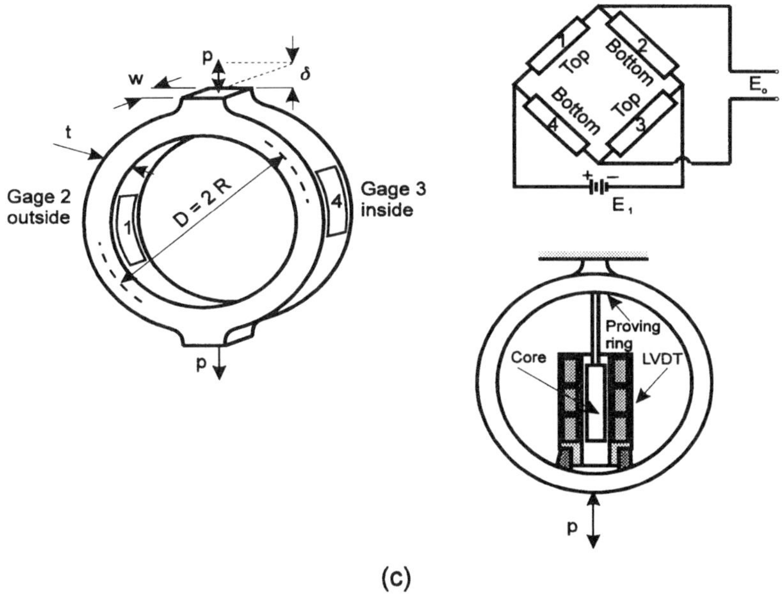

FIGURE 4.7 Ring-type load cells: (*a*) elastic element with strain-gage sensors; (*b*) gage positions in the Wheatstone bridge; and (*c*) elastic element with an LVDT sensor [3].

Piezoelectric Methods

A piezoelectric material exhibits a phenomenon known as the *piezoelectric effect.* This effect states that when asymmetrical, elastic crystals are deformed by a force, an electrical potential will be developed within the distorted crystal lattice. This effect is reversible. That is, if a potential is applied between the surfaces of the crystal, it will change its physical dimensions [4]. Elements exhibiting piezoelectric qualities are sometimes known as electrorestrictive elements.

The magnitude and polarity of the induced surface charges are proportional to the magnitude and direction of the applied force [4]:

$$Q = dF \tag{4.12}$$

where d is the charge sensitivity (a constant for a given crystal) of the crystal in C/N. The force F causes a thickness variation Δt meters of the crystal:

$$F = \frac{aY}{t}\Delta t \tag{4.13}$$

where a is area of crystal, t is thickness of crystal, and Y is Young's modulus.

$$Y = \frac{stress}{strain} = \frac{Ft}{a\Delta t} \tag{4.14}$$

The charge at the electrodes gives rise to a voltage $E_0 = Q/C$, where C is capacitance in farads between the electrodes and $C = \varepsilon a/t$ where ε is the absolute permittivity.

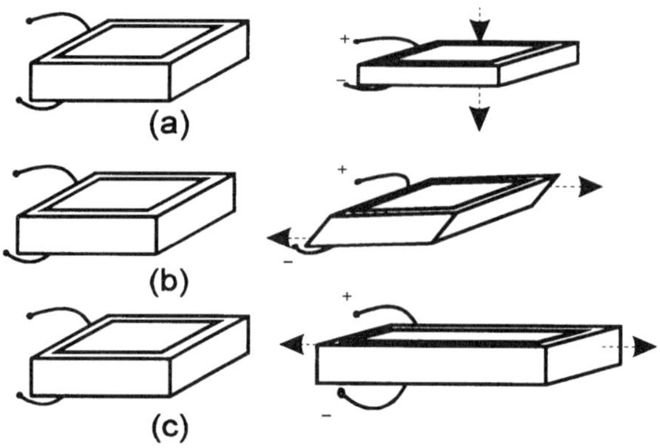

FIGURE 4.8 Modes of operation for a simple plate as a piezoelectric device [4].

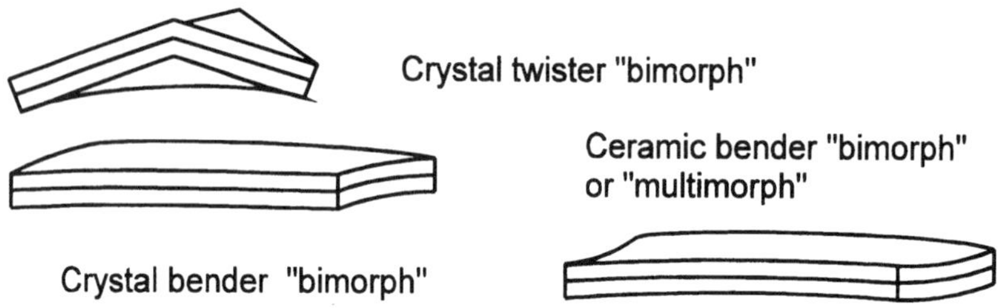

FIGURE 4.9 Curvature of "twister" and "bender" piezoelectric transducers when voltage applied [4].

$$E_o = \frac{dF}{C} = \frac{d}{\varepsilon}\frac{tF}{a} \tag{4.15}$$

The voltage sensitivity = $g = d/\varepsilon$ in volt m/N can be obtained as:

$$E_o = g\frac{t}{a}F = gtP \tag{4.16}$$

The piezoelectric materials used are quartz, tourmaline, Rochelle salt, ammonium dihydrogen phosphate (ADP), lithium sulfate, barium titanate, and lead zirconate titanate (PZT) [4]. Quartz and other earthly piezoelectric crystals are naturally polarized. However, synthetic piezoelectric materials, such as barium titanate ceramic, are made by baking small crystallites under pressure and then placing the resultant material in a strong dc electric field [4]. After that, the crystal is polarized, along the axis on which the force will be applied, to exhibit piezoelectric properties. Artificial piezoelectric elements are free from the limitations imposed by the crystal structure and can be molded into any size and shape. The direction of polarization is designated during their production process.

The different modes of operation of a piezoelectric device for a simple plate are shown in Figure 4.8 [4]. By adhering two crystals together so that their electrical axes are perpendicular, bending moments or torque can be applied to the piezoelectric transducer and a voltage output can be produced (Figure 4.9) [4]. The range of forces that can be measured using piezoelectric transducers are from 1 to 200 kN and at a ratio of 2×10^5.

Piezoelectric crystals can also be used in measuring an instantaneous change in the force (dynamic forces). A thin plate of quartz can be used as an electronic oscillator. The frequency of these oscillations will be dominated by the natural frequency of the thin plate. Any distortion in the shape of the plate caused by an external force, alters the oscillation frequency. Hence, a dynamic force can be measured by the change in frequency of the oscillator.

Resistive Method

The resistive method employs the fact that when the multiple contact area between semiconducting particles (usually carbon) and the distance between the particles are changed, the total resistance is altered. The design of such transducers yields a very small displacement when a force is applied. A transducer might consist of 2 to 60 thin carbon disks mounted between a fixed and a movable electrode. When a force is applied to the movable electrode and the carbon disks move together by 5 to 250 μm per interface, the transfer function of their resistance against the applied force is approximately hyperbolic, that is, highly nonlinear. The device is also subject to large hysteresis and drift together with a high transverse sensitivity.

In order to reduce hysteresis and drift, rings are used instead of disks. The rings are mounted on an insulated rigid core and prestressed. This almost completely eliminates any transverse sensitivity error. The core's resonant frequency is high and can occur at a frequency as high as 10 kHz. The possible measuring range of such a transducer is from 0.1 kg to 10 kg. The accuracy and linear sensitivity of this transducer is very poor.

Inductive Method

The inductive method utilizes the fact that a change in mechanical stress of a ferromagnetic material causes its permeability to alter. The changes in magnetic flux are converted into induced voltages in the pickup coils as the movement takes place. This phenomenon is known as the *Villari effect* or *magnetostriction*. It is known to be particularly strong in nickel–iron alloys.

Transducers utilizing the Villari effect consist of a coil wound on a core of magnetostrictive material. The force to be measured is applied on this core, stressing it and causing a change in its permeability and inductance. This change can be monitored and used for determining the force.

The applicable range for this type of transducer is a function of the cross-sectional area of the core. The accuracy of the device is determined by a calibration process. This transducer has poor linearity and is subject to hysteresis. The permeability of a magnetostrictive material increases when it is subjected to pure torsion, regardless of direction. A flat frequency response is obtained over a wide range from 150 Hz to 15,000 Hz.

Piezotransistor Method

Devices that utilize *anisotropic stress effects* are described as piezotransistors. In this effect, if the upper surface of a *p–n* diode is subjected to a localized stress, a significant reversible change occurs in the current across the junction. These transistors are usually silicon nonplanar type, with an emitter base junction. This junction is mechanically connected to a diaphragm positioned on the upper surface of a typical TO-type can [4]. When a pressure or a force is applied to the diaphragm, an electronic charge is produced. It is advisable to use these force-measuring devices at a constant temperature by virtue of the fact that semiconducting materials also change their electric properties with temperature variations. The attractive characteristic of piezotransistors is that they can withstand a 500% overload.

Multicomponent Dynamometers Using Quartz Crystals As Sensing Elements

The Piezoelectric Effects in Quartz.

For force measurements, the *direct piezoelectric effect* is utilized. The direct longitudinal effect measures compressive force; the direct shear effect measures shear force in one direction. For example, if a disk of crystalline quartz (SiO_2) cut normally to the crystallographic x-axis is loaded by a compression force, it will yield an electric charge, nominally 2.26 pC/N. If a disk of crystalline quartz is cut normally to the

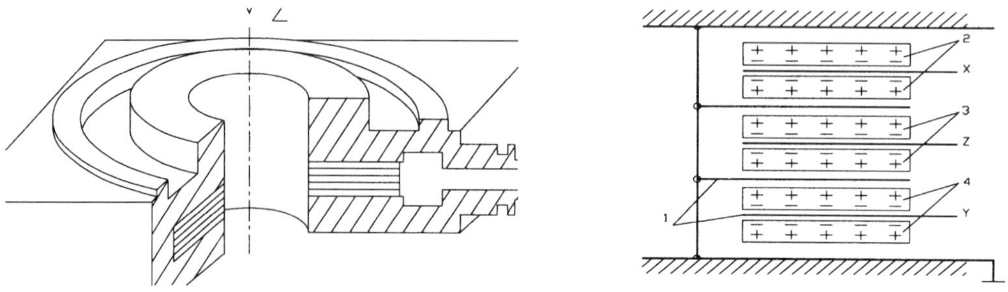

FIGURE 4.10 Three-component force transducer.

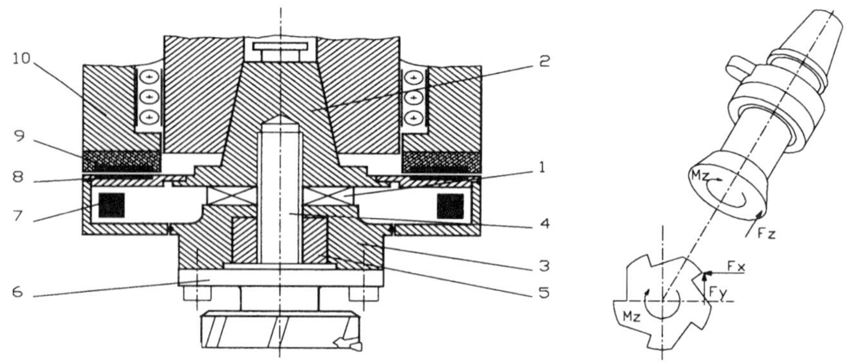

FIGURE 4.11 Force measuring system to determine the tool-related cutting forces in five-axis milling [6].

crystallographic *y*-axis, it will yield an electric charge (4.52 pC/N) if loaded by a shear force in one specific direction. Forces applied in the other directions will not generate any output [5].

A charge amplifier is used to convert the charge yielded by a quartz crystal element into a proportional voltage. The range of a charge amplifier with respect to its conversion factor is determined by a feedback capacitor. Adjustment to mechanical units is obtained by additional operational amplifiers with variable gain.

The Design of Quartz Multicomponent Dynamometers.
The main element for designing multicomponent dynamometers is the three-component force transducer (Figure 4.10). It contains a pair of *X*-cut quartz disks for the normal force component and a pair of *Y*-cut quartz disks (shear-sensitive) for each shear force component.

Three-component dynamometers can be used for measuring cutting forces during machining. Four three-component force transducers sandwiched between a base plate and a top plate are shown in Figure 4.10. The force transducer is subjected to a preload as shear forces are transmitted by friction. The four force transducers experience a drastic change in their load, depending on the type and position of force application. An overhanging introduction of the force develops a tensile force for some transducers, thus reducing the preload. Bending of the dynamometer top plate causes bending and shearing stresses. The measuring ranges of a dynamometer depend not only on the individual forces, but also on the individual bending stresses.

Measuring Signals Transmitted by Telemetry.
Figure 4.11 shows the newly designed force measuring system RCD (rotating cutting force dynamometer). A ring-shaped sensor (1) is fitted in a steep angle taper socket (2) and a base ring (3) allowing sensing of the three force components F_x, F_y and F_z at the cutting edge as well as the moment M_z. The

FIGURE 4.12 Capacitive force transducer [7].

physical operating principle of this measuring cell is based on the piezoelectric effect in quartz plates. The quartz plates incorporated in the sensor are aligned so that the maximum cross-sensitivity between the force components is 1%. As a result of the rigid design of the sensor, the resonant frequencies of the force measuring system range from 1200 Hz to 3000 Hz and the measuring ranges cover a maximum of 10 kN [6].

Force-proportional charges produced at the surfaces of the quartz plates are converted into voltages by four miniature charge amplifiers (7) in hybrid construction. These signals are then filtered by specific electrical circuitry to prevent aliasing effects, and digitized with 8 bit resolution using a high sampling rate (pulse-code modulation). The digitized signals are transmitted by a telemetric unit consisting of a receiver and transmitter module, an antenna at the top of the rotating force measuring system (8), as well as a fixed antenna (9) on the splash cover of the two-axis milling head (10). The electrical components, charge amplifier, and transmitter module are mounted on the circumference of the force measuring system [6].

The cutting forces and the moment measured are digitized with the force measuring system described above. They are modulated on an FM carrier and transmitted by the rotating transmitter to the stationary receiver. The signals transmitted are fed to an external measured-variable conditioning unit.

Measuring Dynamic Forces.

Any mechanical system can be considered in the first approximation as a weakly damped oscillator consisting of a spring and a mass. If a mechanical system has more than one resonant frequency, the lowest one must be taken into consideration. As long as the test frequency remains below 10% of the resonant frequency of the reference transducer (used for calibration), the difference between the dynamic sensitivity obtained from static calibration will be less than 1%. The above considerations assume a sinusoidal force signal. The static calibration of a reference transducer is also valid for dynamic calibration purposes if the test frequency is much lower (at least 10 times lower) than the resonant frequency of the system.

Capacitive Force Transducer

A transducer that uses capacitance variation can be used to measure force. The force is directed onto a membrane whose elastic deflection is detected by a capacitance variation. A highly sensitive force transducer can be constructed because the capacitive transducer senses very small deflections accurately. An electronic circuit converts the capacitance variations into dc-voltage variations [7].

The capacitance sensor illustrated in Figure 4.12 consists of two metal plates separated by an air gap. The capacitance C between terminals is given by the expression:

$$C = \varepsilon_o \varepsilon_r \frac{A}{h} \tag{4.17}$$

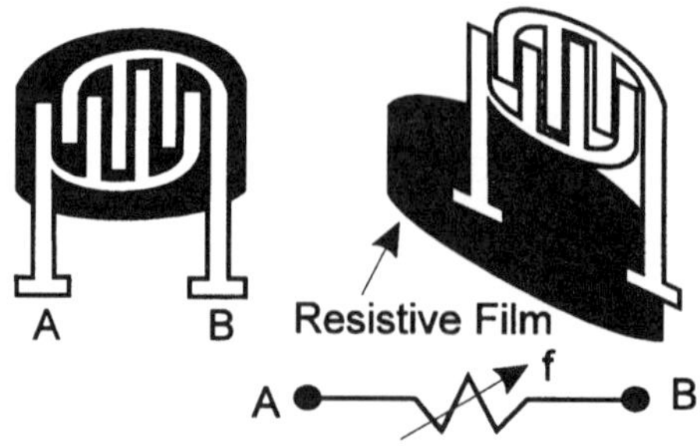

FIGURE 4.13 Diagram of a typical force sensing resistor (FSR).

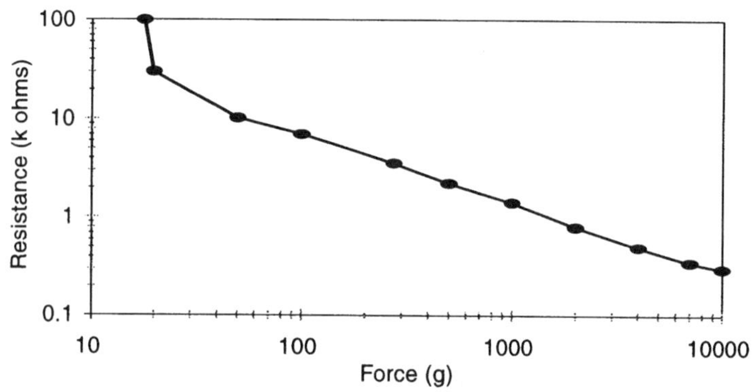

FIGURE 4.14 Resistance as a function of force for a typical force sensing resistor.

where C = Capacitance in farads (F)
ε_0 = Dielectric constant of free space
ε_r = Relative dielectric constant of the insulator
A = Overlapping area for the two plates
h = Thickness of the gap between the two plates

The sensitivity of capacitance-type sensors is inherently low. Theoretically, decreasing the gap h should increase the sensitivity; however, there are practical electrical and mechanical conditions that preclude high sensitivities. One of the main advantages of the capacitive transducer is that moving of one of its plate relative to the other requires an extremely small force to be applied. A second advantage is stability and the sensitivity of the sensor is not influenced by pressure or temperature of the environment.

Force Sensing Resistors (Conductive Polymers)

Force sensing resistors (FSRs) utilize the fact that certain polymer thick-film devices exhibit decreasing resistance with the increase of an applied force. A force sensing resistor is made up of two parts. The first is a resistive material applied to a film. The second is a set of digitating contacts applied to another film. Figure 4.13 shows this configuration. The resistive material completes the electrical circuit between the two sets of conductors on the other film. When a force is applied to this sensor, a better connection is made between the contacts; hence, the conductivity is increased. Over a wide range of forces, it turns out that the conductivity is approximately a linear function of force. Figure 4.14 shows the resistance of the sensor as a function of force. It is important to note that there are three possible regions for the sensor to operate. The first abrupt transition occurs somewhere in the vicinity of 10 g of force. In this region, the resistance changes very rapidly. This behavior is useful when one is designing switches using force sensing resistors.

FSRs should not be used for accurate measurements of force because sensor parts may exhibit 15% to 25% variation in resistance between each other. However, FSRs exhibit little hysteresis and are considered far less costly than other sensing devices. Compared to piezofilm, the FSR is far less sensitive to vibration and heat.

Magnetoresistive Force Sensors

The principle of *magnetoresistive force sensors* is based on the fact that metals, when cooled to low temperatures, show a change of resistivity when subjected to an applied magnetic field. Bismuth, in particular, is quite sensitive in this respect. In practice, these devices are severely limited because of their high sensitivity to ambient temperature changes.

Magnetoelastic Force Sensors

Magnetoelastic transducer devices operate based on the Joule effect; that is, a ferromagnetic material is dimensionally altered when subjected to a magnetic field. The principle of operation is as follows: Initially, a current pulse is applied to the conductor within the waveguide. This sets up a magnetic field circumference-wise around the waveguide over its entire length. There is another magnetic field generated by the permanent magnet that exists only where the magnet is located. This field has a longitudinal component. These two fields join vectorally to form a helical field near the magnet which, in turn, causes the waveguide to experience a minute torsional strain or twist only at the location of the magnet. This twist effect is known as the *Wiedemann effect* [8].

Magnetoelastic force transducers have a high frequency response (on the order of 20 kHz). Some of the materials that exhibit magnetoelastic include Monel metal, Permalloy, Cekas, Alfer, and a number of nickel–iron alloys. Disadvantages of these transducers include: (1) the fact that excessive stress and aging may cause permanent changes, (2) zero drift and sensitivity changes due to temperature sensitivity, and (3) hysteresis errors.

Torsional Balances

Balancing devices that utilize the deflection of a spring may also be used to determine forces. *Torsional balances* are equal arm scale force measuring devices. They are comprised of horizontal steel bands instead of pivots and bearings. The principle of operation is based on force application on one of the arms that will deflect the torsional spring (within its design limits) in proportion to the applied force. This type of instrument is susceptible to hysteresis and temperature errors and therefore is not used for precise measurements.

Tactile Sensors

Tactile sensors are usually interpreted as a touch sensing technique. Tactile sensors cannot be considered as simple touch sensors, where very few discrete force measurements are made. In tactile sensing, a force "distribution" is measured using a closely spaced array of force sensors.

Tactile sensing is important in both grasping and object identification operations. Grasping an object must be done in a stable manner so that the object is not allowed to slip or damaged. Object identification includes recognizing the shape, location, and orientation of a product, as well as identifying surface properties and defects. Ideally, these tasks would require two types of sensing [9]:

1. Continuous sensing of contact forces
2. Sensing of the surface deformation profile

These two types of data are generally related through stress–strain relations of the tactile sensor. As a result, almost continuous variable sensing of tactile forces (the sensing of the tactile deflection profile) is achieved.

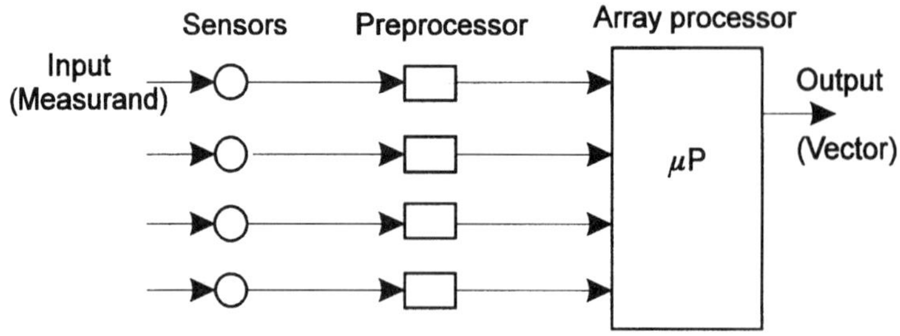

FIGURE 4.15 Tactile array sensor.

Tactile Sensor Requirements.
Significant advances in tactile sensing are taking place in the robotics area. Applications include automated inspection of surface profiles, material handling or parts transfer, parts assembly, and parts identification and gaging in manufacturing applications and fine-manipulation tasks. Some of these applications may need only simple touch (force–torque) sensing if the parts being grasped are properly oriented and if adequate information about the process is already available.

Naturally, the main design objective for tactile sensing devices has been to mimic the capabilities of human fingers [9]. Typical specifications for an industrial tactile sensor include:

1. Spatial resolution of about 2 mm
2. Force resolution (sensitivity) of about 2 g
3. Maximum touch force of about 1 kg
4. Low response time of 5 ms
5. Low hysteresis
6. Durability under extremely difficult working conditions
7. Insensitivity to change in environmental conditions (temperature, dust, humidity, vibration, etc.)
8. Ability to monitor slip

Tactile Array Sensor.
Tactile array sensors (Figure 4.15) consist of a regular pattern of sensing elements to measure the distribution of pressure across the finger tip of a Robot. The 8×8 array of elements at 2 mm spacing in each direction, provides 64 force sensitive elements. Table 4.1 outlines some of the characteristics of early tactile array sensors. The sensor is composed of two crossed layers of copper strips separated by strips of thin silicone rubber. The sensor forms a thin, compliant layer that can be easily attached to a variety of finger-tip shapes and sizes. The entire array is sampled by computer.

A typical tactile sensor array can consist of several sensing elements. Each element or taxel (Figure 4.16) is used to sense the forces present. Since tactile sensors are implemented in applications where sensitivity providing semblance to human touch is desired, an elastomer is utilized to mimic the human skin. The elastomer is generally a conductive material whose electrical conductivity changes locally when pressure is applied. The sensor itself consists of three layers: a protective covering, a sheet of conductive elastomer, and a printed circuit board. The printed circuit board consists of two rows of two "bullseyes," each with conductive inner and outer rings that compromise the taxels of the sensor. The outer rings are connected together and to a column-select transistor. The inner rings are connected to diodes (D) in Figure 4.16. Once the column in the array is selected, the current flows through the diodes, through the elastomer, and thence through a transistor to ground. As such, it is generally not possible to excite just one taxel because the pressure applied causes a local deformation in neighboring taxels. This situation is called *crosstalk* and is eliminated by the diodes [10].

Tactile array sensor signals are used to provide information about the contact kinematics. Several feature parameters, such as contact location, object shape, and the pressure distribution, can be obtained.

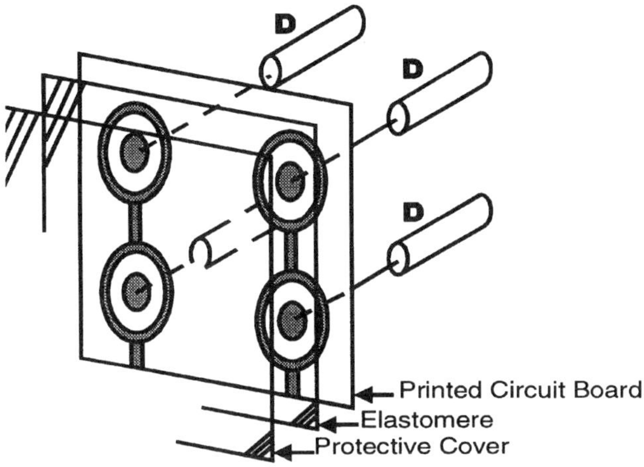

FIGURE 4.16 Typical taxel sensor array.

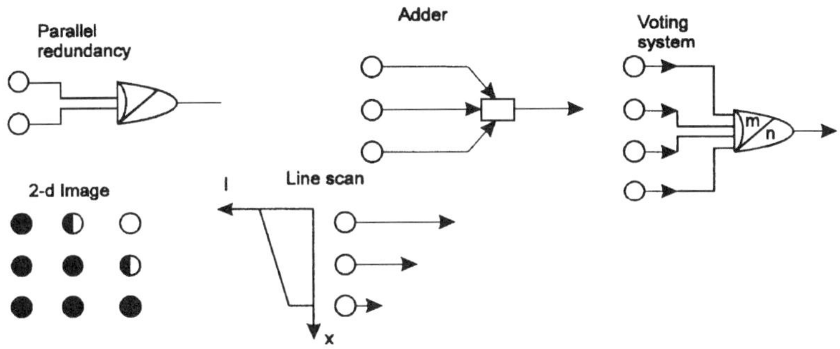

FIGURE 4.17 General arrangement of an intelligent sensor array system [9].

TABLE 4.1 Summary of Some of the Characteristics of Early Tactile Arrays Sensors

Device parameter	Size of array (4 × 4)	(8 × 8)	(16 × 16)
Cell spacing (mm)	4.00	2.00	1.00
Zero-pressure capacitance (fF)	6.48	1.62	0.40
Rupture force (N)	18.90	1.88	0.19
Max. linear capacitance (fF)	4.80	1.20	0.30
Max. output voltage (V)	1.20	0.60	0.30
Max. resolution (bit)	9.00	8.00	8.00
Readout (access) time (μs)	—	<20	—

©IEEE 1985.

The general layout of a sensor array system can be seen in Figure 4.17. An example of this is a contact and force sensing finger. This tactile finger has four contact sensors made of piezoelectric polymer strips on the surface of the fingertip that provide dynamic contact information. A strain gage force sensor provides static grasp force information.

References

1. M. C. Shaw, *Metal Cutting Principles*, Oxford: Oxford Science Publications: Clarendon Press, 1989.
2. E. O. Doebelin, *Measurement Systems, Application and Design*, 4th ed., New York: McGraw-Hill, 1990.
3. J. W. Dally, W. F. Riley, and K. G. McConnel, *Instrumentation for Engineering Measurements*, New York: John Wiley & Sons, 1984.
4. P. H. Mansfield, *Electrical Transducers for Industrial Measurement*, London: The Butterworth Group, 1973.
5. K. H. Martini, Multicomponent dynamometers using quartz crystals as sensing elements, *ISA Trans.*, 22(1), 1983.
6. G. Spur, S. J. Al-Badrawy, and J. Stirnimann, Measuring the Cutting Force in Five-Axis Milling, Translated paper "Zerpankraftmessung bei der funfachsigen Frasbearbeitung", Zeitschrift fur wirtschaftliche Fertigung und Automatisierung 9/93 Carl Hanser, Munchen, Kistler Piezo-Instrumentation, 20.162e 9.94.
7. C. L. Nachtigal, *Instrumentation and Control, Fundamentals and Applications*, Wiley Series in Mechanical Engineering Practice, New York: Wiley Interscience, John Wiley & Sons, 1990.
8. C. W. DeSilva, *Control Sensors and Actuators*, Englewood Cliffs, NJ: Prentice-Hall, 1989.
9. J. W. Gardner, *Microsensors Principles and Applications*, New York: John Wiley & Sons, 1995.
10. W. Stadler, *Analytical Robotics and Mechatronics*, New York: McGraw-Hill, 1995.

Further Information

C. P. Wright, *Applied Measurement Engineering, How to Design Effective Mechanical Measurement Systems*, Englewood Cliffs, NJ: Prentice-Hall, 1995.

E. E. Herceg, *Handbook of Measurement and Control*, Pennsauken, NJ: Schavitz Engineering, 1972.

D. M. Considine, *Encyclopedia of Instrumentation and Control*, New York: McGraw-Hill, 1971.

H. N. Norton, *Sensor and Analyzer Handbook*, Englewood Cliffs, NJ: Prentice Hall, 1982.

S. M. Sze, *Semiconductor Sensors*, New York: John Wiley & Sons, 1994.

B. Lindberg and B. Lindstrom, Measurements of the segmentation frequency in the chip formation process, *Ann. CIRP*, 32(1), 1983.

J. Tlusty and G. C Andrews, A critical review of sensors for unmanned machining, *Ann. CIRP*, 32(2), 1983.

5

Torque and Power Measurement

Ivan J. Garshelis
Magnova, Inc.

Torque, speed, and power are the defining mechanical variables associated with the functional performance of rotating machinery. The ability to accurately measure these quantities is essential for determining a machine's efficiency and for establishing operating regimes that are both safe and conducive to long and reliable services. On-line measurements of these quantities enable real-time control, help to ensure consistency in product quality, and can provide early indications of impending problems. Torque and power measurements are used in testing advanced designs of new machines and in the development of new machine components. Torque measurements also provide a well-established basis for controlling and verifying the tightness of many types of threaded fasteners. This chapter describes the basic concepts as well as the various methods and apparati in current use for the measurement of torque and power; the measurement of speed, or more precisely, angular velocity, is discussed in Reference [1].

5.1 Fundamental Concepts

Angular Displacement, Velocity, and Acceleration

The concept of *rotational* motion is readily formalized: all points within a rotating rigid body move in parallel or coincident planes while remaining at fixed distances from a line called the *axis.* In a perfectly rigid body, all points also remain at fixed distances from each other. Rotation is perceived as a change in the angular position of a reference point on the body, i.e., as its *angular displacement,* $\Delta\theta$, over some time interval, Δt. The motion of that point, and therefore of the whole body, is characterized by its

0-8493-0047-9/00/$0.00+$.50

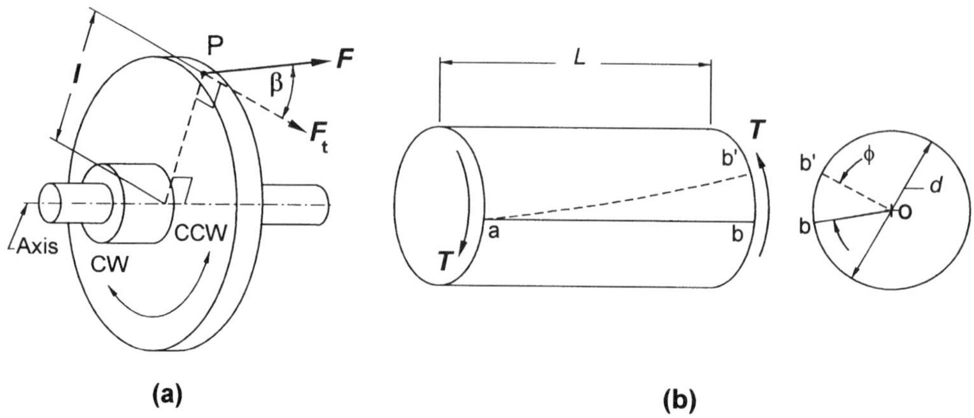

FIGURE 5.1 (*a*) The off-axis force F at P produces a torque $T = (F \cos \beta)l$ tending to rotate the body in the CW direction. (*b*) Transmitting torque T over length L twists the shaft through angle ϕ.

clockwise (CW) or counterclockwise (CCW) *direction* and by its *angular velocity,* $\omega = \Delta\theta/\Delta t$. If during a time interval Δt, the velocity changes by $\Delta\omega$, the body is undergoing an *angular acceleration,* $\alpha = \Delta\omega/\Delta t$. With angles measured in radians, and time in seconds, units of ω become radians per second (rad s^{-1}) and of α, radians per second per second (rad s^{-2}). Angular velocity is often referred to as *rotational speed* and measured in numbers of complete revolutions per minute (rpm) or per second (rps).

Force, Torque, and Equilibrium

Rotational motion, as with motion in general, is controlled by *forces* in accordance with Newton's laws. Because a force directly affects only that component of motion in its line of action, forces or components of forces acting in any plane that includes the axis produce no tendency for rotation about that axis. Rotation can be initiated, altered in velocity, or terminated only by a *tangential force* F_t acting at a finite radial distance l from the axis. The effectiveness of such forces increases with both F_t and l; hence, their product, called a *moment,* is the activating quantity for rotational motion. A moment about the rotational axis constitutes a *torque.* Figure 5.1(*a*) shows a force F acting at an angle β to the tangent at a point P, distant l (the moment arm) from the axis. The torque T is found from the *tangential component* of F as:

$$T = F_t l = \left(F \cos \beta\right) l \tag{5.1}$$

The combined effect, known as the *resultant,* of any number of torques acting at different locations along a body is found from their *algebraic sum,* wherein torques tending to cause rotation in CW and CCW directions are assigned opposite signs. Forces, hence torques, arise from physical contact with other solid bodies, motional interaction with fluids, or via gravitational (including inertial), electric, or magnetic force fields. The *source* of each such torque is subjected to an equal, but oppositely directed, *reaction* torque. With force measured in newtons and distance in meters, Equation 5.1 shows the unit of torque to be a Newton meter (N·m).

A nonzero resultant torque will cause the body to undergo a proportional angular acceleration, found, by application of Newton's second law, from:

$$T_r = I\alpha \tag{5.2}$$

where I, having units of kilogram meter² (kg m²), is the moment of inertia of the body around the axis (i.e., its *polar* moment of inertia). Equation 5.2 is applicable to any body regardless of its state of motion.

When $\alpha = 0$, Equation 5.2 shows that T_r is also zero; the body is said to be in *equilibrium*. For a body to be in equilibrium, there must be either more than one *applied* torque, or none at all.

Stress, Rigidity, and Strain

Any portion of a rigid body in equilibrium is also in equilibrium; hence, as a condition for equilibrium of the portion, any torques applied thereto from *external* sources must be balanced by equal and directionally opposite *internal* torques from adjoining portions of the body. Internal torques are *transmitted* between adjoining portions by the collective action of *stresses* over their common cross-sections. In a solid body having a round cross-section (e.g., a typical shaft), the *shear stress* τ varies linearly from zero at the axis to a maximum value at the surface. The shear stress, τ_m, at the surface of a shaft of diameter, d, transmitting a torque, T, is found from:

$$\tau_m = \frac{16T}{\pi d^3} \tag{5.3}$$

Real materials are not *perfectly* rigid but have instead a *modulus of rigidity, G,* which expresses the finite ratio between τ and *shear strain,* γ. The maximum strain in a solid round shaft therefore also exists at its surface and can be found from:

$$\gamma_m = \frac{\tau_m}{G} = \frac{16T}{\pi d^3 G} \tag{5.4}$$

Figure 5.1(b) shows the manifestation of shear strain as an angular displacement between axially separated cross-sections. Over the length L, the solid round shaft shown will be *twisted* by the torque through an angle ϕ found from:

$$\phi = \frac{32LT}{\pi d^4 G} \tag{5.5}$$

Work, Energy, and Power

If during the time of application of a torque, T, the body rotates through some angle θ, mechanical work:

$$W = T\theta \tag{5.6}$$

is performed. If the torque acts in the same CW or CCW sense as the displacement, the work is said to be done *on* the body, or else it is done *by* the body. Work done *on* the body causes it to accelerate, thereby appearing as an increase in *kinetic energy* (KE = $I\omega^2/2$). Work done *by* the body causes deceleration with a corresponding decrease in kinetic energy. If the body is not accelerating, any work done on it at one location must be done by it at another location. Work and energy are each measured in units called a joule (J). Equation 5.6 shows that 1 J is equivalent to 1 N·m rad, which, since a radian is a dimensionless ratio, ≡ 1 N·m. To avoid confusion with torque, it is preferable to quantify mechanical work in units of m·N, or better yet, in J.

The *rate* at which work is performed is termed *power, P.* If a torque T acts over a small interval of time Δt, during which there is an angular displacement $\Delta\theta$, work equal to $T\Delta\theta$ is performed at the rate $T\Delta\theta/\Delta t$. Replacing $\Delta\theta/\Delta t$ by ω, power is found simply as:

$$P = T\omega \tag{5.7}$$

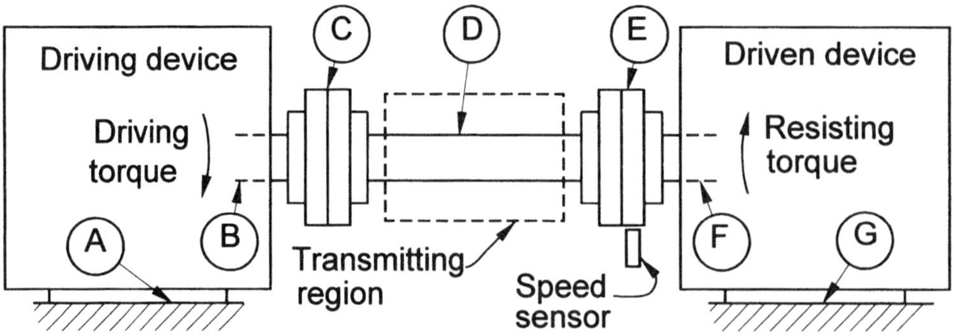

FIGURE 5.2 Schematic arrangement of devices used for the measurement of torque and power.

The unit of power follows from its definition and is given the special name watt (W). 1 W = 1 J s^{-1} = 1 m·N s^{-1}. Historically, power has also been measured in horsepower (Hp), where 1 Hp = 746 W. Rotating bodies effectively transmit power between locations where torques from external sources are applied.

5.2 Arrangements of Apparatus for Torque and Power Measurement

Equations 5.1 through 5.7 express the physical bases for torque and power measurement. Figure 5.2 illustrates a generalized measurement arrangement. The actual apparatus used is selected to fulfill the specific measurement purposes. In general, a driving torque originating within a device at one location (B in Figure 5.2), is resisted by an opposing torque developed by a different device at another location (F). The driving torque (from, e.g., an electric motor, a gasoline engine, a steam turbine, muscular effort, etc.) is coupled through connecting members C, transmitting region D, and additional couplings E, to the driven device (an electric generator, a pump, a machine tool, mated threaded fasteners, etc.) within which the resisting torque is met at F. The torque at B or F is the quantity to be measured. These torques may be *indirectly* determined from a correlated physical quantity, e.g., an electrical current or fluid pressure associated with the operation of the driving or driven device, or more directly by measuring either the *reaction* torque at A or G, or the *transmitted* torque through D. It follows from the cause-and-effect relationship between torque and rotational motion that most interest in transmitted torque will involve rotating bodies.

To the extent that the frames of the driving and driven devices and their mountings to the "Earth" are *perfectly* rigid, the reaction at A will *at every instant* equal the torque at B, as will the reaction at G equal the torque at F. Under equilibrium conditions, these equalities are independent of the compliance of any member. Also under equilibrium conditions, and except for usually minor *parasitic* torques (due, e.g., to bearing friction and air drag over rapidly moving surfaces), the driving torque at B will equal the resisting torque at F.

Reaction torque at A or G is often determined, using Equation 5.1, from measurements of the forces acting at known distances fixed by the apparatus. Transmitted torque is determined from measurements, on a suitable member within region D, of τ_m, γ_m, or ϕ and applying Equations 5.3, 5.4, or 5.5 (or analogous expressions for members having other than solid round cross-sections [2]). *Calibration,* the measurement of the stress, strain, or twist angle resulting from the application of a *known* torque, makes it unnecessary to know any details about the member within D. When $\alpha \neq 0$, and is measurable, T may also be determined from Equation 5.2. Requiring only noninvasive, observational measurements, this method is especially useful for determining transitory torques; for example those associated with firing events in multicylinder internal combustion engines [3].

Equations 5.6 and 5.7 are applicable *only* during rotation because, in the absence of motion, no work is done and power transfer is zero. Equation 5.6 can be used to determine *average* torque from calorimetric measurements of the heat generated (equal to the mechanical work *W*) during a totalized number of revolutions ($\equiv \theta/2\pi$). Equation 5.7 is routinely applied in power measurement, wherein *T* is determined by methods based on Equations 5.1, 5.3, 5.4, or 5.5, and ω is measured by any suitable means [4].

F, *T*, and ϕ are sometimes measured by simple mechanical methods. For example, a "torque wrench" is often used for the controlled tightening of threaded fasteners. In these devices, torque is indicated by the position of a needle moving over a calibrated scale in response to the elastic deflection of a spring member, in the simplest case, the bending of the wrench handle [5]. More generally, instruments, variously called *sensors* or *transducers*, are used to convert the desired (torque or speed related) quantity into a linearly proportional electrical signal. (Force sensors are also known as *load cells*.) The determination of *P* most usually requires multiplication of the two signals from separate sensors of *T* and ω. A transducer, wherein the amplitude of a *single* signal proportional to the power being transmitted along a shaft, has also been described [6].

5.3 Torque Transducer Technologies

Various physical interactions serve to convert *F*, τ, γ, or ϕ into proportional electrical signals. Each requires that some axial portion of the shaft be dedicated to the torque sensing function. Figure 5.3 shows typical features of sensing regions for four sensing technologies in present use.

Surface Strain

Figure 5.3(*a*) illustrates a sensing region configured to convert surface strain (γ_m) into an electric signal proportional to the transmitted torque. Surface strain became the key basis for measuring both force and torque following the invention of bonded wire strain gages by E. E. Simmons, Jr. and Arthur C. Ruge in 1938 [7]. A modern strain gage consists simply of an elongated electrical conductor, generally formed in a serpentine pattern in a very thin foil or film, bonded to a thin insulating carrier. The carrier is attached, usually with an adhesive, to the surface of the load carrying member. Strain is sensed as a change in gage resistance. These changes are generally too small to be accurately measured directly and so it is common to employ two to four gages arranged in a Wheatstone bridge circuit. Independence from axial and bending loads as well as from temperature variations are obtained by using a four-gage bridge comprised of two diametrically opposite pairs of matched strain gages, each aligned along a *principal strain* direction. In round shafts (and other shapes used to transmit torque), tensile and compressive principal strains occur at 45° angles to the axis. Limiting strains, as determined from Equation 5.4 (with τ_m equal to the shear proportional limit of the shaft material), rarely exceed a few parts in 10^3. Typical practice is to increase the compliance of the sensing region (e.g., by reducing its diameter or with hollow or specially shaped sections) in order to attain the limiting strain at the highest value of the torque to be measured. This maximizes the measurement sensitivity.

Twist Angle

If the shaft is *slender* enough (e.g., $L > 5\ d$) ϕ, at limiting values of τ_m for typical shaft materials, can exceed 1°, enough to be resolved with sufficient accuracy for practical torque measurements (ϕ at τ_m can be found by manipulating Equations 5.3, 5.4, and 5.5). Figure 5.3(*b*) shows a common arrangement wherein torque is determined from the difference in tooth-space phasing between two identical "toothed" wheels attached at opposite ends of a compliant "torsion bar." The phase displacement of the periodic electrical signals from the two "pickups" is proportional to the peripheral displacement of salient features on the two wheels, and hence to the twist angle of the torsion bar and thus to the torque. These features are chosen to be sensible by any of a variety of noncontacting magnetic, optical, or capacitive techniques. With more elaborate pickups, the relative angular position of the two wheels appears as the amplitude of a *single* electrical signal, thus providing for the measurement of torque even on a stationary shaft (e.g., [13–15]). In still other constructions, a shaft-mounted variable displacement transformer or a related type of electric device is used to provide speed independent output signals proportional to ϕ.

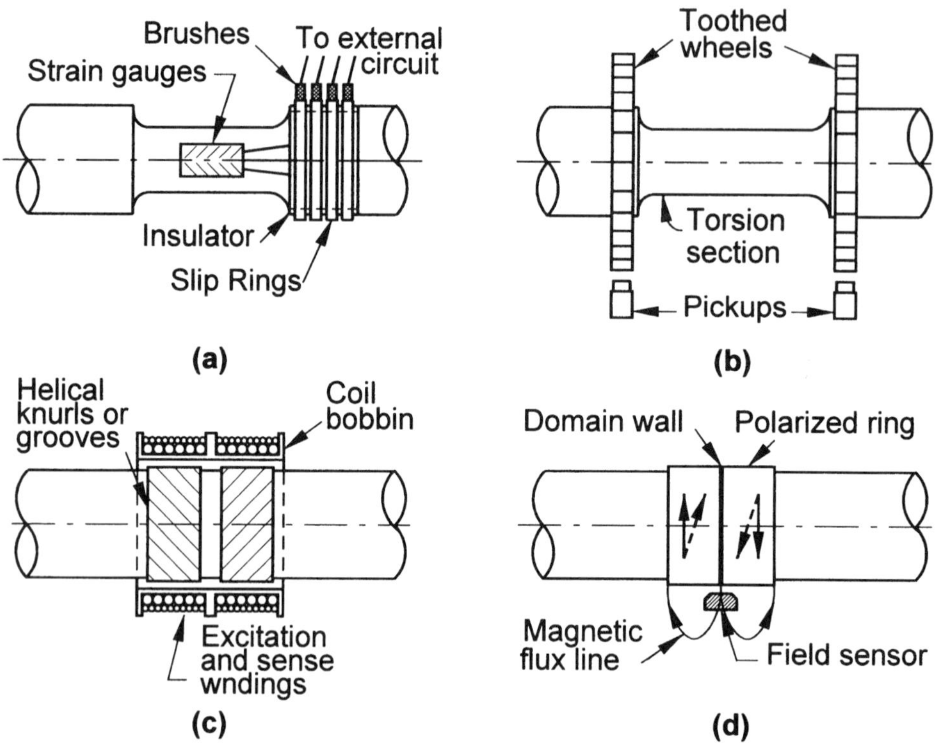

FIGURE 5.3 Four techniques in present use for measuring transmitted torque. (*a*) Torsional strain in the shaft alters the electrical resistance for four strain gages (two not seen) connected in a Wheatstone bridge circuit. In the embodiment shown, electrical connections are made to the bridge through slip rings and brushes. (*b*) Twist of the torsion section causes angular displacement of the surface features on the toothed wheels. This creates a phase difference in the signals from the two pickups. (*c*) The permeabilities of the two grooved regions of the shaft change oppositely with torsional stress. This is sensed as a difference in the output voltages of the two sense windings. (*d*) Torsional stress causes the initially circumferential magnetizations in the ring (solid arrows) to tilt (dashed arrows). These helical magnetizations cause magnetic poles to appear at the domain wall and ring ends. The resulting magnetic field is sensed by the field sensor.

Stress

In addition to elastic strain, the stresses by which torque is transmitted are manifested by changes in the magnetic properties of ferromagnetic shaft materials. This "magnetoelastic interaction" [8] provides an inherently noncontacting basis for measuring torque. Two types of magnetoelastic (sometimes called magnetostrictive) torque transducers are in present use: Type 1 derive output signals from torque-induced variations in magnetic circuit permeances; Type 2 create a magnetic field in response to torque. Type 1 transducers typically employ "branch," "cross," or "solenoidal" constructions [9]. In branch and cross designs, torque is detected as an imbalance in the permeabilities along orthogonal 45° helical paths (the principal stress directions) on the shaft surface or on the surface of an *ad hoc* material attached to the shaft. In solenoidal constructions torque is detected by differences in the *axial* permeabilities of two adjacent surface regions, preendowed with symmetrical magnetic "easy" axes (typically along the 45° principal stress directions). While branch and cross type sensors are readily miniaturized [10], local variations in magnetic properties of typical shaft surfaces limit their accuracy. Solenoidal designs, illustrated in Figure 5.3(*c*), avoid this pitfall by effectively averaging these variations. Type 2 transducers are generally constructed with a ring of magnetoelastically active material rigidly attached to the shaft. The ring is magnetized during manufacture of the transducer, usually with each axial half polarized in an

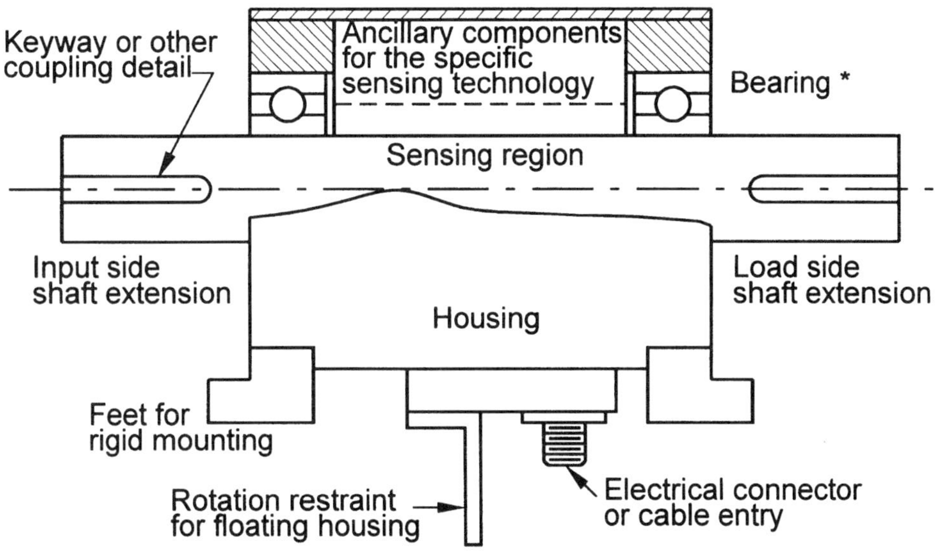

FIGURE 5.4 Modular torque transducer showing generic features and alternative arrangements for free floating or rigid mounting. Bearings* are used only on rotational models. Shaft extensions have keyways or other features to facilitate torque coupling.

opposite circumferential direction as indicated by the solid arrows in Figure 5.3(*d*) [11]. When torque is applied, the magnetizations tilt into helical directions (dashed arrows), causing magnetic poles to develop at the central domain wall and (of opposite polarity) at the ring end faces. Torque is determined from the output signal of one or more magnetic field sensors (e.g., Hall effect, magnetoresistive, or flux gate devices) mounted so as to sense the intensity and polarity of the magnetic field that arises in the space near the ring.

5.4 Torque Transducer Construction, Operation, and Application

Although a torque sensing region can be created directly on a desired shaft, it is more usual to install a preassembled *modular* torque transducer into the driveline. Transducers of this type are available with capacities from 0.001 N·m to 200,000 N·m. Operating principle descriptions and detailed installation and operating instructions can be found in the catalogs and literature of the various manufactures [12–20]. Tradenames often identify specific type of transducers; for example, *Torquemeters* [13] refers to a family of noncontact strain gage models; *Torkducer*® [18] identifies a line of Type 1 magnetoelastic transducers; *Torqstar*™ [12] identifies a line of Type 2 magnetoelastic transducers; *Torquetronic* [16] is a class of transducers using wrap-around twist angle sensors; and *TorXimitor*™ [20] identifies optoelectronic based, noncontact, strain gage transducers. Many of these devices show generic similarities transcending their specific sensing technology as well as their range. Figure 5.4 illustrates many of these common features.

Mechanical Considerations

Maximum operating speeds vary widely; upper limits depend on the size, operating principle, type of bearings, lubrication, and dynamic balance of the rotating assembly. Ball bearings, lubricated by grease, oil, or oil mist, are typical. Parasitic torques associated with bearing lubricants and seals limit the accuracy of low-end torque measurements. (Minute capacity units have no bearings [15]). Forced lubrication can allow operation up to 80,000 rpm [16]. High-speed operation requires careful consideration of the effects of centrifugal stresses on the sensed quantity as well as of critical (vibration inducing) speed ranges. Torsional oscillations associated with resonances of the shaft elasticity (characterized by its spring constant) with the rotational inertia of coupled masses can corrupt the measurement, damage the transducer by dynamic excursions above its rated overload torque, and *even be physically dangerous.*

Housings either *float* on the shaft bearings or are *rigidly mounted.* Free floating housings are restrained from rotating by such "soft" means as a cable, spring, or compliant bracket, or by an eccentric external feature simply resting against a fixed surface. In free floating installations, the axes of the driving and driven shafts must be carefully aligned. Torsionally rigid "flexible" couplings at each shaft end are used to accommodate small angular and/or radial misalignments. Alternatively, the use of dual flexible couplings at one end will allow direct coupling of the other end. Rigidly mounted housings are equipped with mounting feet or lugs similar to those found on the frame of electric motors. Free-floating models are sometimes rigidly mounted using adapter plates fastened to the housing. Rigid mountings are preferred when it is difficult or impractical to align the driving and driven shafts, as for example when driving or driven machines are changed often. Rigidly mounted housings *require* the use of dual flexible couplings at *both* shaft ends.

Modular transducers designed for zero or limited rotation applications have no need for bearings. To ensure that *all* of the torque applied at the ends is sensed, it is important in such "reaction"-type torque transducers to limit attachment of the housing to the shaft to only one side of the sensing region. Whether rotating or stationary, the external shaft ends generally include such torque coupling details as flats, keyways, splines, tapers, flanges, male/female squares drives, etc.

Electrical Considerations

By their very nature, transducers require some electrical input power or *excitation.* The "raw" output signal of the actual sensing device also generally requires "conditioning" into a level and format appropriate for display on a digital or analog meter or to meet the input requirements of data acquisition equipment. Excitation and signal conditioning are supplied by electronic circuits designed to match the characteristics of the specific sensing technology. For example, strain gage bridges are typically powered with 10 V to 20 V (dc or ac) and have outputs in the range of 1.5 mV to 3.0 mV per volt of excitation at the rated load. Raising these millivolt signals to more usable levels requires amplifiers having gains of 100 or more. With ac excitation, oscillators, demodulators (or rectifiers) are also needed. Circuit elements of these types are normal when inductive elements are used either as a necessary part of the sensor or simply to implement noncontact constructions.

Strain gages, differential transformers, and related sensing technologies require that electrical components be mounted *on* the torqued member. Bringing electrical power to and output signals from these components on rotating shafts require special methods. The most direct and common approach is to use conductive means wherein brushes (typically of silver graphite) bear against (silver) slip rings. Useful life is extended by providing means to lift the brushes off the rotating rings when measurements are not being made. Several "noncontacting" methods are also used. For example, power can be supplied via inductive coupling between stationary and rotating transformer windings [12–15], by the illumination of shaft mounted photovoltaic cells [20], or even by batteries strapped to the shaft [21] (limited by centrifugal force to relatively low speeds). Output signals are coupled off the shaft through rotary transformers, by frequency-modulated (infrared) LEDs [19, 20], or by radio-frequency (FM) telemetry [21]. Where shaft rotation is limited to no more than a few full rotations, as in steering gear, valve actuators or oscillating mechanisms, hard wiring both power and signal circuits is often suitable. Flexible cabling minimizes incidental torques and makes for a long and reliable service life. All such wiring considerations are avoided when noncontact technologies or constructions are used.

Costs and Options

Prices of torque transducers reflect the wide range of available capacities, performance ratings, types, styles, optional features, and accessories. In general, prices of any one type increase with increasing capacity. Reaction types cost about half of similarly rated rotating units. A typical foot-mounted, 565 N·m capacity, strain gage transducer with either slip rings or rotary transformers and integral speed sensor, specified nonlinearity and hysteresis each within ±0.1%, costs about $4000 (1997). Compatible instrumentation providing transducer excitation, conditioning, and analog output with digital display of torque and speed costs about $2000. A comparable magnetoelastic transducer with ±0.5% accuracy costs about $1300. High-capacity transducers for extreme speed service with appropriate lubrication options can cost more than $50,000. Type 2 magnetoelastic transducers, mass produced for automotive power steering applications, cost approximately $10.

5.5 Apparatus for Power Measurement

Rotating machinery exists in specific types without limit and can operate at power levels from fractions of a watt to some tens of megawatts, a range spanning more than 10^8. Apparatus for power measurement exists in a similarly wide range of types and sizes. Mechanical power flows from a *driver* to a *load.* This power can be determined *directly* by application of Equation 5.7, simply by measuring, in addition to ω, the output torque of the driver or the input torque to the load, whichever is the device under test (DUT). When the DUT is a driver, measurements are usually required over its full service range of speed and torque. The test apparatus therefore must act as a controllable load and be able to *absorb* the delivered power. Similarly, when the DUT is a pump or fan or other type of load, or one whose function is simply to alter speed and torque (e.g., a gear box), the test apparatus must include a *driver* capable of supplying power over the DUT's full rated range of torque and speed. Mechanical power can also be determined *indirectly* by conversion into (or from) another form of energy (e.g., heat or electricity) and measuring the relevant calorimetric or electrical quantities. In view of the wide range of readily available methods and apparatus for accurately measuring both torque and speed, indirect methods need only be considered when special circumstances make direct methods difficult.

Dynamometer is the special name given to the power-measuring apparatus that includes absorbing or/and driving means and wherein torque is determined by the reaction forces on a stationary part (the *stator*). An effective dynamometer is conveniently assembled by mounting the DUT in such a manner as to allow measurement of the reaction torque on its frame. Figure 5.5 shows a device designed to facilitate such measurements. Commercial models (Torque Table® [12]) rated to support DUTs weighing 222 N to 4900 N are available with torque capacities from 1.3 N·m 226 to N·m. "Torque tubes" [4] or other DUT mounting arrangements are also used. Other than for possible rotational/elastic resonances, these systems have no speed limitations. More generally, and especially for large machinery, dynamometers include a specialized driving or absorbing machine. Such dynamometers are classified according to their function as *absorbing* or *driving* (sometimes *motoring*). A *universal dynamometer* can function as either a driver or an absorber.

Absorption Dynamometers

Absorption dynamometers, often called *brakes* because their operation depends on the creation of a controllable *drag* torque, convert mechanical work into heat. A drag torque, as distinguished from an active torque, can act only to restrain and not to initiate rotational motion. Temperature rise within a dynamometer is controlled by carrying away the heat energy, usually by transfer to a moving fluid, typically air or water. Drag torque is created by inherently dissipative processes such as: friction between rubbing surfaces, shear or turbulence of viscous liquids, the flow of electric current, or magnetic hysteresis. Gaspard Riche de Prony (1755–1839), in 1821 [22], invented a highly useful form of a friction brake to meet the needs for testing the steam engines that were then becoming prevalent. Brakes of this

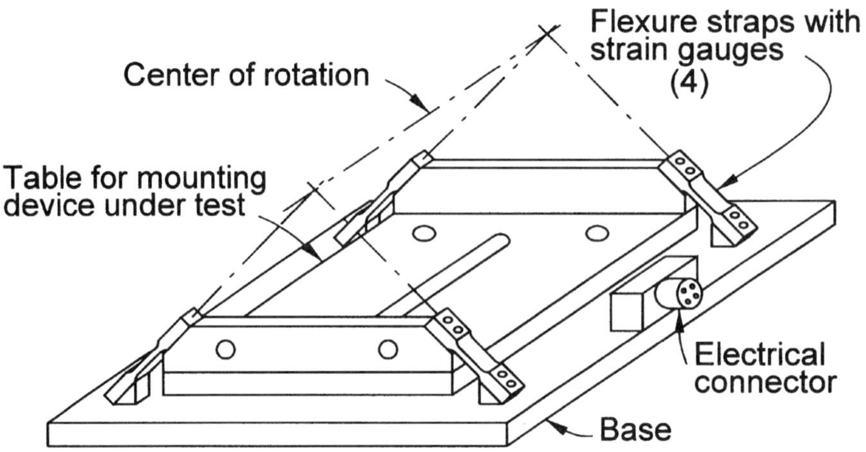

FIGURE 5.5 Support system for measuring the reaction torque of a rotating machine. The axis of the machine must be accurately set on the "center of rotation." The holes and keyway in the table facilitate machine mounting and alignment. Holes in the front upright provide for attaching a lever arm from which calibrating weights may be hung [4, 11].

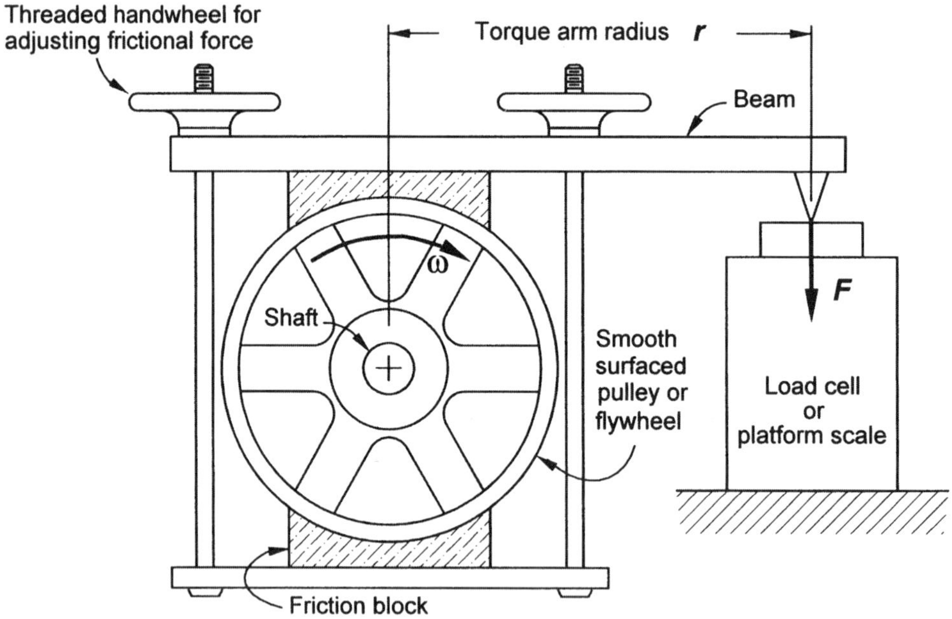

FIGURE 5.6 A classical prony brake. This brake embodies the defining features of all absorbing dynamometers: conversion of mechanical work into heat and determination of power from measured values of reaction torque and rotational velocity.

type are often used for instructional purposes, for they embody the general principles and major operating considerations for all types of absorption dynamometers. Figure 5.6 shows the basic form and constructional features of a *prony brake.* The power that would normally be delivered by the shaft of the driving engine to the driven load is (for measurement purposes) converted instead into heat via the work done by the frictional forces between the friction blocks and the flywheel rim. Adjusting the tightness of the clamping bolts varies the frictional drag torque as required. Heat is removed from the inside surface of the rim by arrangements (not shown) utilizing either a continuous flow or evaporation of water. There is no need to know the magnitude of the frictional forces nor even the radius of the flywheel (facts recognized by Prony), because, while the drag torque tends to rotate the clamped-on apparatus, it is held stationary by the equal but opposite reaction torque Fr. F at the end of the torque arm of radius r (a fixed dimension of the apparatus) is monitored by a scale or load cell. The power is found from Equations 5.1 and 5.7 as $P = Fr\omega = Fr2\pi N/60$ where N is in rpm.

Uneven retarding forces associated with fluctuating coefficients of friction generally make rubbing friction a poor way to generate drag torque. Nevertheless, because they can be easily constructed, *ad hoc* variations of prony brakes, often using only bare ropes or wooden cleats connected by ropes or straps, find use in the laboratory or wherever undemanding or infrequent power measurements are to be made. More sophisticated prony brake constructions are used in standalone dynamometers with self-contained cooling water tanks in sizes up to 746 kW (1000 Hp) for operation up to 3600 rpm with torques to 5400 N·m [23]. Available in stationary and mobile models, they find use in testing large electric motors as well as engines and transmissions on agricultural vehicles. Prony brakes allow full drag torque to be imposed down to zero speed.

William Froude (1810–1879) [24] invented a *water brake* (1877) that does not depend on rubbing friction. Drag torque within a *Froude brake* is developed between the rotor and the stator by the momentum imparted by the rotor to water contained within the brake casing. Rotor rotation forces the water to circulate between cup-like pockets cast into facing surfaces of both rotor and stator. The rotor is supported in the stator by bearings that also fix its axial position. Labyrinth-type seals prevent water leakage while minimizing frictional drag and wear. The stator casing is supported in the dynamometer frame in cradle fashion by *trunnion* bearings. The torque that prevents rotation of the stator is measured by reaction forces in much the same manner as with the prony brake. Drag torque is adjusted by a valve, controlling either the back pressure in the water outlet piping [25] or the inlet flow rate [26] or sometimes (to allow very rapid torque changes) with two valves controlling both [27]. In any case, the absorbed energy is carried away by the continuous water flow. Other types of cradle-mounted water brakes, while externally similar, have substantially different internal constructions and depend on other principles for developing the drag torque (e.g., smooth rotors develop viscous drag by shearing and turbulence). Nevertheless, all *hydraulic dynamometers* purposefully function as *inefficient* centrifugal pumps. Regardless of internal design and valve settings, maximum drag torque is low at low speeds (zero at standstill) but can rise rapidly, typically varying with the square of rotational speed. The irreducible presence of some water, as well as windage, places a speed-dependent lower limit on the *controllable* drag torque. In any one design, wear and vibration caused by cavitation place upper limits on the speed and power level. Hydraulic dynamometers are available in a wide range of capacities between 300 kW and 25,000 kW, with some portable units having capacities as low as 75 kW [26]. The largest ever built [27], absorbing up to about 75,000 kW (100,000 Hp), has been used to test propulsion systems for nuclear submarines. Maximum speeds match the operating speeds of the prime movers that they are built to test and therefore generally decrease with increasing capacity. High-speed gas turbine and aerospace engine test equipment can operate as high as 30,000 rpm [25].

In 1855, Jean B. L. Foucault (1819–1868) [22] demonstrated the conversion of mechanical work into heat by rotating a copper disk between the poles of an electromagnet. This simple means of developing drag torque, based on *eddy currents*, has, since circa 1935, been widely exploited in dynamometers. Figure 5.7 shows the essential features of this type of brake. Rotation of a toothed or spoked steel rotor through a spatially uniform magnetic field, created by direct current through coils in the stator, induces locally circulating (eddy) currents in electrically conductive (copper) portions of the stator. Electromagnetic forces between the rotor, which is magnetized by the uniform field, and the field arising from the eddy currents, create the drag torque. This torque, and hence the mechanical input power, are controlled by adjusting the *excitation* current in the stator coils. Electric input power is less than 1% of the rated capacity. The dynamometer is effectively an internally short-circuited generator because the power associated with the resistive losses from the generated eddy currents is dissipated *within* the machine.

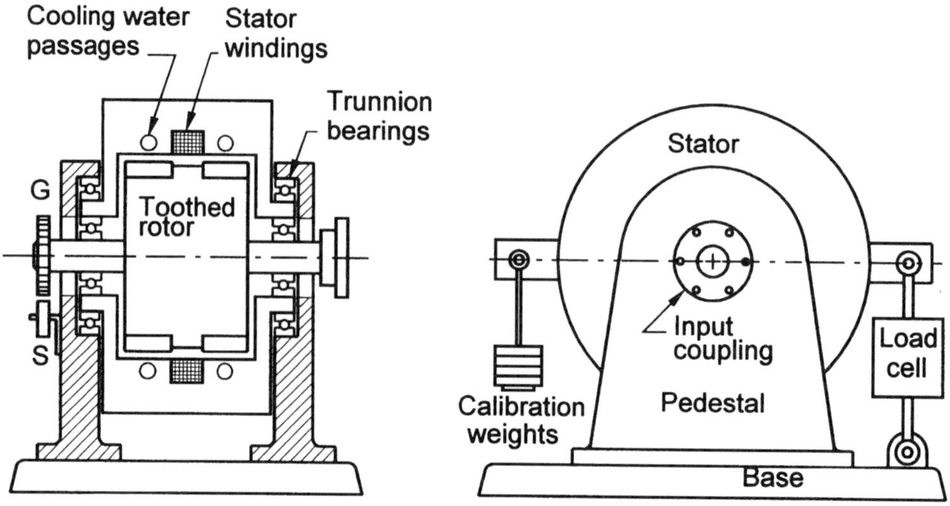

FIGURE 5.7 Cross-section (left) and front view (right) of an eddy current dynamometer. G is a gear wheel and S is a speed sensor. Hoses carrying cooling water and cable carrying electric power to the stator are not shown.

Being heated by the flow of these currents, the stator must be cooled, sometimes (in smaller capacity machines) by air supplied by blowers [23], but more often by the continuous flow of water [25, 27, 28]. In *dry gap* eddy current brakes (the type shown in Figure 5.7), water flow is limited to passages within the stator. Larger machines are often of the *water in gap* type, wherein water also circulates around the rotor [28]. Water in contact with the moving rotor effectively acts as in a water brake, adding a nonelectromagnetic component to the total drag torque, thereby placing a lower limit to the controllable torque. Windage limits the minimum value of controllable torque in dry gap types. Since drag torque is developed by the motion of the rotor, it is zero at standstill for any value of excitation current. Initially rising rapidly, approximately linearly, with speed, torque eventually approaches a current limited saturation value. As in other cradled machines, the torque required to prevent rotation of the stator is measured by the reaction force acting at a fixed known distance from the rotation axis. Standard model eddy current brakes have capacities from less than 1 kW [23, 27] to more than 2000 kW [27, 28], with maximum speeds from 12,000 rpm in the smaller capacity units to 3600 rpm in the largest units. Special units with capacities of 3000 Hp (2238 kW) at speeds to 25,000 rpm have been built [28].

Hysteresis brakes [29] develop drag torque via magnetic attractive/repulsive forces between the magnetic poles established in a reticulated stator structure by a current through the field coil, and those created in a "drag cup" rotor by the stator field gradients. Rotation of the special steel rotor, through the spatial field pattern established by the stator, results in a cyclical reversal of the polarity of its local magnetizations. The energy associated with these reversals (proportional to the area of the hysteresis loop of the rotor material) is converted into heat within the drag cup. Temperature rise is controlled by forced air cooling from a blower or compressed air source. As with eddy current brakes, the drag torque of these devices is controlled by the excitation current. In contrast with eddy current brakes, rated drag torque is available down to zero speed. (Eddy current effects typically add only 1% to the drag torque for each 1000 rpm). As a result of their smooth surfaced rotating parts, hysteresis brakes exhibit low parasitic torques and hence cover a dynamic range as high as 200 to 1. Standard models are available having continuous power capacities up to 6 kW (12 kW with two brakes in tandem cooled by two blowers). Intermittent capacities per unit (for 5 min or less) are 7 kW. Some low-capacity units are convection cooled; the smallest has a continuous rating of just 7 W (35 W for 5 min). Maximum speeds range from 30,000 rpm for the smallest to 10,000 rpm for the largest units. Torque is measured by a strain gage bridge on a moment arm supporting the machine stator.

Driving and Universal Dynamometers

Electric generators, both ac and dc, offer another means for developing a controllable drag torque and they are readily adapted for dynamometer service by cradle mounting their stator structures. Moreover, electric machines of these types can also operate in a motoring mode wherein they can deliver controllable *active* torque. When configured to operate selectively in either driving or absorbing modes, the machine serves as a universal dynamometer. With dc machines in the absorbing mode, the generated power is typically dissipated in a convection-cooled resistor bank. Air cooling the machine with blowers is usually adequate, since *most* of the mechanical power input is dissipated externally. Nevertheless, *all* of the mechanical input power is accounted for by the product of the reaction torque and the rotational speed. In the motoring mode, torque and speed are controlled by adjustment of both field and armature currents. Modern ac machines utilize regenerative input power converters to allow braking power to be returned to the utility power line. In the motoring mode, speed is controlled by high-power, solid-state, adjustable frequency inverters. Internal construction is that of a simple three-phase induction motor, having neither brushes, slip rings, nor commutators. The absence of rotor windings allows for higher speed operation than dc machines. Universal dynamometers are "four-quadrant" machines, a term denoting their ability to produce torque in the same or opposite direction as their rotational velocity. This unique ability allows the effective drag torque to be reduced to zero at any speed. Universal dynamometers [25, 28] are available in a relatively limited range of capacities (56 to 450 kW), with commensurate torque (110 to 1900 N·m) and speed (4500 to 13,500 rpm) ranges, reflecting their principal application in automotive engine development. Special dynamometers for testing transmissions and other vehicular drive train components insert the DUT between a diesel engine or electric motor prime mover and a hydraulic or eddy current brake [30].

Measurement Accuracy

Accuracy of power measurement (see discussion in [4]) is generally limited by the torque measurement ($\pm$0.25% to $\pm$1%) since rotational speed can be measured with almost any desired accuracy. Torque errors can arise from the application of extraneous (i.e., not indicated) torques from hose and cable connections, from windage of external parts, and from miscalibration of the load cell. Undetected friction in the trunnion bearings of cradled dynamometers can compromise the torque measurement accuracy. Ideally, well-lubricated antifriction bearings make no significant contribution to the restraining torque. In practice, however, the unchanging contact region of the balls or other rolling elements on the bearing races makes them prone to brinelling (a form of denting) from forces arising from vibration, unsupported weight of attached devices, or even inadvertently during the alignment of connected machinery. The problem can be alleviated by periodic rotation of the (primarily outer) bearing races. In some bearing-in-bearing constructions, the central races are continuously rotated at low speeds by an electric motor while still others avoid the problem by supporting the stator on hydrostatic oil lift bearings [28].

Costs

The wide range of torque, speed, and power levels, together with the variation in sophistication of associated instrumentation, is reflected in the very wide range of dynamometer prices. Suspension systems of the type illustrated in Figure 5.5 (for which the user must supply the rotating machine) cost $4000 to $6000, increasing with capacity [12]. A 100 Hp (74.6 kW) *portable* water brake equipped with a strain gage load cell and a digital readout instrument for torque, speed, and power costs $4500, or $8950 with more sophisticated data acquisition equipment [26]. Stationary (and some *transportable* [23]) hydraulic dynamometers cost from $113/kW in the smaller sizes [25], down to $35/kW for the very largest [27]. Transportation, installation, and instrumentation can add significantly to these costs. Eddy current dynamometers cost from as little as $57/kW to nearly $700/kW, depending on the rated capacity, type of control system, and instrumentation [24, 25, 28]. Hysteresis brakes with integral speed sensors cost from \$3300 to \$14,000 according to capacity [29]. Compatible controllers, from manual to fully programmable for PC test control and data acquisition via an IEEE-488 interface, vary in price from \$500 to \$4200. The flexibility and high performance of ac universal dynamometers is reflected in their comparatively high prices of \$670 to \$2200/kW [25, 28].

References

1. Pinney, C. P. and Baker, W. E., Velocity Measurement, *The Measurement, Instrumentation and Sensors Handbook,* Webster, J. G., ed., Boca Raton, FL: CRC Press, 1999.
2. S. Timoshenko, *Strength of Materials,* 3rd ed., New York: Robert E. Kreiger, Part I, 281–290; Part II, 235–250, 1956.
3. S. J. Citron, *On-line engine torque and torque fluctuation measurement for engine control utilizing crankshaft speed fluctuations,* U. S. Patent No. 4,697,561, 1987.
4. Supplement to ASME Performance Test Codes, Measurement of Shaft Power, ANSI/ASME PTC 19.7-1980 (Reaffirmed 1988).
5. See, for example, the catalog of torque wrench products of Consolidated Devices, Inc., 19220 San Jose Ave., City of Industry, CA 91748.
6. I. J. Garshelis, C. R. Conto, and W. S. Fiegel, A single transducer for non-contact measurement of the power, torque and speed of a rotating shaft, SAE Paper No. 950536, 1995.
7. C. C. Perry and H. R. Lissner, *The Strain Gage Primer,* 2nd ed., New York: McGraw-Hill, 1962, 9. (This book covers all phases of strain gage technology.)
8. B. D. Cullity, *Introduction to Magnetic Materials,* Reading, MA: Addison-Wesley, 1972, Section 8.5, 266–274.
9. W. J. Fleming, Magnetostrictive torque sensors—comparison of branch, cross and solenoidal designs, SAE Paper No. 900264, 1990.
10. Y. Nonomura, J. Sugiyama, K. Tsukada, M. Takeuchi, K. Itoh, and T. Konomi, Measurements of engine torque with the intra-bearing torque sensor, SAE Paper No. 870472, 1987.
11. I. J. Garshelis, *Circularly magnetized non-contact torque sensor and method for measuring torque using same,* U.S. Patent 5,351,555, 1994 and 5,520,059, 1996.
12. Lebow® Products, Siebe, plc., 1728 Maplelawn Road, Troy, MI 48099, Transducer Design Fundamentals/Product Listings, Load Cell and Torque Sensor Handbook No. 710, 1997, also: Torqstar™ and Torque Table®.
13. S. Himmelstein & Co., 2490 Pembroke, Hoffman Estates, IL 60195, MCRT® Non-Contact Strain Gage Torquemeters and Choosing the Right Torque Sensor.
14. Teledyne Brown Engineering, 513 Mill Street, Marion, MA 02738-0288.
15. Staiger, Mohilo & Co. GmbH, Baumwasenstrasse 5, D-7060 Schorndorf, Germany (In the U.S.: Schlenker Enterprises Ltd., 5143 Electric Ave., Hillside, IL 60162), Torque Measurement.
16. Torquemeters Ltd., Ravensthorpe, Northampton, NN6 8EH, England (In the U.S.: Torquetronics Inc., P.O. Box 100, Allegheny, NY 14707), Power Measurement.
17. Vibrac Corporation, 16 Columbia Drive, Amherst, NH 03031, Torque Measuring Transducer.
18. GSE, Inc., 23640 Research Drive, Farmington Hills, MI 48335-2621, Torkducer®.
19. Sensor Developments Inc., P.O. Box 290, Lake Orion, MI 48361-0290, 1996 Catalog.
20. Bently Nevada Corporation, P.O. Box 157, Minden, NV 89423, TorXimitor™.
21. Binsfield Engineering Inc., 4571 W. MacFarlane, Maple City, MI 49664.
22. C. C. Gillispie (ed.), *Dictionary of Scientific Biography,* Vol. XI, New York: Charles Scribner's Sons, 1975.
23. AW Dynamometer, Inc., P.O. Box 428, Colfax, IL 61728, Traction dynamometers: Portable and stationary dynamometers for motors, engines, vehicle power take-offs.
24. Roy Porter (ed.), *The Biographical Dictionary of Scientists,* 2nd ed., New York: Oxford University Press, 1994.

25. Froude-Consine Inc., 39201 Schoolcraft Rd., Livonia, MI 48150, F Range Hydraulic Dynamometers, AG Range Eddy Current Dynamometers, AC Range Dynamometers.
26. Go-Power Systems, 1419 Upfield Drive, Carrollton, TX 75006, Portable Dynamometer System, Go-Power Portable Dynamometers.
27. Zöllner GmbH, Postfach 6540, D-2300 Kiel 14, Germany (In the U.S. and Canada: Roland Marine Inc., 90 Broad St., New York, NY 10004), Hydraulic Dynamometers Type P, High Dynamic Hydraulic Dynamometers.
28. Dynamatic Corporation, 3122 14th Ave., Kenosha, WI 53141-1412, Eddy Current Dynamometer—Torque Measuring Equipment, Adjustable Frequency Dynamometer.
29. Magtrol, Inc., 70 Gardenville Parkway, Buffalo, NY 14224-1322, Hysteresis Absorption Dynamometers.
30. Hicklin Engineering, 3001 NW 104th St., Des Moines, IA 50322, Transdyne™ (transmission test systems, brake and towed chassis dynamometers).

6

Tactile Sensing

R. E. Saad
University of Toronto

A. Bonen
University of Toronto

K. C. Smith
University of Toronto

B. Benhabib
University of Toronto

Robots in industrial settings perform repetitive tasks, such as machine loading, parts assembly, painting, and welding. Only in rare instances can these autonomous manipulators modify their actions based on sensory information. Although, thus far, a vast majority of research work in the area of robot sensing has concentrated on computer vision, contact sensing is an equally important feature for robots and has received some attention as well. Without tactile-perception capability, a robot cannot be expected to effectively grasp objects. In this context, robotic tactile sensing is the focus of this chapter.

6.1 Sensing Classification

Robotic sensing can be classified as either of the noncontact or contact type [1]. *Noncontact sensing* involves interaction between the robot and its environment by some physical phenomenon, such as acoustic or electromagnetic waves, that interact without contact. The most important types of robotic sensors of the noncontact type are vision and proximity sensors. *Contact sensing*, on the other hand, implies measurement of the general interaction that takes place when the robot's end effector is brought into contact with an object. Contact sensing is further classified into force and tactile sensing.

Force sensing is defined as the measurement of the global mechanical effects of contact, while *tactile sensing* implies the detection of a wide range of local parameters affected by contact. Most significant among those contact-based effects are contact stresses, slippage, heat transfer, and hardness.

The properties of a grasped object that can be derived from tactile sensing can be classified into geometric and dynamometric types [2]. Among the geometric properties are presence, location in relation to the end-effector, shape and dimensions, and surface conditions [3–7]. Among the dynamometric parameters associated with grasping are: force distribution, slippage, elasticity and hardness, and friction [8–12].

Tactile sensing requires sophisticated transducers; yet the availability of these transducers alone is not a sufficient condition for successful tactile sensing. It is also necessary to accurately control the modalities through which the tactile sensor interacts with the explored objects (including contact forces, as well as end-effector position and orientation) [13–15]. This leads to active tactile sensing, which requires a high degree of complexity in the acquisition and processing of the tactile data [16].

0-8493-0047-9/00/$0.00+$.50

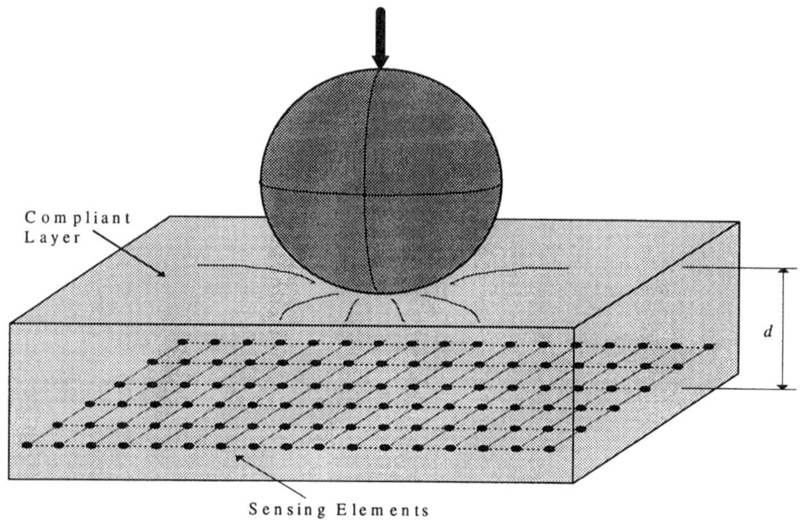

FIGURE 6.1 An object indenting a compliant layer, where an array of force-sensing elements is placed at a distance *d* from the surface.

6.2 Mechanical Effects of Contact

Tactile sensing normally involves a rigid object indenting the compliant cover layer of a tactile sensor array [17], Figure 6.1. The indentation of a compliant layer due to contact can be analyzed from two conceptually different points of view [1]. The first one is the measurement of the actual contact stresses (force distribution) in the layer, which is usually relevant to controlling manipulation tasks. The second one is the deflection profile of the layer, which is usually important for recognizing geometrical object features. Depending on the approach adopted, different processing and control algorithms must be utilized.

There exists a definite relationship between the local shape of a contacting body and a set of subsurface strains (or displacements); however, this relationship is quite complex. Thus, it requires the use of the Theory of Elasticity and Contact Mechanics to model sensor–object interaction [18], and the use of Finite Element Analysis (FEA) as a practical tool for obtaining a more representative model of the sensor [19].

In general, the study of tactile sensors comprises two steps: (1) the *forward analysis*, related to the acquisition of data from the sensor (changes on the stress or strains, induced by the indentation of an object on the compliant surface of the transducer); and, (2) the *inverse problem*, normally related to the recovery of force distribution or, in some cases, the recovery of the indentor's shape.

Simplified Theory for Tactile Sensing

For simplicity, the general two-dimensional tactile problem is reduced herein to a one-dimensional one. Figure 6.2 shows a one-dimensional transducer that consists of a compliant, homogeneous, isotropic, and linear layer subjected to a normal stress $q_v(x)$ created by the indentation of an object.

For modeling purposes, it is assumed that the compliant layer is an elastic half-space. This simplification yields closed-form equations for the analysis and avoids the formation of a more complex problem, in which the effect of the boundary conditions at x_{min} and x_{max} must be taken into account. It has been proven that the modeling of the sensor by an elastic half-space represents a reasonable approximation to the real case [18]. Under these conditions, it can be shown that the normal strain, at a depth $y = d$, due to the normal stress $q_v(y)$ is given by [20]:

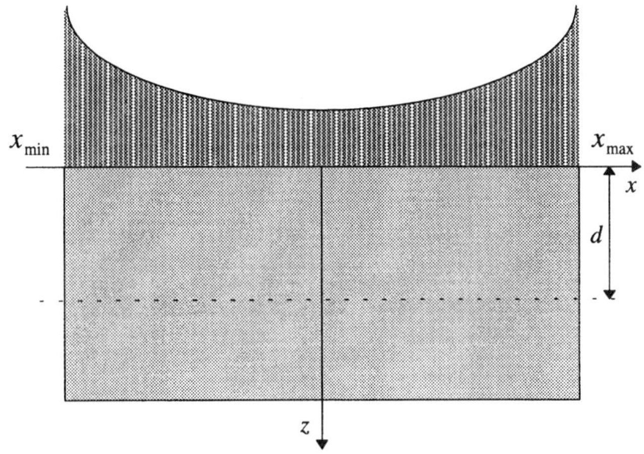

FIGURE 6.2 Ideal one-dimensional transducer subjected to a normal stress.

$$\varepsilon_z(x) = \int_{-\infty}^{\infty} q_v(x - x_0)\, h_z(x_0, d)\, dx_0 \tag{6.1}$$

where ε_z is the strain at x and $z = d$ due to the normal stress on the surface, and

$$h_z(x) = -\frac{2d(1+v)\left[d^2(1-v) - vx^2\right]}{\pi r E\left(x^2 + d^2\right)^2} \tag{6.2}$$

E and v are, respectively, the modulus of elasticity and the Poisson's coefficient of the compliant layer. In obtaining Equation 6.2, it is assumed that the analysis is performed under *planar strain* conditions. It should be noted that a similar analysis can be performed for tangential contact stresses or strains.

The normal displacement at the surface, w, is given by:

$$w(x) = \int_{-\infty}^{\infty} q_v(x - x_0)\, k(x_0)\, dx_0 \tag{6.3}$$

where

$$k(x) = \frac{-2\left(1 - v^2\right)}{\pi E} \log\left|\frac{x}{x_a}\right| \tag{6.4}$$

The singularity at $x = 0$ is expected due to the singularity of stress at that point. Note that, $k(x)$ is the deformation of the surface when a singular load of 1 N is applied at $x = 0$. The constant x_a should be chosen such that at $x = x_a$, the deformation is zero. In this case, zero deformation should occur at $x \to \infty$ (note that it has been assumed that the sensor is modeled by an elastic half space), namely $x_a \to \infty$. This problem is associated with the two-dimensional deformation of an elastic half-space. To eliminate this difficulty, the boundary conditions of the transducer must be taken into account (i.e., a finite transducer must be analyzed), which requires, in general, the use of FEA.

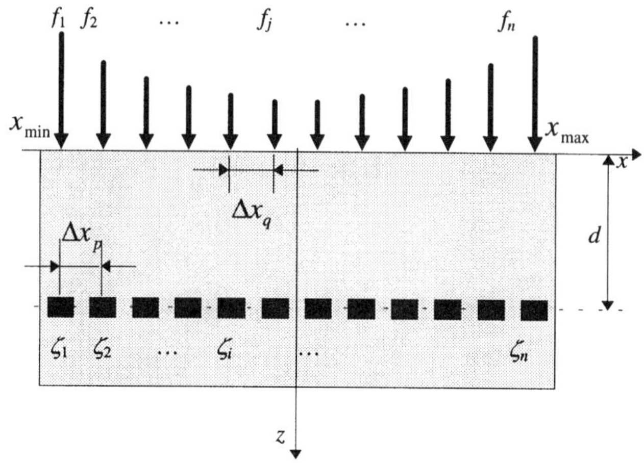

FIGURE 6.3 One-dimensional transducer with discrete sensing elements located at $z = d$.

Since measurements of strain (or stress) are usually done by a discrete number of sensing elements, Equation 6.2 must be discretized (Figure 6.3). Correspondingly, the force distribution must be reconstructed at discrete positions as shown in Figure 6.3. Let Δx_q be the distance between points, where the force distribution must be reconstructed from strain (or stress) measurements carried out by strain (or stress) sensing elements uniformly distributed at intervals Δx_p, at $z = d$. Also assume, even though it is not necessary, that $\Delta x_q = \Delta x_p = \Delta x$ and that the forces are applied at positions immediately above the sensor elements. One can now define the strain (stress)-sample vector, ζ, whose components are given by $\zeta_i = \varepsilon_x(x_i)$, $i = 1, 2, \ldots, n$, and the force distribution vector, F, whose components are given by $f_i = q_v(x_j)$, $j = 1, 2, \ldots$, n. Then, the discrete form of Equation 6.1 is given by:

$$\zeta = \mathbf{TF} \tag{6.5}$$

where the elements of the matrix **T** are given by $T_{ij} = k_v(x_i - x_j)$, i = 1, 2, ..., n and $j = 1, 2, \ldots, n$ [23]. A similar relation to Equation 6.5 can be obtained discretizing Equation 6.3. In the general case, where $\Delta x_q \neq \Delta x_p$, **T** is not square. Furthermore, in the general case, the vector **F** comprises both vertical and tangential components.

Equations 6.1 and 6.3 represent the regular *forward problem*, while Equation 6.5 represents the discretized version of the forward problem. The *inverse problem*, in most cases, consists of recovering the applied force profile from the measurements of strain, stress, or deflection. (Note that the surface displacement can also be used to recover the indentor's shape.)

In [20], it was shown that the inverse problem is ill-posed because the operators h and k, of Equations 6.1 and 6.3, respectively, are ill-conditioned. Consequently, the inverse problem is susceptible to noise. To solve this problem, regularization techniques must be utilized [20].

It has been proven that, in order to avoid aliasing in determining the continuous strain (stress) at a depth d using a discretized transducer, the elements have to be separated by one tenth of the compliant layer's thickness. However, good results were obtained, without much aliasing, by separating the sensing elements by a distance equal to the sensor's depth [18].

Requirements for Tactile Sensors

In 1980, Harmon conducted a survey to determine general specifications for tactile sensors [21]. Those specifications have been used subsequently as guidelines by many tactile sensor designers:

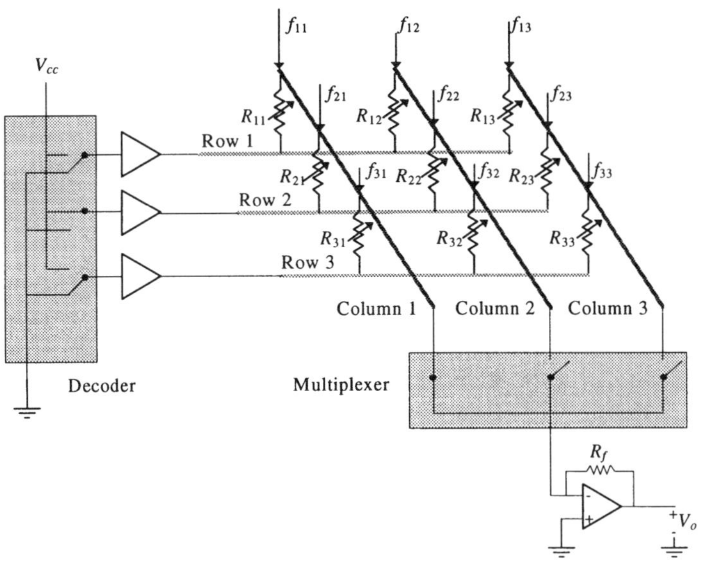

FIGURE 6.4 General configuration of a resistive transducer.

1. Spatial resolution of 1 to 2 mm
2. Array sizes of 5 × 10 to 10 × 20 points
3. Sensitivity of 0.5×10^{-2} to 1×10^{-2} N for each force-sensing element (tactel)
4. Dynamic range of 1000:1
5. Stable behavior and with no hysteresis
6. Sampling rate of 100 Hz to 1 kHz
7. Monotonic response, though not necessarily linear
8. Compliant interface, rugged and inexpensive

While properties (5), (7), and (8) above should apply to any practical sensor, the others are merely suggestions, particularly with respect to the number of array elements and spatial resolution.

Developments on tactile sensing following [21] have identified additional desirable qualities; namely, reliability, modularity, speed, and the availability of multisensor support [16].

6.3 Technologies for Tactile Sensing

The technologies associated with tactile sensing are quite diverse: extensive surveys of the state-of-the-art of robotic-tactile-transduction technologies have been presented in [2, 3, 16, 17]. Some of these technologies will be briefly discussed.

Resistive

The transduction method that has received the most attention in tactile sensor design is concerned with the change in resistance of a conductive material under applied pressure. A basic configuration of a resistive transducer is shown in Figure 6.4. Each resistor, whose value changes with the magnitude of the force, represents a resistive cell of the transducer. Different materials have been utilized to manufacture the basic cell.

Conductive elastomers were among the first resistive materials used for the development of tactile sensors. They are insulating, natural or silicone-based rubbers made conductive by adding particles of conductive or semiconductive materials (e.g., silver or carbon). The changes in resistivity of the elastomers under pressure are produced basically by two different physical mechanisms. In the first approach, the change in resistivity of the elastomer under pressure is associated with deformation that alters the particle density within it. Two typical designs of this kind are given in [22, 23]. In the second approach, while the bulk resistance of the elastomer changes slightly when it is compressed, the design allows the increase of the area of contact between the elastomer and an electrode, and correspondingly a change in the contact resistance. A typical design of this kind is given in [24]. In [25], a newer tactile sensor is reported with both three-axis force sensing and slippage sensing functions. In the former case, the pressure sensing function is achieved utilizing arrays of pressure transducers that measure a change in contact resistance between a specially treated polyimide film and a resistive substrate.

Piezoresistive elements have also been used in several tactile sensors. This technology is specifically attractive at present because, with micromachining, the piezoresistive elements can be integrated together with the signal-processing circuits in a single chip [26]. A 32 × 32-element silicon pressure sensor array incorporating CMOS processing circuits for the detection of a high-resolution pressure distribution was reported in [8]. The sensor array consists of an x–y-matrix-organized array of pressure cells with a cell spacing of 250 μm. CMOS processing circuits are formed around the array on the same chip. Fabrication of the sensor array was carried out using a 3 mm CMOS process combined with silicon micromachining techniques. The associated diaphragm size is 50 μm × 50 μm. The overall sensor-array chip size is 10 mm × 10 mm.

In Figure 6.4, a circuit topology, to scan a 3 × 3 array of piezoresistive elements, is shown. The basic idea was originally proposed in [24] and adapted on several occasions by different researchers. Using this method, the changes in resistance are converted into voltages at the output. With the connections as shown in Figure 6.4, the resistance R_{21} can be determined from:

$$V_0 = \frac{R_f}{R_{21}} V_{cc} \tag{6.6}$$

where V_o is the output voltage, V_{cc} is the bias voltage, and R_f is the feedback resistance of the output amplifier stage.

One problem with the configuration shown in Figure 6.4 is the difficulty in detecting small changes in resistance due to the internal resistance of the multiplexer as well changes in the voltage of power source, which have a great influence at the output. Other methods utilized to scan resistive transducer arrays are summarized in [3].

When piezoresistors and circuits are fabricated on the same silicon substrate, the sensor array can be equipped with a complex switching circuit, next to the sensing elements, that allows a better resolution in the measurements [9].

Capacitive

Tactile sensors within this category are concerned with measuring capacitance, which varies under applied load. The capacitance of a parallel-plate capacitor depends on the separation of the plates and their areas. A sensor using an elastomeric separator between the plates provides compliance such that the capacitance will vary according to the applied normal load, Figure 6.5(*a*).

Figure 6.5(*b*) shows the basic configuration of a capacitive tactile sensor. The intersections of rows and columns of conductor strips form capacitors. Each individual capacitance can be determined by measuring the corresponding output voltage at the selected row and column. To reduce cross-talk and electromagnetic interference, the rows and columns that are not connected are grounded. Figure 6.5(*c*) shows an equivalent circuit when the sensor is configured to measure the capacitance formed at the intersection of row *i* and row *j*, C_{ij}. R_d is the input resistance of the detector and C_d represents the effects of the stray capacitances, including the detector-amplifier input capacitance, the stray capacitance due

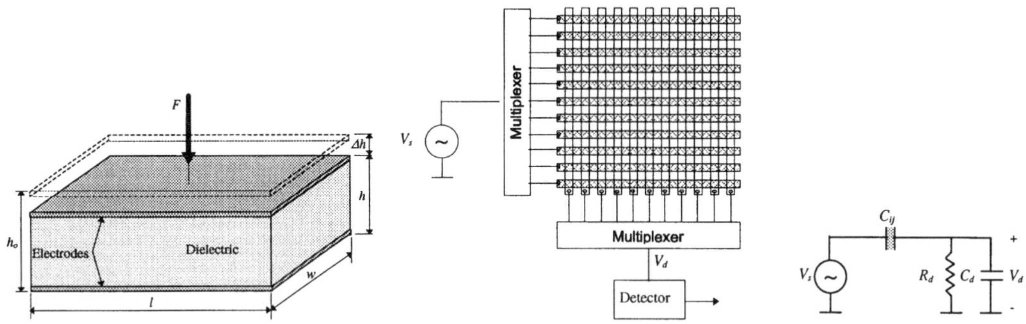

(a) (b) (c)

FIGURE 6.5 (*a*) Basic cell of a capacitor tactile sensor. (*b*) Typical configuration of a capacitive tactile sensor. (*c*) Equivalent circuit for the measurement of the capacitance C_{ij}.

to the unselected rows and columns, and the capacitance contributed by the cable that connects the transducer to the detector. Since the stray capacitance due to the unselected rows and columns changes with the applied forces, the stray capacitance due to the cable is designed to be predominant [18].

The magnitude of voltage at the input of the detector, $|V_d|$ is given by:

$$|V_d| = \frac{C_{ij} R_d \omega}{\sqrt{1+\left[\omega R_d \left(C_{ij}+C_d\right)\right]^2}} |V_s| \tag{6.7}$$

Assuming that $C_d \gg C_{ij}$ and ω is sufficiently large,

$$|V_d| \cong \frac{C_{ij}}{C_d} |V_s| \tag{6.8}$$

When a load is applied to the transducer, the capacitor is deformed as shown in Figure 6.5(*a*). For modeling purposes, it is assumed that the plate capacitor is only under compression. When no load is applied, the capacitance due to the element in the *i*th row and the *j*th column, C^0_{ij}, is given by:

$$C^0_{ij} = \varepsilon \frac{wl}{h_0} \tag{6.9}$$

where ε is the permittivity of the dielectric, w and l are the width and the length of the plate capacitor, respectively, and h_0 is the distance between plates when no load is applied. The voltage at the input of the detector for this particular case is indicated by V_{d0}; then from Equation 6.8, one obtains:

$$|V_{d0}| \cong \frac{C^0_{ij}}{C_d} |V_s| \tag{6.10}$$

When a load is applied, the capacitor is under compression and the capacitance is given by:

$$C_{ij} = \varepsilon \frac{wl}{h_0 - \Delta h} \tag{6.11}$$

The strain in this case is given by:

$$\zeta_z \cong \frac{\Delta h}{h_0} \tag{6.12}$$

where Δh is the displacement of the top metal plate and $\Delta h \ll h_0$. The strain can be measured by:

$$\frac{|V_d| - |V_{d0}|}{|V_d|} = \frac{\dfrac{C_{ij}}{C_d} - \dfrac{C_{ij}^0}{C_d}}{\dfrac{C_{ij}}{C_d}} = 1 - \frac{C_{ij}^0}{C_{ij}} = 1 - \frac{h_0 - \Delta h}{h_0} = \frac{\Delta h}{h_0} = \frac{\Delta h}{h_0} \cong \zeta_z \tag{6.13}$$

Consequently, the strain at each tactel can be determined by measuring the magnitudes of V_d and V_{d0} for each element.

Note that the presence of a tangential force would offset the plates tangentially and change the effective area of the capacitor plates. An ideal capacitive pressure sensor can quantify basic aspects of touch by sensing normal forces, and can detect slippage by measuring tangential forces. However, distinguishing between the two forces at the output of a single sensing element is a difficult task and requires a more complex transducer than the one presented in Figure 6.5(*a*) [27].

Micromachined, silicon-based capacitive devices are especially attractive due to their potential for high accuracy and low drift. A sensor with 1024 elements and a spatial resolution of 0.5 mm was reported in [28]. Several possible structures for implementing capacitive high-density tactile transducers in silicon have been reported in [29]. A cylindrical finger-shaped transducer was reported in [18].

The advantages of capacitive transducers include: wide dynamic range, linear response, and robustness. Their major disadvantages are susceptibility to noise, sensitivity to temperature, and the fact that capacitance decreases with physical size, ultimately limiting the spatial resolution. Research is progressing toward the development of electronic processing circuits for the measurement of small capacitances using charge amplifiers [30], and the development of new capacitive structures [29].

Piezoelectric

A material is called piezoelectric, if, when subjected to a stress or deformation, it produces electricity. Longitudinal piezoelectric effect occurs when the electricity is produced in the same direction of the stress, Figure 6.6. In Figure 6.6(*a*), a normal stress σ (= F/A) is applied along the Direction 3 and the charges are generated on the surfaces perpendicular to Direction 3. A transversal piezoelectric effect occurs when the electricity is produced in the direction perpendicular to the stress.

The voltage V generated across the electrodes by the stress σ is given by:

$$V = d_{33} \frac{h}{\varepsilon} \sigma \tag{6.14}$$

where d_{33} = Piezoelectric constant associated with the longitudinal piezoelectric effect
ε = Permittivity
h = Thickness of the piezoelectric material

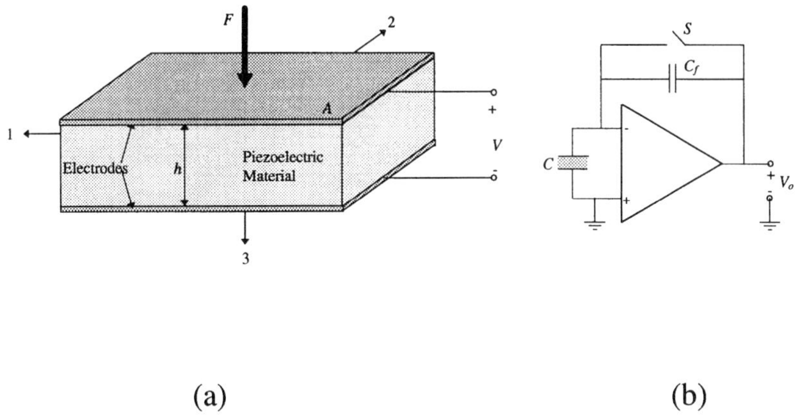

FIGURE 6.6 (*a*) Basic cell of a pizoelectric transducer. (*b*) Charge amplifier utilized for the measurement of the applied force.

Since piezoelectric materials are insulators, the transducer shown in Figure 6.6(*a*), can be considered as a capacitor, from an electrical point of view. Consequently,

$$V = \frac{Q}{C} = \frac{Q}{\varepsilon A} h \tag{6.15}$$

where Q = Charge induced by the stress σ
C = Capacitance of the parallel capacitor
A = Area of each electrode

A comparison of Equations 6.14 and 6.15 leads to:

$$Q = d_{33} A \sigma \tag{6.16}$$

It is concluded that the force applied to the photoelastic material can be determined by finding the charge Q. Charge amplifiers are usually utilized for determining Q. The basic configuration of a charge amplifier is shown in Figure 6.6(*b*). The charge generated in the transducer is transferred to the capacitor C_f and the output voltage, V_o is given by:

$$V_o = -\frac{Q}{C_f} \tag{6.17}$$

The circuit must periodically discharge the feedback capacitor C_f to avoid saturation of the amplifier by stray charges generated by the offset voltages and currents of the operational amplifier. This is achieved by a switch as shown in Figure 6.6(*b*) or by a resistor parallel to C_f.

The piezoelectric material most widely used in the implementation of tactile transducers is PVF2. It shows the largest piezoelectric effect of any known material. Its flexibility, small size, sensitivity, and large electrical output offer many advantages for sensor applications in general, and tactile sensors in particular. Examples of tactile sensors implemented with this technology can be found in [1, 31].

The major advantages of the piezoelectric technology are its wide dynamic range and durability. Unfortunately, the response of available materials does not extend down to dc and therefore steady loads cannot be measured directly. Also, the PVF2 material produces a charge output that is prone to electrical interference and is temperature dependent.

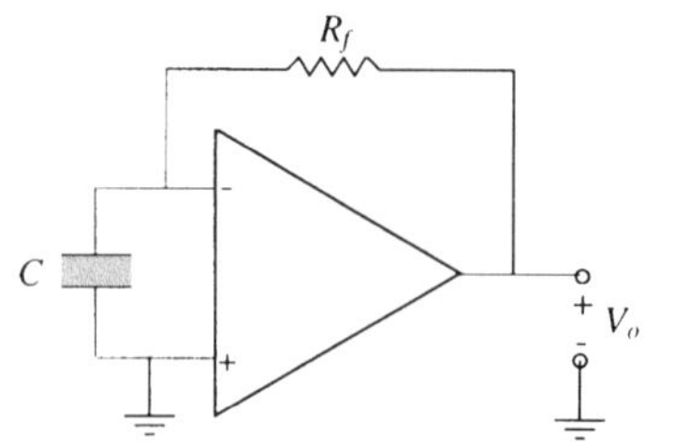

FIGURE 6.7 Current-to-voltage converter.

The possibility of measuring transient phenomenon using piezoelectric material has recently encouraged some researchers to use the piezoelectric effect for detecting vibrations that indicate incipient slip, occurrence of contact, local change in skin curvature, and estimating friction and hardness of the object [7, 10, 11]. If the piezoelectric transducer shown in Figure 6.6(*a*) is connected to an FET-input operational amplifier configured as a current-to-voltage converter as shown in Figure 6.7, the output voltage is given by:

$$V_o = \frac{dQ}{dt} R_f = AR_f d_{33} \frac{d\sigma}{dt} \tag{6.18}$$

where R_f is the feedback resistor. Correspondingly, the circuit configuration provides the mean to measure of changes in the contact stress. A detailed explanation of the behavior of this sensor can be found in [7].

Optical

Recent developments in fiber optic technology and solid-state cameras have led to numerous novel tactile sensor designs [32, 33]. Some of these designs employ flexible membranes incorporating a reflecting surface, Figure 6.8. Light is introduced into the sensor via a fiber optic cable. A wide cone of light propagates out of the fiber, reflects back from the membrane, and is collected by a second fiber. When an external force is applied onto the elastomer, it shortens the distance between the reflective side of the

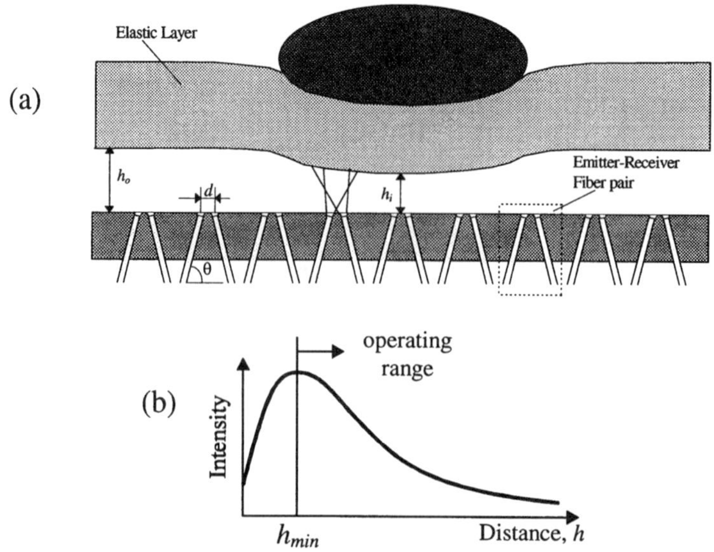

FIGURE 6.8 (*a*) Reflective transducer. (*b*) Light-intensity as a function of the distance *h*.

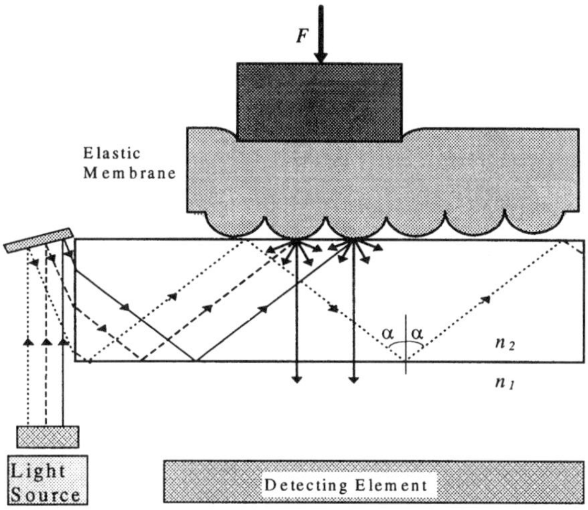

FIGURE 6.9 Tactile transducer based on the principle of internal reflection.

membrane and the fibers, *h*. Consequently, the light gathered by the receiving fiber changes as a function of *h*, Figure 6.8(*b*). To recover univocally the distance from the light intensity, a monotonic function is needed. This can be achieved by designing the transducer such it operates for $h > h_{min}$, where h_{min} is indicated in Figure 6.8(*b*). (The region $h > h_{min}$ is preferred to the $h < h_{min}$ for dynamic range reasons.)

Another optical effect that can be used is that of frustrated total internal reflection [5, 34]. With this technique, an elastic rubber membrane covers, without touching, a glass plate (waveguide); light entering the side edge of the glass is totally reflected by the top and bottom surfaces and propagates along it, Figure 6.9.

The condition for total internal reflection occurs when:

$$n_2 \sin \alpha \leq n_1 \tag{6.19}$$

where n_1 = Index of refraction of the medium surrounding the waveguide (in this case air, $n_1 \cong 1$)
n_2 = Index of refraction of the waveguide
α = Angle of incidence at the interface glass-air

Objects in contact with the elastic membrane deform it and induce contact between the bottom part of the membrane and the top surface of the waveguide, disrupting the total internal reflection. Consequently, the light in the waveguide is scattered at the contact location. Light that escapes through the bottom surface of the waveguide can be detected by an array of photodiodes, a solid-state sensor, or, alternatively, transported away from the transducer by fibers [3]. The detected imaged is stored in a computer for further analysis. A rubber membrane with a flat surface yields a high-resolution binary (contact or noncontact) image [5]. If the rubber sheet is molded with a textured surface (Figure 6.9), then an output proportional to the area of contact is obtained and, consequently, the applied forces can be detected [3]. Shear forces can also be detected using special designs [35]. Sensors based on frustrated internal reflection can be molded into a finger shape [5] and are capable of forming very high-resolution tactile images. Such sensors are commercially available. An improved miniaturized version of a similar sensor was proposed in [34].

Other types of optical transducers use "occluder" devices. One of the few commercially available tactile sensors uses this kind of transducer [36]. In one of the two available designs, the transducer's surface is made of a compliant material, which has on its underside a grid of elongated pins. When force is applied to the compliant surface, the pins on the underside undergo a mechanical motion normal to the surface,

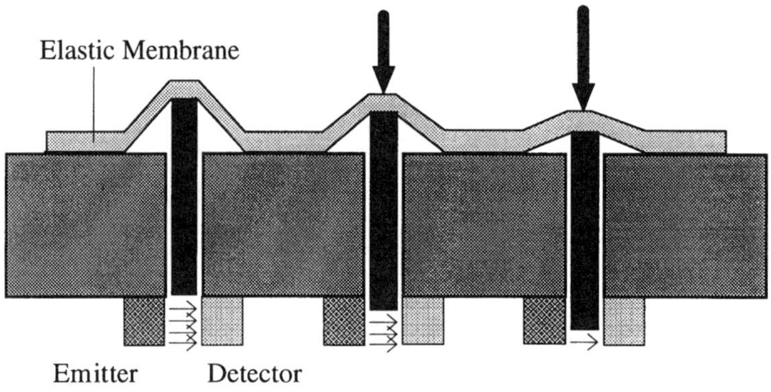

FIGURE 6.10 Principle of operation of an occluder transducer.

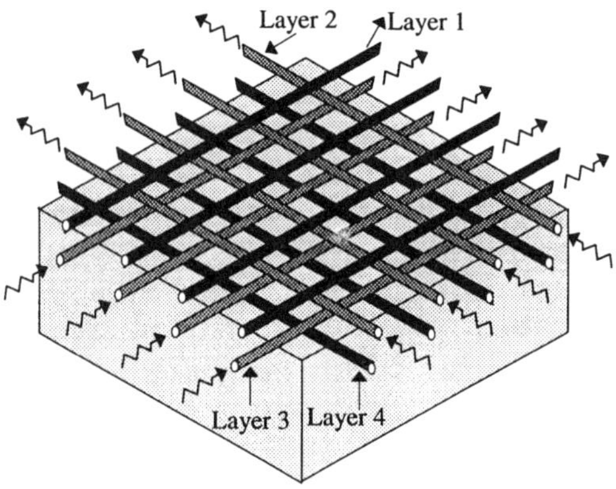

FIGURE 6.11 A four-layer tactile transducer.

blocking the light path of a photoemitter–detector pair. The amount of movement determines the amount of light reaching the photoreceiver. Correspondingly, the more force applied, the less amount of light is collected by the photoreceiver, Figure 6.10. The major problems with this specific device are associated with creep, hysteresis, and temperature variation. This scheme also requires individual calibration of each photoemitter–photodetector pair.

Fibers have also been used directly as transducers in the design of tactile sensors. Their use is based on two properties of fiber optic cables: (1) if a fiber is subjected to a significant amount of bending, then the angle of incidence at the fiber wall can be reduced sufficiently for light to leave the core [37]; and (2) if two fibers pass close to one another and both have roughened surfaces, then light can pass between the fibers. Light coupling between adjacent fibers is a function of their separation [3].

An example of an optical fiber tactile sensor, whose sensing mechanism is based on the increase of light attenuation due to the microbend in the optical fibers, is shown in Figure 6.11 [37]. The transducer consists of a four-layer, two-dimensional fiber optic array constructed by using two layers of optical fibers as a corrugation structure, through which microbends are induced in two orthogonal layers of active fibers. Each active fiber uses an LED as the emitter and a PIN photodiode as a detector. When an object is forced into contact with the transducer, a light distribution is detected at each detector. This light distribution is related to the applied force and the shape of the object. Using complex algorithms and active sensing (moving the object in relation to the transducer), the object position, orientation, size, and contour information can be retrieved [37]. However, the recovery of the applied force profiles was not reported in [37].

Photoelastic

An emerging technology in optical tactile sensing is the development of photoelastic transducers. When a light ray propagates into an optically anisotropic medium, it splits into two rays that are linearly polarized at right angles to each other and propagate at different velocities. This splitting of a ray into two rays that have mutually perpendicular polarizations results from a physical property of crystalline material that is called *optical birefringence* or simply *birefringence*. The direction in which light propagates with the higher velocity is called the *fast axis*; and the one in which it propagates more slowly is called the *slow axis*. Some optically isotropic materials — such as glass, celluloid, bakelite, and transparent plastics in general — become birefringent when they are subjected to a stress field. The birefringent effect lasts only during the application of loads. Thus, this phenomenon is called *temporary* or *artificial birefringence* or, more commonly, the *photoelastic phenomenon*.

Figure 6.12(*a*) shows a photoelastic transducer proposed in [38]. It consists of a fully supported two-layer beam with a mirrored surface sandwiched in between. Normal line forces are applied to the top surface of the beam at discrete tactels, separated by equal distances, *s*, along the beam. The upper compliant layer is for the protection of the mirror, while the lower one is the photoelastic layer.

Circularly polarized monochromatic light, incident along the *z*-axis, illuminates the bottom surface of the transducer. The light propagates parallel to the *z*-axis, passes through the photoelastic layer, and then reflects back from the mirror. If no force is applied to the transducer, the returning light is circularly polarized because unstressed photoelastic material is isotropic. If force is applied, stresses are induced in the photoelastic layer, making the material birefringent. This introduces a certain phase difference between the components of the electric field associated with the light-wave propagation. The two directions of polarization are in the plane perpendicular to the direction of propagation (in this case, the *x*–*y* plane). As a consequence of this effect, the output light is elliptically polarized, creating a phase difference distribution, *p*, between the input light ant the output light at each point in the *x*–*y* plane. The phase difference distribution carries the information of the force distribution applied to the transducer.

A *polariscope* is a practical method to observe the spatial variation on light intensity (fringes) due to the effect of induced phase difference distribution. Polariscopes can be either linear or circular, depending on the required polarization of the light. They can also be characterized as a reflective or a transparent type, depending on whether the photoelastic transducer reflect or transmits the light.

A circular, reflective polariscope, shown in Figure 6.12(*b*), is utilized to illuminate the transducer shown in Figure 6.12(*a*). The input light is linearly polarized and is directed toward the photoelastic transducer by a beam splitter. Before reaching the transducer, the light is circularly polarized by a quarter-wave plate. The output light is elliptically polarized when a force is applied. This light is directed toward a detector passing through the quarter-wave plate, the beam splitter, and an analyzer. Finally, it is detected by a camera linked to a frame grabber connected to a PC, for further data processing. The light that illuminates the camera consists of a set of fringes from where the force distribution applied to the transducer must be recovered. A technique for the recovery of the forces from the fringes is described in [38]. A model of the transducer using FEA is reported in [39].

One of the earlier applications of photoelasticity to tactile sensing dates back to the development phase of the Utah/MIT dexterous hand [40]. The researchers proposed the use of the photoelastic phenomenon as a transduction method for the recovery of the force profile applied to the fingers of the hand. They limited their application to the development of a single-touch transducer, although they claimed that an array of such devices could be implemented. However, the construction of a large array of their devices would be difficult. To overcome this difficulty, another research group proposed a different transducer [41]. Although an analytical model was developed for the sensor, a systematic method for recovering the two-dimensional force profile from the light intensity distribution was not reported. Thus, the sensor was used mainly for the study of the forward analysis, namely, observing the light intensity distribution for different touching objects brought into contact with the sensor. This sensor could eventually be used for determining some simple geometric properties of a touching object.

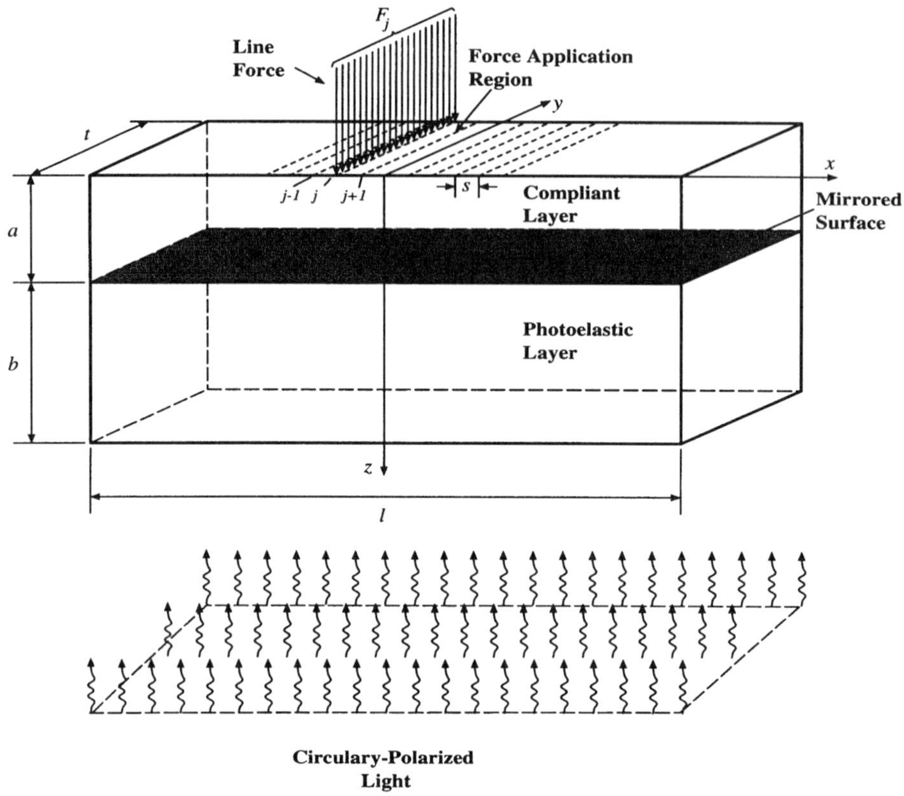

(a)

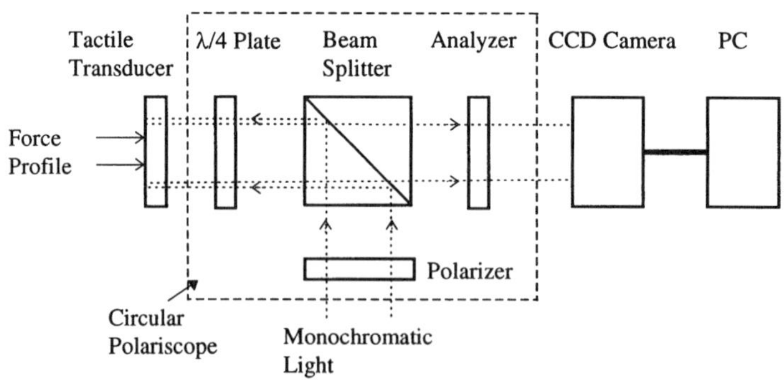

(b)

FIGURE 6.12 (*a*) Photoelastic transducer. (*b*) Circular reflective polariscope.

A tactile sensor reported in [42] is capable of detecting slippage. The output light intensity (the fringe pattern) is captured by a camera interfaced to a PC. When an object moves across the surface of the transducer, the light intensity distribution changes. A direct analysis of the fringes is used to detect movement of the grasped object; a special technique was reported to optimize the comparison process for detecting differences between two fringe patterns occurring due to the slippage of the object in contact with the sensor [42]. It is important to note that such an analysis of the fringes does not require the recovery of the applied force profile.

Photoelasticity offers several attractive properties for the development of tactile sensors: good linearity, compatibility with vision-base sensing technologies, and high spatial resolution associated with the latter, that could lead to the development of high-resolution tactile imagers needed for object recognition and fine manipulation. Also, photoelastic sensors are compatible with fiber optic technology that allows remote location of electronic processing devices and avoidance of interference problems.

Other technologies for tactile sensing include acoustic, magnetic, and microcavity vacuum sensors [43, 44].

References

1. P. Dario, Contact sensing for robot active touch, in *Robotic Science*, M. Brady (ed.), Cambridge, MA: MIT Press, 1989, chap. 3, 138-163.
2. P. P. L. Regtien, Tactile imaging, *Sensors and Actuators, A*, 31, 83-89, 1992.
3. R. A. Russell, *Robot Tactile Sensing*, Brunswick, Australia: Prentice-Hall, 1990.
4. A. D. Berger and P. K. Khosla, Using tactile data for real-time feedback, *Int. J. Robotics Res.*, 10(2), 88-102, 1991.
5. S. Begej, Planar and finger-shaped optical tactile sensors for robotic applications, *IEEE J. Robotics Automation*, 4, 472-484, 1988.
6. R. A. Russell and S. Parkinson, Sensing surface shape by touch, *IEEE Int. Conf. Robotics Automation*, Atlanta, GA, 1993, 423-428.
7. R. D. Howe, A tactile stress rate sensor for perception of fine surface features, *IEEE Int. Conf. Solid-State Sensors Actuators*, San Francisco, CA, 1991, 864-867.
8. S. Sugiyama, K. Kawahata, H. Funabashi, M. Takigawa, and I. Igarashi, A 32 × 32 (1K)-element silicon pressure-sensor array with CMOS processing circuits, *Electron. Commun. Japan*, 75(1), 64-76, 1992.
9. J. S. Son, E. A. Monteverde, and R. D. Howe, A tactile sensor for localizing transient events in manipulation, *IEEE Int. Conf. Robotics Automation*, San Diego, CA, 1994, 471-476.
10. M. R. Tremblay and M. R. Cutkosky, Estimating friction using incipient slip sensing during manipulation task, *IEEE Int. Conf. Robotics Automation*, Atlanta, GA, 1993, 429-434.
11. S. Omata and Y. Terubuna, New tactile sensor like the human hand and its applications, *Sensors and Actuators, A*, 35, 9-15, 1992.
12. R. Bayrleithner and K. Komoriya, Static friction coefficient determination by force sensing and its applications, *IROS'94*, Munich, Germany, 1994, 1639-1646.
13. M. A. Abidi and R. C. Gonzales, The use of multisensor data for robotic applications, *IEEE Trans. Robotics Automation*, 6, 159-177, 1990.
14. A. A. Cole, P. Hsu, and S. S. Sastry, Dynamic control of sliding by robot hands for regrasping, *IEEE Trans. Robotics Automation*, 8, 42-52, 1992.
15. P. K. Allen and P. Michelman, Acquisition and interpretation of 3-D sensor data from touch, *IEEE Trans. Robotics Automation*, 6, 397-404, 1990.
16. H. R. Nicholls (ed.), *Advanced Tactile Sensing for Robotics*, Singapore: World Scientific Publishing, 1992.

17. J. G. Webster (ed.), *Tactile Sensors for Robotics and Medicine*, New York: John Wiley & Sons, 1988.
18. R. S. Fearing, Tactile sensing mechanism, *Int. J. Robotics Res.*, 9(3), 3-23, 1990.
19. T. H. Speeter, Three-dimensional finite element analysis of elastic continua for tactile sensing, *Int. J. Robotics Res.*, 11(1), 1-19, 1992.
20. Y. C. Pati, P. S. Krishnaprasad, and M. C. Peckerar, An analog neural network solution to the inverse problem of early taction, *IEEE Trans. Robotics Automation*, 8(2), 196-212, 1992.
21. L. D. Harmon, Automated tactile sensing, *Int. J. Robotics Res.*, 1(2), 3-32, 1982.
22. W. E. Snyder and J. St. Clair, Conductive elastomers as a sensor for industrial parts handling equipment, *IEEE Trans. Instrum. Meas.*, 27(1), 94-99, 1991.
23. M. Shimojo, M. Ishikawa, and K. Kanaya, A flexible high resolution tactile imager with video signal output, *IEEE Int. Conf. Robotics Automation*, Sacramento, CA, 1991, 384-391
24. W. D. Hillis, A high resolution imaging touch sensor, *Int. J. Robotic Res.*, 1(2), 33-44, 1982.
25. Y. Yamada and M. R. Cutkosky, Tactile sensor with 3-axis force and vibration sensing functions and its applications to detect rotational slip, *IEEE Int. Conf. Robotics Automation*, San Diego, CA, 1994, 3550-3557.
26. K. Njafi and C. H. Mastrangelo, Solid-state microsensors and smart structure, *Ultrasonic Symp.*, Baltimore, MD, 1993, 341-350.
27. F. Zhu and J. W. Spronck, A capacitive tactile sensor for shear and normal force measurements, *Sensors and Actuators, A*, 31, 115-120, 1992.
28. K. Suzuki, K. Najafi, and K. D. Wise, A 1024-element high-performance silicon tactile imager, *IEEE Trans. Electron Devices*, 17(8), 1852-1860, 1990.
29. M. R. Wolffenbuttel and P. L. Regtien, The accurate measurement of a micromechanical force using force-sensitive capacitances, *Conf. Precision Electromagnetic Meas.*, Boulder, CO, 1994, 180-181.
30. M. R. Wolffenbuttel, R. F. Wolffenbuttel, and P. P. L. Regtien, An integrated charge amplifier for a smart tactile sensor, *Sensors and Actuators, A*, 31, 101-109, 1992.
31. E. D. Kolesar, Jr. and C. S. Dyson, Object imaging with piezoelectric robotic tactile sensor, *J. Microelectromechanical Syst.*, 4(2), 87-96, 1995.
32. J. L. Scheiter and T. B. Sheridan, An optical tactile sensor for manipulators, *J. Robot Computer-Integrated Manufacturing*, 1, 65-71, 1989.
33. R. Ristic, B. Benhabib, and A. A. Goldenberg, Analysis and design of a modular electrooptical tactile sensor, *IEEE Trans. Robotics Automation*, 5(3), 362-368, 1989.
34. H. Maekawa, K. K. Tanie, K. Komoriya, M. Kaneko, C. Horiguchi, and T. Sugawara, development of a finger shaped tactile sensor and it evaluation by active touch, *IEEE Int. Conf. Robotics Automation*, Nice, France, 1992, 1327-1334.
35. M. Ohka, Y. Mitsurya, S. Takeuchi, and O. Kamekawa, A three-axis optical tactile sensor (fem contact analyses and sensing experiments using a large-sized tactile sensor), *IEEE Int. Conf. Robotics Automation*, Nagoya, Aichi, Japan, 1995, 817-824.
36. J. Rebman and K. A. Morris, A tactile sensor with electro-optical transduction, in *Robots Sensors, Tactile and Non-Vision*, Vol. 2, A. Pugh (ed.), EFS Publications, 1986, 145-155.
37. S. R. Emge and C. L. Chen, Two dimensional contour imaging with a fiber optic microbend tactile sensor array, *Sensors and Actuators, B*, 3, 31-42, 1991.
38. R. E. Saad, A. Bonen, K. C. Smith, and B. Benhabib, Distributed-force recovery for a planar photoelastic tactile sensor, *IEEE Trans. Instrum. Meas.*, 45, 541-546, 1996.
39. R. E. Saad, A. Bonen, K. C. Smith, and B. Benhabib, Finite-element analysis for photoelastic tactile sensors, *Proc. IEEE Int. Conf. Industrial Electronics, Control, and Instrumentation*, Orlando, FL, 1995, 1202-1207.
40. S. C. Jacobsen, J. E. Wood, D. F. Knutti, and B. Biggers, The Utah/MIT dexterous hand: work in progress, in *Robotics Research: The First International Symposium*, M. Brady and R. Paul (eds.), Cambridge: MIT Press, 1983, 601-653.
41. A. Cameron, R. Daniel, and H. Durrant-Whyte, Touch and motion, *IEEE, Int. Conf. Robotics Automation*, Philadelphia, PA, 1988, 1062-1067.

42. S. H. Hopkins, F. Eghtedari, and D. T. Pham, Algorithms for processing data from a photoelastic slip sensor, *Mechatronics*, 2(1), 15-28, 1992.
43. S. Ando and H. Shinoda, Ultrasonic emission tactile sensing, *IEEE Trans. Control Syst.*, 15(1), 61-69, 1996.
44. J. C. Jiang, V. Faynberg, and R. C. White, Fabrication of micromachined silicon tip transducer for tactile sensing, *J. Vacuum Sci. Technol., B,* 11, 1962-1967, 1993.

II

Mechanical Variables Measurement — Fluid

7

Pressure and Sound Measurement

Kevin H.-L. Chau
Analog Devices, Inc.

Ron Goehner
The Fredericks Company

Emil Drubetsky
The Fredericks Company

Howard M. Brady
The Fredericks Company

William H. Bayles, Jr.
The Fredericks Company

Peder C. Pedersen
Worcester Polytechnic Institute

7.1 Pressure Measurement

Kevin H.-L. Chau

Basic Definitions

Pressure is defined as the normal force per unit area exerted by a fluid (liquid or gas) on any surface. The surface can be either a solid boundary in contact with the fluid or, for purposes of analysis, an imaginary plane drawn through the fluid. Only the component of the force normal to the surface needs to be considered for the determination of pressure. Tangential forces that give rise to shear and fluid motion will not be a relevant subject of discussion here. In the limit that the surface area approaches zero, the ratio of the differential normal force to the differential area represents the pressure at a point on the surface. Furthermore, if there is no shear in the fluid, the pressure at any point can be shown to be independent of the orientation of the imaginary surface under consideration. Finally, it should be noted that pressure is not defined as a vector quantity and is therefore nondirectional.

Three types of pressure measurements are commonly performed:

Absolute pressure is the same as the pressure defined above. It represents the pressure difference between the point of measurement and a perfect vacuum where pressure is zero.

Gage pressure is the pressure difference between the point of measurement and the ambient. In reality, the ambient (atmospheric) pressure can vary, but only the pressure difference is of interest in gage pressure measurements.

0-8493-0047-9/00/$0.00+$.50

TABLE 7.1 Pressure Unit Conversion Table

Units	kPa	psi	in H_2O	cm H_2O	in. Hg	mm Hg	mbar
kPa	1.000	0.1450	4.015	10.20	0.2593	7.501	10.00
psi	6.895	1.000	27.68	70.31	2.036	51.72	68.95
in. H_2O	0.2491	3.613×10^{-2}	1.000	2.540	7.355×10^{-2}	1.868	2.491
cm H_2O	0.09806	1.422×10^{-2}	0.3937	1.000	2.896×10^{-2}	0.7355	0.9806
in. Hg	3.386	0.4912	13.60	34.53	1.000	25.40	33.86
mm Hg	0.1333	1.934×10^{-2}	0.5353	1.360	3.937×10^{-2}	1.000	1.333
mbar	0.1000	0.01450	0.04015	1.020	0.02953	0.7501	1.000

Key:
(1) kPa = kilopascal;
(2) psi = pound force per square inch;
(3) in. H_2O = inch of water at 4°C;
(4) cm H_2O = centimeter of water at 4°C;
(5) in. Hg = inch of mercury at 0°C;
(6) mm Hg = millimeter of mercury at 0°C;
(7) mbar = millibar.

Differential pressure is the pressure difference between two points, one of which is chosen to be the reference. In reality, both pressures can vary, but only the pressure difference is of interest here.

Units of Pressure and Conversion

The SI unit of pressure is the *pascal* (Pa), which is defined as the newton per square meter ($N \cdot m^{-2}$); 1 Pa is a very small unit of pressure. Hence, decimal multiples of the pascal (e.g., kilopascals [kPa] and megapascals [MPa]) are often used for expressing higher pressures. In weather reports, the hectopascal (1 hPa = 100 Pa) has been adopted by many countries to replace the millibar (1 bar = 10^5 Pa; hence, 1 millibar = 10^{-3} bar = 1 hPa) as the unit for atmospheric pressure. In the United States, pressure is commonly expressed in pound force per square inch (psi), which is about 6.90 kPa. In addition, the absolute, gage, and differential pressures are further specified as psia, psig, and psid, respectively. However, no such distinction is made in any pressure units other than the psi. There is another class of units e.g., millimeter of mercury at 0°C (mm Hg, also known as the *torr*) or inch of water at 4°C (in H_2O), which expresses pressure in terms of the height of a static liquid column. The actual pressure p referred to is one that will be developed at the base of the liquid column due to its weight, which is given by Equation 7.1.

$$p = \rho g h \tag{7.1}$$

where ρ is the density of the liquid, g is the acceleration due to gravity, and h is the height of the liquid column. A conversion table for the most popular pressure units is provided in Table 7.1.

Sensing Principles

Sensing Elements

Since pressure is defined as the force per unit area, the most direct way of measuring pressure is to isolate an area on an elastic mechanical element for the force to act on. The deformation of the sensing element produces displacements and strains that can be precisely sensed to give a calibrated measurement of the pressure. This forms the basis for essentially all commercially available pressure sensors today. Specifically, the basic requirements for a pressure-sensing element are a means to isolate two fluidic pressures (one to be measured and the other one as the reference) and an elastic portion to convert the pressure difference into a deformation of the sensing element. Many types of pressure-sensing elements are currently in use. These can be grouped as diaphragms, capsules, bellows, and tubes, as illustrated in Figure 7.1. Diaphragms

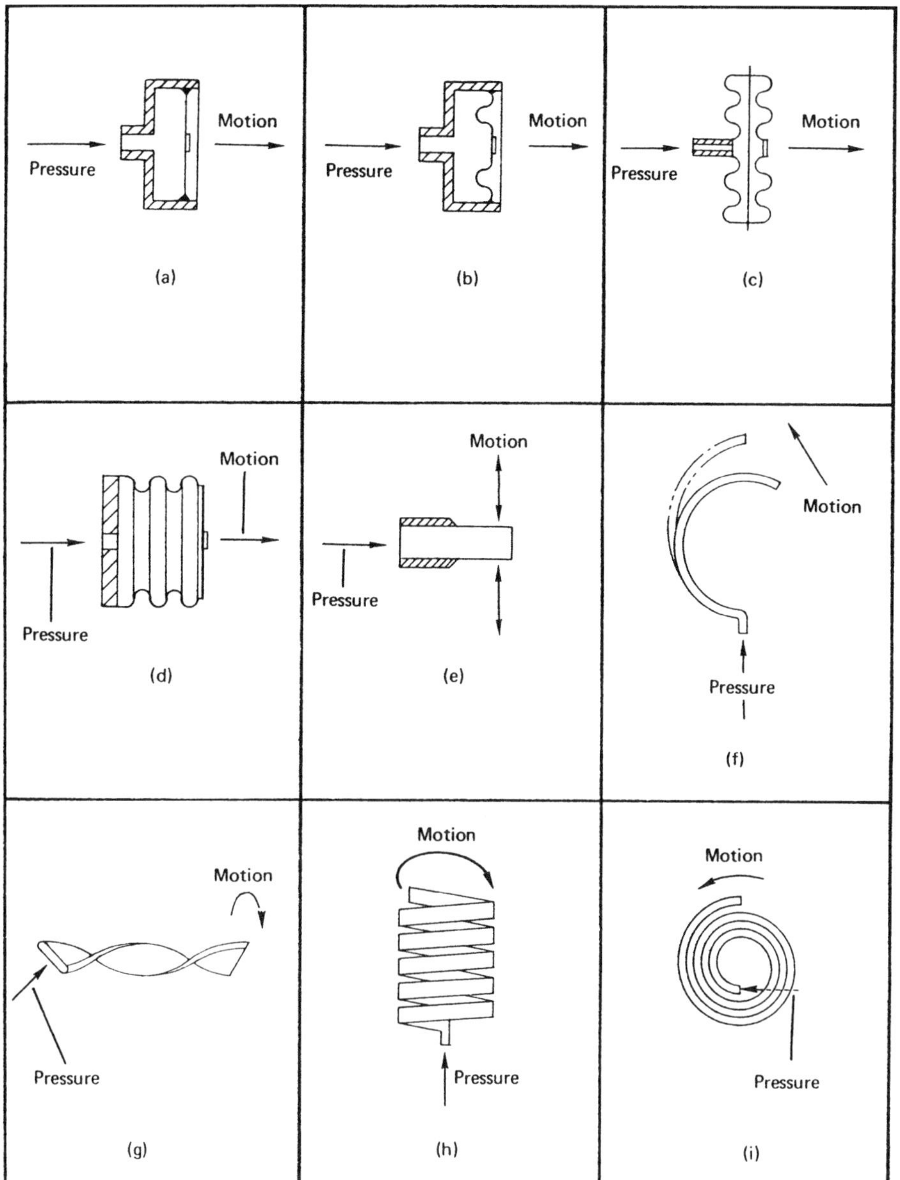

FIGURE 7.1 Pressure-sensing elements: (*a*) flat diaphragm; (*b*) corrugated diaphragm; (*c*) capsule; (*d*) bellows; (*e*) straight tube; (*f*) C-shaped Bourdon tube; (*g*) twisted Bourdon tube; (*h*) helical Bourdon tube; (1) spiral Bourdon tube. (From Norton, H. N., *Handbook of Transducers,* Englewood Cliffs, NJ: Prentice-Hall, 1989, 294-330. Reprinted with permission.)

are by far the most widely used of all sensing elements. A special form of tube, known as the *Bourdon tube,* is curved or twisted along its length and has an oval cross-section. The tube is sealed at one end and tends to unwind or straighten when it is subjected to a pressure applied to the inside. In general, Bourdon tubes are designed for measuring high pressures, while capsules and bellows are usually for measuring low pressures. A detailed description of these sensing elements can be found in [1].

Detection Methods

A detection means is required to convert the deformation of the sensing element into a pressure readout. In the simplest approach, the displacements of a sensing element can be amplified mechanically by lever and flexure linkages to drive a pointer over a graduated scale, for example, in the moving pointer barometers. Some of the earliest pressure sensors employed a Bourdon tube to drive the wiper arm over a potentiometric resistance element. In *linear-variable differential-transformer* (LVDT) pressure sensors, the displacement of a Bourdon tube or capsule is used to move a magnetic core inside a coil assembly to vary its inductance. In *piezoelectric* pressure sensors, the strains associated with the deformation of a sensing element are converted into an electrical charge output by a piezoelectric crystal. Piezoelectric pressure sensors are useful for measuring high-pressure transient events, for example, explosive pressures. In *vibrating-wire* pressure sensors, a metal wire (typically tungsten) is stretched between a fixed anchor and the center of a diaphragm. The wire is located near a permanent magnet and is set into vibration at its resonant frequency by an ac current excitation. A pressure-induced displacement of the diaphragm changes the tension and therefore the resonant frequency of the wire, which is measured by the readout electronics. A detailed description of these and other types of detection methods can be found in [1].

Capacitive Pressure Sensors.
Many highly accurate (better than 0.1%) pressure sensors in use today have been developed using the capacitive detection approach. Capacitive pressure sensors can be designed to cover an extremely wide pressure range. Both high-pressure sensors with full-scale pressures above 10^7 Pa (a few thousand psi) and vacuum sensors (commonly referred to as capacitive *manometers*) usable for pressure measurements below 10^{-3} Pa (10^{-5} torr) are commercially available. The principle of *capacitive pressure sensors* is illustrated in Figure 7.2. A metal or silicon diaphragm serves as the pressure-sensing element and constitutes one electrode of a capacitor. The other electrode, which is stationary, is typically formed by a deposited metal layer on a ceramic or glass substrate. An applied pressure deflects the diaphragm, which in turn changes the gap spacing and the capacitance [2]. In the differential capacitor design, the sensing diaphragm is located in between two stationary electrodes. An applied pressure will cause one capacitance to increase and the other one to decrease, thus resulting in twice the signal while canceling many undesirable common mode effects. Figure 7.3 shows a practical design of a differential capacitive sensing cell that uses two isolating diaphragms and an oil fill to transmit the differential pressure to the sensing diaphragm. The isolating diaphragms are made of special metal alloys that enable them to handle corrosive fluids. The oil is chosen to set a predictable dielectric constant for the capacitor gaps while providing adequate damping to reduce shock and vibration effects. Figure 7.4 shows a rugged capacitive pressure sensor for industrial applications based on the capacitive sensing cell shown in Figure 7.3. The capacitor electrodes are connected to the readout electronics housing at the top. In general, with today's sophisticated electronics and special considerations to minimize stray capacitances (that can degrade the accuracy of measurements), a capacitance change of 10 aF (10^{-18} F) provided by a diaphragm deflection of only a fraction of a nanometer is resolvable.

Piezoresistive Pressure Sensors.
Piezoresistive sensors (also known as *strain-gage* sensors) are the most common type of pressure sensor in use today. *Piezoresistive effect* refers to a change in the electric resistance of a material when stresses or strains are applied. Piezoresistive materials can be used to realize strain gages that, when incorporated into diaphragms, are well suited for sensing the induced strains as the diaphragm is deflected by an applied pressure. The sensitivity of a strain gage is expressed by its *gage factor*, which is defined as the fractional change in resistance, $\Delta R/R$, per unit strain:

$$\text{Gage factor} = \left(\Delta R/R\right)/\varepsilon \tag{7.2}$$

where strain ε is defined as $\Delta L/L$, or the extension per unit length. It is essential to distinguish between two different cases in which: (1) the strain is parallel to the direction of the current flow (along which

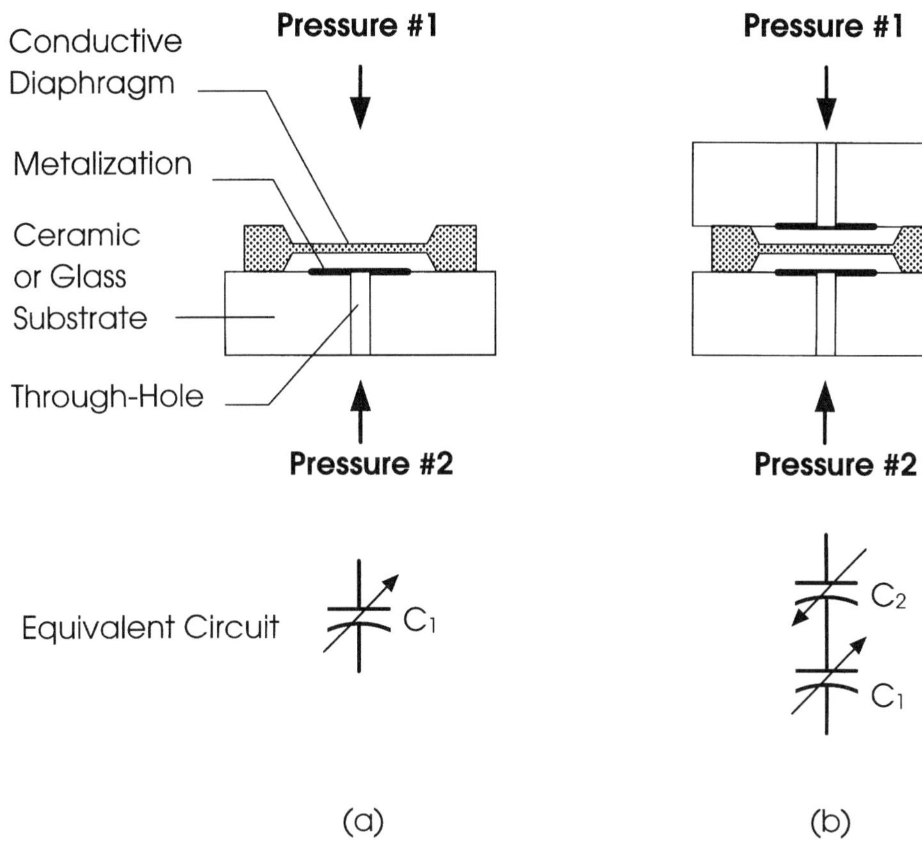

FIGURE 7.2 Operating principle of capacitive pressure sensors. (*a*) Single capacitor design; and (*b*) differential capacitor design.

the resistance change is to be monitored); and (2) the strain is perpendicular to the direction of the current flow. The gage factors associated with these two cases are known as the *longitudinal gage factor* and the *transverse gage factor*, respectively. The two gage factors are generally different in magnitude and often opposite in sign. Typical longitudinal gage factors are ~2 for many useful metals, 10 to 35 for polycrystalline silicon (polysilicon), and 50 to 150 for single-crystalline silicon [3–5]. Because of its large piezoresistive effect, silicon has become the most commonly used material for strain gages. There are several ways to incorporate strain gages into pressure-sensing diaphragms. For example, strain gages can be directly bonded onto a metal diaphragm. However, hysteresis and creep of the bonding agent are potential issues. Alternatively, the strain gage material can be deposited as a thin film on the diaphragm. The adhesion results from strong molecular forces that will not creep, and no additional bonding agent is required. Today, the majority of piezoresistive pressure sensors are realized by integrating the strain gages into the silicon diaphragm using integrated circuit fabrication technology. This important class of silicon pressure sensors will be discussed in detail in the next section.

Silicon Micromachined Pressure Sensors

Silicon micromachined pressure sensors refer to a class of pressure sensors that employ integrated circuit batch processing techniques to realize a thinned-out diaphragm sensing element on a silicon chip. Strain gages made of silicon diffused resistors are typically integrated on the diaphragm to convert the pressure-induced diaphragm deflection into an electric resistance change. Over the past 20 years, silicon micro-machined pressure sensors have gradually replaced their mechanical counterparts and have captured over

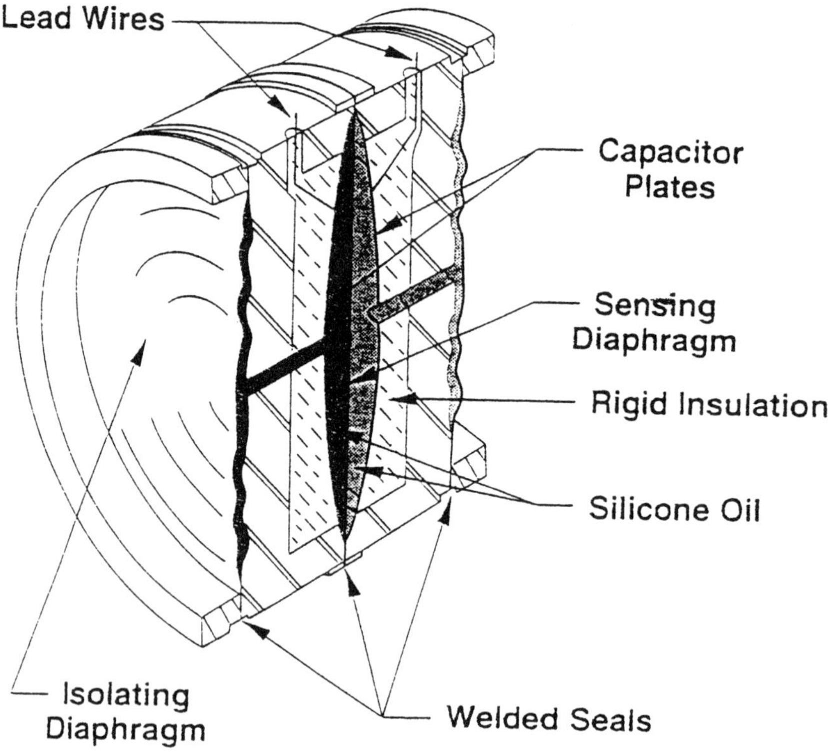

FIGURE 7.3 A differential capacitive sensing cell that is equipped with isolating diaphragms and silicone oil transfer fluid suitable for measuring pressure in corrosive media. (Courtesy of Rosemount, Inc.)

80% of the pressure sensor market. There are several unique advantages that silicon offers. Silicon is an ideal mechanical material that does not display any hysteresis or yield and is elastic up to the fracture limit. It is stronger than steel in yield strength and comparable in Young's modulus [6]. As mentioned in the previous section, the piezoresistive effect in single-crystalline silicon is almost 2 orders of magnitude larger than that of metal strain gages. Silicon has been widely used in integrated circuit manufacturing for which reliable batch fabrication technology and high-precision dimension control techniques have been well developed. A typical silicon wafer yields hundreds of identical pressure sensor chips at very low cost. Further, the necessary signal conditioning circuitry can be integrated on the same sensor chip no more than a few millimeters in size [7]. All these are key factors that contributed to the success of silicon micromachined pressure sensors.

Figure 7.5 shows a typical construction of a silicon piezoresistive pressure sensor. An array of square or rectangular diaphragms is "micromachined" out of a (100) oriented single-crystalline silicon wafer by selectively removing material from the back. An anisotropic silicon etchant (e.g., potassium hydroxide) is typically employed; it etches fastest on (100) surfaces and much slower on (111) surfaces. The result is a pit formed on the backside of the wafer bounded by (111) surfaces and a thinned-out diaphragm section on the front at every sensor site. The diaphragm thickness is controlled by a timed etch or by using suitable etch-stop techniques [6, 8]. To realize strain gages, *p*-type dopant, typically boron, is diffused into the front of the *n*-type silicon diaphragm at stress-sensitive locations to form resistors that are electrically isolated from the diaphragm and from each other by reverse biased *p–n* junctions. The strain gages, the diaphragm, and the rest of the supporting sensor chip all belong to the same single-crystalline silicon. The result is a superb mechanical structure that is free from creep, hysteresis, and thermal expansion coefficient mismatches. However, the sensor die must still be mounted to a sensor housing, which typically has mechanical properties different from that of silicon. It is crucial to ensure

FIGURE 7.4 A rugged capacitive pressure sensor product for industrial applications. It incorporates the sensing cell shown in Figure 7.3. Readout electronics are contained in the housing at the top. (Courtesy of Rosemount, Inc.)

a high degree of stress isolation between the sensor housing and the sensing diaphragm that may otherwise lead to long-term mechanical drifts and undesirable temperature behavior. A common practice is to bond a glass wafer or a second silicon wafer to the back of the sensor wafer to reinforce the overall composite sensor die. This way, the interface stresses generated by the die mount will also be sufficiently remote from the sensing diaphragm and will not seriously affect its stress characteristics. For gage or differential pressure sensing, holes must be provided through the carrier wafer prior to bonding that are aligned to the etch pits of the sensor wafer leading to the back of the sensing diaphragms. No through holes are necessary for absolute pressure sensing. The wafer-to-wafer bonding is performed in a vacuum to achieve a sealed reference vacuum inside the etch pit [6, 9]. Today's silicon pressure sensors are available in a large variety of plastic, ceramic, metal can, and stainless steel packages (some examples are shown in Figure 7.6). Many are suited for printed circuit board mounting. Others have isolating diaphragms and transfer fluids for handling corrosive media. They can be readily designed for a wide range of industrial, medical, automotive, aerospace, and military applications.

Silicon Piezoresistive Pressure Sensor Limitations

Despite the relatively large piezoresistive effects in silicon strain gages, the full-scale resistance change is typically only 1% to 2% of the resistance of the strain gage (which yields an unamplified voltage output of 10 mV/V to 20 mV/V). To achieve an overall accuracy of 0.1% of full scale, for example, the combined effects of mechanical and electrical repeatability, hysteresis, linearity, and stability must be controlled or compensated to within a few parts per million (ppm) of the gage resistance. Furthermore, silicon strain gages are also very temperature sensitive and require careful compensations. There are two primary sources of temperature drifts: (1) the temperature coefficient of resistance of the strain gages (from 0.06%/°C to 0.24%/°C); and (2) the temperature coefficient of the gage factors (from –0.06%/°C to

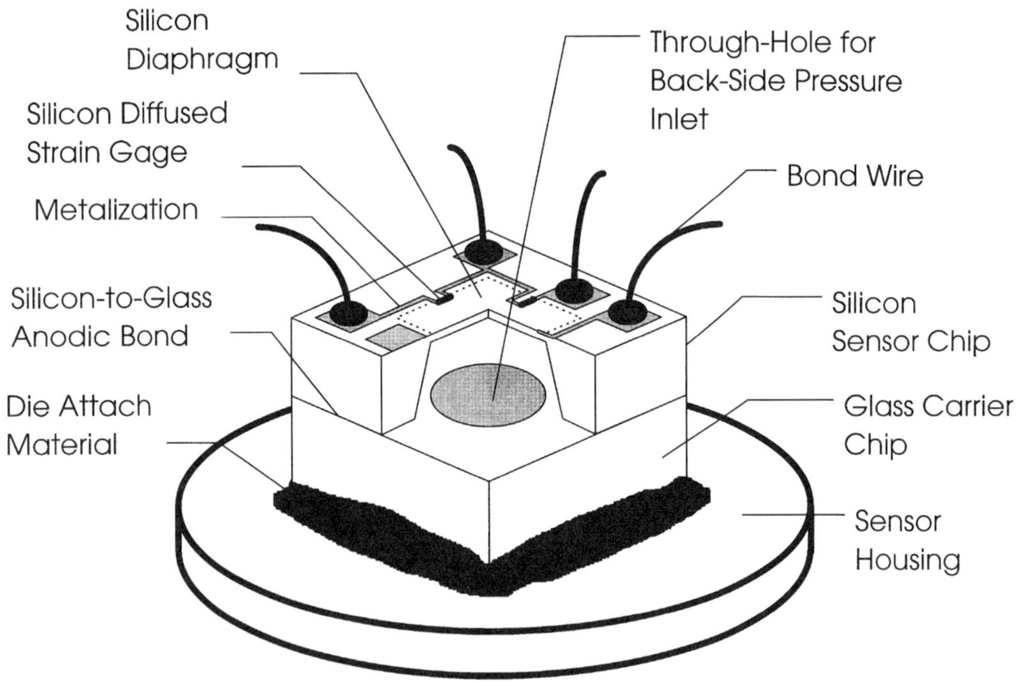

FIGURE 7.5 A cut-away view showing the typical construction of a silicon piezoresistive pressure sensor.

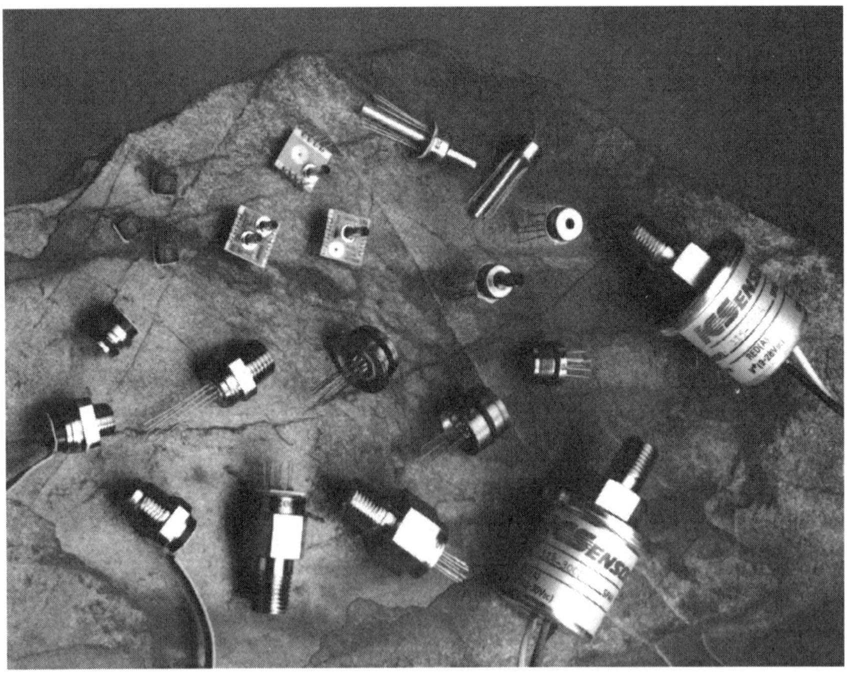

FIGURE 7.6 Examples of commercially available packages for silicon pressure sensors. Shown in the photo are surface-mount units, dual-in-line (DIP) units, TO-8 metal cans, and stainless steel units with isolating diaphragms. (Courtesy of EG&G IC Sensors.)

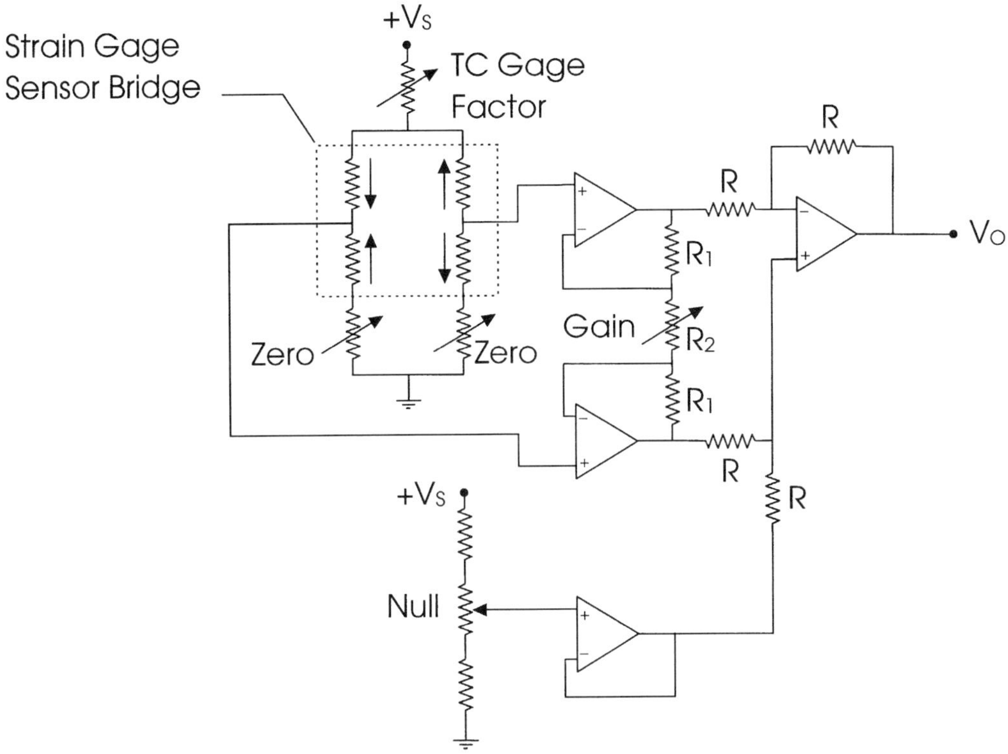

FIGURE 7.7 A signal-conditioning circuit for silicon piezoresistive pressure sensor.

–0.24%/°C), which will cause a decrease in pressure sensitivity as the temperature rises. Figure 7.7 shows a circuit configuration that can be used to achieve offset (resulting from gage resistance mismatch) and temperature compensations as well as providing signal amplification to give a high-level output. Four strain gages that are closely matched in both their resistances and temperature coefficients of resistance are employed to form the four active arms of a Wheatstone bridge. Their resistor geometry on the sensing diaphragm is aligned with the principal strain directions so that two strain gages will produce a resistance increase and the other two a resistance decrease on a given diaphragm deflection. These two pairs of strain gages are configured in the Wheatstone bridge such that an applied pressure will produce a bridge resistance imbalance while the temperature coefficient of resistance will only cause a common mode resistance change in all four gages, keeping the bridge balanced. As for the temperature coefficient of the gage factor, because it is always negative, it is possible (e.g., with the voltage divider circuit in Figure 7.7) to utilize the positive temperature coefficient of the bridge resistance to increase the bridge supply voltage, compensating for the loss in pressure sensitivity as temperature rises. Another major limitation in silicon pressure sensors is the nonlinearity in the pressure response that usually arises from the slight nonlinear behavior in the diaphragm mechanical and the silicon piezoresistive characteristics. The nonlinearity in the pressure response can be compensated by using analog circuit components. However, for the most accurate silicon pressure sensors, digital compensation using a microprocessor with correction coefficients stored in memory is often employed to compensate for all the predictable temperature and nonlinear characteristics. The best silicon pressure sensors today can achieve an accuracy of 0.08% of full scale and a long-term stability of 0.1% of full scale per year. Typical compensated temperature range is from –40°C to 85°C, with the errors of compensation on span and offset both around 1% of full scale. Commercial products are currently available for full-scale pressure ranges from 10 kPa to 70 MPa (1.5 psi to 10,000 psi). The 1998 prices are U.S.$5 to $20 for the most basic uncompensated sensors; $10 to $50 for the compensated (with additional laser trimmed resistors either integrated on-chip or on a ceramic substrate)

TABLE 7.2 Selected Companies That Make Pressure Sensors and Pressure Calibration Systems (This is not intended to be an exhaustive list of all manufacturers.)

(1) Silicon micromachined piezoresistive pressure sensor

Druck Inc.
4 Dunham Drive
New Fairfield, CT 06812
Tel: (203) 746-0400
http://www.druck.com

EG&G IC Sensors
1701 McCarthy Blvd.
Milpitas, CA 95035-7416
Tel: (408) 432-1800

Foxboro ICT
199 River Oaks Pkwy.
San Jose, CA 95134-1996
Tel: (408) 432-1010

Honeywell Inc.
Micro Switch Div.
11 W. Spring St.
Freeport, IL 61032-4353
Tel: (815) 235-5500
http://www.honeywell.com/sensing

Lucas NovaSensor
1055 Mission Ct.
Fremont, CA 94539
Tel: (800) 962-7364
http://www.novasensor.com

Motorola, Inc.
Sensor Products Div.
5005 E. McDowell Rd.
Phoenix, AZ 85008
Tel: (602) 244-3381
http://mot-sps.com/senseon

SenSym, Inc.
1804 McCarthy Blvd.
Milpitas, CA 95035
Tel: (408) 954-1100
http://www.sensym.com

(2) Bonded strain gage pressure sensors

Gefran Inc.
122 Terry Dr.
Newtown, PA 18940
Tel: (215) 968-6238
http://www.gefran.it

(3) Capacitive pressure sensors

Kavlico Corp.
14501 Los Angeles Ave.
Moorpark, CA 93021
Tel: (805) 523-2000

Rosemount Inc.
Measurement Div.
12001 Technology Drive
Eden Prairie, MN 55344
Tel: (800) 999-9307
http://www.rosemount.com

(4) Pressure calibration systems

Mensor Corp.
2230 IH-35 South
San Marcos, TX 78666-5917
Tel: (512) 396-4200
http://www.mensor.com

Ruska Instrument
10311 Westpark Drive
Houston, TX 77042
Tel: (713) 975-0547
http://www.ruska.com

or signal-conditioned (compensated with amplified output) sensors; and \$60 to \$300 for sensors with isolating diaphragms in stainless steel housings. Table 7.2 provides contact information for selected companies making pressure sensors.

References

1. H. N. Norton, *Handbook of Transducers*, Englewood Cliffs, NJ: Prentice-Hall, 1989, 294-330.
2. W. H. Ko, Solid-state capacitive pressure transducers, *Sensors and Actuators*, 10, 303-320, 1986.
3. C. S. Smith, Piezoresistance effect in germanium and silicon, *Phys. Rev.*, 94, 42-49, 1954.
4. O. N. Tufte and E. L. Stelzer, Piezoresistive properties of silicon diffused layers, *J. Appl. Phys.*, 34, 313-318, 1963.

5. D. Schubert, W. Jenschke, T. Uhlig, and F. M. Schmidt, Piezoresistive properties of polycrystalline and crystalline silicon films, *Sensors and Actuators*, 11, 145-155, 1987.
6. K. E. Petersen, Silicon as a mechanical material, *IEEE Proc.*, 70, 420-457, 1982.
7. R. F. Wolffenbuttel (ed.), *Silicon Sensors and Circuits: On-Chip Compatibility*, London: Chapman & Hall, 1996, 171-210.
8. H. Seidel, The mechanism of anisotropic silicon etching and its relevance for micromachining, Tech. Dig., *Transducers '87*, Tokyo, Japan, June 1987, 120-125.
9. E. P. Shankland, Piezoresistive silicon pressure sensors, *Sensors*, 22-26, Aug. 1991.

Further Information

R. S. Muller, R. T. Howe, S. D. Senturia, R. L. Smith, and R. M. White (eds.), *Microsensors*, New York: IEEE Press, 1991, provides an excellent collection of papers on silicon microsensors and silicon micromachining technologies.

R. F. Wolffenbuttel (ed.), *Silicon Sensors and Circuits: On-Chip Compatibility*, London: Chapman & Hall, 1996, provides a thorough discussion on sensor and circuit integration.

ISA Directory of Instrumentation On-Line (http://www.isa.org) from the Instrument Society of America maintains a list of product categories and active links to many sensor manufacturers.

7.2 Vacuum Measurement

Ron Goehner, Emil Drubetsky, Howard M. Brady, and William H. Bayles, Jr.

Background and History of Vacuum Gages

To make measurements in the vacuum region, one must possess a knowledge of the expected pressure range required by the processes taking place in the vacuum chamber as well as the accuracy and/or repeatability of the measurement required for the process. Typical vacuum systems require that many orders of magnitude of pressures must be measured. In many applications, the pressure range may be 8 orders of magnitude, or from atmospheric (1.01×10^5 Pa, 760 torr) to 1×10^{-3} Pa (7.5×10^{-6} torr).

For semiconductor lithography, high-energy physics experiments and surface chemistry, ultimate vacuum of 7.5×10^{-9} torr and much lower are required (a range of 11 orders of magnitude below atmospheric pressure). One gage will not give reasonable measurements over such large pressure ranges. Over the past 50 years, vacuum measuring instruments (commonly called gages) have been developed that used transducers (or sensors) which can be classified as either direct reading (usually mechanical) or indirect reading [1] (usually electronic). Figure 7.8 shows vacuum gages typically in current use. When a force on a surface is used to measure pressure, the gages are mechanical and are called *direct reading* gages, whereas when any property of the gas that changes with density is measured by electronic means, they are called *indirect reading* gages. Figure 7.9 shows the range of operating pressure for various types of vacuum gages.

Direct Reading Gages

A subdivision of direct reading gages can be made by dividing them into those that utilize a liquid wall and those that utilize a solid wall. The force exerted on a surface from the pressure of thermally agitated molecules and atoms is used to measure the pressure.

Liquid Wall Gages

The two common gages that use a liquid wall are the manometer and the McLeod gage. The liquid column *manometer* is the simplest type of vacuum gage. It consists of a straight or U-shaped glass tube

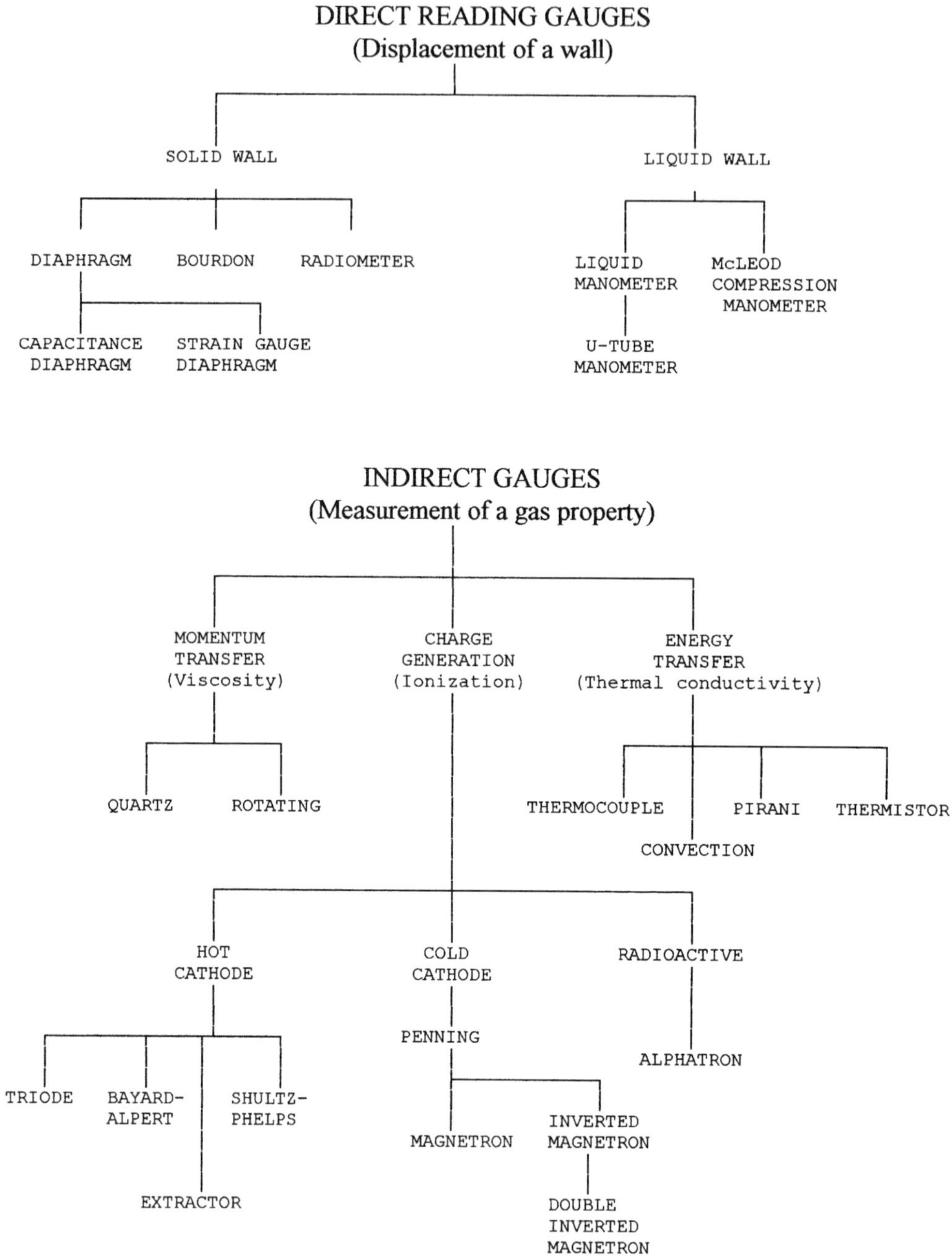

FIGURE 7.8 Classification of pressure gages. (From D.M. Hoffman, B. Singh, and J.H. Thomas, III (eds.), *The Handbook of Vacuum Science and Technology*, Orlando, FL: Academic Press, 1998. With permission.)

evacuated and sealed at one end and filled partly with mercury or a low vapor pressure liquid such as diffusion pump oil (See Figure 7.10). In the straight tube manometer, as the space above the mercury is evacuated, the length of the mercury column decreases. In the case of the U-tube, as the free end is evacuated, the two columns approach equal height. The pressure at the open end is measured by the difference in height of the liquid columns. If the liquid is mercury, the pressure is directly measured in mm of Hg (torr). The manometer is limited to pressures equal to or greater than ~1 torr (133 Pa). If

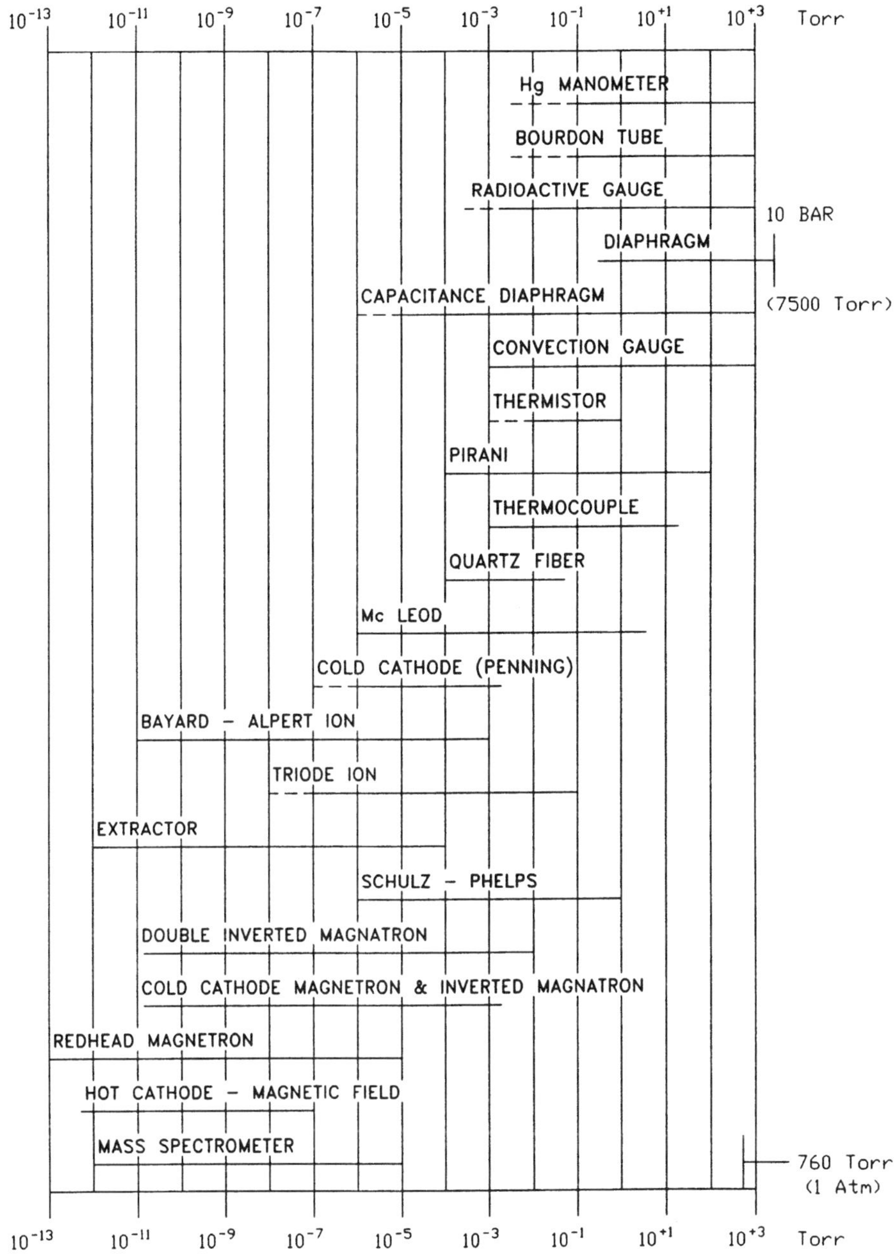

FIGURE 7.9 Pressure ranges for various gages. (From D.M. Hoffman, B. Singh, and J.H. Thomas, III (eds.), *The Handbook of Vacuum Science and Technology,* Orlando, FL: Academic Press, 1998. With permission.)

the liquid is a low density oil, the U-tube is capable of measuring a pressure as low as ~0.1 torr. This is an absolute, direct reading gage but the use of mercury or low density oils that will in time contaminate the vacuum system preclude its use as a permanent vacuum gage.

Due to the pressure limitation of the manometer, the McLeod gage [2, 3] was developed to significantly extend the range of vacuum measurement (see Figure 7.11). This device is essentially a mercury manometer in which a volume of gas is compressed before measurement. This can be used as a primary standard device when a liquid nitrogen trap is used on the vacuum system. Figure 7.11 shows gas at 10^{-6} torr and

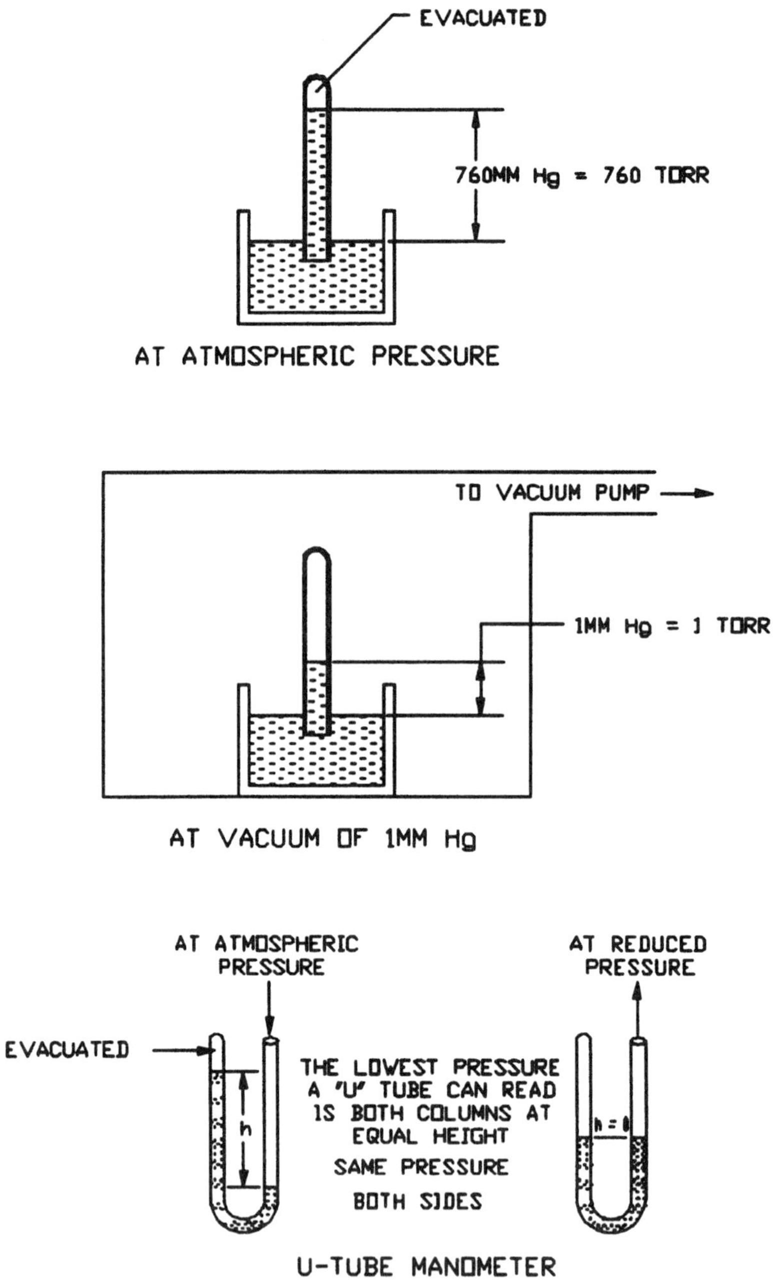

FIGURE 7.10 Mercury manometers. (From W.H. Bayles, Jr., Fundamentals of Vacuum Measurement, Calibration and Certification, Industrial Heating, October 1992. With permission.)

a compression ratio of 10^{+7}. In this example, the difference of the columns will be 10 mm. Extreme care must be taken not to break the glass and expose the surroundings to the mercury. The McLeod Gage is an inexpensive standard but should only be used by skilled and careful technicians. The gage will give a false low reading unless precautions are taken to ensure that any condensible vapors present are removed by liquid nitrogen trapping.

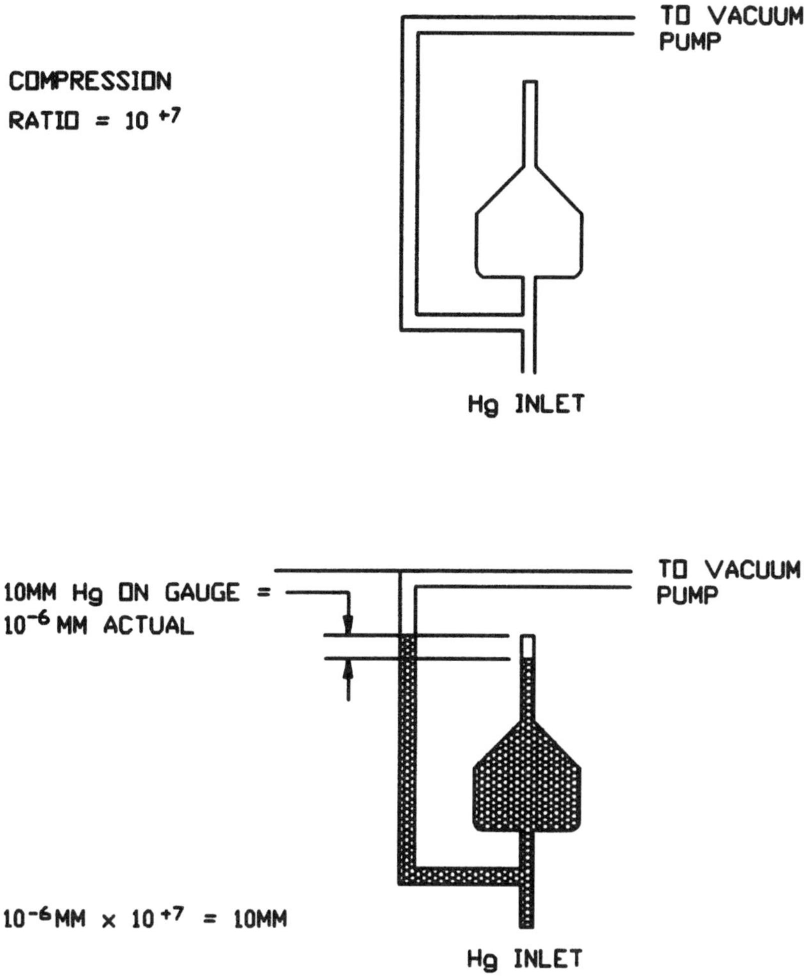

FIGURE 7.11 McLeod gage. (From D.M. Hoffman, B. Singh, and J.H. Thomas, III (eds.), *The Handbook of Vacuum Science and Technology,* Orlando, FL: Academic Press, 1998. With permission.)

Solid Wall Gages

There are two major mechanical solid wall gage types: capsule and diaphragm.

Bourdon Gages.
The capsule-type gages depend on the deformation of the capsule with changing pressure and the resultant deflection of an indicator. Pressure gages using this principle measure pressures above atmospheric to several thousand psi and are commonly used on compressed gas systems. This type of gage is also used at pressures below atmospheric, but the sensitivity is low. The Bourdon gage (Figure 7.12), is used as a moderate vacuum gage. In this case, the capsule is in the form of a thin-walled tube bent in a circle, with the open end attached to the vacuum system with a mechanism and a pointer attached to the other end. The atmospheric pressure deforms the tube; a linear indication of the pressure is given that is independent of the nature of the gas. Certain manufacturers supply capsule gages capable of measuring pressures as low as 1 torr. These gages are rugged, inexpensive, and simple to use and can be made of materials inert to corrosive vapors. Since changing atmospheric pressure causes inaccuracies in the readings, compensated versions of the capsule and Bourdon gage have been developed that improve the accuracy [4].

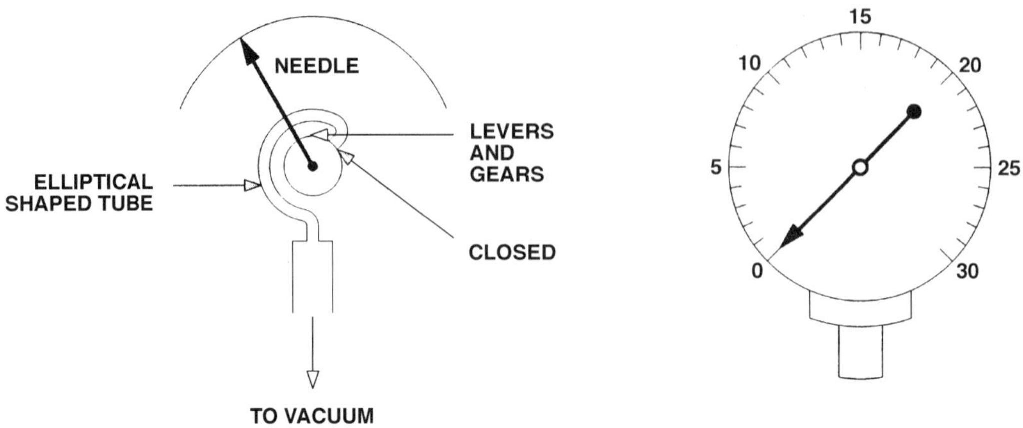

FIGURE 7.12 Bourdon gage. (From Varian Vacuum Technologies, Basic Vacuum Practice, Varian Associates, Inc., Lexington, MA, 1992. With permission.)

Diaphragm Gages.
If compensated capsule or diaphragm mechanisms are combined with sensitive and stable electronic measuring circuits, performance is improved. One such gage is the capacitance diaphragm gage (also referred to as the capacitance manometer).

The capacitance diaphragm gage is shown in Figure 7.13. A flexible diaphragm forms one plate of a capacitor and a fixed probe the other. The flexible diaphragm deforms due to even slight changes in pressure, resulting in a change in the capacitance. The capacitance is converted to a pressure reading. The sensitivity, repeatability, and simplicity of this gage enables this type of direct reading gage to be a standard from 10^{-6} torr to atmospheric pressure, provided multiple heads designed for each pressure range are used. A single head can have a dynamic range of 4 or 5 orders of magnitude [5].

The strain gage type of diaphragm gage is shown in Figure 7.13. In this case, deformation of the diaphragm causes a proportional output from the attached strain gage. Sensitivities and dynamic range tend to be less than those of the capacitance diaphragm gage, but the price of the strain gage type diaphragm gage is usually lower.

Both of these gages are prone to errors caused by small temperature changes due to the inherent high sensitivity of this gage type. Temperature-controlled heads or correction tables built into the electronics have been used to minimize this problem. Other sources of error in all solid wall gages are hysteresis and metal fatigue.

Indirect Reading Gages

Indirect reading gages measure some property of the gas that changes with the density of the gas and usually produces an electric output. Electronic devices amplify and compensate this output to provide a pressure reading.

Thermal Conductivity Gages

Thermal conductivity gages utilize the property of gases in which reduced thermal conductivity corresponds to decreasing density (pressure). The thermal conductivity decreases from a nearly constant value above ~1 torr to essentially 0 at pressures below 10^{-2} torr. The gage controllers are designed to work with a specific sensor tube, and substitutions are limited to those that are truly functionally identical. Heat transfer at various pressures is related to the Knudsen number, as is shown in Figure 7.14 for various heat transfer regimes. The Knudsen number can then be related to pressure through the geometry of the sensor, providing a relationship of heat transfer to pressure for a particular design thermal conductivity gage.

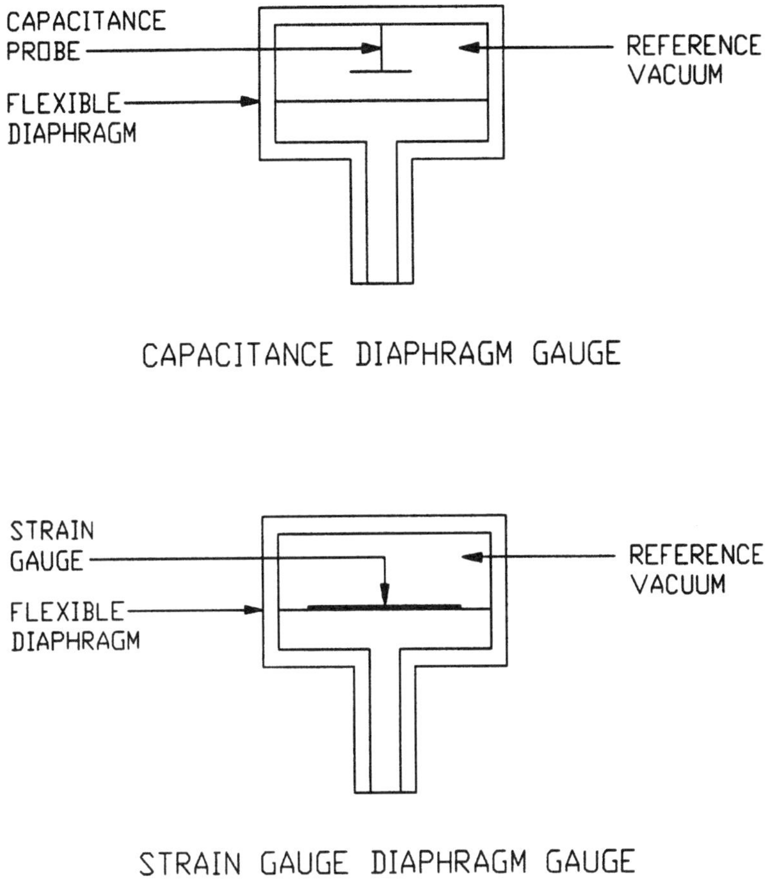

FIGURE 7.13 Diaphragm gage. (From W.H. Bayles, Jr., Fundamentals of Vacuum Measurement, Calibration and Certification, Industrial Heating, October 1992. With permission.)

Pirani Gages.

The Pirani gage is perhaps the oldest indirect gage that is still used today. In operation, a sensing filament carrying current and producing heat is surrounded by the gas to be measured. As the pressure changes, the thermal conductivity changes, thus varying the temperature of the sensing filament. The temperature change causes a change in the resistance of the sensing filament. The sensing filament is usually one leg of a Wheatstone bridge. The bridge can be operated so that the voltage is varied to keep the bridge balanced; that is, the resistance of the sensing filament is kept constant.

This method is called the constant temperature method and is deemed the fastest, most sensitive, and most accurate. To reduce the effect of changing ambient temperature, an identical filament sealed off at very low pressure is placed in the leg adjacent to the sensing filament as a balancing resistor. Because of its high thermal resistance coefficient, the filament material is usually a thin tungsten wire. It has been demonstrated that a 10 W light bulb works quite well [6]. (see Figure 7.15.)

A properly designed, compensated Pirani gage with sensitive circuitry is capable of measuring to 10^{-4} torr. However, the thermal conductivity of gases varies with the gas being measured, causing a variation in gage response. These variations can be as large as a factor of 5 at low pressures and as high as 10 at high pressures (see Figure 7.16). Correction for these variations can be made on the calibration curves supplied by the manufacturer if the composition of the gas is known. Operation in the presence of high partial pressures of organic molecules such as oils is not recommended.

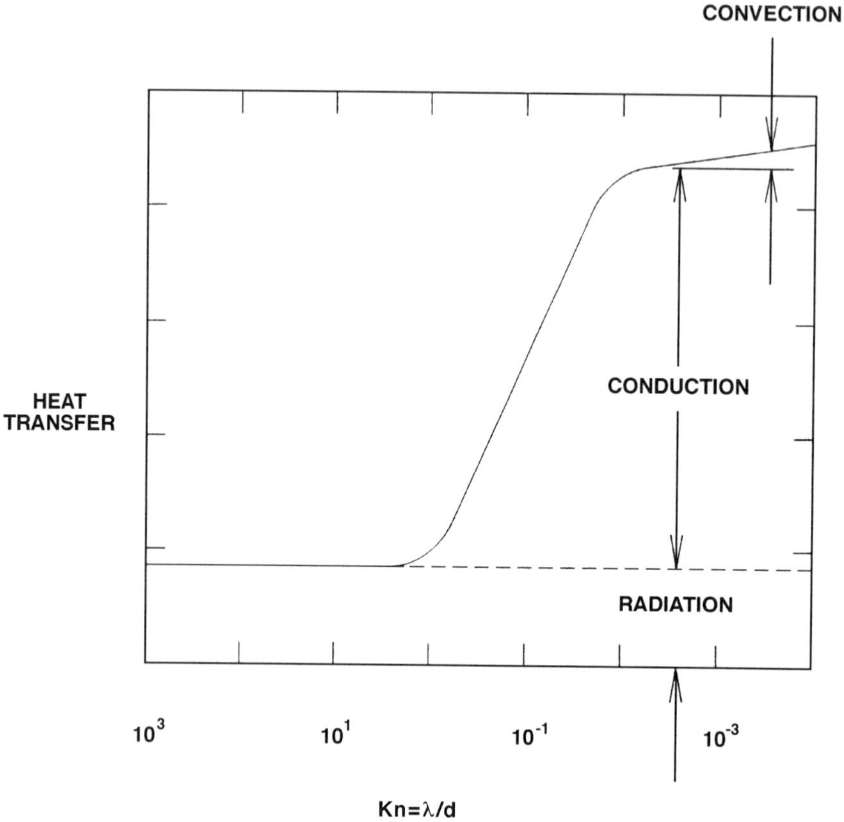

FIGURE 7.14 Heat transfer regimes in a thermal conductivity gage. (From J.F. O'Hanlon, *A User's Guide to Vacuum Technology*, New York: John Wiley & Sons, 1980, 47. With permission.)

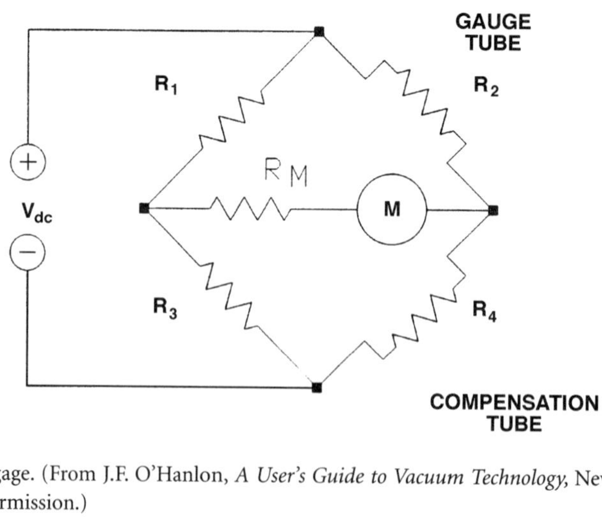

FIGURE 7.15 Pirani gage. (From J.F. O'Hanlon, *A User's Guide to Vacuum Technology*, New York: John Wiley & Sons, 1980, 47. With permission.)

Thermistor Gages.
In the thermistor gage, a thermistor is used as one leg of a bridge circuit. The inverse resistive characteristics of the thermistor element unbalances the bridge as the pressure changes, causing a corresponding change in current. Sensitive electronics measure the current and are calibrated in pressure units. The

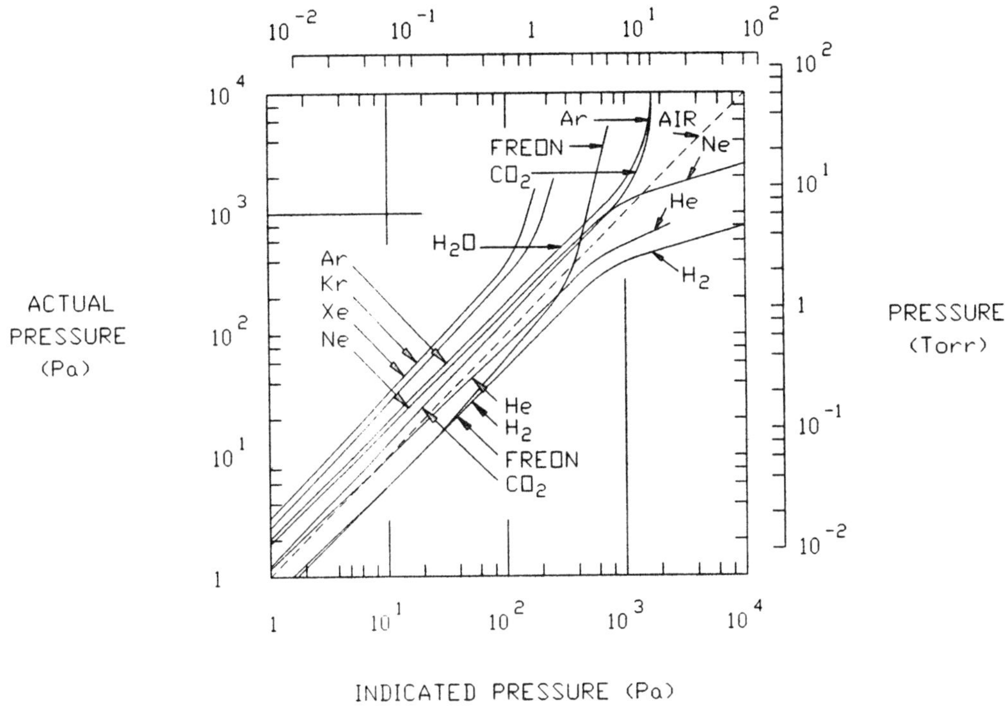

FIGURE 7.16 Calibration curves for the Pirani gage. (Reprinted with permission from Leybold-Herqeus GMblt, Köhn, Germany.)

thermistor gage measures approximately the same pressure range as the thermocouple.The exact calibration depends on the the gas measured. In a well-designed bridge circuit, the plot of current vs. pressure is practically linear in the range 10^{-3} to 1 torr [7]. Modern thermistor gages use constant-temperature techniques.

Thermocouple Gages.

Another example of an indirect reading thermal conductivity gage is the thermocouple gage. This is a relatively inexpensive device with proven reliability and a wide range of applications. In the thermocouple gage, a filament of resistance alloy is heated by the passage of a constant current (see Figure 7.17). A thermocouple is welded to the midpoint of the filament or preferably to a conduction bridge at the center of the heated filament. This provides a means of directly measuring the temperature. With a constant current through the filament, the temperature increases as the pressure decreases as there are fewer molecules surrounding the filament to carry the heat away. The thermocouple output voltage increases as a result of the increased temperature and varies inversely with the pressure. The thermocouple gage can also be operated in the constant-temperature mode.

Gas composition effects apply to all thermal conductivity gages. The calibration curves for a typical thermocouple gage are shown in Figure 7.18. The thermocouple gage can be optimized for operation in various pressure ranges. Operation of the thermocouple gage in high partial pressures of organic molecules such as oils should be avoided. One manufacturer pre-oxidizes the thermocouple sensor for stability in "dirty" environments and for greater interchangeability in clean environments.

Convection Gages.

Below 1 torr a significant change in thermal conductivity occurs as the pressure changes. Thus, the thermal conductivity gage is normally limited to 1 torr.

At pressures above 1 torr, there is, in most gages, a small contribution to heat transfer caused by convection. Manufacturers have developed gages that utilize this convection effect to extend the usable

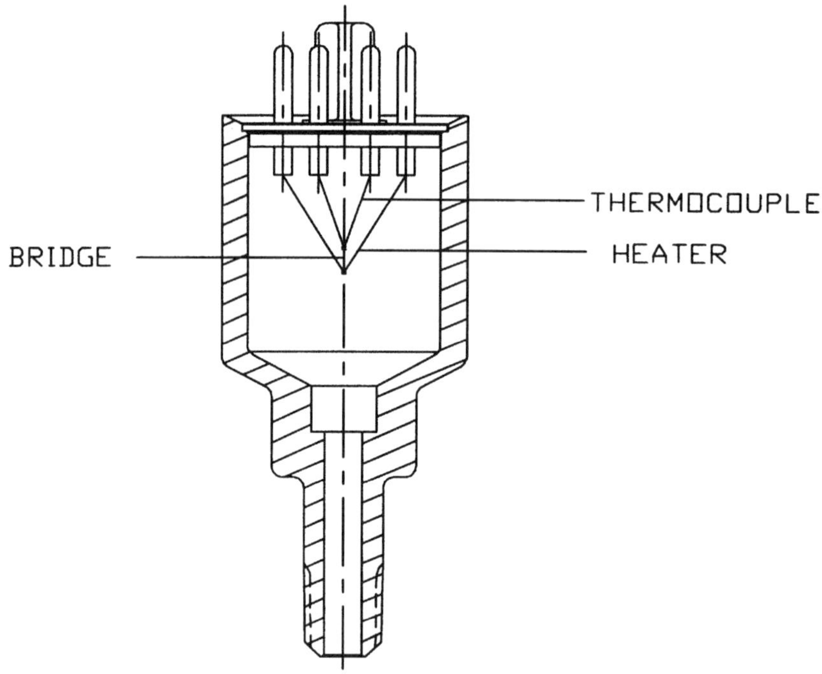

FIGURE 7.17 Thermocouple gage. (Reprinted with permission of Televac Division, The Fredericks Co., Huntingdon Valley, PA.)

range to atmospheric pressure and slightly above [8–12]. Orientation of a convection gage is critical because this convection heat transfer is highly dependent on the orientation of the elements within the gage.

The Convectron™ uses the basic structure of the Pirani with special features to enhance convection cooling in the high-pressure region [13]. To utilize the gage above 1 torr (133 Pa), the sensor tube must be mounted with its major axis in a horizontal position. If the only area of interest is below 1 torr, the tube can be mounted in any position. As mentioned above, the gage controller is designed to be used with a specific model sensor tube; because extensive use is made of calibration curves and look-up tables stored in the controller, no substitution is recommended.

The Televac convection gage uses the basic structure of the thermocouple gage except that two thermocouples are used [14]. As in any thermocouple gage, the convection gage measures the pressure by determining the heat loss from a fine wire maintained at constant temperature. The response of the sensor depends on the gas type. A pair of thermocouples is mounted a fixed distance from each other (see Figure 7.19). The one mounted lower is heated to a constant temperature by a variable current power supply. Power is pulsed to this lower thermocouple and the temperature is measured between heating pulses. The second (upper) thermocouple measures convection effects and also compensates for ambient temperature. At pressures below ~2 torr (270 Pa) the temperature in the upper thermocouple is negligible. The gage tube operates as a typical thermocouple in the constant-temperature mode. Above 2 torr, convective heat transfer causes heating of the upper thermocouple. The voltage output is subtracted from that of the lower thermocouple, thus requiring more current to maintain the wire temperature. Consequently, the range of pressure that can be measured (via current change) is extended to atmospheric pressure (see Figure 7.20). Orientation of the sensor is with the axis vertical.

The use of convection gages with process control electronics allows for automatic pump-down with the assurance that the system will neither open under vacuum nor be subject to over-pressure during backfill to atmospheric pressure. These gages, with their controllers, are relatively inexpensive. In oil-free systems, they afford long life and reproducible results.

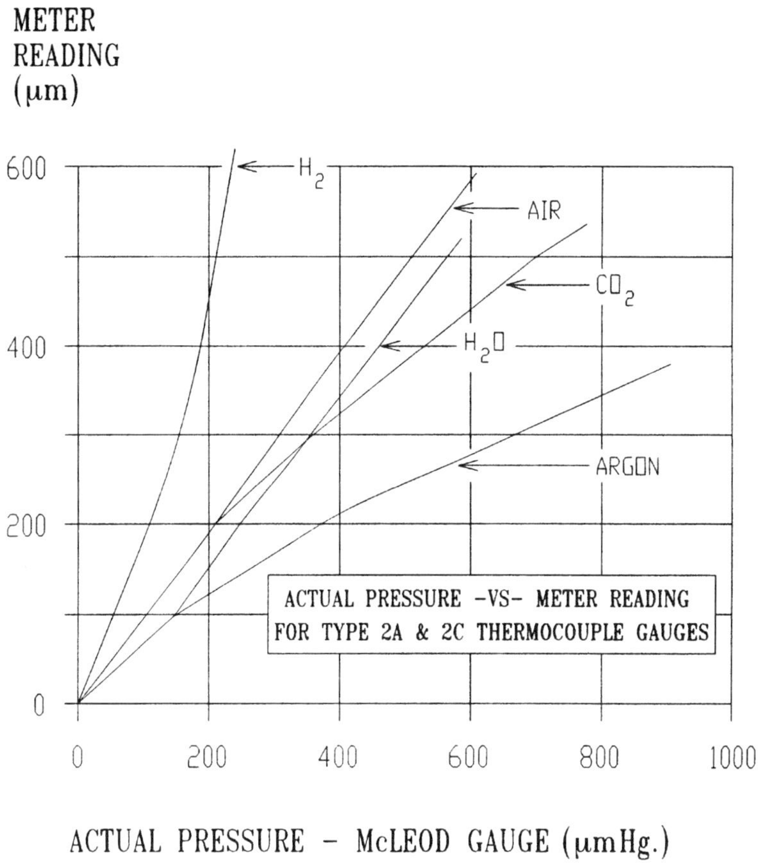

FIGURE 7.18 Calibration curves for the thermocouple gage. (Reprinted with permission from Televac Division, The Fredericks Co., Huntingdon Valley, PA.)

Hot Cathode Ionization Gages

Hot cathode ionization gage designs consist of triode gages, Bayard-Alpert gages, and others.

Triode Hot Cathode Ionization Gages.
For over 80 years, the triode electron tube has been used as an indirect way to measure vacuum [15, 16]. A typical triode connection is as an amplifier, as is shown in Figure 7.21. A brief description of its operation is given here, but more rigorous treatment of triode performance is given in [17–20]. However, if the triode is connected as in Figure 7.22 so that the grid is positive and the plate is negative with respect to the filament, then the ion current collected by the plate for the same electron current to the grid is greatly increased [21].

Today, the triode gage is used in this higher sensitivity mode. Many investigators have shown that a linear change in molecular density (pressure) results in a linear change in ion current [15, 21, 22]. This linearity allows a sensitivity factor S to be defined such that:

$$I_i = S \times I_e \times P \tag{7.3}$$

where I_i = Ion current (A)
I_e = Electron current (A)
P = Pressure
S = Sensitivity (in units of reciprocal pressure)

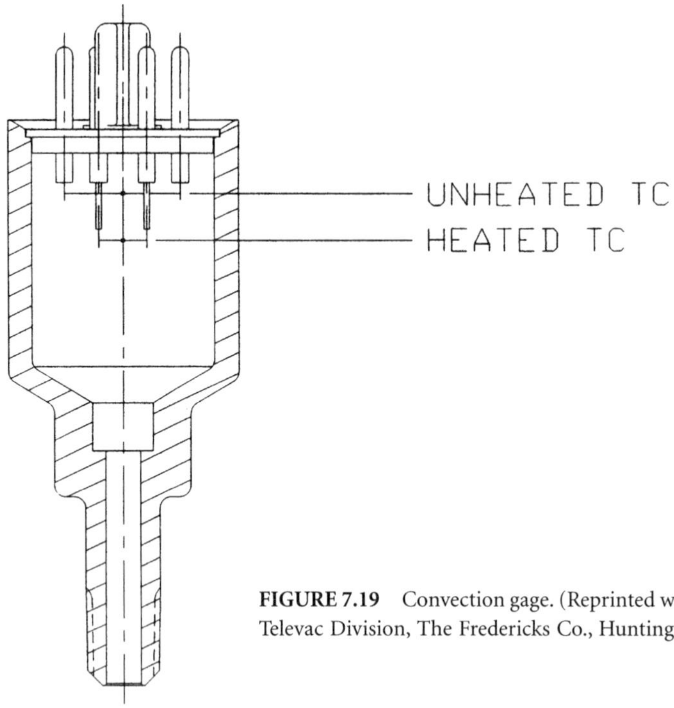

FIGURE 7.19 Convection gage. (Reprinted with permission from Televac Division, The Fredericks Co., Huntingdon Valley, PA.)

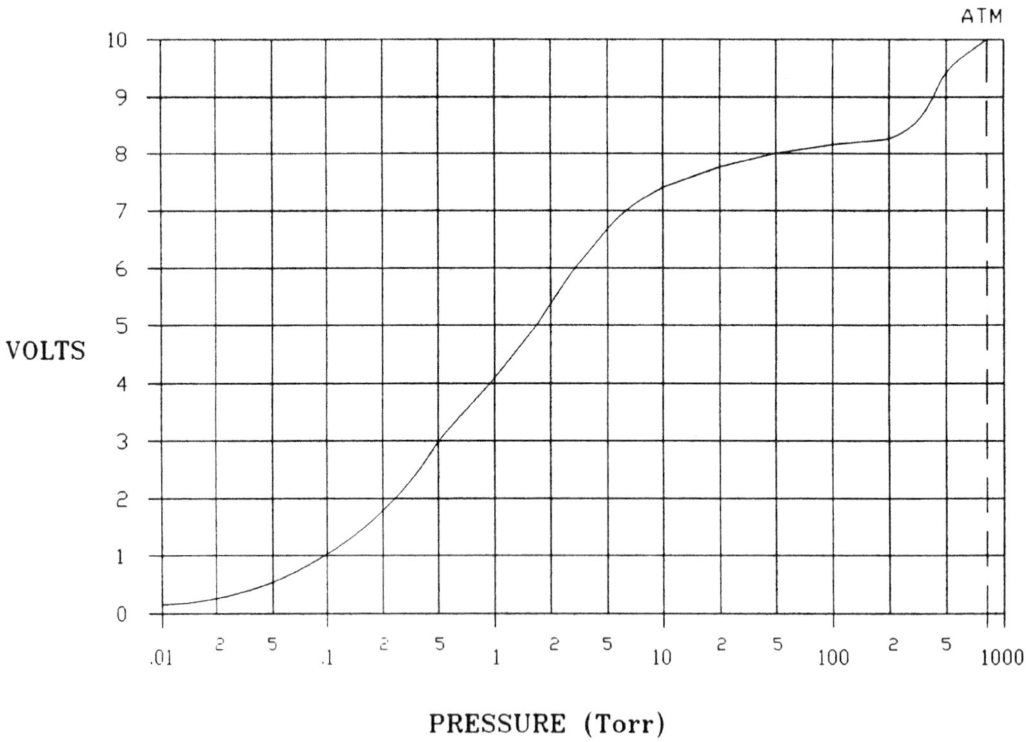

FIGURE 7.20 Output curve for the convection gage. (Reprinted with permission from Televac Division, The Fredericks Co., Huntingdon Valley, PA.)

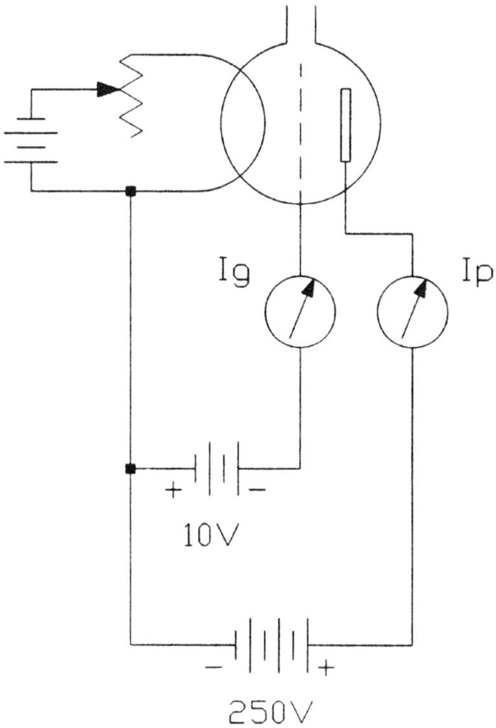

FIGURE 7.21 Typical triode connection.

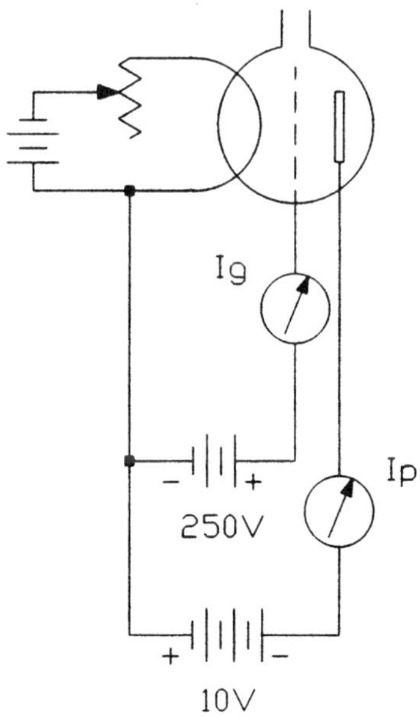

FIGURE 7.22 Alternative triode connection.

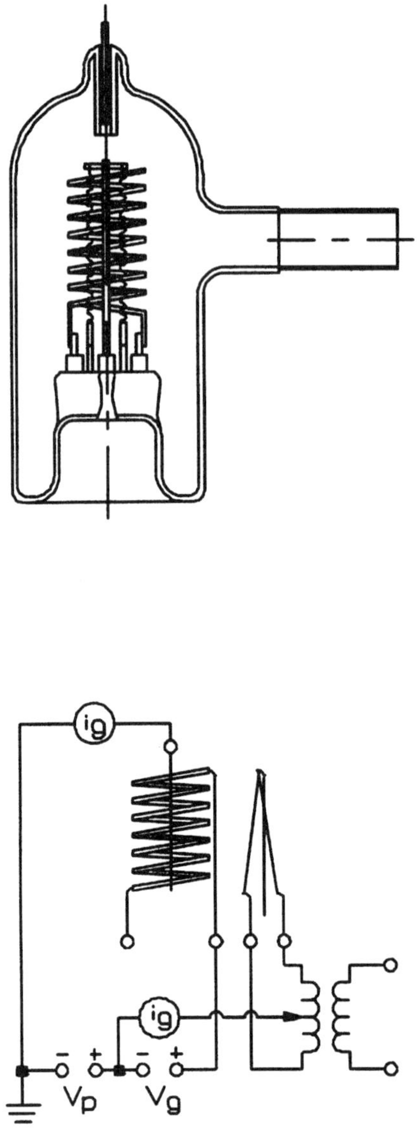

FIGURE 7.23 Bayard–Alpert hot cathode ionization gage.

Additional details are found in [23–25]. In nearly all cases, except at relatively high pressures, the triode gage has been replaced by the Bayard–Alpert gage.

Bayard–Alpert Hot Cathode Ionization Gages.
It became apparent that the pressure barrier observed at 10^{-8} torr was caused by a failure in measurement rather than pumping [26, 27]. A solution to this problem was proposed by Bayard and Alpert [28] that is now the most widely used gage for general UHV measurement.

The Bayard–Alpert gage is similar to a triode gage but has been redesigned so that only a small quantity of the internally generated X-rays strike the collector. The primary features of the Bayard–Alpert gage and its associated circuit are shown in Figures 22.23 and 22.24. The cathode has been replaced by a thin collector located at the center of the grid, and the cathode filament is now outside and several millimeters away from the grid. The Bayard–Alpert design utilizes the same controller as the triode gage, with

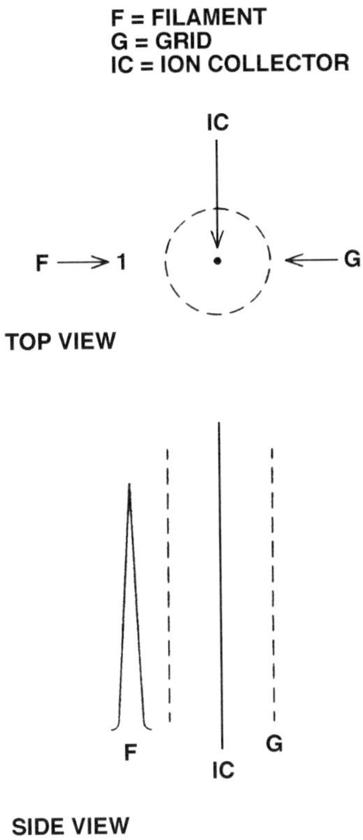

FIGURE 7.24 Bayard–Alpert gage configuration.

corrections for sensitivity differences between the gage designs. When a hot filament gage is exposed to high pressures, burn-out of the tungsten filaments often occurs. To prevent this, platinum metals were coated with refractory oxides to allow the gage to withstand sudden exposure to atmosphere with the filament hot [29, 30]. Typical materials include either thoria or yttria coatings on iridium. Bayard–Alpert and triode gages of identical structure and dimensions but with different filaments (i.e., tungsten vs. thoria iridium)were observed to have different sensitivities, with the tungsten filament versions being 20% to 40% more sensitive than the iridium of the same construction.

The lowest pressure that can be measured is limited by low energy X-rays striking the ion collector and emitting electrons. Several methods to reduce this X-ray limit were developed. Gage designs with very small diameter collectors have been made that extend the high vacuum range down to 10^{-12} torr, but accuracy was lost at the high pressures [31, 32].

The modulated gage was designed by Redhead [33] with an extra electrode near the ion collector. In this configuration, the X-ray current could be subtracted by measuring the ion current at two modulator potentials, thus increasing the range to 5×10^{-12}. Other gages use suppressor electrodes in front of the ion collector [34, 35].

The extractor gage (Figure 7.25) is the most widely used UHV hot cathode gage for those who need to measure 10^{-12} torr [36]. In this gage, the ions are extracted out of the ionizing volume and deflected or focused onto a small collector. More recent designs have been developed [37, 38]. The use of a channel electron multiplier [39] has reduced the low pressure limit to 10^{-15} torr.

The Bayard–Alpert gage suffers from some problems, however. The ion current is geometry dependent. Investigators have reported on the sensitivity variations, inaccuracy, and instability of Bayard-Alpert gages with widely differing results [40–46, 49, 50, 56]. Investigators have developed ways to reduce or eliminate some of these problems [47, 48].

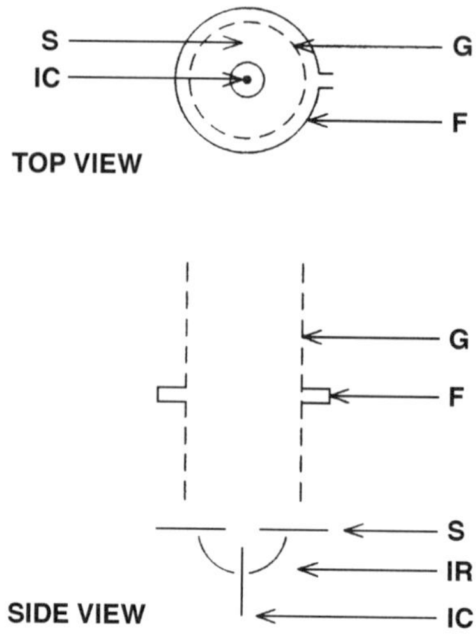

FIGURE 7.25 Extractor ionization gage. F, filament; G, grid; S, shield; IR, iron reflector; IC, ion collector.

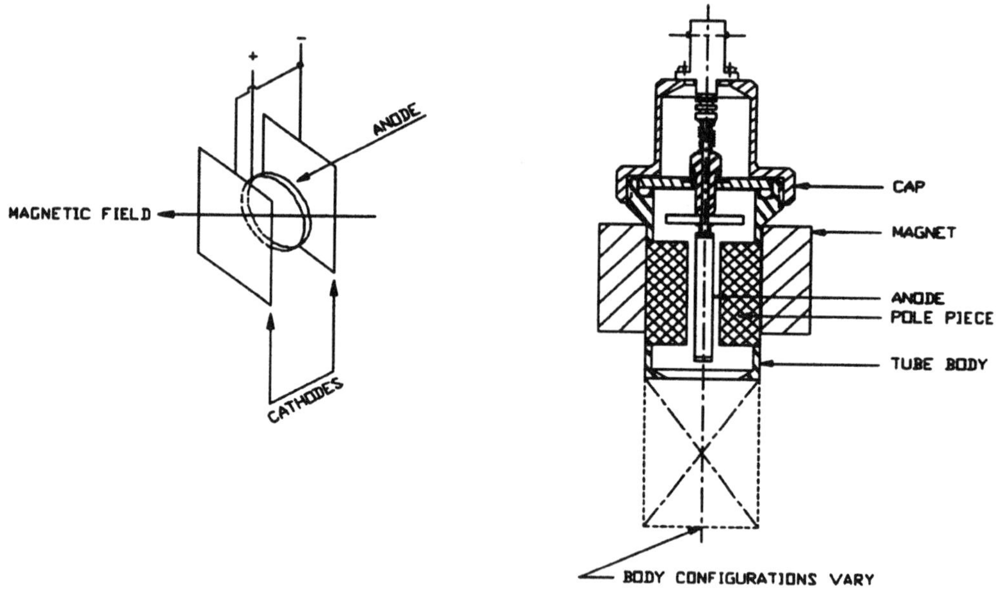

FIGURE 7.26 Penning gages. ([Left] From J.F. O'Hanlon, *A User's Guide to Vacuum Technology*, New York: John Wiley & Sons, 1980, 47. With permission. [Right] Reprinted with permission from Televac Division, The Fredericks Co., Huntingdon Valley, PA.)

Cold Cathode Ionization Gages

To measure pressures below 10^{-3} torr, Penning [51] developed the cold cathode discharge gage. Below 10^{-3} torr, the mean free path is so high that little ionization takes place. The probability of ionization was increased by placing a magnetic field parallel to the paths of ions and electrons to force these particles into helical trajectory.

This gage consists of two parallel cathodes and an anode, which is placed midway between them (see Figure 7.26). The anode is a circular or rectangular loop of metal wire whose plane is parallel to that of the cathodes. A few kilovolts potential difference is maintained between the anode and the cathodes.

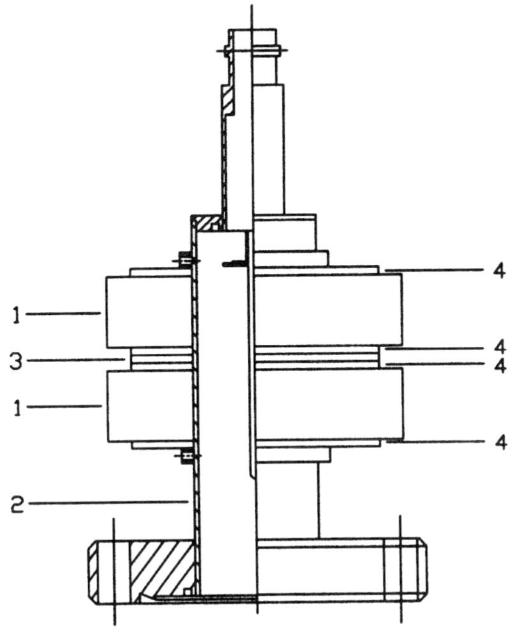

FIGURE 7.27 Double inverted magnetron. (Reprinted with permission from Televac Division, The Fredericks Co., Huntingdon Valley, PA.)

Furthermore, a magnetic field is applied between the cathodes by a permanent magnet usually external to the gage body. Electrons emitted from either of the two cathodes must travel in helical paths due to the magnetic field eventually reaching the anode, which carries a high positive charge. During the travel along this long path, many electrons collide with the molecules of the gas, thus creating positive ions that travel more directly to the cathodes. The ionization current thus produced is read out on a sensitive current meter as pressure.

This is a rugged gage used for industrial applications such as in leak detectors, vacuum furnaces, electron beam welders, and other industrial processes. The Penning gage is rugged, simple, and inexpensive. Its range is typically 10^{-3} torr to 10^{-6} torr, and some instability and lack of accuracy has been observed [52]. The magnetron design [53] and the inverted magnetron design [54] extended the low pressure range to 10^{-12} Torr [55] or better. These improvements produced a better gage but instability, hysterisis, and starting problems remain [57, 58]. Magnetrons currently in use are simpler and do not use an auxiliary cathode.

More recently, a double inverted magnetron was introduced [59]. This gage has greater sensitivity (amp/torr) than the other types (see Figure 7.27). It has been operated successfully at $\sim 1 \times 10^{-11}$ torr. The gage consists of two axially magnetized, annular-shaped magnets (**1**) placed around a cylinder (**2**) so that the north pole of one magnet faces the north pole of the other one. A nonmagnetic spacer (**3**) is placed between the two magnets and thin shims (**4**) are used to focus the magnetic fields. This gage has been operated to date at 10^{-11} torr, and stays ignited and reignites quickly when power is restored at this pressure. Essentially instantaneous reignition has been demonstrated by use of radioactive triggering [60].

Resonance Gages

One example of a resonance-type vacuum gage is the quartz friction vacuum gage [61]. A quartz oscillator can be built to measure pressure by a shift in resonance frequency caused by static pressure of the surrounding gas or by the increased power required to maintain a constant amplitude. Its range is from near atmospheric pressure to about 0.1 torr. A second method is to measure the resonant electrical impedance of a tuning fork oscillator. Test results for this device show an accuracy within ±10% for pressure from 10^{-3} torr to 10^{3} torr. There is little commercial use to date for these devices.

Molecular Drag (Spinning Rotor) Gages

Meyer [62] and Maxwell [63] introduced the idea of measuring pressure by means of the molecular drag of rotating devices in 1875. The rotors of these devices were tethered to a wire or thin filament. The gage was further enhanced by Holmes [64], who introduced the concept of the magnetic rotor suspension, leading to the spinning rotor gage. Nearly 10 years later, Beams et al. [65] disclosed the use of a magnetically levitated, rotating steel ball to measure pressure at high vacuum. Fremerey [66] reported on the historical development of this gage.

The molecular drag gage (MDG), often referred to as the spinning rotor gage, received wider acceptance after its commercial introduction in 1982 [67]. It is claimed to be more stable than other gages at lower pressures [68].

The principle of operation of the modern MDG is based on the fact that the rate of change of the angular velocity of a freely spinning ball is proportional to the gas pressure and inversely proportional to the mean molecular velocity. When the driving force is removed, the angular velocity is determined by measuring the ac voltage induced in the pickup coils by the magnetic moment of the ball (see Figure 7.28).

In current practice, a small rotor (steel ball bearing) about 4.5 mm in diameter is magnetically levitated and spun up to about 400 Hz by induction. The ball, enclosed in a thimble connected to the vacuum system, is allowed to coast by turning off the inductive drive. Then, the time of a revolution of the ball is measured by timing the signal induced in a set of pickup coils by the rotational component of the ball's magnetic moment. Gas molecules will exert a drag on the ball, slowing it at a rate set by the pressure P, its molecular mass m, temperature T, and the coefficient of momentum transfer σ, between the gas and the ball. A perfectly smooth ball would have a value of unity. There is also a pressure-independent residual drag (RD) caused by eddy current losses in the ball and surrounding structure. There will also be temperature effects that will cause the ball diameter and moment of inertia to change.

The pressure in the region of molecular flow is given by:

$$P = \frac{\pi \rho\, a\, \bar{c}}{10\, \sigma_{\text{eff}}} \left(\frac{-\omega' - \text{RD} - 2\, \alpha\, T'}{\omega} \right) \tag{7.4}$$

Note: Some sources include the term $\frac{(8kT')}{(\pi m)}$

where ρ = Density of the rotor
a = Radius of the rotor
ω'/ω = Fractional rate of slowing of the rotor
$\bar{c}$ = Mean gas molecular velocity
α = Linear coefficient of expansion of the ball
T' = Rate of change of the ball's temperature

All of the terms in the first part of the equation can be readily determined except for the accommodation coefficient σ, which depends on the surface of the ball and the molecular adhesion between the gas and the surface of the ball. The accommodation coefficient σ must be determined by calibration of the MDG against a known pressure standard or, if repeatability is more important than the highest accuracy, by assuming a value of 1 for σ. Measurements of σ on many balls over several years have been repeatedly performed by Dittman et al. [68]. The values obtained ranged from 0.97 to 1.06 for 68 visually smooth balls, so using a value of 1 for σ would not introduce a large error and would allow the MDG to be considered a primary standard (Fremerey [66]).

The controller [68] contains the electronics to power and regulate the suspension and drive, detect and amplify the signal from the pickup coils, and then time the rotation of the ball. It also contains a data processor that stores the calibration data and computes the pressure.

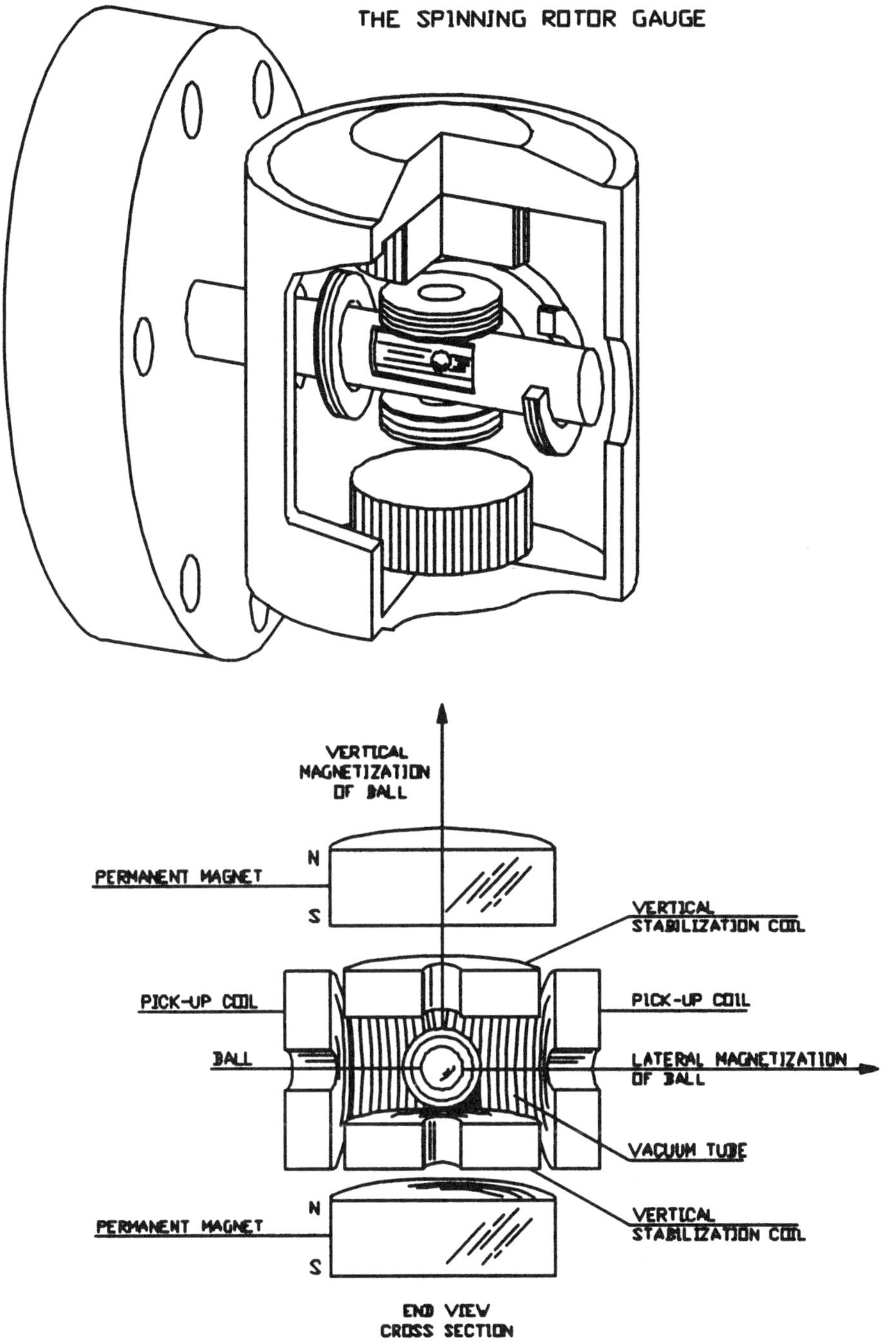

FIGURE 7.28 Molecular drag gage (spinning rotor).

The MDG is perhaps the best available transfer standard for the pressure range of 10^{-2} torr to 10^{-7} torr (1 Pa to 10^{-5} Pa) because it is designed for laboratory use in controlled, relatively vibration-free environments [69, 70].

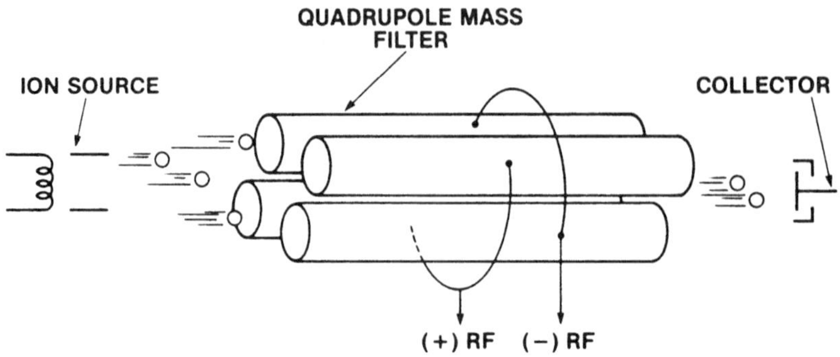

FIGURE 7.29 Quadrupole mass spectrometer. (From Varian Associates, Basic Vacuum Practice, Varian Associates, Inc., Lexington, MA, 1992. With permission.)

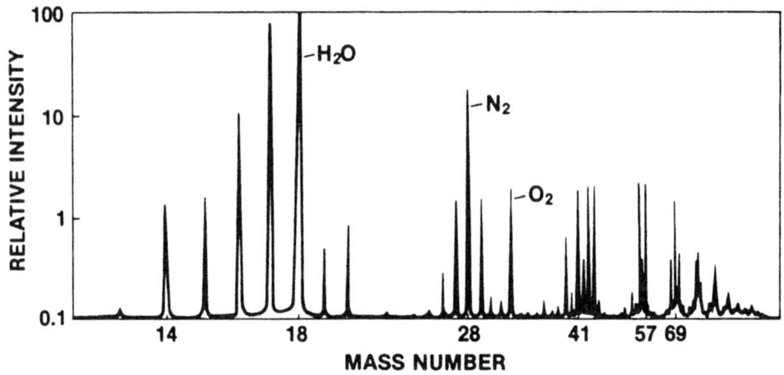

FIGURE 7.30 Relative intensity vs. mass number. (From Varian Vacuum Technologies, Basic Vacuum Practice, Varian Associates, Inc., Lexington, MA, 1992. With permission.)

Partial Pressure Measurements and Mass Spectrometers

The theory and practical applications of partial pressure measurements and mass spectrometers are discussed in detail in the literature [76]; however, an overview is presented herein [78].

A simple device for measuring the partial pressure of nitrogen as well as the total pressure in a vacuum system is the residual nitrogen analyzer (RNA). Operating in high vacuum, it is used to detect leaks in a vacuum system. It is effective because various gases are pumped at different rates and nitrogen is readily pumped, leaving a much lower percentage than is present at atmospheric pressure. Thus, the presense of a significant percentage of nitrogen at high vacuum indicates an air leak. The RNA consists of a cold cathode gage with an optical filter and a photomultiplier tube. Since ionization in the cold cathode tube produces light and the color is determined by the gases present, the RNA filters out all except for that corresponding to nitrogen and is calibrated to give the partial pressure of nitrogen.

A more complex device to measure the partial pressure of many gases in a vacuum chamber is the mass spectrometer-type residual gas analyzer (RGA). This device comes in many forms and several sizes. The quadrupole mass spectrometer is shown in Figure 7.29. The sensing head consists of an ion source, a quadrupole mass filter, and a Faraday cup collector. The quadrupole mass filter consists of two pairs of parallel rods having equal and opposite RF and dc voltages. For each combination of voltages, only ions of a specific mass will pass through the filter. The mass filter is tuned to pass only ions of a specific mass-to-charge ratio at a given time. As the tuning is changed to represent increasing mass numbers, a display such as Figure 7.30 is produced, showing the relative intensity of the signal vs. the mass number. This display can then be compared electronically with similar displays for known gases to determine the composition of the gases in the vacuum chamber. The head can operate only at high vacuum. However, by maintaining the head at high vacuum and using a sampling technique, the partial pressures of gases at higher pressures can be determined.

Calibration of mass spectrometers can be accomplished by equating the integral (the total area under all the peaks, taking into account the scale factors) to the overall pressure as measured by another gage (cold cathode, BA, or MDG). Once calibrated, the mass spectrometer can be used as a sensitive monitor of system pressure. It is also important when monitoring system pressure to know what gases are present. Most gages have vastly different sensitivities to different gas species.

Additional references on the details of the MDG and other types of gages available and on calibration are found in the literature [69–73, 75, 76]. The material in this chapter was summarized from an article on the fundamentals of vacuum measurement, calibration, and certification [77] from the authors' contribution to *The Handbook of Vacuum Technology* [76] and from other referenced sources.

References

1. J.F. O'Hanlon, *A User's Guide to Vacuum Technology,* New York: John Wiley & Sons, 1980, 47.
2. H. McLeod, *Phil. Mag.*, 47, 110, 1874.
3. C. Engleman, Televac Div. The Fredericks Co. 2337 Philmont Ave., Huntingdon Valley, PA, 19006, private communication.
4. Wallace and Tiernan Div., Pennwalt Corp., Bellville, NJ.
5. R.W. Hyland and R.L. Shaffer, Recommended practices of calibration and use of capacitance diaphragm gage for a transfer standard, *J. Vac. Sci. Tech., A,* 9(6), 2843, 1991.
6. K.R. Spangenberg, *Vacuum Tubes,* New York: McGraw-Hill, 1948, 766.
7. S. Dushman, *Scientific Foundations of Vacuum Technique,* 2nd ed., J.M. Lafferty (ed.), New York: John Wiley & Sons, 1962.
8. W. Steckelmacher and B. Fletcher, *J. Physics E.*, 5, 405, 1972.
9. W. Steckelmacher, *Vacuum,* 23, 307, 1973.
10. A. Beiman, *Total Pressure Measurement in Vacuum Technology,* Orlando, FL: Academic Press, 1985.
11. Granville-Phillips, 5675 Arapahoe Ave., Boulder CO, 80303.
12. Televac Div. The Fredericks Co. 2337 Philmont Ave., Huntingdon Valley, PA, 19006.
13. Granville-Phillips Data Sheet 360127, 3/95.
14. Televac U.S. Patent No. 5351551.
15. O.E. Buckley, *Proc. Natl. Acad. Sci.*, 2, 683, 1916.
16. M.D. Sarbey, *Electronics,* 2, 594, 1931.
17. R. Champeix, *Physics and Techniques of Electron Tubes,* Vol. 1, New York: Pergamon Press, 1961, 154-156.
18. N. Morgulis, *Physik Z. Sowjetunion,* 5, 407, 1934.
19. N.B. Reynolds, *Physics,* 1, 182, 1931.
20. J.H. Leck, *Pressure Measurement in Vacuum Systems,* London: Chapman & Hall, 1957, 70-74.
21. S. Dushman and C.G. Found, *Phys. Rev.*, 17, 7, 1921.
22. E.K. Jaycock and H.W. Weinhart, *Rev. Sci. Instr.*, 2, 401, 1931.
23. G.J. Schulz and A.V. Phelps, *Rev. Sci. Instr.*, 28, 1051, 1957.
24. Japanese Industrial Standard (JIS-Z-8570), Method of Calibration for Vacuum Gages,
25. J.W. Leck, op. cit., 69.
26. W.B. Nottingham, *Proc. 7th Annu. Conf. Phys. Electron.*, M.I.T., Cambridge, MA, 1947.
27. H.A. Steinhertz and P.A. Redhead, *Sci. Am.*, March, 2, 1962.
28. R.T. Bayard and D. Alpert, *Rev. Sci. Instr.*, 21, 571, 1950.
29. O.A. Weinreich, *Phys. Rev.*, 82, 573, 1951.
30. O.A. Weinreich and H. Bleecher, *Rev. Sci. Instr.*, 23, 56, 1952.
31. H.C. Hseuh and C. Lanni, *J. Vac. Sci. Technol.*, A 5, 3244, 1987.
32. T.S. Chou and Z.Q. Tang, *J. Vac. Sci. Technol.*, A4, 2280, 1986.
33. P.A. Redhead, *Rev. Sci. Instr.*, 31, 343, 1960.

34. G.H. Metson, *Br. J. Appl. Phys.*, 2, 46, 1951.
35. J.J. Lander, *Rev. Sci. Inst.*, 21, 672, 1950.
36. P.A. Redhead, *J. Vac. Sci. Technol.*, 3, 173, 1966.
37. J. Groszkowski, *Le Vide,* 136, 240, 1968.
38. L.G. Pittaway, *Philips Res. Rept.*, 29, 283, 1974.
39. D. Blechshmidt, *J. Vac Sci. Technol.*, 10, 376, 1973.
40. P.A. Redhead, *J. Vac. Sci. Technol.*, 6, 848, 1969.
41. S.D. Wood and C.R. Tilford, *J. Vac. Sci. Technol.*, A3, 542, 1985.
42. C.R. Tilford, *J. Vac. Sci. Technol.*, A3, 546, 1985.
43. P.C. Arnold and D.G. Bills, *J. Vac. Sci. Technol.*, A2, 159, 1984.
44. P.C. Arnold and J. Borichevsky, *J. Vac. Sci. Technol.*, A12, 568, 1994.
45. D.G. Bills, *J. Vac. Sci. Technol.*, A12, 574, 1994.
46. C.R. Tilford, A.R. Filippelli, et al., *J. Vac. Sci. Technol.*, A13, 485, 1995.
47. P.C. Arnold, D.G. Bills, et al., *J. Vac. Sci.Technol.*, A12, 580, 1994.
48. ETI Division of the Fredericks Co., Gage Type 8184.
49. T.A. Flaim and P.D. Owenby, *J. Vac. Sci. Technol.*, 8, 661, 1971.
50. J.F. O'Hanlon, op. cit., 65.
51. F.M. Pennin, *Physica,* 4, 71, 1937.
52. F.M. Penning and K. Nienhauis, *Philips Tech. Rev.*, 11, 116, 1949.
53. P.A. Redhead, *Can. J. Phys.*, 36, 255, 1958.
54. J.P. Hobson and P.A. Readhead, *Can. J. Phys.*, 33, 271, 1958.
55. NRC type 552 data sheet.
56. N. Ohsako, *J. Vac. Sci.Technol.*, 20, 1153, 1982.
57. D. Pelz and G. Newton, *J. Vac. Sci. Technol.*, 4, 239, 1967.
58. R.N. Peacock, N.T. Peacock, and D.S. Hauschulz, *J. Vac. Sci. Technol.*, A9, 1977 1991.
59. E. Drubetsky, D.R. Taylor, and W.H. Bayles, Jr., *Am. Vac. Soc., New Engl. Chapter, 1993 Symp.*
60. B.R. Kendall and E. Drubetsky, *J. Vac. Sci. Technol.*, A14, 1292, 1996.
61. M. Ono, K. Hirata, et al., Quartz friction vacuum gage for pressure range from 0.001 to 1000 torr, *J. Vac. Sci. Technol.*, A4, 1728, 1986.
62. O.E. Meyer, *Pogg. Ann.*, 125, 177, 1865.
63. J.C. Maxwell, *Phil. Trans. R. Soc.*, 157, 249, 1866.
64. F.T. Holmes, *Rev. Sci. Instrum.*, 8, 444, 1937.
65. J.W. Beams, J.L. Young, and J.W. Moore, *J. Appl. Phys.*, 17, 886, 1946.
66. J.K. Fremery, *Vacuum,* 32, 685, 1946.
67. NIST, Vacuum Calibrations Using the Molecular Drag Gage, Course Notes, April 15-17, 1996.
68. S. Dittman, B.E. Lindenau, and C.R. Tilford, *J. Vac. Sci. Technol.*, A7, 3356, 1989.
69. K.E. McCulloh, S.D. Wood, and C.R. Tilford, *J. Vac. Sci. Technol.*, A3, 1738, 1985.
70. G. Cosma, J.K. Fremerey, B. Lindenau, G. Messer, and P. Rohl, *J. Vac. Sci. Technol.*, 17, 642, 1980.
71. C.R. Tilford, S. Dittman, and K.E. McCulloh, *J. Vac. Sci. Technol.*, A6, 2855, 1988.
72. S. Dittman, NIST Special Publication 250-34, 1989.
73. National Conference of Standards Laboratories, Boulder, CO.
74. M. Hirata, M. Ono, H. Hojo, and K. Nakayama, *J. Vac. Sci. Technol.*, 20(4), 1159, 1982.
75. H. Gantsch, J. Tewes, and G. Messer, *Vacuum,* 35(3), 137, 1985.
76. D.M. Hoffman, B. Singh, and J.H. Thomas, III (eds.), *The Handbook of Vacuum Science and Technology,* Orlando, FL: Academic Press, 1998.
77. W.H. Bayles, Jr., Fundamentals of Vacuum Measurement, Calibration and Certification, Industrial Heating, October 1992.
78. Varian Associates, Basic Vacuum Practice, Varian Associates, Inc., 121 Hartwell Ave., Lexington, MA 02173, 1992.

7.3 Ultrasound Measurement

Peder C. Pedersen

Applications of Ultrasound

Medical

Ultrasound has a broad range of applications in medicine, where it is referred to as *medical ultrasound.* It is widely used in obstetrics to follow the development of the fetus during pregnancy, in cardiology where images can display the dynamics of blood flow and the motion of tissue structures (referred to as real-time imaging), and for locating tumors and cysts. 3-D imaging, surgical applications, imaging from within arteries (intravascular ultrasound), and contrast imaging are among the newer developments.

Industrial

In industry, ultrasound is utilized for examining critical structures, such as pipes and aircraft fuselages, for cracks and fatigue. Manufactured parts can likewise be examined for voids, flaws, and inclusions. Ultrasound has also widespread use in process control. The applications are collectively called *Non-Destructive Testing* (NDT) or *Non-Destructive Evaluation* (NDE). In addition, *acoustic microscopy* refers to microscopic examinations of internal structures that cannot be studied with a light microscope, such as an integrated circuit or biological tissue.

Underwater

Ultrasound is likewise an important tool for locating structures in the ocean, such as wrecks, mines, submarines, or schools of fish; the term SONAR (SOund Navigation And Ranging) is applied to these applications.

There are many other usages of ultrasound that lie outside the scope of this handbook: ultrasound welding, ultrasound cleaning, ultrasound hyperthermia, and ultrasound destruction of kidney stones (lithotripsy).

Definition of Basic Ultrasound Parameters

Ultrasound refers to acoustic waves of frequencies higher than 20,000 cycles per second (20 kHz), equal to the assumed upper limit for sound frequencies detectable by the human ear. As acoustic waves fundamentally are mechanical vibrations, a medium (e.g., water, air, or steel) is required for the waves to travel, or propagate, in. Hence, acoustic waves cannot exist in vacuum, such as outer space. If a single frequency sound wave is produced, also termed a *continuous wave* (CW), the fundamental relationship between frequency, f, in Hz, the sound speed of the medium, c_0, in m s^{-1}, and the wavelength, λ, in meters, is given as:

$$\lambda = \frac{c_0}{f} \tag{7.5}$$

The wavelength λ describes the length, in the direction of propagation, of one period of the sound wave. The wavelength determines, or influences, the behavior of many acoustic functions: The sound field emitted from an acoustic radiator (e.g., a transducer or loudspeaker) is determined by the radiator's size measured in wavelengths; the ability to differentiate between closely spaced reflectors is a function of the separation measured in wavelengths. Even when a sound pulse, rather than a CW sound, is transmitted, the wavelength concept is still useful, as the pulse typically contains a dominant frequency.

The vibrational activity on the surface of the sound source transfers the acoustic energy into the medium. If one were able to observe a very small volume, referred to as a *particle*, of the medium during transmission of sound energy, one would see the particle moving back and forth around a fixed position. Associated with the particle motion is an acoustic pressure, which refers to the pressure variation around the mean pressure (which is typically the atmospheric pressure). This allows the introduction of two important — and closely related — acoustic quantities: the *particle velocity*, $\vec{\mathbf{u}}(\vec{\mathbf{r}},t)$, and the *acoustic pressure*, $p(\vec{\mathbf{r}},t)$. In this notation, the arrow above a symbol in bold indicates a vector. The symbol $\vec{\mathbf{r}}$ represents the position vector, which simply defines a specific location in space. Thus, both particle velocity and pressure are functions of three spatial variables, *x*, *y*, and *z*, and the time variable, *t*.

To characterize a medium acoustically, the most important parameter is the *specific acoustic impedance*, *z*. For a lossless medium, *z* is given as follows:

$$z = \rho_0 c_0 \tag{7.6}$$

In Equation 7.6, ρ_0 is the density of the medium, measured in kg m^{-3}. When a medium absorbs acoustic energy (which all media do to a greater or smaller extent), the expression for acoustic impedance also contains a small imaginary term; this will be ignored in the discussions presented in this chapter. The acoustic impedance relates the particle velocity to the acoustic pressure:

$$z = \frac{p(\vec{\mathbf{r}},t)}{u(\vec{\mathbf{r}},t)} \tag{7.7}$$

Note that the relationship in Equation 7.7 uses the scalar value of the particle velocity (a scalar is a quantity, such as temperature, that does not have a direction associated with it). Equation 7.7 is exact for plane wave fields and a very good approximation for arbitrary acoustic fields. In a plane wave, all points in a plane normal to (i.e., which forms a 90° angle with respect to) the direction of propagation have the same pressure and particle velocity.

The acoustic impedances on either side of an interface (boundary between different media) determine the acoustic pressure reflected from the interface. Let a plane wave traveling in a medium with the acoustic impedance z_1 encounter a planar, smooth interface with another medium having the acoustic impedance z_2. Assume that the plane wave propagates directly toward the interface; this is commonly referred to as insonification under normal incidence. In this case, the pressure and the intensity reflection coefficients, *R* and R_I, respectively, are as follows:

$$\begin{aligned} R &= \frac{p_r}{p_i} = \frac{z_2 - z_1}{z_2 + z_1} \\ R &= \frac{I_r}{I_i} = \left(\frac{z_2 - z_1}{z_2 + z_1}\right)^2 \end{aligned} \tag{7.8}$$

Intensity is a measure of the mean power transmitted through unit area and is measured in watts per square meter. The corresponding pressure and intensity transmission coefficients, *T* and T_I, respectively, are:

$$\begin{aligned} T &= \frac{p_t}{p_i} = \frac{2 z_2}{z_2 + z_1} \\ T_I &= \frac{I_t}{I_i} = \frac{4 z_1 z_2}{\left(z_2 + z_1\right)^2} \end{aligned} \tag{7.9}$$

The subscripts "i," "r," and "t" in Equations 7.8 and 7.9 refer to incident, reflected, and transmitted, respectively. The expressions in these two equations can be considered approximately valid for nonplanar waves under near-normal incidence. However, when the incident angle (angle between direction of propagation and the normal to the surface) becomes large, the reflection and transmission coefficients can change dramatically. In addition, the reflected and transmitted signals will also change if the surface is rough to the extent that the rms (root mean square) height exceeds a few percent of the wavelength.

The wave propagation can take several forms. In fluids and gases, only *longitudinal* (or *compressional*) waves exist, meaning that the direction of wave propagation is equal to the direction of the particle velocity vector. In solids, both longitudinal and *shear* waves exist which propagate in different directions and with different sound speeds. Transverse waves can exist on strings where the particle motion is normal to the direction of propagation. (Strictly speaking, shear waves can propagate a short distance in liquids [fluids] if the viscosity is sufficiently high.)

Conceptual Description of Ultrasound Imaging and Measurements

Most ultrasound measurements are based on the generation of a short ultrasound pulse that propagates in a specified direction and is partly reflected wherever there is an abrupt change in the acoustic properties of the medium and detection of the resulting echoes (pulse-echo ultrasound). A change in properties can be due to a cyst in liver tissue, a crack in a high-pressure pipe, or reflection from layers in the sea bottom. The degree to which a pulse is reflected at an interface is determined by the change in acoustic impedance as described in Equation 7.8. An image is formed by mapping echo strength vs. travel time (proportional to distance) and beam direction, as illustrated in Figure 7.31. This is referred to as *B-mode* imaging (Brightness-mode). Further signal processing can be applied to compensate for attenuation (the damping out of the pressure pulse as it propagates) of the medium or to control focusing. Signal processing can also be applied to analyze echoes for information about the structure of materials or about the surface characteristics of rough surfaces.

A block diagram of a simplified pulse-echo ultrasound measurement system is shown in Figure 7.32. The pulser circuit can generate a large voltage spike for exciting the transducer in B-mode applications, or the arbitrary function source can produce a short burst for Doppler measurements, or a coded waveform, such as a linear sweep. The amplifier brings the driving voltage to a level where the transducer can generate an adequate amount of acoustic energy.

The transducer is made from a piezoelectric ceramic that has the property of producing mechanical vibrations in response to an applied voltage pulse and generating a voltage when subjected to mechanical stress. When an image is required, the transducer can be a mechanical sector probe that produces a fan-shaped image by means of a single, mechanically steered, focused transducer element. Alternatively, the transducer can be a linear array transducer (described later) that produces a rectangular image. In the case of an array transducer, the pulser/amplifier must contain a driving circuit for each element in the array, in addition to delay control. To achieve a short pulse and good sensitivity, the transducer is equipped with backing material and matching layers (to be discussed later).

The receiver block contains a low-noise amplifier with time-varying gain to correct for medium attenuation and often a circuit for logarithmic compression. In the case of array transducers, the receiver circuitry is a complex system of amplifiers, time-varying delay elements, and summing circuits. The signal processing block can be part analog and part digital. The echo signals are envelope detected and digitized; the *envelope* of a signal is a curve that follows the amplitude of the received signal. A scan converter changes the signal into a format suitable for display on a gray-scale monitor. Information about ultrasound pulser-receiver instrumentation for NDE measurements can be found in [1] and for medical imaging in [2].

Not all ultrasound measurement systems are based on the pulse-echo concept. For material characterization, *transmission measurements* are frequently used, as illustrated in Figure 7.33. The main difference between the pulse-echo system and the transmission system is that two transducers are used in the transmission system; the description of the individual blocks for the pulse-echo system applies generally here also. Imaging is generally not possible, although tomographic imaging of either attenuation or

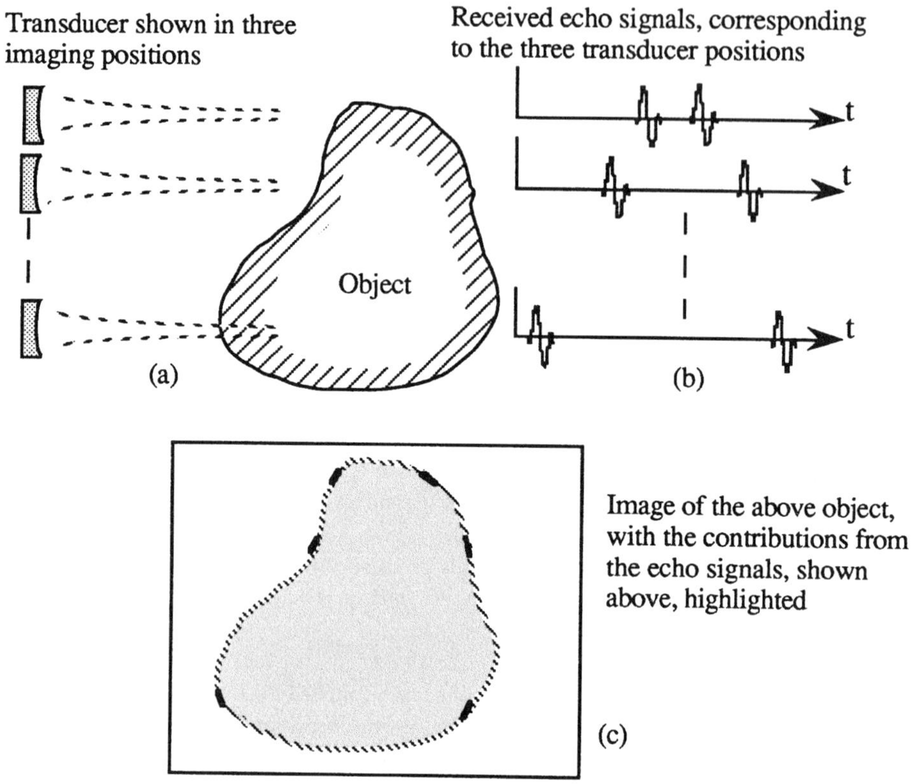

FIGURE 7.31 (*a*) A focused transducer insonifies the irregular object from different positions out of which only three are shown. The different transducer positions can readily be obtained by the use of an array transducer (to be described later). (*b*) Received echoes from the front and back of the object are displayed vs. travel time. It is here assumed that the structure is only weakly reflecting and that the attenuation has only a minimal effect. (*c*) An image is formed, based on the echo strengths and the echo arrival times.

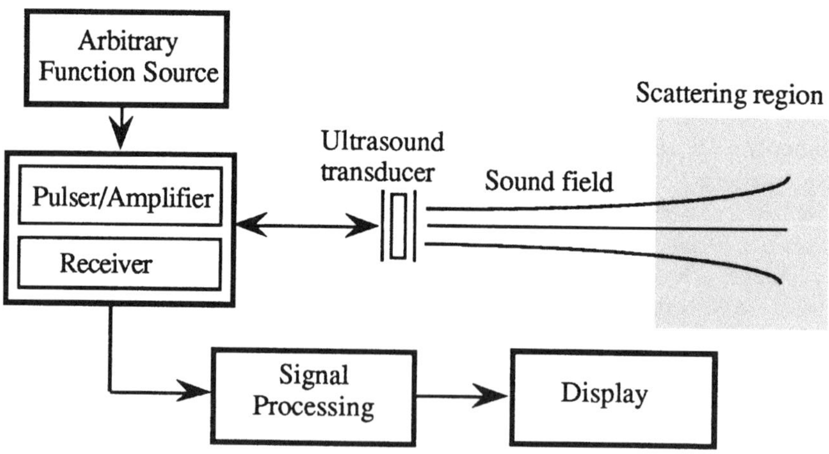

FIGURE 7.32 Simplified block diagram of a pulse-echo ultrasound system.

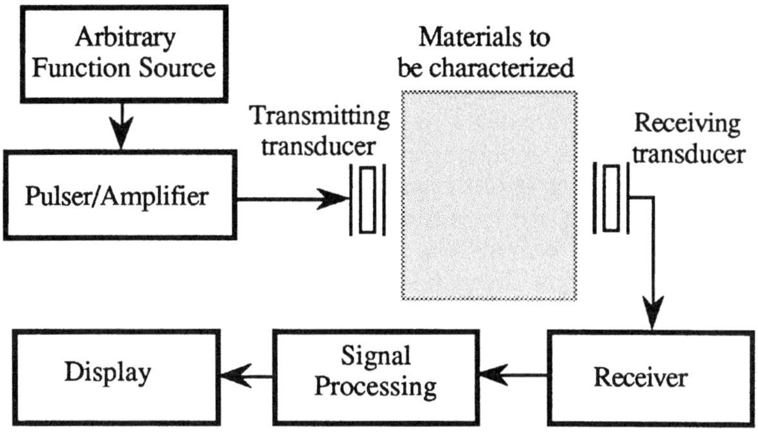

FIGURE 7.33 Simplified block diagram of ultrasound transmission system.

velocity has been attempted. If the two transducers are moved together, voids and inclusions located in the transmission path can be detected.

Transmission measurements require, of course, that the structure or the medium of interest is accessible from opposite sides. This system allows accurate characterization of the attenuation and the sound speed of the medium from which a variety of material properties can be derived. Flow in a pipe or a channel can also be determined with a transmission system, by determining the difference between upstream and downstream propagation times.

Single-Element and Array Transducers

The device that converts electric energy into acoustic energy and vice versa is termed the *transducer* or, more specifically, the ultrasound transducer. Piezoelectric ceramics can be used for all frequencies in the ultrasound range; however, for ultrasound frequencies in the 20 kHz to 200 kHz range, magnetostrictive devices are occasionally used, which are based on materials that exhibit mechanical deformation in response to an applied magnetic field. Ultrasound transducers are commercially available for many applications over a wide range of frequencies.

Single-element transducers are used for basic measurements and material characterization, while array transducers with many individual transducer elements are used for imaging purposes. The former type costs from a few hundreds dollars and up, whereas the latter type costs in the thousands of dollars and requires extensive electronics for beam control and signal processing. A third alternative is the mechanical sector scanner where a single transducer is mechanically rotated over a specified angle. A single-element transducer consists of a piezoelectric element, a backing material that enables the transducer to respond to a fairly wide range of frequencies (but with reduced sensitivity), and a matching layer and faceplate that provide improved coupling into the medium and protect the transducer.

Although broadband transducers are very desirable, the energy conversion characteristics of the practical transducer correspond to that of a bandpass filter, i.e., the ultrasound pulse has most of its energy distributed around one frequency. This is the frequency referred to when, for example, one orders a 3.5 MHz transducer. The 6 dB bandwidth can be from 50% to 100% of the transducer frequency; the 6 dB bandwidth is the frequency range over which the transducer can produce an acoustic pressure that is at least 50% of the acoustic pressure at the most efficient frequency. The radiation characteristics of a transducer are determined by the geometry of the transducer (square, circular, plane, focused, etc.) and by the dimensions measured in wavelengths. Hence, a 10 MHz transducer tends to be smaller than a 5 MHz transducer. Focusing can also be achieved by means of an acoustic lens.

Array transducers exist in three main categories: phased arrays, linear arrays, and annular arrays. A fourth category could exist commercially in a few years: the 2-D array or a sparse 2-D array. Common for these array transducers is the fact that *during transmission*, the excitation time and excitation signal amplitude for each transducer element are controlled independently. This allows the beam to be steered in a given direction, as well as focused at a given point in space. *During reception*, an independently controlled time delay — which may be time varying — can be applied to each element before summation with the signal from other elements. The delay control permits the transducer to have maximal sensitivity to an echo from a specified range, and, moreover, to shift this point in space away from the transducer as echoes from structures further away are received. A limitation of the phased and linear arrays is that beam steering and focusing can only take place within the image plane. In the direction normal to the image plane, a fixed focus is produced by the physical shape of the array elements.

The phased array transducer consists of a fairly small number of elements, such as 16 to 32. Both beam steering and focusing are carried out, producing a pie-shaped image. The phased array is most suitable when only a narrow observation window exists, as is the case when imaging the heart from the space between the ribs. The linear array transducer has far more elements, typically between 128 and 256. It activates only a subset of all the elements for a given measurement, and produces focusing, but generally not beam steering, so that the beam direction is normal to the array surface. Consider a linear array transducer with 128 elements, labeled 1 to 128. If one assumes that the first pulse-echo measurement is made with elements 1 to 16, the next measurement with elements 2 to 17, etc., the beam will have been *linearly* translated along the long dimension of the linear array (hence the name), producing a rectangular image format. The annular array transducer consists of a series of concentric rings. As such, it cannot steer the beam, but can focus both in transmit and receive mode, just as the phased and linear arrays. Thus, the annular array transducer is not suitable for imaging unless the transducer is moved mechanically. Its specific advantage is uniform focusing.

Selection Criteria for Ultrasound Frequencies

The discussion so far has often mentioned *frequency* as an important variable, without providing any guidance as to what ultrasound frequencies should be used. There is generally no technological limitation with respect to choice of ultrasound frequency, as frequencies even in the gigahertz (GHz, 10^9 Hz) range can be produced. The higher the frequency, the better is the control over the direction and the width of the ultrasound beam, leading to improved axial and lateral resolution. However, the attenuation increases with higher frequency. For many media, the attenuation varies with the frequency squared; while for biological soft tissue, the attenuation varies nearly linearly with frequency.

Generally, the ultrasound frequency is chosen as high as possible while still allowing a satisfactory signal-to-noise ratio (SNR) of the received signal. This rule does not always hold; for example, for grain size estimation in metals and for rough surface characterization, the optimal frequency has a specific relationship to the mean grain dimension or the rms roughness.

Basic Parameters in Ultrasound Measurements

Ultrasound technology today has led to a very wide range of applications, and the following overview mentions only the basic parameters. For an in-depth review, see [3]. A discussion of applications is given at the end of the Ultrasound Theory and Applications section.

Reflection (or Transmission) Detected or Not

The most basic measurement consists of determining whether a reflected or transmitted signal is received or not. This can be used to monitor liquid level where the absence of liquid prevents ultrasound transmission from taking place, or to detect the existence of bubbles in a liquid (e.g., for dialysis purposes, where the bubbles are the reflectors) or for counting objects, such as bottles, on a conveyer belt.

Travel Time

Travel time is the elapsed time from transmission of an ultrasound pulse to the detection of a received pulse in either a pulse-echo or a transmission system. If the sound speed is known, the thickness of an object can be found from the time difference between the front surface echo and the back surface echo. Conversely, sound speed can be determined when the thickness is known. By measuring with a broadband pulse (a short pulse containing a broad range of frequencies), velocity dispersion (frequency dependence of velocity) can be found, which has applications for materials characterization. Elastic parameters can also be found from velocity measurements. Imaging applications are generally made based on the assumption of a known, constant velocity; thus, round-trip travel times from the transducer to the reflecting structures and back in a known direction gives the basis for the image formation.

Attenuation

The attenuation and its frequency dependence are important materials parameters, as they can be used to differentiate between normal and pathological biological tissue, measure porosity of ceramics, the grain size of metals, and the structure of composite materials. Attenuation can be found by means of transmission measurements by determining the signal amplitude with the test object first absent and then present; here, transmission losses must also be considered, as quantified in Equation 7.9. Attenuation can be obtained from pulse-echo measurements, by comparing the strength of echoes from the front and back surfaces of the specimen under test. Diffraction effects (described later in the *Advanced Topics in Ultrasound* section) might need to be considered. Attenuation can be considered as a bulk parameter for the medium as a whole, or it can be determined for small regions of the medium.

Reflection Coefficient

The strength of a reflection at an interface between two media is determined by the change in acoustic impedance across the interface, as shown in Equation 7.8. This allows, at least in principle, the impedance of one medium to be determined from the measured reflection coefficient and the impedance of the other medium. In fact, the term *impediography* refers to the determination of the impedance profile of a layered medium by means of this concept. In practice, the measurement is not easy to carry out, as the transducer must be aligned very accurately at normal incidence to the medium surface. By measuring the reflection coefficient vs. incident angle at a planar, smooth surface, a velocity estimate of the reflecting medium can be made; in an alternative application, reflection measurements vs. frequency permit the evaluation of rough surface parameters.

Parameters Obtained Through Signal Processing

A number of object parameters can be obtained from analysis of the received signals. Material properties can be extracted from speckle analysis (analysis of statistical fluctuations in the image) and other forms of statistical analysis. Recognition of objects with complex shapes can, in some cases, be done by extracting specific features of the received signals. Doppler processing for velocity estimation is another example of information that is only attainable by means of signal processing. By combining several "stacked" images, 3-D reconstruction is possible, from which volume estimation and rate of change of volume are possible, such as for determining the dynamics of the left ventricle of the heart.

Ultrasound Theory and Applications

Sound Velocity

The propagation speed, c_0, is generally determined by the density and the elastic properties of the medium. For an ideal gas, c_0 may be expressed as [4]:

$$c_0 = \sqrt{\gamma P_0/\rho_0} = \sqrt{\gamma r T_K} \tag{7.10}$$

where $\gamma = C_P/C_V$, the ratio of specific heats; for air, $\gamma = 1.402$. P_0 is the static pressure of the gas, which at 1 atm is 1.013×10^5 N m^{-2}. ρ_0 is the specific density of the gas, equal to 1.293 kg m^{-3} for air at 1 atm and 0°C. The constant r is the ratio of the universal gas constant and the molecular weight of the gas, and T_K is the temperature in kelvin. Substituting the values for air into the first equation in Equation 7.10 gives $c_0 = 331.6$ m s^{-1} at 0°C. As the ratio P_0/ρ_0 is constant for varying pressure, but constant temperature, c_0 is likewise pressure independent. (For a real gas, c_0 exhibits, in fact, a small dependence on pressure.)

The temperature dependence of an ideal gas can be obtained by rewriting the second expression in Equation 7.10 as follows

$$c_0 = c_{ref} \sqrt{1 + T/273} \tag{7.11}$$

where c_{ref} is the sound speed at 0°C, and T is the temperature in °C.

In liquids, the expression for c_0 is given as

$$c_0 = \sqrt{K_s/\rho_0} \tag{7.12}$$

where K_s is the adiabatic bulk modulus. Expressions for actual liquids as a function of temperature and pressure are not easily derived, and are often empirically determined. For pure water, the sound speed as a function of temperature and pressure can be given as:

$$c_{H_2O} = 1402.4 + 5.01T - 0.055T^2 + 0.00022T^3 + 1.6 \times 10^{-6}\ P_0 \tag{7.13}$$

In Equation 7.13, T is temperature in °C, and P_0 is the static pressure in N m^{-2}; the expression is valid for a pressure range from 1 atm to 100 atm.

Finally, for solids, the sound speed for longitudinal waves, c_L, is [5]:

$$c_L = \left(\frac{c_{11}}{\rho_0}\right)^{1/2} = \left(\frac{\lambda + 2\mu}{\rho_0}\right)^{1/2} \tag{7.14}$$

In Equation 7.14, c_{ij} is the the elastic stiffness constant, such that c_{11} refers to longitudinal stress over longitudinal strain, and $\lambda = c_{12}$ and $\mu = c_{44}$ are the Lamé elastic constants.

Wave Propagation in Homogeneous Media

In order to effectively use ultrasound for measurement purposes, it is essential to be able to describe the behavior of acoustic fields. This includes the general behavior, and maybe even a detailed understanding, of the radiated field from ultrasound transducers; the transmission, reflection, and refraction of an acoustic field at boundaries and layers; and the diffraction of of acoustic field at finite-sized reflectors.

The wave equation is a differential equation that formulates how an acoustic disturbance (acoustic pressure or particle velocity) propagates through a homogeneous medium. For an arbitrary wave field, the wave equation takes the following general form:

$$\nabla^2 p = \frac{1}{c^2}\frac{\partial^2 p}{\partial t^2} \tag{7.15}$$

where ∇^2 is the Laplacian operator. Equation 7.15 is also valid when particle velocity is substituted for pressure. If the condition of a plane wave field at a single frequency (also called a harmonic plane wave field) is imposed, the solution to Equation 7.15 is:

$$p(\vec{\mathbf{r}}, t) = A\exp\left[j\left(\omega t - \vec{\mathbf{k}} \times \vec{\mathbf{r}}\right)\right] + B\exp\left[j\left(\omega t + \vec{\mathbf{k}} \times \vec{\mathbf{r}}\right)\right] \tag{7.16}$$

In Equation 7.16, $p(\vec{\mathbf{r}},t)$ consists of a plane wave propagating in the direction defined by the propagation vector $\vec{\mathbf{k}}$ and a plane wave in the direction $-\vec{\mathbf{k}}$. As before, $\vec{\mathbf{r}}$ is a position vector that, in a Cartesian coordinate system, can be expressed as $\vec{\mathbf{r}} = x\hat{x} + y\hat{y} + z\hat{z}$ where $\hat{x}$, $\hat{y}$, and $\hat{z}$ are unit vectors. The magnitude of $\vec{\mathbf{k}}$ is called the wavenumber:

$$k = \omega / c_0 = 2\pi / \lambda \tag{7.17}$$

Expressing $\vec{\mathbf{k}}$ in the Cartesian coordinate system gives:

$$\vec{\mathbf{k}} = k_x\hat{x} + k_y\hat{y} + k_z\hat{z} \tag{7.18}$$

When k, k_x, and k_y are specified, k_z is defined as well:

$$k_z = \sqrt{\left(\left(\omega / c_0\right)^2 - k_x^2 - k_y^2\right)} \tag{7.19}$$

Applying the expressions for $\vec{\mathbf{r}}$ and $\vec{\mathbf{k}}$ to Equation 7.16 and considering only the plane wave in the direction of $+\vec{\mathbf{k}}$ gives

$$p\left(\vec{\mathbf{r}},t\right) = A \exp\left[j\left(\omega t - \left(k_x x + k_y y + k_z z\right)\right)\right] \tag{7.20}$$

From Equation 7.20, one sees that the amplitude of a plane wave is constant and equal to A, but that the phase varies with both time and space. When the direction of $\vec{\mathbf{k}}$ is specified, k_x, k_y, and k_z are found from projection onto the coordinate axes.

Solving Equation 7.15 under the assumption of spherical waves at a single frequency, with the source placed at the origin of the coordinate system, gives:

$$p\left(r,t\right) = \frac{A}{r}\exp\left[j\left(\omega t - kr\right)\right] + \frac{B}{r}\exp\left[j\left(\omega t + kr\right)\right] \tag{7.21}$$

The first term is a diverging spherical wave, and the second term is a converging spherical wave. The reason that the vector dot product does not appear in Equation 7.21 is that $\vec{\mathbf{k}}$ and $\vec{\mathbf{r}}$ always point in the same direction for spherical waves. While Equation 7.21 gives the complete solution, in most cases only a diverging spherical wave exists. If spherical waves are produced by a spherical source of radius a, then:

$$p\left(r,t\right) = \frac{A}{r}\exp\left[j\left(\omega t - kr\right)\right], \quad r > a \tag{7.22}$$

Acoustic Intensity and Sound Levels

The acoustic intensity, I, describes the mean power transported through a unit area normal to the direction of propagation. A general expression for I for a CW pressure function is:

$$I = \frac{1}{T}\int_0^T p\left(t\right)u\left(t\right)dt \tag{7.23}$$

In Equation 7.23, T represents one full cycle of the pressure function, and $u(t)$ is the particle velocity function. For a plane wave, the intensity is:

$$I = \frac{P^2}{2\rho_0 c_0} = \frac{1}{2} PU \tag{7.24}$$

where P and U are the magnitudes of the pressure and the particle velocity functions, respectively. The expression for I in Equation 7.24 can also be used as an approximation for nonplanar waves, such as spherical waves, as long as (kr) is large.

The intensity levels for pulse-echo measurements are often described by the SPTA (*Spatial Peak, Temporal Average*) value, which therefore refers to the mean intensity at the point in space where the intensity is the highest. Although the guidelines for exposure levels for medical imaging, set by the FDA, vary for different parts of the body, a general upper limit is 100 mW cm^{-2}. Given that the duty cycle (pulse duration/pulse interval) is typically in the order of 0.0001 for ultrasound imaging, the temporal peak intensity is much higher, and is even approaching the level where nonlinear effects can begin to be observed. Intensity levels used for NDE are application dependent, but are generally in the same range as in medical ultrasound.

Sound levels are logarithmic expressions (expressions in dB) of the pressure level or the intensity level. The term SPL stands for sound pressure level and is given as:

$$\mathrm{SPL} = 20\log\left(P_e / P_{ref}\right) \ [\mathrm{dB}] \tag{7.25}$$

In Equation 7.25, P_e is the effective pressure of the acoustic wave and P_{ref} is the reference effective pressure. Correspondingly, IL is the intensity level defined as:

$$\mathrm{IL} = 10\log\left(I / I_{ref}\right) \ [\mathrm{dB}] \tag{7.26}$$

where I_{ref} is the reference acoustic intensity.

Several reference levels are commonly used [6]. In air, P_{ref} = 20 μPa and $I_{ref} = 10^{-12}$ W m^{-2} are nearly equivalent reference levels. For water, typical pressure reference levels are P_{ref} = 1 μPa, P_{ref} = 20 μPa or P_{ref} = 1 μbar = 0.1 Pa, with corresponding intensity reference levels of $I_{ref} = 6.76 \times 10^{-19}$ W m^{-2}, $I_{ref} = 2.70 \times 10^{-16}$ W m^{-2}, and $I_{ref} = 6.76 \times 10^{-9}$ W m^{-2}.

Wave Propagation Across Layers and Boundaries

The reflection and transmission coefficients for a plane wave impinging at normal incidence on the interface between two half spaces were given in Equations 7.8 and 7.9. It is also important to consider the transmission and reflection of waves at non-normal incidence and across a layer.

When a longitudinal wave impinges at a liquid–liquid interface, both the transmitted and the reflected waves are longitudinal waves, as illustrated in Figure 7.34(a). However, when a longitudinal wave propagating in a liquid medium encounters a liquid–solid interface under oblique incidence, both a longitudinal and a shear transmitted wave and a longitudinal reflected wave will result, as shown in Figure 7.34(b). The change of direction of the transmitted waves relative to the incident wave is referred to as *refraction*, which is caused by the difference in sound speed between the two media.

The magnitude and direction of the reflected and transmitted waves are determined by two simple boundary conditions: (1) the *pressure* on both sides of the boundary must be equal at all times, and (2) the *normal particle velocity* on both sides of the boundary must be equal at all times; the normal particle velocity refers to the component of the particle velocity that is normal to the boundary.

The boundary conditions lead to the following two relationships:

$$\sin\theta_i = \sin\theta_r; \quad \frac{\sin\theta_i}{c_1} = \frac{\sin\theta_t}{c_2} \tag{7.27}$$

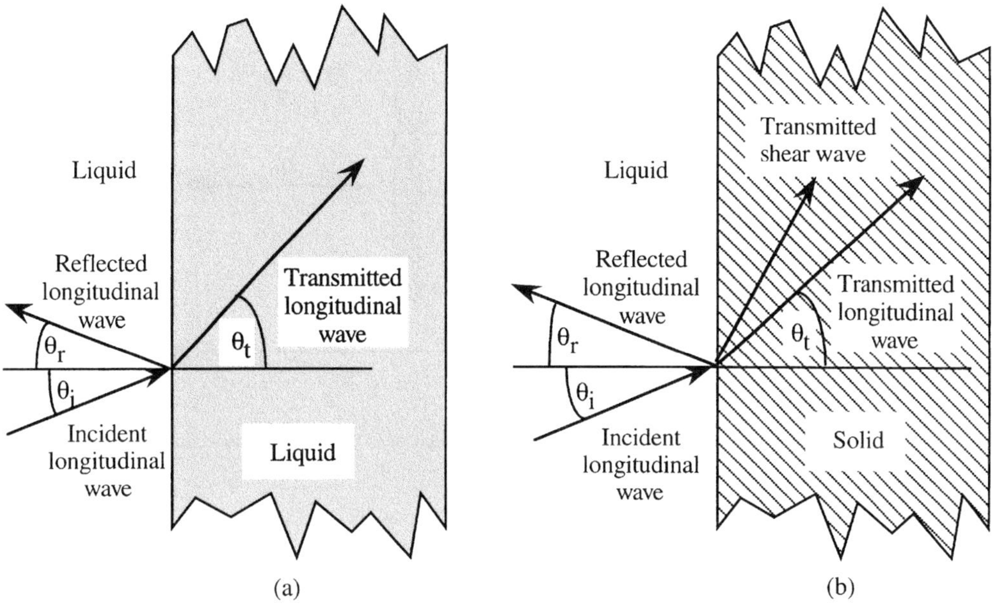

FIGURE 7.34 (*a*). An incident longitudinal wave at a liquid–liquid interface produces a reflected and a transmitted longitudinal wave. (*b*) An incident longitudinal wave at a liquid–solid interface produces a reflected longitudinal wave, transmitted longitudinal, and shear waves. The incident, reflected, and transmitted angles are indicated.

where c_1 and c_2 are the sound speeds of the two media, and θ_i, θ_r, and θ_t are the incident, reflected and transmitted angles, respectively. From Equation 7.27, one sees that

$$\theta_r = \theta_i; \quad \theta_t = \arcsin\left(\frac{c_2}{c_1}\sin\theta_i\right) \tag{7.28}$$

The second expression in Equation 7.28 is referred to as Snell's law and quantifies the degree of refraction. Refraction affects the quality of ultrasound imaging because image formation is based on the assumption that the ultrasound beam travels in a straight path through all layers and inhomogeneities, and that the sound speed is constant throughout the medium. When the actual beam travels along a path that deviates to some extent from a straight path and passes through some parts of the medium faster than it does other parts, the resulting image is a distorted depiction of reality. Correcting this distortion is a very complex problem and, in the near future, one should only expect image improvement in the simplest cases.

It can be seen from Equation 7.28 that angle θ_t is not defined if the argument to the arcsin function exceeds unity. This defines a critical incident angle as follows:

$$\sin\theta_c = \frac{c_1}{c_2} \tag{7.29}$$

at or above which the reflection coefficient is 1 and the transmission coefficient is 0. A critical angle only exists when $c_1 < c_2$. The transmitted shear and longitudinal velocities each correspond to a different critical angle of incidence.

From the boundary conditions, the reflection coefficient as a function of incident angle can be determined [7]:

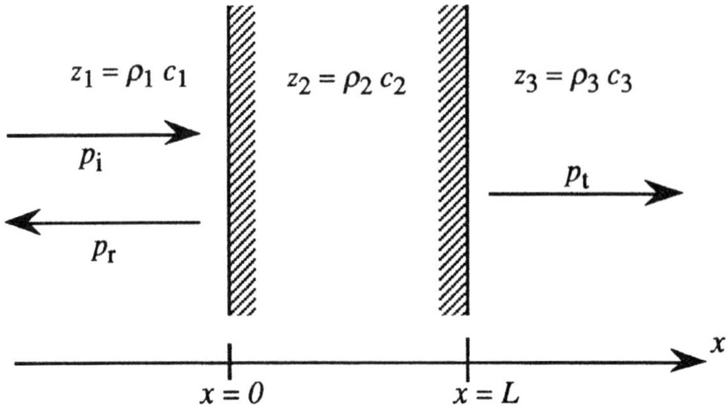

FIGURE 7.35 A pressure wave is incident on a layer of thickness L and with acoustic impedance $z_2 = \rho_2 c_2$.

$$R(\theta_i) = \frac{(z_2/z_1) - (\cos\theta_t/\cos\theta_i)}{(z_2/z_1) + (\cos\theta_t/\cos\theta_i)}, \quad \text{where} \quad \cos\theta_t = \sqrt{1 - (c_2/c_1)^2 \sin^2\theta_i} \tag{7.30}$$

The result in Equation 7.30 is referred to as the Rayleigh reflection coefficient.

The transmission of plane waves under normal incidence through a single layer is often of interest. The dimensions and the medium parameters are defined in Figure 7.35.

By applying the boundary conditions to both interfaces of the layer, both reflection and transmission coefficients are obtained [7], as given in Equations 7.31 and 7.32.

$$R = \frac{p_r}{p_i} = \frac{(1 - z_1/z_3)\cos k_2 L + j(z_2/z_3 - z_1/z_2)\sin k_2 L}{(1 + z_1/z_3)\cos k_2 L + j(z_2/z_3 + z_1/z_2)\sin k_2 L} \tag{7.31}$$

$$T = \frac{p_t}{p_i} = \frac{2}{(1 + z_1/z_3)\cos k_2 L + j(z_2/z_3 + z_1/z_2)\sin k_2 L} \tag{7.32}$$

A number of special conditions for Equation 7.31 can be considered, such as: (1) $z_1 = z_3$, which simplifies the numerator in Equation 7.31; and (2) the layer thickness is only a small fraction of a wavelength, that is, $k_2 L \ll 1$ which makes $\cos k_2 L \approx 1$ and $\sin k_2 L \approx k_2 L$. One case is of particular interest: the choice of thickness and acoustic impedance for the layer that makes $R = 0$, and therefore gives 100% energy transmission. This is fulfilled for:

$$k_2 L = \pi/2 + n\pi; \quad z_2 = \sqrt{z_1 z_3} \tag{7.33}$$

The result in Equation 7.33 states that the layer must be a quarter of a wavelength thick (plus an integer number of half wavelengths) and must have an acoustic impedance that is the geometric mean of the impedances of the media on either side. Among several applications of the *quarter wavelength impedance matching* is the impedance matching between a transducer and the medium, such as water. A drawback with a single matching layer is that it only works effectively over a narrow frequency range, while the actual acoustic pulse contains a fairly broad spectrum of frequencies. A better matching is achieved by using more than one matching layer, and it has been shown that a matching layer that has a continuously varying acoustic impedance across the layer provides broadband impedance matching.

Attenuation: Its Origin, Measurement, and Applications

Attenuation refers to the damping of a signal, here specifically an acoustic signal, with travel time or with travel distance. Attenuation is typically expressed in dB, i.e., on a logarithmic scale. Attenuation is an important parameter to measure in many types of materials characterization, but also the parameter that sets an upper limit for the ultrasound frequency that can be used for a given measurement. In NDE, attenuation is used for grain size estimation [8, 9], for characterization of composite materials, and for determination of porosity [10]. In medical ultrasound, attenuation can be used for tissue characterization [11], such as differentiating between normal and cirrhotic liver tissue and for classification of malignancies. In flowmeters, attenuation caused by vortices can be used to measure the frequency at which they are shed; this frequency is proportional to the flow velocity.

Attenuation represents the combined effect of *absorption* and *scattering*, where absorption refers to the conversion of acoustic energy into heat due to the viscosity of the medium, and scattering refers to the spreading of acoustic energy into other directions due to inhomogeneities in the medium. The absorption can, in part, be due to *classical absorption*, which varies with frequency squared, and *relaxation absorption*, which can result in a complicated frequency dependence of absorption. Gases (except for noble gases), liquids, and biological tissue exhibit mainly relaxation absorption, whereas classical absorption is most prominent in solids, which can also have a significant amount of scattering attenuation.

In general, absorption dominates in homogeneous media (e.g., liquids, gases, fine-grained metals, polymers), whereas scattering dominates in heterogeneous media (e.g., composites, porous ceramics, large-grained materials, bone). The actual attenuation and its frequency dependence can be specified fairly unambiguously for gases and liquids, while for solids it is very dependent on the manufacturing process, which determines the microstructure of the material, such as the grain structure.

Measurement of attenuation can be carried out for at least two purposes: (1) to measure the *bulk attenuation* of a given homogeneous medium; and (2) to measure the *spatial distribution of attenuation* over a plane in an inhomogeneous medium. The former approach is most common in materials characterization, whereas the latter approach is found mainly in medical ultrasound. Bulk attenuation can be performed either with transmission measurements or with pulse-echo measurements, as illustrated in Figure 7.36(*a*) and (*b*), respectively.

For the measurement of attenuation, in dB/cm, of a medium of thickness d, by transmission measurements, define $v_1(t)$ as the received signal without medium present and $v_2(t)$ as the received signal with medium present, as shown in Figure 7.36(*a*). The attenuation is then determined from the ratio of the energies of the two signals, corrected for the transmission losses, as:

$$\text{Att}\left[\text{dB/cm}\right] = \frac{1}{d}\,10\log\left\{\frac{\int_0^\infty \left[v_1(t)\right]^2 dt}{\int_0^\infty \left[v_2(t)\right]^2 dt}\right\} - 20\log\left\{\frac{4\,z_1\,z_2}{\left(z_1 + z_2\right)^2}\right\} \tag{7.34}$$

Correction for transmission losses (2nd term) can be avoided by alternatively measuring the incremental attenuation due to an incremental thickness increase.

For the measurement of attenuation by *pulse-echo measurements*, the front and the back wall echoes are termed $v_f(t)$ and $v_b(t)$, respectively. Based on an *a priori* knowledge of the pulse duration, a time window ΔT is defined. The attenuation is most accurately measured when based on the energies of $v_f(t)$ and $v_b(t)$, but may alternatively be based on the amplitudes of $v_f(t)$ and $v_b(t)$, as stated in Equation 7.35.

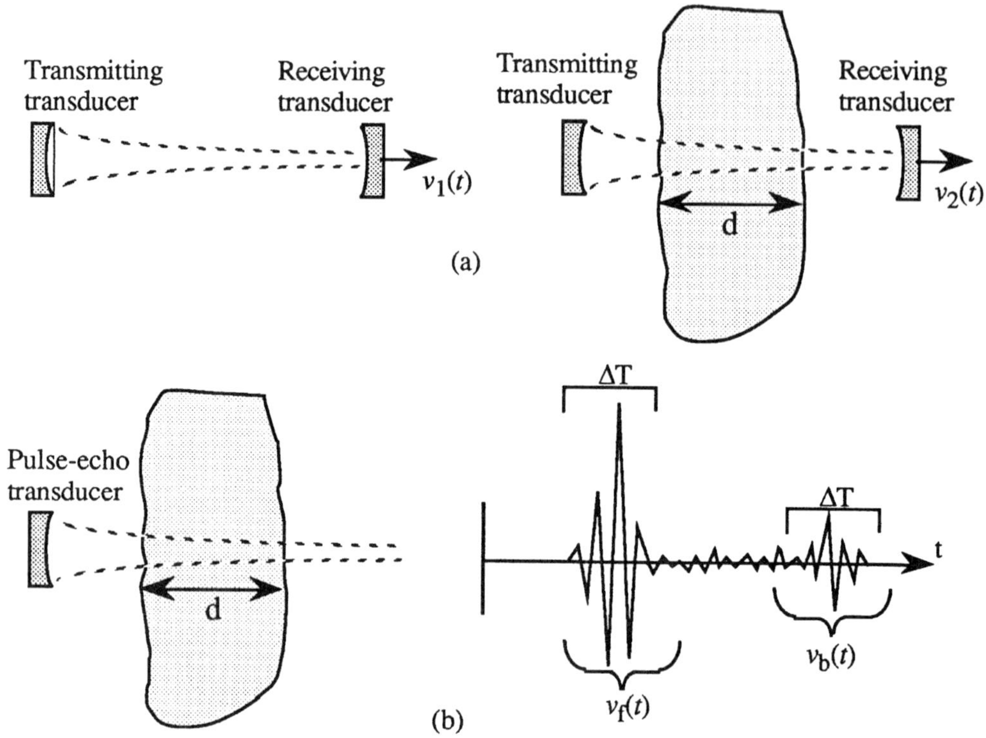

FIGURE 7.36 Measurement of bulk attenuation. (*a*) Measurement of attenuation by transmission measurement. (*b*) Measurement of attenuation by pulse-echo measurements.

$$\text{Att}\left[\text{dB/cm}\right] = \frac{1}{2d}\,10\log\left\{\frac{\int_0^{\Delta t}\left[v_f(t)\right]^2 dt}{\int_0^{\Delta T}\left[v_b(t)\right]^2 dt}\right\} \cong \frac{1}{2d}\,20\log\left\{\frac{\text{peak ampl., } v_f(t)}{\text{peak ampl., } v_b(t)}\right\} \tag{7.35}$$

Accurate attenuation measurements require attention to several potential pitfalls: (1) diffraction effects (even in the absence of attenuation, echoes from different ranges vary in amplitude, due to beam spreading); (2) misalignment effects (if the reflecting surface is not normal to the transducer axis, there is a reduction in detected signal amplitude, due to phase cancellation at the transducer surface); and (3) transmission losses whose magnitude it is not always easy to determine.

The spatial distribution of attenuation must be measured with pulse-echo measurements, where it is assumed that a backscatter signal of sufficient amplitude can be received from all regions of the medium. Use is made of the frequency dependence of attenuation, which has the effect that the shift in mean frequency of a given received echo relative to the mean frequency of the transmitted signal varies proportional to the total absorption. The *rate of shift* in mean frequency is thus proportional to the local attenuation.

CW Fields from Planar Piston Sources

In discussing the acoustic fields generated by acoustic radiators (transducers), a clear distinction must be made between fields due to CW excitation and due to pulse excitation. Although the overall field

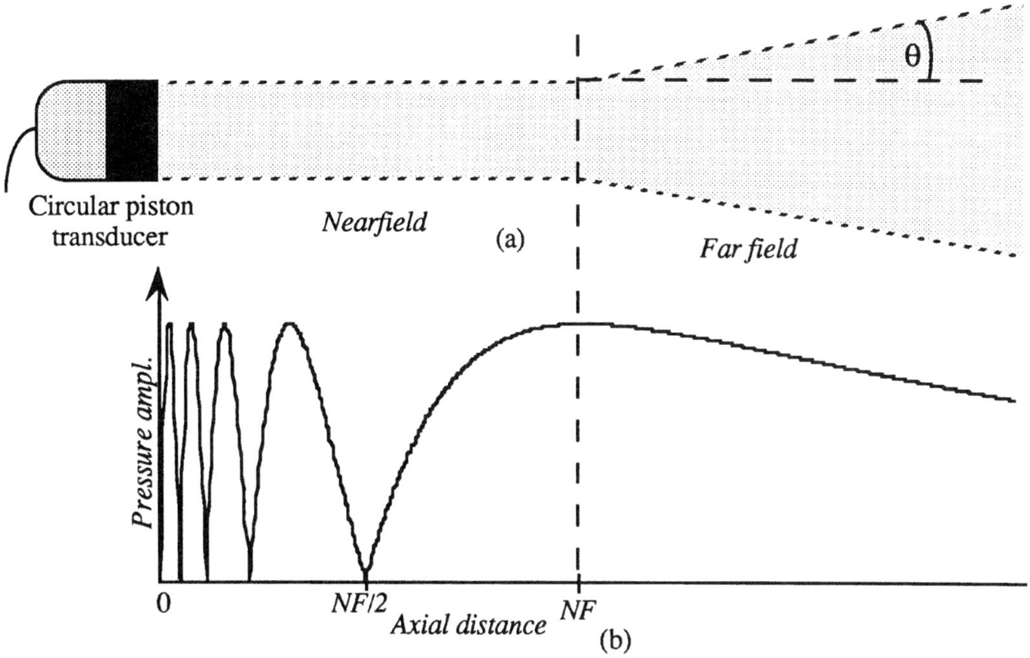

FIGURE 7.37 Pressure fields from a circular planar piston transducer operating with CW excitation. NF = Near Field. (*a*) Approximate field distribution in near field and far field. (*b*) Axial pressure in near field and far field for $(ka) = 31.4$.

patterns for these two cases are quite similar, the detailed field structure is very different. In this section, only CW fields are discussed.

When a CW excitation voltage is applied to a planar piston transducer, the resulting acoustic field can be divided into a *near field* region and a *far field* region. This division is particularly distinct when the transducer has a circular geometry. A piston transducer simply refers to a transducer with the same velocity amplitude at all points on the surface. The length of the near field, NF, is given as:

$$\mathrm{NF} = \frac{a^2}{\lambda} \tag{7.36}$$

where a is the radius of the transducer and λ is the wavelength. As shown in Figure 7.37(*a*), the near field is approximately confined while the far field is diverging. The angle of divergence, θ, is approximately:

$$\theta = \arcsin\left(0.61\frac{\lambda}{a}\right) \tag{7.37}$$

Additional comments to the simplified representation in Figure 7.37(*a*) are in order:

1. The actual field has no sharp boundaries, in contrast to what is shown in Figure 7.37.
2. While the beam diameter is roughly constant in the near field, it is not as regular as shown; and at the same time, the near field structure is very complex.
3. The angle of divergence, θ, is only clearly established well into the far field.
4. The depiction of the far field shows only the main lobe; in addition, there are side lobes, which are directions of significant pressure amplitude, separated by *nulls* (i.e., directions with zero pressure amplitude).

In general, analytical expressions do not exist for the pressure magnitude at an arbitrary field point in the near field, and the pressure must instead be calculated by numerical techniques. However, an exact expression exists for the axial pressure amplitude, $P_a x$, valid in both near and far field [12].

$$P_{ax}(x) = 2\rho_0 c_0 U_0 \left| \sin\left(0.5ka\left[\sqrt{x^2/a^2+1}-(x/a)\right]\right)\right| \tag{7.38}$$

In Equation 7.38, U_0 is the amplitude of the velocity function on the surface of the transducer. The expression is derived for a baffled planar piston transducer where the term "baffled" means that the transducer is mounted in a large rigid surface, called a baffle. Figure 7.37(b) shows the amplitude of the axial field for a planar piston transducer for which $(ka) = 31.4$. For field points located more than 3 to 4 near-field distances away from the transducer, a general expression for the pressure amplitude can be derived [12]:

$$P(r,\theta) = \frac{\rho_0 c_0}{2} U_0 \, ka\left(\frac{a}{r}\right)\left|\left[\frac{2J_1(ka\sin\theta)}{ka\sin\theta}\right]\right| \tag{7.39}$$

where $J_1(\cdot)$ is the Bessel function of the first order. The variable r is the length of the position vector defining the field point, and angle θ is the angle that the position vector makes with the x-axis.

Of interest in evaluating transducers in the far field is the *directivity*, D, defined as follows:

$$D(x) = \frac{I_{ax}(x)\,[\text{given source}]}{I_{ax}(x)\,[\text{simple source}]} \tag{7.40}$$

Thus, the directivity gives the factor with which the axial intensity of the given source is increased over that of a simple source (omnidirectional radiator), radiating the same total energy. For a baffled, circular planar piston transducer, the directivity in the far field can be calculated to be [12]:

$$D = \frac{(ka)^3}{ka - J_1(2ka)} \tag{7.41}$$

When $(2ka) \gg 1$, Equation 7.41 can be approximated to $D = (ka)^2$. Directivity values in the 1000s are common. For example, a 3.5 MHz transducer with 1 cm diameter radiating into water has a (ka) value of 73.3 and a directivity of 5373.

Often, focused transducers are used where the focusing is either created by the curvature of the piezoelectric element or by an acoustic lens in front of the transducer. The degree of focusing is determined by the (ka) value of the transducer.

Generation of Ultrasound: Piezoelectric and Magnetostrictive Phenomena

Ultrasound transducers today are available over a wide frequency range, in many sizes and shapes, and for a wide variety of applications. The behavior of the ultrasound transducer is determined by several parameters: the transduction material, the backing material, the matching layer(s), and the geometry and dimension of the transducer. A good overview of ultrasound transducers is available in [13].

The transduction material is most commonly a piezoelectric material, but can for some applications be a magnetostrictive material instead. These materials are inherently narrowband, meaning that they work efficiently only over a narrow frequency range. This is advantageous for CW applications such as

TABLE 7.3 List of Most Significant Piezoelectric Parameters for Common Piezoelectric Materials

Parameter	Barium titanate ($BaTiO_3$)	Lead zirconate titanate, PZT-5	Lead meta-niobate, $PbNb_2O_6$	Polyvinylidene fluoride (PVDF)
d_{33}	149 (10^{-12} m/V)	374 (10^{-12} m/V)	75 (10^{-12} m/V)	22 (10^{-12} m/V)
g_{33}	14 (10^{-3} Vm/N)	25 (10^{-3} Vm/N)	35 (10^{-3} Vm/N)	339 (10^{-3} Vm/N)
k_{33}	0.50	0.70	0.38	$k_{31} = 0.12$
k_T	0.38	0.68	0.40	0.11
Q_m	600	75	5	19

ultrasound welding and ultrasound hyperthermia, but is a problem for imaging applications, as impulse excitation will produce a long pulse with poor resolving abilities. To overcome this deficiency, a *backing material* is tightly coupled to the back side of the transducer for the purpose of damping the transducer and shortening the pulse. However, the backing material also reduces the sensitivity of the transducer. Some of this reduced sensitivity can be regained by the use of a *matching layer*, specifically selected for the medium of interest. As seen in Equation 7.33, a quarter wavelength matching layer can provide 100% efficient coupling to a medium, albeit only at one frequency. A combination of several matching layers can provide a more broadband impedance matching. The field is determined by both the geometry (planar, focused, etc.) of the transducer and by the frequency content of the velocity function of the surface of the transducer. In this section, unique aspects of the transduction material itself are described.

Piezoelectric Materials.

A piezoelectric material exhibits a mechanical strain (relative deformation) due to the presence of an electric field, and generates an electric field when subjected to a mechanical stress. A detailed review of piezoelectricity is given in [14]. Piezoelectric materials can either be: (1) natural material such as quartz; (2) man-made ceramics (e.g., barium titanate (BaTi), lead zirconate titanate (PZT), or lead meta-niobate); (3) man-made polymers (e.g., polyvinylidene fluoride (PVDF)). The piezoelectric ceramics are the most commonly used materials for ultrasound transducers. These ceramics are made piezoelectric by a so-called poling process in which the material is subjected to a strong electric field while at the same heating it to above the material's Curie temperature.

Several material constants determine the behavior of a given piezoelectric material, the most important of which are listed in Table 7.3 and defined below. An extensive list of parameter values for various piezoelectric materials is available in [15].

d_{33} Transmission constant
g_{33} Receiving constant
k_{33} Piezoelectric coupling coefficient
k_T Piezoelectric coupling coefficient for a transverse clamped material
Q_M Mechanical *Q*

The transmission constant, d_{33}, gives the mechanical deformation of piezoelectric materials for frequencies well below the resonance frequency. A transducer with a large d_{33} value will therefore become an efficient transmitter. If an electric field, E_3, is applied in the polarized direction of a piezoelectric rod or disk, the strain, S_3, in that direction is approximately:

$$S_3 = d_{33} E_3 = d_{33} \frac{V_{appl}}{l} \tag{7.42}$$

where V_{appl} is the applied voltage and l is the thickness. The total deformation, Δl, becomes

$$\Delta l = d_{33} V_{appl} \tag{7.43}$$

The receiving constant, g_{33}, defines the sensitivity of a transducer element as a receiver when the frequency of the applied pressure is well below the resonance frequency of the transducer. If the applied stress (force/area) in the direction of polarization is T_3, the output voltage from a rod or disk is approximately:

$$v_{out} = g_{33} T_3 l \tag{7.44}$$

The coupling coefficient describes the power conversion efficiency of a piezoelectric transducer, *operating at or near resonance.* Specifically, k_{33} is the coupling coefficient for an unclamped rod; that is, the rod is allowed to deform in the directions orthogonal to the direction of applied force or voltage. In contrast, k_T is the coupling coefficient for a *clamped* disk. Finally, Q_M gives the mechanical Q, which is a measure for how narrowband the transducer material inherently is.

Whereas expressions of the type given in Equations 7.42 to 7.44 are adequate for describing static or low-frequency behavior, the behavior near resonance where most transducers operate requires more complex models, which are beyond the scope of this chapter. The Mason model is adequate for narrowband modeling of transducers, whereas the KLM or Redwood models better describe the transducers for broadband applications [13, 15].

Magnetostrictive Materials.
The magnetostrictive phenomenon refers to a magnetically induced contraction or expansion in ferroelectric media, such as in nickel or alfenol, and was discovered by Joule in 1847. Magnetostrictive materials are generally used for ultrasound frequencies below 100 kHz, and are therefore relevant mainly for underwater applications. Eddy current losses influence the performance of magnetostrictive materials, but the losses can be reduced by constructing the magnetostrictive transducer from thin laminations.

Transducer Specifications.
Ultrasound transducers are generally specified by their diameter, center frequency, focal distance, and type of focusing (if applicable). The transducer can be designed as a contact transducer, a submersible transducer, an air transducer, etc.; in addition, the type of connector or cabling can be specified. In many cases, measurement data for the actual transducer can be supplied by the vendor in the form of a measured pressure pulse and the corresponding frequency spectrum, recorded with a hydrophone at a specific field point. Similarly, the beam profile can be measured in the form of pulse amplitude or pulse energy as a function of lateral position.

Detailed information about the acoustic field from a radiating transducer can be obtained with either the Schlieren technique or the optical interferometric technique. Such instruments are quite expensive, falling in the \$50K to \$120K range.

Display Formats for Pulse-Echo Measurements

The basic description of ultrasound imaging was presented earlier in this chapter. Based on the information presented so far in this section, more specific aspects of pulse-echo ultrasound imaging will now be described. Different display formats are used; the simplest of these is the *A-mode* display.

A-mode.
When a pulse-echo transducer has emitted a pulse in a given direction and has been switched to receive mode, an output signal from the transducer is produced based on detected echoes from different ranges, as illustrated conceptually in Figure 7.31(*a*). In this signal, distance to the reflecting structure has been converted into time from the start of signal. This signal is often referred to as the RF signal. Demodulating this signal (i.e., generating the envelope of the signal) produces the A-mode (amplitude mode) display, or the A-line signal. This signal can be the basis for range measurements, attenuation measurements, and measurement of reflection coefficient.

M-mode.
If the A-mode signal from a transducer in a fixed position is used to *intensity* modulate a cathode-ray tube (CRT), such as a monitor or oscilloscope, along a straight vertical line, a line of dots with

brightness according to echo strengths would appear on the screen. Moving the display electronically across the screen results in a set of straight horizontal lines. Now consider the case where the reflecting structures are moving in a direction toward or away from the transducer while pulse-echo measurements were being performed. This results in variations in the arrival time of echoes in the A-line signal, and the resulting lines across the screen are no longer straight, but curved, as determined by their velocity. Such a display is called *M-mode*, or motion mode, display. An application for this would be measurement of the diameter variation of a flexible tube, or blood vessel, due to a varying pressure inside the tube.

B-mode.

If pulse-echo measurements are repeatedly being performed, while the transducer scans the object of interest, an image of the object can be generated, as illustrated in Figure 7.31. Specifically, each A-line signal is used to intensity modulate a line on a CRT corresponding to the location of ultrasound beam, which produces an image that maps the reflectivity of the structures in the object. The resulting image is called a *B-mode* (or brightness mode) image. If the transducer is moved linearly, a rectangular image is produced, whereas a rotated transducer generates a pie-shaped, or sector image. This motion is typically done electronically or electromechanically, as described earlier under single-element and array transducers. When the scanning is done rapidly (say, 30 scans/s), the result is a real-time image.

Many forms of signal processing and image enhancement can be applied in the process of generating the B-mode image. Echoes from deeper lying structures will generally be of smaller amplitude, due to the attenuation of the overlying layers of the medium. Attenuation correction is made especially in medical imaging, so that echoes are displayed approximately with their true strength. This correction consists of a time-varying gain, called *time-gain control*, or TGC, such that the early arriving echoes experience only a low gain and later echoes from deeper lying structures experience a much higher gain. Various forms for signal compressions or nonlinear signal transfer function can selectively enhance the weaker or the stronger echoes.

C-mode.

In a C-mode display, only echoes from a specific depth will be imaged. To generate a complete image, the transducer must therefore be moved in a raster scan fashion over a plane. C-scan imaging is slow and cannot be used for real-time imaging, but has several applications in NDE for examining a given layer, or interface, in a composite structure, or the inner surface of a pipe.

Flow Measurements by Doppler Signal Processing

Flow velocity can be obtained by various ultrasonic methods, e.g., by measuring the Doppler frequency or Doppler spectrum. The Doppler frequency is the difference between the frequency (or pitch) of a moving sound source, as measured by a *stationary* observer, and the actual frequency of the sound source. The change in frequency is determined by the speed and direction of the moving source. The classical example is a moving locomotive with its whistle blowing; the pitch is increased when the train moves toward the observer, and vice versa. With respect to ultrasound measurements, only the reflecting (or scattering) gas, fluid, or structure is moving and not the the sound source, yet the Doppler phenomenon is present here as well. In order for ultrasound Doppler to function, the gas or fluid must contain scatterers that can reflect some of the ultrasound energy back to the receiver. The Doppler frequency, f_d, is given as follows:

$$f_d = \frac{2v}{c_0} \cos \theta \tag{7.45}$$

where v is the velocity of the moving scatterers, and θ is the angle between the velocity vector and the direction of the ultrasound beam. Doppler flowmeters require only access to the moving gas, fluid, or object from one side. For industrial use, when the fluids or gases often do not contain scatterers, *transmission* methods are preferred; these methods, however, are not based on the Doppler principle.

Two main categories of Doppler systems exist: the CW Doppler system and the PW (pulsed wave) Doppler system. The CW Doppler system transmits a continuous signal and does not detect the distance to the moving structure. It is therefore only applicable when there is just one moving structure or cluster of scatterers in the acoustic field. For example, a CW Doppler is appropriate for assessing the pulsatility and nature of blood flow in an arm or leg. CW Doppler systems are small and relatively inexpensive.

The PW Doppler system transmits a short burst at precise time intervals; it is therefore inherently a sampled system, and, as such, is subject to aliasing. It measures changes from one transmitted pulse to the next in the received signal from moving structures at one or several selected ranges, and is able to determine both velocity and range. Assuming that the ultrasound beam is narrow and thus very directional, the PW Doppler can create a *flow image*; that is, it can indicate the magnitude and direction of flow over an image plane. Commonly, color and color saturation are used to indicate direction and speed in a flow image, respectively. In contrast to CW Doppler systems, the PW Doppler systems require much signal processing and are therefore typically expensive and often part of a complete imaging system. When a distribution of velocities, rather than a single velocity, is encountered, as is often the case with fluid flow, a Doppler spectrum rather than a Doppler frequency is determined. An FFT (Fast Fourier Transform) routine is then used to reveal the Doppler frequencies and thus the velocity components present.

Review of Common Applications of Ultrasound and Their Instrumentation

Range Measurements, Air

Ultrasound range measurements are used in cameras, in robotics, for determining dimensions of rooms, etc. Measurement frequencies are typically around 50 kHz to 60 kHz. The measurement concept is pulse-echo, but with burst excitation rather than pulse excitation. Special electronic circuitry and a thin low-acoustic-impedance air transducer is most commonly used. Rugged solid or composite piezoelectric-based transducers, however, can also be used, sometimes up to about 500 kHz.

Thickness Measurement for Testing, Process Control, Etc.

Measurement of thickness is a widely used application of ultrasound. The measurements can be done with direct coupling between the transducer and the object of interest, or — if good surface contact is difficult to establish — with a liquid or another coupling agent between the transducer and the object. Ultrasound measurements of thickness have applications in process control, quality control, measuring build-up of ice on an aircraft wing, detecting wall thickness in pipes, as well as medical applications. The instrumentation involves a broadband transducer, pulser-receiver, and display or, alternatively, echo detecting circuitry and numerical display.

Detection of Defects, such as Flaws, Voids, and Debonds

The main ultrasound application in NDE is inspection for the localization of voids, crack, flaws, debonding, etc. [3]. Such defects can exist immediately after manufacturing, or were formed due to stresses, corrosion, etc. Various types of standard or specialized flaw detection equipment are available from ultrasound NDE vendors.

Doppler Flow Measurements

The flow velocity of a liquid or a moving surface can be determined through Doppler measurements, provided that the liquid or the surface scatters ultrasound back in the direction of the transducer, and that the angle between the flow direction and the ultrasound beam is known. Further details are given in the section about Doppler processing. CW and PW Doppler instruments are commercially available, with CW instrumentation being by far the least expensive.

Upstream/Downstream Volume Flow Measurements

When flow velocity is measured in a pipe with access to one or both sides, an ultrasound transmission technique can be used in which transducers are placed on the same or opposite sides of the pipe, with one transducer placed further upstream than the other transducer. From the measured *difference* in travel time between the upstream direction and the downstream direction, and knowledge about the pipe geometry, the volume flow can be determined. Special clamp-on transducers and instrumentation are available. An overview of flow applications in NDE is given in [16].

Elastic Properties of Solids

Since bulk sound speed varies with the elastic stiffness of the object, as given in Equation 7.15, sound speed measurements can be used to estimate elastic properties of solids under different load conditions and during solidification processes. Such measurements can also be used for measurement of product uniformity and for quality assurance. The measurements can be performed on bulk specimens or on thin rods, using either pulse-echo or transmission instrumentation [17]. Alternatively, measurements of the material's own resonance frequencies can be performed for which commercial instruments, such as the *Grindo-sonics*, are available.

Porosity, Grain Size Estimation

Measurement of ultrasound attenuation can reveal several materials parameters. By observing the attenuation in metals as a function of frequency, the grain size and grain size distribution can be estimated. Attenuation has been used for estimating porosity in composites. In medical ultrasound, attenuation is widely used for tissue characterization, that is, for differentiating between normal and pathological tissues. Pulse-echo instrumentation interfaced with a digitizer and a computer for data analysis is required.

Acoustic Microscopy

The measurement approaches utilized in acoustic microscopy are similar to other ultrasound techniques, in that A-scan, B-scan, and C-scan formats are used. It is in the applications and the frequency ranges where acoustic microscopy differs from conventional pulse-echo techniques. Although acoustic microscopes have been made with transducer frequencies up to 1 GHz, the typical frequency range is 20 MHz to 100 MHz, giving spatial resolutions in the range from 100 μm to 25 μm. Acoustic microscopy is used for component failure analysis, electronic component packaging, and internal delaminations and disbonds in materials, and several types of acoustic microscopes are commercially available.

Medical Ultrasound

Medical imaging is a large and diverse application area of ultrasound, especially in obstetrics, cardiology, vascular studies, and for detecting lesions and abnormalities in organs. The display format is either B-mode, using gray scale image, or a combination of Doppler and B-mode, with flow presented in color and stationary structures in gray scale. A wide variety of instruments and scanners for medical ultrasound are available.

Selected Manufacturers of Ultrasound Products

Table 7.4 contains a representative list of ultrasound equipment and manufacturers.

Advanced Topics in Ultrasound

Overview of Diffraction

The ultrasound theory presented thus far has emphasized basic concepts, and the applications that have been discussed tacitly assume that the field from the transducer is a plane wave field. This simplifying assumption is acceptable for applications such as basic imaging and measurements based on travel time. However, the plane wave assumption introduces errors when materials parameters (e.g., attenuation, surface roughness, and object shape) are sought to be measured with ultrasound. Therefore, to use ultrasound as a quantitative tool, an understanding is needed of the structure of the radiated acoustic field from a given transducer with a given excitation, and — equally important — the ability to calculate the actual radiated field. This leads to the topic of diffraction, which is the effect that accounts for the complex structure of both radiated and scattered fields. Not surprisingly, there are direct parallels between optical diffraction and acoustic diffraction. (As a separate issue, it should be noted that the ultrasound

TABLE 7.4 List of Products for and Manufacturers of Ultrasound Measurements

Product type	Manufacturer
Ultrasound transducers	Panametrics, 221 Crescent St., Waltham, MA 02154. (800) 225-8330
Ultrasound transducers	Krautkramer Branson Inc., 50 Industrial Park Rd., Lewistown, PA 17044. (717) 242-0327
Range measurements, air	Polaroid Corporation, Ultrasonics Components Group, 119 Windsor Street, Cambridge, MA 02139. (800) 225-1618
Pulser-receivers	Panametrics, 221 Crescent St., Waltham, MA 02154. (800) 225-8330
Pulser-receivers	JSR Ultrasonics, 3800 Monroe Ave., Pittsfield, NY 14534. (716) 264-0480
Ultrasound power ampl.	Amplifier Research, 160 School House Rd., Souderton, PA 18964. (800) 254-2677
Ultrasound power ampl.	Ritec, 60 Alhambra Rd., Suite 5, Warwick, RI 02886. (401) 738-3660
NDE instrumentation	Panametrics, 221 Crescent St., Waltham, MA 02154. (800) 225-8330
NDE instrumentation	Krautkramer Branson Inc., 50 Industrial Park Rd., Lewistown, PA 17044. (717) 242-0327
Acoustic microscopy	Sonoscan, 530 E. Green St., Bensenville, IL 60106. (708) 766-4603
Medical Imaging	Hewlett Packard, Andover, MA; ATL, Bothell, WA; Diasonics, Milpitas, CA; Siemens Ultrasound, Issaquah, WA.
Schlieren based imaging of acoustic fields	Optison, 568 Weddell Drive, Suite 6, Sunnyvale, CA 94089. (408) 745-0383
Optical based imaging of acoustic fields	UltraOptec, 27 de Lauzon, Boucherville, Quebec, Canada J4B 1E7. (514) 449-2096

theory presented here assumes that the wave amplitudes [pressure, displacement] are small enough so that nonlinear effects can be disregarded.)

Diffraction is basically an edge effect. Whereas a plane wave incident on a large planar interface is reflected in a specific direction, the plane wave incident on an edge results in waves scattered in many directions. Similar considerations hold for the field produced by a transducer: The surface of the transducer produces a so-called *geometric wave* that has the shape of the transducer itself; the edge of the transducer, however, generates an *edge wave* with the shape of an expanding torus. The actual pressure field is a combination of the two wave fields. Very close to a large transducer, the geometric wave dominates, and diffraction effects might not need to be considered. Over small regions far away from the transducer, the field can be approximated by a plane wave field, and diffraction does not need to be considered. However, in many practical cases, diffraction must be considered if detailed information about the pressure field is desired, and numerical methods must be employed for the calculations.

The structure of the axial field, shown in Figure 7.37(*b*), is a direct result of diffraction. Numerical evaluation of the diffracted field from a transducer can be done in several ways: (1) use of the Fresnel or Fraunhofer integrals (not applicable close to the transducer) to calculate the field at a single frequency at a specified plane normal to the transducer axis; (2) calculation of the pressure function at any point of interest in space, based on a specified velocity function, $u(t)$, on the surface of the transducer, using Rayleigh integral; (3) decomposing the velocity field in the plane of the transducer into its plane wave components, using a 2-D Fourier transform technique, followed by a forward propagation of the plane waves to the plane of interest and an inverse Fourier transform to give the diffracted field; or (4) use of finite element methods to calculate the diffracted field at any point or plane of interest. In the following, methods (1) and (2) will be described.

Fresnel and Fraunhofer Diffraction.

Let the velocity function on the surface of the transducer, $u(x,y)$, be specified for a particular frequency, ω. Assume that the transducer is located in the $(x,y,0)$ plane and that one is interested in the pressure field in the (x_0,y_0,z) plane. The Fresnel diffraction formulation [18] assumes that the paraxial approximation is fulfilled, requiring z to be at least 5 times greater than the transducer radius, in which case the *Fresnel diffraction integral* applies:

$$p\left(x_0, y_0, z, \omega\right) = \frac{A_0}{\lambda_z} \iint_S u\left(x, y, \omega\right) \exp\left[-jk\frac{x^2+y^2}{2z}\right] \exp\left[jk\frac{x_0 x + y_0 y}{z}\right] dx\, dy \qquad (7.46)$$

where S is the surface of the transducer and A_0 is a constant. If one defines the two first terms of the integrand as some complex spatial function, $\Gamma(x,y)$, then Equation 7.46 is a scaled Fourier transform of $\Gamma(x,y)$.

If $k(x^2 + y^2)/2z \ll 1$, or, equivalently, $z > 10\, a^2/\lambda$, the second term in Equation 7.46 can be ignored, and the resulting equation is called the *Fraunhofer diffraction integral*:

$$p\left(x_0, y_0, z, \omega\right) = \frac{A_0}{\lambda z} \iint_S u\left(x, y, \omega\right) \exp\left[jk \frac{x_0 x + y_0 y}{z}\right] dx\, dy \tag{7.47}$$

Thus, one can only use the Fraunhofer integral for calculating the far field diffraction. From Equation 7.47, one can make the interesting observation that the far field of a transducer is a scaled version of the Fourier transform of the source.

Pressure Function at a Given Field Point, Based on Rayleigh Integral.

While the Fresnel and Fraunhofer diffraction methods are CW methods, calculation of pressure from the Rayleigh integral is fundamentally an impulse technique, and is as such better suited for analysis of pulse-echo measurements. The basis for the calculation is the *velocity potential impulse response,* $h(\vec{r},t)$, obtained from the Rayleigh integral:

$$h\left(\vec{\mathbf{r}}, t\right) = \frac{1}{2\pi} \iint_S \frac{\delta\left(t - r'/c\right)}{r'} dS \tag{7.48}$$

In Equation 7.48, r' is the distance from dS on the surface of the transducer to the field point, defined by the position vector $\vec{r}$, and $\delta(t)$ is the Dirac delta function. As can be seen, $h(\vec{r},t)$ is the result of an impulsive velocity excitation on the surface of the transducer and is a function of both time and a spatial location. It is important to note that $h(\vec{r},t)$ exists in analytical form for several transducer geometries, and, by extension, for annular and linear array transducers [19].

For the case of an arbitrary velocity function, $u(t)$, on the transducer surface, the corresponding velocity potential, $\phi(\vec{r},t)$, is obtained as:

$$\phi\left(\vec{\mathbf{r}}, t\right) = u\left(t\right) \otimes h\left(\vec{\mathbf{r}}, t\right) \tag{7.49}$$

where $\otimes$ refers to time domain convolution. Both particle velocity, $u(\vec{r},t)$, and pressure, $p(\vec{r},t)$, can be found from $\phi(\vec{r},t)$, as follows:

$$u\left(\vec{\mathbf{r}}, t\right) = \nabla \phi\left(\vec{\mathbf{r}}, t\right) \tag{7.50}$$

$$p\left(\vec{\mathbf{r}}, t\right) = -\rho_0 \frac{\partial}{\partial t} \phi\left(\vec{\mathbf{r}}, t\right) \tag{7.51}$$

where ∇ is the gradient operator.

Thus, from the expressions above, the pressure can be calculated for any field point, $\vec{r}$, when $u(t)$ and the transducer geometry are defined. In this calculation, all diffraction effects are included. However, given the high frequency content in $h(\vec{r},t)$ and in particular in the time derivative of $h(\vec{r},t)$, care must be taken to avoid aliasing errors, as described in [19].

Received Signal in Pulse-Echo Ultrasound.

The expression in Equation 7.51 allows for quantitative evaluation of the pressure field for an arbitrary point, line, or plane. However, it does not describe the calculation of the received signal in a pulse-echo system. Consider a small planar reflector, placed in a homogeneous medium, and referred to as *dR*. The reflector has the area *dA*. The dimensions of *dR* must be small with respect to the shortest wavelength in the insonifying pulse. The location and the orientation of the planar reflector is given by $\vec{\mathbf{r}}$ and $\vec{\mathbf{n}}$, respectively, where $\hat{\mathbf{n}}$ is a unit normal vector to the small reflector.

The voltage from the receiving transducer in a pulse-echo system due to *dR* is termed $d\,v(\vec{\mathbf{r}},t)$ and can be determined when $u(t)$ is specified. The electro-acoustic transfer function for both the transmitting and the receiving transducer is assumed to be unity for all frequencies. For the case when the acoustic impedance of *dR* is much higher than that of the medium, $d\,v(\vec{\mathbf{r}},t)$ is given as [20]:

$$\begin{aligned} d\,v(\vec{\mathbf{r}},t) &= A_0\,\rho_0\,\frac{\cos\left[\psi(\vec{\mathbf{r}})\right]}{c_0}\left[h(\vec{\mathbf{r}},t)\otimes h(\vec{\mathbf{r}},t)\otimes\frac{\partial^2}{\partial t^2}u(t)\right]dA \\ &= A_0\,\cos\left[\psi(\vec{\mathbf{r}})\right]u(t)\otimes\frac{\rho_0}{c_0}\left[\frac{\partial^2}{\partial t^2}\left(h(\vec{\mathbf{r}},t)\otimes h(\vec{\mathbf{r}},t)\right)\right]dA \end{aligned} \tag{7.52}$$

In Equation 7.52, A_0 is determined by the reflection coefficient of the reflector. The term $\cos[\psi(\vec{\mathbf{r}})]$ is a correction term (obliquity factor) where $\psi(\vec{\mathbf{r}})$ is the angle between $\hat{\mathbf{n}}$ and the propagation direction of the wave field at $\vec{\mathbf{r}}$. For an extended surface, the received voltage can be found by decomposing the surface into small reflectors and calculating the total received signal as the sum of the contributions from all the small reflectors. An efficient numerical technique for this type of integration has been developed [21].

References

1. C. M. Fortunko and D. W. Fitting, Appropriate ultrasonic system components for NDE of thick polymer-composites, *Review of Progress in Quantitative Nondestructive Evaluation, Vol. 10B.* New York: Plenum Press, 1991, 2105-2112.
2. C. R. Hill, Medical imaging and pulse-echo imaging and measurement, in C.R. Hill (ed.) *Physical Principles of Medical Ultrasound,* New York: Halsted Press, 1986, chaps. 7 and 8, 262-304.
3. L. C. Lynnworth, *Ultrasonic Measurements for Process Control,* San Diego: Academic Press, 1989, 53-89.
4. L. E. Kinsler, A. R. Frey, A. B. Coppens, and J. V. Sanders, *Fundamentals of Acoustics,* 3rd. ed., New York: John Wiley, 1982, 106.
5. J. D. Achenbach, *Wave Propagation in Elastic Solids,* 1st ed., New York: Elsevier Science, 1975, 123.
6. L. E. Kinsler, A. R. Frey, A. B. Coppens, and J. V. Sanders, *Fundamentals of Acoustics,* 3rd. ed., New York: John Wiley, 1982, 115-117.
7. L. E. Kinsler, A. R. Frey, A. B. Coppens, and J. V. Sanders, *Fundamentals of Acoustics,* 3rd. ed., New York: John Wiley, 1982, 127-133.
8. J. Saniie and N. M. Bilgutay, Quantitative grain size evaluation using ultrasonic backscattered echoes, *J. Acoust. Soc. Am.,* 80, 1816-1824, 1986.
9. E. P. Papadakis, Scattering in polycrystalline media, in P. D. Edmonds (ed.) *Ultrasonics,* New York: Academic Press, 1981, 237-298.
10. S. M. Handley, M. S. Hughes, J. G. Miller, and E. I. Madaras, Characterization of porosity in graphite/epoxy laminates with polar backscatter and frequency dependent attenuation, *1987 Ultrasonics Symp.,* 1987, 827-830.
11. J. C. Bamber, Attenuation and absorption, in C.R. Hill (ed.), *Physical Principles of Medical Ultrasound,* New York: Halsted Press, 1986, 118-199.
12. L. E. Kinsler, A. R. Frey, A. B. Coppens, and J. V. Sanders, *Fundamentals of Acoustics,* 3rd. ed., New York: John Wiley, 1982, 176-185.
13. M. O'Donnell, L. J. Busse, and J. G. Miller, Piezoelectric transducers, in P. D. Edmonds (ed.), *Ultrasonics,* New York: Academic Press, 1981, 29-65.

14. IEEE Standard on Piezoelectricity, *IEEE Trans. Sonics Ultrasonics*, 31, 8–55, 1984.
15. G.S. Kino, *Acoustic Waves*, Englewood Cliffs, NJ: Prentice-Hall, 1987, 17-83 and 554-557.
16. L. C. Lynnworth, *Ultrasonic Measurements for Process Control*, San Diego, CA: Academic Press, 1989, 245-368.
17. L. C. Lynnworth, *Ultrasonic Measurements for Process Control*, San Diego, CA: Academic Press, 1989, 537-557.
18. V. M. Ristic, *Principles of Acoustic Devices*, New York: John Wiley & Sons, 1983, 316-320.
19. D. P. Orofino and P. C. Pedersen, Multirate digital signal processing algorithm to calculate complex acoustic pressure fields, *J. Acoust. Soc. Am.*, 92, 563-582, 1992.
20. A. Lhemery, Impulse-response method to predict echo responses from targets of complex geometry. I. Theory, *J. Acoust. Soc. Am.*, 90, 2799-2807, 1991.
21. S. K. Jespersen, P. C. Pedersen, and J. E. Wilhjelm, The diffraction response interpolation method, *IEEE Trans. Ultrasonics, Ferroelectrics, and Frequency Control*, 45, Nov. 1998.

Further Information

L. C. Lynnworth, *Ultrasonic Measurements for Process Control*, San Diego, CA: Academic Press, 1989, an excellent overview of industrial applications of ultrasound.

L. E. Kinsler, A. R. Frey, A. B. Coppens, and J. V. Sanders, *Fundamentals of Acoustics*, 3rd. ed., New York: John Wiley & Sons, 1982, a very readable introduction to acoustics.

P. D. Edmonds (ed.), *Ultrasonics*, (Vol. 19 in the series: *Methods of Experimental Physics*). New York: Academic Press, 1981, in-depth description of ultrasound interaction with many types of materials, along with discussion of ultrasound measurement approaches.

J. A. Jensen, *Estimation of Blood Velocities Using Ultrasound*, Cambridge, UK: Cambridge University Press, 1996, a very up-to-date book about ultrasound Doppler measurement of flow and the associated signal processing.

E. P. Papadakis (ed.), *Ultrasonic Instruments and Devices: Reference for Modern Instrumentations, Techniques, and Technology*, in the series *Physical Acoustics*, Vol. 40, New York: Academic Press, 1998.

F. W. Kremkau, Diagnostic Ultrasound: Principles and Instruments, 5th ed., Philadelphia, PA: W. B. Saunders Co., 1998, a very readable and up-to-date introduction to medical ultrasound.

8

Acoustic Measurement

Per Rasmussen
G.R.A.S. Sound and Vibration

Sound is normally defined as vibration of a solid, liquid, or gaseous medium in the frequency range of the human ear, i.e., between about 20 Hz and 20 kHz. Here, the definition is further limited and only vibrations in liquids and gaseous media are considered. In contrast to solid media, a liquid or gaseous medium cannot transmit shear forces, so sound waves are always longitudinal waves, in which the particles moves in the direction of propagation of the wave. The wave propagation in gaseous and liquid media can be described by the three variables: the pressure p, the particle velocity u, and the density ρ. The relation between these is described by the wave equation [1], and this can be derived from three basic equations: the Euler equation (this is essentially Newton's second law applied to a fluid), the Continuity equation, and the State equation. Although the wave equation in principle can be used to describe and calculate all sound waves in all situations, it will in practice often be impossible to perform the necessary calculations. In some special cases, it is possible to get analytical results directly from the wave equation and these cases are therefore of special interest. The cases most often encountered in acoustics is the *free field,* the *diffuse* (or reverberant) *field,* and the *closed coupler.* The free field is, in principle, an infinite, empty (except for the medium and the source) space, with no reflections. Here, the waves are allowed to radiate freely in all directions without reflections. In practice, the free field is implemented in an anechoic chambers, where all walls have been made nearly 100% absorptive. The diffuse field is obtained in a reverberation room where all walls have been made, in principle, 100% reflective. At the same time, the walls are made nonparallel and the result is a sound field with sound waves in all directions. The closed coupler is a small chamber, with dimensions small compared to the wavelength of the sound. A special case of this is the standing wave tube. This is a tube with a diameter smaller than the wavelength and with a sound source in one end. With a suitable loudspeaker as a source, the wave propagation in the tube can be assumed to be one-dimensional. This simplifies the mathematical description so that it is possible to calculate the sound field.

In practice, almost the only parameter measured directly in acoustics is the sound pressure, and all other parameters like sound power, particle velocity, reverberation time, directivity, etc. are derived from pressure measurements. These are performed with measurement microphones in gaseous media and

0-8493-0047-9/00/$0.00+$.50

hydrophones in liquid media. The measurement microphones are all of the condenser type to ensure precision, long-term stability, and sensitivity. Hydrophones are usually made with a rubber coating over a sensitive element of piezoelectric material.

The traditional frequency range from 20 Hz to 20 kHz for acoustic measurements is selected because this is the range audible to the human ear. Sound waves exist outside this range in the form of infrasound (below 20 Hz) and ultrasound (above 20 kHz). As the basic equations (the wave equation) and measurement principles are the same for both infrasound and ultrasound, many of the principles from the frequency range from 20 Hz to 20 kHz can be extended to these ranges.

8.1 The Wave Equation

Sound wave propagation cannot take place in a vacuum, but is always associated with some kind of medium. For simplicity, assume that this medium is air, although the same equations are valid also for all gaseous and fluid media. In this medium, the concept of an air particle can be introduced. An *air particle* is a small volume of air in which the acoustical parameters like pressure, density, etc. can be considered constant. On the other hand, the air volume must be large enough to include a very large number of air molecules, so that the air volume can be considered to be a continuous medium and not a collection of molecules. The *Euler equation* for such an air particle is given by:

$$-\text{grad } p = \rho_0 \left(\frac{\partial \vec{v}}{\partial t} \right) \tag{8.1}$$

where p is the pressure, ρ_0 is the density, and v is the particle velocity. This equation can be considered as Newton's second law ($F = ma$) applied to a fluid. Here, the gradient of the pressure equals the force F acting on the air particle, the density equals the mass m, and the time differentiated particle velocity $\partial v/\partial t$ equals the acceleration.

The second equation necessary to derive the wave equation is the *continuity equation.* This simply states that if you have a small volume of air and you bring in some extra air, the density (or the mass) will increase. Mathematically, this can be formulated as:

$$\text{div } \vec{v} = -\frac{1}{\rho_0} \cdot \frac{\partial \rho}{\partial t} \tag{8.2}$$

where c is the sound velocity. The sound velocity depends on the composition of the air and the temperature. For normal air at 0°C, the velocity is 314 m s^{-1} while at 20°C, the velocity is 340 m s^{-1}.

The third equation is the *state equation,* which relates pressure changes to changes in the density, that is, if a small volume of air is compressed, the density will increase:

$$\frac{\partial \rho}{\partial t} = \frac{1}{c^2} \times \frac{\partial p}{\partial t} \tag{8.3}$$

Now we have three equations relating the three variables: pressure, particle velocity, and density. By eliminating the particle velocity and the density, we obtain one differential equation for the sound pressure:

$$\nabla^2 p = \frac{1}{c^2} \frac{\partial^2 p}{\partial t^2} \tag{8.4}$$

This is the *wave equation* for acoustic waves in gaseous and fluid media. In principle, this allows one to calculate the sound pressure anywhere in a sound field, if some suitable boundary conditions are given. In practice, however, it is only possible to find solutions in a few simple cases.

8.2 Plane Sound Waves

In a free space, at great distance from the sound source, sound waves are approximately plane waves. This means that the wave equation only depends on one of the coordinates in the wave equation. If the direction of propagation of the wave fronts is in the *x*-direction, the solution to the wave equation reduces to:

$$p = A\cos\left[\left(\omega t - k\right)\left(x + \varphi_a\right)\right] \tag{8.5}$$

where ω is the frequency. Similarly, the particle velocity in the free field is given by:

$$v = \frac{A}{\rho c}\cos\left[\left(\omega t - k\right)\left(x + \varphi_a\right)\right] = \frac{p}{\rho c} \tag{8.6}$$

This means that in a plane wave, the particle velocity is equal to the pressure divided by the constant ρc and the pressure and particle velocity are in phase. The constant ρc is the *acoustic impedance;* and for air at 20°C, the density is 1.29 kg m^{-3} and the sound velocity is 340 m s^{-1}, giving an acoustic impedance of 438.6 kg m^{-2} s^{-1}.

The plane wave transmits energy in the direction of propagation. The power transmitted per unit area is the intensity in the direction of the propagation (in many older textbooks, terms like "the intensity of the sound" were mistakenly used for the magnitude of the sound pressure). In general, the intensity is given by the product of the sound pressure and the sound velocity; thus, in the case of the plane wave, the intensity (I) can be calculated from Equations 8.5 and 8.6:

$$I = vp = \frac{p^2}{\rho c} \tag{8.7}$$

Thus, in the plane wave, the intensity transmitted by the wave can be calculated from the sound pressure and, as the sound power is the intensity per unit area, the sound power can be calculated by multiplying the intensity by the area.

8.3 Spherical Waves

Another simple solution to the wave equation can be found for the radiation from a point source into free space. The point source is an infinitely small sphere whose surface is pulsating radially. In practice, for small sound sources (i.e., where the dimensions of the sound source is small compared to the wavelength of the sound), the point source is a good approximation for the real physical source that makes the spherical wave solution of special interest.

The wave equation in Equation 8.5 is transformed into the spherical coordinates r, θ, and ψ. As the point source radiates equally in all directions, the solution depends only on the distances r from the center of the point source:

$$p = \frac{P_0}{r}\cos\left(\omega t - kr\right) \tag{8.8}$$

It can be seen that the sound pressure is inversely proportional to the distance from the sound source. The particle velocity can be divided into a near-field contribution v_n and a far-field contribution v_f:

$$v_f = \frac{P_0}{r\rho c}\cos(\omega t - kr) \tag{8.9}$$

$$v_n = \frac{P_0}{\omega\rho r^2}\sin(\omega t - kr) \tag{8.10}$$

The far-field contribution in Equation 8.9 can be seen to be in-phase with the pressure in Equation 8.8 and also the particle velocity is inversely proportional to the distance r from the point source. The near-field contribution is inversely proportional to the square of the distance to the source and therefore dies away rapidly as the distance to the source increases.

As in the case of the plane wave, the intensity in the spherical wave is the product of the pressure and particle velocity. For the near-field contribution, one obtains:

$$I = v_n p = \frac{P_0}{\omega\rho r^2}\sin(\omega t - kr)\frac{P_0}{r}\cos(\omega t - kr) = 0 \tag{8.11}$$

That is, the near-field part of the particle velocity does not contribute to the radiated power as the particle velocity is 90° out of phase with the pressure.

The far-field contribution is given by:

$$I = v_f p = \frac{P_0}{r\rho c}\cos(\omega t - kr)\frac{P_0}{r}\cos(\omega t - kr) = \frac{P_0^2}{r^2\rho c}\cos^2(\omega t - kr) \tag{8.12}$$

It can be seen that the intensity decreases with the square of the distance to the source and by combining Equations 8.8 and 8.12, one obtains:

$$I = \frac{P_0^2}{r^2\rho c}\cos^2(\omega t - kr) = \frac{p^2}{\rho c} \tag{8.13}$$

which is identical to Equation 8.7 for the plane wave. Thus, as for the plane progressive wave, the intensity in the spherical wave can be calculated from the pressure.

8.4 Acoustic Measurements

As can be seen from the wave equation, the full acoustic field can in principle be described from only pressure measurements. This means that all other acoustic parameters can be derived from pressure measurements and, in practice, pressure is often the only parameter measured. There have been a few attempts to make transducers for particle velocity measurements based on, for example, transmission of ultrasonic waves; but the absolute dominating transducers for acoustic measurements are the condenser-type microphones, Figure 8.1. These have proven to be superior with respect to temperature stability, long-term stability, and insensitivity to rough handling. While measurement microphones are designed and produced to ensure well-defined and accurate measurements, a wide range of other microphones are available for other purposes. These can, for example, be for incorporation in telephones, where price is a very decisive factor, or for studio recordings, where a subjective evaluation is more important than the objective performance.

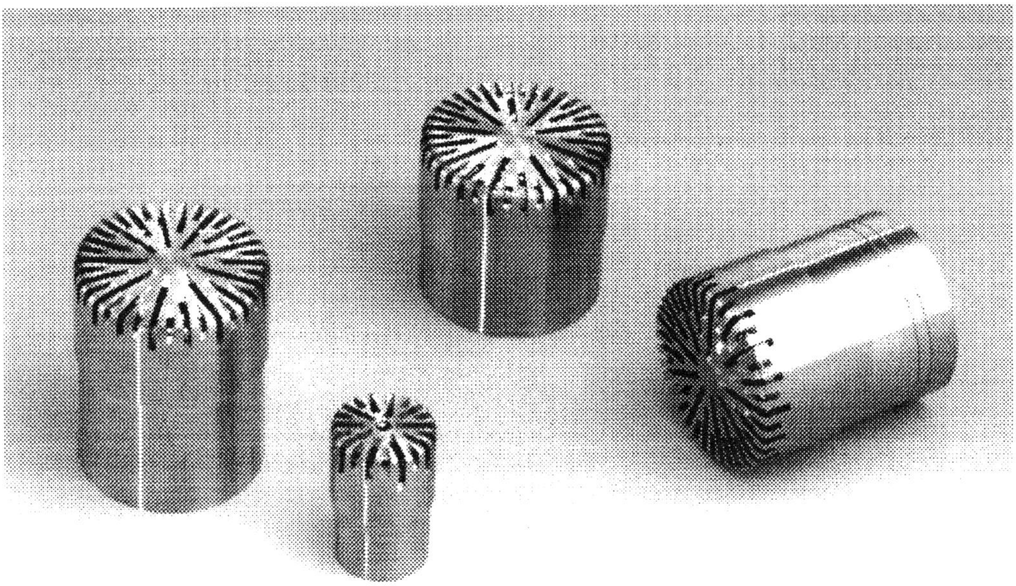

FIGURE 8.1 Measurement microphones: ½″ and ¼″.

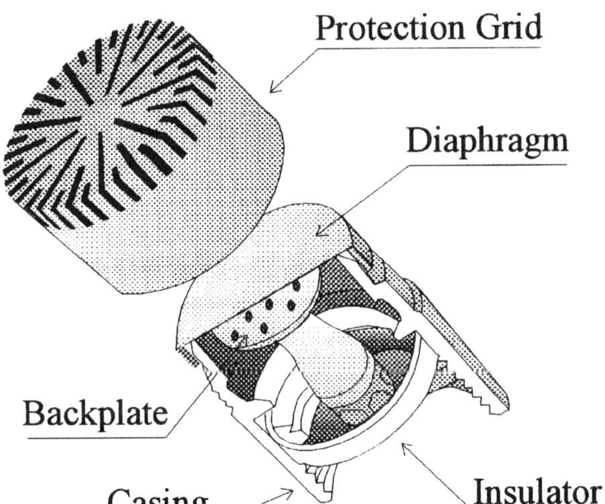

FIGURE 8.2 Basic elements of a measurement microphone.

Condenser Microphones

The condenser microphone consists basically of five elements: protection grid, microphone casing, diaphragm, backplate, and insulator; see Figure 8.2. The diaphragm and the backplate form the parallel plates of an air capacitor. This capacitor is polarized with a charge from an external voltage supply (externally polarized type) or by an electric charge injected directly into an insulating material on the backplate (prepolarized type). When the sound pressure in the sound field fluctuates, the distances between the diaphragm and the backplate will change, and consequently change the capacitance of the diaphragm/backplate capacitor. As the charge on the capacitor is kept constant, the change in capacitance will generate an output voltage on the output terminal of the microphone. The acoustical performance of a microphone is determined by the physical dimensions such as diaphragm area, the distance between the diaphragm and the backplate, the stiffness and mass of the suspended diaphragm, and the internal volume of the microphone casing. These factors will determine the frequency range of the microphone, the sensitivity, and the dynamic range. The sensitivity of the microphone is described as the output voltage of the microphone for a given sound pressure excitation, and is in itself of little interest for the operation of the microphone, except for calibration purposes. However, the sensitivity of the microphone (together with the electric impedance of the cartridge) also determines the lowest sound pressure level that can be measured with the microphone. For example, with a microphone with a sensitivity of 2.5 mV Pa^{-1}, the lowest level that can be measured is around 40 dB (re. 20 μPa), while a microphone with a sensitivity of 50 mV Pa^{-1} can measure levels down to approximately 15 dB (re. 20 μPa).

The size of the microphone is the first parameter determining the sensitivity of the microphone. In general, the larger the diaphragm diameter, the more sensitive the microphone will be. There are, however, limits to how sensitive the microphone can be made by simply making it larger. The polarization voltage between the diaphragm and the backplate will attract the diaphragm and deflect this toward the backplate. As the size of the microphone is increased, the deflection will increase and eventually the diaphragm will be deflected so much that it will touch the backplate. To avoid this, the distance between the diaphragm and the backplate can be increased or the polarization voltage can be decreased. Both of these actions will, however, decrease the sensitivity, so that the optimum size of a practical measurement microphone for use up to 20 kHz is very close to ½″ (12.6 mm).

As the size of the microphone is decreased, the useful frequency range of the microphone is increased. The frequency range, which can be obtained, is determined in part by the size of the microphone. At high frequencies, when the wavelength of the sound waves becomes much smaller than the diameter of the diaphragm, the diaphragm will stop behaving like a rigid piston (the diaphragm "breaks up" — this is not a destructive phenomenon). Different parts of the diaphragm will start to move with different magnitude and phase, and the frequency response of the microphone will change. To avoid this, the upper limiting frequency is placed so that the sensitivity of the microphone drops off before the diaphragm starts to break up. This gives, for a typical 0.5 in. microphone, an upper limiting frequency in the range from 20 kHz to 40 kHz, depending on the diaphragm tension. If the diaphragm is tensioned so that it becomes more stiff, the resonance frequency of the diaphragm will be higher; on the other hand, the sensitivity of the microphone will be reduced as the diaphragm deflection by a certain sound pressure level decreases.

The frequency response of the microphone is determined by the diaphragm tension, the diaphragm mass, and the acoustical damping in the airgap between the diaphragm and the backplate. This system can be represented by the mechanical analogy of a simple mass–spring–damper system as in Figure 8.3. The mass in the analogy represents the mass of the diaphragm and the spring represents the tension in the diaphragm. Thus, if the diaphragm is tensioned to become stiffer, the corresponding spring will become stiffer. The damping element in the analogy represents the acoustical damping between the diaphragm and the backplate. This can be adjusted by, for example, drilling holes in the backplate. This will make it easier for the air to move away from the airgap when the diaphragm is deflected, and therefore decrease damping.

The frequency response of the simple mechanical model of the microphone is given in Figure 8.4, together with the influence of the different parameters. At low frequencies (below the resonant frequency), the response of the microphone is determined by the diaphragm tension, and as described above, the sensitivity will increase if the tension is decreased. The resonant frequency is determined by the diaphragm tension and the diaphragm mass, with an increased tension giving an increased resonant frequency, and an increased mass giving a decreased resonant frequency. The response around the

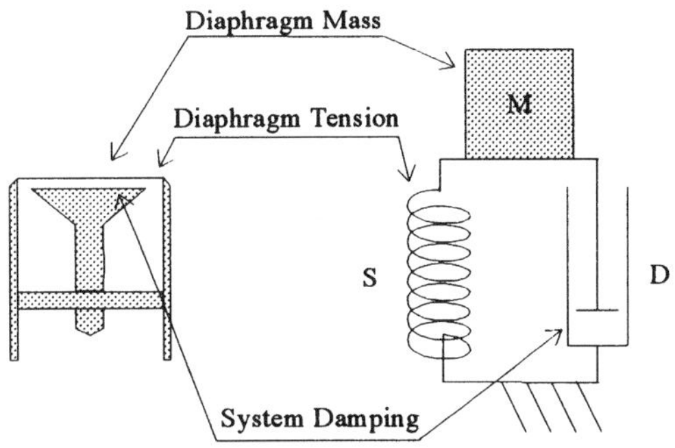

FIGURE 8.3 Mechanical analogy of a microphone.

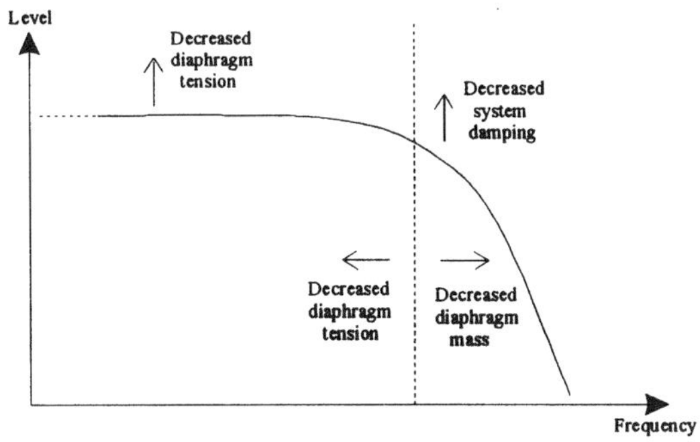

FIGURE 8.4 Influence of microphone parameters on frequency response.

resonant frequency is determined by the acoustical damping, where an increase in the damping will decrease the response.

Although the material selection and assembling techniques have changed during the last few years, the basic types of microphones remain unchanged. The basic types are the free-field microphones, the pressure microphones, and the random incidence microphones. They have been constructed with different frequency characteristics, corresponding to the different requirements.

The *pressure microphone* is meant to measure the actual sound pressure as it exists on the diaphragm. A typical application could be the measurement of the sound pressure in a closed coupler or as in Figure 8.5, the measurement of the sound pressure at a boundary. In this case, the microphone forms part of the wall and measures the sound pressure on the wall itself. The frequency response of this microphone should be flat in a frequency range as wide as possible, taking into account that the sensitivity will decrease as the frequency range is increased. The acoustical damping in the airgap between the diaphragm and the backplate is adjusted so that the frequency response is flat up to and a little beyond the resonant frequency.

The *free-field microphone* is designed to essentially measure the sound pressure as it existed before the microphone was introduced into the sound field. At higher frequencies, the presence of the microphone itself in the sound field will change the sound pressure. In general, the sound pressure around the microphone cartridge will increase due to reflections and diffraction. The free-field microphone is designed so that the frequency characteristics compensate for this pressure increase. The resulting output of the free-field microphone is a signal proportional to the sound pressure as it existed before the microphone was introduced into the sound field. The free-field microphone should always be pointed toward the sound source (0° incidence), as in Figure 8.6. In this situation, the presence of the microphone diaphragm in the sound field will result in a pressure increase in front of the diaphragm, see Figure 8.7(*a*),

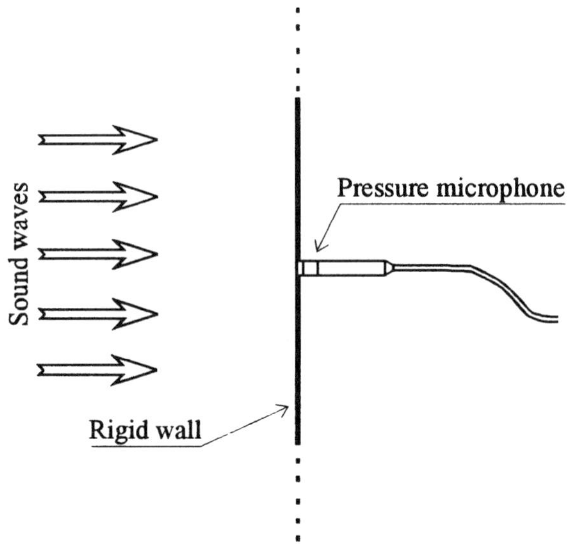

FIGURE 8.5 Application of pressure microphones.

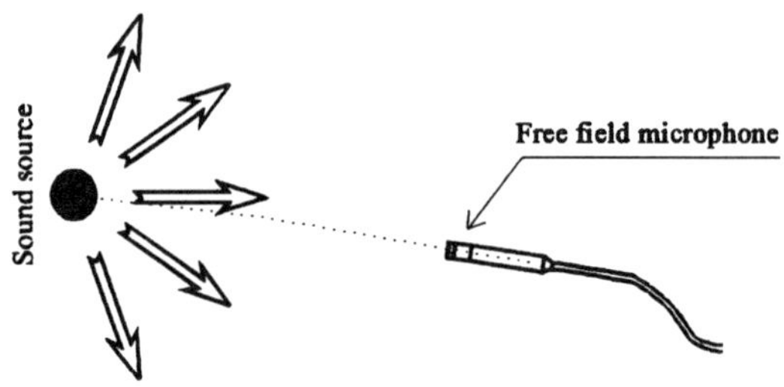

FIGURE 8.6 Application of a free-field microphone.

depending on the wavelength of the sound waves and the microphone diameter. For a typical ½″ microphone, the maximum pressure increase will occur at 26.9 kHz, where the wavelength of the sound (λ = 342 ms^{-1}/26.9 kHz ≈ 12.7 mm ≈ 0.5 in.) coincides with the diameter of the microphone. The microphone is then designed so that the sensitivity of the microphone decreases by the same amount as the acoustical pressure increases in front of the diaphragm. This is obtained by increasing the internal acoustical damping in the microphone cartridge, to obtain a frequency response as in Figure 8.7(*b*). The result is an output from the microphone, Figure 8.7(*c*), which is proportional to the sound pressure as it existed before the microphone was introduced into the sound field. The curve in Figure 8.7(*a*) is also called the "free-field correction curve" for the microphone, as this is the curve that must be added to the frequency response of the microphone cartridge to obtain the acoustical characteristic of the microphone in the free field.

The free-field microphone is required in principle, to be pointed toward the sound source and that the sound waves travel in essentially one direction. In some cases, (e.g., when measuring in a reverberation room or other highly reflecting surroundings), the sound waves will not have a well-defined propagation direction, but will arrive at the microphone from all directions simultaneously. The sound waves arriving at the microphone from the front will cause a pressure increase, as described for the free-field microphone, while the waves arriving from the back of the microphone will be decrease to a certain extent due to the shadowing effects of the microphone cartridge. The combined influence of the waves coming from different directions therefore depends on the distribution of sound waves from different directions. For

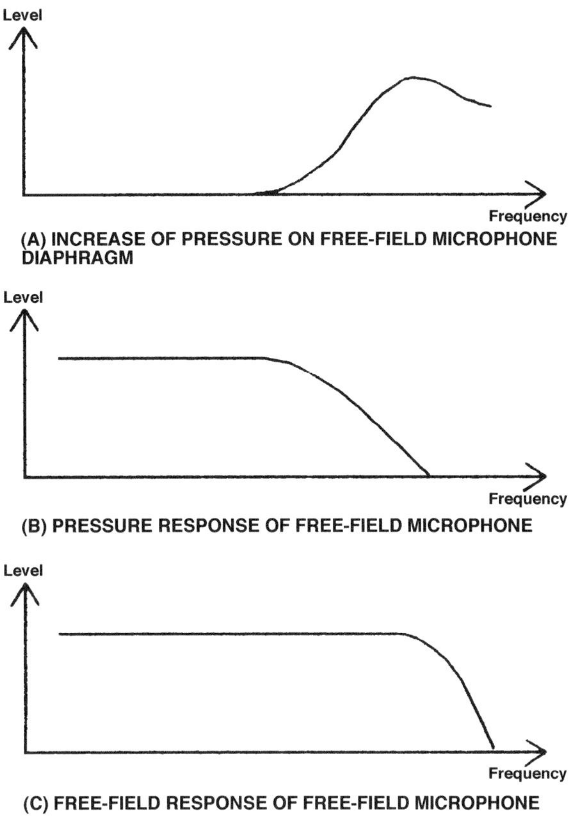

FIGURE 8.7 Frequency response of free-field microphone: (*a*) pressure increase in front of diaphragm; (*b*) microphone pressure response; (*c*) resulting microphone output.

measurement microphones, a standard distribution has been defined, based on statistical considerations, resulting in a standardized random incidence microphone.

As mentioned previously, measurement microphones can be either of the externally polarized type or the prepolarized type. The externally polarized types are by far the most stable and accurate microphones and should be preferred for precision measurements. The prepolarized microphones are, however, preferred in some cases, in that they do not require the external polarization voltage source. This is typically the case when the microphone will be used on small hand-held devices like sound level meters, where a power supply for polarization voltage would add excessively to cost, weight, and battery consumption. Still, it should be realized that prepolarized microphones in general are much less stable to environmental changes than externally polarized microphones.

8.5 Sound Pressure Level Measurements

The human ear basically hears the sound pressure, but the sensitivity varies with the frequency. The human ear is most sensitive to sound in the frequency range from 1 kHz to 5 kHz, while the sensitivity drops at higher and lower frequencies. This has led to the development of several frequency weighting functions, which attempt to replicate the sensitivity of the human ear. Also, the response of the human ear to time-varying signals and impulses has led to the development of instruments with well-defined time weighting functions. The resulting measurement instrumentation is the *sound level meter,* as defined in for example by the IEC International Standard 651, "Sound Level Meters" [2]. The standard defines four classes of sound level meters for different accuracy's (Table 8.1). Type 0 is the most accurate, intended

TABLE 8.1 IEC 651 Sound Level Meter Requirements

Frequency (Hz)	Type 0 (dB)	Type 1 (dB)	Type 2 (dB)	Type 3 (dB)
10	+2; –∞	+3; –∞	+5; –∞	+5; –∞
12.5	+2; –∞	+3; –∞	+5; –∞	+5; –∞
16	+2; –∞	+3; –∞	+5; –∞	+5; –∞
20	±2	±3	±3	+5; –∞
25	±1.5	±2	±3	+5; –∞
31.5	±1	±1.5	±3	±4
40	±1	±1.5	±2	±4
50	±1	±1.5	±2	±3
63	±1	±1.5	±2	±3
80	±1	±1.5	±2	±3
100	±0.7	±1	±1.5	±3
125	±0.7	±1	±1.5	±2
160	±0.7	±1	±1.5	±2
200	±0.7	±1	±1.5	±2
250	±0.7	±1	±1.5	±2
315	±0.7	±1	±1.5	±2
400	±0.7	±1	±1.5	±2
500	±0.7	±1	±1.5	±2
630	±0.7	±1	±1.5	±2
800	±0.7	±1	±1.5	±2
1000	±0.7	±1	±1.5	±2
1250	±0.7	±1	±1.5	±2.5
1600	±0.7	±1	±2	±3
2000	±0.7	±1	±2	±3
2500	±0.7	±1	±2.5	±4
3125	±0.7	±1	±2.5	±4.5
4000	±0.7	±1	±3	±5
5000	±1	±1.5	±3.5	±6
6300	+1; –1.5	+1.5; –2	±4.5	±6
8000	+1; –2	+1.5; –3	±5	±6
10000	+2; –3	+2; –4	+5; –∞	+6; –∞
12500	+2; –3	+3; –6	+5; –∞	+6; –∞
16000	+2; –3	+3; –∞	+5; –∞	+6; –∞
16000	+2; –3	+3; –∞	+5; –∞	+6; –∞

for precision laboratory measurements, while Type 1 is most widely used for general-purpose measurements, see Figure 8.8. Type 2 is used where low price is of importance, while Type 3 is not used in practice because of the wide tolerances, making the results too unreliable. The output of the sound level meter is, in principle, assumed to be an approximate measure of the impression perceived by the human ear.

The sound level meter can be functionally divided into four parts: microphone and preamplifier, A-weighting filter, rms detector and display (Figure 8.9). The microphone should ensure the correct measurement of the sound pressure within the frequency range for the given class. Also, the standard gives requirements for the directionality of the microphone. The frequency response of the instrument, including the weighting filter, is given for sound waves arriving at the microphone along the reference direction. For sound waves arriving from other directions, the standard allows wider tolerances at higher frequencies, taking into account the inevitable reflections and diffraction occurring at higher frequencies.

The preamplifier converts the high-impedance output signal from the microphone to a low-impedance signal, but has in itself no or even negative voltage amplification. The signal from the preamplifier is then passed through an A-weighting filter. This is a standardized filter which, in principle, resembles the sensitivity of the human ear, so that a measurement utilizing this filter will give a result which correlates with the subjective response of an average listener. The filter, with the filter characteristic as in Figure 8.10, attenuates low and high frequencies and slightly amplifies frequencies in the mid-frequency range from

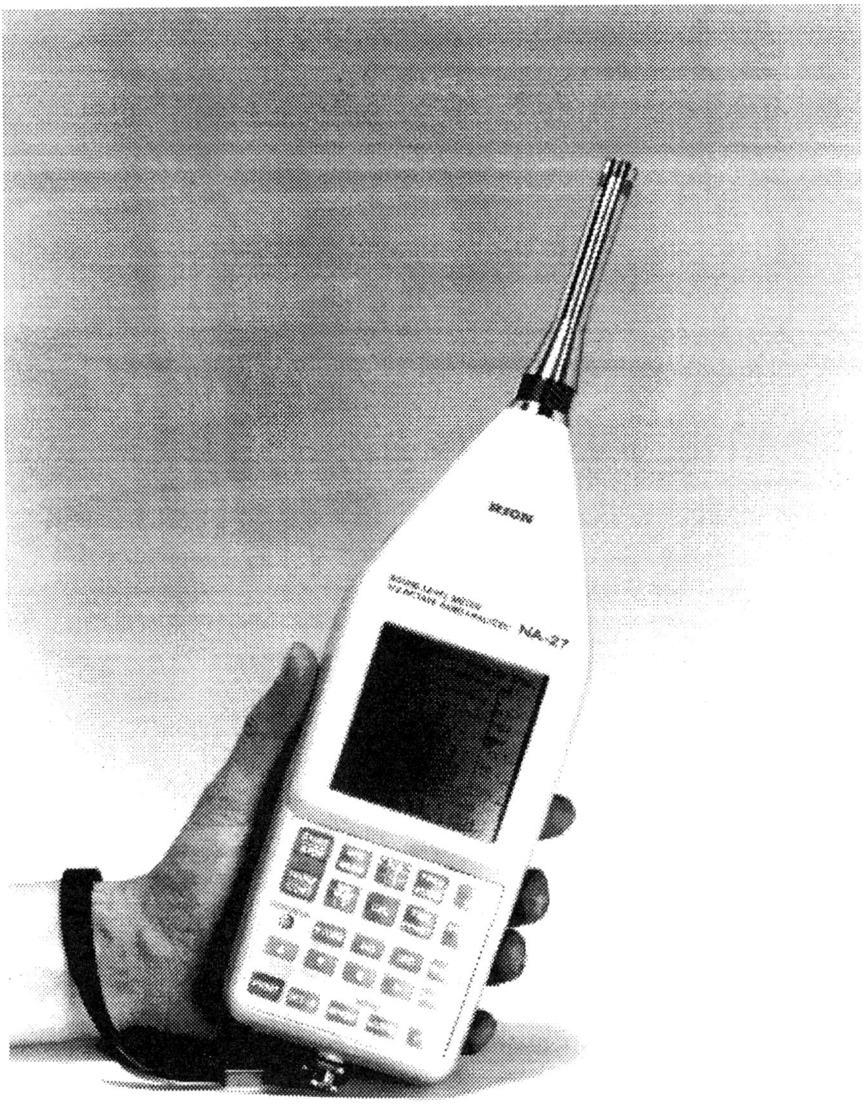

FIGURE 8.8 Modern Type 1 sound level meter with built-in frequency analyzer.

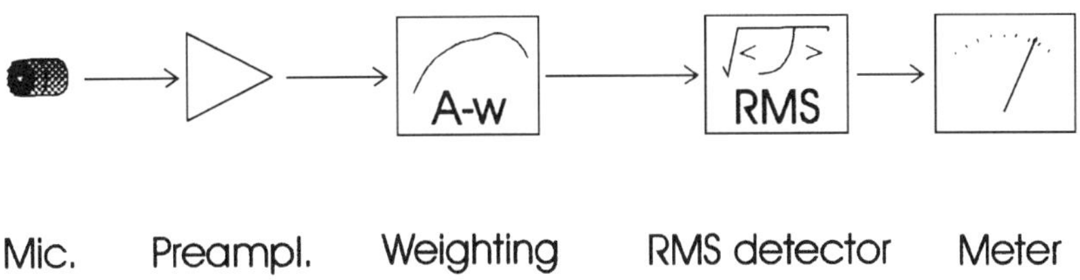

FIGURE 8.9 Functional parts of a sound level meter.

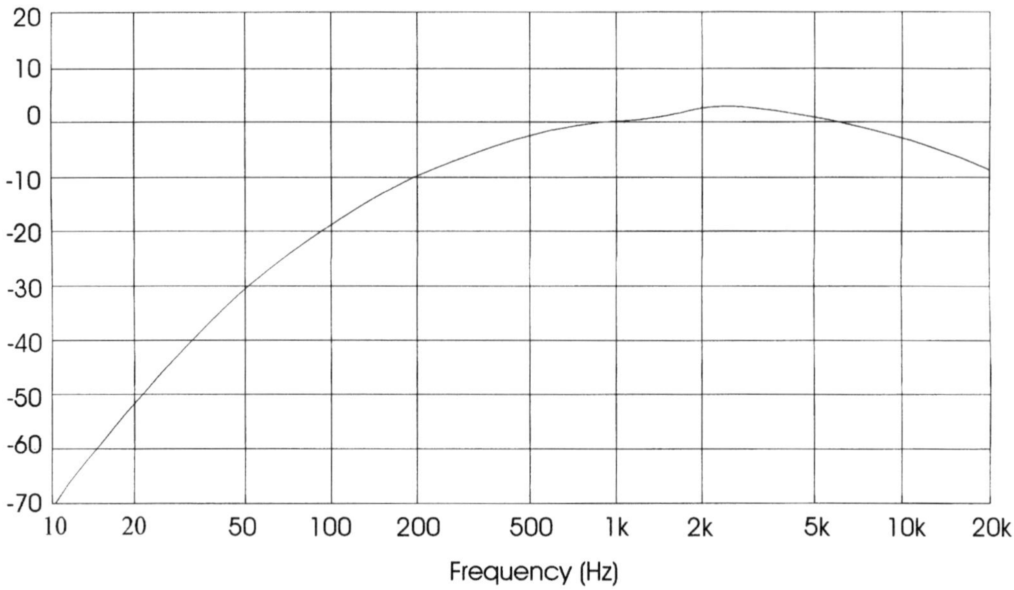

FIGURE 8.10 A-weighting curve.

1 kHz to 5 kHz. There are a number of other weighting curves, denoted B-weighting, C-weighting and D-weighting, which may give better correlation with subjective responses in special cases, such as for very high or very low levels, or for aircraft noise.

The signal from the A-weighting filter is subsequently passed through an exponential rms detector, with a time constant of either 125 ms ("fast") or 1 s ("slow"). These time constants simulate the behavior of the human ear when subjected to time-varying signals. Especially when the duration of the sound stimuli to the human ear becomes shorts (e.g., around 200 ms), the sound is subjectively judged as being lower compared to the same sound heard continuously. The same effect is obtained using the "fast" averaging time. This will however give a higher statistical uncertainty on the level estimate than when using the time constant "slow," so this should be chosen if the sound signal is continuously. Other standards describe special sound level meters such as integrating sound level meters or impulse sound level meters intended for special purposes.

8.6 Frequency Analyzers

While the sound level meter gives a single reading for the sound level in the frequency range from 20 Hz to 20 kHz, it is often desirable to have more detailed information about the frequency content of the signal. Two types of frequency analyzers are commonly used in acoustic measurements: FFT analyzers and real-time filter analyzers. The *FFT analyzers,* with very high frequency resolutions, can give a wealth of frequency information and make it possible to separate closely space harmonics (e.g., from a gear box). In contrast to this, the *real-time filter analyzers,* Figure 8.11, uses a much broader frequency resolution, usually in 1/3-octave bands. The frequency analysis is performed by a bank of filters (nowadays mostly digital filters) with well-defined frequency and time responses. The filter responses have been internationally standardized, with the center frequencies and passbands as shown in Table 8.2. In the frequency range from 20 Hz to 20 kHz, the 1/3-octave filterbank consists of 31 filters, simultaneously measuring the input signal. The resulting 1/3-octave spectrum resembles the subjective response of the human ear.

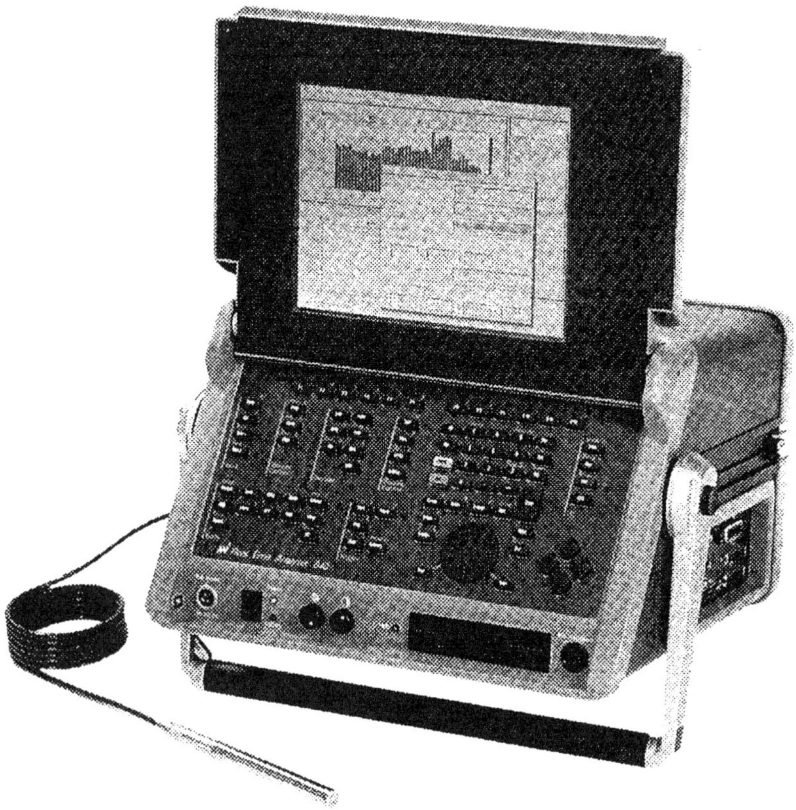

FIGURE 8.11 Real-time frequency analyzer for acoustic measurements.

8.7 Pressure-Based Measurements

The result of sound pressure measurements will be influenced by many factors: source, source operating conditions, surroundings, measurement position, etc. Depending on the goal of the measurement, these parameters can be controlled in different manners. If the goal of the measurement is to quantify the noise exposure to an operator's ear in a noisy environment, it is important that the microphone is in the same position as the operator's ear would normally be in, and that the environment is equal to the normal operating environment. If, on the other hand, the task is to describe the sound source as a noise-emitting machine, it is important to minimize the influence of the environment on the measurement result.

If the aim is to describe the measuring object as a noise source, it is customary to state the radiated sound power for the source. This is a global parameter quantifying the total noise radiation from the source, and to a certain extent independent of the environment. The sound power can be measured in a number of different ways: in a free field, in a reverberation room, using a substitution technique, or using sound intensity technique. A free field is a sound field in which the sound is radiated freely in all directions, with no restricting walls or reflections. This is most often obtained in a semi-anechoic chamber, where all walls and ceiling have been covered by nearly 100% absorptive material, with only the floor made of reflecting material. When the sound source is placed in the semi-anechoic chamber, the emitted sound waves will radiate freely away from the source and, in the far field, the waves can be considered to be plane waves or spherical waves. Therefore, the sound intensity can be calculated from pressure measurements using Equation 8.7.

TABLE 8.2 1/3-Octave Analysis Frequencies

Nominal center frequency (Hz)	Exact center frequency (Hz)	Passband (Hz)
20	19.95	17.8–22.4
25	25.12	22.4–28.2
31.5	31.62	28.2–35.5
40	39.81	35.5–44.7
50	50.12	44.7–56.2
63	63.1	56.2–70.8
80	79.43	70.8–89.1
100	100.0	89.1–112
125	125.89	112–141
160	158.49	141–178
200	199.53	178–224
250	251.19	224–282
315	316.23	282–355
400	398.11	355–447
500	501.19	447–562
630	630.96	562–708
800	794.33	708–891
1000	1000.0	891–1120
1250	1258.9	1120–1410
1600	1584.9	1410–1780
2000	1995.3	1780–2240
2500	2511.9	2240–2820
3150	3162.3	2820–3550
4000	3981.1	3550–4470
5000	5011.9	4470–5620
6300	6309.6	5620–7080
8000	7943.3	7080–8910
10000	10000.0	8910–11200
12500	12589.3	11200–14100
16000	15848.9	14100–17800
20000	19952.6	17800–22400

As real sound sources seldom radiate equally in all directions, a number of measurements around the test object are averaged. ISO Standard 3745 "Acoustics—Determination of sound power levels of noise sources—Precision method for anechoic and semi-anechoic rooms" [3] specifies an array of microphone positions on a hemisphere over the test object, as in Figure 8.12, with the coordinates as in Table 8.3. As all points are associated with the same area, and as the sound power is intensity times the area, the total radiated sound power can be calculated as:

$$P = A\sum I_{\mathrm{n}} = \frac{2\pi r}{\rho c}\sum p_{\mathrm{n}}^{2} \tag{8.14}$$

where A is the area of the test hemisphere with radius r, and p_{n} is the pressure measured in point number n.

8.8 Sound Intensity Measurements

The calculation in Equation 8.14 of the sound power from sound pressure measurements is based on Equation 8.7. This equation, which gives the intensity based on a pressure measurement, is however only valid in a free field, in the direction of propagation. In general, in the presence of background noise or with reflections from walls, etc., it is not possible to calculate the sound intensity from a single pressure

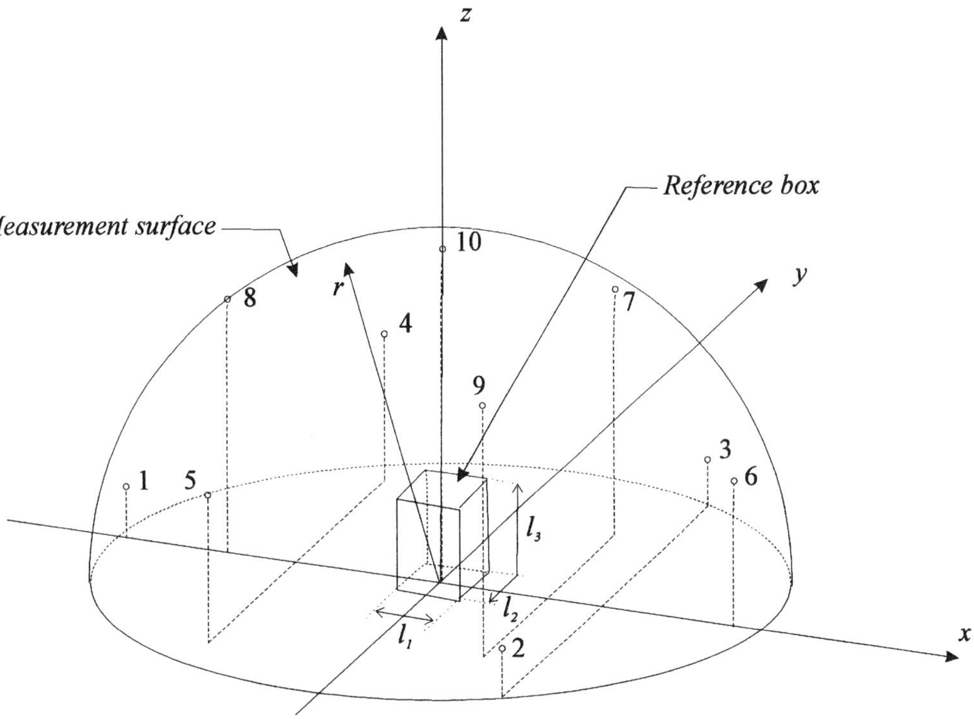

FIGURE 8.12 Measurement points for sound power determination.

TABLE 8.3. Coordinates of Measurement Points for Hemisphere with Radius *r*

Measurement point no.	x/r	y/r	z/r
1	–0.99	0	0.15
2	0.5	–0.86	0.15
3	0.5	0.86	0.15
4	–0.45	0.77	0.45
5	–0.45	–0.77	0.45
6	0.89	0	0.45
7	0.33	0.57	0.75
8	–0.66	0	0.75
9	0.33	–0.57	0.75
10	0	0	1.0

measurement. In these cases, it is however possible to measure directly the sound intensity with a two-microphone intensity probe, Figure 8.13.

Sound intensity I is the product of the pressure and the particle velocity:

$$I = pv \tag{8.15}$$

While the pressure p is a scalar and independent of the direction, the particle velocity is a vector quantity and directionally dependent. When the particle velocity is stated as in Equation 8.15, it is implicit that the velocity is in a certain direction and that the resulting intensity is calculated in the

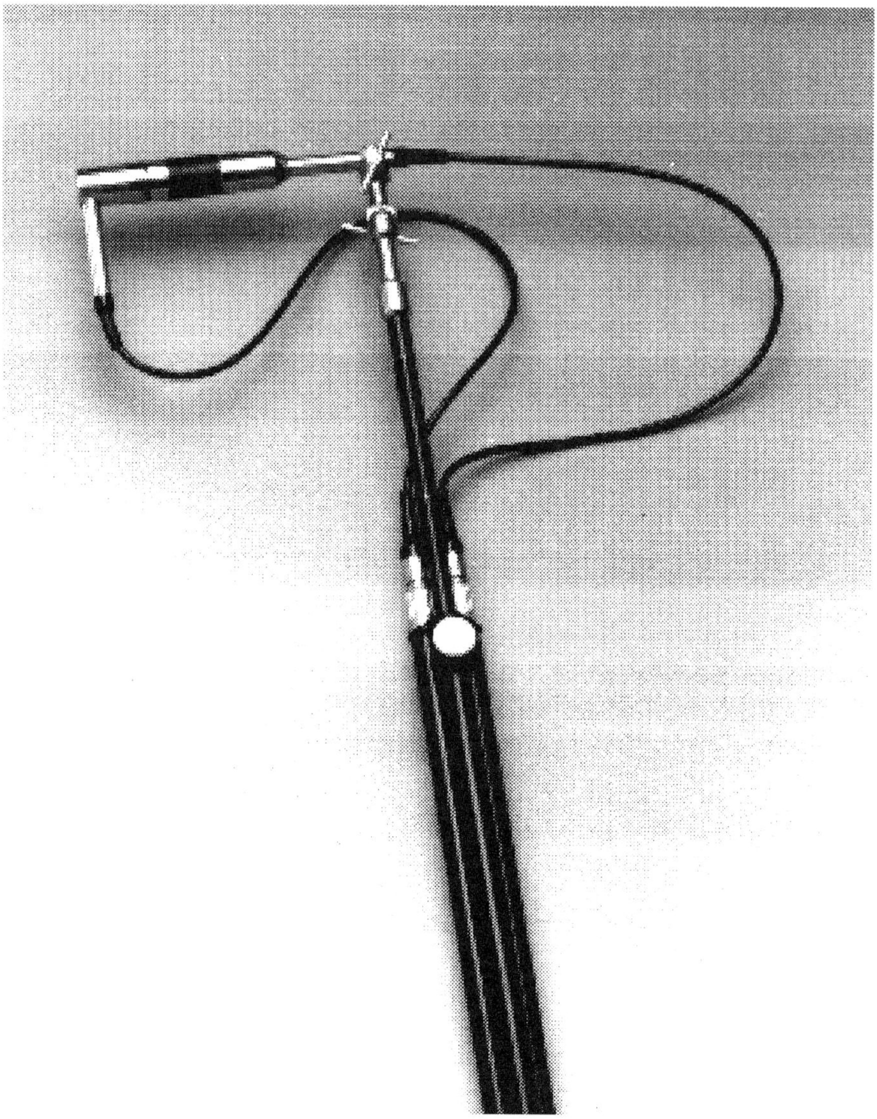

FIGURE 8.13 Two-microphone sound intensity probe.

same direction. For example, the particle velocity v in the direction of propagation, Figure 8.14(a), gives the intensity radiation away from the point source, while the particle velocity perpendicular to the propagation direction, Figure 8.14(b), is zero. The intensity calculated from Equation 8.15 will therefore be zero in the direction perpendicular to the propagation direction even though the sound pressure is the same. This means that the sound energy flows away radially from the point source and no energy is flowing tangentially.

The measurement of the sound intensity according to Equation 8.15 requires the measurement of the sound pressure and the particle velocity. With the two-microphone intensity probe, the pressure in a position in between the two microphones is calculated as the mean pressure measured by the two microphones:

$$p = \frac{p_1 + p_2}{2} \tag{8.16}$$

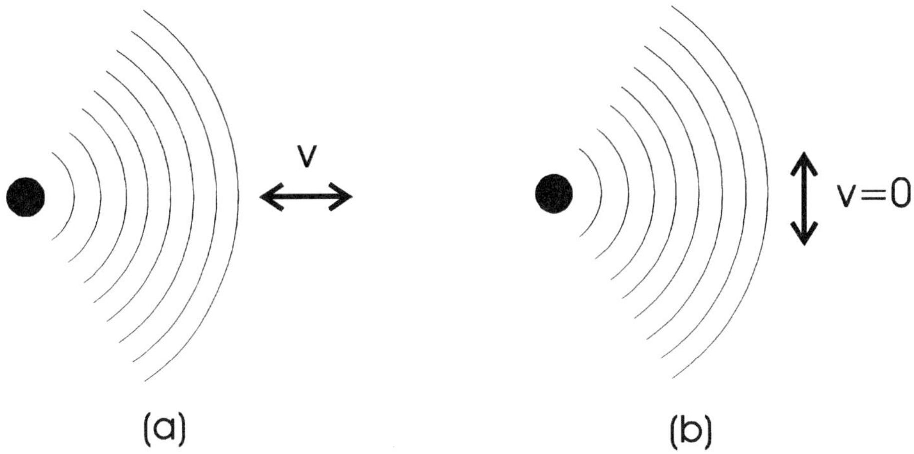

FIGURE 8.14 Particle velocity (*a*) along direction of propagation, and (*b*) perpendicular to direction of propagation.

The air particle velocity *v*, in the direction of the intensity probe, can be calculated from the pressure differences between the two microphone measurements:

$$v = \int \frac{(p_2 - p_1)}{\rho \Delta r} \partial \tau \tag{8.17}$$

where ρ is the density of the air and Δr is the distance between the microphones. The intensity *I* is then obtained by multiplying the pressure and the velocity:

$$I = pv = \frac{p_1 + p_2}{2} \int \frac{(p_2 - p_1)}{\rho \Delta r} \partial \tau \tag{8.18}$$

The intensity measurement technique is a powerful tool to localize acoustical noise source and to determine the sound power radiated from a sound source, even in the presence of other strong sound sources.

8.9 Near-Field Acoustic Holography Measurements

The term *acoustic holography* comes from the analogy to optical holographs. It is well known how holography, as opposed to a normal photo, enables one to reconstruct the full image of an object. This is obtained by "recording" information about both the magnitude and the phase of the light, while a normal photo only "records" the magnitude of the light. Similarly, with *acoustic holography*, both the magnitude and the phase of the sound field are measured over a plane surface. These measurements result in a complete description of the sound field where both magnitude and phase are known at all points. It is then possible to calculate acoustic quantities, including sound intensity distribution, particle velocity, sound power, radiation pattern, etc.

The basic assumption behind *near-field acoustic holography* (NAH) is that the sound field can be decomposed into two simple wave types: plane waves and evanescent waves. The *plane waves* describe the part of the sound field that is propagated away from the near field toward the far field, and the *evanescent waves* describe the complicated sound field existing in the near field. Any sound field can be described as a combination of plane waves and evanescent waves with different magnitude and directions.

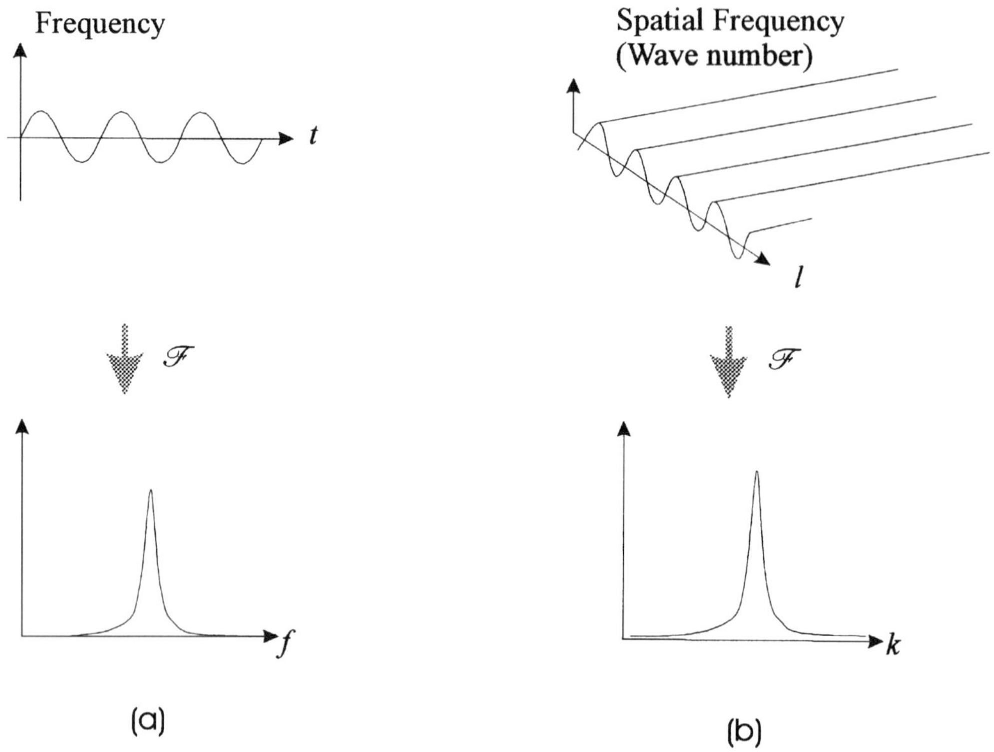

FIGURE 8.15 Temporal and spatial frequency of plane wave.

The magnitude and direction of the individual waves can be described by their spatial frequencies or wave numbers. For a simple plane wave propagating in a certain direction, this can be described in terms of its temporal frequency as well as by its spatial frequency. The temporal frequency, Figure 8.15(*a*), is obtained by looking at the pressure changes with time at a certain point in the sound field. This gives the temporal frequency in hertz or radians per second. Similarly, the spatial frequency, Figure 8.15(*b*), is obtained by looking at the pressure changes at a certain time. At that instant in time, the pressure will be different in different positions in space. If one moves in a certain direction in space, one will see a certain change in the pressure, corresponding to a spatial frequency, measured with the unit cycles per meter or radians per meter. As the temporal frequency gives information about how often the pressure changes with time at a certain point, the spatial frequency gives information about how often the pressure changes with position at a certain time. In the example of Figure 8.15(*b*), the propagation direction of the plane wave was identical to the direction of the axis along which the spatial frequency was measured. In this case, shown again in Figure 8.16(*a*), the relationship between the spatial frequency k_0 (i.e., the wave number) and the temporal frequency f is given by the speed of sound c:

$$k_0 = \frac{2\pi f}{c} = \frac{\omega}{c} = \frac{2\pi}{\lambda} \tag{8.19}$$

where λ is the wavelength. If, however, the axis along which the spatial frequency is measured is not the same as the propagation direction, see the example in Figure 8.16(*b*), this simple relationship is not valid. In this case, although the temporal frequency is the same as in Figure 8.16(*a*), the spatial frequency is lower. For one particular temporal frequency, the spatial frequencies will thus give information about the propagation directions. Therefore, if the sound field is made up of several plane waves with the same temporal frequency, but with different propagation directions, this will be shown in the spatial spectrum

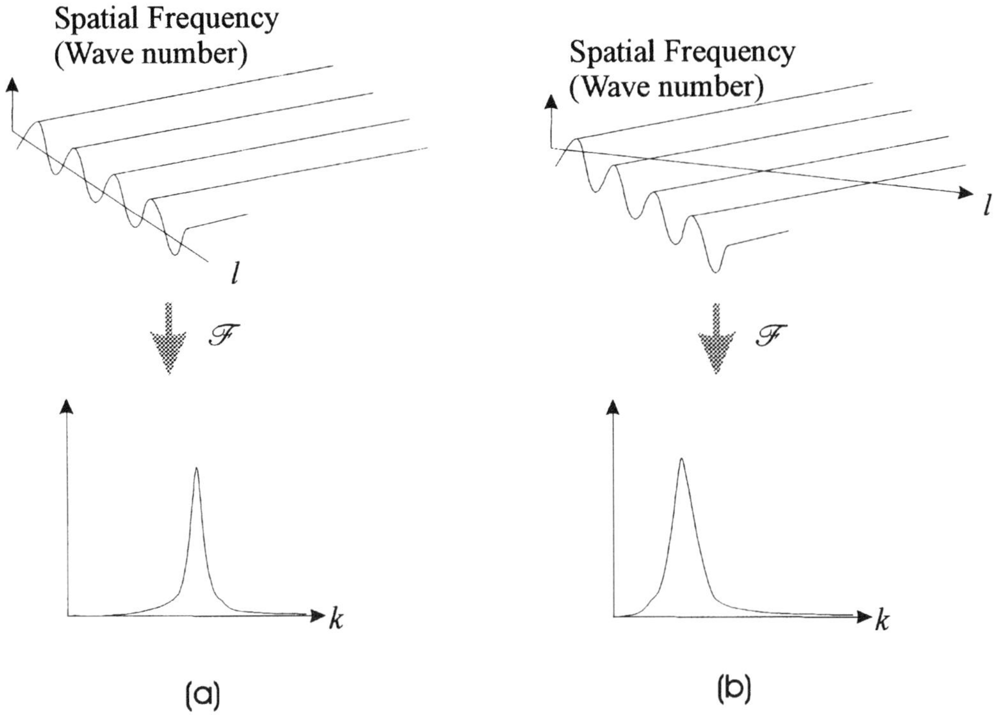

FIGURE 8.16 Spatial frequencies of waves propagating in different directions.

as several spatial frequency components. If, for example, the sound field is a sum of two waves, Figure 8.17, where one wave is traveling along the axis of measurement and the other at an angle of 45° relative to the first wave, the spatial spectrum will contain two spatial frequencies. One spatial frequency will be k, corresponding to a wave in the direction along the axis, and the other frequency will be k^* cos (45°). Thus far, the spatial frequencies have been defined along a single axis corresponding to a one-dimensional Fourier transformation. In the NAH technique, the sound field is sampled not only along a single axis, but over a plane. Therefore, a two-dimensional Fourier transformation is used instead. This gives as a result a two-dimensional spatial frequency spectrum, but otherwise the information is the same as before: namely, information about the direction and magnitude of the simple wave types.

The sound field from a point source cannot be explained by simple plane waves such as those in Figures 8.16 and 8.17, as the amplitude decreases with the distance from the origin. The plane waves retain the same magnitude over the full plane. Thus, to described the near-field phenomenon, one must introduce evanescent waves. In the one-dimensional Fourier spectrum, the evanescent waves can be identified as spatial frequencies higher than $k_0 = 2\pi f/c$. Similarly, in the two-dimensional spatial frequency spectrum, the evanescent waves can be identified as having spatial frequencies or wavenumbers higher than k_0.

The individual spatial frequencies in the two-dimensional spatial frequency spectrum correspond to simple plane waves or evanescent waves in the scan plane (i.e., the measurement plane). For each of these simple wave types, it is easy to calculate the pressure in other planes, see Figure 8.18. For the plane waves, a simple phase shift of the wave is required to calculate the result in a new plane. For the evanescent waves, the changes in amplitude must be taken into account; but in principle, this is also a simple transfer function applied to the two-dimensional spatial frequency spectrum. In this way, the two-dimensional spatial frequency spectrum in a new plane can be calculated from the original data by applying simple transfer function operations. The new two-dimensional spatial frequency spectrum is then an inverse Fourier transform (in two dimensions) to get the sound field in the new plane.

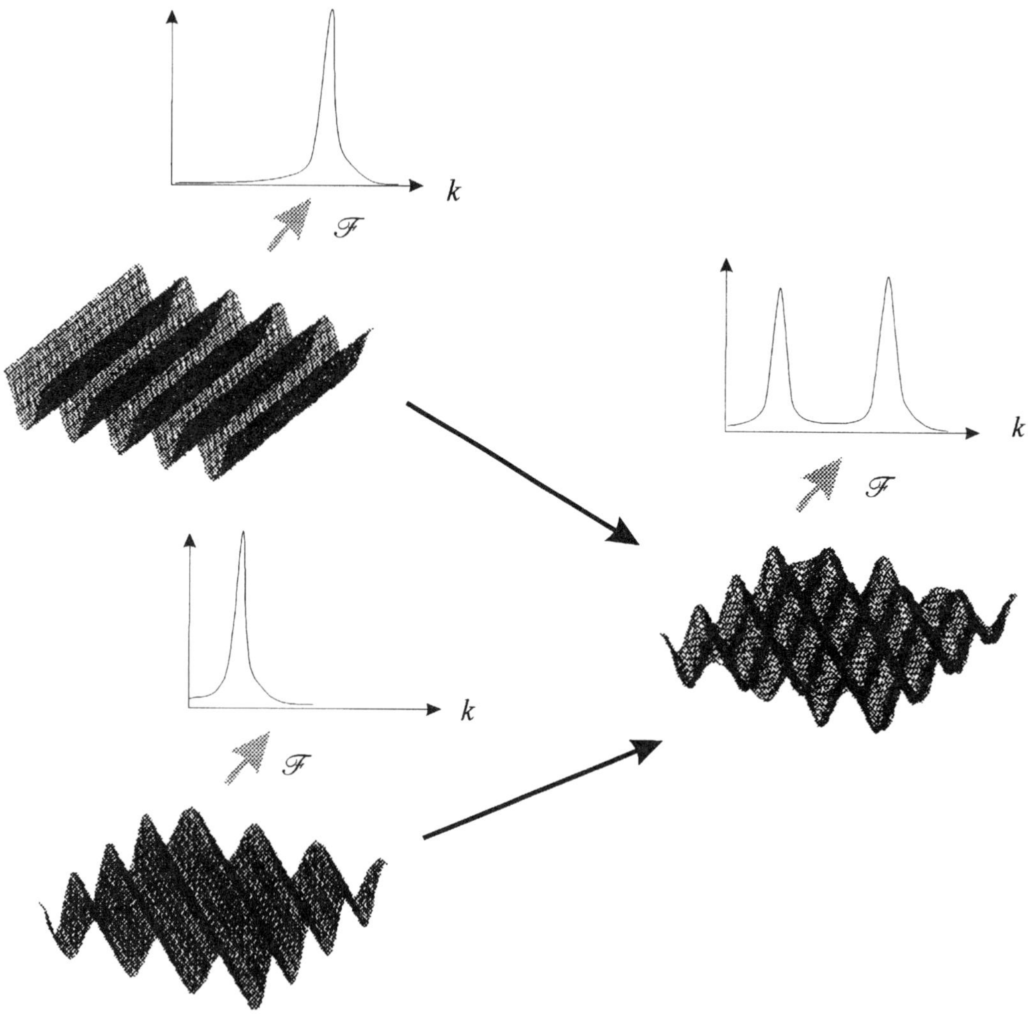

FIGURE 8.17 Spatial spectrum of two waves propagating in different directions.

The overall principle of near-field acoustic holography can be simplified as in Figure 8.19. The sound field is scanned in a plane close to the measuring object. This gives an array of temporal spectra, one for each scan position. Looking at one temporal frequency at a time, one takes out the information from each of the spectra corresponding to the actual frequency of interest. This generates a new array with information about only one temporal frequency. A Fourier transform (in two dimensions) is then applied to the array to generate a two-dimensional spatial frequency spectrum. This can then be transformed to new planes using simple transfer function operations. When the two-dimensional spatial frequency spectrum in the new plane has been calculated, an inverse Fourier transform is used to obtain the new pressure distribution in the new plane. In principle, the NAH technique requires that all cross-spectra between all the scan positions are given; that is, in each of the scan positions, all the cross-spectra to all other scan positions must be determined. A simple scan of a sound field with 2540 scan positions, defining $N = 1000$ scan positions, would result in $\frac{1}{2}N(N + 1) = 500{,}500$ cross-spectra. Instead of measuring all these cross-spectra, the system uses a set of reference transducers to reduce the amount of cross-spectra. The number of necessary reference transducers to give a complete description of the sound field without measuring the full amount of data is determined by the complexity of the sound field. A measurement

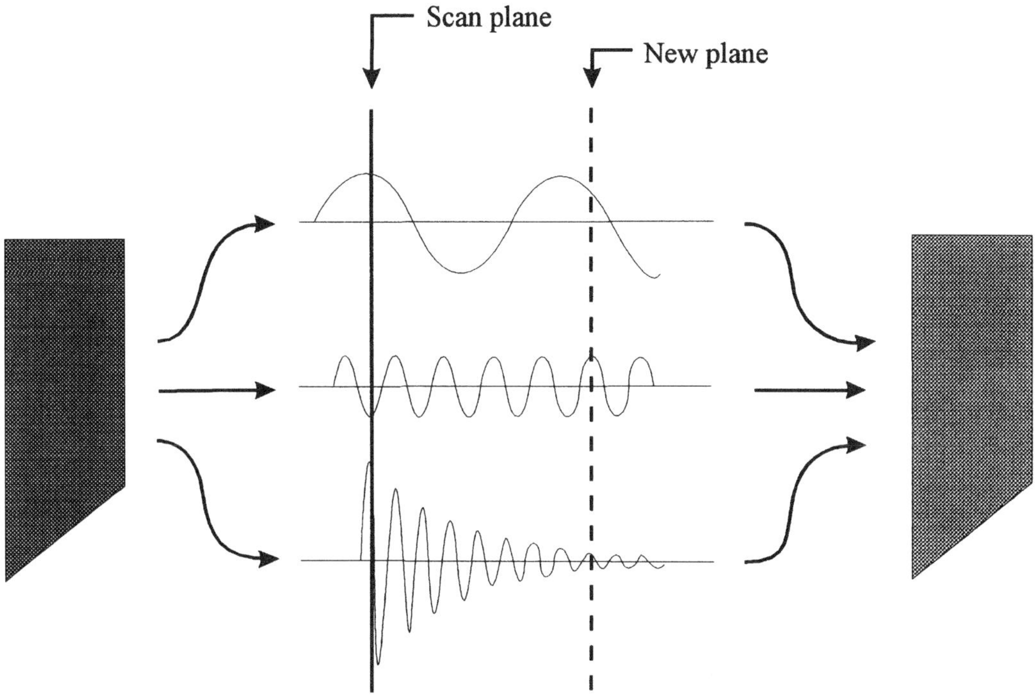

FIGURE 8.18 Transformation of simple wave types in a spatial spectrum, from one measurement plane to another plane.

with, for example, four reference transducers and 2540 scan positions will then be reduced to $4N = 4000$ cross-spectrum measurements.

8.10 Calibration

In order to make accurate and reliable measurements, the microphone and connected instruments must be properly calibrated. The calibration of measurement microphones can be divided into two parts: a level calibration and a frequency response calibration. The *level calibration* establishes the output signal of the microphone for a given acoustic input signal at a given frequency, while the *frequency response* gives the output at other frequencies relative to the level calibration frequency. The level calibration can be performed by a number of different methods with different accuracies.

The most accurate method is the *reciprocity calibration method.* This method utilizes the fact that a condenser microphone is a reciprocal transducer; that is, it can be used as a microphone (to convert an acoustical signal to a voltage signal) and as a loudspeaker (to convert a voltage signal into an acoustical signal). By measuring the relationship between three test microphones driven as both transmitters and receivers, one obtains a set of three equations with the three microphone sensitivities as the unknowns. By solving these three equations, one obtains the sensitivity of the three microphones. The reciprocity calibration method is very accurate but rather tedious and requires well-controlled environmental conditions and is therefore seldom used in practical situations.

The comparison or substitution methods are essentially identical in that they are based on measuring the differences between the test microphone and a reference microphone with known sensitivity. In this case, the reference microphone is often calibrated at an accredited national acoustical laboratory like NIST, NPL, or PTB, whereby the traceability is ensured. In the substitution method, the acoustical output of a sound source is measured with the reference microphone. Afterward, the reference microphone is replaced with a test microphone and the output is measured again. Provided that the sound source has

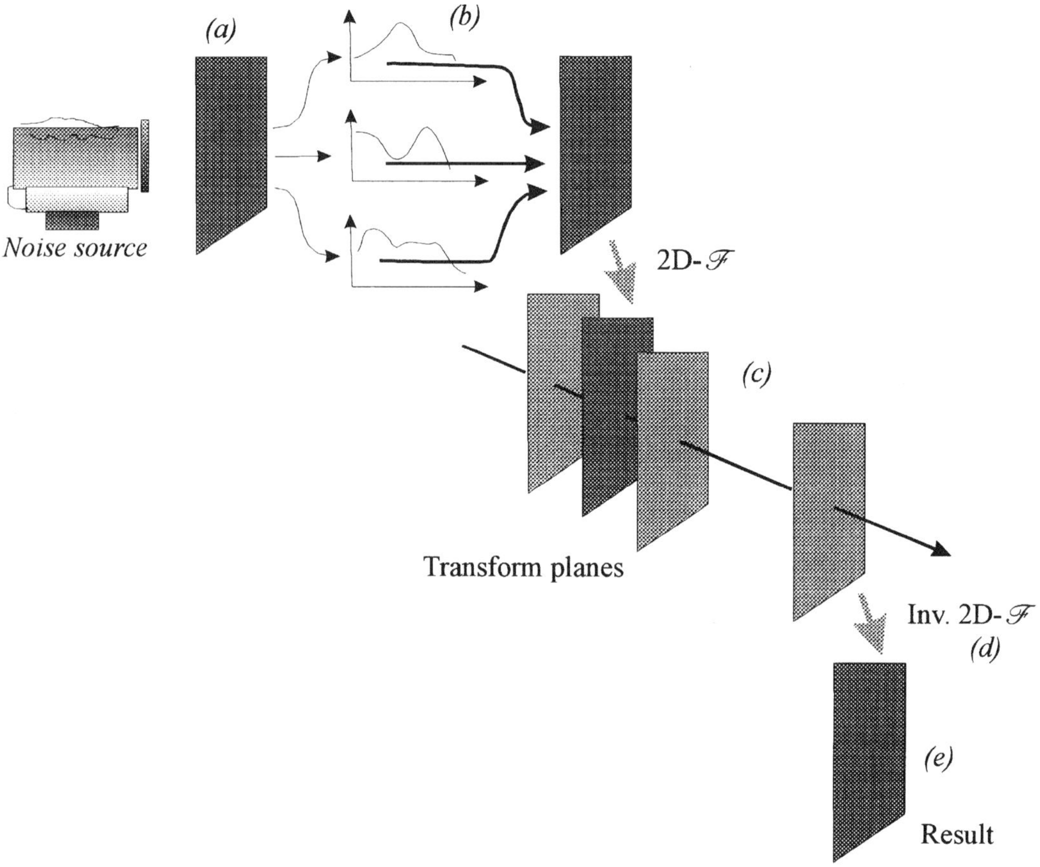

FIGURE 8.19 Overall principle of NAH: (*a*) measurement of cross-spectra in the scan plane; (*b*) calculation for one temporal frequency at a time; (*c*) 2-D spatial Fourier transformation; (*d*) transformation of simple wave types; (*e*) inverse 2-D transformation; (*f*) to obtain the sound field in the new plane.

been stable, the sensitivity of the test microphone can then be calculated. The comparison method is similar to the substitution method, except that the reference microphone and the test microphone are subjected to the same sound pressure simultaneously and therefore the requirements to the sound source stability are less important.

An often-used method for microphone calibration is the *pistonphone method.* The pistonphone, Figure 8.20, is a very stable sound source, which produces a well-defined sound pressure level inside a closed coupler. It works by volume displacements, Figure 8.21, with a well-defined velocity, usually at 250 Hz. As the piston is moving in and out, the volume of the closed coupler is changed and this will result in pressure variations. The actual pressure level obtained in the pistonphone depends on the volume of the coupler, the volume displacement of the pistons, the barometric pressure, and—to a lesser degree—on other factors such as humidity, heat dissipation, etc. As the pistonphone is based on a relatively simple mechanical system, it is very reliable and easy to use in practice, with an accuracy around 0.1 dB. Also, the pistonphone is often used as the stable sound source for calibrations using comparisons or substitution methods.

A *sound pressure calibrator* is basically a small self-contained comparison calibration device. The test microphone is inserted into a small, closed volume and a small loudspeaker produces a single frequency signal, usually at 1 kHz. The output level of the loudspeaker is controlled in a feedback system with a signal from a reference microphone. Provided that the reference microphone and the feedback gain are

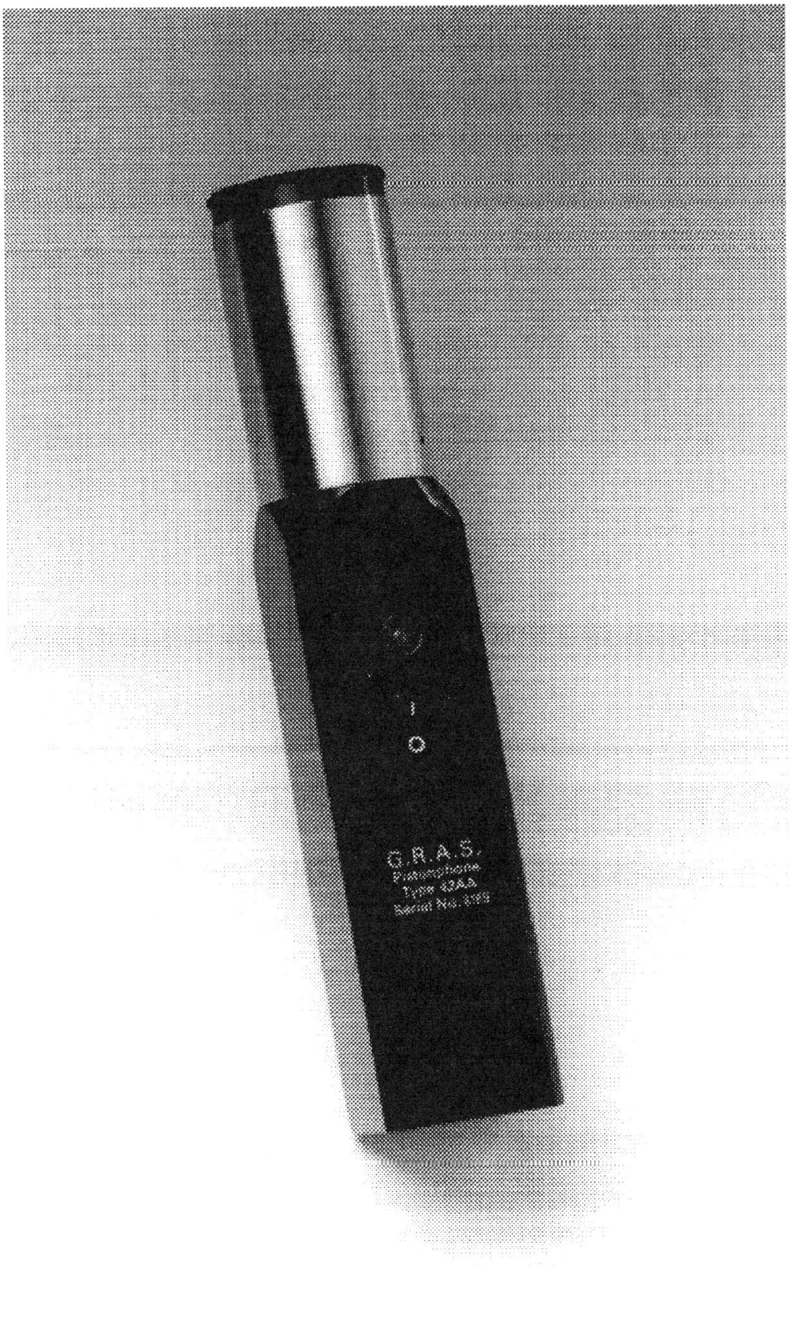

FIGURE 8.20 Pistonphone for microphone calibration.

stable, the sound level at the test microphone will be well-defined and the sensitivity can be determined. The sound level calibrators are normally not used to make accurate microphone calibrations, but rather to make field checks of the integrity of a complete measurement system.

The frequency response of a microphone is most often determined by the *electrostatic actuator method.* A conducting grid is placed close to and parallel to the microphone diaphragm. An electric field is established between the actuator and the diaphragm by applying 800 V dc to the actuator. A test signal of 50 to 150 V ac is superimposed on the dc signal, and the electrostatic forces will push and pull the

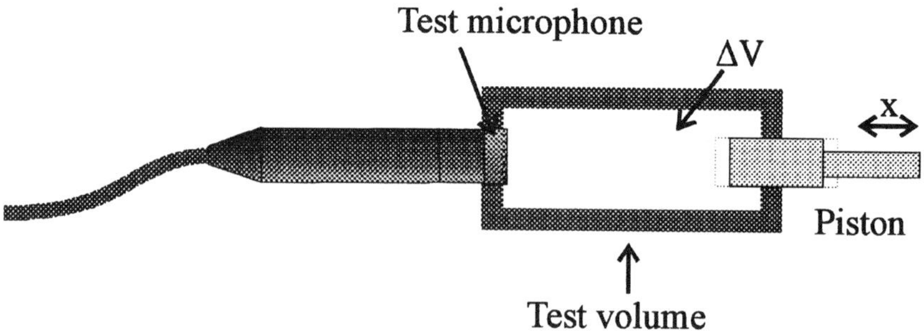

FIGURE 8.21 Principle of a pistonphone.

diaphragm, similar to a sound pressure of 1 to 10 Pa. By sweeping the test signal through the frequencies of interest, the pressure response of the test microphone can be recorded. The electrostatic actuator technique is widely used as a convenient and accurate test method, both during production and final calibration of measurement microphones.

Available instrumentation and manufacturers are given in Tables 8.4 and 8.5.

TABLE 8.4

Instrumentation	Types available	Approx. price	Manufacturers
Meas. microphones	½″ Free field ½″ Pressure ¼″ Free field ¼″ Pressure	\$750–\$825	GRAS Sound & Vibration ACO Pacific B&K The Modal Shop Larson Davies
Preamplifiers	½″ and 0.25 in.	\$600–\$850	GRAS Sound & Vibration ACO Pacific B&K The Modal Shop Larson Davies
Sound level meters	Simple type 1 SLM	\$800–\$2000	Rion CEL B&K
Sound level meters	Advanced SLM with freq. analysis and data storage	\$2000–\$10,000	Rion CEL Larson Davies B&K
Frequency analyzers	Real-time frequency analyzers/FFT	\$5000–\$50,000	Hewlett Packard Norsonic Data Physics 01dB
Near-field acoustical holography	Complete system with 16–64 channel acquisition and postprocessing	\$100,000–\$200,000	LMS B&K

TABLE 8.5 Companies That Makes Acoustical Measurement Instruments

G.R.A.S. Sound & Vibration Skelstedet 10B 2950 Vedbaek Denmark Tel: +45 45 66 40 46	Larson Davies Inc. 1681 West 820 North Provo, UT 84601 Tel: (801) 375 0177
LMS International Interleuvenlaan 68 B-3001 Leuven Belgium Tel: +32 16 384 571	Brüel & Kjær Spectris Technologies Inc. 2364 Park Central Blvd. Decatur, GA 30035-3987 Tel: (800) 332 2040
Hewlett-Packard Co. P.O. Box 95052-8059 Santa Clara, CA 95052 Tel: (206) 335 2000	Rion Scantek, Inc. 916 Gist Avenue Silver Springs, MD 20910 Tel: (301) 495 7738
Norsonic AS P.O. Box 24 N-3420 Lierskogen Norway Tel: +47 32 85 20 80	ACO Pacific, Inc. 2604 Read Avenue Belmont, CA 94002 Tel: (415) 595 8588
The Modal Shop Inc. 1776 Mentor Avenue, Suite 170 Cincinnati, OH 45212-3521 Tel: (513) 351 9919	CEL Instruments 1 Westchester Drive Milford, NH 03055 Tel: (800) 366 2966
01dB 111 rue du 1er Mars F69100 Villeurbanne France Tel: +33 4 78 53 96 96	

References

1. E. Skudrzyk, *The Foundation of Acoustics,* New York: Springer-Verlag, 1971.
2. International Electrotechnical Commission, *Publication 651: Sound Level Meters,* Genève, Switzerland, IEC, 1971.
3. International Organization for Standardization, *Standard 3745 "Acoustics—Determination of sound power levels of noise sources—Precision method for anechoic and semi-anechoic rooms,* Genève, Switzerland, ISO, 1981.

9

Flow Measurement

Richard Thorn
University of Derby

Adrian Melling
Universitaet Erlangen-Nuember

Herbert Köchner
Universitaet Erlangen-Nuember

Reinhard Haak
Universitaet Erlangen-Nuember

Zaki D. Husain
Daniel Flow Products, Inc.

Donald J. Wass
Daniel Flow Products, Inc.

David Wadlow
Sensors Research Consulting, Inc.

Harold M. Miller
Data Industrial Corporation

Halit Eren
Curtin University of Technology

Hans-Peter Vaterlaus
Rittmeyer Ltd.

Thomas Hossle
Rittmeyer Ltd.

Paolo Giordano
Rittmeyer Ltd.

Christophe Bruttin
Rittmeyer Ltd.

Wade M. Mattar
The Foxboro Company

James H. Vignos
The Foxboro Company

Nam-Trung Nguyen
University of California at Berkeley

Jesse Yoder
Automation Research Corporation

Rekha Philip-Chandy
Liverpool John Moores University

Roger Morgan
Liverpool John Moores University

Patricia J. Scully
Liverpool John Moores University

0-8493-0047-9/00/$0.00+$.50

9.1 Differential Pressure Flowmeters

Richard Thorn

Flow measurement is an everyday event. Whether you are filling up a car with petrol (gasoline) or wanting to know how much water the garden sprinkler is consuming, a flowmeter is required. Similarly, it is also difficult to think of a sector of industry in which a flowmeter of one type or another does not play a part. The world market in flowmeters was estimated to be worth $2500 million in 1995, and is expected to grow steadily for the foreseeable future. The value of product being measured by these meters is also very large. For example, in the U.K. alone, it was estimated that in 1994 the value of crude oil produced was worth $15 billion.

Given the size of the flowmeter market, and the value of product being measured, it is somewhat surprising that both the accuracy and capability of many flowmeters are poor in comparison to those instruments used for measurement of other common process variables such as pressure and temperature. For example, the orifice plate flowmeter, which was first used commercially in the early 1900s and has a typical accuracy of ±2% of reading, is still the only flowmeter approved by most countries for the fiscal measurement of natural gas. Although newer techniques such as Coriolis flowmeters have become increasingly popular in recent years, the flow measurement industry is by nature conservative and still dominated by traditional measurement techniques. For a review of recent flowmeter developments, refer to [1].

Over 40% of all liquid, gas, and steam measurements made in industry are still accomplished using common types of *differential pressure flowmeter*; that is, the orifice plate, Venturi tube, and nozzle. The operation of these flowmeters is based on the observation made by Bernoulli that if an annular restriction is placed in a pipeline, then the velocity of the fluid through the restriction is increased. The increase in velocity at the restriction causes the static pressure to decrease at this section, and a pressure difference is created across the element. The difference between the pressure upstream and pressure downstream of this obstruction is related to the rate of fluid flowing through the restriction and therefore through the pipe. A differential pressure flowmeter consists of two basic elements: an obstruction to cause a pressure drop in the flow (a *differential producer*) and a method of measuring the pressure drop across this obstruction (a *differential pressure transducer*).

One of the major advantages of the orifice plate, Venturi tube, or nozzle is that the measurement uncertainty can be predicted without the need for calibration, if it is manufactured and installed in accordance with one of the international standards covering these devices. In addition, this type of differential pressure flowmeter is simple, has no moving parts, and is therefore reliable. The main disadvantages of these devices are their limited range (typically 3:1), the permanent pressure drop they produce in the pipeline (which can result in higher pumping costs), and their sensitivity to installation effects (which can be minimized using straight lengths of pipe before and after the flowmeter). The combined advantages of this type of flowmeter are still quite hard to beat, and although it has limitations, these have been well investigated and can be compensated for in most circumstances. Unless very high accuracy is required, or unless the application makes a nonintrusive device essential, the differential flowmeter should be considered. Despite the predictions of its demise, there is little doubt that the differential pressure flowmeter will remain a common method of flow measurement for many years to come.

Important Principles of Fluid Flow in Pipes

There are a number of important principles relating to the flow of fluid in a pipe that should be understood before a differential pressure flowmeter can be used with confidence. These are the difference between laminar and turbulent flow, the meaning of Reynolds number, and the importance of the flow's velocity profile.

Fluid motion in a pipe can be characterized as one of three types: laminar, transitional, or turbulent. In *laminar flow*, the fluid travels as parallel layers (known as streamlines) that do not mix as they move in the direction of the flow. If the flow is turbulent, the fluid does not travel in parallel layers, but moves in a haphazard manner with only the average motion of the fluid being parallel to the axis of the pipe. If the flow is *transitional*, then both types may be present at different points along the pipeline or the flow may switch between the two.

In 1883, Osborne Reynolds performed a classic set of experiments at the University of Manchester that showed that the flow characteristic can be predicted using a dimensionless number, now known as the Reynolds number. The Reynolds number Re is the ratio of the inertia forces in the flow ($\rho\bar{v}D$) to the viscous forces in the flow (η) and can be calculated using:

$$\mathrm{Re} = \frac{\rho\bar{v}D}{\eta} \tag{9.1}$$

where ρ = Density of the fluid
$\bar{v}$ = Mean velocity of the fluid
D = Pipe diameter
η = Dynamic viscosity of the fluid

If Re is less than 2000, viscous forces in the flow dominate and the flow will be laminar. If Re is greater than 4000, inertia forces in the flow dominate and the flow will be turbulent. If Re is between 2000 and 4000, the flow is transitional and either mode can be present. The Reynolds number is calculated using mainly properties of the fluid and does not take into account factors such as pipe roughness, bends, and valves that also affect the flow characteristic. Nevertheless, the Reynolds number is a good guide to the type of flow that can be expected in most situations.

The fluid velocity across a pipe cross-section is not constant, and depends on the type of flow present. In laminar flow, the velocity profile is parabolic since the large viscous forces present cause the fluid to move more slowly near the pipe walls. Under these conditions, the velocity at the center of the pipe is twice the average velocity across the pipe cross-section. The laminar flow profile is unaffected by the roughness of the pipe wall. In turbulent flow, inertia forces dominate, pipe wall effects are less, and the flow's velocity profile is flatter, with the velocity at the center being about 1.2 times the mean velocity. The exact flow profile in a turbulent flow depends on pipe wall roughness and Reynolds number. Figure 9.1 shows the "fully developed" flow profiles for laminar and turbulent flow. These are the flow profiles that would be obtained at the end of a very long pipe, thus ensuring that any changes to the flow profile due to pipe bends and fittings are no longer present. To have confidence in the performance of a differential pressure flowmeter, both the characteristic and velocity profile of the flow passing through the flowmeter should be stable and known.

Bernoulli's Equation

The Bernoulli equation defines the relationship between fluid velocity (v), fluid pressure (p), and height (h) above some fixed point for a fluid flowing through a pipe of varying cross-section, and is the starting point for understanding the principle of the differential pressure flowmeter. For the inclined, tapered pipe shown in Figure 9.2, Bernoulli's equation states that:

$$\frac{p_1}{\rho g} + \frac{v_1^2}{2g} + h_1 = \frac{p_2}{\rho g} + \frac{v_2^2}{2g} + h_2 \tag{9.2}$$

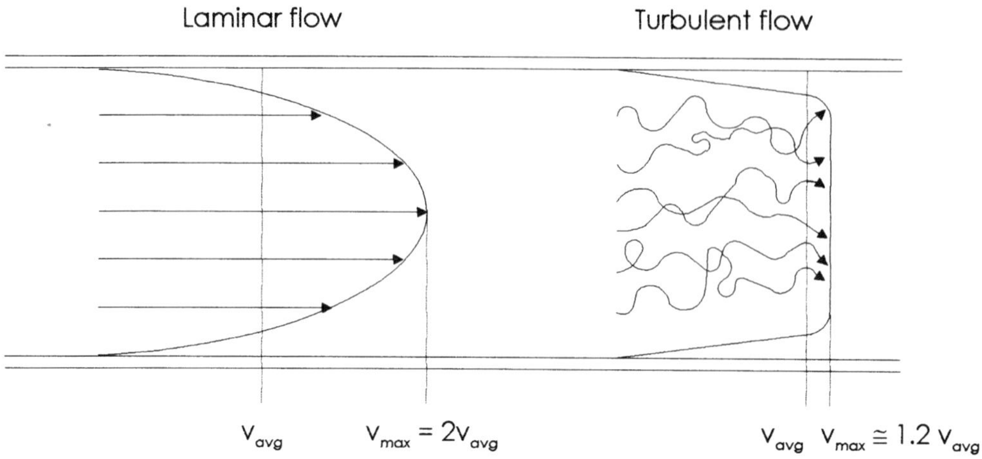

FIGURE 9.1 Velocity profiles in laminar and turbulent flow.

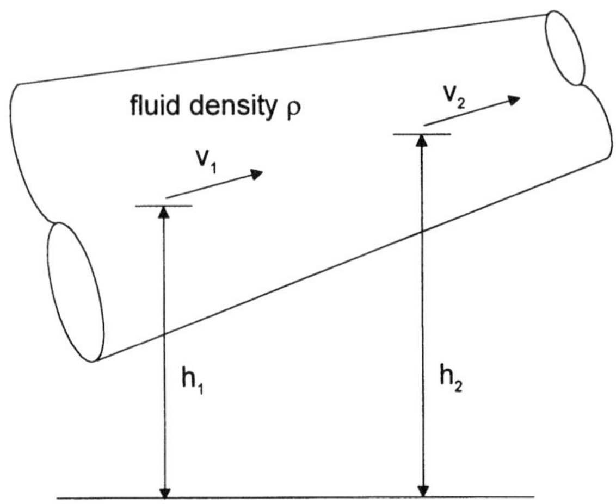

FIGURE 9.2 Flow through an inclined, tapered pipe.

Thus, the sum of the pressure head ($p/\rho g$), the velocity head ($v/2g$), and potential head (h) is constant along a flow streamline. The term "head" is commonly used because each of these terms has the unit of meters. Equation 9.2 assumes that the fluid is frictionless (zero viscosity) and of constant density (incompressible). Further details on the derivation and significance of Bernoulli's equation can be found in most undergraduate fluid dynamics textbooks (e.g., [2]).

Bernoulli's equation can be used to show how a restriction in a pipe can be used to measure flow rate. Consider the pipe section shown in Figure 9.3. Since the pipe is horizontal, $h_1 = h_2$, and Equation 9.2 reduces to:

$$\frac{p_1 - p_2}{\rho} = \frac{v_1^2 - v_2^2}{2} \tag{9.3}$$

The conservation of mass principle requires that:

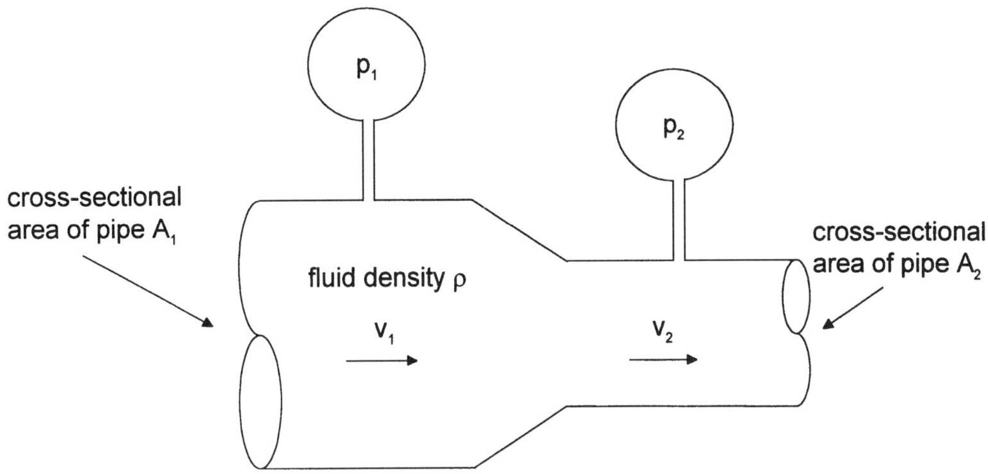

FIGURE 9.3 Using a restriction in a pipe to measure fluid flow rate.

$$v_1 A_1 \rho = v_2 A_2 \rho \tag{9.4}$$

Rearranging Equation 9.4 and substituting for v_2 in Equation 9.3 gives:

$$Q = v_1 A_1 = \frac{A_2}{\sqrt{1 - \left(\frac{A_2}{A_1}\right)^2}} \sqrt{\frac{2(p_1 - p_2)}{\rho}} \tag{9.5}$$

This shows that the volumetric flow rate of fluid Q can be determined by measuring the drop in pressure ($p_1 - p_2$) across the restriction in the pipeline — the basic principle of all differential pressure flowmeters. Equation 9.5 has limitations, the main ones being that it is assumed that the fluid is incompressible (a reasonable assumption for most liquids), and that the fluid has no viscosity (resulting in a flat velocity profile). These assumptions need to be compensated for when Equation 9.5 is used for practical flow measurement.

Common Differential Pressure Flowmeters

The Orifice Plate

The orifice plate is the simplest and cheapest type of differential pressure flowmeter. It is simply a plate with a hole of specified size and position cut in it, which can then clamped between flanges in a pipeline (Figure 9.4). The increase that occurs in the velocity of a fluid as it passes through the hole in the plate results in a pressure drop being developed across the plate. After passing through this restriction, the fluid flow jet continues to contract until a minimum diameter known as the vena contracta is reached. If Equation 9.5 is used to calculate volumetric flow rate from a measurement of the pressure drop across the orifice plate, then an error would result. This is because A_2 should strictly be the area of the vena contracta, which of course is unknown. In addition, turbulence between the vena contracta and the pipe wall results in an energy loss that is not accounted for in this equation.

To overcome the problems caused by the practical application of Equation 9.5, two empirically determined correction factors are added. After some reorganization Equation 9.5 can be written as:

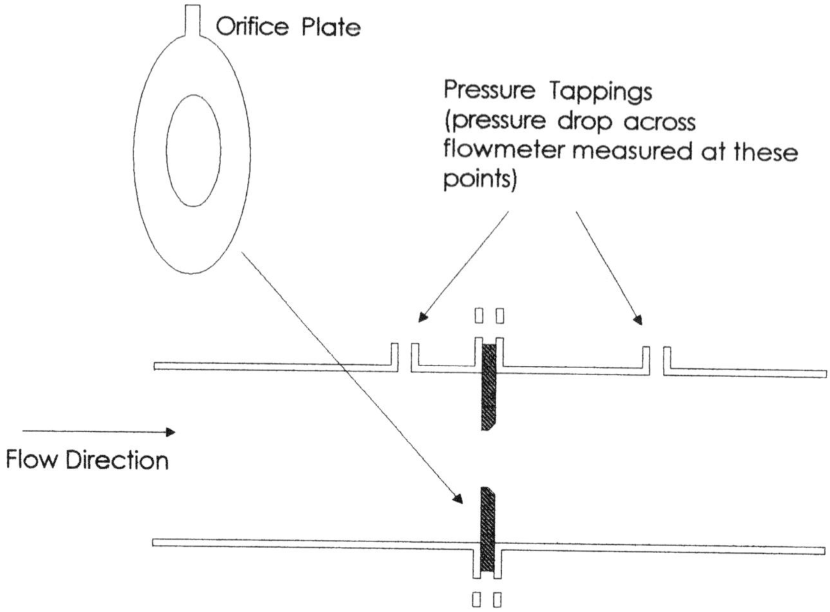

FIGURE 9.4 A square-edged orifice plate flowmeter.

$$Q = \frac{C}{\sqrt{1-\beta^4}} \varepsilon \frac{\pi}{4} d^2 \sqrt{\frac{2(p_1 - p_2)}{\rho}} \tag{9.6}$$

where ρ = Density of the fluid upstream of the orifice plate
d = Diameter of the hole in the orifice plate
β = Diameter ratio d/D, where D is the upstream internal pipe diameter

The two empirically determined correction factors are C the discharge coefficient, and ε the expansibility factor. C is affected by changes in the diameter ratio, Reynolds number, pipe roughness, the sharpness of the leading edge of the orifice, and the points at which the differential pressure across the plate are measured. However, for a fixed geometry, it has been shown that C is only dependent on Reynolds number and so this coefficient can be determined for a particular application. ε is used to account for the compressibility of the fluid being monitored. Both C and ε can be determined from equations and tables in a number of internationally recognized documents known as standards. These standards not only specify C and ε, but also the geometry and installation conditions for the square-edged orifice plate, Venturi tube, and nozzle, and are essentially a design guide for the use of the most commonly used types of differential pressure flowmeter. Installation recommendations are intended to ensure that fully developed turbulent flow conditions exist within the measurement section of the flowmeter. The most commonly used standard in Europe is ISO 5167-1 [3], while in the U.S., API 2530 is the most popular [4]. There are differences between some of the recommendations in these two standards (e.g., the minimum recommended length of straight pipe upstream of the flowmeter), but work is underway to resolve these.

Equation 9.6 illustrates perhaps the greatest strength of the orifice plate, which is that measurement performance can be confidently predicted without the need for calibration if the device is manufactured, installed, and operated in accordance with one of the international standards. In addition, the device is cheap to manufacture, has no moving parts, is reliable, and can be used for metering most clean gases, liquids, and steam.

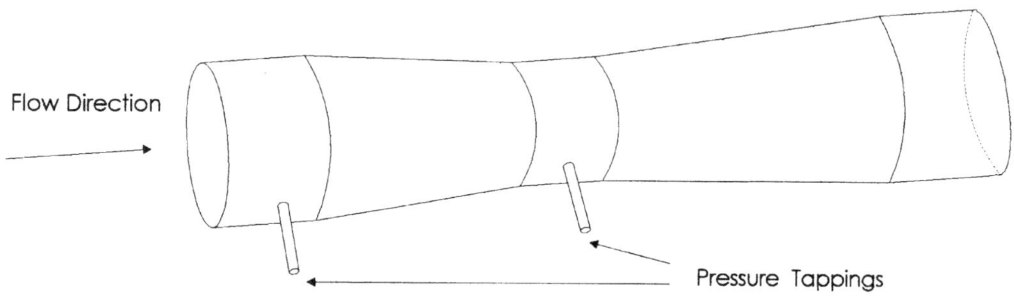

FIGURE 9.5 A Venturi tube flowmeter.

The major disadvantages of the orifice plate are its limited range and sensitivity to flow disturbances. The fact that fluid flow rate is proportional to the square root of the measured differential pressure limits the range of a one plate/one differential pressure transmitter combination to about 3:1. The required diameter ratio (also known as beta ratio) of the plate depends on the maximum flow rate to be measured and the range of the differential pressure transducer available. Sizing of the orifice plate is covered in most of the books in the further reading list, and nowadays computer programs are also available to help perform this task. The flow measurement range can be increased by switching; ways in which this may be achieved are described in [5]. Equation 9.6 assumes a fully developed and stable flow profile, and so installation of the device is critical, particularly the need for sufficient straight pipework upstream of the meter. Wear of the leading edge of the orifice plate can severely alter measurement accuracy; thus; this device is normally only used with clean fluids.

Only one type of orifice plate, the square-edged concentric, is covered by the standards. However, other types exist, having been designed for specific applications. One example is the eccentric orifice plate, which is suited for use with dirty fluids. Details of these other types of orifice plate can be found in [6].

The Venturi Tube

The classical or Herschel Venturi tube is the oldest type of differential pressure flowmeter, having first been used in 1887. As Figure 9.5 shows, a restriction is introduced into the flow in a more gradual way than for the orifice plate. The resulting flow through a Venturi tube is closer to that predicted in theory by Equation 9.5 and so the discharge coefficient C is much nearer unity, being typically 0.95. In addition, the permanent pressure loss caused by the Venturi tube is lower, but the differential pressure is also lower than for an orifice plate of the same diameter ratio. The smooth design of the Venturi tube means that it is less sensitive to erosion than the orifice plate, and thus more suitable for use with dirty gases or liquids. The Venturi tube is also less sensitive to upstream disturbances, and therefore needs shorter lengths of straight pipework upstream of the meter than the equivalent orifice plate or nozzle. Like the orifice plate and nozzle, the design, installation, and use of the Venturi tube is covered by a number of international standards.

The major disadvantages of the Venturi tube flowmeter are its size and cost. It is more difficult, and therefore more expensive to manufacture than the orifice plate. Since a Venturi tube can be typically 6 diameters long, it can become cumbersome to use with larger pipe sizes, with associated maintenance of upstream and downstream pipe lengths also becoming a problem.

The Nozzle

The nozzle (Figure 9.6) combines some of the best features of the orifice plate and Venturi tube. It is compact and yet, because of its curved inlet, has a discharge coefficient close to unity. There are a number of designs of nozzle, but one of the most commonly used in Europe is the ISA-1932 nozzle, while in the U.S., the ASME long radius nozzle is more popular. Both of these nozzles are covered by international standards.

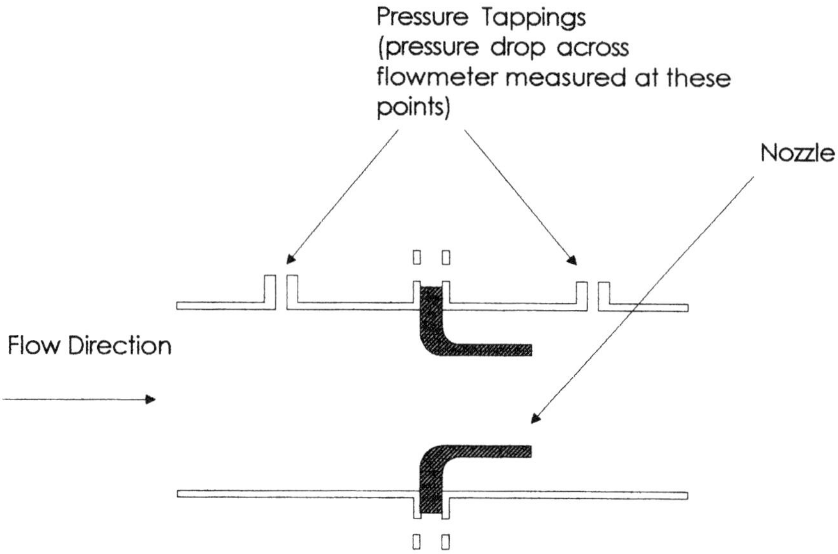

FIGURE 9.6 A nozzle flowmeter.

The smooth inlet of the nozzle means that it is more expensive to manufacture than the orifice plate as the curvature of the inlet changes with diameter ratio, although it is cheaper than the Venturi tube. The device has no sharp edges to erode and cause changes in calibration, and thus is well suited for use with dirty and abrasive fluids. The nozzle is also commonly used for high-velocity, high-temperature applications such as steam metering.

A variation of the nozzle is the sonic (or critical flow Venturi) nozzle, which has been used both as a calibration standard for testing gas meters and a transfer standard in interlaboratory comparisons [7].

Other Differential Pressure Flowmeters

There are many other types of differential pressure flowmeter, including the segmental wedge, V-cone, elbow, and Dall tube. Each of these has advantages over the orifice plate, Venturi tube, and nozzle for specific applications. For example, the segmental wedge can be used with flows having a low Reynolds number, and a Dall tube has a lower permanent pressure loss than a Venturi tube. However, none of these instruments are yet covered by international standards and, thus, calibration is needed to determine their accuracy. Further information on these, and other less-common types of differential pressure flowmeter, can be found in [8].

Performance and Applications

Table 9.1 shows the performance characteristics and main application areas of the square-edged orifice plate, Venturi tube, and nozzle flowmeters. Compared to other types of flowmeters on the market, these differential pressure flowmeters only have moderate accuracy, typically ±2% of reading; but of course, this can be improved if the device is calibrated after installation. Although in some circumstances these flowmeters can be used with dirty gases or liquids, usually only small amounts of a second component can be tolerated before large measurement errors occur. When calculating the cost and performance of a differential flowmeter, both the primary element and the differential pressure transducer should be taken into account. Although the orifice plate is the cheapest of the primary elements, the cost of the fitting needed to mount it in the pipeline, particularly if on-line removal is required, can be significant.

Choosing which flowmeter is best for a particular application can be very difficult. The main factors that influence this choice are the required performance, the properties of the fluid to be metered, the

TABLE 9.1 The Performance and Application Areas of Common Differential Pressure Flowmeters

	Performance					Applications				
	Typical uncalibrated accuracy	Typical range	Typical pipe diameter (mm)	Permanent pressure loss	Comparative cost	Clean gas	Dirty gas	Clean liquid	Slurry	Steam
Orifice plate	±2%	3:1	10–1000	High	Low	Yes	No	Yes	No	Yes
Venturi tube	±2%	3:1	25–500	Low	High	Yes	Maybe	Yes	Maybe	Maybe
Nozzle	±2%	3:1	25–250	High	Medium	Yes	Maybe	Yes	No	Yes

TABLE 9.2 A Selection of Companies That Supply Differential Pressure Flowmeters

ABB Kent-Taylor Oldens Lane, Stonehouse Gloucestershire, GL10 3TA England Tel: + 44 1453 826661 Fax: + 44 1453 826358	ISA Controls Ltd. Hackworth Industrial Park Shildon County Durham DL4 1LH England Tel: + 44 1388 773065 Fax: + 44 1388 774888
Daniel Industries Inc. 9720 Katy Road P.O. Box 19097 Houston, TX 77224 Tel: (713) 467-6000 Fax: (713) 827-3880	Perry Equipment Corporation Wolters Industrial Park P.O. Box 640 Mineral Wells, TX 76067 Tel: (817) 325-2575 Fax: (817) 325-4622
Hartmann & Braun (UK) Ltd. Bush Beach Engineering Division Stanley Green Trading Estate, Cheadle Hulme Cheshire SK8 6RN England Tel: + 44 161 4858151 Fax: + 44 161 4884048	

installation requirements, the environment in which the instrument is to be used, and, of course, cost. There are two standards that can be used to help select a flowmeter: BS 1042: Section 1.4, which is a guide to the use of the standard differential pressure flowmeters [9]; and BS 7405, which is concerned with the wider principles of flowmeter selection [10].

Because all three flowmeters have similar accuracy, one strategy for selecting the most appropriate instrument is to decide if there are any good reasons for not using the cheapest flowmeter that can be used over the widest range of pipe sizes: the orifice plate. Where permanent pressure loss is important, the Venturi tube should be considered, although the high cost of this meter can only usually be justified where large quantities of fluid are being metered. For high-temperature or high-velocity applications, the nozzle should be considered because under these conditions, it is more predictable than the orifice plate. For metering dirty fluids, either the Venturi tube or the nozzle should be considered in preference to the orifice plate, the choice between the two depending on cost and pressure loss requirements. Table 9.2 lists some suppliers of differential pressure flowmeters.

Installation

Correct installation is essential for successful use of a differential pressure flowmeter because the predicted uncertainty in the flow rate/differential pressure relationship in Equation 9.6 assumes a steady flow, with

TABLE 9.3 The Minimum Straight Lengths of Pipe Required between Various Fittings and an Orifice Plate or Venturi Tube (as recommended in ISO 5167-1) to Ensure That a Fully Developed Flow Profile Exists in the Measurement Section. All Lengths Are Multiples of the Pipe Diameter

	Upstream of the flowmeter				Downstream of the flowmeter
Diameter Ratio β	Single 90° bend	Two 90° bends in the same plane	Two 90° bends in different planes	Globe valve fully open	For any of the fittings shown to the left
0.2	10	14	34	18	4
0.4	14	18	36	20	6
0.6	18	26	48	26	7
0.8	46	50	80	44	8

a fully developed turbulent velocity profile, is passing through the flowmeter. Standards contain detailed recommendations for the minimum straight lengths of pipe required before and after the flowmeter, in order to ensure a fully developed flow profile. Straight lengths of pipe are required after the flowmeter because disturbances caused by a valve or bend can travel upstream and thus also affect the installed flowmeter. Table 9.3 gives examples of installation requirements taken from ISO 5167-1. If it is not possible to fit the recommended lengths of straight pipe before and after the flowmeter, then the flowmeter must be calibrated once it has been installed.

The other major problem one faces during installation is the presence of a rotating flow or swirl. This condition distorts the flow velocity profile in a very unpredictable way, and is obviously not desirable. Situations that create swirl, such as two 90° bends in different planes, should preferably be avoided. However, if this is not possible, then swirl can be removed by placing a flow conditioner (also known as a flow straightener) between the source of the swirl and the flowmeter. There are a wide range of flow conditioner designs, some of which can be used to both remove swirl and correct a distorted velocity profile [11]. Because they obstruct the flow, all flow conditioners produce an unrecoverable pressure loss, which in general increases with their capability (and complexity).

Differential Pressure Measurement

Apart from the differential producer, the other main element of a differential pressure flowmeter is the *transducer* needed to measure the pressure drop across the producer. The correct selection and installation of the differential pressure transducer plays an important part in determining the accuracy of the flow rate measurement.

The main factors that should be considered when choosing a differential pressure transducer for a flow measurement application are the differential pressure range to be covered, the accuracy required, the maximum pipeline pressure, and the type and temperature range of the fluid being metered.

Most modern differential pressure transducers consist of a pressure capsule in which either capacitance, strain gage, or resonant wire techniques are used to detect the movement of a diaphragm. Using these techniques, a typical accuracy of ±0.1% of full scale is possible. The transducer is usually part of a unit known as a transmitter, which converts differential pressure, static pressure, and ambient temperature measurements into a standardized analog or digital output signal. "Smart" transmitters use a local, dedicated microprocessor to condition signals from the individual sensors and compute volumetric or mass flow rate. These devices can be remotely configured, and a wide range of diagnostic and maintenance functions are possible using their built-in "intelligence."

As far as installation is concerned, the transmitter should be located as close to the differential producer as possible. This helps ensure a fast dynamic response and reduces problems caused by vibration of the connecting tubes. The position of the pressure tappings is also important. If liquid flow in a horizontal pipe is being measured, then the pressure tappings should be located at the side of the pipe so that they cannot be blocked with dirt or filled with air bubbles. For horizontal gas flows, if the gas is clean, the pressure tappings should be vertical; if steam or dirty gas is being metered, then the tappings should be located at the side of the pipe. These general guidelines show that

TABLE 9.4 Standards Related to Differential Pressure Flow Measurement

	American National Standards Institute, New York
ANSI/ASHRAE 41.8	Standard methods of measurement of flow of liquids in pipes using orifice flowmeters.
ANSI/ASME MFC-7M	Measurement of gas flow by means of critical flow Venturi nozzles.
ANSI/ASME MFC-14M	Measurement of fluid flow using small bore precision orifice meters.
	American Petroleum Institute, Washington, D.C.
API 2530	Manual of Petroleum Measurement Standards, Chapter 14 — Natural gas fluids measurement, Section 3 — Orifice metering of natural gas and other related hydrocarbon fluids.
	American Society of Mechanical Engineers, New York
ASME MFC-3M	Measurement of fluid flow in pipes using orifice, nozzle and Venturi.
ASME MFC-8M	Fluid flow in closed conduits — Connections for pressure signal transmissions between primary and secondary devices.
	British Standards Institution, London
BS 1042	Measurement of fluid flow in closed conduits.
	International Organization for Standardization, Geneva
ISO 2186	Fluid flow in closed conduits — connections for pressure transmissions between primary and secondary elements.
ISO TR 3313	Measurement of pulsating fluid flow in a pipe by means of orifice plates, nozzles or Venturi tubes.
ISO 5167-1	Measurement of fluid flow by means of pressure differential devices.
ISO 9300	Measurement of gas flow by means of critical flow Venturi nozzles.

considerable care must be taken with the installation of the differential pressure transmitter if large measurement errors are to be avoided. For further details on the installation of differential pressure transmittters, see ISO 2186 [12].

Standards

International standards that specify the design, installation, and use of the orifice plate, Venturi tube, and nozzle, and allow their accuracy to be calculated without the need for calibration, are probably the main reason for the continuing use of this type of flowmeter. Table 9.4 gives details of the most common standards related to differential pressure flowmeters. There are still inconsistencies in the various standards. For example, ISO 5167-1 states that any flow conditioner should be preceded by at least 22 diameters of straight pipe and followed by at least 20 diameters of straight pipe. This would seem to contradict one application of a flow conditioner, which is to reduce the length of straight pipe required upstream of a flowmeter.

Despite the occasional inconsistancy and difference between the standards, they are the internationally accepted rules for the installation and use of the square-edged orifice plate, Venturi tube, and nozzle.

Future Developments

In spite of the vast amount of published data available on differential pressure flowmeters, continued research is needed to improve the understanding of the effect of flow conditions, and flowmeter geometry, on the uncalibrated accuracy of these devices. For example, work has been recently undertaken to derive an improved equation for the discharge coefficient of an orifice plate [13].

The metering of multiphase flow is an area of increasing importance. In addition to developing new measurement techniques, many people are investigating ways in which traditional flowmeters can be used to meter multiphase flows. A good review of the use of differential pressure flowmeters for multiphase flow measurement can be found in [14].

The development of "smart" differential pressure transmitters has overcome the limitations of differential pressure flowmeters in some applications. For example, these devices are being used to linearize and extend the range of differential pressure flowmeters.

The above developments should help to ensure the continued popularity of the differential pressure flowmeter for the forseeable future, despite increasing competition from newer types of instrument.

Defining Terms

Differential pressure flowmeter: A flowmeter in which the pressure drop across an annular restriction placed in the pipeline is used to measure fluid flow rate. The most common types use an orifice plate, Venturi tube, or nozzle as the primary device.

Orifice plate: Primary device consisting of a thin plate in which a circular aperture has been cut.

Venturi tube: Primary device consisting of a converging inlet, cylindrical mid-section, and diverging outlet.

Nozzle: Primary device consisting of a convergent inlet connected to a cylindrical section.

Differential pressure transmitter: Secondary device that measures the differential pressure across the primary device and converts it into an electrical signal.

References

1. R. A. Furness, Flowmetering: evolution or revolution, *Measurement and Control*, 27 (8), 15-18, 1994.
2. B. S. Massey, *Mechanics of Fluids*, 6th ed., London: Chapman and Hall, 1989.
3. International Organization for Standardization, ISO 5167-1, Measurement of Fluid Flow by Means of Pressure Differential Devices — Part 1 Orifice plates, nozzles and Venturi tubes inserted in circular cross-section conduits running full, Geneva, Switzerland, 1991.
4. American Petroleum Institute, API 2530, Manual of Petroleum Measurement Standards Chapter 14 — Natural Gas Fluids Measurement, Section 3 — Orifice Metering of Natural Gas and Other Related Hydrocarbon Fluids, Washington, 1985.
5. E. L. Upp, *Fluid Flow Measurement*, Houston: Gulf Publishing, 1993.
6. H. S. Bean, *Fluid Meters Their Theory and Application*, 6th ed., New York: American Society of Mechanical Engineers, 1983.
7. P. H. Wright, The application of sonic (critical flow) nozzles in the gas industry, *Flow Meas. Instrum.*, 4 (2), 67-71, 1993.
8. D. W. Spitzer, *Industrial Flow Measurement*, 2nd ed., Research Triangle Park, NC: ISA, 1990.
9. British Standards Institution, BS 1042, Measurement of Fluid Flow in Closed Conduits — Part 1 Pressure differential devices — Section 1.4 Guide to the use of devices specified in Sections 1.1 and 1.2, London, 1992.
10. British Standards Institution, BS7405, Guide to the Selection and Application of Flowmeters for Measurement of Fluid Flow in Closed Conduits, London, 1991.
11. E. M. Laws and A. K. Ouazzane, Compact installations for differential pressure flow measurement, *Flow Meas. Instrum.*, 5 (2), 79-85, 1994.
12. International Organization for Standardization, ISO 2186, Fluid Flow in Closed Conduits — Connections for Pressure Signal Transmissions Between Primary and Secondary Elements, Geneva, Switzerland, 1973.
13. M. J. Reader-Harris, J. A. Slattery, and E. P. Spearman, The orifice plate discharge coefficient equation — further work, *Flow. Meas. Instrum.*, 6 (2), 101-114, 1995.
14. F. C. Kinghorn, Two-phase flow measurement using differential pressure meters, *Multi- Phase Flow Measurement Short Course*, London, 17th-18th June 1985.

Further Information

R. C. Baker, *An Introductory Guide to Flow Measurement*, London: Mechanical Engineering Publications, 1989, a good pocket sized guide on the choice and use of flowmeters commonly used in industry.

A. T. J. Hayward, *Flowmeters — A Basic Guide and Source-Book for Users*, London: Macmillan, 1979, an overview of the important areas of flow measurement which is a joy to read.

R. W. Miller, *Flow Measurement Engineering Handbook*, 3rd ed., New York: McGraw-Hill, 1996, a thorough reference book particularly on differential pressure flowmeters, covers European and U.S. standards with calculations using both U.S. and SI units.

D. W. Spitzer, *Flow Measurement: Practical Guides for Measurement and Control*, Research Triangle Park, NC: ISA, 1991, intended for practicing engineers, this book covers most aspects of industrial flow measurement.

E. L. Upp, *Fluid Flow Measurement*, Houston: Gulf Publishing, 1993, contains a lot of practical advice on the use of differential pressure flowmeters.

Flow Measurement and Instrumentation, UK: Butterworth-Heinemann, a quarterly journal covering all aspects of flowmeters and their applications. A good source of information on current research activity.

9.2 Variable Area Flowmeters

Adrian Melling, Herbert Köchner, and Reinhard Haak

The term *variable area flowmeters* refers to those meters in which the minimum cross-sectional area available to the flow through the meter varies with the flow rate. Meters of this type that are discussed in this section include the rotameter and the movable vane meter used in pipe flows, and the weir or flume used in open-channel flows. The measure of the flow rate is a geometrical quantity such as the height of a bob in the rotameter, the angle of the vane, or the change in height of the free surface of the liquid flowing over the weir or through the flume.

Most of the discussion here is devoted to the rotameter, firstly because the number of installed rotameters and movable vane meters is large relative to the number of weirs and flumes, and secondly because the movable vane is often used simply as a flow indicator rather than as a meter.

The following section includes basic information describing the main constructional features and applications of each type of meter. In the third section, the principles of measurement and design of rotameters and open-channel meters are described in some detail; the movable vane meter is not considered further because most design aspects are similar to those of rotameters. Then, the contribution of modern computational and experimental methods of fluid mechanics to flowmeter design is discussed, using the results of a detailed investigation of the internal flow.

Details of manufacturers of rotameters and movable vane meters, together with approximate costs of these meters, are also tabulated in Tables 9.5 through 9.7.

General Description of Variable Area Flowmeters

Rotameter

The *rotameter* is a robust and simple flowmeter for gases and liquids, and holds a large share of the market for pipe diameters smaller than about 100 mm. In its basic form, the rotameter consists of a conical transparent vertical glass tube containing a "bob" (Figure 9.7), which rises in the tube with increasing flow rate until a balance is reached between gravitational, buoyancy, and drag forces on the bob. Within the range of a particular flowmeter (depending on the bob shape and density, the tube shape and the fluid density and viscosity), the flow rate is linearly proportional to the height of the bob in the tube and is determined simply by reading the level of the upper edge of the bob. The rotameter is also

TABLE 9.5 Approximate Rotameter Prices

Type	Size	Price (plastic)	Price (stainless steel, borosilicate glass)
Glass	< 1/2 in. (12.7 mm)	$50	$200
Plastic	1/2 in. (12.7 mm)	$50	$200
	1 in. (25.4 mm)	$70	$150
	2 in. (50.8 mm)	$200	$500
Metal	1/2 in. (12.7 mm)		$500
	1 in. (25.4 mm)		$600
	2 in. (50.8 mm)		$700
	3 in. (76.2 mm)		$1100
	4 in. (101.6 mm)		$1400
Additional costs			
Electric limit switch	$100		
Current transducer	$500		
Pneumatic	$1200		

TABLE 9.6 Approximate Movable Vane Meter Prices

Type	Price
Plastic versions (small scale)	From $10
Direct coupled pointer versions incl. electric flow switch: Metal, screw connections 1/2 in. (12.7 mm) to 2 in. (50.8 mm)	$400 (brass) to $600 (stainless steel)
Flanged versions	Add $200 (1/2 in., 12.7 mm) to $400 (2 in., 50.8 mm)
Magnetic coupled indicator, stainless steel	$1200 (1/2 in., 12.7 mm) to $2100 (8 in., 203 mm)

known as “floating element flowmeter,” although the buoyancy force on the “floating element” is not sufficient to make it float.

The bob is commonly formed of a combination of cylindrical and conical sections, but a spherical bob is often used in small diameter tubes (see Figures 9.8). More complicated bob geometries can reduce the sensitivity to viscosity of the fluid. Frequently, the bob has several shallow inclined grooves around the upper rim that induce a slow rotation (frequency about 1 Hz) of the bob, which helps to maintain a stable position of the bob. In larger rotameters where the additional friction is acceptable, the bob can be allowed to slide up and down a rod on the tube axis to prevent any sideways motion.

Various tube geometries are in use. The basic requirement for an increase in the cross-sectional area up the height of the tube leads to the conical tube. An alternative form uses three ribs arranged circumferentially around the tube to guide the bob; the flow area between the ribs increases along the tube height. For spherical bobs, a triangular tube in which the bob remains in contact with three surfaces over the whole height of the tube prevents lateral movement of the bob with minimum friction. Commonly, the tube is made of glass to facilitate the reading of the flow rate from a scale engraved on the tube. For general-purpose applications, the scale can be marked in millimeters; the flow rate is then determined by a conversion factor depending on the tube dimensions, the mass of the bob, the pressure and temperature, and the properties of the fluid. For specific application to a single fluid under controlled conditions, it is more convenient to have the flow rate directly marked on the tube.

For laboratory use, medical equipment, and other applications with small flow rates, the glass tube is almost universal. Laboratory meters are often equipped with a flow valve that allows direct flow setting and reading. For many small rotameters, integrated flow controllers are offered; these hold the pressure drop and the flow setting constant, even with changing upstream or downstream pressure.

For most rotameters, electric limit switches or analog electric signal transducers are available. For glass and plastic meters, inductive or optical switches are used that can be positioned to the desired level of

TABLE 9.7 Manufacturers of Rotameters and Movable Vane Meters

Tokyo Keiso Co. Ltd. Shiba Toho Building 1-7-24, Shibakoen Minato-ku Tokyo 105, Japan Tel: +81-3-3431-1625 Fax: +81-3-3433-4922	Porter Instruments Co. 245 Township Line Road Hatfield, PA 19440 Tel: (215) 723-4000 Fax: (215) 723-2199
Brooks Instrument Division Emerson Electric 407 W. Vine St. Hatfield, PA 19440 Tel: (215) 362-3500 Fax: (215) 362-3745	Wallace & Tiernan, Inc. 25 Main St. St. Belleville, NJ 07109 Tel: (201) 759-8000 Fax: (201) 759-9333
Krohne America Inc. 7 Dearborn Rd. Peabody, MA 01960 Tel: (508) 535-6060, (800) 356-9464 Fax: (508) 535-1720	KDG-Mobrey Ltd. Crompton Way Crawley West Sussex RH10 2YZ, Great Britain Tel: +44-1293-518632 Fax: +44-1293-533095
Bailey Fischer & Porter Co. County Line Rd. Warminster, PA 18974 Tel: (215) 674-6000 Fax: (215) 674-7183	CT Platon Jays Close Viables Basingstoke, Hampshire RG22 4BS, Great Britain Tel: +44-1256-470456 Fax: +44-1256-363345
Bopp & Reuther Heinrichs Messtechnik GmbH Stolberger Str. 393 D-50933 Köln, Germany Tel: +49-221-49708-0 Fax: +49-221-497088	Kobold Messring GmbH Nordring 22-24 D-65719 Hofheim/Taunus, Germany Tel: +49 6192 2990 Fax: +49 6192 23398
Rota Yokogawa GmbH & Co. KG Rheinstrasse 8 D-79664 Wehr, Germany Tel: +49-7761-567-0 Fax: +49-7761-567-126	

the bob. While some units just detect the bob within their active area, other switches have bistable operation, wherein the switch is toggled by the float passing the switch.

For industrial applications, the metering of corrosive fluids or fluids at high temperature or pressure leads to the use of stainless steel tubes or a wide range of other materials chosen to suit the application requirements (e.g., temperature, pressure, corrosion). Most metal rotameters use a permanent magnetic coupling between the float and the pointer of the indicator, enabling a direct analog flow reading without electric supply. Electric switches can, however, be added to the indicator, and electric and pneumatic transmitters for the flow reading are offered by most producers. Some units are equipped with options such as flow totalizers, mass flow calculation units for gas applications in conjunction with temperature and pressure sensors, or energy calculation units. Due to the frequent application of the rotameter in the chemical and petrochemical industry, electric options are often designed for use in hazardous areas.

The rotameter is characterized by:

- Simple and robust construction
- High reliability
- Low pressure drop

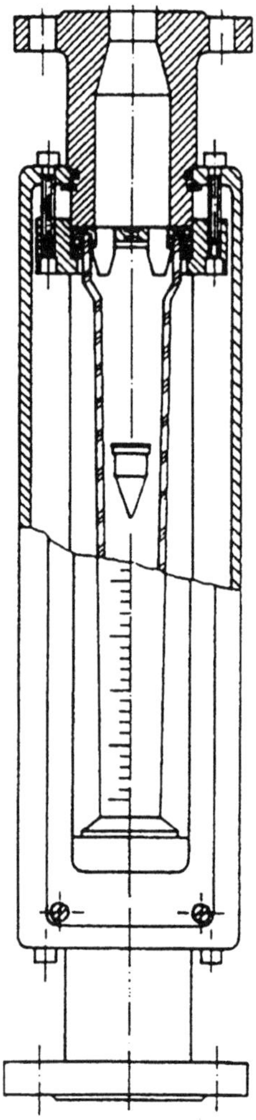

FIGURE 9.7 Cross-section of a rotameter. The level of the bob rises linearly with increasing flow rate.

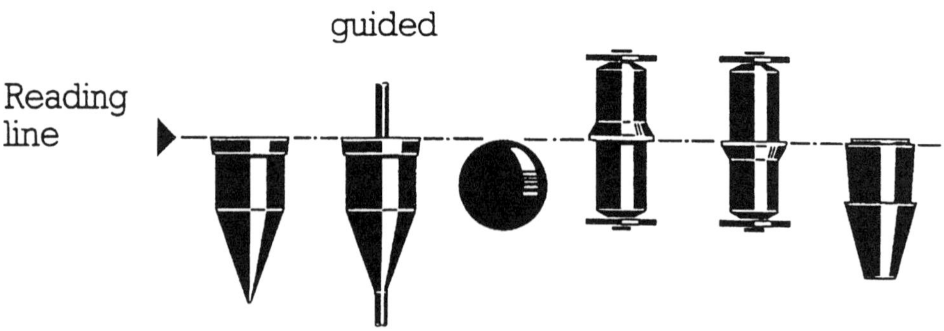

FIGURE 9.8 Typical rotameter bob geometries.

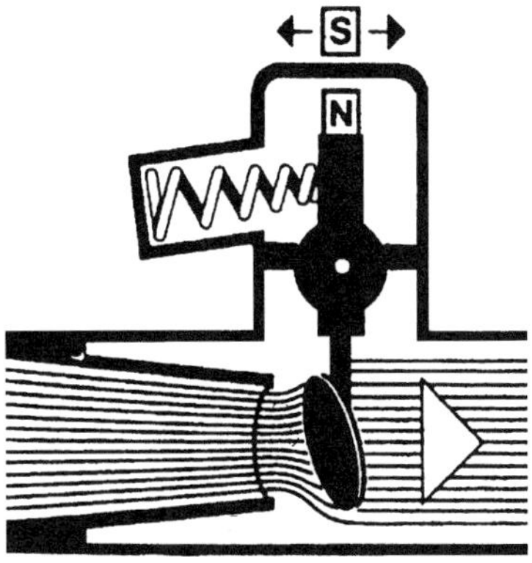

FIGURE 9.9 Movable vane meter. The magnet (N,S) transmits the vane position to an indicator.

- Applicable to a wide variety of gases and liquids
- Flow range typically 0.04 L h^{-1} to 150 m^3 h^{-1} for water
- Flow range typically 0.5 L h^{-1} to 3000 m^3 h^{-1} for air
- 10:1 flow range for given bob-tube combination
- Uncertainty 0.4% to 4% of maximum flow
- Insensitivity to nonuniformity in the inflow (no upstream straight piping needed)
- Typical maximum temperature 400°C
- Typical maximum pressure 4 MPa (40 bar)
- Low investment cost
- Low installation cost

Movable Vane Meter

The movable vane meter is a robust device suitable for the measurement of high flow rates where only moderate requirements on the measurement accuracy are made. Dirty fluids can also be metered. It contains a flap that at zero flow is held closed by a weight or a spring (Figure 9.9). A flow forces the vane open until the dynamic force of the flow is in balance with the restoring force of the weight or the spring. The angle of the vane is thus a measure of the flow rate, which can be directly indicated by a pointer attached to the shaft of the vane on a calibrated scale. The resistance provided by the vane depends on the vane position and hence on the flow rate or Reynolds number; a recalibration is therefore necessary when the fluid is changed. An important application is the metering of the air flow in automotive engines with fuel injection.

In low-cost flow indicators, a glass or plastic window allows a direct view of the flap (Figure 9.10). Sometimes, optical or reed switches in combination with a permanent magnet attached to the flap are used as electric flow switches. All-metal units use a magnetic coupling between the vane and the pointer of the indicator, thus avoiding most of the friction and material selection problems. Many applications use the movable vane meter as a flow switch with appropriate mechanical, magnetic, inductive, or optical switches. The setting of the switch is done once for all during the installation. Since a continuous flow indication is not needed, flow reading uncertainties are not of concern.

Most aspects of design and application of rotameters can be applied to movable vane meters, although the latter are characterized by a higher uncertainty of the flow reading. General features of both types of variable area flowmeter are summarized, for example, in reference [1]. Since the basic construction of the two meter types is very similar, the unit costs also lie in the same range.

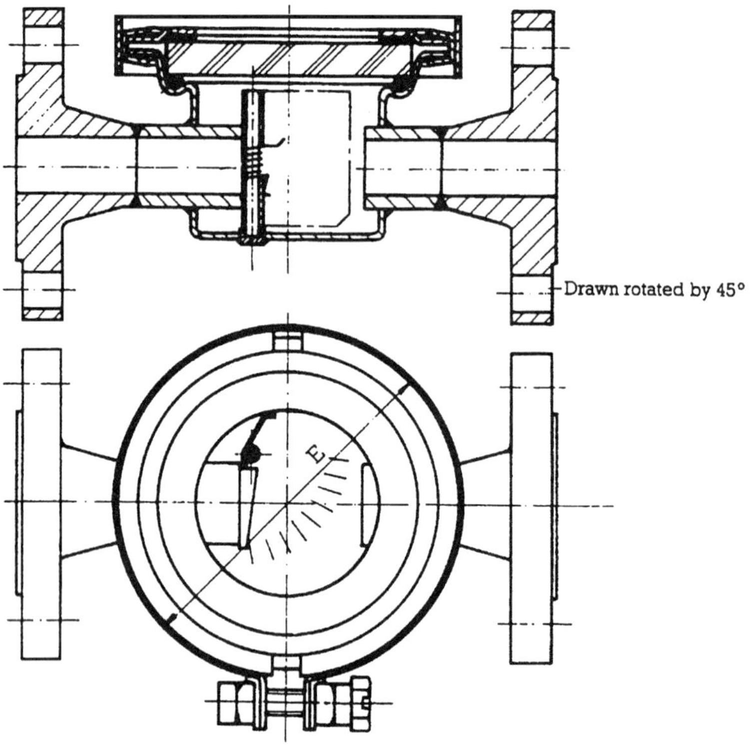

FIGURE 9.10 Flow indicator. The angle of the vane provides a measure of the flow rate.

Weir, Flume

Of the methods available for the metering of a liquid (generally water) in open-channel flow, the weir and the flume fall within the scope of this discussion on variable area flowmeters. In each case, the flow metering depends on measurement of the difference in height h of the water surface over an obstruction across the channel and the surface sufficiently far upstream. There is a wide variety of geometries in use (see reference [2]), but most of these can be described as variants of three basic types. In the sharp crested weir, also known as the thin plate weir (Figure 9.11), the sill or crest is only about 1 mm to 2 mm thick. The sheet of liquid flowing over the weir, called the nappe or vein, separates from the weir body after passing over the crest. An air-filled zone at atmospheric pressure is formed underneath the outflowing jet, and the streamlines above the weir are strongly curved. If the width of the weir is less than that of the upstream channel (Figures 9.12 and 9.13), the term "notch" is frequently used. A broad crested weir (Figure 9.14) has a sill which is long enough for straight, parallel streamlines to form above the crest. In this review, the term "weir" is applied generally to both the weir and the notch.

The resistance to flow introduced by the weir causes the water level upstream to rise. If it is assumed that the flow velocity some distance upstream of the weir is zero, then a simple measure of the upstream water level suffices to determine the discharge over a weir of known geometry. As in the case of an orifice in pipe flow, there will be a contraction of the nappe and a frictional resistance at the sides as water flows over a weir. The actual discharge is less than the theoretical discharge according to an empirically determined coefficient of discharge C_d. In the analysis of rectangular notches and weirs, it is frequently assumed that the channel width remains constant so that a contraction of the nappe occurs only in the vertical direction as the water accelerates over the obstacle. When the weir is narrower than the upstream channel, there is an additional horizontal contraction. The effect of a significant upstream velocity is accounted for empirically.

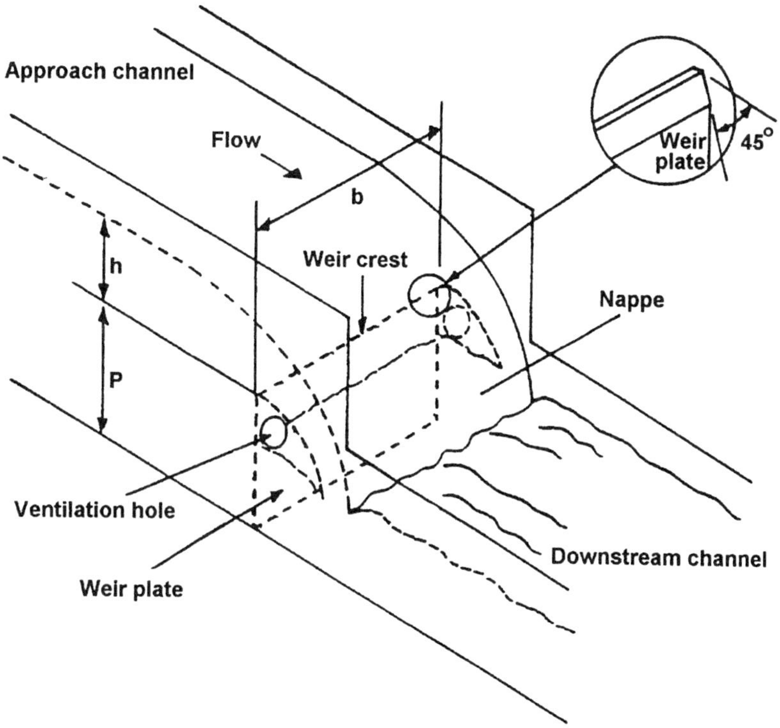

FIGURE 9.11 Full-width thin plate weir. Flow rate $\dot{Q}$ varies with water level h above the weir crest ($\dot{Q} \sim h^{3/2}$) and with a flow-dependent discharge coefficient C_d.

As an alternative to the weir, the flume (Figure 9.15) provides a method for flow metering with relatively low pressure loss. By restricting the channel width, analogous to the Venturi tube used in pipe flow, the flow velocity in the narrow portion of the channel is increased and the water level sinks accordingly. Most of the head of water is recovered in the diffusing section of the weir. The water levels upstream and in the throat of the weir can be determined by simple floats and recorded on a chart by pens driven mechanically from the floats. For remote monitoring, the use of echo sounders is advantageous.

The weir is preferentially used in natural river beds and the flume in canalized water courses. The flume must be used in streams with sediment transport to avoid the accumulation of deposits that would occur at the approach to a weir. Weirs and flumes are characterized by

- Simple measurement of the water level
- Simple maintenance
- Reliable measurement of large flow rates at low stream velocity
- Limited measurement accuracy (at best about 2%)
- High installation costs, particularly for flumes

Recommendations for installation of weirs and flumes, for the location of the head measuring station upstream, and for determining the discharge from the measured head are given, for example, in [2] and [3] as well as in appropriate standards (e.g., [4–6]).

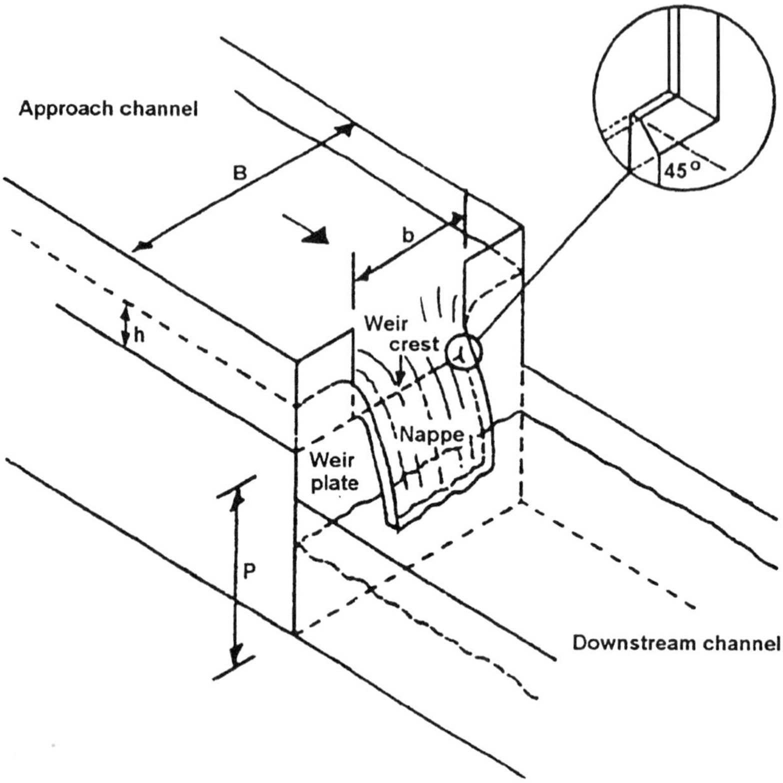

FIGURE 9.12 Thin plate rectangular notch weir: discharge coefficient C_d is flow dependent.

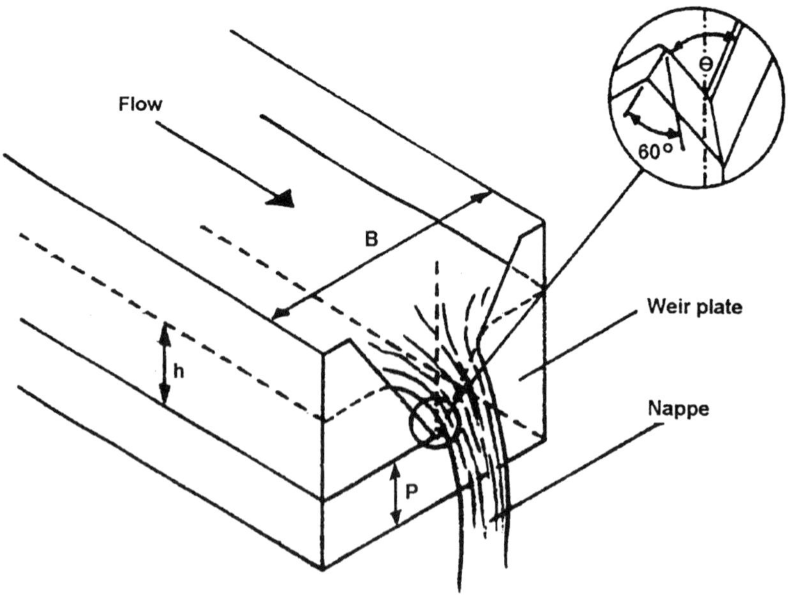

FIGURE 9.13 Thin plate V-notch weir: discharge coefficient C_d is almost independent of the flow.

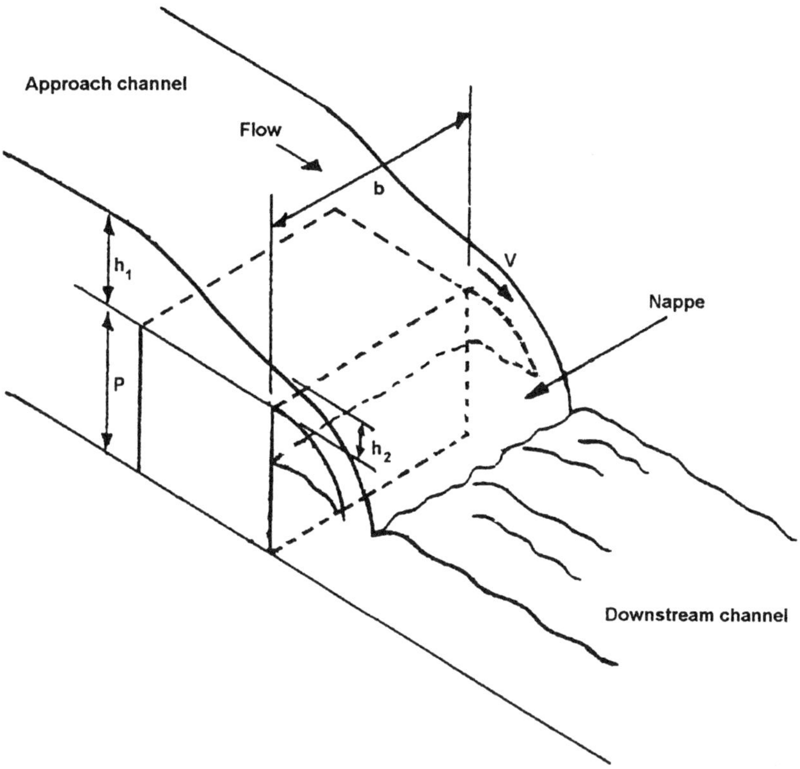

FIGURE 9.14 Broad-crested weir: maximum discharge when $h_2 = 2h_1/3$.

Measuring Principles of Variable Area Flowmeters

Rotameter

Flow Rate Analysis.
The forces acting on the bob lead to equilibrium between the weight of the bob $\rho_b g V_b$ acting downwards and the buoyancy force $\rho g V_b$ and the drag force F_d acting upwards, where V_b is the volume and ρ_b is the density of the bob, ρ is the density of the fluid, and g is the gravitational acceleration:

$$\rho_b g V_b = \rho g V_b + F_d \tag{9.7}$$

The drag force results from the flow field surrounding the bob and particularly from the wake of the bob. In flow analyses based on similarity principles, these influences are accounted for by empirical coefficient C_L or C_T in the drag law for:

$$\text{Laminar flow} \quad F_d = C_L \mu D_b U \tag{9.8}$$

$$\text{Turbulent flow} \quad F_d = C_T \rho D_b^2 U^2 \tag{9.9}$$

where μ = Fluid viscosity
D_b = Maximum bob diameter
U = Velocity in the annular gap around the bob at the minimum cross-section

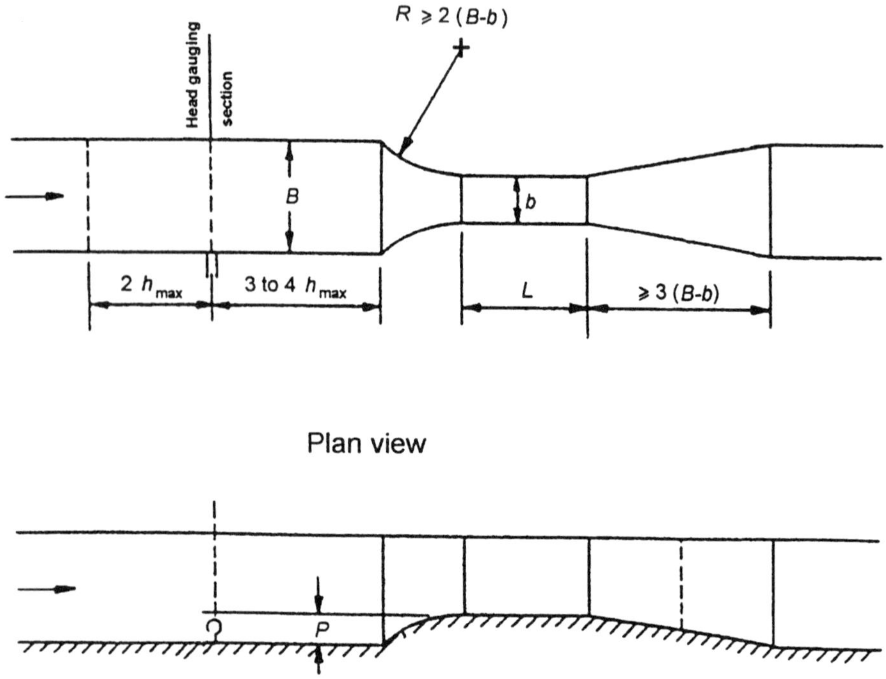

FIGURE 9.15 Venturi flume: maximum discharge when throat depth = 2/3 × total head.

The volume flow rate through the rotameter is:

$$\dot{Q} = \frac{\pi}{4}\left(D^2 - D_b^2\right)U \tag{9.10}$$

or

$$\dot{Q} = m\frac{\pi}{4}D_b^2 U \tag{9.11}$$

where m is the open area ratio, defined as:

$$m = \frac{D^2 - D_b^2}{D_b^2} \tag{9.12}$$

and D is the tube diameter at the height of the bob.

Combining Equations 9.7, 9.8, and 9.10 gives for laminar flow:

$$\dot{Q}_L = \alpha D_b^4 \frac{\left(\rho_b - \rho\right)g}{\mu} \tag{9.13}$$

where the parameter α is defined in terms of a constant $K = V_b/D_b^3$ characteristic of the shape of the bob:

$$\alpha = \frac{\pi m K}{4C_L} \tag{9.14}$$

Using Equation 9.9 instead of Equation 9.8 yields for turbulent flow:

$$\dot{Q}_T = \beta D_b^{5/2} \sqrt{\frac{(\rho_b - \rho)g}{\rho}} \tag{9.15}$$

where

$$\beta = \frac{\pi m}{4} \sqrt{\frac{K}{C_T}} \tag{9.16}$$

With either laminar or turbulent flow through the rotameter, it is clear from Equations 9.13 and 9.15 that the flow rate is proportional to m. If the cross-sectional area of the tube is made to increase linearly with length, i.e.,

$$D = D_b(1 + h \tan \phi) \tag{9.17}$$

then since the cone angle ϕ of the tube is small, Equation 9.12 can be written as:

$$m = 2h \tan\phi \tag{9.18}$$

and the flow rate is directly proportional to the height h of the bob.

Similarity Analysis.

In early studies of floating element flowmeters, Ruppel and Umpfenbach [7] proposed the introduction of characteristic dimensionless quantities, to permit the use of experimentally determined flow coefficients in flowmeter analysis. Lutz [8] extended these ideas by showing that the transfer of flow coefficients from one flowmeter to another is possible if geometrical similarity exists. More recent works [9–12] have used these principles to produce graphical or computer-based design schemes and have proposed general guidelines for laying out practical flow metering systems.

The basic scaling parameter for flow is the Reynolds number, defined as:

$$Re = \frac{\rho U_{IN} D_b}{\mu} \tag{9.19}$$

where U_{IN} is the velocity at the rotameter inlet, and the tube diameter D is represented by its value at the inlet, equal to the bob diameter D_b. Through the Reynolds number regimes of laminar or turbulent flow, and particularly important for the rotameter flow regimes with strong or weak viscosity dependence, can be distinguished. Originating in the work [7], it has been found to be practical for rotameters to use an alternative characteristic number, the Ruppel number, defined as:

$$Ru = \frac{\mu}{\sqrt{m_b g \rho (1 - \rho/\rho_b)}} \tag{9.20}$$

where $m_b = \rho_b D_b^3$ is the mass of the bob. By combining Equations 9.15, 9.16, and 9.20, the mass flow $\dot{m}$ through the rotameter can be written as:

$$\dot{m} = \frac{\pi}{4} \frac{mD_b\mu}{\sqrt{C_T\,Ru}} \tag{9.21}$$

Alternatively, from the definition:

$$\dot{m} = \rho\dot{Q} = \rho U_{IN} \frac{\pi}{4} D_b^2 \tag{9.22}$$

and Equation 9.19, the flow rate is:

$$\dot{m} = \frac{\pi}{4} D_b \mu Re \tag{9.23}$$

Equations 9.21 and 9.23 give the following relationship between the Ruppel number and the Reynolds number:

$$Ru = \frac{m}{\sqrt{C_T\,Re}} \tag{9.24}$$

An analysis for laminar flow leads to the relationship:

$$Ru = \sqrt{\frac{m}{C_L\,Re}} \tag{9.25}$$

The advantage of the Ruppel number is its independence of the flow rate. Since the Ruppel number contains only fluid properties and the mass and the density of the bob, it is a constant for a particular instrument.

At low Ruppel numbers the linear resistance law assumed in Equation 9.8 applies and α is a constant for given m, as shown in Figure 9.16. At higher Ruppel numbers, the flow is transitional or turbulent, and log α decreases linearly with log Ru. In Figure 9.17, curves for β against Ruppel number show a linear increase of log β with log Ru in the laminar region, followed by a gradual transition to horizontal curves in fully turbulent flow.

Similarity analysis of rotameters allows the easy calculation of the flow reading with changing density and viscosity of the fluid. For lower viscosities, only the density effect has to be taken into account; hence, for most gas measurements, simple conversion factors can be used. With higher viscosity, manufacturers offer either Ruppel number-related conversion factors or two-dimensional tables of conversion factors to be applied to different heights of the bob. Some manufacturers offer recalibration services, allowing the user to order new scales for instruments to be used in changed applications and environments. For large-order users, simple computer programs are available from some manufacturers for scale calculations.

Theories based on similarity considerations can predict the variation of the flow coefficient with the Ruppel number in the laminar and turbulent flow regimes but not in laminar-turbulent transitional flow. Detailed experimental and computational studies can assist the flow analysis of floating element flowmeters. An example is given in the section "Rotameter Internal Flow Analysis."

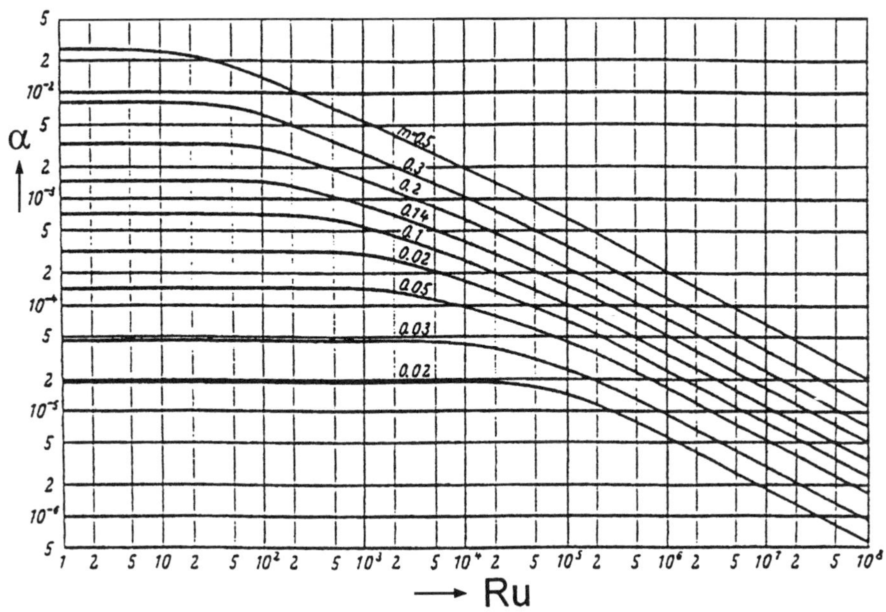

FIGURE 9.16 Rotameter flow coefficient α as a function of Ruppel number *Ru* and open area ratio *m*. For a given geometry, α is constant for low *Ru* (laminar flow).

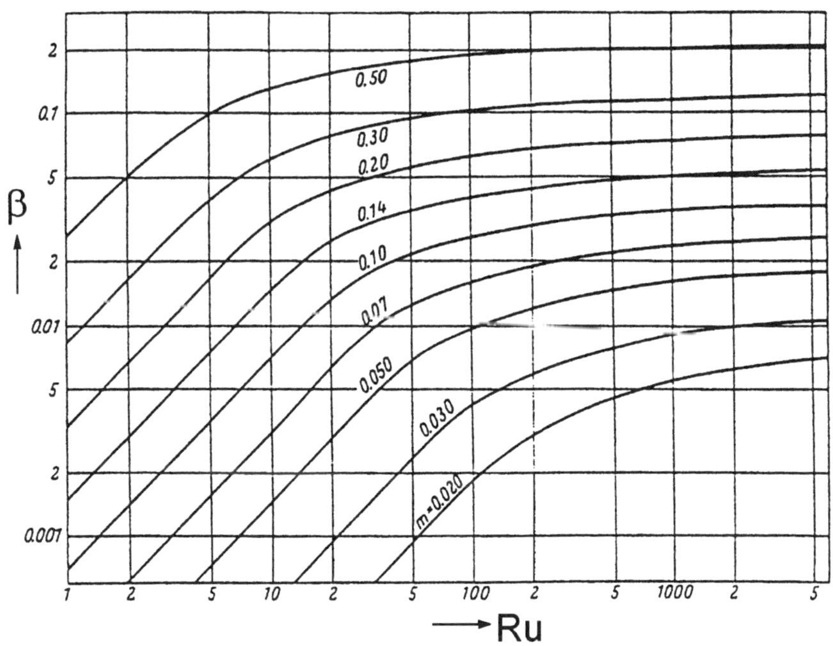

FIGURE 9.17 Rotameter flow coefficient β as a function of Ruppel number *Ru* and open area ratio *m*. For a given geometry, β is constant for high *Ru* (turbulent flow).

Weir, Flume

Flow Rate Analysis for Weirs.

Although the determination of the discharge across a weir or a flume generally requires the knowledge of one or more empirical coefficients, the basic equations describing the head-discharge characteristics of the various geometrical forms are readily derived from simple fluid mechanics considerations.

The discharge over a sharp crested weir of width b (Figure 9.11) when the water level over the sill is h is analyzed by considering a horizontal strip of water of thickness δy at a depth y below the water surface. Since the velocity of the water through this strip is $\sqrt{2gy}$, the discharge is:

$$\delta\dot{Q} = b\delta y\sqrt{2gy} \tag{9.26}$$

The total discharge is obtained by integration, introducing a coefficient of discharge C_d:

$$\dot{Q} = C_d\sqrt{2g}b\int_0^h \sqrt{y}dy \tag{9.27}$$

$$\dot{Q} = \frac{2}{3}C_d\sqrt{2g}bh^{3\ 2} \tag{9.28}$$

Empirical corrections to Equation 9.28 to account for the weir height p are given, for example, in [3], but provided h/p does not exceed 0.5, it is adequate to use Equation 9.28 with a discharge coefficient of 0.63 to achieve a flow rate tolerance of 3%.

For the rectangular notch, Equation 9.28 also applies with a discharge coefficient of 0.59 and a flow rate tolerance of 3%. Since in practice the discharge coefficient varies slightly with the head of water for both weirs and notches, it is sometimes preferred to replace Equation 9.28 by $\dot{Q} = Kbh^n$, where K and n are determined by calibration.

The triangular notch (Figure 9.13) offers the advantage that the length of the wetted edge varies with the head of water, in contrast with the rectangular notch where it is constant. Consequently, the discharge coefficient of the triangular notch is constant for all heads. Furthermore, the achievable range (ratio of maximum to minimum flow rates) is higher for the triangular geometry. If h is the height of the water surface and θ is the angle of the notch, then a horizontal strip of the notch of thickness δy at depth y has a width $2(h - y)\tan(\theta/2)$. The discharge through this strip is

$$\delta\dot{Q} = 2(h-y)\tan\frac{\theta}{2}\delta y\sqrt{2gy}C_d \tag{9.29}$$

giving a total discharge through the notch:

$$\dot{Q} = 2C_d\sqrt{2g}\tan\frac{\theta}{2}\int_0^h (h-y)\sqrt{y}dy \tag{9.30}$$

$$\dot{Q} = \frac{8}{15}C_d\sqrt{2g}\tan\frac{\theta}{2}h^{5\ 2} \tag{9.31}$$

The discharge coefficient is about 0.58.

Commonly used notch angles are 90° ($\tan\theta/2 = 1$, $C_d = 0.578$), 53.13° ($\tan\theta/2 = 0.5$, $C_d = 0.577$), and 28.07° ($\tan\theta/2 = 0.25$, $C_d = 0.587$). Notches with $\theta = 53.13°$ and $\theta = 28.07°$ deliver, respectively, one half and one quarter of the discharge of the 90° notch at the same head.

The above derivations assume that the water is discharging from a reservoir with cross-sectional area far exceeding the flow area at the weir, so that the velocity upstream is negligible. When the weir is built into a channel of cross-sectional area A, the water will have a finite velocity of approach $v_1 = \dot{Q}/A$. As a first approximation, $\dot{Q}$ is obtained from Equation 9.28 assuming zero velocity of approach. Assuming further that v_1 is uniform over the weir, there will be an additional head $v_1^2/2g$ acting over the entire weir. Referring to Figure 9.11, the total discharge is then:

$$\dot{Q} = C_d \sqrt{2g}\, b \int_{v_1^2/2g}^{h+v_1^2/2g} \sqrt{y}\, dy \tag{9.32}$$

$$\dot{Q} = \frac{2}{3} C_d \sqrt{2g}\, b \left[\left(h + v_1^2\ 2g\right)^{3/2} - \left(v_1^2/2g\right)^{3/2} \right] \tag{9.33}$$

From Equation 9.33, a corrected value of v_1 can be determined. Further iterations converge rapidly to the final value of the discharge $\dot{Q}$. Equation 9.33 can be written as:

$$\dot{Q} = \frac{2}{3} C_v C_d \sqrt{2g}\, b h^{3\ 2} \tag{9.34}$$

where

$$C_v = \left(1 + \frac{v_1^2}{2gh}\right)^{3/2} - \left(\frac{v_1^2}{2gh}\right)^{3/2} \tag{9.35}$$

is the coefficient of velocity, which is frequently determined empirically.

Over a broad-crested weir (Figure 9.14), the discharge depends on the head h_1, the width b, and the length l of the sill. There is also a dependence on the roughness of the sill surface and the viscosity. Consequently, there is a loss of head as the water flows over the sill. It is assumed that the sill is of sufficient length to allow the velocity to be uniform at a value v throughout the depth h_2 of the water at the downstream edge of the sill. Neglecting losses

$$v = \sqrt{2g\left(h_1 - h_2\right)}$$

and the discharge is:

$$\dot{Q} = C_d b h_2 v = C_d b \sqrt{2g\left(h_1 h_2^2 - h_2^3\right)} \tag{9.36}$$

From Equation 9.36, the discharge is a maximum when $(h_1 h_2^2 - h_2^3)$ reaches a maximum, i.e., when $h_2 = 2h_1/3$. This discharge is then:

$$\dot{Q} = \left(\frac{2}{3}\right)^{3/2} C_d b \sqrt{g}\, h_1^{3/2} \tag{9.37}$$

The stable condition of the weir lies at the maximum discharge.

Flow Rate Analysis for Flumes.
The discharge through a flume depends on the water level *h*, the channel width *b*, and velocity *v* at each of the stations 1 (upstream) and 2 (at the minimum cross-section) in Figure 9.15. Applying Bernoulli's equation to the inlet and the throat, neglecting all losses, one obtains:

$$\rho g H = \rho g h_1 + \frac{1}{2}\rho v_1^2 = \rho g h_2 + \frac{1}{2}\rho v_2^2 \tag{9.38}$$

where H is the total head. The analysis is then formally the same as that for the broad-crested weir, giving the same expression for the discharge as in Equation 9.37. The flow through the flume is a maximum when the depth at the throat is two thirds of the total head. Normally, a coefficient of velocity is introduced so that the discharge equation is:

$$\dot{Q} = \left(\frac{2}{3}\right)^{3/2} C_v C_d b \sqrt{g} h_1^{3/2} \tag{9.39}$$

The combined coefficient is determined empirically. In [3], a value $C_v C_d = 1.061 \pm 0.085$ is quoted.

The validity of the empirical coefficients quoted above, or of other coefficients given in the various standards for metering of open-channel flows, can only be guaranteed with a given confidence level for meter installations satisfying certain geometrical constraints. Limits on quantities such as b, h, h/b, etc. are given in [3] and the standards.

Rotameter: Internal Flow Analysis

Computation of Internal Flow.
Improvement of rotameter design could be assisted by detailed knowledge of the internal flow field, which is characterized by steep velocity gradients and regions of separated flow. Measurements of the internal flow field are complicated by the small dimensions of the gap around the bob and the strongly curved glass tube. Bückle et al. [13] successfully used laser Doppler anemometry [14] for velocity measurements in a rotameter. The working fluid was a glycerine solution with an index of refraction (1.455) close to that of glass (1.476). Problems with refraction of the laser beams at the curved tube wall were thus avoided, but the high viscosity of glycerine restricted the experiments to laminar flow at Reynolds numbers $Re = \rho U_{IN} D_b / \mu < 400$.

The application of computational fluid dynamics to the flow in a rotameter [13] involves the finite volume solution of the conservation equations for mass and momentum. For a two-dimensional laminar flow, these equations can be written in a cylindrical polar coordinate system as:

$$\frac{\partial \rho}{\partial t} + \frac{\partial(\rho u)}{\partial z} + \frac{1}{r}\frac{\partial(\rho r v)}{\partial r} = 0 \tag{9.40}$$

$$\frac{\partial(\rho u)}{\partial t} + \frac{\partial}{\partial z}\left(\rho u u - 2\mu \frac{\partial u}{\partial z}\right) + \frac{1}{r}\frac{\partial}{\partial r}\left(\rho r u v - \mu r\left(\frac{\partial u}{\partial r} + \frac{\partial v}{\partial z}\right)\right) = -\frac{\partial p}{\partial z} \tag{9.41}$$

$$\frac{\partial(\rho v)}{\partial t} + \frac{\partial}{\partial z}\left(\rho u v - \mu r\left(\frac{\partial u}{\partial r} + \frac{\partial v}{\partial z}\right)\right) + \frac{1}{r}\frac{\partial}{\partial r}\left(\rho r v v - 2\mu r \frac{\partial v}{\partial r}\right) = -\frac{\partial p}{\partial r} \tag{9.42}$$

where ρ is the density, u, v and r, z are the velocity components and coordinate directions in the axial and radial directions, respectively, μ is the dynamic viscosity, and p is the pressure.

For the numerical solution, the governing equations for a generalized transport variable ϕ (i.e., u or v) are formally integrated over each control volume (CV) of the computational grid. The resulting flux balance equation can in turn be discretized as an algebraic equation for ϕ at the center P of each CV in terms of the values ϕ_{nb} at the four nearest neighbors of point P and known functions A:

$$A_{p\phi p} + \sum_{nb} A_{nb}\,\phi_{nb} = S_{\phi} \tag{9.43}$$

For the whole solution domain, a system of equations results that can be solved by a suitable algorithm (e.g. [15]).

For a rotameter, the symmetry of the problem allows the computational solution domain to be chosen as one half diametral plane of the rotameter. The boundary conditions are:

$$u = v = 0 \tag{9.44}$$

along the walls, and

$$\frac{\partial u}{\partial r} = v = 0 \tag{9.45}$$

along the axis of symmetry. At the input boundary, the initial profile for the u-velocity is taken from the experiment. At the outlet boundary, zero gradient is assumed for all dependent variables.

Computed Flow Field.
Representative calculations for Reynolds number 220 are shown as velocity vectors (Figure 9.18) and streamlines (Figure 9.19). Experimental and computed velocity profiles over a radius of the flowmeter tube are shown in Figure 9.20.

At z = 45 mm, the computations indicate a stagnation point at the bob tip, but the measurements show a low but finite axial velocity. The deviation from zero is attributable to slight unsteadiness in the position of the bob and to smearing of the steep radial velocity gradient along the measuring volume. This effect is also apparent at radii up to about 5 mm, where the computations show a notably steeper velocity gradient than the measurements. The measured and computed peak velocities agree well, although the measured peak is located at a larger radius.

In the strongly converging annulus between the bob and the tube wall (z = 65 mm) and in the plane z = 88 mm, the computations reproduce well the trend of the measured results, particularly near the flowmeter wall. Discrepancies in the region adjacent to the bob are attributable to asymmetry in measured profiles arising from the piping upstream of the rotameter tube.

Above the bob (z = 110 mm), there is a strong upward flow in an annular region near the tube wall and a recirculation region around the axis occupying almost half the tube cross-section. The forward velocities show a good match between computations and measurements, but there is a very marked discrepancy in the recirculation zone. At z = 135 mm, the recirculation zone is finished, but the wake of the bob is still very evident. Except for a discrepancy of about 10% on the axis, the results from the two methods agree remarkably well.

The effect of rotation of the bob on the flow field was considered by computations, including an equation for the azimuthal velocity component w.

$$\frac{\partial(\rho w)}{\partial t} + \frac{\partial}{\partial z}\left(\rho u w - \mu\frac{\partial w}{\partial z}\right) + \frac{1}{r}\frac{\partial}{\partial r}\left(\rho r v w - \mu r\frac{\partial w}{\partial r}\right) = -\rho\frac{vw}{r} - \mu\frac{w}{r^2} \tag{9.46}$$

Calculations for 1 Hz rotation frequency showed the highest swirl velocities close to the axis above the bob. There was no observable influence of the rotation on the u- and v-velocity components [16].

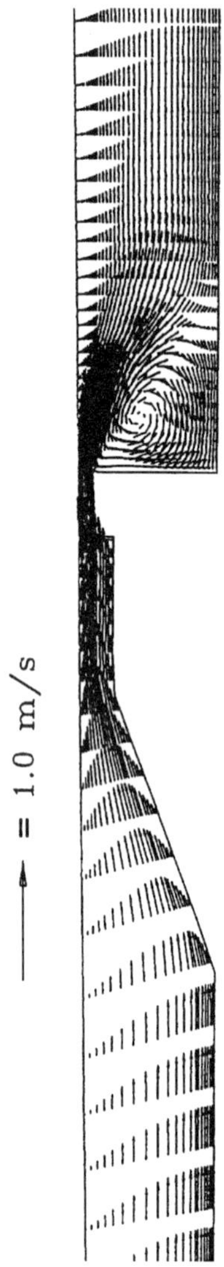

FIGURE 9.18 Computed velocity vectors for laminar flow through a rotameter. Computations in a half diametral plane for axisymmetric flow at Reynolds number 220.

Summary

For pipe flows, variable area flowmeters are most suitable for low flow rates of gases or liquids at moderate temperatures and pressures. Favorable features include rugged construction, high reliability, low pressure drop, easy installation, and low cost. Disadvantages include measurement uncertainty of 1% or more, limited range (10:1), slow response, and restrictions on the meter orientation. A generally good price/performance ratio has led to widespread use of these meters in numerous scientific and medical instruments and in many industrial applications for flow monitoring.

Variable area flowmeters in open-channel flows have applications for flow measurements in waste water plants, waterworks, rivers and streams, irrigation, and drainage canals. Hydrological applications of weirs with adjustable sill or crest height include flow regulation, flow measurement, upstream water level control, and discharge of excess flow in streams, rivers, and canals. Ruggedness, simplicity, and low maintenance costs are favorable characteristics for field applications of these meters.

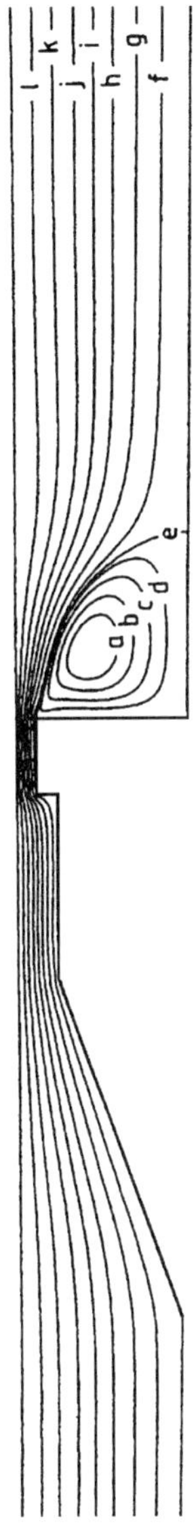

FIGURE 9.19 Computed streamlines for laminar flow through a rotameter (Reynolds number 220).

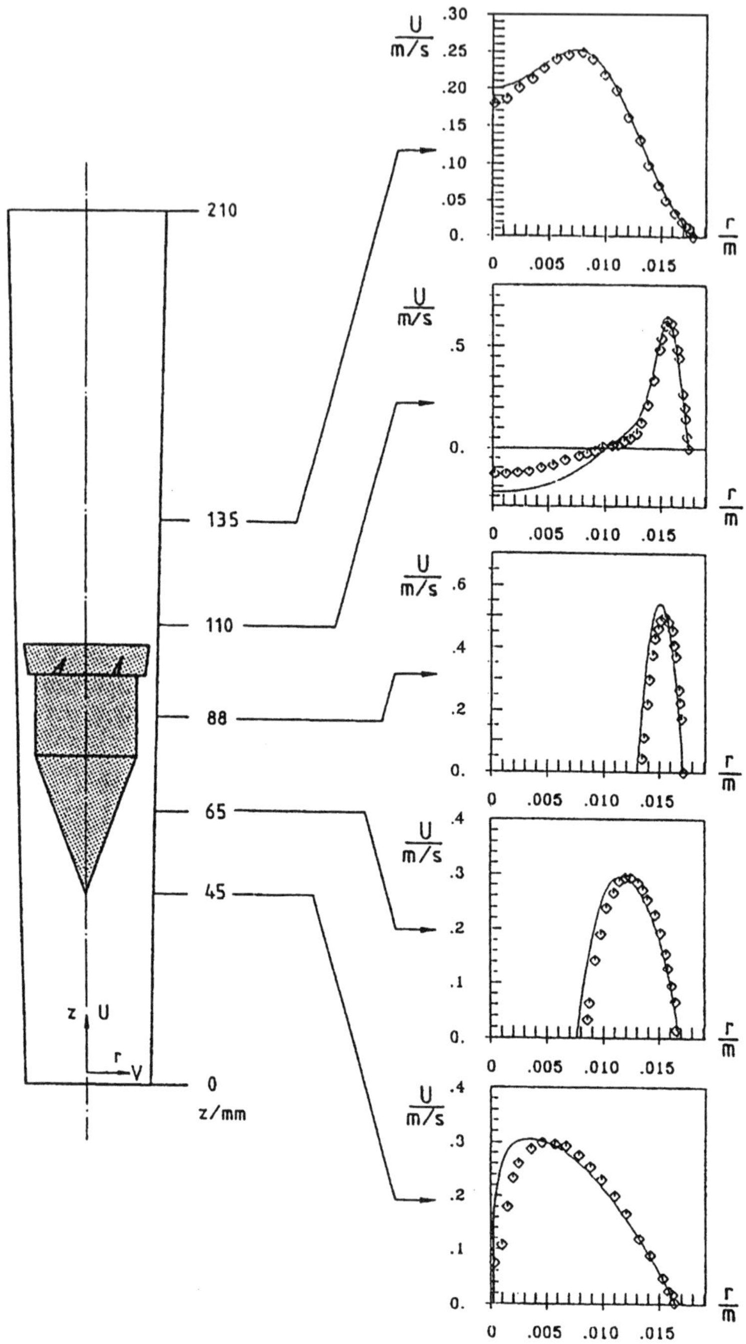

FIGURE 9.20 Comparison of measured and computed velocity profiles for laminar flow through a rotameter (Reynolds number 220). U = axial component (m/s), z = axial position (mm), r = radial position (m).

References

1. R. A. Furness, BS7045: the principles of flowmeter selection, *Flow Meas. Instrum.*, 2, 233–242, 1991.
2. W. Boiten, Flow-measuring structures, *Flow Meas. Instrum.*, 4, 17–24, 1993.
3. R. Hershey, General purpose flow measurement equations for flumes and thin plate weirs, *Flow Meas. Instrum.*, 6, 283–293, 1995.
4. ISO 1438, Thin Plate Weirs, International Standards Organisation, Geneva, 1980.
5. ISO 4359, Rectangular, Trapezoidal and U-shaped Flumes, International Standards Organisation, Geneva, 1983.
6. ISO 8368, Guidelines for the Selection of Flow Gauging Structures, International Standards Organisation, Geneva, 1985.
7. G. Ruppel and K. J. Umpfenbach, Strömungstechnische Untersuchungen an Schwimmermessern, *Technische Mechanik und Thermodynamik,* 1, 225–233, 257–267, 290–296, 1930.
8. K. Lutz, Die Berechnung des Schwebekörper-Durchflußmessers, *Regelungstechnik,* 10, 355–360, 1959.
9. D. Bender, Ähnlichkeitsparameter und Durchflußgleichungen für Schwebekörperdurchflußmesser, *ATM-Archiv für Technisches Messen,* 391, 97–102, 1968.
10. VDE/VDI-Fachgruppe Meßtechnik 3513, *Schwebekörperdurchflußmesser, Berechnungsverfahren,* VDE/VDI 3513, 1971.
11. H. Nikolaus, Berechnungsverfahren für Schwebekörperdurchflußmesser, theoretische Grundlagen und Aufbereitung für die Datenverarbeitungsanlage, *ATM-Archiv für Technisches Messen,* 435, 49–55, 1972.
12. H. Nikolaus and M. Feck, Graphische Verfahren zur Bestimmung von Durchflußkennlinien für Schwebekörperdurchflußmeßgeräte, *ATM-Archiv für Technisches Messen,* 1247-5, 171–176, 1974.
13. U. Bückle, F. Durst, B. Howe, and A. Melling, Investigation of a floating element flow meter, *Flow Meas. Instrum.*, 3, 215–225, 1992.
14. F. Durst, A. Melling, and J. H. Whitelaw, *Principles and Practice of Laser-Doppler Anemometry,* 2nd ed., London: Academic Press, 1981.
15. M. Perić, M. Schäfer, and E. Schreck, Computation of fluid flow with a parallel multigrid solver, *Proc. Conf. Parallel Computational Fluid Dynamics,* Stuttgart, 1991.
16. U. Bückle, F. Durst, H. Köchner, and A. Melling, Further investigation of a floating element flowmeter, *Flow Meas. Instrum.*, 6, 75–78, 1995.

9.3 Positive Displacement Flowmeters

Zaki D. Husain and Donald J. Wass

A *positive displacement flowmeter,* commonly called a PD meter, measures the volume flow rate of a continuous flow stream by momentarily entrapping a segment of the fluid into a chamber of known volume and releasing that fluid back into the flow stream on the discharge side of the meter. By monitoring the number of entrapments for a known period of time or number of entrapments per unit time, the total volume of flow or the flow rate of the stream can be ascertained. The total volume and the flow rate can then be displayed locally or transmitted to a remote monitoring station.

The positive displacement flowmeter has been in use for many decades to measure both liquid and gas flows. A PD meter can be viewed as a hydraulic motor with high volumetric efficiency that generally absorbs a small amount of energy from the flowing fluid. The energy absorption is primarily to overcome the internal resistance of the moving parts of the meter and its accessories. This loss of energy is observed as the pressure drop across the meter. The differential pressure across the meter is the driving force for the internals of the PD meter.

Design and Construction

A positive displacement meter has three basic components: an outer housing, the internal mechanism that creates the dividing chamber through its repetitive motion, and the display or counter accessories that determines the number of entrapments of fluid in the dividing chamber and infers the flow rate and the total volume of flow through the meter.

The external housing acts as the holding chamber of the flowing fluid before and after the entrapments by the internals. For low line pressures, the housing is usually single walled while, for higher operating pressures, the housing is double walled where the inner wall is the containment wall for the entrapment chamber and the outer wall is the pressure vessel. For the double-walled housing, the entrapment chamber walls are subjected to the differential pressure across the meter while the external housing is subjected to the total line pressure. This allows fabrication of a thin-walled inner chamber that can retain its precise shape and dimensions independent of line pressure.

The measuring mechanism consists of precise metering elements that require tight tolerances on the mating parts. The metering elements consist of the containment wall of the metering chamber and the moving components of the meter that forms the entrapment volume of the flowing fluid by cyclic or repetitive motion of those elements. The most common types of positive displacement meters are the oscillating piston, nutating disk, oval gear, sliding vane, birotor, trirotor, and diaphragm designs.

The counter or output mechanism converts the motion of the internal measuring chamber of a PD meter and displays the flow rate or total flow by correlating the number of entrapments and each entrapped volume. Many positive displacement meters have a mechanical gear train that requires seals and packing glands to transmit the motion of the inner mechanism to the outside counters. This type of display requires more driving power to overcome resistance of the moving parts and the seals, which results in an additional pressure drop for the meter. Many PD meters transmit the motion of the inner mechanism to the counters through switch output utilizing electromechanical, magnetic, optical, or purely electronic techniques in counting the entrapments and displaying the flow rate and total flow volume. The latter processing techniques normally have less pressure drop than the all-mechanical or part-mechanical transmission methods. All-mechanical drive counters do not require external power and, through proper selection of gear train, can display the actual flow volume or the flow rate. Thus, meters can be installed at remote locations devoid of any external power source. Meters with mechanical display cannot easily correct for the changes in volume due to thermal expansion or contraction of the measuring chamber due to flow temperature variation and, in the case of single-walled housing, the changes in the entrapment volume due to variations of the line pressure. An electronically processed output device could monitor both the pressure and temperature of the meter and provide necessary corrections to the meter output. With constantly changing and improving electronics technology, many PD meters are now installed with solar or battery-powered electronic output for installations at remote locations with no external power source. Many electronic displays allow access to meter data from a central monitoring station via radio or satellite communication.

Some Commercially Available PD Meter Designs

Commercially available PD meters have many noticeably different working mechanisms and a few designs are unique and proprietary. Although each design and working mechanism can be noticeably different from another, all positive displacement meters have a stationary fluid retaining wall and a mechanism that momentarily entraps inlet fluid into a partitioned chamber before releasing it to the downstream side of the meter. This entrapment and release of the flowing fluid occur with such repetitive and sweeping motion that, for most practical purposes, the flow rate appears to be uniform and steady, even though, in reality, the exit flow does have some pulsation. The flow pulsation out of the meter may be more pronounced for some designs of positive displacement meters than others. These flow pulsations are more pronounced at the lower flow rates for all designs. Some designs are more suitable for either liquid or gas flows, while some designs can measure both gas and liquid. For liquid applications, PD meters

FIGURE 9.21 Sliding-vane type PD meter. (Courtesy of Daniel Industries, Inc.)

FIGURE 9.22 Trirotor type PD meter. (Courtesy of Liquid Controls LLC.)

work best for liquids with heavy viscosities. Almost all PD meters require precisely machined, high-tolerance mating parts; thus, measured fluid must be clean for longevity of the meter and to maintain the measurement precision.

Sliding-Vane Type Positive Displacement Meter

Figure 9.21 shows a working cycle of a sliding-vane type PD meter where vanes are designed to move in and out of the rotating inner mechanism. The position of each sliding vane relative to specific angular rotation of the rotor is usually maintained by a mechanical cam. In the design shown in Figure 9.21, each blade is independently retained in the slots. There are designs with even numbers of blades where diametrically opposite blades are one integral unit. High-pressure application would utilize dual wall construction.

Tri-Rotor Type PD Meter

This design has three rotating parts that entrap fluid between the rotors and an outer wall. The working cycle of this type of meter is shown Figure 9.22. In this design, for one rotation of the top mechanism, two blades rotate twice. The driving mechanism and rotation of each blade with respect to the others is maintained by a three-gear assembly where each rotating blade shaft is connected to one gear of the three-gear assembly to maintain relative rotational speed. This design is used to measure liquid flows.

Birotor PD Meter

The measuring unit of the birotor PD meter has two spiral type rotors kept in perfect timing by a set of precision gears and is used to measure liquid flows. Flow can enter the meter either perpendicular or parallel to the axis of rotation of the mating rotors, as shown in Figure 9.23. The axial design birotor is identical to the standard model in component parts and in principle; however, it utilizes a measuring unit mounted parallel, rather than perpendicular, to flow. Meter output is registered mechanically through a gear train located outside the measuring chamber or, electronically, using a pickup assembly mounted with a timing gear. Axial orientation results in compact installation, improved accuracy, and low pressure loss. The axial design is ideal for high flow rate applications.

Piston Type PD Meter

A typical design of a piston type PD meter is shown in Figure 9.24. A centrally located rotating part generates the reciprocating motion for each of the four pistons of the meter. The reciprocating motion of each of the piston is timed such that the discharge from an individual piston cylinder occurs in a cycle to generate a semicontinuous discharge from the meter. These PD meters are used for very low flow rates of liquid flows. A piston-cylinder design can withstand large differential pressures across the meter, so high viscous liquids can be measured by this type of meters and

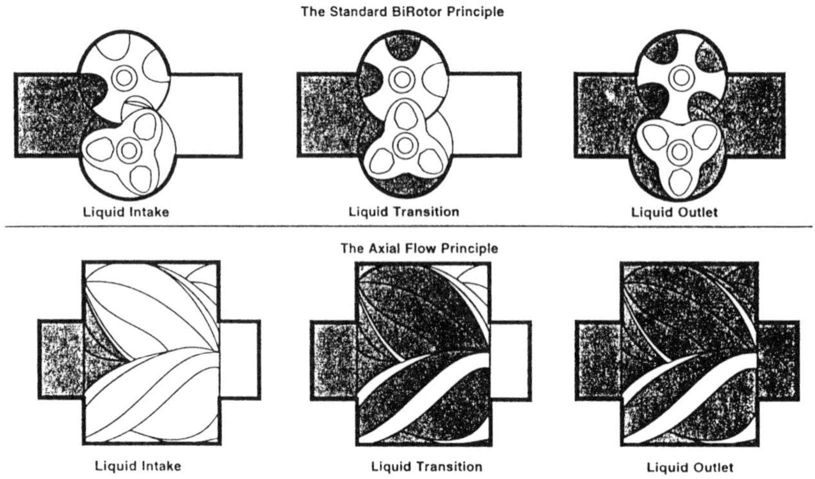

FIGURE 9.23 Birotor type PD meter. (Courtesy of Brooks Instruments Division, Emerson Electric Company.)

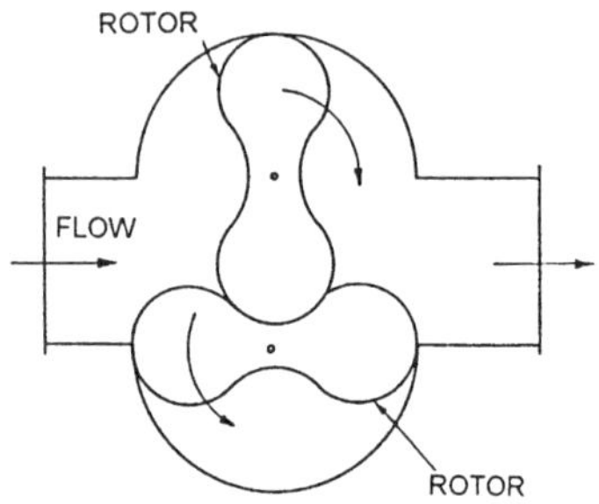

FIGURE 9.24 Piston-type PD meter. (Courtesy of Pierburg Instruments, Inc.)

FIGURE 9.25 Oval Gear Meter. (Courtesy of Daniel Industries, Inc.)

measurements are very precise. Mechanical components of this type of meter require very precise mechanical tolerances, which increase the product cost.

Oval Gear PD Meter

The measurement of volumetric flow of an oval gear meter is obtained by separating partial volumes formed between oval gears and the measuring chamber wall, as shown in Figure 9.25. The rotation of the oval gears results from the differential pressure across the flowmeter. During one revolution of the oval gears, four partial volumes are transferred. The rotation of the gears is transmitted from the measuring chamber to the output shaft directly with mechanical seals or via a magnetic coupling. This type of meter is used to measure liquids having a wide range of viscosity.

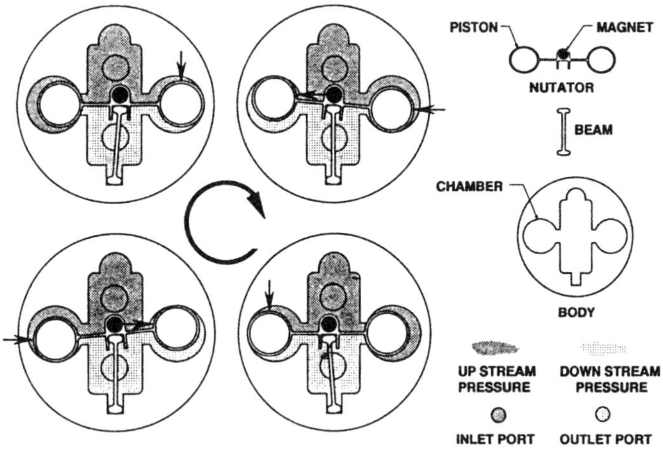

FIGURE 9.26 Nutating-disk type PD meter. (Courtesy of DEA Engineering Company.)

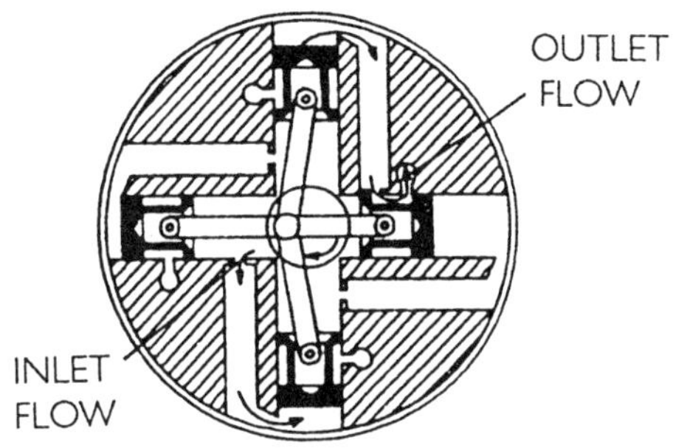

FIGURE 9.27 Schematic diagram of roots meter.

Nutating-Disk Type PD Meters

A disk placed within the confines of a boundary wall at a specific orientation can induce a flow instability that can generate a wobbling or nutating motion to the disk. The operating cycle of one such design is shown in Figure 9.26. The entrapment and discharge of the fluid to and from the two chambers occur during different phases of the repetitive cycle of the nutating disk. This type of meter is used to measure liquids. The design provides economic flow measurement where accuracy is not of great importance and is often used in water meters.

Meter designs described hereafter are for gas flows. PD meters for gas flows require very tight tolerances of the moving parts to reduce leak paths and, therefore, the gas must be clean because the meters are less tolerant to solid particles.

Roots PD Meter

The roots meter design is shown in Figure 9.27. This is the most commonly used PD meter to measure gas flows. This is one of the oldest designs of a positive displacement meter. The mechanical clearance between the rotors and the housing requires precisely machined parts. Roots meters are adversely affected if inlet flow to the meter has a relatively high level of pulsation. The main disadvantage of this meter is that it introduces cyclic pulsations into the flow. This meter cannot tolerate dirt and requires filtering upstream of the meter. A roots meter in single-case design is used at near-ambient line pressures. This

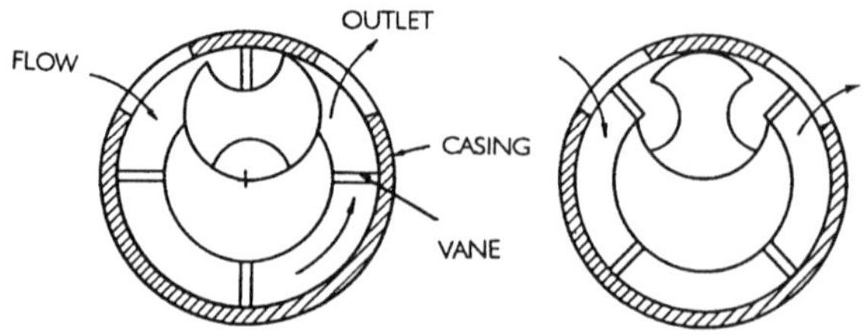

FIGURE 9.28 Operation of CVM meter.

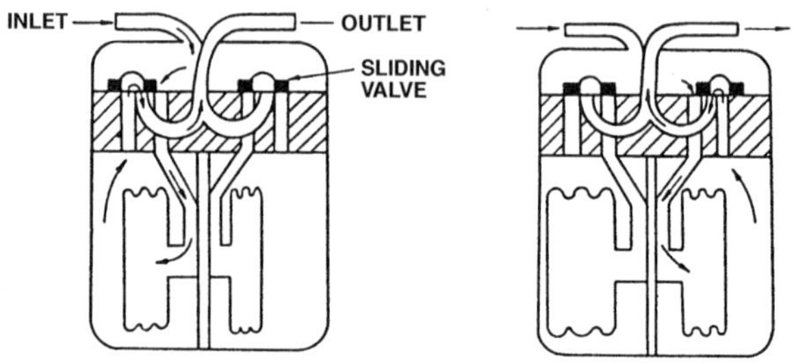

FIGURE 9.29 Operation of a diaphragm meter.

meter design is very widely used in the low-pressure transmission and distribution market of natural gas. If this meter ever seizes up at any position, the flow is completely blocked.

The CVM Meter

The CVM meter (Figure 9.28) is a proprietary design of a positive displacement meter used to measure gas flows. This meter has a set of four vanes rotating in an annular space about the center of the circular housing. A gate similar in shape to the center piece of the trirotor design (Figure 9.22) allows the vanes to pass — but not the gas. The measurement accuracy of the CVM meter is similar to the roots meter, but the amplitude of exit pulsation is lower than the roots meter for similar meter size and flow rate. This design is also not as sensitive to inlet flow fluctuations, and is widely used in the low-pressure, natural gas distribution market in Europe.

Diaphragm Meter

The diaphragm meter, also known as a bellows-type meter, is simple, relatively inexpensive, has reliable metering accuracy, and is widely used as a domestic gas meter. This design, shown in Figure 9.29, acts like a reciprocating piston meter, where the bellows act as the piston. The operation is controlled by a double slide valve. In one position, the slide valve connects the inlet to the inside of the bellow, while the outside of the bellow is open to the outlet. In the other position, the outside is connected to the inlet while the inside of the bellow is open to the outlet. This design can measure extremely low flow rates (e.g., pilot light of a gas burner).

Advantages and Disadvantages of PD Meters

High-quality, positive displacement meters will measure with high accuracy over a wide range of flow rates, and are very reliable over long periods. Unlike most flowmeters, a positive displacement meter is insensitive to inlet flow profile distortions. Thus, PD meters can be installed in close proximity to any upstream or downstream piping installations without any loss of accuracy. In general, PD meters have minimal pressure drop across the meter; hence, they can be installed in a pipeline with very low line pressures. Until the introduction of electronic correctors and flow controls on other types of meters, PD meters were most widely used in batch loading and dispensing applications. All mechanical units can be installed in remote locations.

Positive displacement meters are generally bulky, especially in the larger sizes. Due to the tight clearance necessary between mating parts, the fluid must be clean for measurement accuracy and longevity of the meter. More accurate PD meters are quite expensive. If a PD meter ever seizes up, it would completely block the flow. Many PD meters have high inertia of the moving parts; therefore, a sudden change in the flow rate can damage the meter. PD meters are normally suitable over only limited ranges of pressure and temperature. Some designs will introduce noticeably high pulsations into the flow. Most PD meters require a good maintenance schedule and are high repair and maintenance meters. Recurring costs in maintaining a positive displacement flowmeter can be a significant factor in overall flowmeter cost.

Applications

Liquid PD meters are capable of measuring fluids with a wide range of viscosity and density. Minor changes in viscosity of the fluid have minimal influence on the accuracy of a PD meter. However, no one specific design of PD meter is capable of such a broad application range. The physical properties of the fluid, especially the fluid viscosity, must be reviewed for proper selection of a meter for the application. Many PD meter designs accurately measure flow velocities over a wide range.

Careful dimensional control is required to produce well-regulated clearances around the moving parts of the PD meter. The clearances provide a leakage path around the flow measurement mechanism, and the amount of leakage depends on the relative velocity of the moving parts, the viscosity of the fluid, and the clearances between parts. In general, the leakage flow rate follows the Poiseuille equation; that is, leakage is directly proportional to the differential pressure and inversely proportional to the absolute viscosity and the length of the leakage path.

The leakage flow, in percentage of the total flow rate, is useful information for error measurement. At low flow rates, percent leakage error can be significant; while with modest flow rates, leakage can be insignificant. However, the differential pressure across the PD meter increases exponentially with increasing flow rates. Therefore, at very high flow rates, leakage flow can again be a significant portion of the total flow. As a result, PD meters tend to dispense more fluid than is indicated by the register at the very low and very high flow rates. The amount of error due to leakage depends on the meter design and on the viscosity of the fluid.

PD meters are driven by the differential pressure across the meter. The primary losses within the meter can be attributed to the friction losses and the viscous drag of the fluid on the moving parts. At very low flow rates, the bearing losses are predominant, while at high flow rates, the viscous drags predominate. The viscous drag is directly proportional to the fluid viscosity and the relative velocity of the moving parts and inversely proportional to the clearance. Within each design, the differential pressure is limited to a predetermined value. Excessive differential pressures can damage the meter. High differential pressure can be avoided by limiting the maximum flow rate, lowering viscosity by heating the fluid, increasing clearances, or by various combinations of these techniques. In general, if the differential pressure limit can be addressed, PD meters measuring high-viscosity fluids have less measurement error over a wide flow rate range.

The viscosity of most gases is too low to cause application problems relating to viscous drag of the form discussed for liquid PD meters. Gas PD meters utilize smaller clearances than liquid PD meters; therefore, gas must be very clean. Even small particles of foreign material can damage gas PD meters. PD meters used in gas flow measurement are primarily in the low line pressure application.

Accessories

Many accessories are available for PD meters. Among the most common and useful accessories for liquid PD meters are the automatic air eliminator and the direct-coupled shut-off valve for batch loading operations. The automatic air eliminator consists of a pressure-containing case surrounding a float valve. When gas in the liquid flow enters the chamber, the float descends by gravity and allows the float valve to open and purge the gas. An air eliminator improves the precision of flow measurement in many applications and is an absolute necessity in applications where air or gas is trapped in the flow stream. The direct-coupled shut-off valve is directly linked to the mechanical register. A predetermined fluid quantity can be entered into the register. When the dispensed amount equals the amount entered in the register, the mechanical linkage closes the valve. Two-stage shut-off valves are often used to avoid line shock in high flow rate dispensing applications. Similarly, all-electronic units can control batch operations. The all-electronic units provide great convenience in the handling of data. However, the register-driven mechanical valves have advantages in simplicity of design, field maintenance or repair, and initial cost.

Gas PD meters can be equipped with mechanical registers that correct the totals for pressure and temperature. The pressure and temperature compensation range is limited in the mechanical units but with appropriate application; the range limitation is not a problem. Electronic totalizers for gas PD meters can correct for broad ranges of temperature and pressure. Electronic accessories can provide various data storage, logging, and remote data access capabilities.

Price and Performance

In general, the accuracy of the PD meter is reflected in the sales price. However, several additional factors influence the cost. The intended application, the materials of construction, and the pressure rating have a strong influence on the cost. Although they provide excellent accuracy, small PD meters for residential water or residential natural gas service are very inexpensive. Meters produced for the general industrial market provide good accuracy at a reasonable cost, and meters manufactured for the pipeline or petrochemical markets provide excellent accuracy and are expensive.

Very inexpensive PD meters are available with plastic cases or plastic internal mechanisms. Aluminum, brass, ductile iron, or cast iron PD meters are moderately priced. Steel PD meters are relatively expensive, especially for high-pressure applications. If stainless steel or unusual materials are necessary for corrosive fluids or special service requirements, the costs are very high. PD meters are manufactured in a wide range of sizes, pressure ratings, and materials, with a multitude of flow rate ranges, and with an accuracy to match most requirements. The cost is directly related to the performance requirements.

Further Information

D. W. Spitzer, Ed., *Flow Measurement: Practical Guides for Measurement and Control*, Research Triangle Park, NC: ISA, 1991.

R. W. Miller, *Flow Measurement Engineering Handbook*, New York: McGraw-Hill, 1989.

Fluid Meters: Their Theory and Application, 6th ed., Report of ASME Research Committee on Fluid Meters, American Society of Mechanical Engineers, 1971.

V. L. Streeter, *Fluid Mechanics, 4th ed.*, New York: McGraw-Hill, 1966.

V. L. Streeter, *Handbook of Fluid Dynamics*, New York: McGraw-Hill, 1961.

Measurement of Liquid Hydrocarbons by Displacement Meters, Manual of Petroleum Measurement Standard, Chapter 5.2, American Petroleum Institute, 1992.

9.4 Turbine and Vane Flowmeters

David Wadlow

This section describes a range of closed-conduit flowmeters that utilize rotating vaned transduction elements, with particular emphasis on axial turbine flowmeters. Single jet and insertion tangential turbines, also known as paddlewheel flowmeters, are described in another section. The various vaned flowmeters used for open-channel and free-field flow measurement are not included in this section.

Axial Turbine Flowmeters

The modern axial turbine flowmeter, when properly installed and calibrated, is a reliable device capable of providing the highest accuracies attainable by any currently available flow sensor for both liquid and gas volumetric flow measurement. It is the product of decades of intensive innovation and refinements to the original axial vaned flowmeter principle first credited to Woltman in 1790, and at that time applied to measuring water flow. The initial impetus for the modern development activity was largely the increasing needs of the U.S. natural gas industry in the late 1940s and 1950s for a means to accurately measure the flow in large-diameter, high-pressure, interstate natural gas lines. Today, due to the tremendous success of this principle, axial turbine flowmeters of different and often proprietary designs are used for a variety of applications where accuracy, reliability, and rangeability are required in numerous major industries besides water and natural gas, including oil, petrochemical, chemical process, cryogenics, milk and beverage, aerospace, biomedical, and others.

Figure 9.30 is a schematic longitudinal section through the axis of symmetry depicting the key components of a typical meter. As one can see, the meter is an in-line sensor comprising a single turbine rotor, concentrically mounted on a shaft within a cylindrical housing through which the flow passes. The shaft or shaft bearings are located by end supports inside suspended upstream and downstream aerodynamic structures called diffusers, stators, or simply cones. The flow thus passes through an annular region occupied by the rotor blades. The blades, which are usually flat but can be slightly twisted, are inclined at an angle to the incident flow velocity and hence experience a torque that drives the rotor. The rate of rotation, which can be up to several $\times 10^4$ rpm for smaller meters, is detected by a pickup, which is usually a magnetic type, and registration of each rotor blade passing infers the passage of a fixed volume of fluid.

General Performance Characteristics

Axial turbines perform best when measuring clean, conditioned, steady flows of gases and liquids with low kinematic viscosities (below about 10^{-5} m^{2}s^{-1}, 10 cSt, although they are used up to 10^{-4} m^{2}s^{-1}, 100 cSt), and are linear for subsonic, turbulent flows. Under these conditions, the inherent mechanical stability of the meter design gives rise to excellent repeatability performance. Not including the special case of water meters, which are described later, the main performance characteristics are:

- Sizes (internal diameter) range from 6 mm to 760 mm, (1/4 in. to 30 in.).
- Maximum measurement capacities range from 0.025 Am^3 h^{-1} to 25,500 Am^3 h^{-1}, (0.015 ACFM to 15,000 ACFM), for gases and 0.036 m^3 h^{-1} to 13,000 m^3 h^{-1}, (0.16 gpm to 57,000 gpm or 82,000 barrels per hour), for liquids, where A denotes actual.
- Typical measurement repeatability is ±0.1% of reading for liquids and ±0.25% for gases with up to ±0.02% for high-accuracy meters. Typical linearities (before electronic linearization) are between ±0.25% and ±0.5% of reading for liquids, and ±0.5% and ±1.0% for gases. High-accuracy meters have linearities of ±0.15% for liquids and ±0.25% for gases, usually specified over a 10:1 dynamic range below maximum rated flow. Traceability to NIST (National Institute of Standards and Technology) is frequently available, allowing one to estimate the overall absolute accuracy performance of a flowmeter under specified conditions. Under ideal conditions, absolute accuracies for optimum designs and installations can come close to the accuracy capabilities at the NIST, which are stated as ±0.13% for liquid flows and ±0.25% for air.
- Rangeability, when defined as the ratio of flow rates over which the linearity specification applies, is typically between 10:1 and 100:1.
- Operating temperature ranges span –270°C to 650°C, (–450°F to 1200°F).
- Operating pressure ranges span coarse vacuum to 414 MPa (60,000 psi).
- Pressure drop at the maximum rated flow rate ranges from around 0.3 kPa (0.05 psi) for gases to in the region of 70 kPa (10 psi) for liquids.

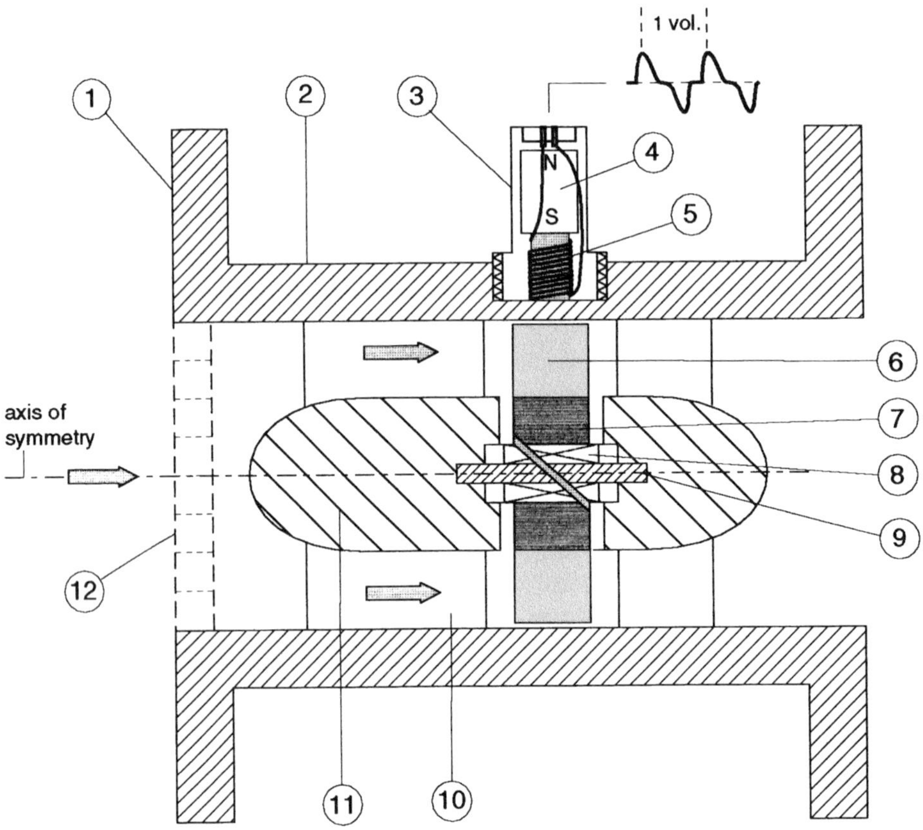

FIGURE 9.30 Longitudinal section of an axial turbine flowmeter depicting the key components. The flowmeter body is usually a magnetically transparent stainless steel such as 304. Common end-fittings include face flanges (depicted), various threaded fittings and tri-clover fittings. The upstream and downstream diffusers are the same in bidirectional meters, and generally supported by three or more flat plates, or sometimes tubular structures, aligned with the body, which also act as flow straighteners. The relative size of the annular flow passage at the rotor varies among different designs. Journal rotor bearings are frequently used for liquids, while ball bearings are often used for gases. Magnetic reluctance pickups (depicted) are frequently used. Others types include mechanical and modulated carrier pickups. (**1**) End fitting — flange shown; (**2**) flowmeter body; (**3**) rotation pickup — magnetic, reluctance type shown; (**4**) permanent magnet; (**5**) pickup cold wound on pole piece; (**6**) rotor blade; (**7**) rotor hub; (**8**) rotor shaft bearing — journal type shown; (**9**) rotor shaft; (**10**) diffuser support and flow straightener; (**11**) diffuser; (**12**) flow conditioning plate (dotted) — optional with some meters.

Theory

There are two approaches described in the current literature for analyzing axial turbine performance. The first approach describes the fluid driving torque in terms of momentum exchange, while the second describes it in terms of aerodynamic lift via airfoil theory. The former approach has the advantage that it readily produces analytical results describing basic operation, some of which have not appeared via airfoil analysis. The latter approach has the advantage that it allows more complete descriptions using fewer approximations. However, it is mathematically intensive and leads rapidly into computer-generated solutions. One prominent pioneer of the momentum approach is Lee [1] who, using this approach, later went on to invent one of the few, currently successful, dual rotor turbine flowmeters, while Thompson and Grey [2] provided one of the most comprehensive models currently available using the airfoil

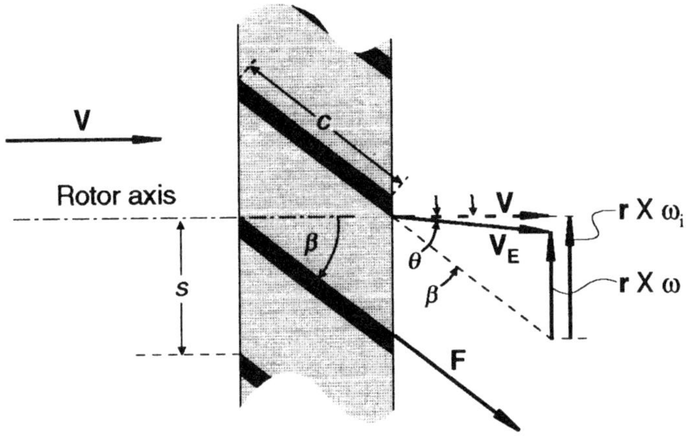

FIGURE 9.31 Vector diagram for a flat-bladed axial turbine rotor. The difference between the ideal (subscript i) and actual tangential velocity vectors is the rotor slip velocity and is caused by the net effect of the rotor retarding torques. This gives rise to linearity errors and creates swirl in the exit flow. **V**, incident fluid velocity vector; $\mathbf{V_E}$, exit fluid velocity vector; θ, exit flow swirl angle due to rotor retarding torques; β, blade pitch angle, same as angle of attack for parallel flow; ω, rotor angular velocity vector; **r**, rotor radius vector; **F**, flow-induced drag force acting on each blade surface; *c*, blade chord; *s*, blade spacing along the hub; *c*/*s*, rotor solidity factor.

approach, which for example, took into account blade interference effects. In the following, the momentum exchange approach is used to highlight the basic concepts of the axial turbine flowmeter.

In a hypothetical situation, where there are no forces acting to slow down the rotor, it will rotate at a speed that exactly maintains the fluid flow velocity vector at the blade surfaces. Figure 9.31 is a vector diagram for a flat-bladed rotor with a blade pitch angle equal to β. Assuming that the rotor blades are flat and that the velocity is everywhere uniform and parallel to the rotor axis, then referring to Figure 9.31, one obtains:

$$r\omega_i = \frac{\tan\beta}{V} \tag{9.47}$$

When one introduces the total flow rate, this becomes:

$$\frac{\omega_i}{Q} = \frac{\tan\beta}{\bar{r}A} \tag{9.48}$$

where ω_i = "Ideal" rotational speed
Q = Volumetric flow rate
A = Area of the annular flow cross-section
$\bar{r}$ = Root-mean-square of the inner and outer blade radii, (R, a)

Eliminating the time dimension from the left-hand-side quantity reduces it to the number of rotor rotations per unit fluid volume, which is essentially the flowmeter *K* factor specified by most manufacturers. Hence, according to Equation 9.48, in the ideal situation, the meter response is perfectly linear and determined only by geometry. (In some flowmeter designs, the rotor blades are helically twisted to improve efficiency. This is especially true of blades with large radius ratios, (R/a). If the flow velocity profile is assumed to be flat, then the blade angle in this case can be described by tan β = constant × *r*. This is sometimes called the "ideal" helical blade.) In practice, there are instead a number of rotor retarding torques of varying relative magnitudes. Under steady flow, the rotor assumes a speed that satisfies the following equilibrium:

$$\begin{aligned}\text{Fluid driving torque} = {} & \text{rotor blade surfaces fluid drag torque} + \text{rotor hub} \\ & \text{and tip clearance fluid drag torque} + \text{rotation sensor} \\ & \text{drag torque} + \text{bearing friction retarding torque}\end{aligned} \tag{9.49}$$

Referring again to Figure 9.31, the difference between the actual rotor speed, $r\omega$, and the ideal rotor speed, $r\omega_{\mathrm{i}}$, is the rotor slip velocity due to the combined effect of all the rotor retarding torques as described in Equation 9.49, and as a result of which the fluid velocity vector is deflected through an exit or swirl angle, θ. Denoting the radius variable by r, and equating the total rate of change of angular momentum of the fluid passing through the rotor to the retarding torque, one obtains:

$$\int_a^R \frac{\rho Q 2\pi r^2 \left(r\omega_{\mathrm{i}} - r\omega\right)}{\pi\left(R^2 - a^2\right)} dr = N_{\mathrm{T}} \tag{9.50}$$

which yields:

$$\bar{r}^2 \rho Q\left(\omega_{\mathrm{i}} - \omega\right) = N_{\mathrm{T}} \tag{9.51}$$

where ρ is the fluid density and N_{T} is the total retarding torque. Combining Equations 9.47 and 9.51 and rearranging, yields:

$$\frac{\omega}{Q} = \frac{\tan\beta}{\bar{r} A} - \frac{N_{\mathrm{T}}}{\bar{r}^2 \rho Q^2} \tag{9.52}$$

The trends evident in Equation 9.52 reflect the characteristic decline in meter response at very low flows and why lower friction bearings and lower drag pickups tend to be used in gas vs. liquid applications and small diameter meters. In most flowmeter designs, especially for liquids, the latter three of the four retarding torques described in Equation 9.49 are small under normal operating conditions compared with the torque due to induced drag across the blade surfaces. As shown in Figure 9.31, the force, F, due to this effect acts in a direction along the blade surface and has a magnitude given by:

$$F = \frac{\rho V^2}{2} C_{\mathrm{D}} S \tag{9.53}$$

where C_{D} is the drag coefficient and S is the blade surface area per side. Using the expression for drag coefficient corresponding to turbulent flow, selected by Pate et al. [3] and others, this force can be estimated by:

$$F = \rho V^2 0.074\ \mathrm{Re}^{-0.2}\ S \tag{9.54}$$

where Re is the flow Reynolds number based on the blade chord shown as dimension c in Figure 9.31. Assuming θ is small compared with β, then after integration, the magnitude of the retarding torque due to the induced drag along the blade surfaces of a rotor with n blades is found to be:

$$N_{\mathrm{D}} = n\left(R + a\right)\rho V^2\ 0.037\ \mathrm{Re}^{-0.2}\ S \sin\beta \tag{9.55}$$

Combining Equations 9.55 and 9.52, and rearranging yields:

$$\frac{\omega}{Q} = \frac{\tan\beta}{\bar{r}A} - \frac{0.036n\left(R+a\right)SA^2\ \mathrm{Re}^{-0.2}\ \sin\beta}{\bar{r}^2} \qquad (9.56)$$

Equation 9.56 is an approximate expression for the K factor because it neglects the effects of several of the rotor retarding torques, and a number of important detailed meter design and aerodynamic factors, such as rotor solidity and flow velocity profile. Nevertheless, it reveals that linearity variations under normal, specified operating conditions are a function of certain basic geometric factors and Reynolds number. These results reflect general trends that influence design and calibration. Additionally, the marked departure from an approximate ρV^2 (actually $\rho^{0.8}V^{1.8}\mu^{-0.2}$ via Re in Equation 9.54) dependence of the fluid drag retarding torque on flow properties under turbulent flow, to other relationships under transitional and laminar flow, gives rise to major variations in the K factor vs. flow rate and media properties for low-flow Reynolds numbers. This is the key reason why axial turbine flowmeters are generally recommended for turbulent flow measurement.

Calibration, Installation, and Maintenance

Axial turbine flowmeters have a working dynamic range of at least 10:1 over which the linearity is specified. The maximum flow rate is determined by design factors related to size vs. maximum pressure drop and maximum rotor speed. The minimum of the range is determined by the linearity specification itself. Due to small, unavoidable, manufacturing variances, linearity error curves are unique to individual meters and are normally provided by the manufacturer. However, although recommended where possible, the conditions of the application cannot usually and need not necessarily duplicate those of the initial or even subsequent calibrations. This has pivotal importance in applications where actual operating conditions are extreme or the medium is expensive or difficult to handle. Figure 9.32 depicts a typically shaped calibration curve of linearity vs. flow rate expressed in terms of multiple alternative measures, various combinations of which can be found in current use. The vertical axis thus represents either the linearity error as a percentage of flow rate, a K factor expressed in terms of the number of pulses from the rotation sensor output per volume of fluid, or the deviation from 100% registration; the latter only applies to flowmeters with mechanical pickups. The horizontal axis can be expressed in terms of flow rate in volume units/time, Reynolds number (Re), or pulse frequency (from the rotation sensor for nonmechanical) divided by kinematic viscosity, (f/v), in units of Hz per m^2s^{-1}, (Hz/cSt or Hz/SSU; 10^{-6} m^2s^{-1} = 1 centistoke = 31.0 seconds, Saybolt Universal), and where kinematic viscosity is the ratio of absolute viscosity (μ) to density. Calibrations are preferably expressed vs. Re or f/v, which is proportional to Re. The hump shown in the curve is a characteristic frequently observed at lower Re and is due to velocity profile effects. K factor vs. f/v calibration curves are specifically called universal viscosity curves (UVC) and, for most meters, are available from the manufacturer for an extra charge. A key utility of UVC is that where media type and properties differ significantly from those of the original calibration, accuracies much greater than the overall linearity error can still readily be obtained via the flowmeter UVC if the kinematic viscosity of the application is known. An alternative, advanced calibration technique [4] is to provide response in terms of Strouhal number vs. Re or Roshko number. This approach is not widely adopted, but it is particularly relevant to high-accuracy and extreme temperature applications because it further allows correct compensation for flowmeter thermal expansion errors.

The accuracy of axial turbine flowmeters is reduced by unconditioned flow, especially swirl. An installation incorporating flow conditioners along with specific upstream and downstream straight pipe lengths is generally recommended [5]. Some axial turbine flowmeters can be purchased with additional large flow straighteners that mount directly ahead of the flowmeter body or conditioning plates that are integral to the body. The manufacturer is the first source of information regarding installation. Errors due to flow velocity pulsations are another concern, particularly in certain gas installations. However, no standard technique for effectively counteracting this source of error has yet been adopted. Periodic

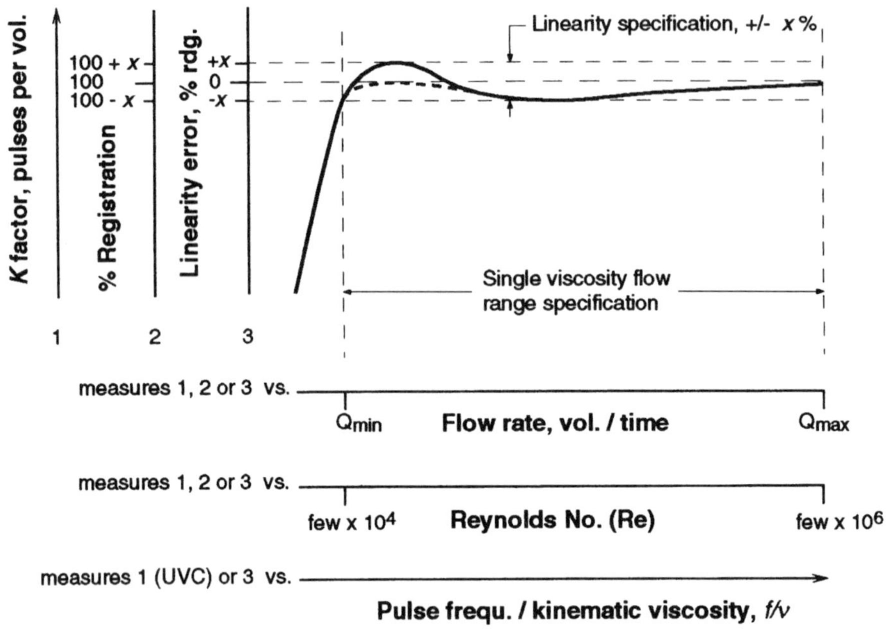

FIGURE 9.32 A typical single rotor axial turbine linearity error, or calibration, curve for a low-viscosity fluid showing the main alternative presentations in current use. Higher accuracy specifications usually correspond to a 10:1 flow range down from Q_{max}, while extended operating ranges usually correspond to reduced accuracies. The hump in the depicted curve is a characteristic feature caused by flow velocity profile changes as Re approaches the laminar region. This feature varies in magnitude between meters. Sensitivity and repeatability performance degrade at low Re. Percent registration is only used with meters that have mechanical pickups. All other meters have a *K* factor. Universal viscosity curve (UVC) and Re calibrations remain in effect at different known media viscosities provided Re or f/v stays within the specified range. Re is referenced to the connecting conduit diameter and is less within the flowmeter. The Re range shown is therefore approximate and can vary by an order of magnitude, depending on the meter. Linearity error can also be expressed in terms of Strouhal number (fD/V) vs. Re (VD/υ) or Roshko number (fD^2/υ), when instead *D* is a flowmeter reference diameter [4]. UVC, Universal Viscosity Curve; – – –, the effect of a rotor shroud in a viscosity compensated flowmeter.

maintenance, testing, and recalibration is required because the calibration will shift over time due to wear, damage, or contamination. For certain applications, especially those involving custody transfer of oil and natural gas, national standards, international standards, and other recommendations exist that specify the minimum requirements for turbine meters with respect to these aspects [6–10].

Design and Construction

There are numerous, often proprietary, designs incorporating variations in rotors, bearings, pickups, and other components in format and materials that are tailored to different applications. Meter bodies are available with a wide range of standard end-fittings. Within application constraints, the primary objective is usually to optimize the overall mechanical stability and fit in order to achieve good repeatability performance. Design for performance, application, and manufacture considerations impacts every internal component, but most of all the rotor with respect to blade shape and pitch, blade count, balance and rigidity vs. drag, stress, and inertia, bearings with respect to precision vs. friction, speed rating and durability, and rotation pickup vs. performance and drag.

Most low-radius ratio blades are machined flat, while high-ratio blades tend to be twisted. The blade count varies from about 6 to 20 or more, depending on the pitch angle and blade radius ratio so that the required rotor solidity is achieved. Rotor solidity is a measure of the "openness" to the flow such that higher solidity rotors are more highly coupled to the flow and achieve a better dynamic range. The pitch angle, which primarily determines the rotor speed, is typically 30° to 45° but can be lower in flowmeters designed for low-density, gas applications. Rotor assemblies are usually a close fit to the inside of the housing. In large-diameter meters, the rotor often incorporates a shroud around the outer perimeter for enhanced stability. Also, since large meters are often used for heavy petroleum products, via selection of a suitable wall clearance, the fluid drag resulting from this clearance gap is often designed to offset the tendency at high media viscosities for the meter to speed up at lower Reynolds numbers. The materials of construction range from nonmagnetic to magnetic steels to plastics.

Stainless steel ball bearings tend to be used for gas meters and low lubricity liquids such as cryogenic liquids and freon, while combination tungsten carbide or ceramic journal and thrust bearings are often considered best for many other liquid meters, depending on the medium lubricity. Fluid bearings (sometimes called "bearingless" designs) are often used in conjunction with the latter, but also sometimes with gases, for reducing the drag. They operate by various designs that use flow-induced forces to balance the rotor away from the shaft ends. Bearing lubrication is either derived from the metered medium or an internal or external system is provided. The more fragile, jeweled pivot bearings are also used in certain gas applications and small meters. Sanitary meters can incorporate flush holes in the bearing assembly to meet 3A crack and crevice standards.

The most common types of rotation sensor are magnetic, modulated carrier, and mechanical, while optical, capacitive, and electric resistance are also used. In research, a modulated nuclear radiation flux rotation sensor for use in certain nuclear reactors has also been reported [11, 12]. Mechanical pickups, which sometimes incorporate a magnetic coupling, are traditional in some applications and can have high resolution; one advantage is that they require no electric power. However, the pickup drag tends to be high. The magnetic and modulated carrier types utilize at least a coil in a pickup assembly that screws into the meter housing near the rotor. In magnetic inductance types, which are now less common, the blades or shroud carry magnetized inserts, and signals are induced in the coil by the traversing magnetic fields. In the more prevalent magnetic reluctance type, an example of which is schematically depicted in Figure 9.30, the coil is wrapped around a permanent magnet or magnet pole piece in the pickup assembly which is mounted next to a high magnetic permeability bladed rotor (or machined shroud). The latter is then typically made of a magnetic grade of stainless steel such as 416, 430, or 17-4Ph. As the rotor turns, the reluctance of the magnetic circuit varies, producing signals at the coil. In the more expensive modulated carrier types, the rotor need only be electrically conductive. The coil is part of a radio frequency (RF) oscillator circuit and the proximity of the rotor blades changes the circuit impedance, giving rise to modulation at a lower frequency that is recovered. The RF types have much lower drag, higher signal levels at low flow, and can operate at temperatures above the Curie point of typical ferromagnetic materials. They are preferred for wide dynamic range and high-temperature applications. Bidirectional flowmeters usually have two magnetic pickups to determine flow direction. This is useful, for example, in the monitoring of container filling and emptying operations often encountered in sanitary applications. Multiple magnetic pickups are also used in some designs to provide increased measurement resolution. Regarding output, various pulse amplifiers, totalizers, flow computers for gas pressure and temperature correction, along with 4–20 mA and other standard interface protocols, are available to suit particular applications. As an example of advanced transmitters, at least one manufacturer (EG&G Flow Technology, Inc.) provides a real-time, miniature, reprogrammable, "smart" transmitter that is integrated into the pickup housing along with a meter body temperature sensor, for full viscosity compensation and UVC linearization. These are for use in dedicated applications, such as airborne fuel management, where the medium viscosity–temperature relationship is known.

Certain applications have uniquely different design requirements and solutions, and two are discussed separately in the following.

Propeller Meters.

Propeller meters are used in either municipal, irrigation, or wastewater measurement. Although in some designs, propeller and turbine meters look almost identical and operate on the same axial rotor principle, this type of flowmeter is currently commercially and officially [13, 14] distinguished as a separate category distinct from the axial turbine. Diameters up to 2440 mm (96 in.) are available. The flow rate capacity of a 1800 mm (72 in.) diameter propeller meter is up to about 25,000 $m^3 h^{-1}$, (110,000 gpm). Typical accuracies are ±2% of reading. A primary requirement is ruggedness, and it is in the designs most suited to harsh environments that the formats are most distinctive. Rotor and pickup assemblies are generally flanged to the housing and removable. The rotors have large clearances, are often cantilevered into the flow, and supported via a sealed bearing without stators. The rotors are typically made of plastic or rubber and carry as few as three highly twisted, high radius ratio blades. Pickups are always mechanical and frequently have magnetic couplings.

Spirometers.
Monitoring spirometers measure the volumes of gas flows entering and leaving the lungs and can also be incorporated in ventilator circuits. Diagnostic spirometers are used to monitor the degree and nature of respiration. With these, a clinician can determine patient respiratory condition by various measures and clinical maneuvers. Low cost, light weight, speed of response, and patient safety are major considerations. Measurement capabilities include the gas volume of a single exhalation and also the peak expiratory flow for diagnostic types, measured in liters and liters per second, respectively. Various technologies are used. However, the Wright respirometer, named after the original inventor [15], today refers to a type of hand-held monitoring spirometer that utilizes a special type of tangential turbine transducer with a two-bladed rotor connected to a mechanical pickup and a dial readout for the volume. These particular spirometers are routinely used by respiratory therapists for patient weaning and ventilator checking. Other axial turbine-based flowmeters are available for ventilation measurements involving, for example, patient metabolics measurements. One axial turbine-based diagnostic spirometer made by Micro Medical, Ltd., currently claims most of the European market. This device utilizes an infrared, optical pickup and has a battery-powered microprocessor-controlled display. In these medical devices, rotors tend to be plastic with a large blade radius ratio. Flow conditioning is minimal or absent. The meters are typically accurate to ±a few percent of reading. In the U.S., spirometers are designated as class 2 medical devices and as such certain FDA approvals are required concerning manufacture and marketing. In the EU they are class IIb medical devices under a different system, and other approvals are required.

Dual-Rotor Axial Turbines

Dual-rotor axial turbines have performance features not found in single rotor designs. In 1981, Lee et al. [16] were issued a U.S. patent for a self-correcting, self-checking dual-rotor turbine flowmeter that is currently manufactured exclusively by Equimeter, Inc. and sold as the Auto-Adjust. This is a high-accuracy flowmeter primarily intended for use on large natural gas lines where even small undetected flow measurement errors can be costly. It incorporates two closely coupled turbine rotors that rotate in the same direction. The upstream rotor is the main rotor and the second rotor, which has a much shallower blade angle, is the sensor rotor. Continuous and automatic correction of measurement errors due to varying bearing friction is achieved by calculating the flow rate based on the difference between the rotor speeds. As shown in Figure 9.31 and discussed in the theory section, the flow exit angle is due to the net rotor retarding torque. If this torque increases in the main rotor, thereby reducing its speed, the exit angle increases and the speed of the sensor rotor is then also reduced. The meter is also insensitive to inlet swirl angle because the swirl affects both rotor speeds in the same sense and the effect is then subtracted in the flow calculation. The meter also checks itself for wear and faults by monitoring the ratio of the two rotor speeds and comparing this number with the installation value [17].

A dual-rotor liquid flowmeter, invented by Ruffner et al. [18], was recently introduced by Exact Flow, LLC. It is being offered as a high-accuracy flowmeter (up to ±0.1% accuracy and ±0.02% repeatability), which has an extraordinarily wide dynamic range of 500:1 with a single-viscosity liquid. This flowmeter has had early commercial success in fuel flow measurement in large jet engine test stands where the wide dynamic range is particularly useful [19]. The meter comprises two, closely and hydraulically coupled rotors that rotate in opposite directions. Due to the exit angle generated by the first rotor, the second rotor continues to rotate to much lower flow rates compared with the first.

Two-Phase Flow Measurement Using Axial Turbines

A differential pressure producing flowmeter such as a venturimeter in series with a turbine is known to be a technically appropriate and straightforward method for measuring the volumetric and mass flow rates of some fine, solid aerosols. However, this section highlights a current research area in the application of axial rotor turbine meters to a range of industrial flow measurement problems where gas/liquid, two-phase flows are encountered. Customarily, turbine meters are not designed for and cannot measure such flows accurately. Errors of the order 10% arise in metering liquids with void fractions of around 20%. Such flows are normally measured after gas separators. Although this problem is not restricted to these industries, the current main impetuses for research are the direct measurement of crude oil in offshore multiphase production systems, the measurement of water/steam mixtures in the cooling loops of nuclear reactors, and the measurement of freon liquid–vapor flows in refrigeration and air conditioning equipment. Several techniques investigated thus far use an auxiliary sensor. This can either be a void fraction sensor or a pressure drop device such as a venturimeter or drag disk, of which the pressure drop approach appears to be technically more promising [20, 21]. Also, from a practical standpoint, gamma densitometers for measuring void fraction are additional and expensive equipment and not, for example, well adapted for use in undersea oil fields. Two techniques currently being studied do not require an auxiliary in-line sensor. The first uses the turbine meter itself as the drag body and combines the output of the turbine with that of a small differential pressure sensor connected across the inlet and outlet regions. This technique requires a homogenizer ahead of the turbine, and measurement accuracies of ±3% for the volumetric flow rates of both phases have recently been reported for air/water mixtures up to a void fraction of 80% [22]. The second technique is based entirely on analysis of the turbine output signal and has provided significant correlations of the signal fluctuations with void fraction. Accuracies of water volumetric flow rate measurement of ±2% have been reported when using air/water mixtures with void fractions of up to 25% [23].

Insertion Axial Turbine Flowmeters

These flowmeters comprise a small axial rotor mounted on a stem that is inserted radially through the conduit wall, often through a shut-off valve. They measure the flow velocity at the rotor position from which the volumetric flow rate is inferred. They are an economical solution to flow measurement problems where pipe diameters are high and accuracy requirements are moderate, and also may be technically preferred where negligible pressure drop is an advantage, as in high-speed flows. They are typically more linear than insertion tangential turbine flowmeters and compete also with magnetic and vortex shedding insertion flowmeters. They are available for the measurement of a range of liquids and gases, including steam, similar to the media range of full bore axial turbines, and have a similarly linear response. Flow Automation, Inc., is currently the leading manufacturer. The rotors, which are usually metal but can be plastic, typically have diameters of 25 mm to 51 mm (1 in. to 2 in.). They can be inserted into pipes with diameters ranging from 51 mm to 2032 mm (2 in. to 80 in.). Velocity measurement ranges cover 0.046 ms^{-1} to 91 ms^{-1}, (9 fpm to 18,000 fpm) for gases and 0.03 ms^{-1} to 30 ms^{-1} (6 fpm to 6000 fpm) for liquids. Dynamic ranges vary between 10:1 and 100:1. The maximum flow rate measurement capacity in a 1836 mm (72 in.) diameter pipe can be as high as nearly 56,500 $m^3 h^{-1}$, (about 250,000 gpm). Since these devices are local velocity sensors, calculating the volumetric flow rate requires a knowledge of the area velocity profile and the actual flow area. Flow conditioning is therefore particularly important for accurate volumetric measurements, while radial positioning, which is a further responsibility of the user, must be according to the manufacturer's recommendation, which can either be centerline, one third of the diameter, 12% of the diameter, or determined by "profiling." Quick [24] discusses operation and installation for natural gas measurement. Although linearities or "accuracies" can be quoted up to ±1% of velocity, achieving the same accuracy for the volumetric flow rate, although possible, can be difficult or impractical. In this respect, a unique dual rotor design, exclusive to Onicon, Inc., and primarily used for chilled-water flow measurement in HVAC systems, requires less flow conditioning than single rotor designs. An Onicon dual rotor turbine assembly is depicted in Figure 9.33. It comprises two rotors that

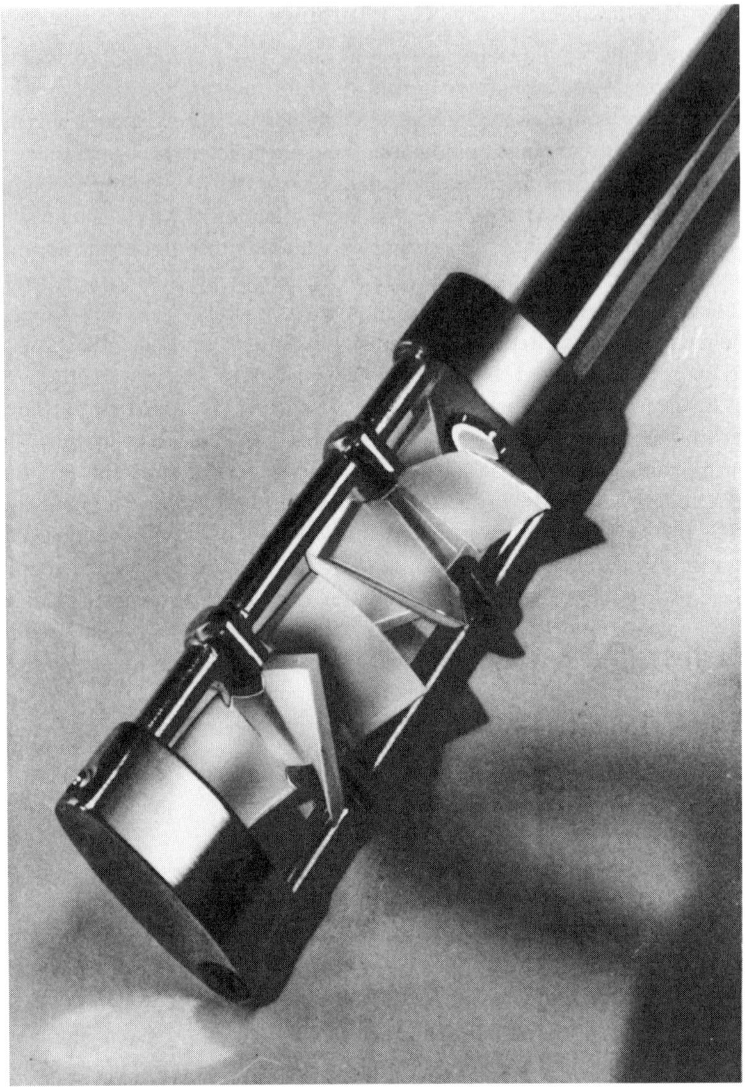

FIGURE 9.33 The rotor assembly of a dual rotor, insertion axial turbine flowmeter for water flow measurement. This patented design renders the flowmeter insensitive to errors due to flow swirl; an important source of potential error in single rotor axial turbine flowmeters. The rotations are sensed by two separate, electric impedance sensors. (Courtesy of Onicon Incorporated.)

rotate in opposite directions. The output is based on the average rotor speed. Any flow swirl present due to poor flow conditioning changes the speed of rotation of each rotor by the same but opposite amounts. Swirl-induced error is thus virtually absent in the averaged output. Also, flow profile sampling is improved over that of a single rotor. The devices are calibrated using a volumetric prover and the specified accuracy of ±2% of reading is for volumetric flow rate rather than velocity. This is the total error and includes an allowance for dimensional variations in industry standard pipes.

Angular Momentum Flowmeters

These are accurate, linear, liquid mass flowmeters that utilize vaned components and are used in aerospace applications. They are currently the instrument of choice for airborne, jet engine fuel flow measurement for large commercial aircraft and some military aircraft for afterburner applications. They are more expensive than the equivalent turbine flowmeters for this application, but they provide mass flow rate measurements directly and are unaffected by fuel density variations. Typical accuracies lie between ±0.5% and ±1.0% over a 40:1 or greater dynamic range. Measurement ranges are available for fuel flows from about 0.01 to 6 kg s^{-1} (70 to 46,000 PPH).

The principle of operation is long established and based on imparting angular momentum to the fluid flow using a driven, flat-bladed impeller. The force required to drive the impeller at constant speed is monitored as a proportional indication of mass flow rate as this quantity varies. Some designs use an electric motor to drive the impeller. However, the current trend in design is motorless. In such a device, a constant driving speed in the region of 100 rpm to 200 rpm is provided by one or more turbine rotors driven by the flow. A variable shunt metering valve assembly adjacent to the turbine rotor mechanically opens and closes in response to flow dynamic pressure and thereby automatically maintains a constant speed of rotation provided that the flow rate is above the minimum of the range. The driven impeller carries vanes that are parallel to the flow, and resides on a common axis with the turbine rotor. A flow straightener ensures parallel flow past the impeller. There is a carefully engineered constant rate spring connection between the turbine shaft and the impeller so that the angular deflection between the two is proportional to the applied torque, and this quantity is directly proportional to the mass flow rate. Two pickup coils sense the rotations of the turbine and the vaned impeller, and only the time difference between the pulses in these two signals is measured and used to calculate the flow rate. Chiles et al. [25] provide a detailed illustration and explanation of the intricate mechanism.

Multijet Turbine Flowmeters

These are linear, volumetric flowmeters designed for liquids measurements and comprise a single, radial-vaned impeller, vertically mounted on a shaft bearing within a vertically divided flow chamber, sometimes called a distributor. The impeller is often plastic and can even be neutrally buoyant in water. There are various designs, but typically, both chambers access a series of radially distributed and angled jets. The lower chamber jets connect to the flowmeter input port and distribute the flow tangentially onto the lower region of the impeller blades, while the upper series, which is angled oppositely, allow the flow to exit. The flow pattern within the flow chamber is thus a vertical spiral and the dynamic pressure drives the impeller to track the flow. This design gives the meters good sensitivity at low flow rates. Due to the distribution principle, the meters are also insensitive to upstream flow condition. Impeller rotation pickups are always mechanical, often magnetically coupled, and frequently also connect with electric contact transmitters. They are primarily used in water measurement, including potable water measurement for domestic and business billing purposes and in conjunction with energy management systems such as hot water building or district heating, and to a much lesser extent in some chemical and pharmaceutical industries for dosing and filling systems involving solvents, refrigerants, acids and alkalis with absolute viscosities less than 4.5 mPa s, (0.045 Poise). Available sizes range from 15 mm to 50 mm. Dynamic ranges lie between 25:1 and 130:1 and flow measurement ranges cover 0.03 to 30 m^3 h^{-1}, (0.13 to 130 gpm). Measurement linearities range between ±1% to ±2%, with typical repeatabilities of ±0.3%. Operating temperatures range from normal to 90°C (200°F) and maximum operating pressures are available up to 6.9 MPa (1000 psi). A number of potable water measurement systems come with sophisticated telemetry options that allow remote interrogation by radio or telephone. For potable water applications in the U.S., these meters normally comply with the applicable AWWA standard [26], while in Europe EEC, DIN, and other national standards apply.

In the author's opinion, there is also another type of vaned flowmeter that could be classified as a type of multijet turbine. This type comprises an axially mounted, vaned impeller with an upstream element that imparts a helical swirl to the flow. The transducer is typically a small, low-cost, sometimes disposable, plastic component, and is usually designed for liquids (but also to lower accuracies, gases), low-flow rate measurements (down to 50 mL min^{-1}). The dynamic range is high and accuracies range up to ±0.5%.

Goss [27] describes one particular design in current use. Specialized applications cover the pharmaceutical, medical, and beverage industries.

Cylindrical Rotor Flowmeter

Instead of coupling to a turbulent flow using the fluid dynamic pressure or momentum flux, as in most axial and tangential turbines, a demonstrated research device due to Wadlow et al. [28] provides a linear, volumetric, low gas flow rate measurement using a single, low inertia, smooth cylindrical tangential rotor that couples to a laminar flow via surface friction. The geometry is conceptually that of a single-surfaced rotating vane. A plane Poiseuille flow is created that passes azimuthally over most of the curved rotor surface in a narrow annular passage between the rotor and a concentric housing. The rotor is motor driven via a feedback control loop connected to an error signal producing, differential pressure sensor connected across the gas ports so as to maintain the meter pressure drop equal to zero. Under this condition, the response, as indicated by motor shaft speed, is exactly linear, determined only by geometry and is independent of gas density and viscosity. The demonstration device reported has a 40 mm diameter rotor and 15 mm diameter gas ports. It measures up to a maximum of 25 Lpm bidirectionally, has a linear dynamic range greater than 100:1, is insensitive to upstream flow condition, and has a $1/e$ step response time of 42 ms, limited only by the external motor torque and combined motor and rotor inertia.

Manufacturers

There is significant and dynamic competition among the numerous manufacturers of the different types of flowmeters described in this section. In all flowmeter types, most manufacturers have exclusive patent rights concerning one or more detailed design aspects that make their products perform differently from those of the competition. Every few years, one or more major turbine flowmeter company can be identified that has changed ownership and name or formed a new partnership. Identifying the competition and selecting the manufacturer are important and sometimes time-consuming parts of the flowmeter selection and specification process. To assist with this, Table 9.8 gives a few selected examples from different manufacturers, of all of the different flowmeter types described in this section. Table 9.9 gives the corresponding contact information for those selected manufacturers, along with the general types offered by each.

Conclusion

However anachronistic intricate mechanical sensors might appear amid current everyday high technology, there are fundamental reasons why axial turbines are likely to experience continued support and development rather than obsolescence, especially for in-line applications requiring in the region of tenth percent volumetric accuracy. Mechanical coupling is the most direct *volume* interaction for a flowing fluid, which is why mechanical meters historically developed first and continue to be the most accurate and reliable types of flowmeter for so many different fluids. Other nonmechanical, perhaps higher technology, or newer approaches thus face high demands for accurate compensation to render such less directly volume-coupled techniques as generally accurate, or more accurate. This is because the error in each corrected factor or assumption in an indirect technique contributes to the overall error. The technology of high-accuracy flowmeters continues to be driven by applications, such as the custody transfer of valuable oil and natural gas, which demand high accuracy and reliability. There is a continuing demand for accurate and reliable water flowmeters. By reason of long proven field experience, turbine and other vane type devices have become one of a few broadly accepted techniques in many major applications such as these where the demands for flow sensors is significant or growing.

TABLE 9.8 Examples of Turbine and Vaned Flowmeters

Type	Size(s), in.	Description	Example application(s)	Manufacturer	Approx. price range
Axial	0.5	FT 4-8, with SIL smart transmitter	Helicopters — fuel	EG&G Flow Technology	$2,500
Axial	1	6700 series, type 60	Raw milk, de-ionized water	Flow Automation	$1,500
Axial	16	Parity series	Custody transfer oil	Fisher-Rosemount Petroleum	$30,000–$38,000
Axial dual rotor	12	Auto-Adjust Turbo-Meter	Custody transfer gas	Equimeter	$42,000
Axial dual rotor	0.25–2	DR Series	Fuel — large jet engine test stands	Exact Flow	$1,300–$1,800
Propeller	48	FM182 (150 PSI)	Municipal water	Sparling	$6,500
Propeller	6	FM102 (150 PSI)	Irrigation	Sparling	$850
Special tangential	—	Wright Mark 8	Monitoring spirometer	Ferraris Medical	$800
Axial	—	MS03/MS04 MicroPlus	Diagnostic spirometer	Micro Medical	$600
Axial	4	WTX802	Chemical dosing for water treatment	SeaMetrics	$450–$550
Axial insertion	1.5–10	TX101	Municipal water, water — HVAC	SeaMetrics	$450
Axial insertion	2+	VTS-300	High pressure steam	Flow Automation	$2,900
Axial insertion	3+	VL-150-LP	Flare stack control	Flow Automation	$2,500
Axial insertion dual rotor	2.5–72	F-1200	Water — HVAC	Onicon	$900
Angular momentum	—	Model 9-217 True Mass Fuel Flowmeter	Large jet aircraft, e.g., Airbus A320, A321	Eldec	Not available
Multijet	1	1720	Domestic water	ABB Kent Messtechnik	$170
Multijet	1.5	AMD3000	Chemical liquid filling and dosing	Aquametro AG	$950
Multijet axial	9/32	DFS-2W	Pharmaceutical filling lines, kidney dialysis dialyte, beverage dispensers, OEM	Digiflow Systems	$42 or less, transducer; $235 electronics

References

1. W. F. Z. Lee and H. J. Evans, Density effect and Reynolds number effect on gas turbine flowmeters, *Trans. ASME, J. Basic Eng.*, 87 (4): 1043-1057, 1965.
2. R. E. Thompson and J. Grey, Turbine flowmeter performance model, *Trans. ASME, J. Basic Eng.*, 92(4), 712-723, 1970.
3. M. B. Pate, A. Myklebust, and J. H. Cole, A computer simulation of the turbine flow meter rotor as a drag body, *Proc. Int. Comput. in Eng. Conf. and Exhibit 1984,* Las Vegas: 184-191, New York: ASME, 1984.
4. P. D. Olivier and D. Ruffner, Improved turbine meter accuracy by utilization of dimensionless data, *Proc. 1992 National Conf. Standards Labs.* (NCSL) Workshop and Symp., Boulder, CO: NCSL, 1992, 595-607.
5. ISA-RP 31.1, *Specification, Installation and Calibration of Turbine Flowmeters*, Research Triangle Park, NC: ISA, 1977.

TABLE 9.9 Manufacturer Contact Information

Manufacturer	U.S. distributor	Relevant types
EG&G Flow Technology, Inc. 4250E Broadway Road Phoenix, AZ 85040 Tel: (602) 437-1315	Not applicable	Axial turbines — liquid and gas, including 3A sanitary Insertion axial turbines — liquid and gas
Flow Automation, Inc. 9303 W. Sam Houston Pkwy S. Houston, TX 77099 Tel: (713) 272-0404	Not applicable	Insertion axial turbines — liquid and gas Axial turbines — liquid and gas, including 3A sanitary
Fisher-Rosemount Petroleum Highway 301 North P.O. Box 450 Statesboro, GA 30459 Tel: (912) 489-0200	Not applicable	Axial turbines — liquid
Equimeter, Inc. 805 Liberty Blvd. DuBois, PA 15801 Tel: (814) 371-8000	Not applicable	Axial turbines — gas Dual rotor axial turbines — gas
Exact Flow, LLC P.O. Box 14515 Scottsdale, AZ 85267-4545 Tel: (602) 922-7446	Not applicable	Dual rotor axial turbines — liquid
Sparling Instruments Co., Inc. 4097 North Temple City Blvd. P.O. Box 5988 El Monte, CA 91734 Tel: (818) 444-0571	Not applicable	Propeller meters
Ferraris Medical, Ltd Ferraris House Aden Road Enfield Middlesex EN3 7SE U.K. Tel: 44 (0)1818059055	Ferraris Medical, Inc. P.O. Box 344 9681 Wagner Road Holland, NY 14080 (716) 537-2391	Tangential turbine monitoring spirometers
Micro Medical, Ltd. The Admiral's Offices The Chatham Historic Dockyard Chatham, Kent ME4 4TQ U.K. 44 (0)163 843383	Micro Direct, Inc. P.O. Box 239 Auburn, ME 04212 (800) 588-3381	Axial turbine diagnostic spirometers
SeaMetrics Inc. P.O. Box 1589 Kent, WA 98035 Tel: (206) 872-0284	Not applicable	Axial turbines — liquid Insertion axial turbines — liquid Multijet turbines
Onicon, Inc. 2161 Logan St. Clearwater, FL 34625 Tel: (813) 447-6140	Not applicable	Axial turbines — liquid Insertion axial turbines, single and dual rotor — liquid

TABLE 9.9 (continued) Manufacturer Contact Information

Manufacturer	U.S. distributor	Relevant types
Eldec Corporation 16700 13th Avenue W. P.O. Box 97027 Lynnwood, WA 98046 Tel: (206) 743-8499	Not applicable	Angular momentum flowmeters
ABB Kent Messtechnik GmbH Otto-Hahn-Strasse 25 D-68623 Lampertheim Germany 49 62069330	ISTEC Corporation 415 Hope Avenue Roselle, NJ 07203 (908) 241-8880	Multijet turbines
Aquametro AG Ringstrasse 75 CH-4106 Therwil 061 725 11 22	ISTEC Corporation 415 Hope Avenue Roselle, NJ 07203 (908) 241-8880	Multijet turbines
Digiflow Systems B.V. Postbus 46 6580 AA Malden The Netherlands 31 243 582929	Digiflow Systems 781 Clifton Blvd. Mansfield, OH 44907 (419) 756-1746	'Multijet' axial turbines — liquid

6. ANSI/ASME MFC-4M-1986 (R1990), *Measurement of Gas Flow by Turbine Meters*, New York: ASME.
7. API MPM, *Measurement of Liquid Hydrocarbons by Turbine Meters*, 3rd ed., Washington, D.C.: API (Amer. Petroleum Inst.), 1995, chap. 5.3.
8. AGA Transmission Meas. Committee Rep. No. 7, Measurement of fuel gas by turbine meters, Arlington, VA: AGA (Amer. Gas Assoc.), 1981.
9. Int. Recommendation R32, Rotary piston gas meters and turbine gas meters, Paris: OIML (Int. Organization of Legal Metrology), 1989.
10. ISO 9951:1993, Measurement of gas flow in closed conduits — turbine meters, Geneva, Switzerland: Int. Organization for Standardization, (also available ANSI), 1993.
11. T. H. J. J. Van Der Hagen, Proof of principle of a nuclear turbine flowmeter, *Nucl. Technol.*, 102(2), 167-176, 1993.
12. K. Termaat, W. J. Oosterkamp, and W. Nissen, *Nuclear turbine coolant flow meter,* U.S. Patent No. 5,425,064, 1995.
13. AWWA C704-92, Propeller-type meters for waterworks applications, Denver, CO: Amer. Water Works Assoc., 1992.
14. ANSI/AWWA C701-88, Cold water meters — turbine type, for customer service, Denver, CO: Amer. Water Works Assoc., 1988.
15. B. M. Wright and C. B. McKerrow, Maximum forced expiratory flow rate as a measure of ventilatory capacity, *Br. Med. J.*, 1041-1047, 1959.
16. W. F. Z. Lee, R. V. White, F. M. Sciulli, and A. Charwat, *Self-correcting self-checking turbine meter,* U.S. Patent No. 4,305,281, 1981.
17. W. F. Z. Lee, D. C. Blakeslee, and R. V. White, A self-correcting and self-checking gas turbine meter, Trans. ASME, *J. Fluids Eng.*, 104, 143-149, 1982.
18. D. F. Ruffner, and P. D. Olivier, *Wide range, high accuracy flow meter,* U.S. Patent No. 5,689,071, 1997.
19. D. F. Ruffner, Private communication, 1996.

20. A. Abdul-Razzak, M. Shoukri, and J. S. Chang, Measurement of two-phase refrigerant liquid-vapor mass flow rate. III. Combined turbine and venturi meters and comparison with other methods, *ASHRAE Trans.: Research,* 101(2), 532-538, 1995.
21. W. J. Shim, T. J. Dougherty, and H. Y. Cheh, Turbine meter response in two-phase flows, *Proc. Int. Conf. Nucl. Eng. — 4,* 1 part B: 943-953, New York: ASME, 1996.
22. K. Minemura, K. Egashira, K. Ihara, H. Furuta, and K. Yamamoto, Simultaneous measuring method for both volumetric flow rates of air-water mixture using a turbine flowmeter, *Trans. ASME, J. Energy Resources Technol.,* 118, 29-35, 1996.
23. M. W. Johnson and S. Farroll, Development of a turbine meter for two-phase flow measurement in vertical pipes, *Flow Meas. Instrum.,* 6(4), 279-282, 1995.
24. L. A. Quick, Gas measurement by insertion turbine meter, *Proc. 70th Int. School Hydrocarbon Meas.,* OK, 1995. (Available E. Blanchard, Arrangements Chair, Shreveport, LA, (318) 868–0603.)
25. W. E. Chiles, L. E. Vetsch, and J. V. Peterson, *Shrouded flowmeter turbine and improved fluid flowmeter using the same,* U.S. Patent No. 4,012,957, 1977.
26. ANSI/AWWA C708-91, *Cold-water meters, multi-jet-type,* Denver, CO: Amer. Water Works Assoc., 1991.
27. J. Goss, *Flow meter,* U.S. Patent No. 5,337,615, 1994.
28. D. Wadlow, and L. M. Layden, *Controlled flow volumetric flowmeter,* U.S. Patent No. 5,284,053, 1994.
29. C. R. Sparks, private communication, 1996.

Further Information

D. W. Spitzer (ed.), *Flow measurement,* Research Triangle Park, NC: ISA, 1991, is a popular 646 page practical engineering guide of which four chapters concern turbine flowmeters, sanitary flowmeters, insertion flowmeters and custody transfer issues.

A. J. Nicholl, Factors affecting the performance of turbine meters, *Brown Bov. Rev.,* 64(11), 684-692, 1977, describes some basic design factors not commonly discussed elsewhere.

J. W. DeFeo, Turbine flowmeters for measuring cryogenic liquids, *Adv. Instrum. Proc.,* 47 pt.1, 465-472, ISA, 1992, provides guidance for a less frequently discussed application in which axial turbines perform well.

ATS Standardization of spirometry, 1994 Update, *Am. J. Respir. Care Med.,* 152, 1107-1136, 1995, is the latest version of the official U.S. guideline for spirometry generated by the American Thoracic Society.

M. D. Lebowitz, The use of expiratory flow rate measurements in respiratory disease, *Ped. Pulmonol.,* 11, 166-174, 1991, provides a review of the diagnostic usefulness of PEFR using portable spirometers.

W. M. Jungowski and M. H. Weiss, Effects of flow pulsation on a single-rotor turbine meter, *Trans. ASME, J. Fluids Eng.,* 118(1), 198-201, 1996.

C. R. Sparks and R. J. McKee, *Method and apparatus for assessing and quantifying pulsation induced error in gas turbine flow meters,* U.S. Patent No. 5,481,924, 1996, assigned to the Gas Research Institute, Chicago, is a potential solution to the accurate measurement of pulsating gas flows which will require engineering development and a high performance rotation sensor [29].

K. Ogawa, S. Ito, and C. Kuroda, Laminar-turbulent velocity profile transition for flows in concentric annuli, parallel plates and pipes, *J. Chem. Eng. Japan,* 13(3), 183-188, 1990, provides mathematical descriptions of velocity profiles.

J. Lui and B. Huan, Turbine meter for the measurement of bulk solids flow rate, *Powder Technol.,* 82, 145-151, 1995, describes theory and experiments relating to a very simple design for a unique application, namely the volumetric measurement of plug flows of sands in a pipe.

9.5 Impeller Flowmeters

Harold M. Miller

Impeller flowmeters, sometimes called *paddlewheel* meters, are one of the most commonly used flowmeter variety. The impeller flow sensor is a direct offshoot of the old-fashioned undershot waterwheel. They are a cost-effective alternative to turbine meters, and can be used in applications that are difficult to handle with other types of flow metering instruments. In their mechanical construction, they are usually very simple, with any sophistication residing in the electronics used to detect the rotation rate of the impeller, and in the choice of materials of construction for chemical corrosion attributes of the metered fluids. They are related to turbine meters in that both types use a rotating mechanical element to produce the output signal. They differ in the fact that the impeller meter has its rotary axis transverse to the flow stream, as opposed to the turbine meter axis, which is parallel thereto.

These devices are available in two basic types. The insertion style is the most common. (See Figures 9.34(*b*) and 9.35.) The sensor is directly installed into a hole in a pipe, with saddle or welded fitting installed at the entry to seal the sensor to the pipe. The sensor can also be preinstalled in appropriate Tees in the pipeline. Most suppliers have designs that can be installed in operating pipe systems, with little, if any, loss of the fluid during the installation process. The impeller design is also supplied in in-line (through-flow) sensors for those applications where such use is desirable (Figure 9.34(*a*)). The in-line meters are commonly of somewhat higher accuracy than the insertion style.

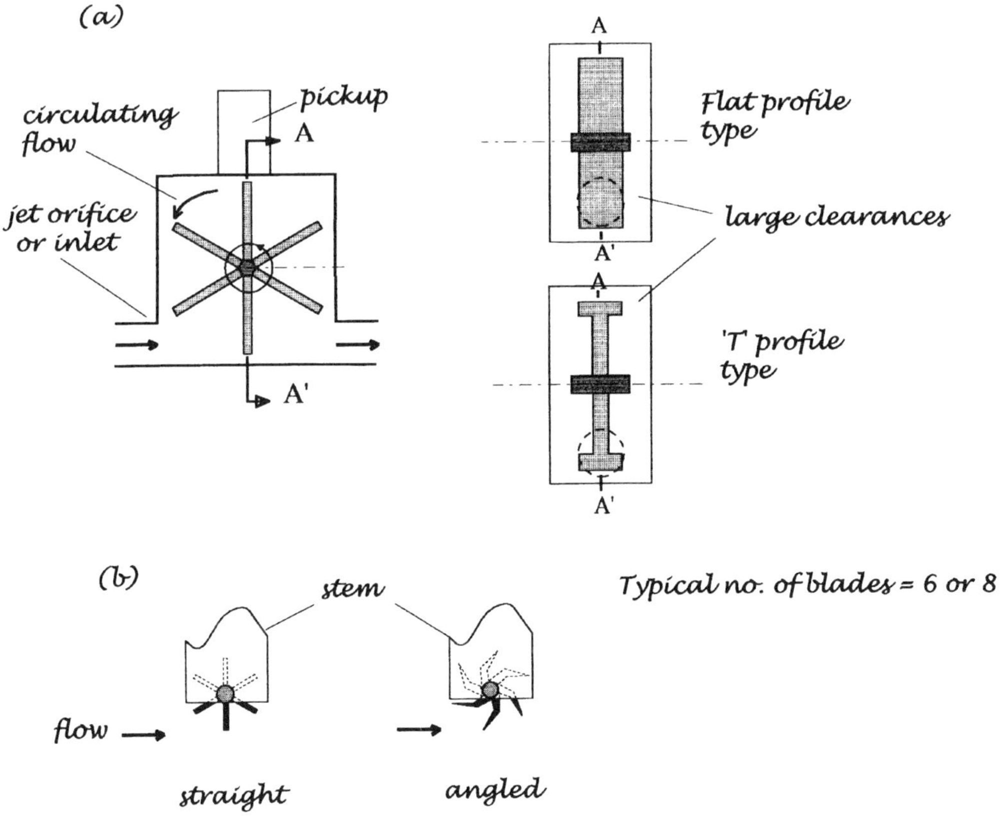

FIGURE 9.34 Impeller flowmeter rotor design variations (*a*) in-line meters; (*b*) insertion meters. (Figure courtesy of David Wadlow.)

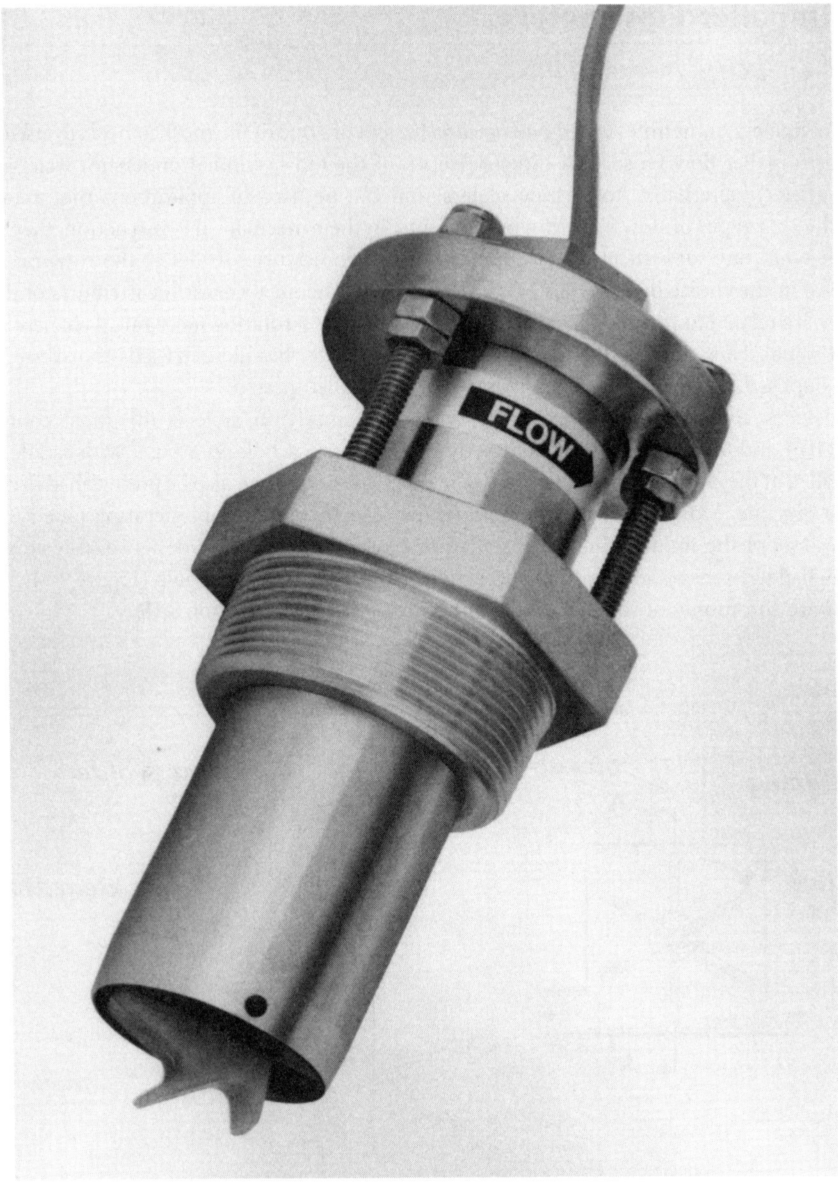

FIGURE 9.35 A typical insertion type impeller flow sensor.

In-line meters are *linear, inferential, volumetric* flowmeters for measuring *liquid and gas flows* and are more *sensitive at lower flow rates* compared with, for example, axial turbine flowmeters. (This is of course because the flow blade incidence angle is much greater.) Insertion impeller flowmeters instead measure the flow velocity in a small region within a flow conduit.

Impeller flowmeters are generally suited to much lower flow rate ranges than the same size axial turbines and hence often find applications where axial turbines cannot be used. This, and a tendency for lower cost and high reliability, are key strengths. Competition with in-line axial turbines can only occur in overlapping design flow ranges.

Some in-line meters have interchangeable orifice sizes, allowing the same body to be used over different flow ranges. The orificed in-line meters are typically insensitive to flow condition. Insertion meters are, of course, very sensitive to flow condition if volumetric measurements are inferred. This is an important distinction and can extend to differences between requirements for specific in-line versions. Installation is important. It affects the user when he or she is designing an installation, selecting a flowmeter, and deciding on how much confidence to place in the measurement reading.

Paddlewheel flowmeters are *never* used in liquid hydrocarbon custody transfer applications. The flow measurement capacity is insufficient to compete in monitoring large volume transfers, and accuracy is not sufficient for valuable fluids.

These meters go by a variety of names, depending on which name the manufacturer selects and also the application. One often-quoted, historical root is also the impulse turbine invented by Pelton in the 19th century and has given rise to the "Pelton wheel turbine" description currently used by some suppliers. A device like this was originally used to drive milling wheels directly about the vertical axis, rather than through a right angle gear as in the "undershot" wheel described. Another common name, besides paddlewheel and impeller-type, based at least on suppliers' descriptions for the in-line variety, is single-jet tangential, or simply tangential, turbine. The former description distinguishes this design from that of multijet tangential turbine flowmeters and therefore deserves mention. There are thus a confusing variety of names in current use, all relating to essentially the same impeller-like, vaned flowmetering principle:

1. Impeller type: insertion and in-line
2. Paddlewheel type: insertion and in-line
3. Pelton turbine wheel type: in-line
4. Tangential turbine flowmeter: insertion and in-line
5. Single-jet tangential turbine flowmeter: in-line
6. Impulse turbine flowmeter: in-line

Historically, the impeller sensor is based on early electronic speedometers used in pleasure boating. Signet, followed quickly by Data Industrial, moved to modify the basic design of such marine instruments to meet the significantly more demanding service life requirements of industrial flow measurement. The resulting flowmeters, and their descendants, have been widely used since 1975. Significant engineering efforts have resulted in highly reliable instruments used in many extremely demanding applications.

Sensing Principles

All impeller flow sensors must detect the rotation of the impeller, and in their usual form transmit a pulse train, at a frequency related to the rotational velocity of the impeller. Being essentially a digital output, impeller sensors can typically transmit signals over quite long distances, up to 1 km when so required. Detection principles used include:

1. One or more magnets retained in the impeller or mechanically connected thereto, using the zero crossing of an induced ac field to generate the pulse train.
2. One or more coupling devices contained within the impeller, modulating a transmitted frequency that is processed to produce the pulse train.
3. One or more metallic targets installed within the sensor, sensed by any of the proximity pickup techniques commercially available, to produce output pulses.
4. One or more magnets retained in the impeller, used to switch a Hall effect device, producing the output pulse train.
5. Optical devices, both transmissive and reflective, have been used to sense the passage of the impeller blades to produce the output pulse train.
6. Measurement of the change in electric reactance due to the passage of impeller vanes through the measurement field area, conditioned to produce the output pulse train.

Any given supplier can produce several types of impeller flowmeter, each type using a different detection method depending on market requirements. Since the impeller can operate to rotational velocities of 4000 rpm or higher, output frequencies can be as high as 500 Hz. At low flow rates, the frequencies can be as low as 0.2 Hz. This factor should be addressed early in the selection procedure because the chosen output device must be capable of the frequency output range of the sensor.

Most sensor manufacturers can either supply or specify an appropriate meter to relate sensor output to flow rate in Engineering units, either U.S. Customary or SI. These meters are usually capable of displaying both flow rate and accumulated flow for the sensor to which they are connected. Additional outputs are available, either stand-alone or in combination with the meter, providing periodic pulse outputs at definable flow increments or analog outputs scaled to flow rate. In addition, certain control functions, alarms, and other special features required by the various markets served are often incorporated in these meters.

Flow Ranges

The insertion style of impeller meter, even when Tee-installed, is a local sensing device, measuring the flow velocity in only a part of the flow stream. The manufacturer calibrates the meter for average flow across the entire cross-section of the pipe. The paddlewheel location is usually close to the inner diameter of the pipe in a region of flow with a velocity significantly below the average flow velocity in the pipe. Proper calibration practices by the manufacturer make the meters effective at flow rates as low as 10 cm s^{-1} average velocity in spite of the low local velocities at the impeller. Generally speaking, flows at Reynolds numbers as low as 5000 can be run with no requirement for special calibration, and frequently lower numbers can be handled with no difficulty. The usual range specification is a flow rate equivalent to 0.3 to 10 m s^{-1} average flow velocity. The diversity of products available allows operation to velocities considerably lower than the 0.3 m s^{-1} range, indeed to as low as 0.07 m s^{-1} in certain specialty impeller flowmeters.

Installation

Pipe sizes: Pipe sizes in which these sensors have been installed run the gamut from small bore tubing to 2.3 m outside diameter. The larger pipe sizes are those that show the greatest installed cost savings over alternative metering systems. The impeller meter, in fact, can be cost-effective in any flowmeter application that is consistent with the accuracy of the instrument, particularly if the application involves pump or valve control.

Piping system restrictions: Most suppliers require at least 10 pipe diameter lengths of straight pipe upstream of the installed meter, and 5 downstream. These conditions are required to minimize the asymmetry of the flow stream in the neighborhood of the impeller, in the installed piping, which can be caused by elbows, tees, and valves. More is better; no supplier is likely to complain that there is too great a straight pipe length upstream. Less can adversely affect the calibration of the sensor due to the local variations of velocity resulting from flow disturbances.

Operating pressure: Manufacturers' standard offerings are usually consistent with the pressure limitations of the materials of construction of the piping system with plastic piping systems, and are commonly as high as 2.7 MPa with steel or brass piping. Higher pressures are available, but are usually somewhat more expensive. Pressure drop generated by the installed flowmeter is usually low. Manufacturers can usually supply information on the anticipated pressure drop.

Calibration: Calibration of the sensors is usually specified by the manufacturer. Some manufacturers provide a calibration factor in terms of gallons per pulse or in pulses per gallon. Others, such as Data Industrial, provide data relating frequency to flow rate in GPM or other volumetric rate units. For insertion-style sensors, any such instrument must be field calibratable to accommodate the variation in the relation of impeller rotational velocity with average flow rate with pipe size variation.

Accuracy: Accuracies for in-line impeller flowmeters vary considerably but can be high, ranging from ±0.2% reading for liquids for one manufacturer to several percent of full scale for several others. The difference between accuracies specified as % full scale and % reading should be particularly noted.

Manufacturers use both, particularly with this class of meter. A ±1% full-scale device is often far less accurate than, for example, a ±2% reading device. For example, it is in error by 10% at the minimum of its flow range if the turndown is 10:1, whereas the ±2% reading device is still only in error by ±2%.

Output accuracy is usually specified as ±1% of full rated flow. This specification is broad enough to handle the dimensional and frictional variations in the sensors as manufactured. Some producers will custom-calibrate the sensors to meet special needs, but this task should be performed only when the mating pipe entry and exit can be shipped to, and accommodated by, the manufacturer. Alternatively, the manufacturer can usually provide new calibration values if the meter used is identified, the meter reading is known, and the actual flow is known for at least two points on the flow curve. When calibration in place is required, the anticipated accuracy is in the range of ±0.5% of full scale or 1% of indicated flow, whichever is greater.

Repeatability of readings is usually on the order of 0.5%. Linearity, except at the extremes of the flow range, is also expected to be no worse than 0.5%, and over the full design range of the meter is normally accurate within the ±1% of full rated flow.

To achieve the accuracies noted above, careful attention must be paid to proper installation, particularly with regard to insertion depth.

Design features: These flow sensors are offered in materials of construction compatible with a broad range of aqueous solutions, of both high and low pH and with deionized water. For this latter service, the history of the sensor, particularly for particle generation, should be reviewed. Flow streams with high concentrations of solids, and particularly of fibrous solids, should be carefully reviewed when the impeller sensor is under consideration. Construction is reasonably forgiving with particulates, but caution in such applications is warranted. Heat transfer fluids are also compatible with sensors from at least two of the suppliers. Basic application limits should be discussed with the supplier, who should have a broad history of successful applications to back up recommendations, as well as sensor materials with test results specific to the measured fluid. So-called "hot tap" or "wet tap" versions are available for installation in operating pipelines and in-line or on-line service. A submersible version capable of metering in flooded well pits for extended periods of time is available from some suppliers. At least two manufacturers provide "intrinsically safe" sensors for application in hazardous environments.

Areas of application: Impeller flow sensors are widely used in the following fields with considerable success:

- Agricultural and horticultural irrigation
- Deionized water systems, including silicon wafer fabrication
- Heating, ventilating, and air conditioning, energy management
- Industrial waste treatment
- Industrial filtration systems
- Chemical reagent metering and batching
- Municipal water systems (potable water)

The sensors can be supplied with analog output for control applications, if desired.

Some suppliers have developed specialty impeller meters adapted to provide long life, ruggedness, and maintainability required in more demanding applications. The history of acceptable operation in applications similar to that proposed for the metering system should be reviewed with the supplier. Custodial transfer applications are normally unsuitable for the impeller sensor.

Manufacturers

The following is a selection of U.S. manufacturers. Their locations and phone numbers are listed in Table 9.10.

TABLE 9.10 U.S. Manufacturers of Impeller Flowmeters

Manufacturer	City, State	Phone number
Data Industrial Corp.	Mattapoisett, MA	(508) 758-6390
G. Fischer (Signet)	Tustin, CA	(714) 731-8800
SeaMetrics	Kent, WA	(206) 872-0284
Roper Flow Technology, Inc.	Phoenix, AZ	(602) 437-1315
Blancett	Altus, OK	(405) 482-0036
Hoffer Flow Controls, Inc.	Elizabeth City, NC	(800) 628-4584; (919) 331
McMillan Co.	Georgetown, TX	(512) 863-0231
Flowmetrics, Inc.	Canoga Park, CA	(818) 346-4492
Proteus Industries, Inc.	Mountain View, CA	(415) 964-4163

Data Industrial

In-line and Insertion Meters: 0.07 to 18 m/s (0.25 to 60 fps) fluid velocity. Available materials to accommodate most industrial fluids. Temperatures from –29°C to 152°C (–20°F to 305°F). Energy monitoring. Digital and/or analog meter outputs. NEMA 4, 4X, and 6P constructions. FM, CSA Approvals.

G. Fischer (Signet)

In-line and Insertion Meters: 0.1 to 7 m/s (0.3 to 20 fps) fluid velocity. Available materials to accommodate most industrial fluids. Temperatures to 149°C (300°F). Energy monitoring. Digital and/or analog meter ouputs. NEMA 4, 4X constructions. FM, CE Approvals.

SeaMetrics

In-line meter. Sizes 3/8 in. to 2 in. Plastic, brass. Clean water applications. Flows from 0.05 to 40.0 gpm. Accuracy ±1% of full scale, (FS). Insertion. Range of IP probes. Plastic, brass, stainless. Sapphire bearings and carbide shafts. Up to 250°F. 0.3 to 30 fps. Accuracy ±1% to ±2% FS.

Roper Flow Technology

In-line meters: Omniflow®, liquids and gases, highly precise, range 7.6 to 5677 mL min^{-1} (liquid), 0.0025 to 0.68 Am^3h^{-1} (gases). Cryogenics to 593°C, pressure to 60,000 psi, viscosity compensated.

Optiflo®, plastic construction for liquids only with some reduction in performance capabilities.

Hoffer

MF Series (Miniflow). High-performance, in-line low flowmeters using a "Pelton rotor" for liquids and gases. Liquid flow ranges from 57 mL min^{-1} to 11.5 rpm. Linearity ±1% reading over 10:1 range. Repeatability ±0.1% typical. Gas flow ranges depending on density from 0.02 ACFM to 1.0 ACFM. Linearity ±1.5% reading. Repeatability ±0.2%. Various bearings available. Ball bearings used for high accuracy applications. UVC curves available. "Smart" transmitter available for temperature–viscosity correction + linearization.

McMillan

Low viscosity liquids: In-line meters. Ranges from 13 mL min^{-1} to 10 L min^{-1}. Accuracies ±1% to ±3% full scale. IR rotation sensors. Have a Teflon version for corrosives.

Gases: Teflon version for chlorine, fluorine, etc. Accuracy ±3% full scale. Ranges: 25 AmL min^{-1} to 5 AmL min^{-1}.

Flowmetrics

Series 600: Stainless steel, in-line tangential turbine for gases and liquids. Ranges: 0.001 to 2.0 GPM liquids; 0.001 to 2.0 ACFM gases. Linearities vary according to range from ±1% FS to ±5% FS. Repeatabilities ± 0.05%, traceable NIST.

Proteus

In-line paddlewheel flowmeters and flow switches — liquids only. Typically used for water cooling lines on electrical or vacuum equipment. Quoted for use with liquids up to 120 cSt. Media include: water, treated water, deionized water, ethylene glycol, light oils, etc. Range 0.08 to 60 gpm. Accuracies vary between models in the ±3% to ±4% FS range. Meters bodies can incorporate optional integral temperature and pressure transducers. These meters are also available with a FluidTalk™ computer interface protocol and PC software for constructing embedded control systems for flow, temperature, and pressure.

References

1. N. P. Cheremisinoff, *Applied Fluid Flow Measurement,* New York: Marcel Dekker, 1979, 106.
2. H. S. Bean (ed.), *Fluid Meters, 6th ed.,* New York: American Society of Mechanical Engineers, 1971, 99.
3. Flow Meter Finder™ Impeller-Type Flow Meters, http://www.seametrics.com/flowmeter-finder/impeller.htm.
4. R. Koch and D. Palmer, Revisiting Measurement Technologies, Multijet vs. Positive Displacement Meters in 1996, http://www.wateronline.com/companies/mastermeterinc/tech.html.
5. L. C. Kjelstrom, Methods of measuring pumpage through closed conduit irrigation systems, *J. Irrigation Drainage Eng.-ASCE,* 117, 748-757. 1991.

9.6 Electromagnetic Flowmeters

Halit Eren

Magnetic flowmeters have been widely used in industry for many years. Unlike many other types of flowmeters, they offer true noninvasive measurements. They are easy to install and use to the extent that existing pipes in a process can be turned into meters simply by adding external electrodes and suitable magnets. They can measure reverse flows and are insensitive to viscosity, density, and flow disturbances. *Electromagnetic flowmeters* can rapidly respond to flow changes and they are linear devices for a wide range of measurements. In recent years, technological refinements have resulted in much more economical, accurate, and smaller instruments than the previous versions.

As in the case of many electric devices, the underlying principle of the electromagnetic flowmeter is Faraday's law of electromagnetic induction. The induced voltages in an electromagnetic flowmeter are linearly proportional to the mean velocity of liquids or to the volumetric flow rates. As is the case in many applications, if the pipe walls are made from nonconducting elements, then the induced voltage is independent of the properties of the fluid.

The accuracy of these meters can be as low as 0.25% and, in most applications, an accuracy of 1% is used. At worst, 5% accuracy is obtained in some difficult applications where impurities of liquids and the contact resistances of the electrodes are inferior as in the case of low-purity sodium liquid solutions.

Faraday's Law of Induction

This law states that if a conductor of length l (m) is moving with a velocity v (m s^{-1}), perpendicular to a magnetic field of flux density B (Tesla), then the induced voltage e across the ends of conductor can be expressed by:

$$e = B\,l\,v \tag{9.57}$$

The principle of application of Faraday's law to an electromagnetic flowmeter is given in Figure 9.36. The magnetic field, the direction of the movement of the conductor, and the induced emf are all perpendicular to each other.

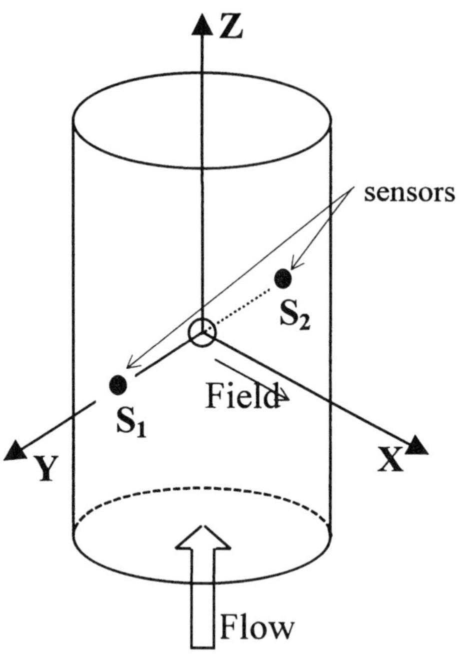

FIGURE 9.36 Operational principle of electromagnetic flowmeters. Faraday's law states that a voltage is induced in a conductor moving in a magnetic field. In electromagnetic flowmeters, the direction of movement of the conductor, the magnetic field, and the induced emf are perpendicular to each other on *x*, *y*, and *z* axes. Sensors S1 and S2 experience a virtual conductor due to liquid in the pipe.

Figure 9.37 illustrates a simplified electromagnetic flowmeter in greater detail. Externally located electromagnets create a homogeneous magnetic field passing through the pipe and the liquid inside it. When a conducting flowing liquid cuts through the magnetic field, a voltage is generated along the liquid path between two electrodes positioned on the opposite sides of the pipe.

In the case of electromagnetic flowmeters, the conductor is the liquid flowing through the pipe, and the length of the conductor is the distance between the two electrodes, which is equal to the tube diameter. The velocity of the conductor is proportional to the mean flow velocity of the liquid. Hence, the induced voltage becomes:

$$e = B\,D\,v \tag{9.58}$$

where D (m) is the diameter of pipe. If the magnetic field is constant and the diameter of the pipe is fixed, the magnitude of the induced voltage will only be proportional to the velocity of the liquid. If the ends of the conductor, in this case the sensors, are connected to an external circuit, the induced voltage causes a current, i, to flow, which can be processed suitably as a measure of the flow rate. The resistance of the moving conductor can be represented by R to give the terminal voltage v_T of the moving conductor as $v_T = e - iR$.

Electromagnetic flowmeters are often calibrated to determine the volumetric flow of the liquid. The volume of liquid flow, Q (L s^{-1}), can be related to the average fluid velocity as:

$$Q = A\,v \tag{9.59}$$

Writing the area, A (m^2), of the pipe as:

$$A = \pi D^2/4 \tag{9.60}$$

gives the induced voltage as a function of the flow rate.

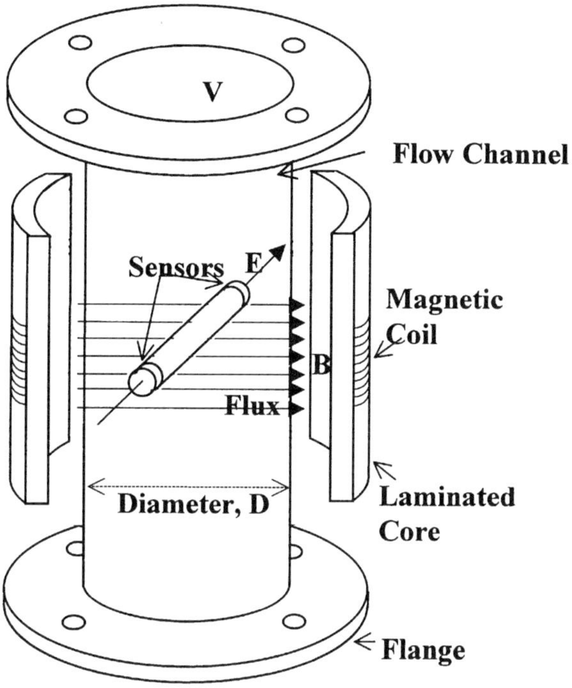

FIGURE 9.37 Construction of practical flowmeters. External electromagnets create a homogeneous magnetic field that passes through the pipe and the liquid inside. Sensors are located 90° to the magnetic field and the direction of the flow. Sensors are insulated from the pipe walls. Flanges are provided for fixing the flowmeter to external pipes. Usually, manufacturers supply information about the minimum lengths of the straight portions of external pipes.

$$e = B4/\pi DQ \tag{9.61}$$

Equation 9.61 indicates that in a carefully designed flowmeter, if all other parameters are kept constant, then the induced voltage is linearly proportional to the liquid flow only.

Based on Faraday's law of induction, there are many different types of electromagnetic flowmeters available, such as ac, dc, and permanent magnets. The two most commonly used ones are the ac and dc types. This section concentrates mainly on ac and dc type flowmeters.

Although the induced voltage is directly proportional to the mean value of the liquid flow, the main difficulty in the use of electromagnetic flowmeters is that the amplitude of the induced voltage is small relative to extraneous voltages and noise. Noise sources include:

- Stray voltage in the process liquid
- Capacitive coupling between signal and power circuits
- Capacitive coupling in connection leads
- Electromechanical emf induced in the electrodes and the process fluid
- Inductive coupling of the magnets within the flowmeter

Construction and Operation of Electromagnetic Flowmeters

Common to both ac and dc electromagnetic flowmeters, the magnetic coils create a magnetic field that passes through the flow tube and process fluid. As the conductive fluid flows through the flowmeter, a voltage is induced between the electrodes in contact with the process liquid. The electrodes are placed at positions where maximum potential differences occur. The electrodes are electrically isolated from the

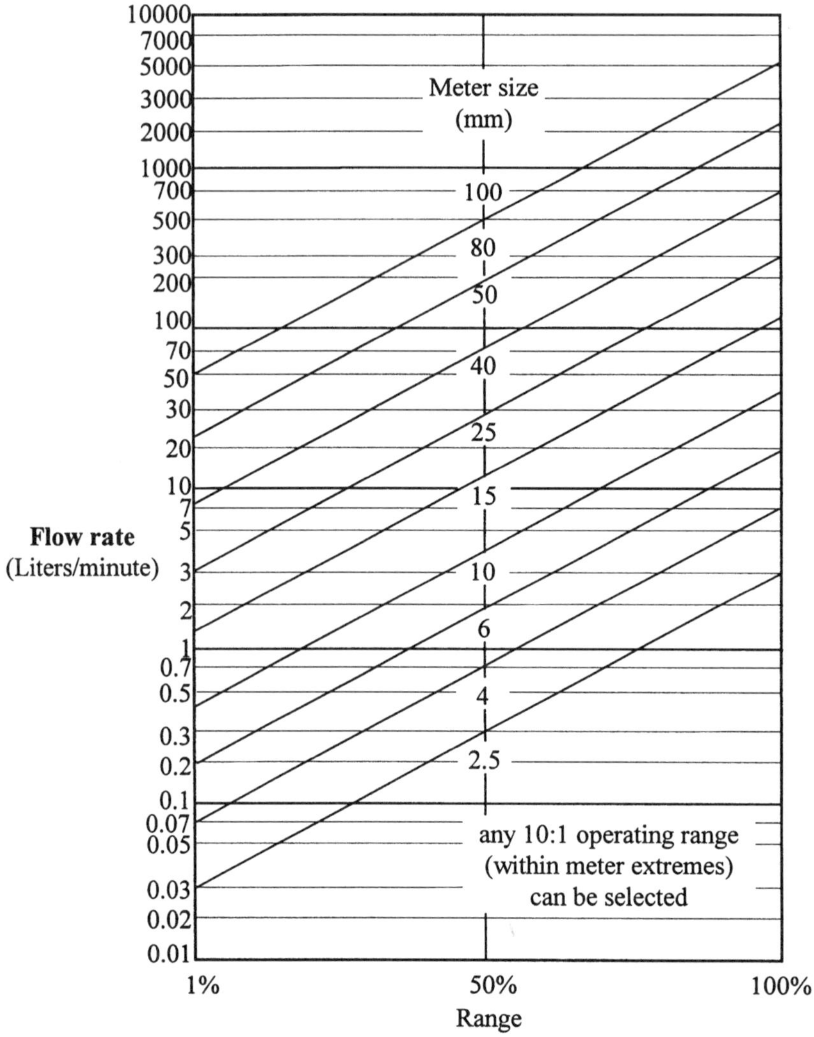

FIGURE 9.38 Selection of flowmeters. In the selection of a suitable flowmeter for a particular application, care must be exercised in handling the anticipated liquid velocities. The velocity of liquid must be within the linear range of the device. For example, a flowmeter with 100 mm internal diameter can handle flows between 50 L min^{-1} to 4000 L min^{-1}. An optimum operation will be achieved at a flow rate of 500 L min^{-1}.

pipe walls by nonconductive liners to prevent short-circuiting of electrode signals. The liner also serves as protection to the flow tube to eliminate galvanic action and possible corrosion due to metal contacts. Electrodes are held in place by holders that also act as sealing.

Dimensionally, magnetic flowmeters are manufactured from 2 mm to 1.2 m in diameter. In a particular application, the determination of the size of the flowmeter is a matter of selecting the one that can handle the anticipated liquid velocities. The anticipated velocity of the liquid must be within the linear range of the device. As an example of a typical guide for selection, the capacities of various size flowmeters are given in Figure 9.38.

Some electromagnetic flowmeters are made from replaceable flow tubes whereby the field coils are located external to the tubes. In these flowmeters, the flanges are located far apart in order to reduce their adverse effects on the accuracy of measurements; hence, they are relatively larger in dimensions. Whereas in others, the field coils are located closer to the flow tube or even totally integrated together.

In this case, the flanges could be located closer to the magnets and the electrodes, thus giving relatively smaller dimensions. On the other hand, the miniature and electrodeless magnetic flowmeters are so compact in size that face-to-face dimensions are short enough to allow them to be installed between two flanges.

The wetted parts of a magnetic flowmeter include the liners, electrodes, and electrode holders. Many different materials such as rubber, teflon, polyurethane, polyethylene, etc. are used in the construction to suit process corrosivity, abrasiveness, and temperature constraints. The main body of a flowmeter and electrodes can be manufactured from stainless steel, tantalum, titanium, and various other alloys. Liners are selected mainly to withstand the abrasive and corrosive properties of the liquid. The electrodes must be selected such that they cannot be coated with insulating deposits of the process liquid during long periods of operation.

The pipe between the electromagnets of a flowmeter must be made from nonmagnetic materials to allow the field to penetrate the fluid without any distortion. Therefore, the flow tubes are usually constructed of stainless steel or plastics. The use of steel is a better option because it adds strength to the construction. Flanges are protected with appropriate liners and do not make contact with the process fluid.

In some electromagnetic flowmeters, electrodes are cleaned continuously or periodically by ultrasonic or electric means. Ultrasonics are specified for ac and dc type magnetic flowmeters when frequent severe insulating coating is expected on the electrodes that might cause the flowmeter to cease to operate in an anticipated manner.

The operation of a magnetic flowmeter is generally limited by factors such as linear characteristics, pressure ratings of flanges, and temperatures of the process fluids. The maximum temperature limit is largely dependant on the liner material selection and usually is set to around 200°C. For example, ceramic liners can withstand high temperatures, but are subject to cracking in case of sudden changes in temperatures of the process fluid.

During the selection of electromagnetic flowmeters, the velocity constraints should be evaluated carefully to secure accurate performance over the expected range. The full-scale velocity of the flowmeter is typically 0.3 m s^{-1} to 10 m s^{-1}. Some flowmeters can measure lower velocities with somewhat poorer accuracy. Generally, employment of electromagnetic flowmeters over a velocity of 5 m s^{-1} should be considered carefully because erosion of the pipe and damages to liners can be significant.

The accuracy of a conventional magnetic flowmeter is usually expressed as a function of full scale (FS), typically 0.5% to 1% FS. However, dc flowmeters have a well-defined zero due to an automatic zeroing capabilities; therefore, they have a percentage rate of accuracy better than ac types, typically 0.5% to 2%.

Types of Electromagnetic Flowmeters

AC Magnetic Flowmeters

In many commercial electromagnetic flowmeters, an alternating current of 50 Hz to 60 Hz in coils creates the magnetic field to excite the liquid flowing within the pipe. A voltage is induced in the liquid due to Faraday's law of induction, as explained above. A typical value of the induced emf in an ac flowmeter fixed on a 50 mm internal diameter pipe carrying 500 L min^{-1} is about 2.5 mV.

Historically, ac magnetic flowmeters were the most commonly used types because they reduced polarization effects at the electrodes. In general, they are less affected by the flow profiles of the liquid inside the pipes. They allow the use of high Z_{in} amplifiers with low drift and high pass filters to eliminate slow and spurious voltage drifts emanating mainly from thermocouple and galvanic actions. These flowmeters find many applications as diverse as the measurement of blood flow in living specimens. Miniaturized sensors allow measurements on pipes and vessels as small as 2 mm in diameter. In these applications, the excitation frequencies are higher than industrial types, 200 Hz to 1000 Hz.

A major disadvantage of the ac flowmeter is that the powerful ac field induces spurious ac signals in the measurement circuits. This necessitates periodic adjustment of zero output at zero velocity conditions — more frequently than for dc counterparts. Also, in some harsh industrial applications, currents in the magnetic field can vary, due to voltage fluctuations and frequency variations in the

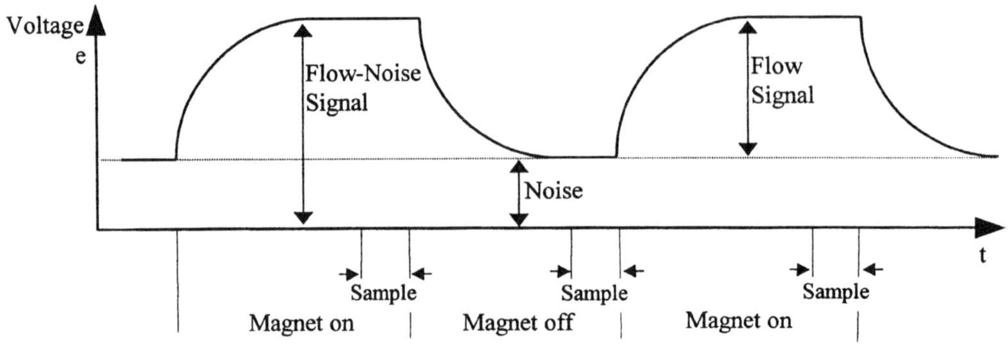

FIGURE 9.39 The signals observed at the electrodes represent the sum of the induced voltage and the noise. When the current in the magnetic coils is turned off, the signal across the electrodes represents only the noise. Subtracting the measurement of the flowmeter when no current flows through the magnet from the measurement when current flows through the magnet effectively cancels out the effect of noise.

power lines. The effect of fluctuations in the magnetic field can be minimized by the use of a reference voltage proportional to the strength of the magnetic field to compensate for these variations. To avoid the effects of noise and fluctuations, special cabling and calibration practices recommended by the manufacturers must be used to ensure accurate operation. Usually, the use of two conduits is required — one for signals and one for power. The cable lengths should also be set to certain levels to minimize noise and sensitivity problems.

Ac flowmeters operating at 50, 60, or 400 Hz are readily available. In general, ac flowmeters can operate from 10 Hz to 5000 Hz. High frequencies are preferred in determining the instantaneous behavior of transients and pulsating flows. Nevertheless, in applications where extremely good conducting fluids and liquid metals are used, the frequency must be kept low to avoid skin effects. On the other hand, if the fluid is a poor conductor, the frequency must not be so high such that dielectric relaxation is not instantaneous.

Dc Magnetic Flowmeters

Unlike ac magnetic flowmeters, direct current or pulsed magnetic flowmeters excite the flowing liquid with a field operating at 3 Hz to 8 Hz. As the current to the magnet is turned on, a dc voltage is induced at the electrodes. The signals observed at the electrodes represent the sum of the induced voltage and the noise, as illustrated in Figure 9.39. When the current in the magnetic coils is turned off, the signal represents only the noise. Subtracting the measurement of the flowmeter when no current flows through the magnet from the measurement when current flows through the magnet effectively cancels out the effect of noise.

If the magnetic field coils are energized by normal direct current, then several problems can occur: polarization, which is the formation of a layer of gas around the measured electrodes, as well as electrochemical and electromechanical effects. Some of these problems can be overcome by energizing the field coils at higher frequencies or ac. However, higher frequencies and ac generate transformer action in the signal leads and fluid path. Therefore, the coils are excited by dc pulses at low repetition rates to eliminate the transformer action. In some flowmeters, by appropriate sampling and digital signal processing techniques, the zero errors and the noise can be rejected substantially.

The zero compensation inherent in dc magnetic flowmeters eliminates the necessity of zero adjustment. This allows the extraction of flow signals regardless of zero shifts due to spurious noise or electrode coating. Unlike ac flowmeters, a larger insulating electrode coating can be tolerated that could shift the effective conductivity significantly without affecting performance. If the effective conductivity remains high enough, a dc flowmeter will operate satisfactorily. Therefore, dc flowmeters are less susceptible to drifts, electrode coatings, and changes in process conditions in comparison to conventional ac flowmeters.

TABLE 9.11 List of Manufacturers of Electromagnetic Flowmeters

ABB K-Flow Inc. P.O. Box 849 45 Reese Rd. Millville, NJ 08332 Tel: (800) 294-8116 Fax: (609) 825-1678	Marsh-McBirney, Inc. 4539 Metropolitan Court Frederick, MD 21704 Tel: (301) 879-5599 Fax: (301) 874-2172
Control Warehouse Shores Industrial park Ocala, FL 34472 Tel: (800) 633-0319 Fax: (352) 687-8925	Nusonics Inc. 11391 E. Tecumseh St. Tulsa, OK 74116-1606 Tel: (918) 438-1010 Fax: (918) 438-6420
Davis Instruments 4701 Mount Hope Dr. Baltimore, MD 21215 Tel: (410) 358-3900 Fax: (410) 358-0252	Rosemount Inc. Dept. MCA 15 12001 Technology Dr. Eden Prairie, MN 55344 Tel: (612) 828-3006 Fax: (612) 828-3088
Fischer Porter 50 Northwestern Dr. P.O. Box 1167T Salem, NH 03079-1137 Tel: (603) 893-9181 Fax: (603) 893-7690	Sparling Instruments Co., Inc. 4097 Temple City Blvd. P.O. Box 5988 El Monte, CA 91734-1988 Tel: (800) 423-4539
Johnson Yokogawa Dept. P, Dart Rd. Newman, GA 30265 Tel: (800) 394-9134 Fax: (770) 251-6427	Universal Flow Monitors, Inc. 1751 E. Nine Mile Rd. Hazel Park, MI 48030 Tel: (313) 542-9635 Fax: (313) 398-4274

Dc magnetic flowmeters do not have good response times due to the slow pulsed nature of operations. However, as long as there are not rapid variations in the flow patterns, zero to full-scale response times of a few seconds do not create problems in the majority of applications. Power requirements are also much less as the magnet is energized part of the time. This gives an advantage in power saving of up to 75%.

If the dc current to the magnet is constant, the proportional magnetic field can be kept steady. Therefore, the amplitudes of the dc voltages generated at the electrodes will be linearly proportional to the flow. However, in practice, the current to the magnet varies slightly due to line voltage and frequency variations. As in the case of ac flowmeters, voltage and frequency variations could necessitate the use of a reference voltage. Because the effect of noise can be eliminated more easily, the cabling requirements are not as stringent.

To avoid electrolytic polarization of the electrodes, bipolar pulsed dc flowmeters have been designed. Also, modification of dc flowmeters led to the development of miniature dc magnetic flowmeters with wafer design for a limited range of applications. The wafer design reduces the weights as well as the power requirements.

Table 9.11 provides a listing of several manufacturers of electromagnetic flowmeters.

Installation and Practical Applications of Electromagnetic Flowmeters

Conventional ac and dc magnetic flowmeters have flanges at the inlet and the outlet that need to be bolted to the flanges of the pipe. Face-to-face dimensions of magnetic flowmeters differ between manufacturers; therefore, once the flowmeter is installed, new piping arrangements could be necessary if a flowmeter is replaced by one from a different manufacturer.

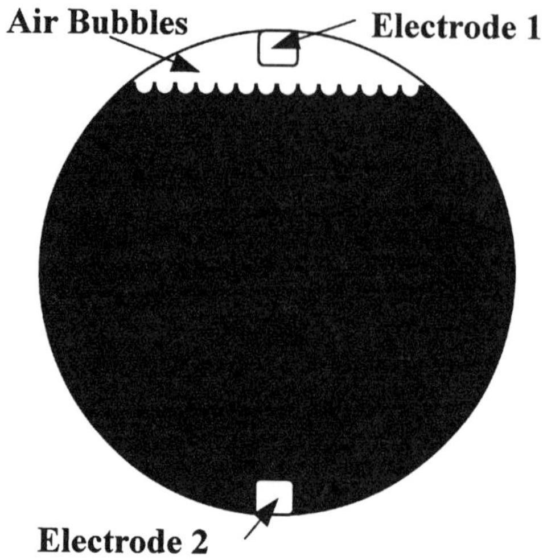

FIGURE 9.40 The pipes of electromagnetic flowmeters must be full of liquid at all times for accurate measurement. If the liquid does not make full contact with electrodes, the high impedance prevents the current flow; hence, measurements cannot be taken. Also, if the pipe is not full, even if contact is maintained between the liquid and electrodes, the empty portions of the pipe will lead to miscalculated flow rates.

Upstream and downstream straight piping requirements can vary from one flowmeter to another, depending on the manufacturer's specifications. As a rule of thumb, the straight portion of the pipe should be at least 5D/2D from the electrodes and 5D/5D from the face of the flowmeter in upstream and downstream directions, respectively. For good accuracy, one should adhere carefully to the recommendations of manufacturers for piping requirements. In some magnetic flowmeters, coils are used in such a way that the magnetic field is distributed in the coil to minimize the piping effect.

For accurate measurements, magnetic flowmeters must be kept full of liquid at all times. If the liquid does not contact the electrodes, measurements cannot be taken. Figure 9.40 illustrates this point. If the measurements are made in other than vertical flows, the electrodes should be located in horizontal directions to eliminate the possible adverse effect of the air bubbles, because the air bubbles tend to concentrate on the top vertical part of the liquid.

In the selection of magnetic flowmeters, a number of considerations must be taken into account, such as:

- Cost, simplicity, precision, and reproducibility
- Metallurgical aspects
- Velocity profiles and upstream disturbances

Most processes employ circular piping that adds simplicity to the construction of the system. The flowmeters connected to circular pipes give relatively better results compared to rectangular or square-shaped pipes, and velocity profiles of the liquid are not affected by the asymmetry. However, in circular pipes, the fringing of the magnetic field can be significant, making it necessary to employ empirical calibrations.

Selection of materials for constructing the channel of the magnetic flowmeter demands care. If the fluid is nonmetallic, a nonconducting or an insulated channel should be sufficient. In this case, wetted electrodes must be used. Electrodes must be designed to have sufficiently large dimensions to keep the output impedance at acceptable levels. Also, careful handling of electrode signals must be observed because, in many cases, malfunctioning of reference signal electronics is the main cause of flowmeter failure.

Magnetic flowmeters do not require continuous maintenance, except for periodic calibrations. Nevertheless, electrode coating, damage to the liners, and electronic failures can occur. Any modification or repair must be treated carefully because, when installed again, some accuracy can be lost. After each modification or repair, recalibration is usually necessary.

Often, magnetic flowmeter liners are damaged by the presence of debris and solids in the process liquid. Also, the use of incompatible liquid with the liners, wear due to abrasion, excess temperature, and installation and removals can contribute to the damage of liners. The corrosion in the electrodes can also be a contributing factor for the damage. In some cases, magnetic flowmeters can be repaired on-site even if severe damage occurs; in other cases, they must be shipped to the manufacturer for repairs. Usually, manufacturers supply spare parts for electrodes, liners, flow tubes and electronic components.

Calibration of electromagnetic flowmeters is achieved with a magnetic flowmeter calibrator or by electronic means. The magnetic flowmeter calibrators are precision instruments that inject simulated output signals of the primary flowmeter into the transmitter. Effectively, this signal is used to check correct operation of electronic components and make adjustments to the electronic circuits. Alternatively, calibrations can also be made by injecting suitable test signals to discrete electronic components. In some cases, empirical calibrations must be performed at zero flow while the flowmeter is filled with the stationary process liquid.

Application of magnetic flowmeters can only be realized with conductive liquids such as acids, bases, slurries, foods, dyes, polymers, emulsions, and suitable mixtures that have conductivities greater than the minimum conductivity requirements. Generally, magnetic flowmeters are not suitable for liquids containing organic materials and hydrocarbons. As a rule of thumb, magnetic flowmeters can be applied if the process liquids constitute a minimum of about 10% conductive liquid in the mixture.

The lack of any direct Reynolds number constraints and the obstructionless design of magnetic flowmeters make it practical for applications that involve conductive liquids that have high viscosity which could plug other flowmeters. They also measure bidirectional flows.

Despite the contrary belief, magnetic flowmeters demonstrate a certain degree of sensitivity to flow profiles. Another important aspect is the effect of turbulence. Unfortunately, there is very little information available on the behavior of turbulent flows when they are in transverse magnetic fields. Figure 9.41 shows an example of a flow profile in which the velocity profile is perturbed. The fluid is being retarded near the center of the channel, and accelerated at the top and bottom near the electrodes.

An important point in electromagnetic flowmeters is the effect of magneto-hydrodynamics, especially prominent in fluids with magnetic properties. Hydrodynamics is the ability of a magnetic field to modify the flow pattern. In some applications, the velocity perturbation due to the magneto-hydrodynamic effect can be serious enough to influence the accuracy of operations, as in the case of liquid sodium and its solutions.

Effects of Electric Conductivity of Fluid

For electromagnetic flowmeters to operate accurately, the process liquid must have minimum conductivity of 1 μS cm^{-1} to 5 μS cm^{-1}. Most common applications involve liquids with conductivities greater than 5 μS cm^{-1}. Nevertheless, for accurate operation, the requirement for the minimum conductivity of liquid can be affected by length of leads from sensors to transmitter electronics.

Ac flowmeters are sensitive to the nonconductive coating of the electrodes that may result in calibration shift or complete loss of signals. The dc flowmeters, on the other hand, should not be affected by nonconductive coating to such a great extent, unless the conductivity between electrodes is less than the minimum required value. In some flowmeters, electrodes can be replaced easily; while in others, electrodes can be cleaned by suitable methods. If coating is a continual problem, ultrasonic and other cleaning methods should be considered.

Zero adjustment of ac magnetic flowmeters requires compensation for noise. If the zero adjustment is performed with any fluid other than the process fluid, serious errors can result because of possible

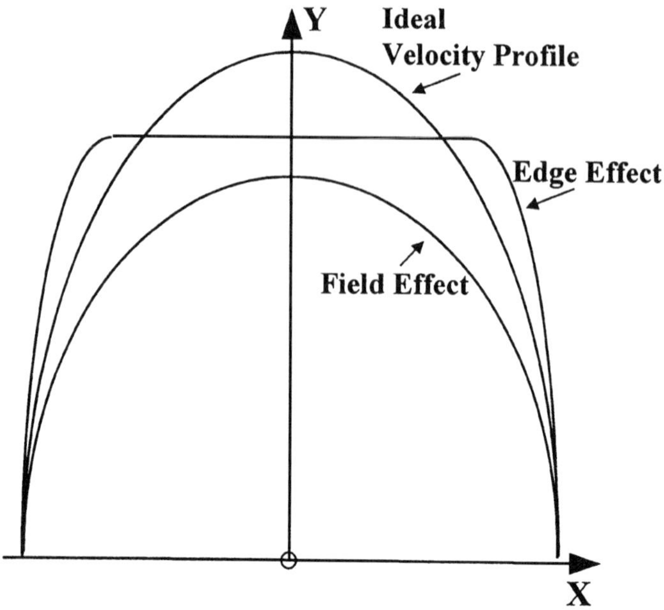

FIGURE 9.41 Flow profiles in the pipes. Magnetic flowmeters demonstrate a certain degree of sensitivity to flow profiles. The ideal velocity profile can be distorted due to edge effects and also field effects known as magneto-hydrodynamics. In some applications, the velocity perturbation due to the magnetohydrodynamic effect can be serious enough to severely influence the accuracy of operations.

differences in conductivities. Similarly, if the electrodes are coated with an insulating substance, the effective conductivity of the electrodes can be altered, thereby causing a calibration shift. If the coating changes in time, the flowmeter can continually require calibration for repeatable readings.

The resistance between electrodes can be approximated by $R = 1/\delta d$, where δ is the fluid conductivity and d is the electrode diameter. For tap water, $\delta = 200$ μS cm^{-1}; for gasoline, $\delta = 0.01$ μS cm^{-1}; and for alcohol, $\delta = 0.2$ μS cm^{-1}. A typical electrode with a 0.74 cm diameter in contact with tap water results in a resistance of 6756 Ω..

Signal Pickup and Demodulation Techniques

Magnetic flowmeters are four-wire devices that require an external power source for operations. Particularly in ac magnetic flowmeters, the high-voltage power cables and low-voltage signal cables must run separately, preferably in different conduits; whereas, in dc magnetic flowmeters, the power and signal cables can be run in one conduit. This is because in dc-type magnetic flowmeters, the voltage and the frequency of excitation of the electromagnets are relatively much lower. Some manufacturers supply special cables along with their flowmeters.

In ac flowmeters, the electrode signals can be amplified much more readily compared to their dc counterparts. That is the reason why ac flowmeters have been used successfully to measure very low flow rates, as well as the flow of very weakly conducting fluids. Nevertheless, ac flowmeters tend to be more complicated, bulky, expensive, and they require electromagnets with laminated yokes together with stabilized power supplies. In some magnetic flowmeters, it is feasible to obtain sufficiently large flow signal outputs without the use of a yoke by means of producing a magnetic field by naked coils. In this case, the transformer action to the connecting leads can be reduced considerably.

One of the main drawbacks of ac-type flowmeters is that it is difficult to eliminate the signals due to transformer action from the useful signals. The separation of the two signals is achieved by exploiting the fact that the flow-dependent signal and the transformer signal are in quadrature. That is, the useful signal is proportional to the field strength, and the transformer action is proportional to the time derivative of the field strength. The total voltage v_T can be expressed as:

$$v_T = v_F + v_t = V_F \sin(\omega t) + V_t \cos(\omega t) \tag{9.62}$$

where v_F = Induced voltage due to liquid flow
v_t = Voltage due to transformer action on wires, etc.

Phase-sensitive demodulation techniques can be employed to eliminate the transformer action voltage. The coil magnetizing current, $i_m = I_m \sin(\omega t)$, is sensed and multiplied by the total voltage v_T, giving:

$$v_T\, i_m = \left[V_F \sin(\omega t) + V_t \cos(\omega t)\right] I_m \sin(\omega t) \tag{9.63}$$

Integration of Equation 9.63 over one period between 0 and 2π eliminates the transformer voltage, yielding only the voltage that is proportional to the flow.

$$V_f = V_F\, I_m\, \pi \tag{9.64}$$

Where V_f is the voltage after integration. This voltage is proportional to the induced voltage modified by constants I_m and π.

In reality, this situation can be much more complicated because of phase shift due to eddy currents in nearby solids and conductors. Other reasons for complexity include: the harmonics because of non-linearity such as hysteresis, and capacitive pickup.

A good electrical grounding of magnetic flowmeters, as illustrated in Figure 9.42, is required to isolate relatively high common mode potential. The sources of ground potential can be in the liquid or in the pipes. In practice, if the pipe is conductive and makes contact with the liquid, the flowmeter should be grounded to the pipe. If the pipe is made from nonconductive materials, the ground rings should be installed to maintain contact with the process liquid.

If the flowmeter is not grounded carefully relative to the potential of the fluid in the pipe, then the flowmeter electrodes could be exposed to excessive common mode voltages that can severely limit the accuracy. In some cases, excessive ground potential can damage the electronics because the least-resistance path to the ground for any stray voltage in the liquid would be via the electrodes.

Some commercial magnetic flowmeters have been developed that can operate on saw-tooth or square waveforms. Universally standardized magnetic flowmeters and generalized calibration procedures still do not exist, and manufacturers use their own particular design of flow channels, electromagnets, coils, and signal processors. Most manufacturers provide their own calibration data.

Further Information

J. P. Bentley, *Principles of Measurement Systems,* 2nd ed., New York: Longman Scientific and Technical, 1988.

E. O. Doebelin, *Measurement Systems: Application and Design,* 4th ed., New York: McGraw-Hill, 1990.

J. P. Holman, *Experimental Methods for Engineers,* 5th ed., New York: McGraw-Hill, 1989.

J. A. Shercliff, *Electromagnetic Flow-Measurements,* New York: Cambridge University Press, 1987.

D. W. Spitzer, *Industrial Flow Measurement,* Research Triangle Park, NC: Instrument Society of America, 1990.

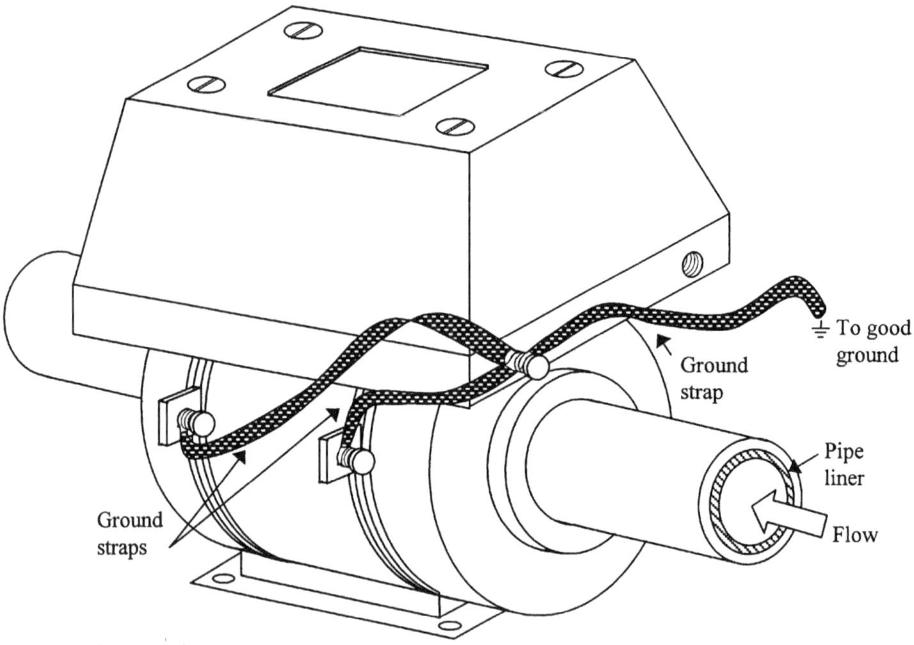

FIGURE 9.42 Grounding of electromagnetic flowmeters. A good grounding is absolutely essential to isolate noise and high common mode potential. If the pipe is conductive and makes contact with the liquid, the flowmeter should be grounded to the pipe. If the pipe is made from nonconductive materials, ground rings should be installed to maintain contact with the process liquid. Improper grounding results in excessive common mode voltages that can severely limit the accuracy and also damage the processing electronics.

9.7 Ultrasonic Flowmeters

Hans-Peter Vaterlaus, Thomas Hossle, Paolo Giordano, and Christophe Bruttin

Flow is one of the most important physical parameters measured in industry and water management. There are various kinds of flowmeters available, depending on the requirements defined by the different market segments. For many years, differential pressure types of flowmeters have been the most widely applied flow measuring device for fluid flows in pipes and open channels that require accurate measurement at reasonable cost. In markets like waterpower, water supply, irrigation, etc., however, flow must be measured without any head losses or any pressure drop. This means no moving parts, no secondary devices, nor are any restrictions allowed. Two types of flowmeters presently fulfill this requirement: Electromagnetic and ultrasonic flowmeters. Whereas *ultrasonic flowmeters* can be applied in nearly any kind of flowing liquid, *electromagnetic* flowmeters require a minimum electric conductivity of the liquid for operation. In addition, the cost of ultrasonic flowmeters is nearly independent of pipe diameter, whereas the price of electromagnetic flowmeters increases drastically with pipe diameter.

There are various types of ultrasonic flowmeters in use for discharge measurement: (1) *Transit time:* This is today's state-of-the-art technology and most widely used type, and will be discussed in this chapter section. This type of ultrasonic flowmeter makes use of the difference in the time for a sonic pulse to travel a fixed distance, first against the flow and then in the direction of flow. Transmit time flowmeters are sensitive to suspended solids or air bubbles in the fluid. (2) *Doppler:* This type is more popular and less expensive, but is not considered as accurate as the transit time flowmeter. It makes use of the Doppler

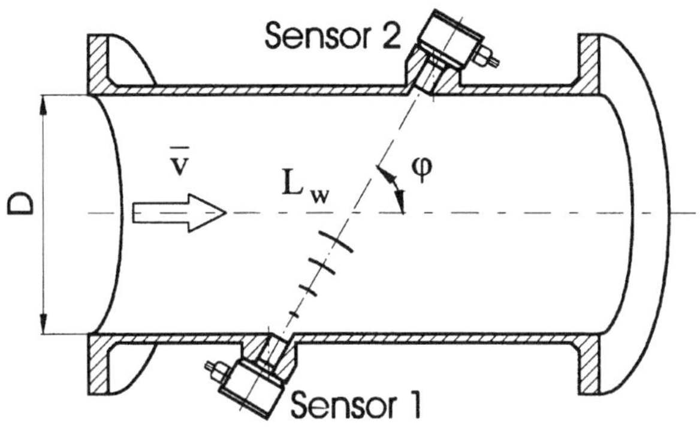

FIGURE 9.43 Principle of transit time flowmeters. Transmitting an ultrasonic pulse upstream and downstream across the flow: the liquid is moving with velocity $\bar{v}_a$ and with angle φ to the ultrasonic pulse.

frequency shift caused by sound reflected or scattered from suspensions in the flow path and is therefore more complementary than competitive to transit time flowmeters. (3) *Cross-correlation:* Two measuring sections are installed with a certain distance to each other. Both measure the energy absorption of the ultrasonic signal. A cross-correlation calculates the flow velocity. (4) *Phase shift:* The phase position of the transmitting and receiving signal is measured in the direction of the flow and against it. The resulting phase shift angle is directly proportional to the flow velocity. (5) *Drift:* The drift of an ultrasonic signal crossing the flow is measured by signal attenuation.

Transit Time Flowmeter

Principle of Operation

The acoustic method of discharge measurement is based on the fact that the propagation velocity of an acoustic wave and the flow velocity are summed vectorially. This type of flowmeter measures the difference in transit times between two ultrasonic pulses transmitted upstream t_{21} and downstream t_{12} across the flow, as shown in Figure 9.43. If there are no transverse flow components in the conduit, these two transmit times of acoustic pulses are given by:

$$t_{12} = \frac{L_w}{c + v_a \cos\varphi} \quad \text{and} \quad t_{21} = \frac{L_w}{c - v_a \cos\varphi} \tag{9.65}$$

where L_w = Distance in the fluid between the two transducers
c = Speed of sound at the operating conditions
ϕ = Angle between the axis of the conduit and the acoustic path
$\bar{v}_a$ = Axial low velocity averaged along the distance L_w

Since the transducers are generally used both as transmitters and receivers, the difference in travel time can be determined with the same pair of transducers. Thus, the mean axial velocity $\bar{v}_a$ along the path is given by:

$$\bar{v}_a = \frac{L_w}{2\cos\varphi}\left(\frac{1}{t_{21}} - \frac{1}{t_{12}}\right) = \frac{D}{2\cos\varphi\sin\varphi}\left(\frac{1}{t_{21}} - \frac{1}{t_{12}}\right) \tag{9.66}$$

The following example shows the demands on the time measurement technique: assuming a closed conduit with diameter D = 150 mm, angle ϕ = 60°, flow velocity $\bar{v}_a$ = 1 m· s^{-1}, and water temperature = 20°C. This results in transmit times of about 116 s and a time difference Δt ($\Delta t = t_{12} - t_{21}$) on the order

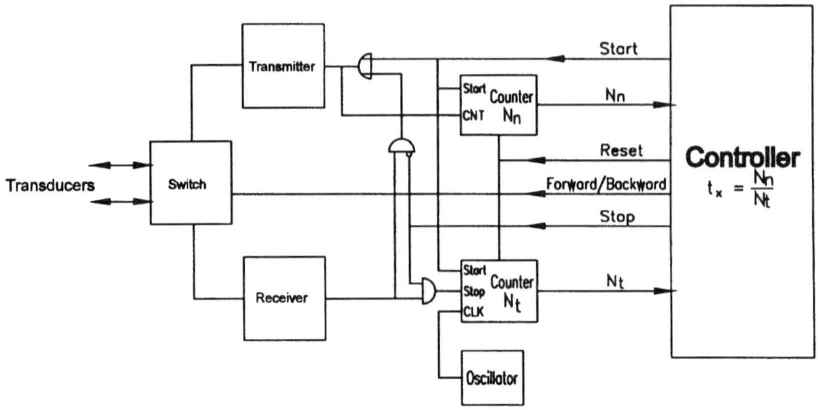

FIGURE 9.44 Block diagram of a transit time ultrasonic flowmeter using oversampling for higher resolution.

of 78 ns. To achieve an accuracy of 1% of the corresponding full-scale range, Δt has to be measured with a resolution of at least 100 ps (1×10^{-10} s).

Standard time measurement techniques are not able to meet such requirements so that special techniques must be applied. The advantage of the approach of state-of-the-art real digital measurement is to process the measured value directly by a microcomputer. The most difficult problem is to reach the required resolution and to cope with the jitter of digital logic gates. It is well known in the measurement technique that, if signals are sampled multiple times (N_n), the resolution increases with the number of samples ("oversampling") [1]. This knowledge is not only applied for analog signals but also for transit time signals and is used in today's technology of flowmeters. The propagation time t (t_{12} t_{21}), depending on the distance the sound pulse has to travel through the fluid, is measured several times. Due to this fact, one obtains the following relation [2]:

$$t = \frac{1}{N_n} \int_0^{\tau} \frac{1}{f} \mathrm{d}t \tag{9.67}$$

where τ is integration time, and f is the frequency.

According to the block diagram in Figure 9.44, two counters are used. One counter N_t is clocked during the measuring period by a stable quartz oscillator; the other one counts the number of samples N_n. The measurement stops after a certain period of some milliseconds and after having reached an integer value for N_n. The two counts of N_t and N_n are used to calculate propagation time t_{12} or t_{21} by dividing N_t by N_n. The resolution r is calculated by:

$$r = \frac{1}{N_n \cdot f} \tag{9.68}$$

Advantages of this measurement method for transit time flowmeters are:

1. The resolution of the velocity measurement is constant (typical value 0.8 mm s^{-1}).
2. The accuracy depends almost only on the stability and the temperature coefficient of the quartz oscillator.
3. Due to the multiple sampling, the jitter of the digital logic is averaged.

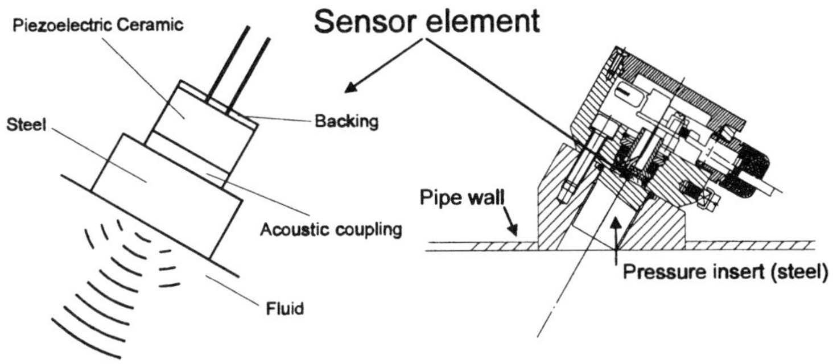

FIGURE 9.45 Ultrasonic transducer in principle and as an example of an existing version. The sensor element can be changed even under pressure.

FIGURE 9.46 Axial sensor type. The ultrasonic pulse passes directly down the axis of the pipe.

Sensors.
The transducer comprises the piezoelectric element that converts electric to acoustic energy and the basic structure for supporting the piezoceramic and providing electric connections. Transducer design entails choice of the piezoelectric element, determination of suitable dimensions and resonant frequency, and construction to withstand thermal and mechanical stress, see Figure 9.45. Considering the transmission line model [3, 4], the optimal electroacoustic response and the best matching of acoustic impedances $Z = \rho c$ of the different transducer elements is important. Not only must suitable waveforms to be detected by the electronics be obtained, but also the energy loss must be minimized when crossing several interface boundaries. Because of the wide versatility of pipe sizes and flow conditions, there are a number of different sensor configurations for transit time flowmeters.

Axial.
Due to the small differences in small diameter pipes, it is necessary to pass the sonic pulse directly down the axis of the pipe to ensure that there is sufficient path length. Figure 9.46 shows an example of an axial sensor pipe section. The upper pipe size limit for this kind of sensor is about 0.075 m [5].

Radial.
Many manufacturers supply complete metering sections with built-in sonic transducers on either side of a spool section. Such sensors, shown in Figure 9.47, are generally called "radial" because of the transducer placement. A lower pipe size limit of radial type sensors is about 100 mm.

Field Mounting.
Radial-type sensors are often used to instrument an existing line, where it is desirable to make the installation without cutting out a section of the pipe. To meet this requirement, field-mountable transducers, cemented, drilled, or welded into an existing pipe section, can be used. Metering sections with diameters up to 13 m or more can be achieved in this way.

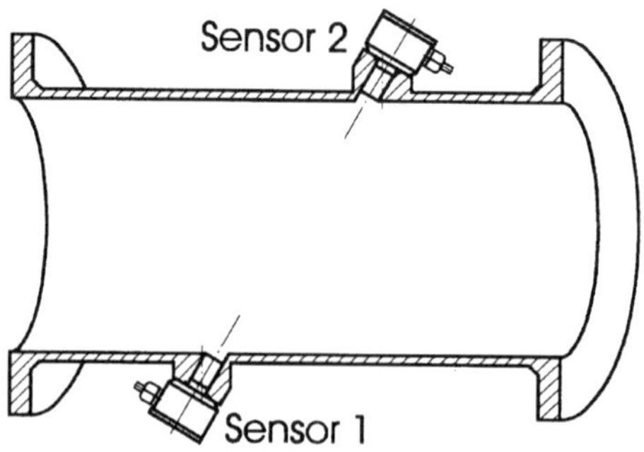

FIGURE 9.47 Radial sensor type. Manufacturers provide them as complete metering sections or as field-mounting sensors for existing conduits.

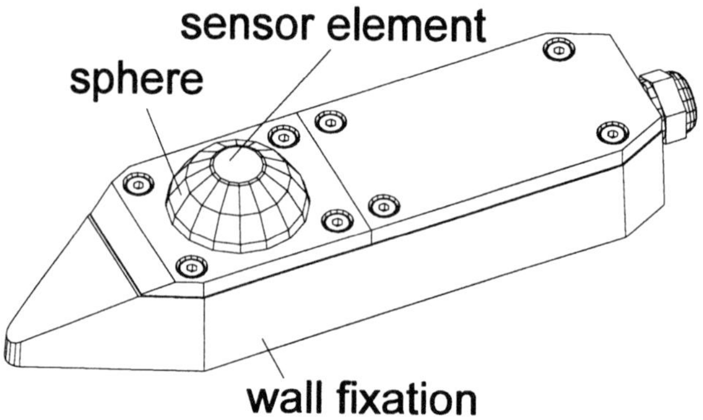

FIGURE 9.48 Open-channel sensor with sensor element placed on a movable sphere. Alignment of two sensors can be executed by laser equipment.

Clamp-on.
If there is a need for an installation where the pipe wall it not penetrated by the transducers, clamp-on systems are the right choice. Achieving a lower accuracy and being somewhat more complex to calibrate, clamp-on systems have their entitlement in applications where an easy movement of the metering section is an important requirement or where an existing process cannot be interrupted. The transducers are mounted on a calibration device and acoustically coupled to the pipe wall with grease and/or epoxy.

Open Channels and Special Applications.
In open channels, the transducers are normally mounted on or dug into the channel walls. Figure 9.48 shows an example of an open-channel sensor. The piezoceramic element is placed on a sphere to achieve a wide range of mounting possibilities. Sometimes, existing pipe sections are completely enclosed with rock or concrete. In these cases, the transducers can be fixed in the wall of the conduit. For small diameters, the resulting protrusion of the transducers into the metering section must be taken into account.

Measurement of Flow in Closed Conduits.
The most important issue in applying ultrasonic flowmeters is an understanding of the effects of the velocity profile of the flowing fluid within the conduit. The flow profile depends on the fluid, the Reynolds number Re, the relative roughness and shape of the conduit, upstream and downstream disturbances, and other factors. Transit time flowmeters give an average flow velocity v along the sonic path. The acoustic flow rate Q_{ADM} is therefore calculated through $Q = \bar{v}A$, with $\bar{v}$ the area-averaged flow velocity and A the cross-section of the conduit. In order to obtain the area-averaged flow velocity $\bar{v}$, the measured velocity $\bar{v}_a$ must be corrected by a hydraulic coefficient k_h that depends on the type of the conduit and

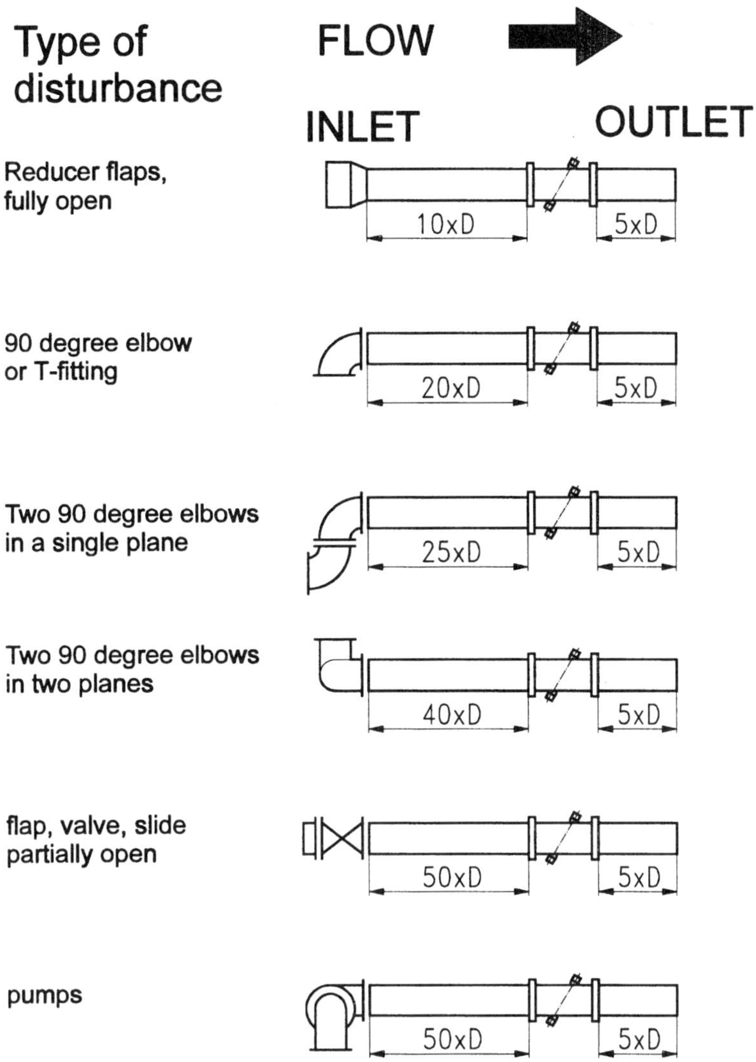

FIGURE 9.49 Minimum straight run requirements for a 1% accuracy of a single-path transit time flowmeter.

the Reynolds number. In order to achieve maximum performance and accuracy of ultrasonic flowmeters, one has to keep to sufficient straight run requirements as shown for some examples in Figure 9.49. By doing so, a typical accuracy of 1% of reading or better can be achieved, even when applying a single-path measurement system. Reduced straight runs lead to reduced accuracy. In some applications, this reduced accuracy is acceptable; if not, a multipath ultrasonic flowmeter must be installed. These flowmeters provide averaging of the various error-producing flow components. Accuracy of 0.5% of reading can be achieved even under nonideal conditions or insufficient straight runs. Figure 9.50 shows four examples of possible sonic path arrangements in a closed conduit.

Single Path with Circular and Rectangular Cross-sections.
The Reynolds number Re is given by:

$$\mathrm{Re} = \frac{\bar{v} \cdot D}{\nu} \tag{9.69}$$

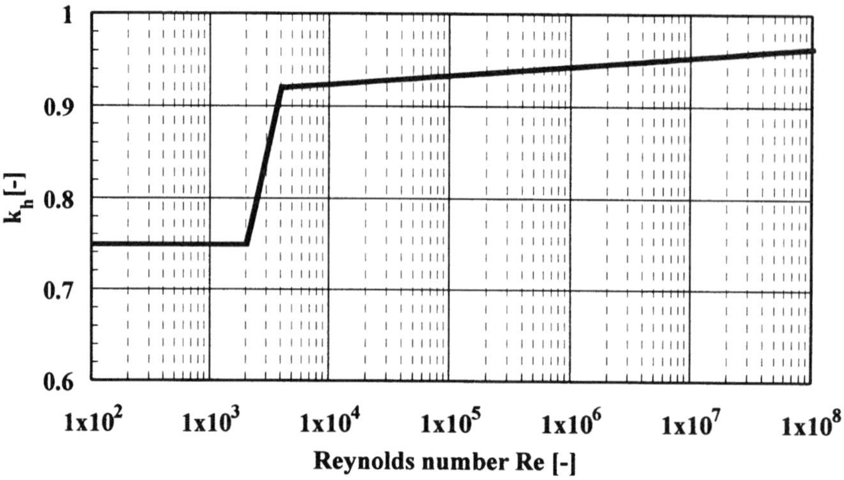

FIGURE 9.50 Dependence of the k_h factor for single-path measurement on the Reynolds number. Between a Reynolds number of 2000 and 4000, the transitional flow regime occurs.

where $\bar{v}$ = Mean velocity over the cross section $[m \cdot s^{-1}]$
D = Pipe diameter
ν = Temperature-dependent cinematic viscosity

In normal piping, a laminar flow exists as long as the Reynolds number is below about 2600. The shape of the velocity profile conforms to a parabola, and the velocity of a point on the profile is given by:

$$v(r) = v_{max}\left(1 - \left(\frac{r}{R}\right)^2\right) \tag{9.70}$$

where r = Variable radius
R = Pipe radius

Between a Reynolds number Re of 2600 and 4000, the transitional flow regime with continuous switching between laminar and turbulent velocity profile comes into existence. When the Reynolds number exceeds 4000, the velocity profile enters the turbulent flow regime. Nikuradse [6] showed that the turbulent velocity profile of an axis-symmetrical flow in a closed conduit without swirl and sufficient inlet and outlet sections and smooth walls can be expressed by:

$$v(r) = v_{max}\left(\frac{R - r}{R}\right)^{\frac{1}{n}} \tag{9.71}$$

where, according to Nikuradse, n is a Reynolds number-dependent exponent given by:

$$n = \frac{1}{\left(0.2525 - 0.0229 \times \log(\text{Re})\right)} \tag{9.72}$$

With the definition of a hydraulic corrective coefficient k_h given by:

$$k_h = \frac{\bar{v}}{\bar{v}_a} \tag{9.73}$$

where $\bar{v}$ = Mean velocity over the cross-section
$\bar{v}_a$ = Mean velocity along the sonic path

and integrating over the cross-section according to:

$$k_h = \frac{\frac{1}{A}\int_0^R\int_0^{2\pi} v(r) r \, dr \, d\theta}{\frac{1}{2R}\int_{-R}^{R} v(r) \, dr} \tag{9.74}$$

one obtains for laminar flow a hydraulic corrective coefficient of $k_h = 0.75$.

In the turbulent flow regime, according to Nikuradse [6], the hydraulic corrective coefficient k_h, dependent on the Reynolds number Re, can be expressed by:

$$k_h = \frac{1}{1.125 - 0.011 \times \log(\mathrm{Re})} \tag{9.75}$$

for a circular cross-section, and

$$k_h = 0.79 + 0.02 \times \log(\mathrm{Re}) \tag{9.76}$$

for a rectangular cross-section.

In hydropower applications, the Reynolds number Re generally exceeds the value of 4000. Figure 9.50 shows the dependence of the k_h factor in closed conduits with circular cross-section on the Reynolds number. The Reynolds number not only changes its value as a function of the flow velocity v for a given diameter D, but is also strongly dependent on the temperature-dependent cinematic viscosity $\nu(T)$. Not taking into account for correct value of Re can easily lead to errors of the k_h factor and thus to the flow-rate Q on the order of 2% to 3%. Modern transit time meters with microprocessors update the k_h factor at a rate of 4 times a second and by measuring the temperature T of the fluid at the same rate. A correct k_h factor is obtained and hence a temperature and Reynolds number-compensated flow rate Q.

Multipath Integration in Circular Cross-sections.

In reality, however, the straight run requirements as defined in Figure 9.49 cannot always be kept. In addition, cross flow errors occur when nonaxial components of velocity in a pipe alter the transit times of a pulse between the sensors. Nonaxial velocities are caused by such disturbances in closed conduits as bends, asymmetric intake flows, and discontinuities in the pipe wall or pumps. It has become accepted practice to eliminate the sensitivity of an acoustic flowmeter to velocity distributions by increasing the number of acoustic paths n. Additional paths not only decrease substantially the velocity distribution error, but also reduce the numerical integration error as well as errors due to path misalignment. This suggests that for accurate measurements for which a few tenths of a percent error are significant, the cost of installing more sensors for a four-path arrangement according to Figure 9.51 can be justified. In a

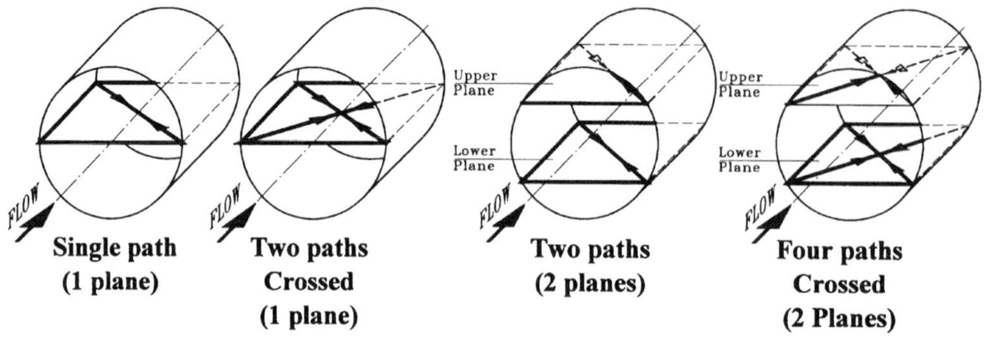

FIGURE 9.51 Possible sonic path arrangements for transit time flowmeters in closed conduits. Using multiple paths leads to reduced straight run requirements.

multipath measurement system, according to the integration method described by IEC 41, Appendix J1 [7], the flowrate Q_{ADM} can be expressed by:

$$Q_{ADM} = k\frac{D}{2}\sum_{i=1}^{n} \bar{v}_{ai} W_i L_{wi} \sin \varphi_i \tag{9.77}$$

where $\bar{v}_{ai}$ = Velocities along the acoustic path i
W_i = Corresponding weighting factor
L_{wi} = Corresponding path length
φ_i = Corresponding path angle

The method described in IEC 41 needs a very accurate transducer positioning due to the fact that the weighting factors W_i obtained by mathematical analysis are only valid when the sensors are positioned at the correct locations. Misalignment of the acoustic paths in conjunction with fixed weights can lead to considerable errors.

Measurement of Flow in Open Channels

Open-channel flow measurement is used in many applications, including water supply networks, hydrography, allocation of water for irrigation and agricultural purposes, sewage treatment plants, etc. Discharge measurements in rivers and open channels are often computed by means of a rating curve, used to convert records of water level readings into flow rates. The rating curve is developed using a set of discharge measurements and water level in the stream, and must be checked periodically to ensure that the level-discharge relationship has remained constant; many phenomena can cause the rating curve to change so that the same recorded water level produces a different discharge. This is the case of open channels under changeable hydraulic conditions due, for example, to backwater effects, gates, and where an univocal stage-discharge relationship does not exist. In this context, acoustic flowmeter application is extremely interesting and is currently experiencing wide success in water management. In fact, while different flow rate values can correspond to a given water level in relation to the hydraulic characteristics, there is always an univocal relationship between the acoustic wave propagation velocity in a flowing fluid and the flowing fluid itself.

By means of the ultrasonic technique, discharge through open channels can be determined using single- and multipath technology. In a single-path configuration, particular attention must be paid to define the vertical velocity distribution (Figure 9.52) in order to achieve a good level of accuracy by a single "line" velocity reading $\bar{v}_{az}$ at the distance z above the bottom. On the other hand, in a multipath configuration, the mean line velocity profile is well described. In this case, special attention must be paid to the integration method used to determine the flow rate from the acoustic path readings.

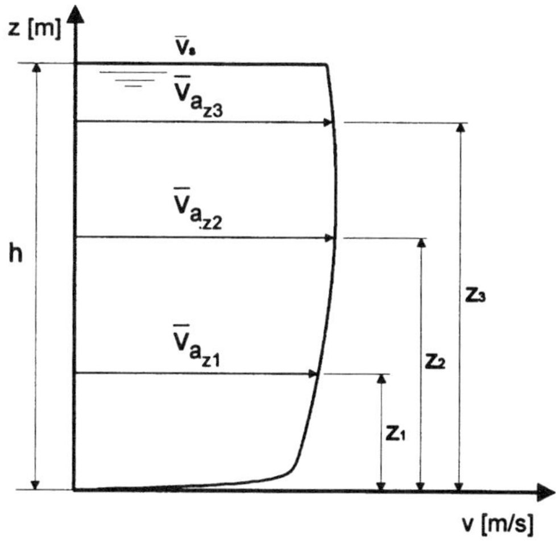

FIGURE 9.52 Open channel: possible vertical velocity distribution.

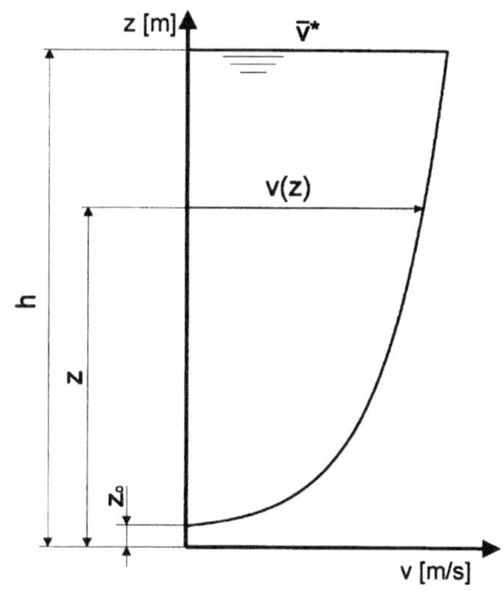

FIGURE 9.53 Logarithmic law describing open-channel velocity profile.

Single Path.

In single-path measurements, using the area-velocity method, the flow rate Q is calculated through $Q = \bar{v}A$ with $\bar{v}$ (m s^{-1}) the area-averaged flow velocity and A the cross-section. To obtain the velocity average $\bar{v}$ over the entire cross-section, the mean path velocity $\bar{v}_{az}$, measured by the acoustic flowmeter at a given depth z, must be corrected by a dimensionless hydraulic corrective coefficient k_h according to the relation $\bar{v} = k_h\bar{v}_{az}$. In general, the coefficient k_h reflects the influence of the horizontal and vertical velocity profile. It mainly depends on the water level, on the cross-section shape, and on the boundary roughness. The mean vertical velocity profile can be described by the well-known logarithmic law (Figure 9.53) given by [8].

$$\bar{v}(z) = \left(\frac{\bar{v}^*}{k}\right) \ln \frac{z}{z_0} \tag{9.78}$$

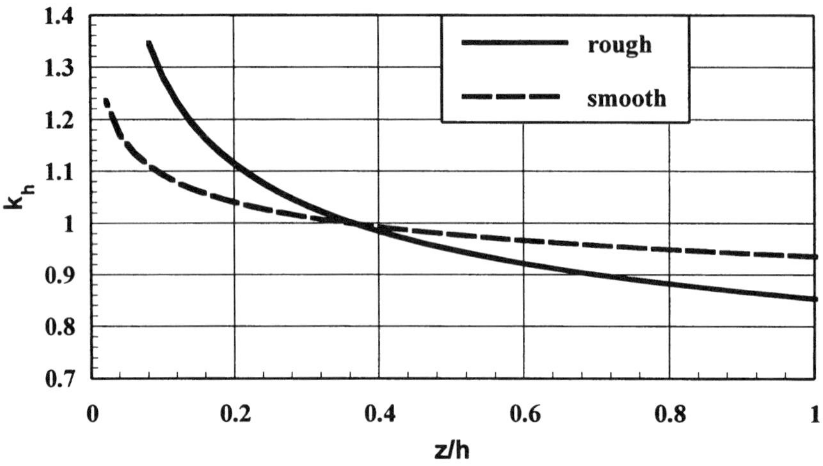

FIGURE 9.54 Logarithmic k_h factor in open channels for two different roughnesses k_s.

where $\bar{v}(z)$ = Mean flow velocity at a distance z above the bottom
k = Von Kàrmàn's turbulence constant
z = Distance above the bottom
$\bar{v}^*$ = Shear velocity
z_0 = Constant of integration, dependent on the boundary roughness

When the boundary surface is hydraulically rough, z_0 has been found to depend solely on the roughness height k_s according to the relation $z_0 = k_s/30$. Integrating Equation 9.78 over the total water height h and substituting for $\bar{v}$, the following equation is obtained:

$$k_h = \frac{\ln\left(\frac{h}{k_s}\right) - 1}{\ln\frac{z}{k_s}} \tag{9.79}$$

assuming z_0 negligible with respect to h.

In Figure 9.54, the logarithmic k_h factor is represented for two different bottom roughness. However, in many practical applications, it is often difficult to define correct values for k_s. For this reason, another k_h model has been developed on the basis of the power law [9, 10]. In this formula, the exponent $1/m$ is not constant anymore, but depends on the roughness and the hydraulic radius to take into account the influence of the channel shape. The mean vertical velocity profile $\bar{v}(z)$, according to the power law, can be expressed by the following formula (Figure 9.55):

$$\bar{v}(z) = \bar{v}_s\left(\frac{z}{h}\right)^{\frac{1}{m}} \tag{9.80}$$

where h = Current water level
$\bar{v}$ = Mean line velocity at the free surface (maximum value)
$1/m$ = Exponent

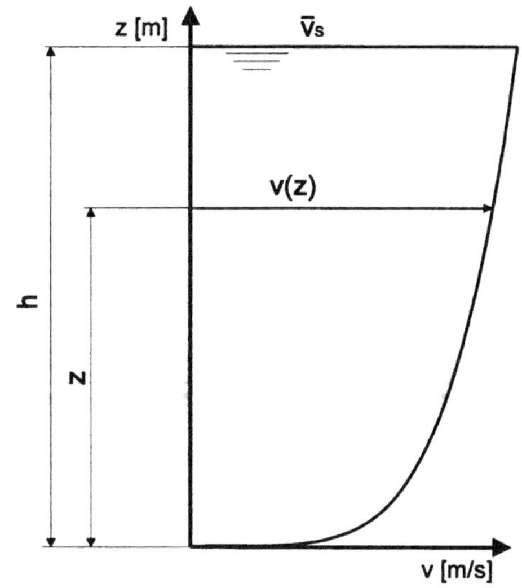

FIGURE 9.55 Power law describing open-channel velocity profile.

From the integration of Equation 9.80 over the total water depth, one obtains the following expression for the k_h factor:

$$k_h = \frac{m}{m+1}\left(\frac{h}{z}\right)^{\frac{1}{m}} \tag{9.81}$$

The value of *m*, depending on the roughness, can be expressed using the a dimensional friction factor *f* of the Darcy–Weisbach formula [8], according to:

$$m = k\sqrt{\frac{8}{f}} \tag{9.82}$$

where k = Von Kàrmàn constant, varying from 0.2 to 0.4

k depends on the suspended load (low values for high turbidity), or can be expressed using the Manning formula by the relation [11]:

$$m = \left(\frac{k}{\sqrt{g}}\right)\frac{R_h^{\frac{1}{6}}}{n} \tag{9.83}$$

where g = Gravity acceleration
n = Manning's roughness coefficient
R_h = Hydraulic radius

In Figure 9.56, the corrective factor is represented for smooth and rough surfaces. By means of this relation, an easier field application has been obtained due to wide familiarity with the Manning formula in open-channel flow computation because of its simplicity of form, high versatility, and satisfactory results.

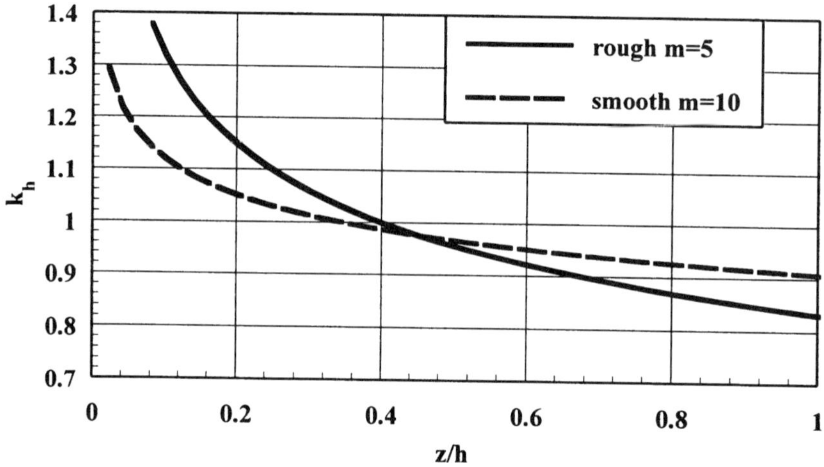

FIGURE 9.56 Power law k_h factor in open channels for two different roughnesses.

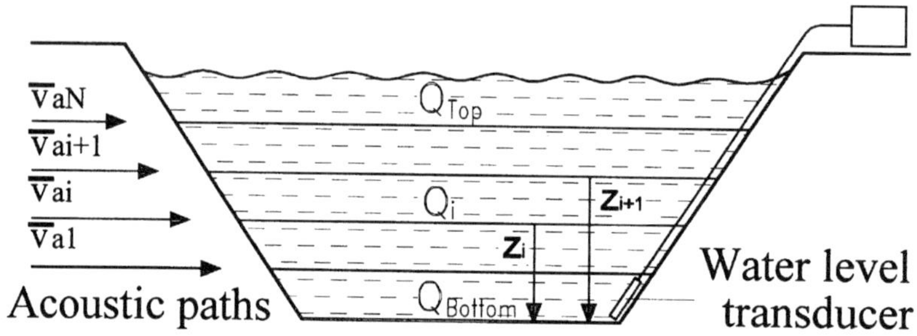

FIGURE 9.57 Multipath measurement in open channels by the mean section method. The flow velocity is measured by several levels.

Multipath.

The foregoing equations for the mean vertical velocity profile in open channels predict that the maximum mean velocity occurs at the free surface. Field and laboratory measurements, however, demonstrate that the maximum mean velocity occurs below the free surface, strongly depending on the ratio B/h, with B being the channel width. These observations show that a one-dimensional velocity distribution law cannot always completely describe flow profiles in open channels. Therefore, to reduce uncertainties in velocity profile description and to achieve high accuracies even under unfavorable hydraulic conditions, a multipath configuration must be used.

In multipath measurement, using the "mean section method," the flow velocity is measured at several levels between the free surface and the channel bottom (Figure 9.57). The total discharge Q_{ADM} is performed by the relation [12]:

$$Q_{ADM} = Q_b + Q_t + \sum_{i=1}^{n} \left(\frac{\bar{v}_{ai} + \bar{v}_{ai+1}}{2} \left[A\left(z_{i+1}\right) - A\left(z_i\right) \right] \right) \tag{9.84}$$

where Q_b = Flow rate in the bottom section with the bottom velocity obtained from the lowest path velocity by correction for bottom friction

Q_t = Flow rate at the highest active section with velocity v_{top}, interpolated from the velocity profile

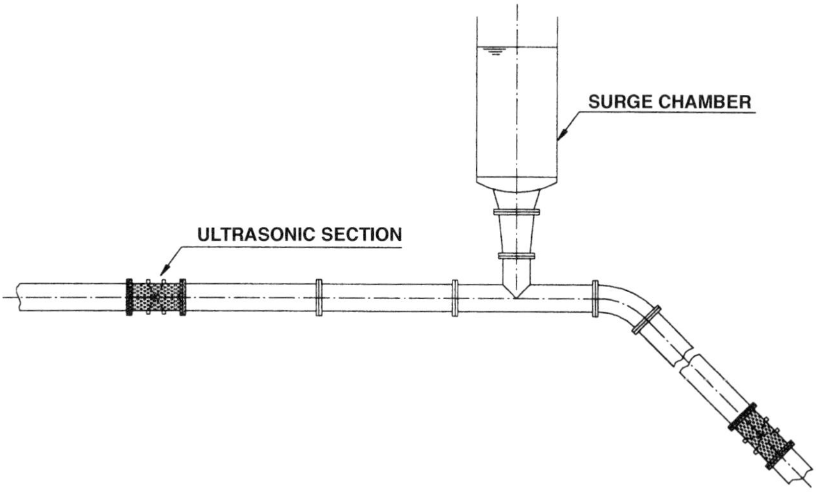

FIGURE 9.58 Penstock leak detection by transit time flowmeter. The penstock contains a surge chamber that causes mass oscillation in case of load changes of the turbine.

$\bar{v}_{ai}$ = Mean velocity along the ith acoustic path
$A(z_i)$ = Cross-section below the ith path

Modern microprocessor-controlled ultrasonic flowmeters can cope with either single-path or multi-path measurement in open channels, using the logarithmic or the power law for single-path measurement. In addition, up-to-date completely modular systems are able to use the major part of the equipment for measuring both in open channels and closed conduits.

Application: Penstock Leak Detection with Surge Chambers

Penstock leak detection is a typical acoustic flowmeter application [13]. It is used for immediate recognition of pipe rupture and leak losses in penstocks. Two flowmeters are installed at opposite ends of the penstock: one for measuring Q_{up} as near as possible to the intake, the other for measuring Q_{down} at the powerhouse entrance. In this description, the upstream flow is compared to the downstream flow and the flow difference. $\Delta Q = Q_{up} - Q_{down}$ is calculated and supervised. If the difference ΔQ exceeds a given threshold, the system enunciates alarms or valve closure contacts.

In hydropower applications another problem arises due to the presence of surge chambers (Figure 9.58). This hydraulic structure causes mass oscillations due to load changes of the turbine. Up to now, if there was a surge tank in a penstock, one had two possibilities. On the one hand, one could divide the penstock into two parts: one before the surge chamber, one after it, and protect them separately. This solution is expensive and very often impractical because the surge chamber is built into the rock and inaccessible. On the other hand, one could develop something like a "leak detection algorithm," which includes the oscillatory behavior of a penstock. Such a system, however, needs extensive field tests and the knowledge for exact settings of resonant frequency, damping factor, and thresholds. The latest technology in ultrasonic flowmeters, combined with accurate water level sensors, can offer a solution to this problem. By applying the formula:

$$\Delta Q = Q_{up} - Q_{down} - \frac{\Delta V_{chamber}}{\Delta t} \tag{9.85}$$

changes of the volume of the surge chamber due to mass oscillations are taken into account, leading to a better and more realistic behavior of penstock leak detection in the presence of surge tanks.

TABLE 9.12 Companies Manufacturing and Distributing Ultrasonic Transit Time Flowmeters

Rittmeyer AG P.O. Box 2143 CH-6302 Zug Switzerland Tel: (+4141)-767-1000 Fax: (+4141)-767-1075 instrumentation@rittmeyer.ch http://www.rittmeyer.com/	Accusonic Division, ORE International, Inc. P.O. Box 709 Falmouth, MA 02541 Tel: (508) 548-5800 http://www.ore.com/
Krohne Messtechnik GmbH&Co.KG Postfach 10 08 62 Ludwig-Krohne-Strasse 5 4100 Duisburg 1 Germany http://www.krohne.com/	Fuji Electric Co., Ltd. 12-1 Yurakucho 1-chome Chiyoda-ku Tokyo 100, Japan Tel: Tokyo 211-7111
Danfoss A/S DK-6430 Nordborg Denmark Tel: (+45) 74 88 22 22	Crouzet SA Division "Aérospatial" 25, rue Jules-Védrines 26027 Valence Cedex, France Tel: 75 79 85 11
Panametrics, Inc. 221 Crescent Street Waltham, MA 02254 Tel: (617) 899-2719 http://www.panametrics.com/	Nusonics Inc. 11391 E. Tecumseh St. Tulsa, OK 74116-1602 Tel: (918) 438-1010
	Ultraflux le technoparc 17, rue Charles Edouard Jeanneret 78306 Poissy Cedex, France Tel: 33(1)39 79 26 40

Instrumentation and Manufacturers/Distributors

Table 9.12 gives an overview of some companies manufacturing and distributing transit time flowmeters. Prices of transit time flowmeters have a very wide range due to the wide versatility of different applications and are therefore difficult to list accurately. Generally, ultrasonic flowmeters are seldom sold as a "pre-packed" instrument. For this reason, the price for a metering section, including installation, varies from a few $1000 to nearly $100,000.

References

1. M. Barmettler and P. Gruber, Anwendung von Oversampling-Verfahren zur Erhöhung der Auflösung digital erfasster Signale, *Technisches Messen,* Oldenbourg Verlag, 1992.
2. D. Hoppe, Kombinierte Zählung und Abstandsbestimmung von Impulssignalen, *Technisches Messen,* Oldenbourg Verlag, 10, 1991.
3. R. Krimholtz, D. Leedom, and G. Matthaei, New equivalent circuits for elementary piezoelectric transducers, *Electronics Lett.,* 6, 398, 1970.
4. P. D. Edmonds, *Methods of Experimental Physics, Ultrasonics,* New York: Academic Press, 1981.
5. D. W. Spitzer, *Flow Measurement, Practical Guides for Measurement and Control,* Research Triangle Park, NC: Instrument Society of America, 1991.
6. J. Nikuradse, Gesetzmässigkeiten der turbulenten Strömung in glatten Rohren, *VDI Verlag GmbH,* 1932.
7. F. L. Brand, Akustische Verfahren zur Durchflussmessung, *Messen, Prüfen Automatisieren,* April 1987.
8. *International Standard IEC 41,* 3rd ed., 1991.
9. R. H. French, *Open Channel Hydraulics,* New York: McGraw-Hill, 1985, 30.

10. Chen-Iung Chen, *J. Hydraulic Eng.*, 379, 117, 1990.
11. M. F. Karim and J. F. Kennedy, *J. Hydraulic Eng.*, 162, 113, 1987.
12. G. Grego, M. Baldin, et al., *Application of an Acoustic Flowmeter for Discharge Measurement in the Po.*
13. G. Grego and M. Baldin, *Energia Elettrica,* 1, 52, 72, 1995.
14. H. P. Vaterlaus and H. Gabler, A new intelligent ultrasonic flowmeter for hydropower applications, *Int. Water Power & Dam Construction,* 1994.

9.8 Vortex Shedding Flowmeters

Wade M. Mattar and James H. Vignos

The *vortex shedding flowmeter* first emerged 25 to 30 years ago and has steadily grown in acceptance since then to be a major flow measurement technique. Its appeal is due, in part, to the fact that it has no moving parts yet produces a frequency output that varies linearly with flow rate over a wide range of Reynolds numbers. The vortex meter has a very simple construction, provides accuracy (1% or better) comparable to higher priced and/or more maintenance-intensive techniques, and works equally well on liquids and gases. In addition, it is powered primarily by the fluid and lends itself more readily than other linear flow devices to two-wire operation. Comparing the vortex shedding flowmeter to an orifice plate, the former has higher accuracy and rangeability, does not require complex pressure impulse lines, is less sensitive to wear and, for volumetric flow measurement, does not require the need to compensate for fluid density.

Industrial vortex shedding flowmeters are normally available in pipe sizes ranging from 15 mm to 300 mm (1/2 in. to 12 in.), with some manufacturers offering sizes up to 400 mm (16 in.). Flow ranges covered depend on fluid properties and meter design. Typical ranges for a 15 mm meter are:

- Water at 21°C (70°F); 0.06 to 2.2 L s^{-1} (1 to 35 gallons per minute)
- Air at 16°C (60°F) and 101 kPa (14.7 psia); 1.1 to 15.7 L s^{-1} (140 to 2000 cubic feet per hour)
- Dry saturated steam at 689 kPa (100 psig); 4.5 to 225 kg h^{-1} (10 to 500 pounds per hour)

Typical ranges for a 300 mm (12 in.) meter are:

- Water at 21°C (70°F); 5.4 to 5400 L s^{-1} (85 to 8500 gallons per minute)
- Air at 16°C (60°F) and 101 kPa (14.7 psia); 157 to 12500 L s^{-1} (20,000 to 1,600,000 cubic feet per hour)
- Dry saturated steam at 689 kPa (100 psig); 1240 to 124000 kg h^{-1} (2750 to 275,000 pounds per hour)

Temperature capability ranges from cryogenic temperatures up to 427°C (800°F). Pressure capability as high as 20.7 MPa (3000 psig) is available.

Principle of Operation

Probably the first time, ages ago, that anyone placed a blunt obstacle in a flowing fluid, he or she observed the whirlpools or vortices that naturally form and shed downstream. In everyday life, examples of vortex shedding are numerous. The undulation of a flag is due to vortex shedding from the pole, and the singing of telephone wires in a strong wind is due to shedding from the wires. Analysis by Theodore von Karman in 1911 described the stability criterion for the array of shed vortices. Consequently, when a stable array of vortices form downstream from an obstacle, it is often referred to as the von Karman vortex street (Figure 9.59).

Very early on, it was noted that, for a large class of obstacles, as the velocity increased, the number of vortices shed in a given time (or frequency of vortex shedding) increased in direct proportion to the velocity. The dimensionless Strouhal number, St, is used to describe the relationship between vortex shedding frequency and fluid velocity and is given by:

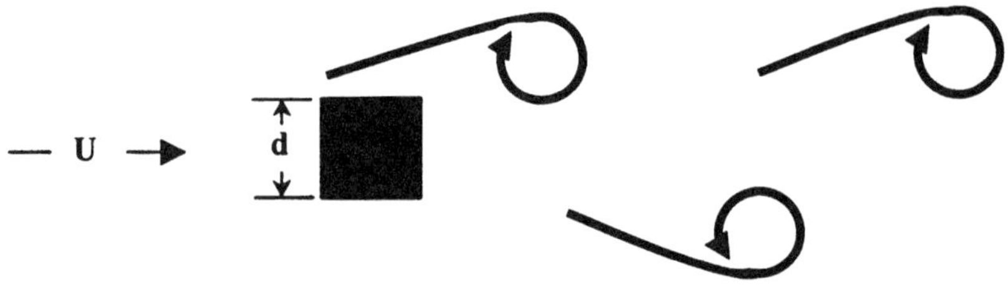

FIGURE 9.59 Von Karman vortex street.

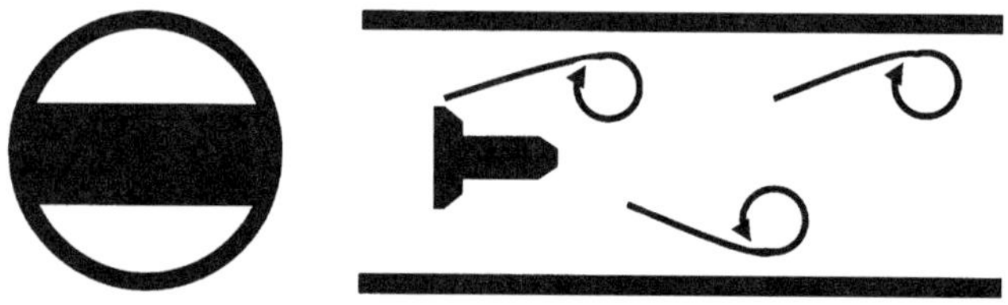

FIGURE 9.60 Vortex shedding in a pipe.

$$\mathrm{St} = \frac{f \times d}{U} \tag{9.86}$$

where f = Vortex shedding frequency
d = Width of shedding body
U = Fluid velocity

Alternatively,

$$U = \frac{f \times d}{\mathrm{St}} \tag{9.87}$$

Although early studies were conducted in unconfined flow, it was later observed that vortex shedding also occurred in confined flow, such as exists in a pipe (see Figure 9.60). For this case, the average fluid velocity, $\overline{U}$, and the meter Strouhal number, St′, replace the fluid velocity and Strouhal number, respectively, in Equation 9.87 to give:

$$\overline{U} = \frac{f \times d}{\mathrm{St}'} \tag{9.88}$$

Since the cross-sectional area, A, of the pipe is fixed, it is possible to define a flowmeter K factor, K, that relates the volumetric flow rate (Q) to the vortex shedding frequency. Given that:

$$Q = A \times \overline{U} \tag{9.89}$$

From Equation 9.88, one obtains:

$$Q = \left(\frac{(A \times d)}{\mathrm{St}'} \right) \times f \tag{9.90}$$

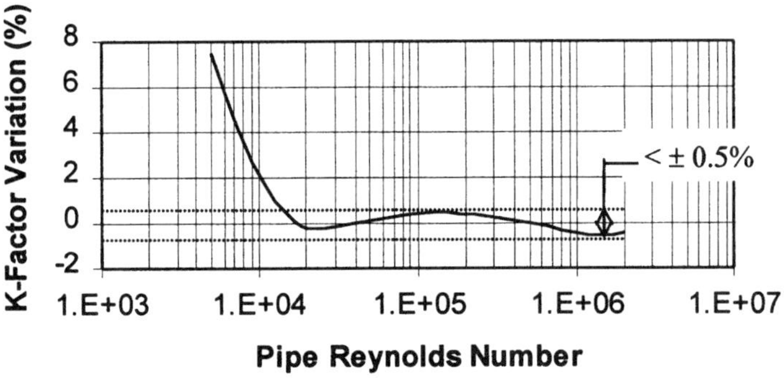

FIGURE 9.61 Typical *K* factor curve.

Defining:

$$K = \frac{\mathrm{St}'}{(A \times d)} \tag{9.91}$$

results in:

$$Q = \frac{f}{K} \tag{9.92}$$

Vortex shedding frequencies range from less than 1 Hz to greater than 3000 Hz, the former being for large meters at low velocities and the latter for small meters at high velocities.

For a vortex shedding flowmeter, an obstacle is chosen that will produce a constant *K* factor over a wide range of pipe Reynolds numbers. Thus, simply counting the vortices that are shed in a given amount of time and dividing by the *K* factor will give a measurement of the total volume of fluid that has passed through the meter. A typical *K* factor vs. Reynolds number curve is shown in Figure 9.61.

The variation in *K* factor over a specified Reynolds number range is sometimes referred to as *linearity.* For the example in Figure 9.61, it can be seen that between Reynolds numbers from 15,000 to 2,000,000, the *K* factor is the most linear. This is referred to as the *linear range* of the shedder. The wider the linear range a shedder exhibits, the more suitable the device is as a flowmeter.

At Reynolds numbers below the linear range, linearization is possible but flowmeter uncertainty can increase.

Calculation of Mass Flow and Standard Volume

Although the vortex flowmeter is a volumetric flowmeter, it is often combined with additional measurements to calculate or infer mass flow or standard volume.

To determine mass flow, $\dot{M}$:

$$\dot{M} = \rho_f \times Q = \rho_f \times \frac{f}{K} \tag{9.93}$$

where ρ_f = Fluid density at flowing conditions.

It is often desirable to know what the volumetric flow rate would be at standard process conditions with respect to pressure and temperature. This is referred to as *standard volume.* In different parts of the world and for different industries, the standard temperature and pressure can be different. The fluid density at standard conditions is referred to as the *base density,* ρ_b.

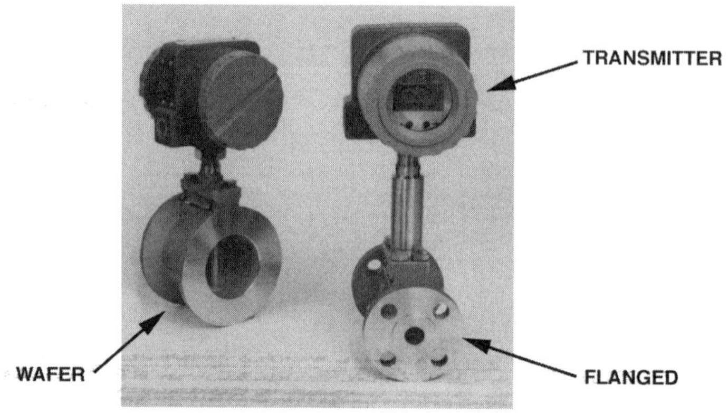

FIGURE 9.62 Vortex flowmeter construction.

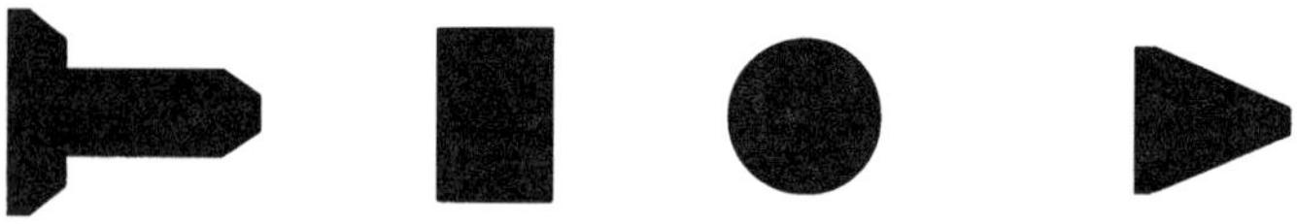

FIGURE 9.63 Shedder cross-sections.

To calculate *standard volume*, Q_V:

$$Q_V = \left(\frac{\rho_f}{\rho_b}\right) \times \frac{f}{K} \tag{9.94}$$

Flowmeter Construction

The vortex shedding flowmeter can be described as having two major components: the *flow tube* and the *transmitter.* Both are described below.

Flow Tube

The flow tube is composed of three functional parts: the flowmeter *body*, which contains the fluid and acts as a housing for the hydraulic components; the *shedder*, which generates the vortices when the fluid passes by; and the *sensor(s)*, which by some transducing means detects the vortices and produces a usable electric signal.

Flowmeter Body.

The pressure-containing portion of the vortex flowmeter, sometimes referred to as the *flowmeter body*, is available in two forms: wafer or flanged (see Figure 9.62). The wafer design, which is sandwiched between flanges of adjacent pipe, is generally lower in cost than the flanged design, but often its use is limited by piping codes.

Shedder.

The *shedder* spans the flowmeter body along the diameter and has a constant cross-section along its length. Typical cross-sections are shown in Figure 9.63.

Sensors.

When shedding is present, both the pressure and velocity fields in the vicinity of the shedder will oscillate at the vortex shedding frequency. *Pressure or velocity sensors* are used to transform the oscillating fields to an electric signal, current or voltage, from which the vortex shedding frequency can be extracted.

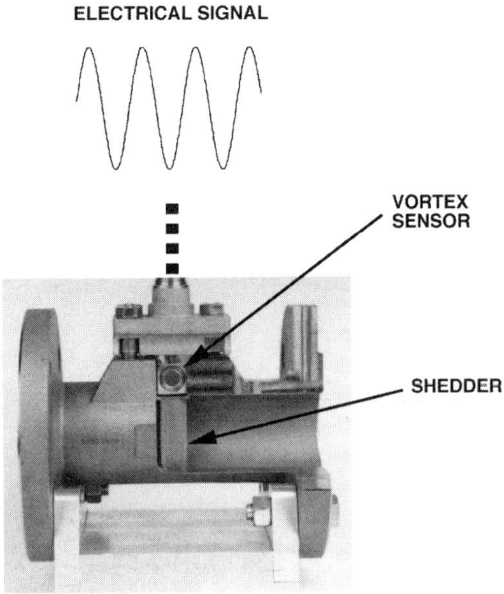

FIGURE 9.64 Example of vortex sensor.

FIGURE 9.65 Isolation manifold for sensor replacement.

Figure 9.64 shows an example of a piezoelectric differential pressure sensor located in the upper portion of the shedder, which converts the oscillating differential pressure that exists between the two sides of the shedder into an electric signal. Some other sensing means utilized to detect vortex shedding are capacitive, thermal, and ultrasonic. Often times the flowmeter electronics are mounted remotely from the flow tube. This might require a local preamplifier to power the sensor or boost its signal strength.

Since the sensor is the most likely mechanical component in a vortex flowmeter to fail, most designs have a provision for replacement of the sensors. In some cases, they can be replaced without removing the flowmeter from the pipeline. An example of an isolation manifold for sensor replacement under process conditions is shown in Figure 9.65.

Transmitter

Vortex flowmeter *transmitters* can be classified into two broad groups: analog and intelligent (or smart). Communication with the analog transmitter is carried out via local manual means, including switches

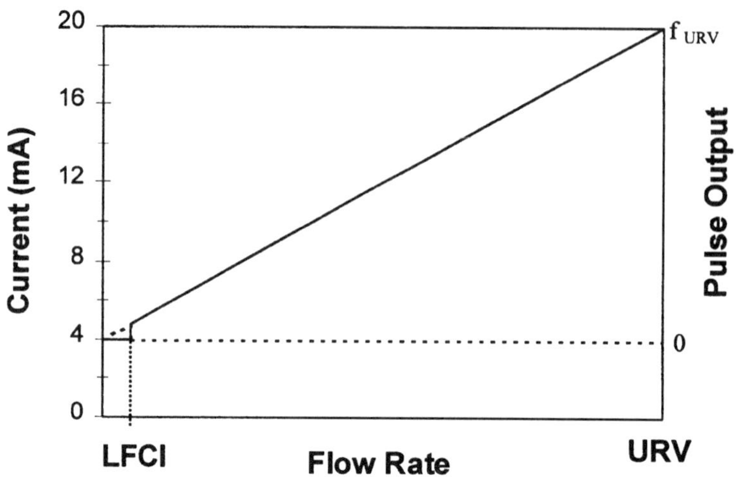

FIGURE 9.66 4 mA to 20 mA and pulse outputs.

and jumper wires. Communication with the newer intelligent device is carried out via digital electronic techniques. Both types of transmitters are two-wire devices. One of the more important features of the intelligent transmitter is that it allows application-specific information to be loaded into the transmitter. This information is then used internally to automatically tailor the transmitter to the application, including calibration of the 4 mA to 20 mA output. Before describing these devices, it is useful to consider the three most common forms of flow measurement signals provided by vortex transmitters.

Measurement Output Signals.

The three most common ways for a transmitter to communicate measurement information to the outside world are 4 mA to 20 mA, digital, and pulse signals.

The *4 mA to 20 mA signal* is the dc current flowing in the power leads to the transmitter. This current is directly proportional to the vortex shedding frequency and, hence, is also linear with flow (see Figure 9.66). 4 mA corresponds to zero flow and 20 mA to the maximum flow rate, i.e., the upper range value (URV) of the meter. A frequency-to-analog converter in the analog meter and a digital-to-analog converter in the intelligent meter produce this output.

The *digital signal* is a digitized numeric value of the measured flow rate in engineering units transmitted over the two wires powering the meter.

The *pulse signal* is a squared-up version of the raw vortex signal coming from the sensor (see Figure 9.67), and is accessible via a pair of electric terminals inside the transmitter housing. The frequency of the pulse signal is either identical to the vortex shedding frequency (raw pulse) or some multiple thereof (scaled pulse). As discussed in the Principle of Operation section, in either case, the frequency of the pulse signal is linearly proportional to flow rate, going from zero to the frequency at the URV, f_{URV} (see Figure 9.66).

As shown in Figure 9.66, at a low but nonzero flow rate, the frequency and mA signals drop to 0 Hz and 4 mA, respectively. The flow rate at which this abrupt change takes place is normally referred to as the low flow cut-in, LFCI, or cut-off. The reason for this forced zero is to avoid erroneous flow measurements at low flow, which result from process noise, including hydrodynamic fluctuations, mechanical vibration, and electrical interference. The digital signal also drops to zero below the LFCI flow rate.

Analog Transmitter.

Originally, *analog transmitters* were constructed entirely of analog electronic components. Today, they are built around a combination of analog and digital electronic components. In either case, the measurement output is in the form of a raw pulse and/or a 4 mA to 20 mA signal. Depending on the particular

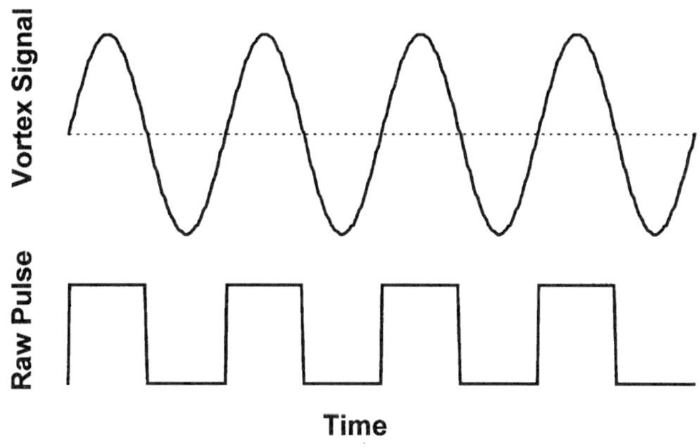

FIGURE 9.67 Raw pulse output.

transmitter, one or more of the following functions are available for tailoring the device via mechanical means to the specific application.

1. Signal output selection: If the transmitter provides both raw pulse and 4 mA to 20 mA signals, but not simultaneously, a means is available for selecting the one desired.
2. 4 mA to 20 mA calibration: Use of the this signal requires that 20 mA correspond to the desired URV. This is accomplished by inputting, via a signal or pulse generator, a periodic signal whose frequency corresponds to the upper range frequency (URF), and adjusting the output current until it reads 20 mA. The URF, which is the frequency corresponding to the vortex shedding frequency at the desired URV, is calculated using the equation URF = $K \times$ URV.

   In order to achieve the accuracy specified by the manufacturer, the K used in the above calculation must be corrected for process temperature and piping effects according to the manufacturer's instructions. The temperature effect is a result of thermal expansion of the flow tube, and is described by:

   $$\Delta K(\%) = -300 \times \alpha \times (T - T_0) \tag{9.95}$$

   where α is the thermal expansion coefficient of the flow tube material, and T_0 is the fluid temperature at which the meter was calibrated. If the shedder and meter body materials are different, α must be replaced by $(2\alpha_1 + \alpha_2)/3$, where α_1 is the thermal expansion coefficient of the meter body material and α_2 that of the shedder.

   Piping disturbances also affect the K factor because they alter the flow profile within the flow tube. This will be discussed in more detail in the section entitled "Adjacent Piping."
3. LFCI: For optimum measurement performance, the low flow cut-in should be set to fit the specific application. The goal is to set it as low as possible, while at the same time avoiding an erroneous flow measurement output.
4. Filter settings: To reduce noise present on the signal from the sensor, electronic filters are built into the transmitter. Normally, means are provided for adjusting these filters, that is, setting the frequencies at which they become active. By attenuating frequencies outside the range of the vortex shedding frequency, which varies from one application to another, better measurement performance is achieved.

Intelligent Transmitters.

Intelligent transmitters, which are microprocessor-based digital electronic devices, have measurement outputs that usually include two or more of the following: raw pulse, scaled pulse, 4 mA to 20 mA and digital. With regard to the digital output, there is at present no single, universally accepted protocol for digital communication; however, a number of proprietary and nonproprietary protocols exist.

The presence of a microprocessor in the intelligent transmitter allows for improved functionality and features compared to the analog transmitter, including:

- Elimination of the need for 4 mA to 20 mA calibration
- Automatic setting of low flow cut-in
- Automatic setting of filters
- Adaptive filtering (active filters that track the vortex frequency)
- Digital signal conditioning
- *K* factor correction for temperature and piping disturbances
- Correction for nonlinearity of *K* factor curve, including the pronounced nonlinearity at low Reynolds numbers (see Figure 9.61)
- Integral flow totalization
- Digital measurement output in desired engineering units

Configuring — that is, tailoring the transmitter to a specific application — is carried out by one or more of the following digital communicators:

- Local configurator: a configurator, built into a transmitter, that has a display and keypad
- Hand-held terminal: a palm-size digital device programmed for configuration purposes
- PC configurator: a personal computer containing configuration software
- System configurator: a digital measurement and control system with imbedded configuration software

Using one of these configurators, the dataset of parameters that defines the configuration can be modified to fit the application in question. The details of this dataset vary, depending on the specific transmitter; however, the general categories of information listed below apply.

- Flowtube parameters (e.g., tube bore, *K* factor, serial no.)
- User identification parameters (e.g., tag no., location)
- Transmitter options (e.g., measurement units, function selections)
- Process fluid parameters (e.g., fluid density and viscosity, process temperature)
- Application parameters (e.g., *K* factor corrections, URV, LFCI level)
- Output options (e.g., measurement output modes, damping, fail-safe state)

Application Considerations

Meter Selection

From a safety viewpoint, it is essential that the vortex flowmeter satisfy the appropriate electrical safety requirements and be compatible with the process, that is, be able to withstand the temperature, pressure, and chemical nature of the process fluid. From a mechanical viewpoint, it must have the proper end connections and, if required for critical applications, have a sensor that can be replaced without shutting down the process. Meter size and measurement output signal type are also very important selection factors.

Size.

Contrary to what one might expect, the required *meter size* is not always the same as the nominal size of the piping in which it is to be installed. In some applications, selecting the size based on adjacent piping will not allow the low end of the required flow range to be measured. The appropriate criteria for selecting meter size is that the meter provides a reliable and accurate measurement over the entire required flow range. This could dictate a meter size that is less than the adjacent piping.

Pressure drop is a competing sizing criteria to that described above. This drop is given by:

$$\Delta P = C \times \rho_f \times Q^2 / D^4 \qquad (9.96)$$

where C is a constant dependent on meter design, and D is the bore diameter of the flow tube. The tendency is to pick a flow tube with the same nominal diameter as the adjacent piping to eliminate the extra pressure drop introduced by a smaller-sized meter. However, in the majority of cases, this added drop is of little consequence.

The meter manufacturer can provide the needed information for making the proper selection. In some cases, sizing programs from manufacturers are available on the Internet in either an interactive or downloadable form.

Measurement Output Options.
As mentioned previously, three types of *measurement outputs* are in current use: a 4 mA to 20 mA analog signal, a pulse train, and a digital signal. Some vortex meters will provide all three of these outputs, but not always simultaneously. It is essential that the meter has the output(s) required by the application.

Meter Installation

The performance specifications of a vortex flowmeter are normally established under the following conditions: (1) the flow tube is installed in a pipeline running full with a single-phase fluid; (2) the piping adjacent to the flow tube consists of straight sections of specified schedule pipe (normally Schedule 40), typically a minimum of 30 PD (pipe diameters) in length upstream and 5 PD downstream of the flow tube with no flow-disturbing elements located within these sections; and (3) the meter is located in a vibration free and electrical interference free environment. As a consequence, certain constraints are placed on where and how the meter is installed in process piping if published performance specifications are to be achieved. These constraints are discussed below. Because meters from different suppliers differ in their sensitivity to the above influences, the statements made are of a qualitative nature. The specific manufacturer should be consulted for more quantitative information.

Location.
The flowmeter should be located in a place where vibration and electrical interference levels are low. Both of these influences can decrease the signal-to-noise ratio at the input to the transmitter. This reduction can degrade the ability of the meter to measure low flows.

The meter should not be installed in a vertical line in which the fluid is flowing down because there is a good possibility that the pipe will not be full.

Adjacent Piping.
Recommended practice is to mount the flowmeter in the process piping according to the manufacturer's stated upstream and downstream minimum straight-length piping requirements. These are typically 15 to 30 PD and 5 PD, respectively. Piping elements such as elbows or reducers upstream of the meter normally affect its K factor, but not its linearity. This allows a bias correction to be applied to the K factor. Many manufacturers provide bias factors for common upstream piping arrangements. Some who offer intelligent flowmeters make the corrections internally once the user has selected the appropriate configuration from a picklist. For piping elements and arrangements where the bias correction is not available, an *in situ* calibration should be run if the manufacturer's specified uncertainty is to be achieved. If this is not possible, calibration in a test facility with an identical configuration should be run.

The same situation as above applies if the pipe schedule adjacent to the meter differs from that under which the meter was calibrated.

To avoid disturbance to the flow, flange gaskets should never protrude into the process fluid.

The following recommendations apply if a control valve is to be situated near a vortex flowmeter. In liquid applications, the control valve should be located a minimum of 5 PD downstream of the flowmeter. This not only prevents disturbance to the flow profile in the flow tube, but also aids in preventing flashing and cavitation (see below). In gas applications, the control valve should be installed upstream of the meter, typically a minimum of 30 PD upstream of the meter to ensure an undisturbed flow profile. Having the pressure drop across the valve upstream of the meter results in a decreased density and subsequent increased velocity at the flowmeter. This helps in achieving good measurements at low flows.

For condensable gases, such as steam, it also helps to reduce the amount of condensate that might otherwise be present at the flowmeter.

Orientation.
In general, meter orientation is not an issue for vortex flowmeters, particularly for vertical pipe installations. However, for meters having electronics at the flow tube, it is recommended in high-temperature horizontal pipe applications that the flow tube be oriented with the electronics beneath the meter. Although vortex flowmeters are not recommended for multiphase applications, they do operate with somewhat degraded performance with dirty fluids (i.e., small amounts of gas bubbles in liquid, solid particles in liquid, or liquid droplets in gas). The degree of degradation in horizontal pipe applications depends to some extent on the specific meter design. Orienting the flow tube according to manufacturer's recommendations for the dirty fluid in question can help to alleviate this problem.

Pressure and Temperature Taps.
The placement of pressure and temperature taps for determining gas densities, if required, is also an important consideration. Recommendations for location of these taps vary, depending on the manufacturer. The temperature probe is inserted typically 6 PD downstream of the flow tube. This prevents any flow disturbance in the meter, and at the same time gets the probe as close to the meter as possible. The pressure tap is made typically 4 PD downstream of the meter. Although a pressure tap does not significantly affect the flow, its placement is critical for achieving an accurate density measurement.

Process Conditions.
Flashing and cavitation can occur in a liquid application just downstream of the shedder if the pressure drop across the meter results in the downstream pressure being below the vapor pressure of the liquid. These phenomena lead to undefined measurement errors and possibly to structural damage, and hence should be avoided. This is usually accomplished by increasing the inlet pressure or inserting a back-pressure valve downstream of the meter. To avoid flashing and cavitation, the downstream pressure after recovery (approximately 5 PD downstream) must be equal to or greater than P_{dmin}, where:

$$P_{\text{dmin}} = c_1 \times \Delta P + c_2 \times P_{\text{vap}} \tag{9.97}$$

where P_{dmin} = Minimum absolute downstream pressure after recovery
P_{vap} = Vapor pressure of the liquid at the flowing temperature
ΔP = Overall pressure drop
c_1, c_2 = Empirical constants for a specific meter (normally available from the meter manufacturer)

Pulsating flow can also, in some circumstances, lead to measurement errors. It is best to avoid placing the meter in process lines where noticeable pulsation exists.

Meter Configuration

It is important when installing an analog or intelligent vortex flowmeter that it be configured for the specific application (see section above on Flowmeter construction). This is often done by the supplier prior to shipping if the user supplies the relevant information at the time the order is placed. If this is not the case, the user must carry out the configuration procedures provided by the manufacturer.

Recent Developments

Recent efforts have been made to make the vortex flowmeter into a real-time mass flow measurement device. As was demonstrated in the "Principle of Operation" section, the output of the meter, based on the frequency of vortex shedding, is related to actual volumetric flow (see Equation 9.92). In intelligent transmitters, the flowing density (the density at flowing conditions) and the base density can be entered into the transmitter's database. Based on these values, mass flow or standard volumetric flow can be computed (see Equations 9.93 and 9.94). This procedure is valid if the flowing density does not vary in time. If this is not the case, an on-line, real-time measure of the density must be provided. Two different approaches have been used. One (multisensor) employs sensors in addition to the vortex sensor; the other (single sensor) relies on additional information being extracted from the vortex shedding signal.

Multisensor

In this method, temperature and pressure measurements are made in addition to the vortex frequency. This approach is similar to that used in orifice-d/p mass flowmetering, in which case temperature and pressure ports are located in the pipe normally downstream of the orifice plate. However, for the multisensor vortex, the temperature and pressure sensors are incorporated into the flowmeter rather than located in the adjacent piping. Using these two additional measurements, the flowing density is calculated from the equation of state for the process fluid.

Single Sensor

This approach takes advantage of the fact that, in principle, for a force- or pressure-based vortex shedding sensor, the amplitude of the vortex shedding signal is directly proportional to the density times the square of the fluid velocity; that is:

$$\text{Signal amplitude} \propto \rho_f \times U^2 \tag{9.98}$$

The fluid velocity can be determined from the vortex frequency; that is:

$$\text{Frequency} \propto U \tag{9.99}$$

Hence,

$$\frac{\text{Signal amplitude}}{\text{Frequency}} \propto \rho_f \times U \propto \text{Mass flow} \tag{9.100}$$

This approach, in principle, is independent of the process fluid, and requires no additional sensors.

Further Information

R. W. Miller, *Flow Measurement Engineering Handbook,* 3rd ed., New York: McGraw-Hill, 1996, chap. 14.

W. C. Gotthardt, Oscillatory flowmeters, in *Practical Guides for Measurement and Control: Flow Measurement,* D. W. Spitzer (ed.), Research Triangle Park, NC: Instrument Society of America, 1991, chap. 12.

J. P. DeCarlo, *Fundamentals of Flow Measurement,* Research Triangle Park, NC: Instrument Society of America, 1984, chap. 8.

ASME MFC-6M, Measurement of Fluid Flow in Closed Conduits Using Vortex Flowmeters, American Society of Mechanical Engineeers, 1998.

9.9 Thermal Mass Flow Sensors

Nam-Trung Nguyen

This chapter section deals with thermal mass flowmeters. The obvious question that arises is, what is actually meant by mass flow and thermal mass flowmeter? To answer this question, one should first understand what is the flow and what is the physical quantity measured by the meter. The flow can be understood here by the motion of a continuum (fluid) in a closed structure (channel, orifice), and is the

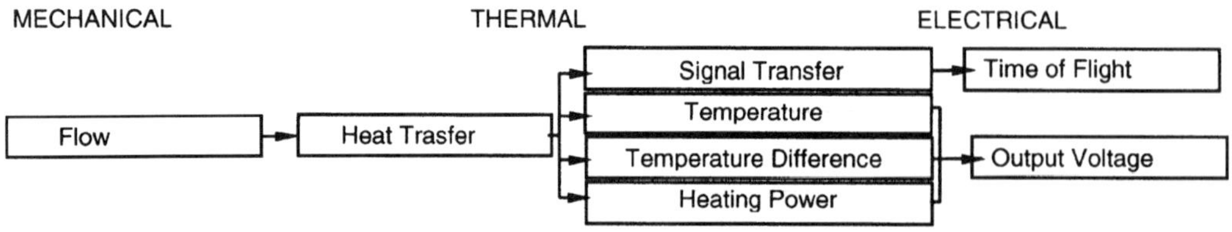

FIGURE 9.68 The three signal domains and the signal transfer process of a thermal flow sensor.

measured object. The associated physical quantity measured by the meter is the mass flux that flows through a unit cross-section. The equation for the volume flow rate Φ_v is given by:

$$\Phi_v = \mathrm{d}V/\mathrm{d}t = vA \tag{9.101}$$

where V = Volume through in the time t
v = Average velocity over the cross-section area A of the channel

With the relation between volume V, mass M, and the density of fluid ρ:

$$V = M/\rho \tag{9.102}$$

the mass flow rate Φ_m can be derived by using Equations 9.101 and 9.102 to obtain:

$$\Phi_m = (\Phi_v \rho) + \left(V \frac{\mathrm{d}\rho}{\mathrm{d}t} \right) \tag{9.103}$$

For time invariable fluid density, one obtains:

$$\Phi_m = \Phi_v \rho = Av\rho \tag{9.104}$$

A thermal mass flow sensor will generally output a signal related to the mass flux:

$$\phi_m = \Phi_m / A = v\rho \tag{9.105}$$

and convert the mechanical variable (mass flow) via a thermal variable (heat transfer) into an electrical signal (current or voltage) that can be processed by, for example, a microcontroller. Figure 9.68 illustrates this working principle. The working range for any mass flux sensor is somewhat dependent on the fluid properties, such as thermal conductivity, specific heat, and density, but not on the physical state (gas or liquid) of the fluid.

Principles of Conventional Thermal Mass Flowmeters

With two heater control modes and two evaluation modes, there are six operational modes shown in Table 9.13 and three types of thermal mass flowmeters:

- Thermal mass flowmeters that measure the effect of the flowing fluid on a hot body (increase of heating power with constant heater temperature, decrease of heater temperature with constant heating power). They are usually called hot-wire, hot-film sensors, or hot-element sensors.
- Thermal mass flowmeters that measure the displacement of temperature profile around the heater, which is modulated by the fluid flow. These are called calorimetric sensors.
- Thermal mass flowmeters that measure the passage time of a heat pulse over a known distance. They are usually called time-of-flight sensors.

TABLE 9.13 Operational Modes of Thermal Mass Flow Sensors

Heater controls	Constant heating power		Constant heater temperature	
Evaluation	Heater temperature	Temperature difference	Heating power	Temperature difference
Operational Modes	Hot-wire and hot-film type Time-of-flight type	Calorimetric type	Hot-wire and hot-film type Time-of-flight type	Calorimetric type

TABLE 9.14 Typical Arrangements of Flow Channel

NOZZLE TYPE	VENTURI TYPE
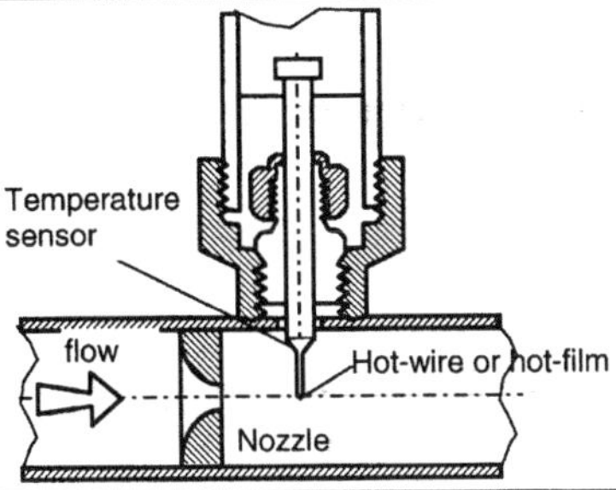	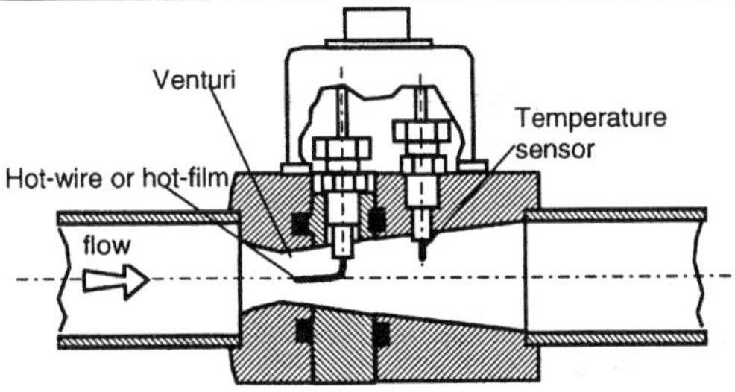

Hot-Wire and Hot-Film Sensors

The dependence of the heat loss between a fine wire as well as a thin film and the surrounding fluid has traditionally been the most accepted method for measuring a fluid flow: the hot-wire method. The hot-film method uses film sensors for detecting the flow. The basic elements of this sensor type are discussed.

Flow channel: In contrast to the thermal anemometer described in Chapter 10 for point measurement, thermal mass flow sensors of the hot-wire and hot-film type have a flow channel defining the mass flow. Table 9.14 shows two typical arrangements.

Sensor element: Hot-wire sensors are fabricated from platinum, platinum-coated tungsten, or a platinum–iridium alloy. Since the wire sensor is extremely fragile, hot-wire sensors are usually used only for clean air or gas applications. On the other hand, hot-film sensors are extremely rugged; therefore, they can be used in both liquid and contaminated-gas environments. In the hot-film sensor, the high-purity platinum film is bonded to the rod. The thin film is protected by a thin coating of alumina if the sensor will be used in a gas, or of quartz if the sensor will be used in a liquid. The alumina coatings have a high abrasion resistance and high thermal conductivity. Quartz coatings are less porous and can be used in heavier layers for electrical insulation. Typical hot-wire and hot-film sensors are shown in Table 9.15.

The sensor element, whether it is a wire or a film, should be a resistor that has a resistance with a high temperature coefficient α. For most sensor materials, the temperature dependence can simply be expressed by a first-order function:

$$R = R_r\left[1 + \alpha\left(T - T_r\right)\right] \tag{9.106}$$

where R = Resistance at operating temperature T
R_r = Resistance at reference temperature T_r
α = Temperature coefficient

For research applications, cylindrical sensors are most common, either a fine wire (typical diameters from 1 to 15 μm) or a cylindrical film (typical diameters from 25 to 150 μm). Industrial sensors are often a resistance wire wrapped around a ceramic substrate that has typical diameters from 0.02 mm to 2 mm. Table 9.16 shows some typical parameters of industrial hot-wire and hot-film sensors.

TABLE 9.15 Typical Hot-Wire and Hot-Film Sensors (**1**, hot-wire; **2**, sensor supports; **3**, electric leads; **4**, hot-film; **5**, contact caps; **6**, quartz rod)

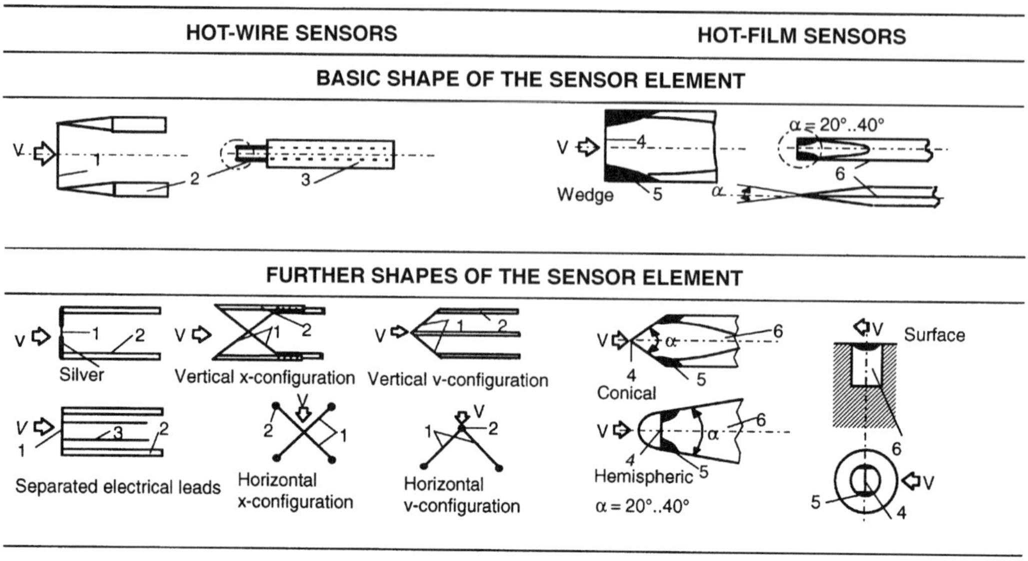

TABLE 9.16 Typical Parameters of Hot-Wire and Hot-Film Sensors [3]

Parameter	Hot-wire	Hot-film
Sensor element	Platinum hot-wire (diameter 70 μm)	Platinum hot-film (alumina coated)
Operational mode	Constant heater temperature in air	
Working temperature range	–30 to 200°C	
Characteristics	Nonlinear	
Accuracy in %	±4	±2
Time response in ms	<5	12
Sensitivity in mV kg^{-1} h^{-1}	1	5

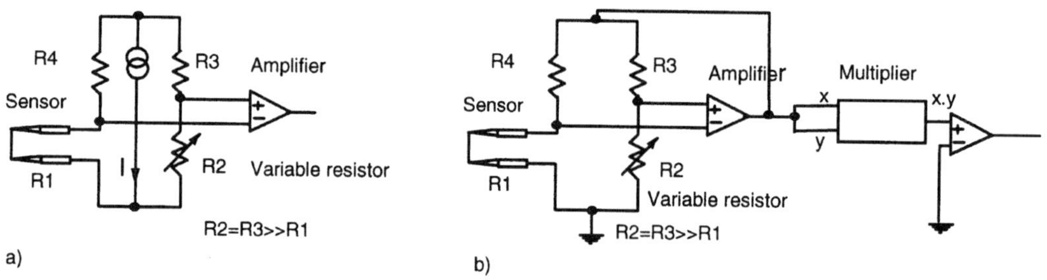

FIGURE 9.69 Control and evaluation circuit of heat-wire and heat-film sensors: (*a*) constant-current bridge; (*b*) constant-temperature bridge.

Control and evaluation circuit: The constant-current and constant-temperature bridge are conventional circuits for control and evaluation of heat-wire or heat-film sensors.

A constant-current Wheatstone bridge with a hot-wire sensor is shown schematically in Figure 9.69(*a*). In this circuit, resistors *R*3 and *R*4 are much larger than sensor resistor *R*1. Therefore, current through *R*1 is essentially independent of changes in the sensor resistor *R*1. Any flow in the channel cools the hot wire, decreases its resistance as given by Equation 9.106, and unbalances the bridge. The unbalanced

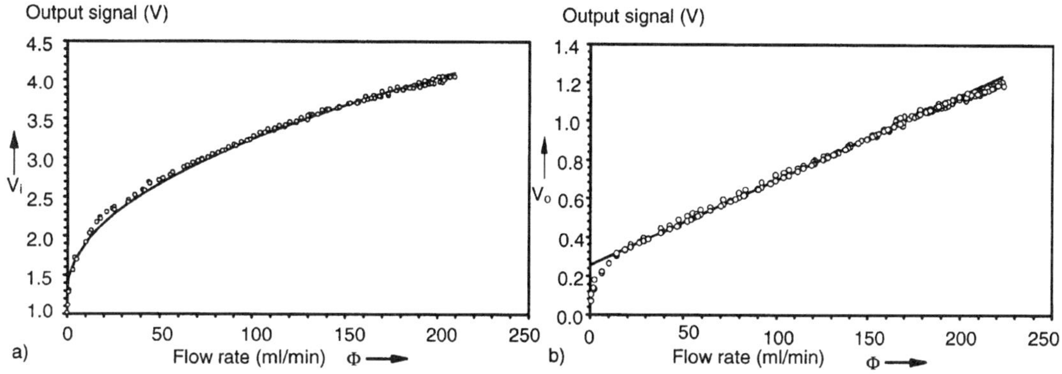

FIGURE 9.70 Sensor characteristics of the constant-temperature mode before (*a*) and after (*b*) linearization.

TABLE 9.17 Typical Calorimetric Sensors

INTRUSIVE TYPE	NON-INTRUSIVE TYPE

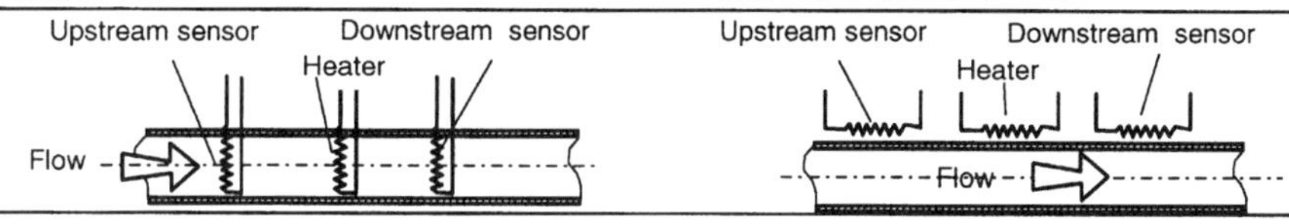

bridge produces an output voltage V_o, which is related to the mass flow. Because the output voltage V_o from the bridge is small, it must be amplified before it is recorded. The value of the thermal coefficient α for *R*1 and *R*2 should be equal in order to eliminate signal errors due to changes in ambient temperature. Similarly, thermal coefficients for *R*3 and *R*4 should also be equal.

The constant-temperature Wheatstone bridge is shown in Figure 9.69(*b*). The bridge is balanced under no-flow conditions with the variable resistor *R*2. The flow cools the hot wire, and its resistance decreases and unbalances the bridge. A differential amplifier balances the bridge with the feedback voltage. The output signal can be linearized before recording.

Because the output signal has a square-root-like characteristic, the linearizer can be realized easily using a multiplier (i.e., AD534 of Analog Devices) with two equal input signals (a squarer). Figure 9.70 illustrates the results. With this method, there is linearization error in the low flow range.

Calorimetric Sensors

The displacement of the temperature profile caused by the fluid flow around a heating element can be used for measuring very small mass flow. Depending on the location of the heating and sensing elements, there are two types of calorimetric sensors: the intrusive sensors that lie in the fluid, and the nonintrusive sensors that are located outside the flow. Table 9.17 illustrates the two typical calorimetric sensors.

The intrusive type has many limitations. The heater and the temperature sensors must protrude into the fluid. Therefore, corrosion and erosion damage these elements easily. Furthermore, the integrity of the piping is sacrificed by the protrusions into the flow, thus increasing the danger of leakage.

In the nonintrusive sensor type, the heater and the temperature sensors essentially surround the flow by being located on the outside of the tube that contains the flow. The major advantage of this sensor type is the fact that no sensor is exposed to the flowing fluid, which can be very corrosive. This technique is generally applied to flows in the range of 1 mL min^{-1} to 500 L min^{-1}. The larger flows are measured using the bypass arrangement. Figure 9.71(*a*) shows the measured shift of the temperature distribution around the heater. The asymmetricity of the temperature profile increases with flow. The

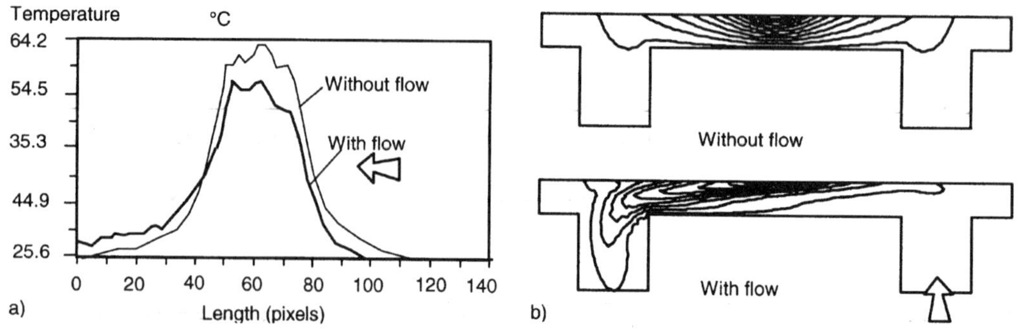

FIGURE 9.71 Temperature distribution around the heater: (*a*) measurement (*b*) numerical simulation.

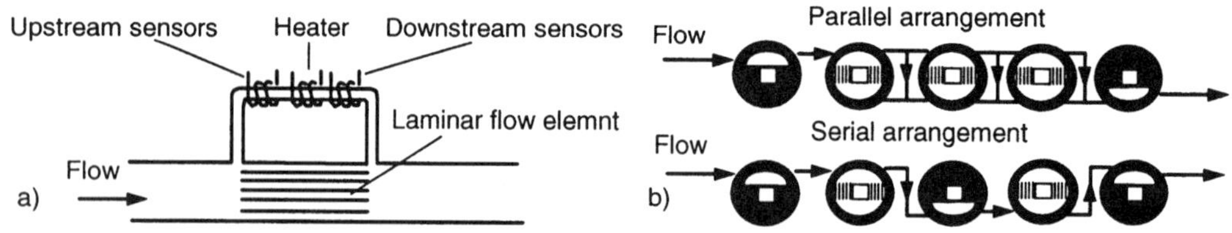

FIGURE 9.72 Bypass arrangement for large flow range: (*a*) principle (*b*) a solution for the laminar flow element [4].

TABLE 9.18 Typical Parameters of Calorimetric Sensors [4]

Parameter	Gases	Liquids
Working temperature	0°C to 70°C	0°C to 70°C
Accuracy in %	±1%	±1%
Linearity	±0.2%	±0.2%
Flow range 1:50	min. 5 mL min^{-1}	min. 5 g h^{-1}
	max. 100 L min^{-1}	max. 1000 g h^{-1}

measurement was carried out using a thermography system [10]. To understand the working principle, the effect of fluid mechanics and heat transfer should be reviewed. The mathematical theory for this problem is discussed later in this chapter. Figure 9.71(*b*) illustrates the influence of the flow over the temperature distribution.

Because the calorimetric mass flow sensors are sensitive in low-flow ranges, bypass designs have been introduced in order to make the sensors suitable for the measurement of larger flow ranges. The sensor element is a small capillary tube (usually less than 3 mm in diameter). They ensure laminar flow over the full measurement range. The laminar flow elements are located parallel to the sensor element as a bypass (Figure 9.72(*a*)). They are usually a small tube bundle, a stack of disks with etched capillary channels [4] (Figure 9.72(*b*)), or a machined annular channel.

Table 9.18 shows typical parameters of calorimetric flow sensors. Compared to the hot-wire or hot-film sensors, this sensor type has good linearity and is only limited by signal noise at low flows and saturation at high flows. While the linear range may exceed a 100:1 ratio, the measurable range may be as large as 10,000:1 [21]. The small size of the capillary sensor tube is advantageous in minimizing the electric power requirement and also in increasing the time of response. Because of the small size of the tube, it necessitates the use of upstream filters to protect against plugging of dust particles. With the bypass arrangement, a relatively wide flow range is possible.

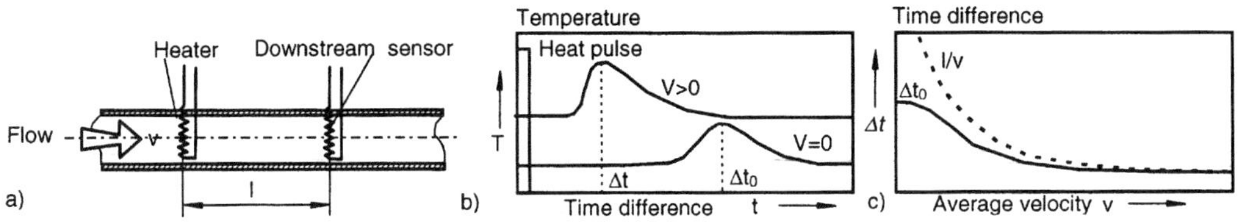

FIGURE 9.73 Time-of-flight sensors: (*a*) principle; (*b*) temperature at downstream sensor; (*c*) sensor characteristic.

Time-of-Flight Sensors

The time-of-flight sensor consists of a heater and one or more temperature sensors downstream, Figure 9.73(*a*). The heater is activated by a current pulse. The transport of the generated heat is a combination of diffusion and forced convection. The resulting temperature field can be detected by temperature sensors located downstream. The detected temperature output signal of the temperature sensor is a function of time and flow velocity. The sensor output is the time difference between the starting point of the generated heat pulse and the point in time at which a maximum temperature at the downstream sensor is reached, Figure 9.73(*b*). At the relatively low flow rates, the time difference depends mainly on the diffusivity of the fluid medium. At relatively high flow rates, the time difference tends to relate to the ratio of the heater–sensor distance and the average flow velocity [5].

Because of the arrangement shown in Figure 9.73(*a*), the time-of-flight sensors have the same limitations as the intrusive type of calorimetric sensors: corrosion, erosion, and leakage. Since the signal processing needs a while to measure the time difference, this sensor type is not suitable for dynamic measurement. The advantage of this type of volumetric flow sensor is the independence of fluid properties as well as fluid temperature in the higher flow range. The influence of fluid properties on the mass flow sensor output is described in [21], as well as an approach to compensate for changes in these properties, which is valid for both hot-element and calorimetric sensors.

Mass and Heat Transfer

The most important signal of the transfer process shown in Figure 9.68 is the thermal signal. There are different kinds of thermal signals: temperature, heat, heat capacity, and thermal resistance. In the following, the transfer of heat and the interaction between heat and temperature will be explained by three mean heat transfer processes: conduction, convection, and radiation. The first two processes can be described by the general equation of a transfer process. The transfer variables in the equation can be the momentum (momentum equation), the temperature (energy equation), or the mass (mass equation).

A transfer process consists of four elements: accumulation, conduction, induction, and convection. The *accumulation process* describes the time dependence of the transfer variable. The *conduction* presents the molecular transfer. The *convection* is the result of the interaction between the flow field and the field of the transfer variable. The *induction* describes the influence of external fields and sources.

Conduction

When there is a temperature gradient in a substance, the heat will flow from the hotter to the colder region, and this heat flow q (in W m^{-2})will be directly proportional to the value of the temperature gradient:

$$q = -\lambda \frac{\mathrm{d}T}{\mathrm{d}x} \tag{9.107}$$

where T is the temperature. The above expression is called the Fourier's law of heat conduction and defines the material constant λ (in W K^{-1} m^{-1}), the thermal conductivity. Figure 9.74 shows the order of the thermal conductivity λ of different materials.

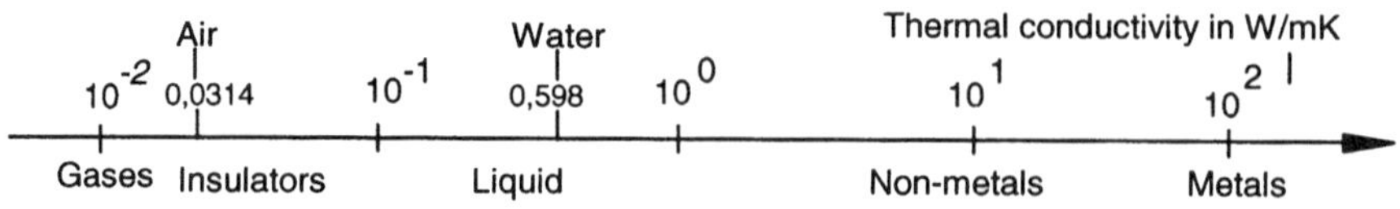

FIGURE 9.74 The order of thermal conductivity of different materials.

The differential form of the heat-conduction equation is a special case of the energy equation (see next subsection on convection). The transfer equation only consists of the accumulative, conductive, and inductive terms:

$$\frac{\partial T}{\partial t} = \frac{\lambda}{\rho c}\left(\frac{\partial^2 T}{\partial x^2} + \frac{\partial^2 T}{\partial y^2} + \frac{\partial^2 T}{\partial z^2}\right) + \frac{q'}{\rho c} \tag{9.108}$$

where ρ is the density, c is the specific heat at constant pressure and q' (in W m^{-3}) is the amount of heat (in joules) per unit of volume and time that can be generated inside the material itself, either through the action of a separate heat source, or through a change in phase of matter.

In the steady state without internal heat sources, the equation of conduction reduces to:

$$\frac{\partial^2 T}{\partial x^2} + \frac{\partial^2 T}{\partial y^2} + \frac{\partial^2 T}{\partial z^2} = 0 \tag{9.109}$$

Convection

In general, there are two kinds of convection: forced convection and natural convection. The first one is caused by a fluid flow, the other one by itself because of the temperature dependency of fluid density and the buoyancy forces. To describe convection, three conservation equations are required:

- Conservation of mass: continuity equation

$$\frac{\partial \rho}{\partial t} + \nabla(\rho v) = 0 \tag{9.110}$$

- Conservation of momentum: Navier–Stokes equation

$$\frac{\rho \partial v}{\partial t} + v\nabla v = -\nabla p + \eta \nabla^2 v + \rho g \tag{9.111}$$

- Conservation of energy: energy equation

$$\frac{\partial T}{\partial t} + v\nabla T = \left(\frac{\lambda}{\rho c}\right)\nabla^2 T + \frac{q'}{\rho c} \tag{9.112}$$

where η is the dynamic viscosity of the fluid. The temperature field and the heat power can be found by solving these three equations. For designing and understanding the thermal flow sensor, the convective heat transfer can be expressed in the simplest form:

$$Q = \varepsilon A \Delta T \tag{9.113}$$

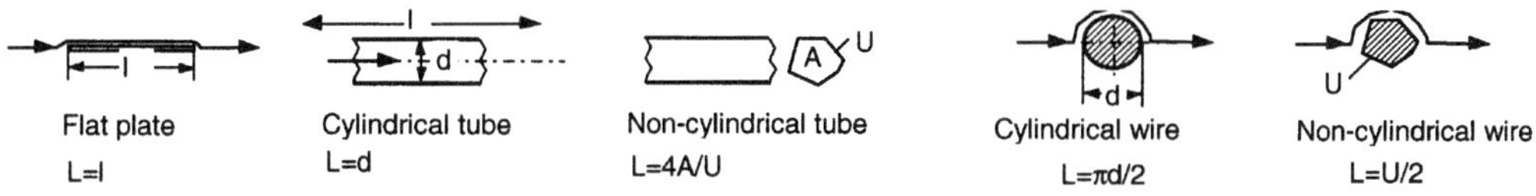

FIGURE 9.75 Definition of the characteristic length of different objects [6].

where Q (in W) = heat transfer rate (or the heat power)
ε = heat transfer coefficient between the heated surface A and the fluid
ΔT = temperature difference between the heated body and ambient

The dimensionless Nusselt number describes the heat transfer. The relationship between the heat transfer coefficient ε and the Nusselt number Nu can be expressed as follows:

$$\varepsilon = \mathrm{Nu}\frac{\lambda}{L} \tag{9.114}$$

where L is the characteristic length (the length L of a flat plate, the hydraulic diameter D_h of a tube, and the half of the perimeter of a wire, Figure 9.75). The hydraulic diameter D_h can be calculated using the wetted perimeter U and the cross-sectional area A of the tube:

$$D_h = \frac{4A}{U} \tag{9.115}$$

The relevant dimensionless number which describes the flow is the Reynolds number Re:

$$Re = \frac{vL}{\nu} \tag{9.116}$$

where v is the average flow velocity and ν the kinematic viscosity of the fluid, which is defined by the density ρ and the dynamic viscosity η:

$$\nu = \frac{\eta}{\rho} \tag{9.117}$$

Table 9.19 shows a collection of formulae for calculating the Nusselt number. The fluid properties (kinematic viscosity ν and Prandtl number Pr) should be chosen at the average temperature T_{av} between the heater temperature $T + \Delta T$ and the fluid temperature T:

$$T_{av} = T + \frac{\Delta T}{2} \tag{2.118}$$

In the case of natural convection, the Nusselt number depends on the Grashof number Gr, which describes the influence of buoyancy forces:

$$\mathrm{Gr} = \frac{g\beta L^3 \Delta T}{\nu^2} \tag{9.119}$$

TABLE 9.19 Nusselt Number (Nu) of Forced Convection [6]

Object	Nu_{lam} for laminar regime	Nu_{turb} for turbulent regime	Average Nusselt number Nu
Flat plate	$Nu_{lam} = 0.664\sqrt{Re^3}\sqrt{Pr}$	$Nu_{turb} = \dfrac{0.037\ Re^{0.8}\ Pr}{1+2.443\ Re^{-0.1}\left(Pr^{2/3}-1\right)}$	$Nu = \sqrt{Nu_{lam}^2 + Nu_{turb}^2}$
	Re<10⁵; 0.6<Pr<2000 → $Re<10^5$; $0.6<Pr<2000$	$5.10^5<Re<10^7$; $0.6<Pr<2000$	$10<Re<10^7$; $0.6<Pr<2000$
Cylindrical tube	$Nu_{lam} = 3.65 + \dfrac{0.19\left(Re\ Pr\ d/l\right)^{0.8}}{1+0.117\left(Re\ Pr\ d/l\right)^{0.467}}$	$Nu_{turb} = \dfrac{\xi/8\left(Re-1000\right)Pr}{1+12.7\sqrt{\xi/8}\left(Pr^{2/3}-1\right)}\left[1+\left(\dfrac{d}{l}\right)^{2/3}\right]$	
	$Re<2300$; $0.1<(Re\ Pr\ d/l)<10^4$	where $\xi = (1.28\log_{10} Re - 1.64)^{-2}$	
Short cylindrical tube $0.1<d/l<1$	$Nu_{lam} = 0.664\sqrt[3]{Pr}\sqrt{Re\ d/l}$ $Re<2300$; $0.1<(Re\ Pr\ d/l)<10^4$	Thermal entrance, fully developed flow, $2300<Re<10^4$ $Nu_{turb} = \sqrt[3]{3.66^3 + 1.61^3\ Re\ Pr\ d/l}$ Thermal entrance, developing flow, $2300<Re<10^6$	
Wire	$Nu_{lam} = 0.664\sqrt{Re^3}\sqrt{Pr}$ $10<Re<10^7$; $0.6<Pr<1000$	$Nu_{turb} = \dfrac{0.037\ Re^{0.8}\ Pr}{1+2.443\ Re^{-0.1}\left(Pr^{2/3}-1\right)}$	$Nu = 0.3 + \sqrt{Nu_{lam}^2 + Nu_{turb}^2}$

TABLE 9.20 The Average Nusselt Number of Free Convection in Some Special Cases

Cases	Equation	Ref.
Vertical flat plate or wire	Nu = 0.55(Gr Pr)$^{1/4}$	[7]
Horizontal flat plate	For the upper surface: Nu = 0.76(Gr Pr)$^{1/4}$	[8]
	For the lower surface: Nu = 0.38(Gr Pr)$^{1/4}$	

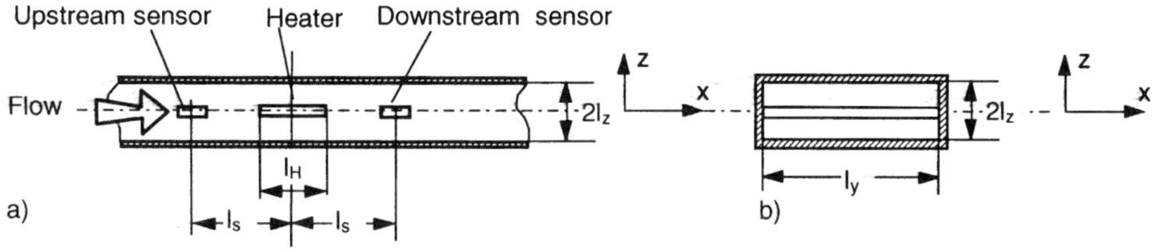

FIGURE 9.76 Analytical model for the intrusive type of calorimetric sensors: (*a*) length cut, (*b*) cross-section.

where g is the acceleration due to the gravity (9.81 m s^{-2}), β the thermal expansion coefficient of the fluid, and ΔT is the temperature difference between the hot fluid and the ambient. The average Nusselt number can be calculated for a laminar flow ($10^4 < \mathrm{GrPr} < 10^8$) in Table 9.20.

Radiation

A body can either emit or absorb thermal radiation. Radiation is not important for the operational principle of thermal mass flow because of its relatively low magnitude.

Following, the physical and mathematical backgrounds of these three sensor types are discussed in detail. The working principle and the influence of fluid properties on the sensor signal can be determined using these mathematical models. However, the mathematical models only describe the relationship between thermal variables (heat power, temperature) and the average velocity. Further relationships between mass flow, mass flux, thermal, and electrical variables can be derived using Equations 9.101 to 9.106.

Analytical Models for Calorimetric Flowmeters

Model for the Intrusive Type of Calorimetric Sensors

In a quasi-static situation, the incoming heat at a certain point in the fluid must be equal to the outgoing heat. The heat is transported either by conduction in the fluid and/or supporting beams, or by convection through the thermal mass of the fluid. Ultimately, the heat is transported to the walls of the tube; see Figure 9.76 and Table 9.18. A heat balance equation results in a differential equation for T in x. The temperature profile in the y and z directions is assumed to be constant and linear, respectively [9].

Referring to Figure 9.76 and using A as the cross-section area of the flow channel ($A = l_y 2l_z$), ρ as the fluid density, c as the fluid heat capacity (at constant pressure), v as the average fluid velocity, and λ as the fluid thermal conductivity, one finds that:

$$\frac{\partial^2 T}{\partial x^2} - v\left(\frac{\rho c}{\lambda}\right)\frac{\partial T}{\partial x} - \left(\frac{T}{l_z^2}\right) = 0 \tag{9.120}$$

or

$$\frac{\partial^2 T}{\partial x^2} - \left(\frac{v}{a}\right)\frac{\partial T}{\partial x} - \frac{T}{l_z^2} = 0 \tag{9.121}$$

where $a = \lambda/\rho c$ is the thermal diffusivity of the fluid. Equation 9.121 is linear in T. Solving the differential equation using a heater length l_H, a heater power Q, and the boundary condition:

$$\lim_{x \to \pm\infty} T(x) = 0 \tag{9.122}$$

the following temperature distribution results:

$$x < \frac{-l_H}{2} \quad \text{for} \quad T(x) = T_0 \exp\left[\gamma_1\left(x + \frac{l_H}{2}\right)\right] \tag{9.123}$$

$$\frac{l_H}{2} < x < \frac{l_H}{2} \quad \text{for} \quad T(x) = T_0 \tag{9.124}$$

$$x > \frac{l_H}{2} \quad \text{for} \quad T(x) = T_0 \exp\left[\gamma_2\left(x - \frac{l_H}{2}\right)\right] \tag{9.125}$$

where

$$\gamma_{1,2} = \frac{\left(v \pm \sqrt{v^2 + 2a^2/l_z^2}\right)}{(2a)} \tag{9.126}$$

$$T_0 = \frac{P}{\left[\left(\frac{2\lambda l_y l_H}{l_z}\right) + A\lambda(\gamma_1 - \gamma_2)\right]} \tag{9.127}$$

The temperature difference between the two sides, upstream (at $x = l_s$) and downstream (at $x = -l_s$) can be then calculated as:

$$\Delta T(v) = T_0\left\{\exp\left[\gamma_2\left(\frac{l_s - l_H}{2}\right)\right] - \exp\left[\gamma_1\left(\frac{-l_s + l_H}{2}\right)\right]\right\} \tag{9.128}$$

Model for the Nonintrusive Type of Calorimetric Sensors

A simple, one-dimensional model is used to show the working principle of the nonintrusive type with capillary-tube and heater wire winding around it. Geometric parameters and assumptions are given in Figure 9.77. Because of the symmetry, only half of the capillary-tube will be considered for the calculation model. The conservation of thermal energy in a lumped element (Figure 9.77(*c*)) can be given in the following equation:

$$Q_{\text{cond.,x,fluid}} + Q_{\text{conv.,x,fluid}} + Q_{\text{cond.,x,wall}} = Q_{\text{cond.,y,fluid}} \tag{9.129}$$

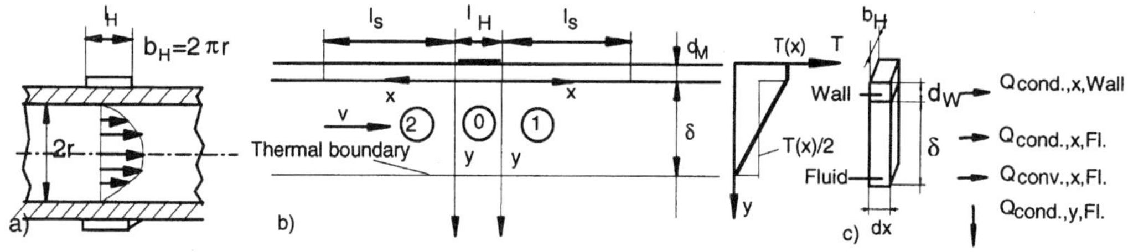

FIGURE 9.77 Analytical model for the nonintrusive type of calorimetric sensors: (*a*) heater and channel geometry, (*b*) model geometry, and (*c*) model of a lumped element.

The indices define the conduction or convection in the *x*- or *y*-axis in the fluid as well as in the heated wall. Defining the temperature along the *x*-axis as $T(x)$, the average flow velocity as *v*, the thermal conductivities of wall material as λ_w and of fluid as λ, the thermal diffusivity of fluid $a = \lambda/(\rho c)$, and the thickness of the average thermal boundary layer as δ one finds the heat balance equation:

$$\left[\frac{1}{2}+\left(\frac{\lambda_W d_W}{\lambda\delta}\right)\right]\frac{\partial^2 T}{\partial x^2}-\frac{v}{(2a)}\frac{\partial T}{\partial x}-\frac{T(x)}{\partial^2}=0 \tag{9.130}$$

The thickness of the average thermal boundary layer δ depends on the flow velocity [15]. For gases with a small Prandtl number (Pr < 1) or liquids with a low Reynolds number, one can assume that:

$$\delta = r \tag{9.131}$$

After solving Equation 9.130 in the local coordinate systems 1 and 2 (Figure 9.77(*b*)), one obtains the temperature difference $\Delta T(v)$ between the temperature sensors:

$$\Delta T(v) = \vartheta_0\left[\exp(\gamma_2 l_s) - \exp(\gamma_1 l_s)\right] \tag{9.132}$$

with:

$$\gamma_{1,2} = \frac{\left(v \pm \sqrt{v^2 + 16a^2\kappa/\delta^2}\right)}{(4a\kappa)} \tag{9.133}$$

The dimensionless factor:

$$\kappa = \frac{1}{2} + \frac{(\lambda_W d_W)}{(\lambda\delta)} \tag{9.134}$$

describes the influence of the wall on the heat balance. If the wall is neglected, we get $\kappa = 1/2$ as in the similar case of Equation 9.121. The heater temperature T_0 can be calculated for the constant heat power *Q*:

$$T_0 = \frac{Q}{\left\{2\pi r\lambda\left[\frac{l_H}{\delta} + \sqrt{\frac{(v^2\delta^2)}{(4a^2)+4\kappa}}\right]\right\}} \tag{9.135}$$

Model for the Time-of-Flight Type

The transport of the heat generated in a line source through a fluid is governed by the energy equation (112). The analytical solution of this differential equation for a pulse signal with input strength q'_0 (W m^{-1}) is given in [11] as:

$$T(x, y, t) = \left(\frac{q'_0}{4\pi\lambda t} \right) \exp\left\{ -\frac{\left[(x - vt)^2 + y^2 \right]}{4at} \right\} \tag{9.136}$$

where a denotes the thermal diffusivity. By measuring the top time τ at which the signal passes the detection element ($y = 0$), in other words differential Equation 9.136 with respect to time, one can obtain the basic equation for the so-called "time-of-flight" of the heat pulse:

$$v = \frac{x}{t} \tag{9.137}$$

For Equation 9.136 to be valid, the term $4at$ must be much smaller than the heater-sensor distance x. This assumes that forced convection by the flow is dominating over the diffusive component. In other words, Equation 9.136 is true at high flow velocities. When the diffusive effect is taken into account, the time-of-flight is given by:

$$\Delta t = \tau = \frac{\left[-2a + \left(4a^2 + v^2 x^2 \right)^{1/2} \right]}{v^2} \qquad v \neq 0 \tag{9.138}$$

$$\tau = \frac{x^2}{4a} \qquad v = 0 \tag{9.139}$$

Principles of Microflow Sensors

In research papers, the first reference to thermal mass flow sensor normally cited is that of King in 1914. Since then, microsystems technology has been developed. The development of thermal flow sensor can be realized in micron-size using the three current technologies: bulk micromachining, surface micromachining, epimicromachining and LIGA-techniques (LIGA: German description of "Lithographie, Galvanoformung, Abformung"). These fabrication techniques (except LIGA) are compatible with conventional microelectronic processing technology. Thermal flow sensors developed using these technologies will be called "microflow sensors" in this section. The operational modes are similar to the conventional thermal flow sensor. In Table 9.21, the microflow sensors are classified after their transducing principle. With these new sensors, very small flows in the nanoliter and microliter range can be measured. Table 9.22 shows some realized examples of microflow sensors.

Smart Thermal Flow Sensors and Advanced Evaluation Circuits

Conventional sensors usually have separate electronics, which causes high cost and prevents large serial production. An integrated smart thermal flow sensor is defined as a chip that contains one or more sensors, signal conditioning, A/D conversion, and a bus output [15]. Therefore, there is a need for advanced evaluation methods that convert the thermal signal directly into a frequency and duty-cycle output.

TABLE 9.21 The Transducing Principle of Microflow Sensors

Transducing principle	Realization	Application
Thermoresistive	Metal film (platinum), polycrystalline silicon, single crystalline silicon or metal alloys.	Measurement of temperature, temperature difference, and heat power
Thermoelectric	*p*Si-Al (bipolar-technology), polySi-Al (CMOS-technology) or *p*PolySi-*n*PolySi thermopiles	Measurement of temperature and temperature difference
Thermoelectronic	Transistors, diodes	Measurement of temperature and temperature difference
Pyroelectric	Pyroelectric materials ($LiTaO_3$) with metal or silicon resistors as heater and electrodes	Measurement of heat power
Frequency analog	SAW oscillators	Measurement of temperature

The Duty-Cycle Modulation for the Hot-Wire Sensors

The constant heater temperature can be controlled by modulation of the amplitude of the heat voltage (conventional principle) or by modulation of the duty-cycle. The heater is activated when the output signal is high. This heater controlling output goes low when the temperature level $T_0 + \Delta T$ is reached. During low output, the heating temperature decreases to the temperature level $T_0 + \Delta T$, where the output goes high again and restarts the heating cycle. The temperature level is determined by a reference resistor and a variable resistor. An increase in the flow rate increases the convective cooling of the heater, and it needs a longer time to reach the temperature level. That results in the higher output time t_H. Figure 9.78 shows the working principle of the duty-cycle modulation, where t_d is the time delay in the switching action. Defining the maximum heating power as Q_{max} (the output signal is always high), one obtains the relation:

$$Q = Q_{max}\left(\frac{t_H}{t_{total}}\right) \tag{9.140}$$

The Electrical Sigma–Delta Modulation for Calorimetric Sensors

Conventional signal conditioning circuits use an analog-to-digital converter (ADC) to get digital sensor readout, which can be regarded as an amplifier or a voltmeter. They read the information signal from the sensor, but do not interact with the sensor. In contrast, sigma–delta converters are a part of the sensor function since they act as a feedback amplifier for the sensor output. Hence, sigma–delta conversion normally results in a much more robust sensor signal than is provided by conventional ADCs. Following, the principle of sigma–delta conversion applied to calorimetric flow sensors is explained.

The transistors T_1 and T_2 represent two switchers that feed the constant current I_0 into the *RC* network. It is assumed in this example that the temperature on the resistor R_{s2} is higher than the temperature on R_{s1}. Therefore, the resistance of R_{s2} is larger than R_{s1}. It results in a larger time constant for the charging of C_2. With the help of the comparators and the D-flip-flop, the constant current I_0 can be switched on and off. The switching signals *f*1 and *f*2 have, in the same time period, different pulse numbers $N + S$ and N:

$$\frac{R_{s2}}{R_{s1}} = \frac{(N+S)}{N} = 1 + \frac{\Delta R}{R_{s1}} \tag{9.141}$$

Thus:

$$\frac{\Delta R}{R_{s1}} = \frac{S}{N} \tag{9.142}$$

TABLE 9.22 Examples of Microthermal Mass Flow Sensors

TRANSDUCING PRINCIPLE	HOT-WIRE OR HOT-FILM	CALORIMETRIC
Thermoresistive [13, 14]	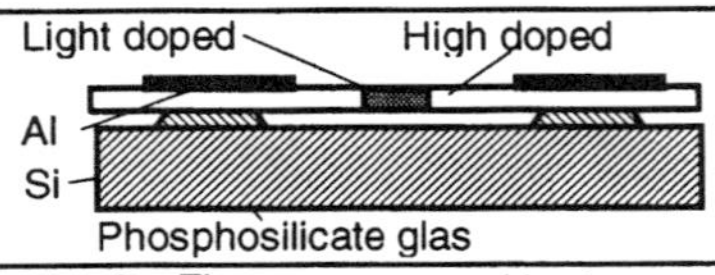	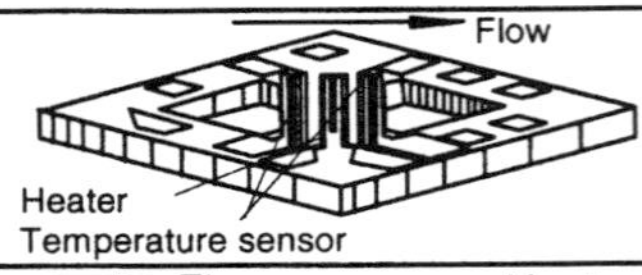
Thermoelectric [15,16]	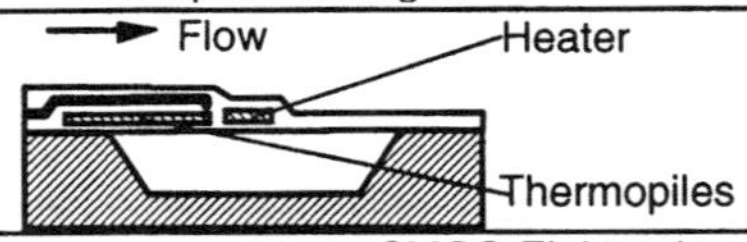	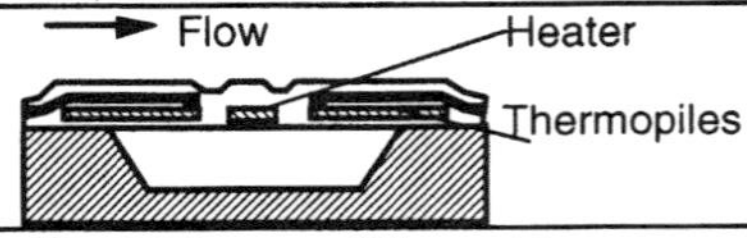
Thermoelectronic [17, 18]	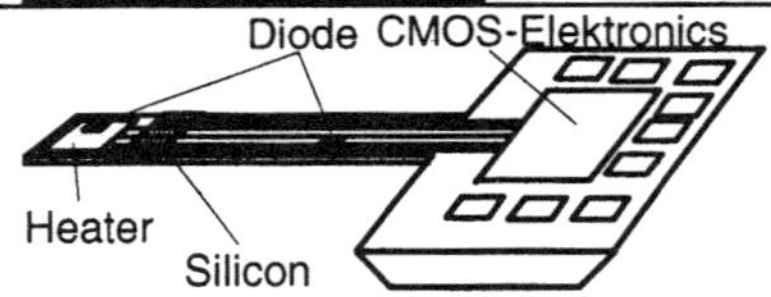	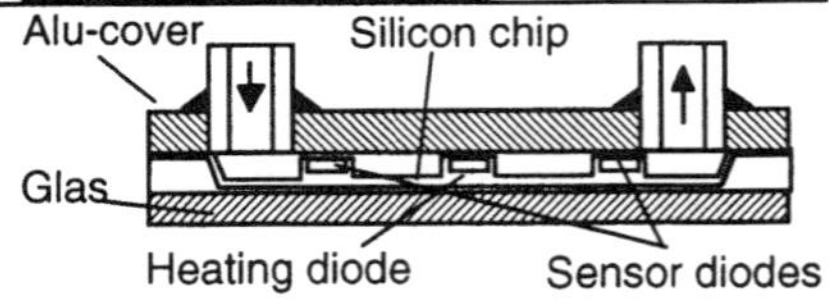
Pyroelectric [19]	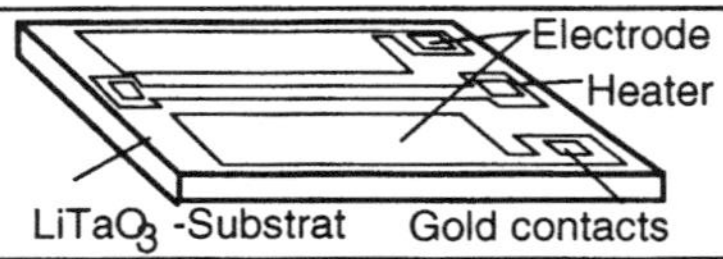	
Frequency analog [20]	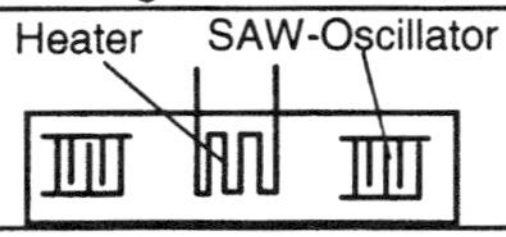	

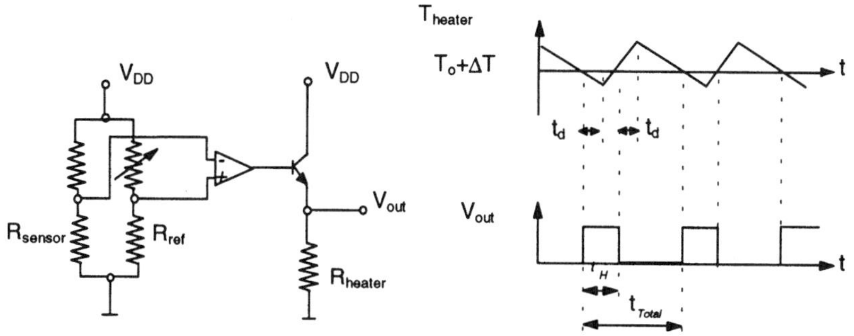

FIGURE 9.78 The duty-cycle modulation for hot wire sensors: (*a*) basic circuit; (*b*) the heater temperature detected by R_{sensor} and the output voltage V_{out} vs. the time.

with:

$$R_{s1} = \frac{\left(V_{ref} N\right)}{\left(I_0 G\right)} \tag{9.143}$$

The resistance difference as well as the temperature difference can be calculated:

$$\Delta R = \frac{\left(S V_{ref}\right)}{G I_0} \tag{9.144}$$

The counting and recording of S and G are shown in Figure 9.79.

The Thermal Sigma–Delta Modulation

The principle of thermal sigma–delta modulation is based on the conventional electrical sigma–delta: the thermal sigma-delta converter uses a thermal integrator (thermal *R*/*C*-network) instead of an electric integrator (electric *R*/*C*-network). Figure 9.80 shows the principle of thermal sigma–delta modulation. The comparator output modulates the flip-flop. The flip-flop synchronizes the heating signal by its clock. Therefore, the heating periods are chopped into further small pulses that depend on the clock frequency. The heater is actuated step-by-step until the comparator switches again, and one obtains a frequency analog output at the flip-flop [22].

Calibration Conditions

Calibration of Hot-Wire and Hot-Film Sensors

The hot-wire and hot-film sensors are based on the point velocity measurement (see Table 9.14). Therefore, the measurement results depend on the velocity distribution inside the flow channel. Achieving a high signal-to-noise ratio can thus require spatial arrays of hot-wire and hot-film sensors that give more information about the velocity field and thus more accurate results of the mass flow in channel. With the use of a nozzle, a Venturi, or a flow conditioner, the flow profile is preconditioned, which leads to an acceptable accuracy.

Temperature Dependence of Fluid Properties

Most fluid properties depend on the working temperature. The heat transfer process depends on the fluid properties. The measurement of fluid temperature (see Table 9.14) keeps the heater on a constant

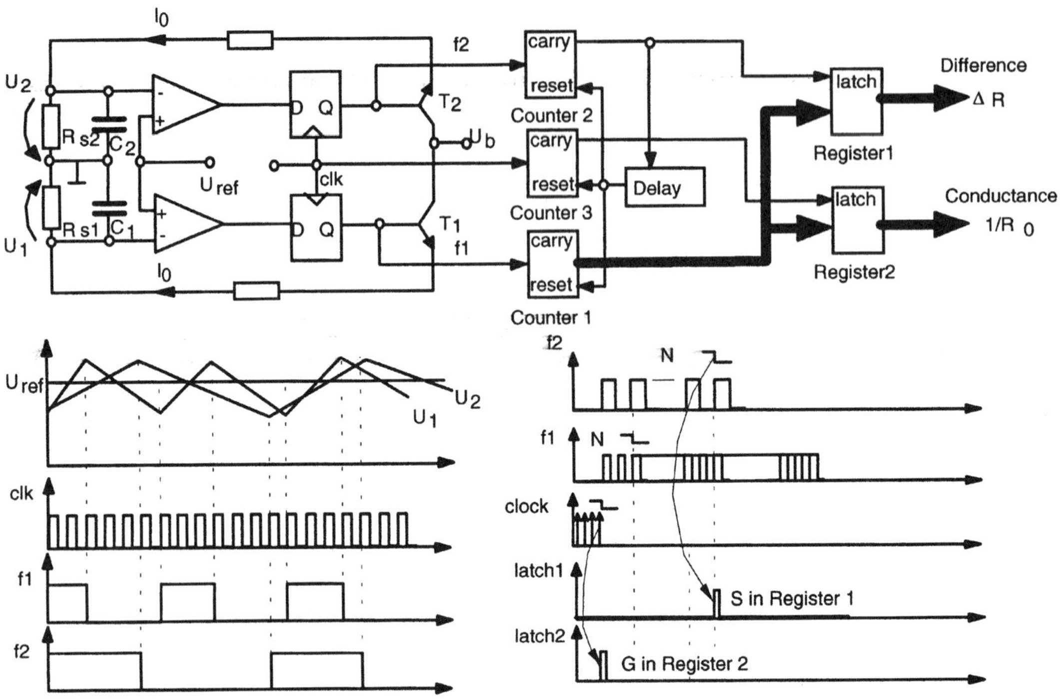

FIGURE 9.79 Sigma–delta conversion for calorimetric sensors.

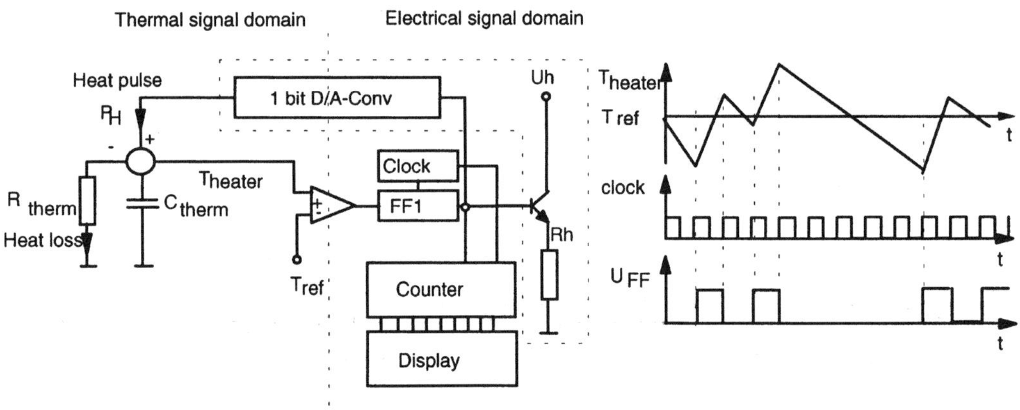

FIGURE 9.80 Principle of thermal sigma–delta modulation.

temperature difference to the fluid and can also be used for compensation of variations in temperature. For these reasons, accurate thermal mass flow sensors require both flow calibration and calibration for temperature compensation. Furthermore, the influence of temperature and/or fluid composition can be derived utilizing the relationships developed in the sections on modeling (Equations 9.120–9.135).

Instrumentation and Components

Table 9.23 lists some companies that manufacture and market thermal mass flowmeters.

TABLE 9.23 Manufactures of Thermal Mass Flow Sensors

Manufacturers	Data	Approximate price
KOBOLD Instruments Inc. 1801 Parkway View Drive Pittsburg, PA 15205	Calorimetric type: MAS-Series; air; min. range 0–10 mL min; max. range 0–40 L min. T max. 50°C; max. pressure 10 bar; accuracy ±2%	~ US$850
	Hot wire type: ANE-Series; air; range 0–20 m/s; working temperature 20–70°C	~ US$1200
HÖNTZSCH GmbH Box 1324, Robert-Bosch Str. 8 D-7050 Waiblingen, Germany	Hot wire type: range 0.05–20 m/s	
BRONKHORST HI-TEC Nijverheidstraat 1A 7261 AK Ruurlo, Netherlands	Calorimetric type: gases and liquids; min. range (gas) 0–5 mL min; max. range (liquid) 0–1000 ml/min; max. pressure 400 bar	~ US$1800
HONEYWELL, MicroSwitch, Freeport, IL	Calorimetric type: Gases; range 0–1000 ml/min; max. pressure 1.75 bar	
SIERRA INSTRUMENTS INC. 5 Harris Ct., Bldg. L Monterey, CA 93924	Calorimetric type: all gases from 1 ml/min–10,000 l/min; –40–100°C; 30 bar max.; 1% accuracy	~ US$400–1200
	Hot-wire type in stainless-steel sheath: gases from 0–100 m/s; –40–400°C; 100 bar max.; 2% accuracy	~ US$60–2000
BROOKS INSTRUMENT DIVISION Emerson Electric Co. 407 W. Vine Street Hatfield, PA 19440	Calorimetric type: all gases from 1 ml/min to 10,000 l/min; –40–100°C; 100 bar max.; 1% accuracy	~ US$400–1200

References

1. Béla G. Lipták, *Flow Measurement,* Radnor, PA: Chilton Book, 1993.
2. H. Strickert, *Hitzdraht und Hitzfilmanemometrie* (Hot-wire and hot-film anemometry), Berlin: Verlag Technik, 1974.
3. G. Schnell, Sensoren in der Automatisierungstechnik (Sensors in the automation techniques), Brauschweing: Verlag Viehweg, 1993.
4. *Mass Flow and Pressure Meters/Controllers,* The Netherlands: Bronkhorst Hi-Tec B. V., 1994.
5. T. S. J. Lammerink, F. Dijkstra, Z. Houkes, and J. van Kuijk, Intelligent gas-mixture flow sensor, *Sensors and Actuators A,* 46/47, 380-384, 1995.
6. VDI-Wärmeatlas, VDI-Verlag, 1994.
7. H. Schlichting, *Boundary Layer Theory,* 7th ed., New York: McGraw-Hill, 1979.
8. A. J. Chapman, *Heat Transfer,* 4th ed., New York: Macmillan, 1984.
9. T. S. J. Lammerink, Niels R. Tas, Miko Elwenspoek, and J. H. J. Fluitman, Micro-liquid flow sensor, *Sensors and Actuators A,* 37/38, 45-50, 1993.
10. N.T. Nguyen and W. Dötzel, A novel method for designing multi-range electrocaloric mass flow sensors: asymmetrical locating with heater- and sensor arrays, *Sensors and Actuators A,* 62, 506-512, 1997.
11. J. van Kuijk, T. S. T. Lammerink, H.-E. de Bree, M. Elwenspoek, and J. H. J. Fluitman, Multi-parameter detection in fluid flows, *Sensors and Actuators A,* 46/47, 380- 384, 1995.
12. S. M. Sze, *Semiconductor Sensors,* New York: John Wiley & Sons, 1994.
13. You-Chong Tai and Richard S. Muller, Lightly-doped polysilicon bridge as a flow meter, *Sensors and Actuators A,* 15, 63-75, 1988.
14. R. G. Johnson and R. E. Higashi, A highly sensitive silicon chip microtransducer for air flow and differential pressure sensing applications, *Sensors and Actuators A,* 11, 63-67, 1987.
15. H. J. Verhoeven and J. H. Huijsing, An integrated gas flow sensor with high sensitivity, low response time and pulse-rate output, *Sensors and Actuators A,* 41/42, 217-220, 1994.

16. F. Mayer, G. Salis, J. Funk, O. Paul, and H. Baltes, Scaling of thermal CMOS gas flow microsensors experiment and simulation, *MEMS '96*, San Diego, 1996, 116-121.
17. Göran Stemme, A CMOS integrated silicon gas-flow sensor with pulse-modulated output, *Sensors and Actuators A*, 14, 293-303, 1988.
18. Canqian Yang and Heinrik Soeberg, Monolithic flow sensor for measuring millilitre per minute liquid flow, *Sensors and Actuators A*, 33, 143-153, 1992.
19. Dun Yu, H. Y. Hsieh, and J. N. Zemel, Microchannel pyroelectric anemometer, *Sensors and Actuators A*, 39, 29-35, 1993.
20. S. G. Joshi, Flow sensors based on surface acoustic waves, *Sensors and Actuators A*, 44, 63-72, 1994.
21. U. Bonne, Fully compensated flow microsensor for electronic gas metering, *Proc. Int. Gas Research Conf.*, 16-19 Nov. '92, Orlando, FL, 3, 859, 1992.
22. J. H. Huijsing, F. R. Riedijk, and G. van der Horn, Developments in integrated smart sensors, *Proc. of Transducer 93*, Yokohama, Japan, 1993, 320-326.

9.10 Coriolis Effect Mass Flowmeters

Jesse Yoder

Coriolis flowmeters were developed in the 1980s to fill the need for a flowmeter that measures mass directly, as opposed to those that measure velocity or volume. Because they are independent of changing fluid parameters, Coriolis meters have found wide application. Many velocity and volumetric meters are affected by changes in fluid pressure, temperature, viscosity, and density. Coriolis meters, on the other hand, are virtually unaffected by these types of changes. By measuring mass directly as it passes through the meter, Coriolis meters make a highly accurate measurement that is virtually independent of changing process conditions. As a result, Coriolis meters can be used on a variety of process fluids without recalibration and without compensating for parameters specific to a particular type of fluid. Coriolis flowmeters are named after Gaspard G. Coriolis (1792–1843), a French civil engineer and physicist for whom the Coriolis force is named.

Coriolis meters have become widely used in industrial environments because they have the highest accuracy of all types of flowmeters. They measure mass directly, rather than inferentially. Coriolis meters do not have moving parts like turbine and positive displacement meters, which have parts that are subject to wear. Maintenance requirements for Coriolis meters are low, and they do not require frequent calibration. Wetted parts can be made from a variety of materials to make these meters adaptable to many types of fluids. Coriolis meters can handle corrosive fluids and fluids that contain solids or particulate matter. While these meters were used mainly for liquids when they were first introduced, they have recently become adaptable for gas applications.

Theory of Operation

Coriolis meters typically consist of one or two vibrating tubes with an inlet and an outlet. While some are U-shaped, most Coriolis meters have some type of complex geometric shape that is proprietary to the manufacturer. Fluid enters the meter in the inlet, and mass flow is determined based on the action of the fluid on the vibrating tubes. Figure 9.81 shows flow tube response to Coriolis acceleration.

Common to Coriolis meters is a central point that serves as the axis of rotation. This point is also the peak amplitude of vibration. What is distinctive about this point is that fluid behaves differently, depending on which side of the axis of rotation, or point of peak amplitude, it is on. As fluid flows toward this central point, the fluid takes on acceleration due to the vibration of the tube. As the fluid flows away from the amplitude of peak vibration, it decelerates as it moves toward the tube outlet. On the inlet side of the tube, the accelerating force of the flowing fluid causes the tube to lag behind its no-flow position. On the outlet side of the tube, the decelerating force of the flowing fluid causes the tube to lead ahead of its no-flow position. As a result of these forces, the tube takes on a twisting motion as it passes through

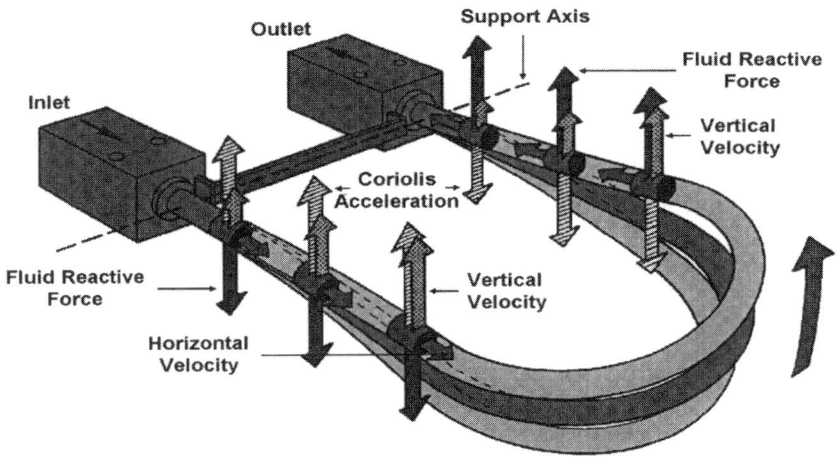

FIGURE 9.81 Flow tube response to Coriolis acceleration.

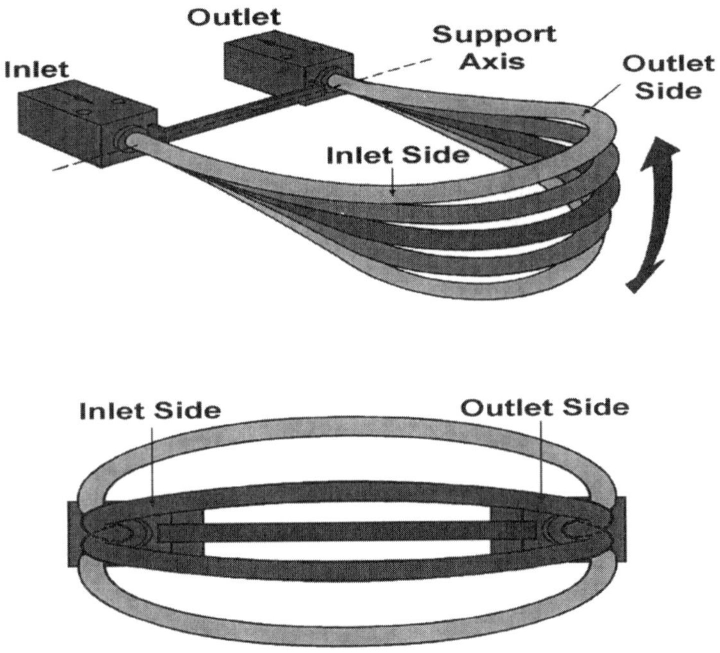

FIGURE 9.82 Two views of an oscillating flow tube with no flow.

each vibrational cycle; the amount of twist is directly proportional to the mass flow through the tube. Figure 9.82 shows the Coriolis flow tube in a no flow situation, and Figure 9.83 shows Coriolis tube response to flow.

The Coriolis tube (or tubes, for multitube devices) is vibrated through the use of electromagnetic devices. The tube has a drive assembly, and has a predictable vibratory profile in the no-flow position. As flow occurs and the tube twists in response to the flow, it departs from this predictable profile. The degree of tube twisting is sensed by the Coriolis meter's detector system. At any point on the tube, tube motion represents a sine wave. As mass flow occurs, there is a phase shift between the inlet side and the outlet side. This is shown in Figure 9.84.

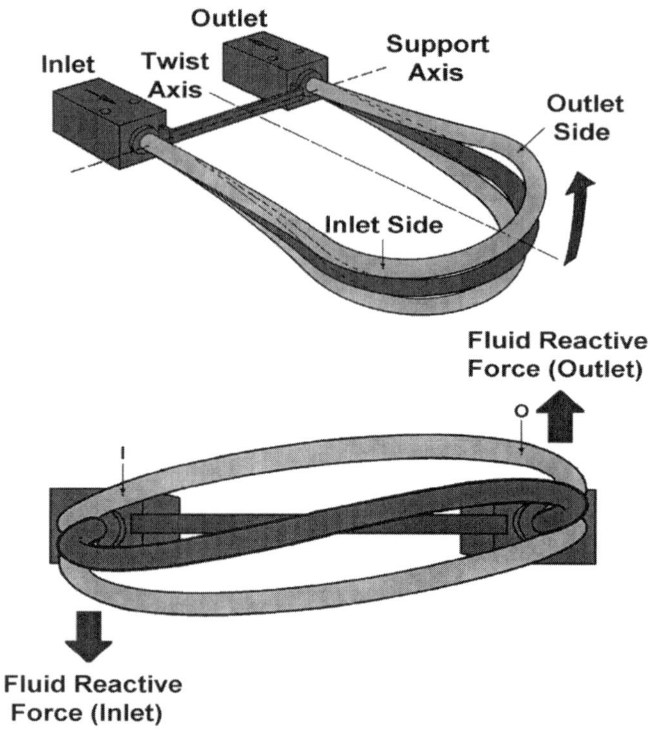

FIGURE 9.83 Two views of an oscillating flow tube in response to flow.

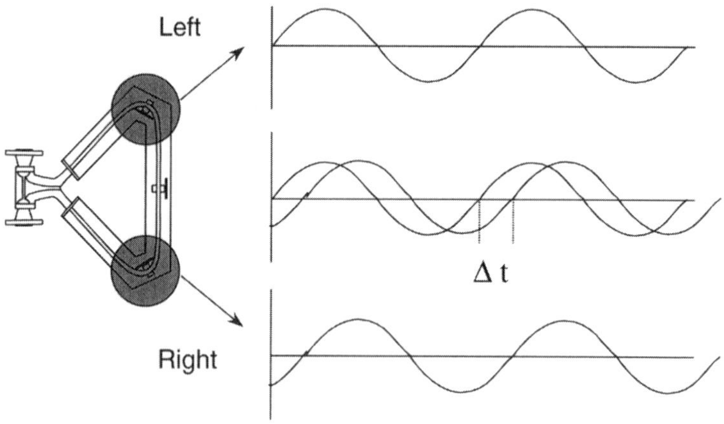

FIGURE 9.84 Phase shift between inlet side and outlet side.

The Coriolis force induced by flow is described by an equation that is equivalent to Newton's second law for rotational motion. This equation is as follows:

$$F = 2m\omega\bar{v} \tag{9.145}$$

In Equation 9.145, F is force, m is the mass to be applied to a known point at a distance L from the axis 0-0, ω is a vector representing angular motion, and $\bar{v}$ is a vector that represents average velocity.

Construction

The internal part of the Coriolis tube is the only part of the meter that is wetted. A typical material of construction is stainless steel. Other corrosion-resistant metals such as Hastelloy are used for tube construction. Some meters are lined with Teflon.

Some designs have thin-wall as well as standard tubes. A thin-wall design makes the meter more useful for gas and low-velocity liquid applications, where the amount of twist by the tube is reduced. It is important to be aware of the extent to which the fluid degrades or attacks the tube wall or lining. If the fluid eats away at the wall, this can reduce the accuracy of the meter.

Advantages

The most significant advantage of Coriolis meters is high accuracy under wide flow ranges and conditions. Because Coriolis meters measure mass flow directly, they have fewer sources of errors. Coriolis meters have a high turndown, which makes them applicable over a wide flow range. This gives them a strong advantage over orifice plate meters, which typically have low turndown. Coriolis meters are also insensitive to swirl effects, making flow conditioning unnecessary. Flow conditioners are placed upstream from some flowmeters to reduce swirl and turbulence for flowmeters whose accuracy or reliability is affected by these factors.

Coriolis meters have a low cost of ownership. Unlike turbine and positive displacement meters, Coriolis meters have no moving parts to wear down over time. The only motion is due to the vibration of the tube, and the motion of the fluid flowing inside the tube. Because Coriolis flowmeters are designed not to be affected by fluid parameters such as viscosity, pressure, temperature, and density, they do not have to be recalibrated for different fluids. Installation is simpler than installation for many other flowmeters, especially orifice plate meters, because Coriolis meters have fewer components.

Coriolis meters can measure more than one process variable. Besides mass flow, they can also measure density, temperature, and viscosity. This makes them especially valuable in process applications where information about these variables reduces costs. It also makes it unnecessary to have a separate instrument to measure these additional variables.

Disadvantages

The chief disadvantage of Coriolis meters is their initial cost. While some small meters have prices as low as $4000, the base price for most Coriolis meters is $6000 and up. The cost of Coriolis meters rises significantly as line sizes increase. The physical size of Coriolis meters increases substantially with the increase in line size, making 150 mm (6 in.) the upper line size limit on Coriolis meters today. The large size of some Coriolis meters makes them difficult to handle, and can also make installation difficult in some cases.

The lack of an established body of knowledge about Coriolis meters is a substantial disadvantage. Because Coriolis meters were recently invented, not nearly as much data are available about them as are for differential pressure-based flowmeters. This has made it difficult for Coriolis meters to gain approvals from industry associations such as the American Petroleum Institute. This will change with time, as more manufacturers enter the market and users build up a larger base of experience.

Applications

Coriolis meters have no Reynolds number constraints, and can be applied to almost any liquid or gas flowing at a sufficient mass flow to affect vibration of the flowmeter. Typical liquid applications include foods, slurries, harsh chemicals, and blending systems. The versatility of Coriolis meters in handling multiple fluids makes them very useful for plants where the flow of multiple fluid types must be measured.

There are an increasing number of gas applications for Coriolis meters. While gas applications are still very much in the minority, the use of this meter to measure gas is likely to increase as more is learned about its use for this purpose.

A Look Ahead

There are some important areas of research for Coriolis meters. While most Coriolis meters have been bent, several manufacturers have recently introduced straight-tube designs. Manufacturers will continue to fine-tune the single-tube vs. double-tube design, and to work on tube geometry. As noted above, the use of Coriolis meters for gas and also for steam applications is another area for future development.

References

E. O. Doebelin, *Measurement Systems: Application and Design,* 4th ed., New York: McGraw Hill, 1990, 603–605.

R. S. Figliola and D. E. Beasley, *Theory and Design for Mechanical Measurements,* 2nd ed., New York: John Wiley & Sons, 1995, 475–478.

K. O. Plache, Coriolis/Gyroscopic Flowmeter, *Mechanical Engineering,* MicroMotion Inc., Boulder, CO, March 1979, 36–41.

L. Smith and J. R. Ruesch, *Flow Measurement,* D. W. Spitzer (ed.), Research Triangle Park, NC: Instrument Society of America, 1996, 221-247.

9.11 Drag Force Flowmeters

Rekha Philip-Chandy, Roger Morgan, and Patricia J. Scully

In a *target flowmeter* a solid object known as a *drag element* is exposed to the flow of fluid that is to be measured. The force exerted by the fluid on the drag element is measured and converted to a value for speed of flow.

The flow-sensing element has no rotating parts, and this makes the instrument suitable for conditions where abrasion, contamination, or corrosion make more conventional instruments unsuitable. An important application of such flowmeters involves environmental monitoring in areas such as meteorology, hydrology, and maritime studies to measure speeds of air or water flow and turbulence close to the surface. In these applications, the fluid flows are sporadic and multidirectional.

A further advantage of the instrument is that it can be made to generate a measurement of flow *direction* in two dimensions, or even in three dimensions, as well as of flow speed. To implement this feature, the drag element must be symmetrical in the appropriate number of dimensions, and it is necessary to measure the force on it vectorially, again in the appropriate number of dimensions. Provided that the deflecting forces are independent in the sensing directions, the resulting outputs can be added vectorially to generate independent values for flow speed and direction. Target flowmeters using strain gage technology have been used by industry, utilities, aerospace, and research laboratories. They have been used to successfully measure the flow of uni- and bidirectional liquids (including cryogenic), gases, and steam (both saturated and superheated) for almost half a century.

Despite these advantages, the target flowmeter appears to have been neglected in favor of more complex and sophisticated devices. The authors have sought to remedy this neglect by developing a sensor suitable for measuring multidirectional flows in two dimensions, instead of measuring only bidirectional flows in a single dimension.

Design of the Flow Sensor

The sensor described in this chapter section is ideally suited for environmental flow measurement. The operation is based on strain measurement of deformation of an elastic rubber cantilever, to which a force is applied by a spherically symmetrical drag element (Figure 9.85). This sensor has many advantages, including compactness and a simple construction requiring no infrastructure other than a rigid support, and it can cope with fluids containing solid matter such as sludge and slurries provided that they do not tangle with the drag element or the rubber beam.

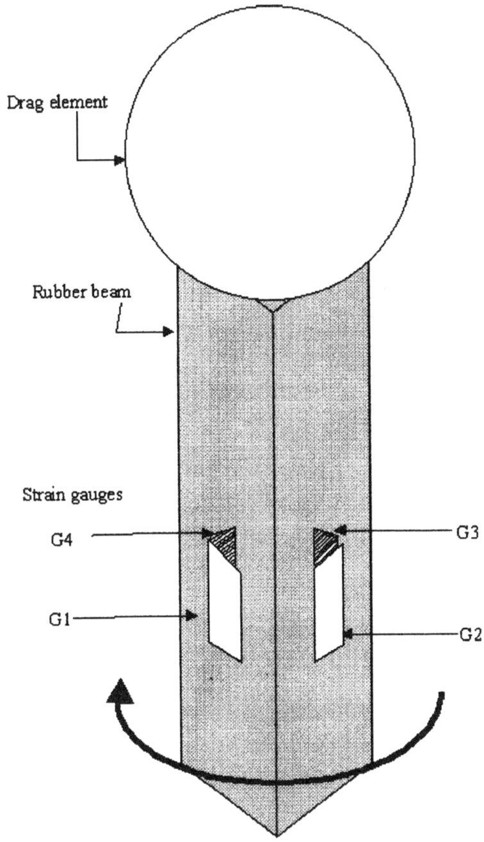

FIGURE 9.85 Schematic of the electric resistance strain gage drag force flow sensor.

According to Clarke [1], the ideal drag element is a flat disk, because this configuration gives a drag coefficient independent of flow rate. Using a spherical drag element, which departs from the ideal of a flat disk [1], the drag coefficient is likely to vary with flow speed, and therefore the gage must be calibrated and optimized for the conditions of intended use. In this discussion, a gage is developed for air flows in the range normally encountered in the natural environment.

The strain measurement can be performed with conventional strain gages, but this limits the applications of the device to conditions where corrosion of the metal-resistive track of the strain gage can be avoided. Therefore, an optical fiber strain gage has been developed as an alternative.

Principle of Fluid Flow Measurement

The drag force F_D exerted by a fluid on a solid object exposed to it is given by the *drag equation*, which from incompressible fluid dynamics is:

$$F_D = \frac{C_D \rho A V^2}{2} \tag{9.146}$$

where ρ = Fluid density
V = Fluid's velocity at the point of measurement
A = Projected area of the body normal to the flow
C_D = Overall drag coefficient

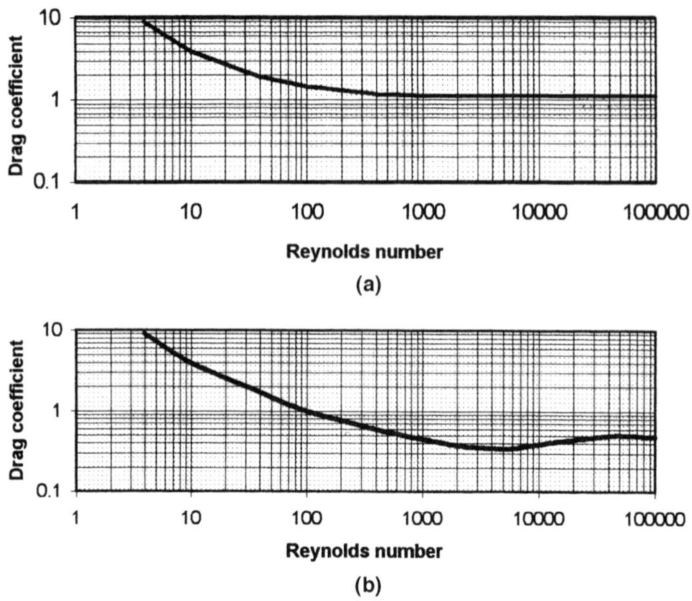

FIGURE 9.86 Graph shows drag coefficient C_D plotted against Reynolds number Re for (*a*) a flat disk and (*b*) a sphere.

C_D is a dimensionless factor, whose magnitude depends primarily on the physical shape of the object and its orientation relative to the fluid stream. The drag coefficient C_D for a sphere is related to the Reynolds number (Re), another dimensionless factor, given by:

$$\mathrm{Re} = \rho\, V A / \eta \tag{9.147}$$

where η = viscosity of the fluid.

A graph of C_D against Reynolds number (derived from Clarke [1]) is shown in Figure 9.86(*a*) for a flat disk and Figure 9.86(*b*) for a sphere, from which it is evident that the value of C_D, although not constant, is not subject to wide variation over the range of the graph.

Now consider the effect of the force F_D on an elastic beam to which the drag element is attached (as in Figure 9.85). If the mass of the beam can be ignored, the deflection of the beam will be due only to force exerted on the drag element by the fluid. From the theory of elasticity for a cantilever beam of length L with point load at its end [2], the shear force P will be constant along the beam. The bending moment, M at a point x along the beam will be $P(L–x)$ so that it varies linearly from PL at $x = 0$ to 0 at $x = L$. The distance y is measured from the neutral plane, which for a rectangular section is the mid-plane.

The inverse of the radius of curvature is given by:

$$\frac{1}{\rho} = \frac{M}{EI} \tag{9.148}$$

where E is the modulus of elasticity or the Young's modulus and I is the moment of inertia of the cross-section where it is assumed that the force is applied perpendicular to the broad face of width b.

$$I = \frac{b\, a^3}{12} \tag{9.149}$$

Through the thickness a of the beam, the strain is ε_x.

$$\varepsilon_x = \frac{\sigma_x}{E} = \frac{Y}{\rho} \tag{9.150}$$

Substituting Equation 9.148 in Equation 9.150 gives:

$$\varepsilon_x = \frac{Y M}{E I} \tag{9.151}$$

The strain at the surfaces is:

$$\varepsilon_x = \frac{a}{2\rho} = \frac{Y M}{2 E I} \tag{9.152}$$

Substituting Equation 9.149 and the bending moment into Equation 9.151 gives:

$$\varepsilon_x = \frac{6\, P\left(L - x\right)}{E\, a^2\, b} \tag{9.153}$$

or, the shear force P is:

$$P = \frac{\varepsilon\, E\, a^2\, b}{6\left(L - x\right)} \tag{9.154}$$

Consequently,

$$\frac{C_D\, \rho\, A V^2}{2} = \frac{\varepsilon E a^2 b}{6\left(L - x\right)} \tag{9.155}$$

Therefore, the strain is:

$$\varepsilon = \frac{3 C_D \rho\, A V^2 \left(L - x\right)}{E a^2 b} \tag{9.156}$$

From Equation 9.156, the strain is a square law function of fluid speed.

For most fluid flows in the natural environment, a two-dimensional measurement is necessary as the flow in the natural environment is almost always two-dimensional. For a measurement of flow speed and direction, it is necessary to relate wind speed to strain measured in two orthogonal directions, on the assumption (which has been justified by experiment) that the velocity vector has a zero component along (parallel to) the rubber beam support (i.e., $U_z = 0$). The wind speed U has orthogonal components U_x and U_y that are proportional to the strain measured in the x (strain_x) and y directions (strain_y), respectively. Since U_x is proportional to $\sqrt{\text{strain}_x}$ and U_y is proportional to $\sqrt{\text{strain}_y}$, then the velocity magnitude $|U|$ can be written as

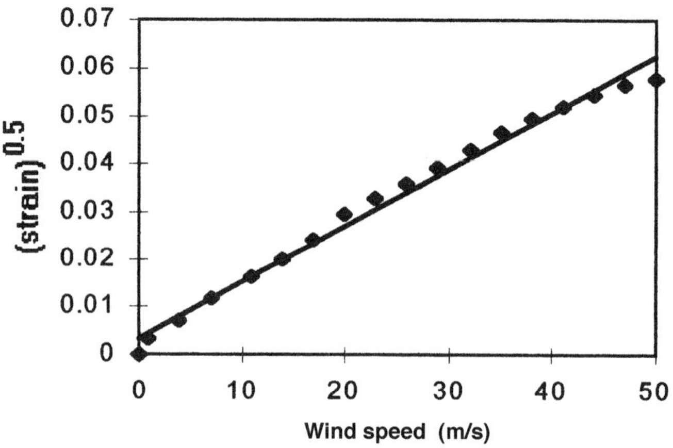

FIGURE 9.87 Root strain vs. wind speed for one dimensional air flow measurement.

$$\left|U\right| = \sqrt{U_x^2 + U_y^2} \tag{9.157}$$

thus,

$$\left|U\right| \alpha \sqrt{\text{strain}_x + \text{strain}_y} \tag{9.158}$$

Therefore, the magnitude of the velocity, $|U|$, is proportional to the square root of the sum of the orthogonal strain components, strain_x and strain_y. The velocity direction, θ, is calculated using the relation:

$$\theta = \tan^{-1}\left[\frac{\sqrt{\text{strain}_y}}{\sqrt{\text{strain}_x}}\right] \tag{9.159}$$

Implementation Using Resistive Strain Gages

A sensor was constructed for measuring one-dimensional flows by bonding two strain gages onto the opposite sides of a square-sectioned elastic beam, 165 mm long and 13.5 mm square, made from polybutadiene polymer (unfilled vulcanized rubber) supplied by the Malaysian Rubber Products Research Association. The modulus of the beam, at approximately 1.2×10^6 N m^{-2}, was relatively low in order to achieve good sensitivity. The drag element was a table tennis ball of diameter 20 mm, glued to the end of the rubber beam. The strain gages were cupro-nickel alloy on a polyimide film, and were bonded to the rubber beam with epoxy resin. The two gage elements were connected in a half-bridge configuration, and the output was taken to a digital strain-gage indicator reading directly in microstrain units.

The sensor was mounted in a calibrated wind tunnel, in such a way that one gage underwent compression and the other gage underwent extension. The wind speed was set at different values and the corresponding strain readings were noted. Results were plotted as wind speed against square root of strain (Figure 9.87). The graph obtained was linear with a correlation co-efficient of 0.98. The linearity shows that the strain is a square-law function of speed, which confirms the theory above.

In this first prototype of the sensor, the range was limited by beam oscillations at speeds greater than 23 m s^{-1}, causing spurious signals. In principle, however, much greater speeds up to 50 m s^{-1} can be monitored by this sensor, if the construction is suitable.

A second sensor was constructed for measuring two-dimensional flows, in which a square-sectioned rubber beam similar to the previous one was set up with four strain gages attached to the four longitudinal

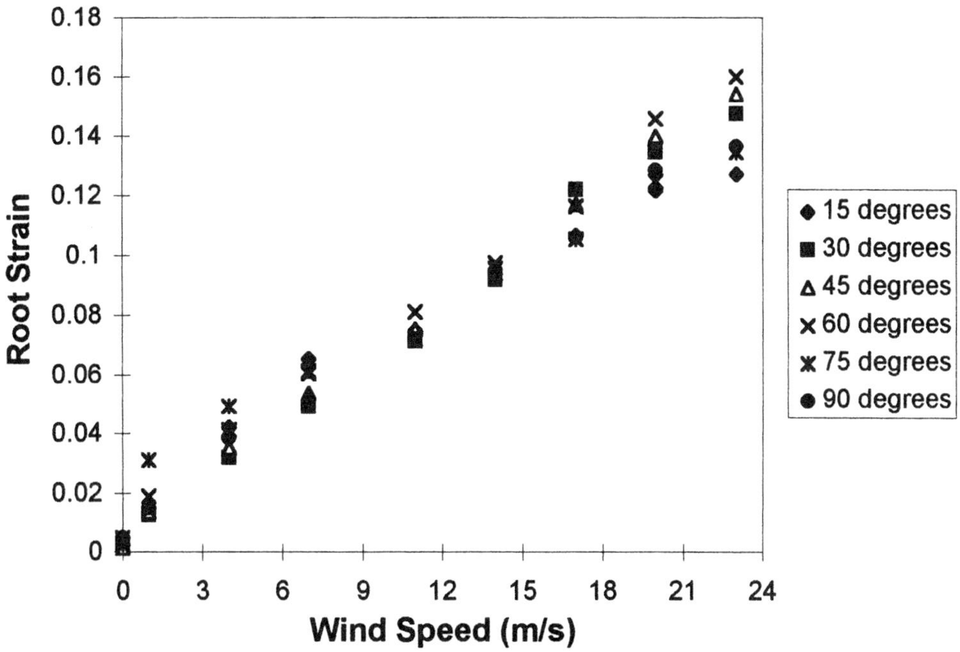

FIGURE 9.88 Root strain plotted against wind speed with the flowmeter oriented at various angles to wind flow.

surfaces of the beam. Each opposite pair of strain gage elements was connected in a half-bridge configuration as before. The two outputs were taken to instrumentation amplifiers and then to a data acquisition board interfaced to the LabView instrumentation software package (National Instruments Corporation). After appropriate signal processing by the software, the values of $strain_x$ and $strain_y$ from both the channels are added and then the square root of the absolute value of this sum is found.

The device was clamped on a turntable and this was rotated about its longitudinal axis from 0° to 90° at 15° intervals. At each angle, the *x* and *y* strain gage outputs were recorded as a function of velocity, over a range from 0 to 23 m s^{-1}. The graph of speed vs. $(strain_x + strain_y)^{1/2}$ for different angles is shown in Figure 9.88. The data presented in Figure 9.88 can be used to obtain a speed calibration curve for the sensor, which will be valid for multidirectional air flow.

The direction of wind flow was calculated according to Equation 9.159 for the different wind speeds. Figure 9.89 shows the plots of the wind flow direction at different wind speeds for various sensor positions. Several authors have published the inability of calculating the wind direction accurately at low wind speeds. The situation is similar in this case and the readings get more accurate as the wind speed increases, especially beyond 11 m s^{-1}.

For the measurement of gusts, the time response of the sensor becomes a critical parameter. Experiments were performed to measure the response time by dropping known weights on the free end of the sensor and using a specially written program in Labview to acquire and record the response. The sensor indicated a 95% response time of 50 ms and a time constant of 30 ms.

Optical Fiber Strain Gage Drag Force Flowmeter

Optical fibers have been applied to measurement of fluid flow, and strain measurement, using interferometric techniques and bending losses [3–5]. The instrument described in the previous sections can be adapted to use optical strain measurement instead of resistive strain gages. In this way, the advantages of the instrument can be preserved; in addition, the flowmeter has the usual benefits of optical fiber sensors such as immunity to electromagnetic interference and intrinsic safety in hazardous environments.

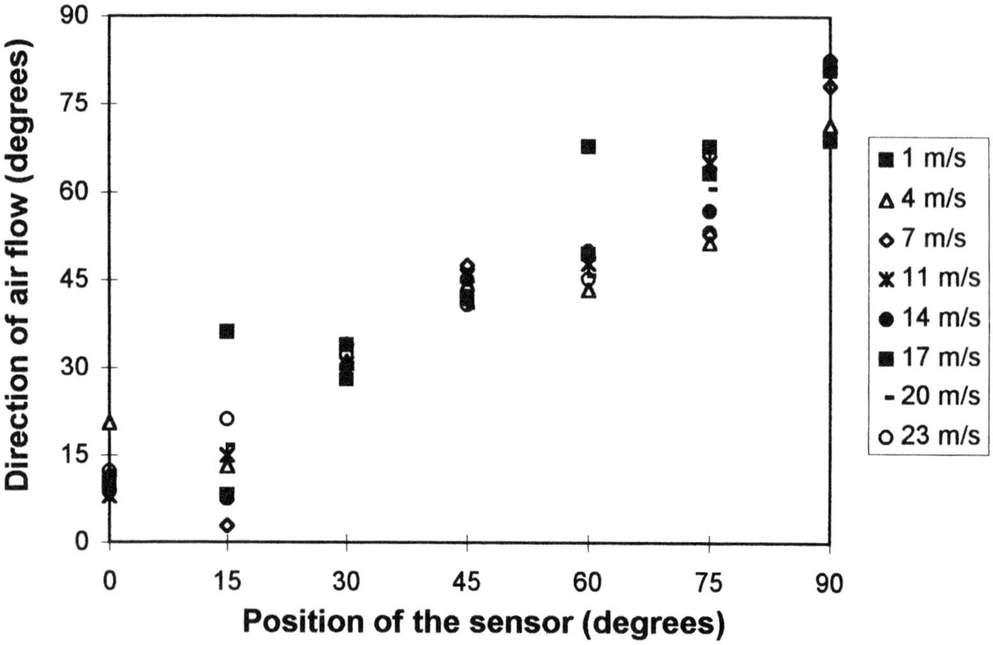

FIGURE 9.89 Calculated direction of wind velocity at different wind speeds.

The principle of an optical fiber strain gage is based on the effect of one or more grooves inserted radially into a 1 mm diameter PMMA fiber, which cause a loss in light transmission. As the optical fiber is bent, as in a cantilever, the angle of the grooves varies. These changes of angle cause light to be attenuated at each groove. The intensity variation can be related to the change of the angle of the groove caused by the bending of the cantilever.

To develop an analogy between the optical strain measurement and the resistive strain measurement, the optical strain was calculated using the formula:

$$\text{Strain}_{\text{opt}} = \frac{\text{Change in power output}}{\text{Original power output}} = \frac{\Delta P}{P} \tag{9.160}$$

Hence,

$$Strain_x = \frac{\Delta P_x}{P_x} \quad \text{and} \quad strain_y = \frac{\Delta P_y}{P_y} \tag{9.161}$$

Therefore, the magnitude of the velocity, $|U|$ is given by:

$$|U| = \sqrt{U_x^2 + U_y^2} = \sqrt{strain_x + strain_y}$$

Substituting Equation 9.161 in $|U|$, the optical root strain magnitude is:

$$|\text{U}| = \sqrt{\left(\frac{\Delta P_x}{P_x} + \frac{\Delta P_y}{P_y}\right)} \tag{9.162}$$

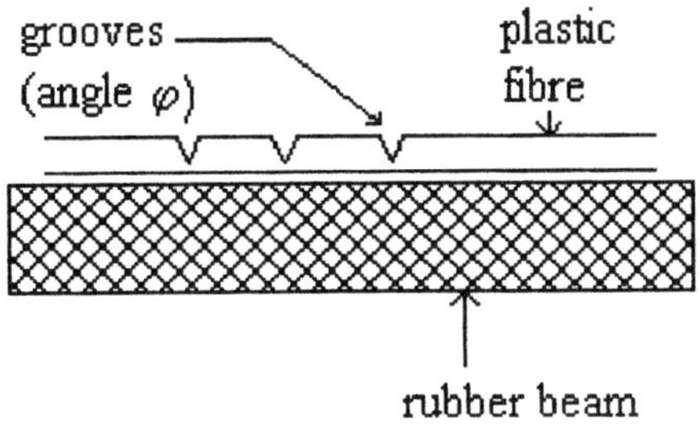

FIGURE 9.90 Principle of operation of the fiber optic drag force flow sensor.

and the direction of wind velocity, θ:

$$\theta = \tan^{-1}\left[U_y / U_x\right] = \tan^{-1}\sqrt{\frac{\text{Strain}_y}{\text{Strain}_x}} \tag{9.163}$$

Substituting Equation 9.161 in θ, the optical strain direction is:

$$\theta = \tan^{-1}\frac{\sqrt{\left(\dfrac{\Delta P_y}{P_y}\right)}}{\sqrt{\left(\dfrac{\Delta P_x}{P_x}\right)}} \tag{9.164}$$

Two types of the optical fiber flow sensor are described here. In the first type, the rubber beam is used as the deflected device with optical fiber strain gages attached to the deflected beam. The second version of this flow sensor uses an unsupported sensitised 1 mm diameter plastic optical fiber that undergoes deflection in the airflow.

Fiber Optic Flow Sensor WITH Rubber Beam

The sensor has been built using a rectangular cross-sectioned rubber beam and a spherical drag element with a 1 mm plastic optical fiber glued to the beam with epoxy. The fiber is looped inside the drag element to enhance the sensitivity of the device. Grooves have been made in the fiber surface that extends into the core of the fiber to increase the losses as a function of the bending of the rubber beam (Figure 9.90).

The grooves are normal to the rubber beam on its outside face, and their depths affect only the cladding of the fiber. As the cantilever bends due to the force exerted by the air flow, the angle of the grooves vary. The groove angle increases when the air flow is facing the sensor, and vice versa. These changes of groove angle cause an intensity modulation of the light transmitted through the fiber because light is lost at each groove. Changes in intensity can be related to changes in the angle caused by the force inducing the bending of the cantilever, and therefore to the velocity of the fluid.

The orthogonal components of strain were measured by attaching two grooved optical fibers on adjacent sides of the rubber beam, orthogonal to each other, as illustrated in Figure 9.91. These two fibers measured the x and y components of optical strain.

Light from a 1 mW helium-neon laser of peak wavelength 633 nm was split into equal components using a cubic beam splitter, and coupled into each fiber. The transmitted intensity through each of the two fibers was monitored using a power meter. The signals from the power meter were sent to a 486 DX2 laptop computer via a data acquisition card to be acquired and processed by Labview. Experiments indicated that this version of the flow sensor could measure wind speeds up to 30 m s^{-1} with a resolution of 1.3 m s^{-1}.

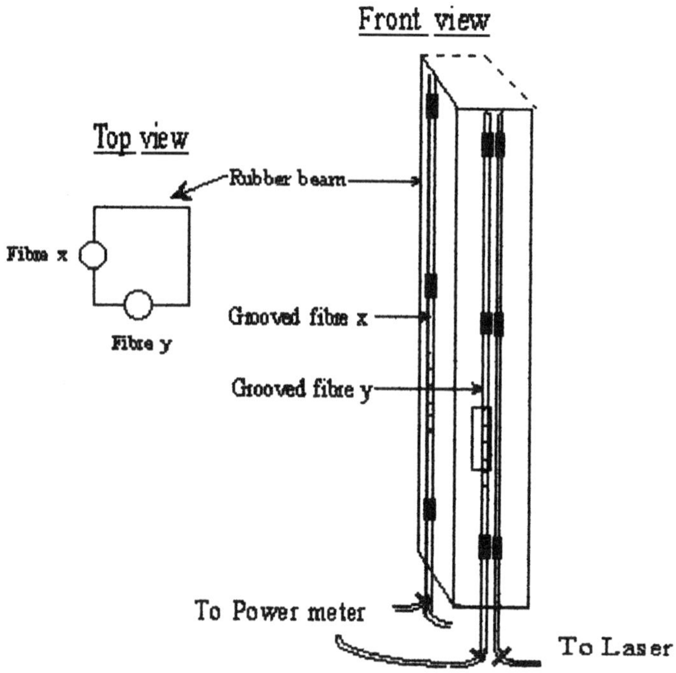

FIGURE 9.91 Front and top views of a section of the fiber optic drag force flow sensor used to measure the two dimensional fluid flow.

Fiber Optic Flow Sensor WITHOUT Rubber Beam

This version of the sensor uses 1 mm diameter polymer fiber as the deflection element, with a core diameter of 0.980 mm, and a thin cladding layer of approximately 20 μm [6]. Multiple grooves were etched radially into the fiber surface to a depth of 0.5 mm, extending into the core of the fiber, using a hot scalpel, and a manufactured V-groove template to ensure uniformity of grooves. Six grooves were determined as the optimum number to achieve a compromise between insertion losses and strain sensitivity, spaced 0.4 cm apart, over a length of 2.5 cm.

In order to measure strain in two orthogonal directions, perpendicular to the longitudinal axis of the fiber, two fibers were used, as shown in Figure 9.92, with the grooves oriented at 90° to each other. This was achieved by positioning the fibers on two adjacent faces of a beam of square cross-section. This beam was short enough to prevent any restriction to the deflection and yet long enough to hold and support the optical fibers. The fibers were looped around so that the looped ends acted as drag elements. The grooved portion of the fiber was unsupported and free to deflect in the air stream. Wind tunnel calibration indicated that the sensor could measure two-dimensional flow up to 35 m s^{-1} with a resolution of 0.96 m s^{-1}.

Conclusion

The measurement of one- and two-dimensional fluid velocity, using both optical fiber and conventional strain gages on a deflected beam, for a range of 0 to 30 m s^{-1}, has been demonstrated experimentally. Although flow visualization and modeling techniques are well advanced in engineering, there is still a need for real measurements, especially in the natural environment. The sensors described in this study are particularly suited for such measurements, which are almost always two-dimensional. The outputs

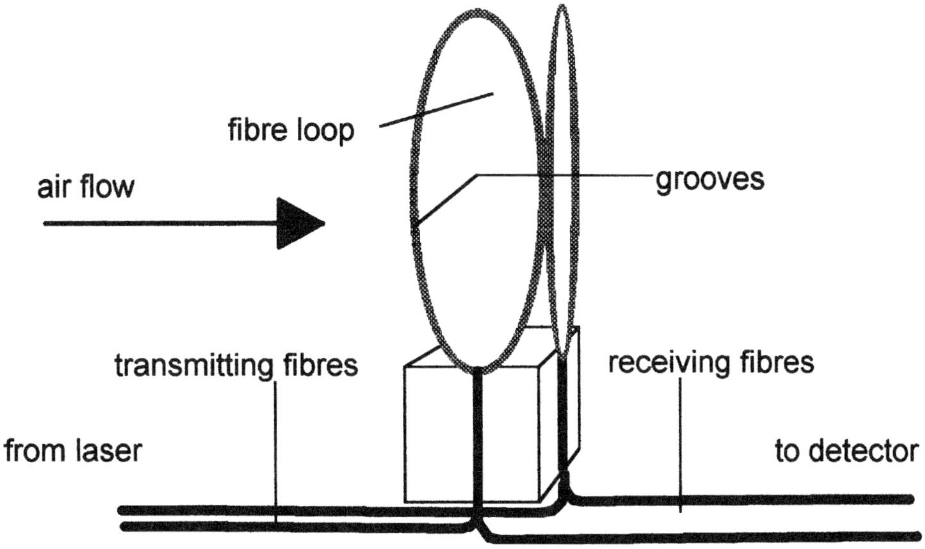

FIGURE 9.92 Optical fiber drag force flow sensor without rubber beam.

of the sensor representing speed and direction of fluid flow are independent of each other, so the sensor is suitable for environmental applications such as wind measurement or river flow, where the fluid forms gusts and can change direction as well as speed. One noteworthy feature of this sensor is its quick response time of 50 ms, which easily enables the measurement of gusts. The dimensions and materials of the sensor must be chosen to suit the fluid. In terms of sensitivity and resolution, the resistive strain gage sensor is a better option, but replacing the conventional strain gages with optical fiber strain gages ensures electrical noise immunity and intrinsic safety for use in hazardous environments.

References

1. T. Clarke, Design and operation of target flowmeters, *Encyclopedia of Fluid Mechanics*, Vol 1, Houston, TX: Gulf Publishing Company, 1986.
2. F. H. Newman and V. H. L. Searle, The general properties of matter. Edward Arnold, 1948.
3. S. Webster, R. McBride, J. S. Barton, and J. D. C. Jones, Air flow measurement by vortex shedding from multimode and monomode fibres. *Measurement, Science and Technology*, 3, 210-216, 1992.
4. J. S. Barton and M. Saudi, A fibre optic vortex flowmeter. *J. Phys. E: Sci. Instrum.* 19, 64-66, 1986.
5. N. Narendran, A. Shukla, and S. Letcher, Optical fibre interferometric strain sensor using a single fibre, *Experimental Techniques*, 16(2), 33-36, 1992.
6. R. Philip-Chandy, Ph.D. thesis, *Fluid flow measurement using electrical and optical fibre strain gauges*, Liverpool John Moores University, UK, 1997.

10

Point Velocity Measurement

John A. Kleppe
University of Nevada

John G. Olin
Sierra Instruments, Inc.

Rajan K. Menon
TSI Inc.

10.1 Pitot Probe Anemometry

John A. Kleppe

Theory

It is instructive to review briefly the principles of fluid dynamics in order to understand *Pitot tube theory* and applications. Consider, for example, a constant-density fluid flowing steadily without friction through the simple device shown in Figure 10.1. If it is assumed that there is no heat being added and no shaft work being produced by the fluid, a simple expression can be developed to describe this flow:

$$\frac{p_1}{w}+\frac{v_1^2}{2g}+z_1=\frac{p_2}{w}+\frac{v_2^2}{2g}+z_2 \tag{10.1}$$

where p_1, v_1, z_1 = Pressure, velocity, and elevation at the inlet
p_2, v_2, z_2 = Pressure, velocity, and elevation at the outlet
w = ρg, the specific weight of the fluid
ρ = Density
g = 9.80665 m s^{-2}

Equation 10.1 is the well known Bernoulli equation. The following example will demonstrate the use of Equation 10.1 and lead to a discussion of the theory of Pitot tubes.

0-8493-0047-9/00/$0.00+$.50

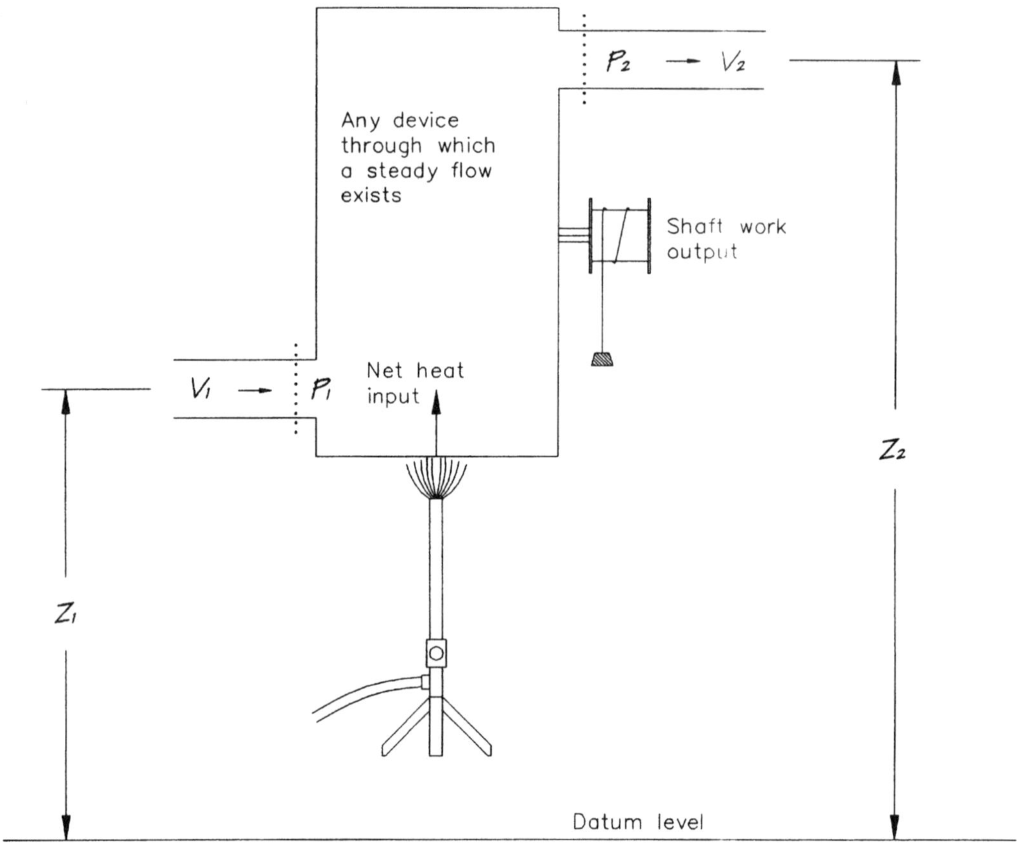

FIGURE 10.1 A device demonstrating Bernoulli's equation for steady flow, neglecting losses. (From [1].)

Example

A manometer [2] is used to measure the dynamic pressure of the tube assembly shown in Figure 10.2 [3]. The manometer fluid is mercury with a density of 13,600 kg m^{-3}. For a measured elevation change, Δh, of 2.5 cm, calculate the flow rate in the tube if the flowing fluids is (a) water, (b) air. Neglect all losses and assume STP conditions for the air flowing in the tube and $g = 9.81$ m s^{-2}.

Solution

Begin by writing expressions for the pressure at point 3.

$$p_3 = h_1\, w_{Hg} + \left(h_3 - h_1\right) w + p_1 \tag{10.2}$$

and

$$p_3 = h_2\, w_{Hg} + \left(h_3 - h_2\right) w + p_2 \tag{10.3}$$

Subtracting these equations and rearrangement yields an expression for the pressure difference.

$$p_2 - p_1 = \Delta h \left(w_{Hg} - w\right) \tag{10.4}$$

where w is the specific weight for water or air, etc.

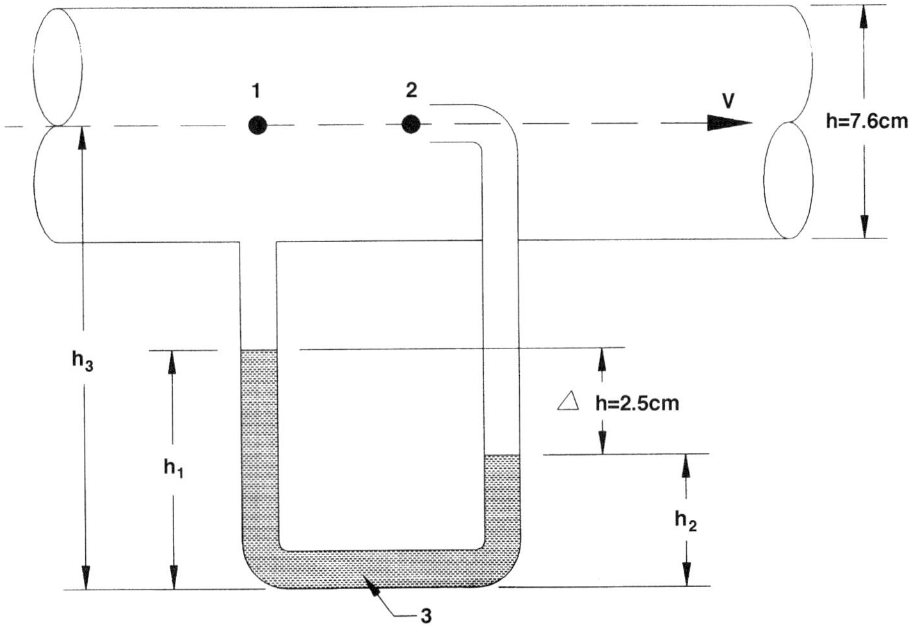

FIGURE 10.2 Using a manometer to measure a Pitot-static tube type assembly - Example (1).

Also using Equation 10.1, one can show that for $z_1 = z_2$ and $v_2 = 0$:

$$p_2 - p_1 = \frac{w\,v_1^2}{2g} \tag{10.5}$$

(a) For water,

$$\begin{aligned} p_2 - p_1 &= \Delta h\left(w_{\text{Hg}} - w_{\text{H}_2\text{O}}\right) \\ &= 0.025\left[13{,}600\left(9.81\right) - 998\left(9.81\right)\right] \\ &= 3090.6 \text{ Pa} \end{aligned} \tag{10.6}$$

Then,

$$3090.6 = \frac{\left(998\right)\left(9.81\right)v_1^2}{2\left(9.81\right)} \tag{10.7}$$

or

$$v_1 = 2.5 \text{ m s}^{-1} \tag{10.8}$$

The flow Q is then calculated to be:

$$Q = A_1 v_1 = \frac{\pi d^2 v_1}{4} = \frac{\pi\left(.076\right)^2\left(2.5\right)}{4} = 0.011 \text{ m}^3 \text{ s}^{-1} \tag{10.9}$$

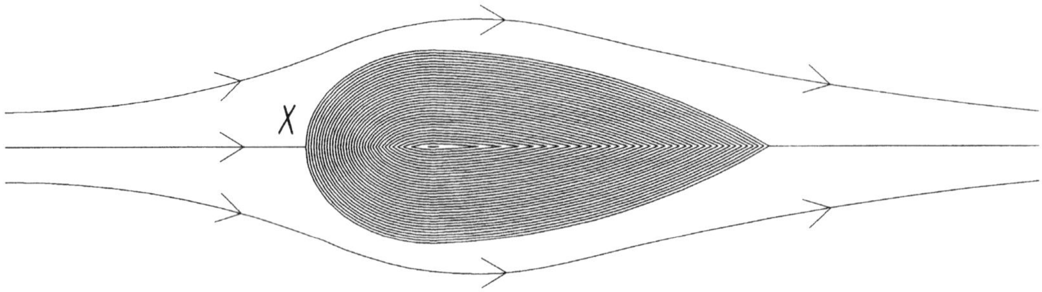

FIGURE 10.3 Flow around a nonrotating solid body.

(b) For air, one can use these same methods to show that:

$$Q = 0.34 \text{ m}^3 \text{ s}^{-1} \tag{10.10}$$

A point in a fluid stream where the velocity is reduced to zero is known as a stagnation point [1]. Any nonrotating object placed in the fluid stream will produce a stagnation point, x, as seen in Figure 10.3. A manometer connected to point x would record the stagnation pressure of the fluid. From Bernoulli's equation (Equation 10.1), the quantity $p + ½\rho v^2 + \rho gz$ is constant along a streamline for the steady flow of a fluid of constant density. Consequently, if the velocity v at a particular point is brought to zero, the pressure there is increased from p to $p + ½\rho v^2$. For a constant-density fluid, the quantity $p + ½\rho v^2$ is known as the stagnation pressure p_0 of that streamline, while the term $½\rho v^2$ — that part of the stagnation pressure due to the motion — is termed the dynamic pressure. A manometer connected to point x would measure the stagnation pressure and, if the static pressure p were also known, then $½\rho v^2$ could be obtained. One can show that:

$$p_t = p + p_v \tag{10.11}$$

where p_t = Total pressure, which is the sum of the static and dynamic pressures which can be sensed by a probe that is at rest with respect to the system boundaries when it locally stagnates the fluid isentropically

p = The actual pressure of the fluid whether in motion or at rest and can be sensed by a probe that is at rest with respect to the fluid and does not disturb the fluid in any way

p_v = The dynamic or velocity pressure equivalent of the directed kinetic energy of the fluid

Using Equation 10.11, one can develop an expression that relates to the velocity of the fluid:

$$p_t = p + 1/2\rho v^2 \tag{10.12}$$

or, solving for v:

$$v = \sqrt{\frac{2(p_t - p)}{\rho}} \tag{10.13}$$

Consider as an example the tube arrangement shown in Figure 10.4. A right-angled tube, large enough to neglect capillary effects, has one end A facing the flow. When equilibrium is attained, the fluid at A is stationary and the pressure in the tube exceeds that of the surrounding stream by $½\rho v^2$. The liquid is forced up the vertical part of the tube to a height:

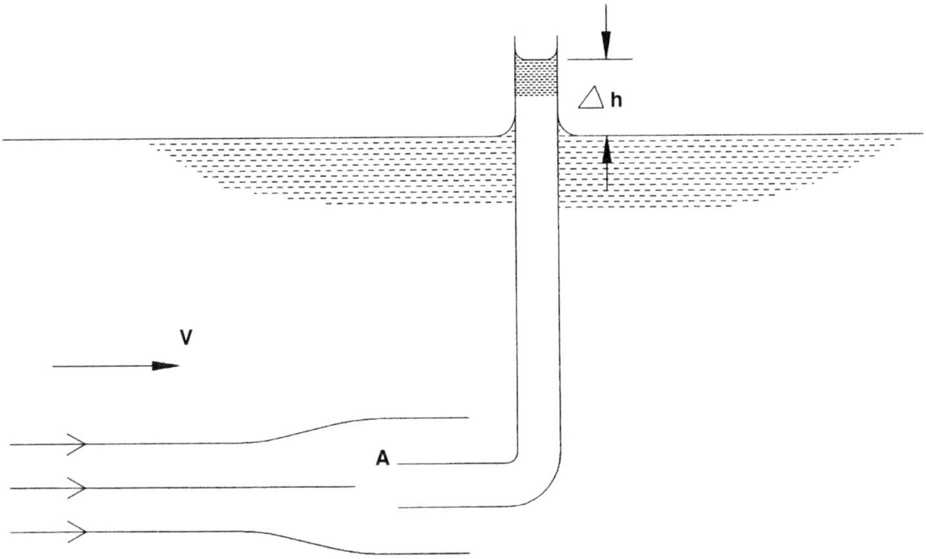

FIGURE 10.4 Right-angle tube in a flow system.

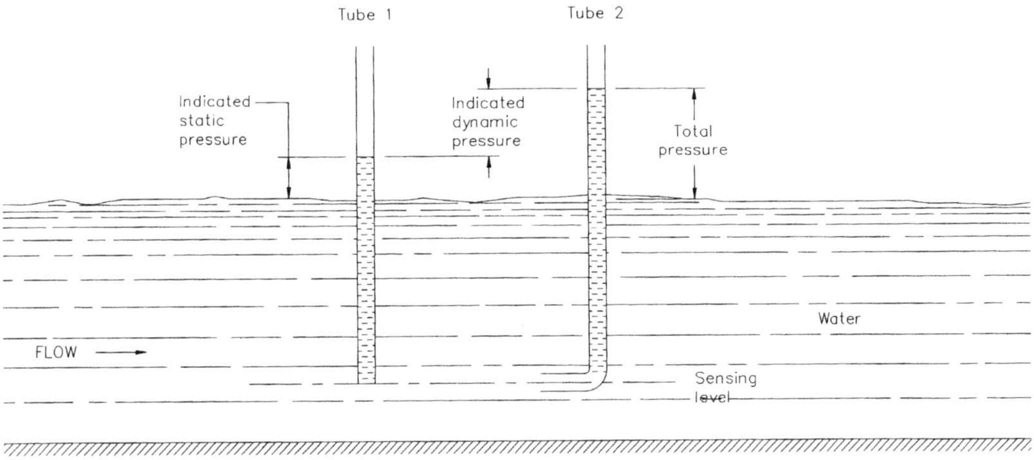

FIGURE 10.5 Basic Pitot tube method of sensing static, dynamic, and total pressure. (From R. P. Benedict, *Fundamentals of Temperature, Pressure and Flow Measurements,* 3rd ed., New York: John Wiley & Sons, 1984. With permission.)

$$\Delta h = \frac{\Delta p}{w} = \frac{v^2}{2g} \tag{10.14}$$

This relationship was used in the example given earlier to solve for v. It must be remembered that the total pressure in a fluid can be sensed only by stagnating the flow isentropically; that is, when its entropy is identical at all points in the flow. Such stagnation can be accomplished by a Pitot tube, as first developed by Henri de Pitot in 1732 [4]. In order to obtain a velocity measurement in the River Seine (in France),

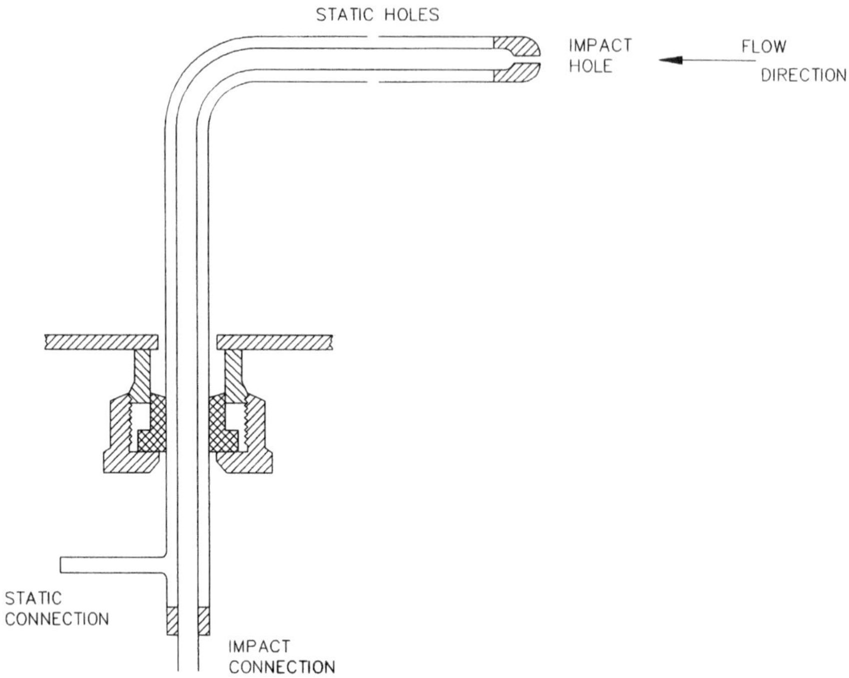

FIGURE 10.6 A modern Pitot-static tube assembly. (From ASME/ANSI PTC 19.2-1987, Instruments and Apparatus, Part 2, Pressure Measurements, 1987. With permission.)

Pitot made use of two tubes immersed in water. Figure 10.5 shows his basic Pitot tube method. The lower opening in one of the tubes was taken to be a measurement of the static pressure. The rise of fluid in the 90° tube was used as an indication of the velocity of the flow. For reasons to be discussed later, Pitot's method for measuring the static pressure was highly inadequate and would be considered incorrect today [4].

A modern-day Pitot-static tube assembly is shown in Figure 10.6 [5]. The static pressure is measured using "static holes" or pressure taps in the boundary. A pressure tap usually takes the form of a hole drilled in the side of a flow passage and is assumed to sense the "true" static pressure. When the fluid is moving past in the tap, which is usually the case, the tap will not indicate the true static pressure. The streamlines are deflected into the holes as shown in Figure 10.7, setting up a system of eddies. The streamline curvature results in a pressure at the tap "mouth" different from the true fluid pressure. These factors in combination result in a higher pressure at the tap mouth than the true fluid pressure, a positive pressure error. The magnitude of this pressure error is a function of the Reynolds number based on the shear velocity and the tap diameter [5]. Larger tap diameters and high velocities give larger errors [5]. The effect of compressibility on tap errors is not well understood or demonstrated, although correlations for this effect have been suggested [5]. It is possible to reduce tap errors by moving the location of the tap to a nonaccelerating flow location, or use pressure taps of smaller diameter. The effect of edge burrs is also noteworthy. All burrs must be removed. There is also an error that results with the angle of attack of the Pitot tube with the flow direction. Figure 10.8 shows the variation of total pressure indications as a function of the angle of attack. It can be seen that little error results if the angle of attack is less than $\pm 10°$.

A widely used variation of the Pitot-static tube is the type S Pitot tube assembly shown in Figure 10.9. It must be carefully designed and fabricated to ensure it will properly measure the static pressure. The "static" tube faces backwards into the wake behind the probe where the pressure is usually somewhat lower than the undisturbed static pressure. The type S Pitot tube therefore requires the application of a

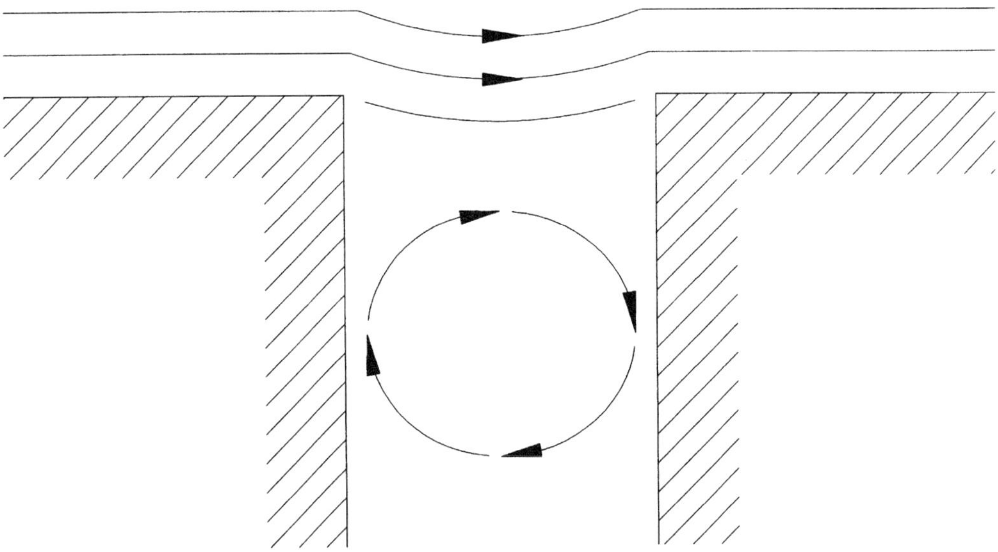

FIGURE 10.7 Pressure tap flow field.

correction factor (usually in the range of 0.84). This correction factor will be valid only over a limited range of velocity measurement. The type S Pitot tube does, however, have the advantage of being compact and relatively inexpensive. A type S Pitot tube can be traversed across a duct or stack to determine the velocity profile and hence total volumetric flow. This is discussed later.

The Pitot Tube in Flow with Variable Density

When a Pitot-static tube is used to determined the velocity of a constant-density fluid, the stagnation pressure and static pressure need not be separately measured: It is sufficient to measure their difference. A high-velocity gas stream, however, can undergo an appreciable change of density in being brought to rest at the front of the Pitot-static tube; under these circumstances, stagnation and static pressures must be separately measured. Moreover, if the flow is initially supersonic, a shock wave is formed ahead of the tube, and, thus, results for supersonic flow differ essentially from those for subsonic flow. Consider first the Pitot-static tube in uniform subsonic flow, as in Figure 10.10.

The process by which the fluid is brought to rest at the nose of the tube is assumed to be frictionless and adiabatic. From the energy equation for a perfect gas, it can be shown that [1]:

$$\frac{v^2}{2} = C_p\left(T_0 - T\right) = C_p T_0 \left\{1 - \left(\frac{p}{p_0}\right)^{(\gamma-1)/\gamma}\right\} \tag{10.15}$$

where v = Velocity
C_p = Specific heat at constant pressure
T = Absolute temperature of the gas
T_0 = Absolute temperature at stagnation conditions
p = Total pressure
γ = Ratio of specific heats

For measuring T_0, it is usual to incorporate in the instrument a small thermocouple surrounded by an open-ended jacket. If T_0 and the ratio of static to stagnation pressure are known, the velocity of the stream can then be determined from Equation 10.15.

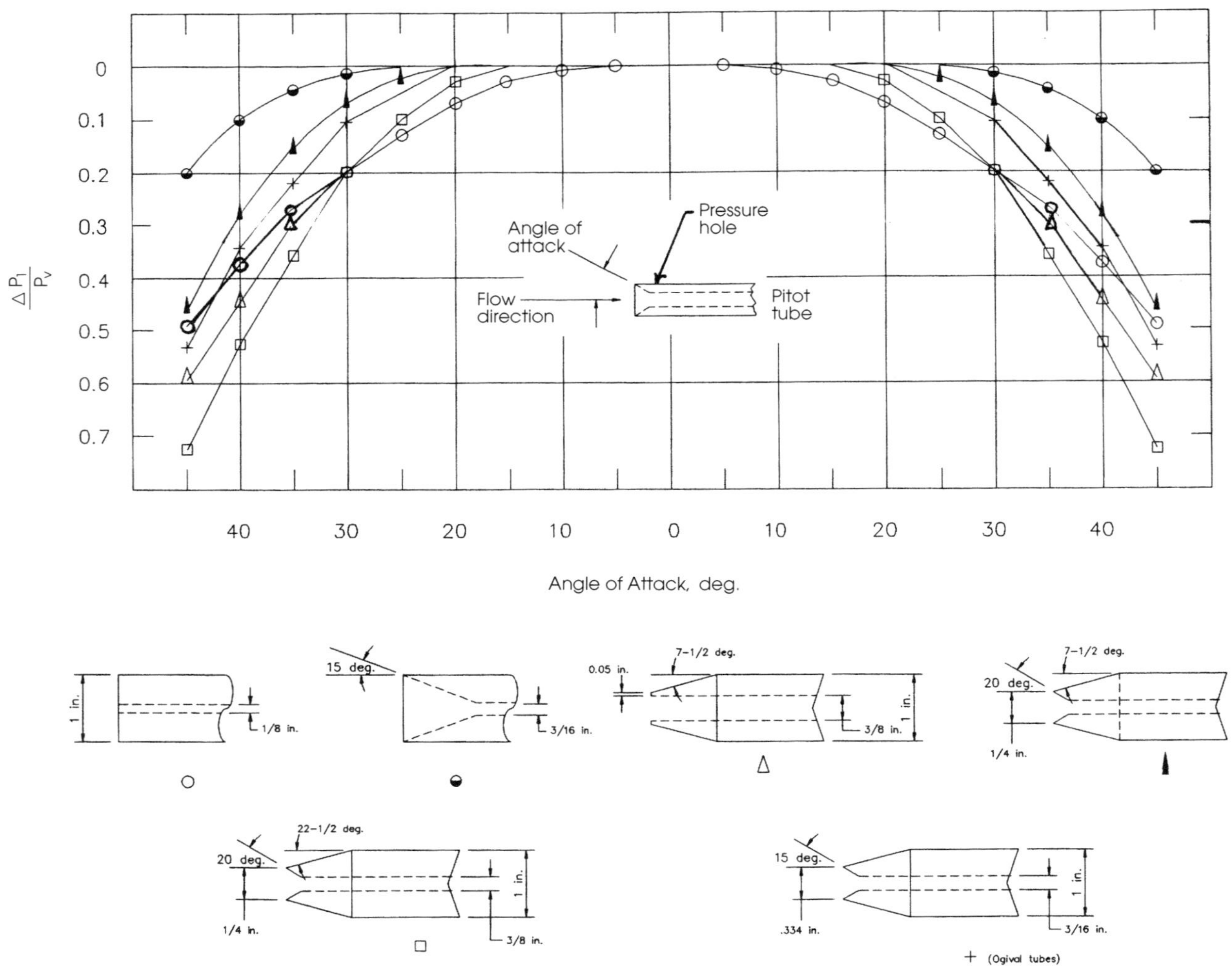

FIGURE 10.8 Variation of total pressure indication with angle of attach and geometry for Pitot tubes. (From ASME/ANSI PTC 19.2-1987, Instruments and Apparatus, Part 2, Pressure Measurements, 1987. With permission.)

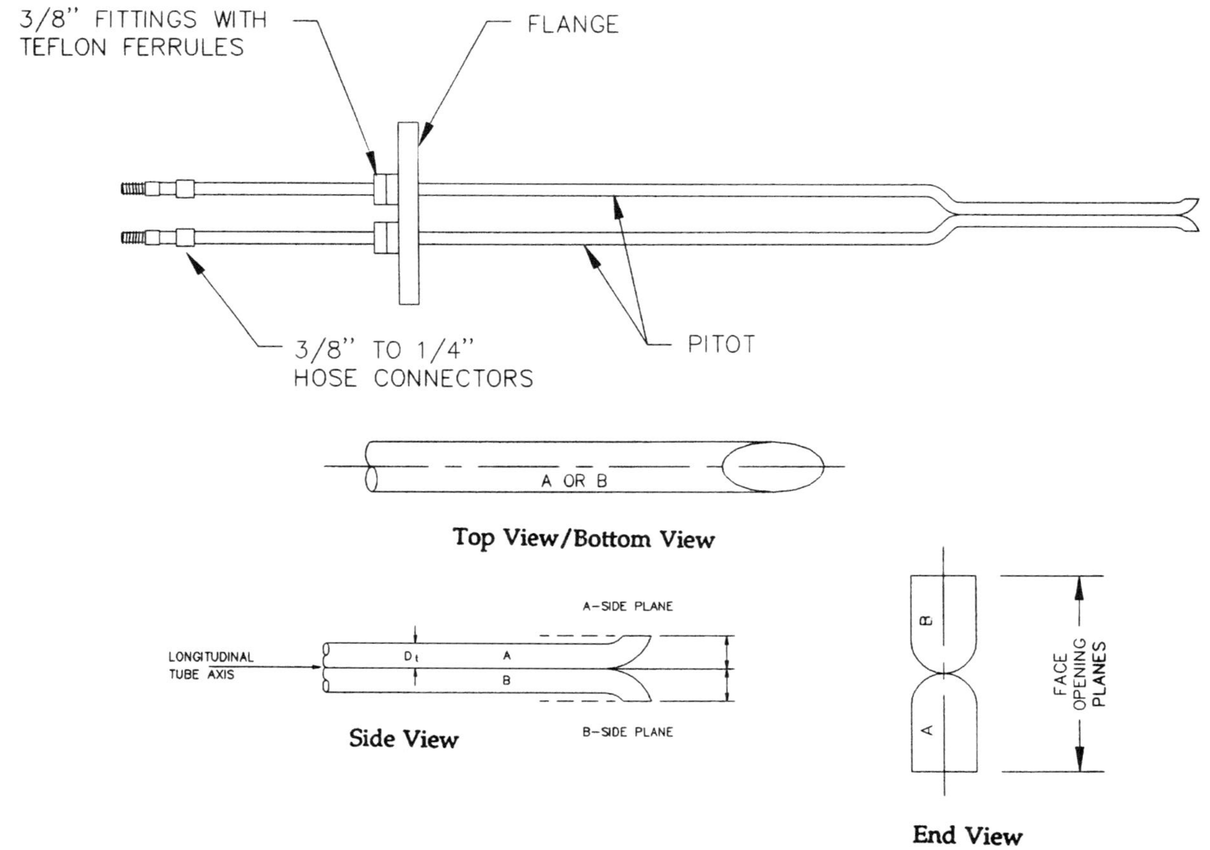

FIGURE 10.9 An S type Pitot tube for use in gas flow measurement will have specific design parameters. For example, the diameter of the tubing D_t, a gas probe will be between 0.48 and 0.95 cm. There should be equal distances from the base of each leg of the Pitot tube to its face opening plane, dimensions d_1, d_2. This distance should be between 1.05 and 1.50 times the external tubing diameter, D_t. The face openings of the Pitot tube should be aligned as shown. This configuration of the type S Pitot tube results in a correction coefficient of approximately 0.84. (From EPA, CFR 40 Part 60, Appendix A—Test Methods, 1 July 1995.)

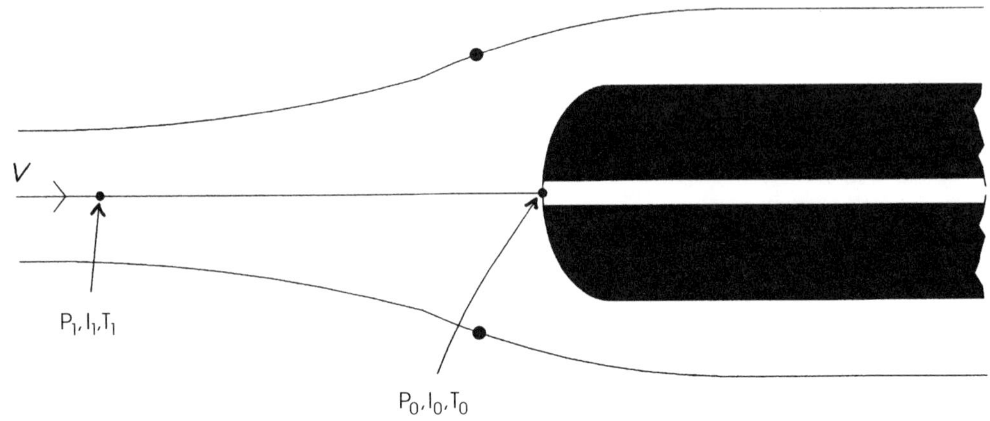

FIGURE 10.10

The influence of compressibility is best illustrated using the Mach number, M. It can be shown that: [1]

$$\frac{p_0}{p} = \left(1 + \frac{\gamma - 1}{2} M^2\right)^{\gamma/(\gamma-1)} \tag{10.16}$$

For subsonic flow, $[(\gamma - 1)/2]M^2 < 1$ and so the right side of Equation 10.16 can be expanded by the binomial theorem to give:

$$\frac{p_0}{p} = 1 + \frac{\gamma}{2} M^2 + \frac{\gamma}{8} M^4 + \frac{\gamma(2-\gamma)}{48} M^6 + \ldots \tag{10.17}$$

$$p_0 - p = \frac{p\gamma M^2}{2}\left\{1 + \frac{M^2}{4} + \left(\frac{2-\gamma}{24}\right) M^4 + \ldots\right\} \tag{10.18}$$

$$= 1/2\ \rho v^2 \left\{1 + \frac{M^2}{4} + \left(\frac{2-\gamma}{24}\right) M^4 + \ldots\right\} \tag{10.19}$$

The bracketed quantity is the compressibility factor and represents the effect of compressibility. Table 10.1 indicates the variation of the compressibility factor with M for air with $\gamma = 1.4$

It is seen that for $M < 0.2$, compressibility affects the pressure difference by less than 1%, and the simple formula for flow at constant density is then sufficiently accurate. For larger values of M, however, the compressibility must be taken into account.

For supersonic flow, Equation 10.16 is not valid because a shock wave forms ahead of the Pitot tube, as shown in Figure 10.11 and, thus, the fluid is not brought to rest isentropically. The nose of the tube must be shaped so that the shock wave is detached, i.e., the semiangle must be greater than 45.6° [1].

If the axis of the tube is parallel to the oncoming flow, the wave can be assumed normal to the streamline leading to the stagnation point. The pressure rise across the shock can therefore be given by:

$$\frac{p_2}{p_1} = \frac{1 + \gamma M_1^2}{1 + \gamma M_2^2} \tag{10.20}$$

TABLE 10.1 Variation of "Compressibility Factor" for Air

M	$\frac{p_0 - p}{\frac{1}{2}\rho v^2}$
0.1	1.003
0.2	1.010
0.3	1.023
0.4	1.041
0.5	1.064
0.6	1.093
0.7	1.129
0.8	1.170
0.9	1.219
1.0	1.276

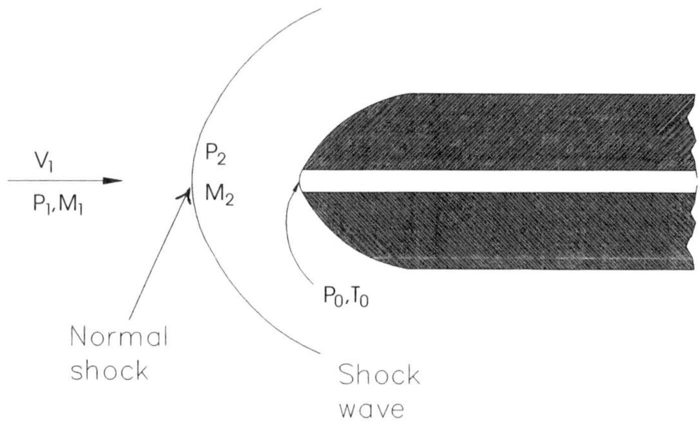

FIGURE 10.11

In the subsonic region downstream of the shock, there is a gradual isentropic pressure rise that can be represented as:

$$\frac{p_0}{p_1} = \frac{p_0}{p_2}\frac{p_2}{p_1} = \left(1 + \frac{\gamma - 1}{2} M_2^2\right)^{\gamma/(\gamma-1)} \frac{1 + \gamma M_1^2}{1 + \gamma M_2^2} \quad (10.21)$$

Finally, one obtains Rayleigh's formula:

$$\frac{p_0}{p_1} = \left\{ \frac{(\gamma + 1)^{\gamma+1}}{2\gamma M_1^2 - \gamma + 1} \left(\frac{M_1^2}{2}\right)^{\gamma} \right\}^{1/(\gamma-1)} \quad (10.22)$$

This expression for air reduces to:

$$\frac{p_0}{p_1} = \frac{166.9 M_1^7}{\left(7M_1^2 - 1\right)^{2.5}} \quad \text{when } \gamma = 1.4 \quad (10.23)$$

Although a conventional Pitot-static tube gives satisfactory results at Mach numbers low enough for no shock waves to form, it is unsuitable in supersonic flow because its "static holes" or "pressure taps", being in the region downstream of the shock, do not then register p_1; nor do they register p_2 since this is found only on the central streamline, immediately behind the normal part of the shock wave. Consequently, p_1 is best determined independently — for example, through an orifice in a boundary wall well upstream of the shock. Where independent measurement of p_1 is not possible, a special Pitot-static tube can be used, in which the static holes are much further back (about 10 times the outside diameter of the tube) from the nose. The oblique shock wave on each side of the tube has by then degenerated into a Mach wave across which the pressure rise is very small.

When $M_1 = 1$, the pressure rise across the shock is infinitesimal and, thus, Equations 10.16 and 10.22 both give:

$$\frac{p_0}{p_1} = \left\{(\gamma+1)/2\right\}^{\gamma/(\gamma-1)} = 1.893 \text{ (for air)} \tag{10.24}$$

A small value of p_0/p therefore indicates subsonic flow, a larger value supersonic flow.

Notice that Equation 10.22 enables the upstream Mach number to be calculated from the ratio of stagnation to static pressure. Since the stagnation temperature does not change across a shock wave:

$$C_p T_0 = C_p T_1 + \frac{v_1^2}{2} = C_p \frac{v_1^2}{\gamma R M_1^2} + \frac{v_1^2}{2} \tag{10.25}$$

Thus, v_1 can also be calculated if T_0 is determined.

Volumetric Flow Measurements

The currently accepted method for measuring volumetric gas flow in ducts and stacks involves the use of Pitot tubes to obtain the velocity at points of equal area of the cross-sectional areas of the stack [7]. For example, Figure 10.12 shows a case where the circular stack of cross-sectional area A has been divided into twelve (12) equal areas. An estimate of the average volumetric flow velocity is determined using the following relationship:

$$\bar{v}_n \approx \frac{\Sigma\, v_n A_n}{A} = \frac{A_i\, \Sigma\, v_n}{N A_i} = \frac{1}{N} \Sigma\, v_n \tag{10.26}$$

where A_i = One segment of the equal area segments
N = Number of equal area segments
v_n = Velocity measured at each point of equal area segment

This relationship shows that one can estimate the average volumetric flow velocity by taking velocity measurements at each point of equal area and then calculate the arithmetic mean of these measurements. It is clearly seen that a different result would be obtained if one were to simply take velocity measurements at equidistant points across the measurement plane and then take the arithmetic mean of these measurements. What would result in this case would be the path-averaged velocity, $\bar{v}_p$, which would be in error.

The sampling site and the number of traverse points designated will affect the quality of the volumetric flow measurement. The acceptability of the sampling procedure is generally determined by the distances from the nearest upstream and downstream disturbances (obstruction or change in direction) to gas flow. The minimum requirements for an acceptable sampling procedure can be found in the literature [7].

An automated system for accomplishing this measurement is shown in Figure 10.13 [8].

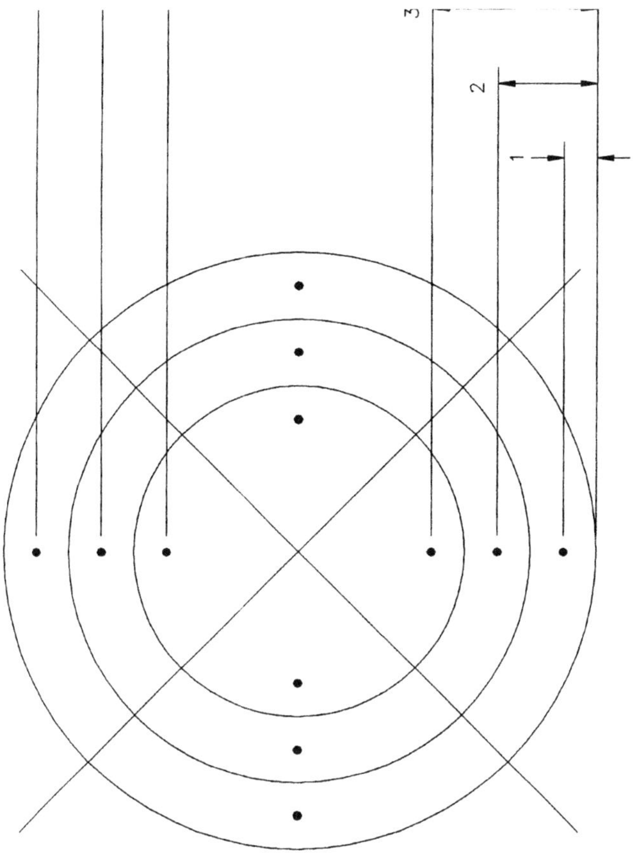

Traverse point	Distance % of diameter
1	4.4
2	14.6
3	29.6
4	70.4
5	85.4
6	95.6

FIGURE 10.12 Example showing circular stack cross-section divided into 12 equal areas, with location of traverse points indicated.

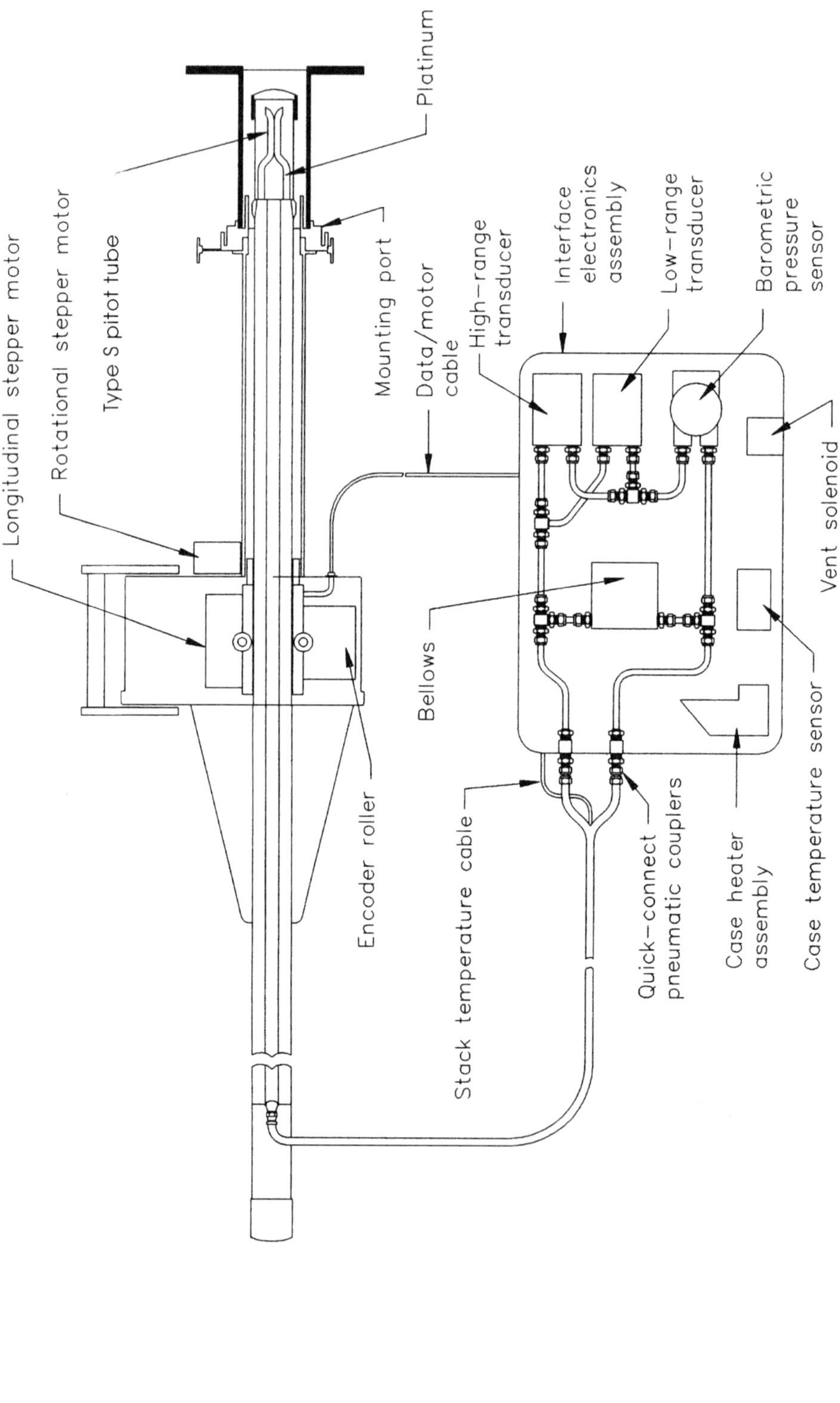

FIGURE 10.13 Automated probe consists of a type S Pitot tube and a platinum RTD mounted onto a type 319 stainless steel probe. (From T. C. Elliott, CEM System: Lynchpin Holding CAA Compliance Together, *Power*, May 1995, 31–40. With permission.)

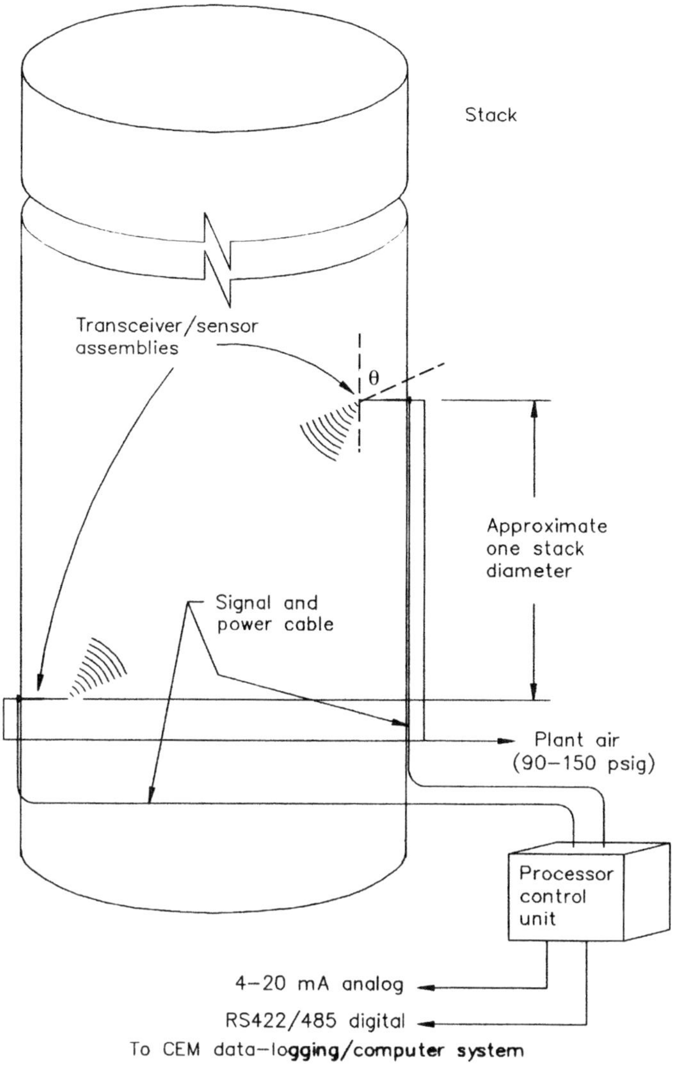

FIGURE 10.14 Block diagram of the hybrid system.

A Hybrid System

A hybrid system that combines sonic (acoustic) and Pitot tube technology has been developed to measure volumetric flow in large ducts and stacks [9–12]. A block diagram of this system is shown in Figure 10.14. The sensors (Figure 10.15) are mounted on opposite sides of the stack or duct at an angle θ to the flow direction. The acoustic portion of the sensor measures the flight time of the sound waves with and against the gas flow. It can easily be shown [9] that by transmitting and receiving the sound waves in opposite directions, the path average velocity of the gaseous medium can be determined from:

$$\bar{v}_{\mathrm{P}} = \frac{d}{2\cos\theta}\left(\frac{\tau_2 - \tau_1}{\tau_1\,\tau_2}\right)\ \mathrm{m\,s^{-1}} \qquad (10.27)$$

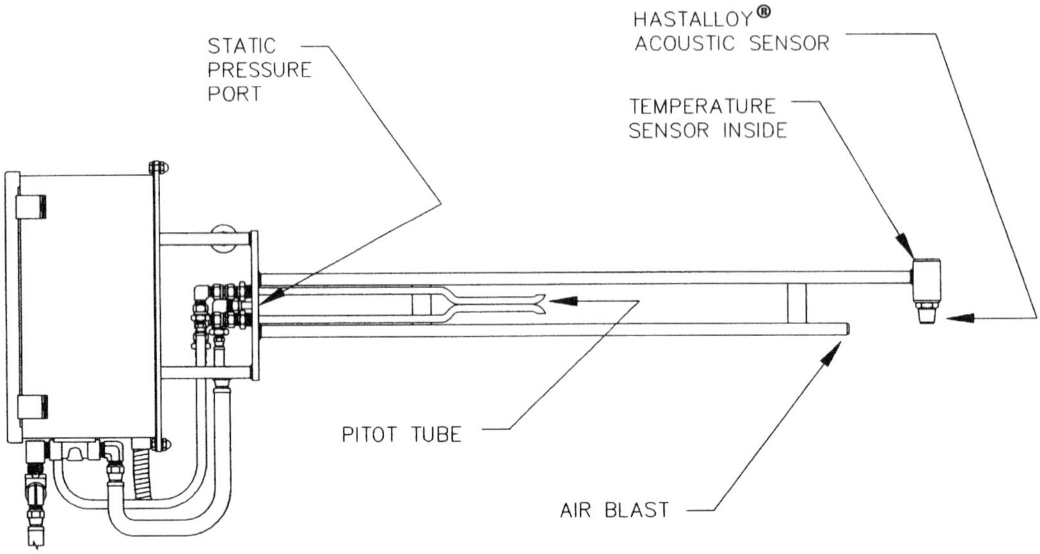

FIGURE 10.15 Acoustic probe contains acoustic, Pitot, and temperature sensors.

where $\bar{v}_p$ = Path average velocity of the gas m s^{-1}
d = Distance between the transceivers (m)
θ = Angle, in degrees, of the path of the transducers with the vertical
τ_1 = Flight time of the sound with the gas flow (s)
τ_2 = Flight time of the sound against gas flow (s)

The result of this part of the total measurement is the area under the velocity curve plotted in Figure 10.16. The Pitot tubes provide differential pressure measurements at two points within the stack. The differential pressure is converted to velocity in a unique manner. The flight times of the acoustic wave, when properly combined with the temperature sensor reading, provide a measurement of the molecular weight of the wet flue gas. This value is then used to obtain the point velocity measurements shown as V_2 and V_3 in Figure 10.16. The actual flow profile curve is then estimated using the values V_1, V_2, V_3, and V_4 and the area under the flow profile curve generated by the acoustic portion of the system. The final part of the measurement involves using the static pressure measurements and the stack temperature measurements to calculate the total standard volumetric flow in scfh (wet).

Commercial Availability

There are a variety of material used to construct Pitot tubes. The reasons for this are that Pitot tubes are used to measure a wide range of fluids. For example, to use a type S Pitot tube in a large power plant stack with a wet scrubber where the environment is extremely hostile and corrosive, stainless steel 316 or C276 (Hastaloy®) must be used. This, of course, makes the price of the Pitot tube as varied as its application. Many of the basic type S Pitot tube probes themselves are manufactured by a few small companies who, in turn, supply them on an OEM basis to others.

A typical type S Pitot tube assembly, such as that shown in Figure 10.9, constructed using stainless steel 316 can be purchased (in small quantities) for $310 each. They are available from:

EEMC/EMRC
3730 North Pellegrino Drive
Tucson, AZ 85749
Tel: (520) 749-2167
Fax: (520) 749-3582

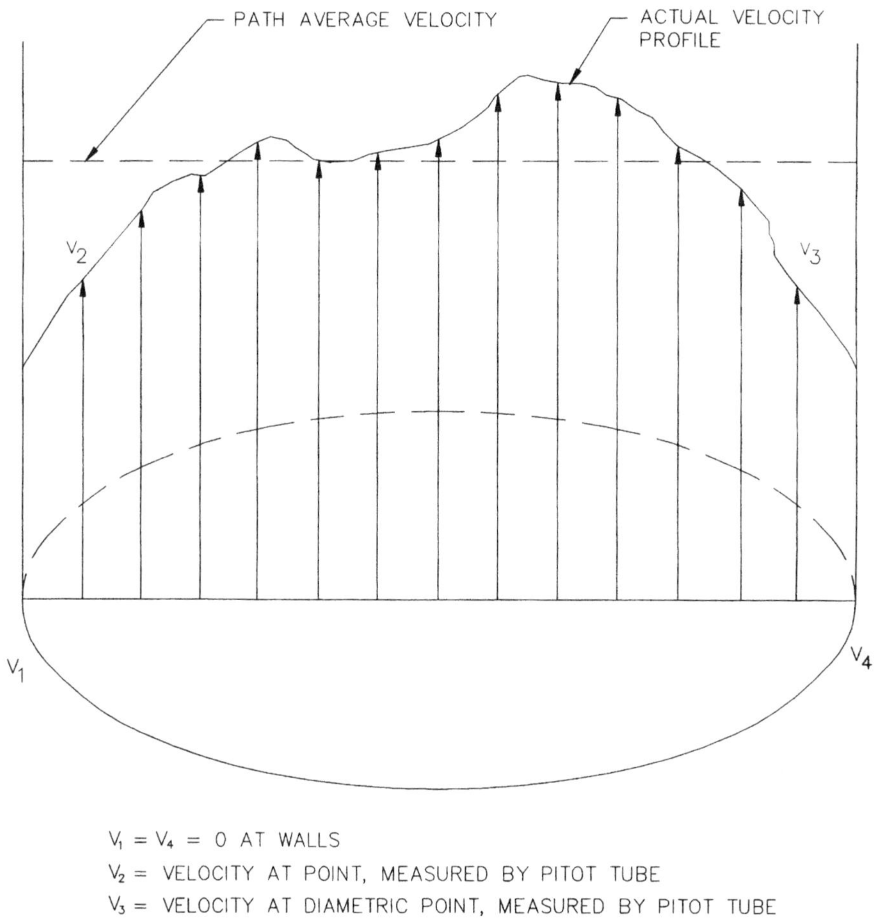

FIGURE 10.16 The velocity profile in a typical large duct or stack can vary greatly, thus changing the total volumetric flow. The hybrid system assumes V_1 and V_4 to be zero; measures V_2 and V_3 using the Pitot tubes; and provides the path average (area under the curve) using the acoustic portions of this sensor.

A typical modern Pitot-static assembly, such as that shown in Figure 10.6, can be purchased (in small quantities) for $34 and are available from:

Dwyer Instruments, Inc.
P.O. Box 373
Michigan City, IN 46361
Tel: (219) 879-8000
Fax: (219) 872-9057

More complex, custom-designed and fabricated Pitot-static probes for use on aircraft are available from:

Rosemount Aerospace Inc.
14300 Judicial Road
Burnsville, MN 55306-4898
Tel: (612) 892-4300
Fax: (612) 892-4430

Table 10.2 lists a number of manufactures/vendors that sell Pitot tube and general differential pressure measurement instrumentation.

TABLE 10.2 A Sample of Manufacturers/Vendors

Name	Address	Telephone/Fax	Probe Type
EEMC/EMRC	3730 North Pellegrino Dr. Tucson, AZ 85749	Tel: (520) 749-2167 Fax: (520) 749-3582	Type S Pitot probe
Dwyer Instruments, Inc.	P.O. Box 373 Michigan City, IN 46361	Tel: (219) 879-8000 Fax: (219) 872-9057	Pitot-static tubes and type S Pitot probe
Rosemount Aerospace, Inc.	14300 Judicial Rd. Burnsville, MN 55306-4898	Tel: (612) 892-4300 Fax: (612) 892-4430	Flow angle sensors, Pitot/Pitot-static tubes, vane angle of attack sensors, temperature sensors, ice detectors, and pressure transducers
Dieterich Standard	P.O. Box 9000 Boulder, CO 80301	Tel: (303) 530-9600 Fax: (303) 530-7064	Multipoint, self-averaging ANNUBAR®
Air Monitor Corporation	P.O. Box 6358 Santa Rosa, CA 95406	Tel: (707) 544-2706 (800) AIRFLOW Fax: (707) 526-9970	Multipoint, self-averaging
United Sciences, Inc.	5310 North Pioneer Rd. Gibsonia, PA 15044	Tel: (412) 443-8610 Fax: (412) 443-7180	Auto-PROBE 2000® automated Method 2 Testing
Scientific Engineering Instruments, Inc.	1275 Kleppe Lane, Suite 14 Sparks, NV 89431-6499	Tel: (702) 358-0937 Fax: (702) 358-0956	STACKWATCH® Hybrid System for volumetric flow sensing in large ducts and stacks (CEMS)

References

1. B. S. Massey, *Mechanics of Fluids,* Princeton, NJ: Van Nostrand, 1968.
2. W. F. Hughes and J. A. Brighton, *Theory and Problems of Fluid Dynamics,* New York: McGraw-Hill, 1967.
3. J. B. Evett and C. Liu, *Fluid Mechanics and Hydraulics: 2500 Solved Problems,* New York: McGraw-Hill, 1989.
4. R. P. Benedict, *Fundamentals of Temperature, Pressure and Flow Measurements,* 3rd ed., New York: John Wiley & Sons, 1984.
5. ASME/ANSI PTC 19.2 - 1987, *Instruments and Apparatus, Part 2, Pressure Measurement,* 1987.
6. S. P. Parker, *Fluid Mechanics Source Book,* New York: McGraw-Hill, 1988.
7. EPA, *CFR 40 Part 60, Appendix A—Test Methods,* 1 July 1995.
8. T. C. Elliott, CEM System: Lynchpin Holding CAA Compliance Together, *Power,* May 1995, 31–40.
9. J. A. Kleppe, Principles and Applications of Acoustic Sensors Used for Gas Temperature and Flow Measurement, *Proc. SENSOR EXPO,* Boston, May 1995, 337–374.
10. J. A. Kleppe, Acoustic Gas Flow Measurement in Large Ducts and Stacks, *Sensors J.,* 12(5), 18–24 and 85–87, 1995.
11. *Guidelines for Flue Gas Flow Rate Monitoring,* EPRI TR-104527, Project 1961-13 Final Report, June 1995.
12. A. Mann and J. A. Kleppe, A Report on the Performance of a Hybrid Flow Monitor Used for CEMS and Heat Rate Applications, *Proc. EPRI 1996 Heat Rate Improvement Conf.,* Dallas, TX, May 1996, Part 33, 1–13.

10.2 Thermal Anemometry

John G. Olin

General Description

A thermal anemometer measures the velocity at a point in a flowing fluid — a liquid or a gas. Figure 10.17 shows a typical industrial thermal anemometer used to monitor velocity in gas flows. It has two sensors — a velocity sensor and a temperature sensor — that automatically correct for changes in gas temperature.

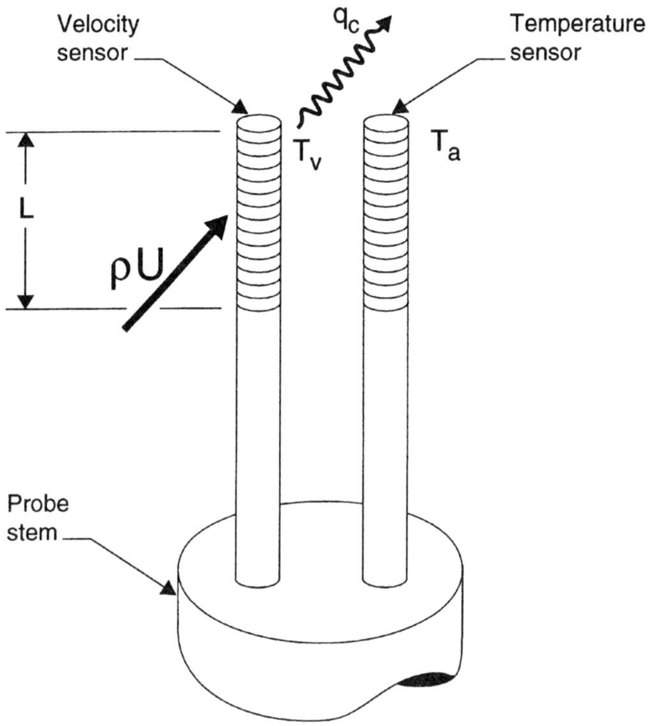

FIGURE 10.17 The principle of operation of a typical industrial thermal anemometer. T_v is the temperature of the heated velocity sensor; T_a is the gas temperature measured by the temperature sensor; ρ is the gas mass density; U is the gas velocity; q_c is the heat carried away by the flowing gas stream; and L is the length of the heated tip of the sensor. (Reprinted with the permission of Sierra Instruments, Inc.)

Both sensors are reference-grade platinum resistance temperature detectors (RTDs). The electric resistance of RTDs increases as temperature increases. For this reason, they are one of the most commonly used sensors for accurate temperature measurements. The electronics circuit passes current through the velocity sensor, thereby heating it to a constant temperature differential ($T_v - T_a$) above the gas temperature T_a and measures the heat q_c carried away by the cooler gas as it flows past the sensor. Hence, it is called a "constant-temperature thermal anemometer."

Because the heat is carried away by the gas molecules, the heated sensor directly measures gas mass velocity (mass flow rate per unit area) ρU. The mass velocity is typically expressed as U_s in engineering units of normal meters per second, or *normal* m s^{-1}, referenced to normal conditions of 0°C or 20°C temperature and 1 atm pressure. If the fluid's temperature and pressure are constant, then the anemometer's measurement can be expressed as *actual* meters per second, or m s^{-1}. When the mass velocity is multiplied by the cross-sectional area of a flow channel, the mass flow rate through the channel is obtained. Mass flow rate, rather than volumetric flow rate, is the direct quantity of interest in most practical and industrial applications, such as any chemical reaction, combustion, heating, cooling, drying, mixing, fluid power, human respiration, meteorology, and natural convection.

The thermal anemometer is often called an *immersible* thermal mass flowmeter because it is immersed in the flow stream, in contrast to the *capillary-tube* thermal mass flowmeter, another thermal methodology commonly configured as an in-line mass flowmeter for low gas flows. The thermal anemometer has some advantages and disadvantages when compared with the two other common point-velocity instruments — Pitot tubes and laser Doppler anemometers. Compared with Pitot tubes, the thermal anemometer measures lower velocities, has much wider rangeability, and can be made smaller, but it generally has a higher cost and is not recommended for nonresearch liquid flows. When thermal anemometers are compared with laser Doppler anemometers, they have a much lower cost, do not require seeding the flow with particles, can have a faster time response, can be made to have better spatial resolution, and can have a higher signal-to-noise ratio. On the other hand, in nonfluctuating flows, laser Doppler anemometers provide a fundamental measurement of velocity, independent of temperature and fluid properties. For this reason, they are often used to calibrate thermal anemometers.

Thermal anemometers are subdivided into two categories: industrial and research. Figure 10.18 shows typical sensors of industrial and research thermal anemometers.

Industrial Thermal Anemometers

Industrial thermal anemometers measure the point velocity or point mass velocity of gases in most practical and industrial applications. They seldom are used to monitor liquid flows because avoidance of cavitation problems limits the temperature T_v of the velocity sensor to only 10°C to 20°C above the liquid temperature, resulting in reduced velocity sensitivity and increased dependence on small changes in liquid temperature. Additionally, industrial liquid flows can cause sensor contamination and fouling. Typical gases monitored by industrial thermal anemometers include air, nitrogen, oxygen, carbon dioxide, methane, natural gas, propane, hydrogen, argon, helium, and stack gases. Common applications are: combustion air; preheated air; fuel gas; stack gas; natural gas distribution; semiconductor manufacturing gas distribution; heating, ventilation, and air conditioning; multipoint traversals of large ducts and stacks; drying; aeration and digester gas; occupational safety and health monitoring; environmental, natural convection, and solar studies; fermentors; and human inhalation monitoring. Industrial thermal anemometers have become the most commonly used instrument for monitoring the point velocity of gases.

The velocity sensor of an industrial thermal anemometer is a reference-grade platinum wire (approximately 25 μm in diameter and 20 Ω in resistance) wound around a cylindrical ceramic mandrel, such as alumina. Alternatively, the sensor is a thin platinum film deposited on a glass or ceramic substrate. To withstand the harsh environment encountered in many industrial applications, the cylindrical platinum RTD is tightly cemented into the tip of a thin-walled, stainless-steel, Hastelloy, or Inconel tube (typically 3 mm outside diameter and 2 cm to 6 cm long). Because the gas temperature usually varies in industrial applications, industrial thermal anemometer probes almost always have a separate, but integrally mounted, unheated platinum RTD sensor for measuring the local gas temperature T_a. When operated in the constant-temperature anemometer mode, the temperature difference ($T_v - T_a$) is usually in the 30°C to 100°C range. The temperature sensor is constructed just like the velocity sensor, but has a resistance in the 300 Ω to 1000 Ω range. As shown in Figure 10.17, the dual-sensor probe has the velocity and temperature sensor mounted side-by-side on a cylindrical probe stem (usually 6 mm to 25 mm in diameter and 0.1 m to 3 m long). A shield usually is provided to prevent breakage of the sensing head. The spatial resolution of this industrial thermal anemometer is 1 cm to 2 cm. The electronics for the industrial thermal anemometer is usually mounted directly on the probe stem in an explosion-proof housing. Industrial thermal anemometer systems like this measure gas velocity over the range of 0.5 normal m s^{-1} to 150 normal m s^{-1}.

In use, the industrial thermal anemometer probe is inserted through a sealed compression fitting or flanged stub in the wall of a duct, pipe, stack, or other flow passage. In this case, it is usually called an *insertion* thermal mass flowmeter. In another common configuration, the dual-sensor probe is permanently fitted into a pipe or tube (typically 8 mm to 300 mm in diameter) with either threaded or flanged gas connections. This configuration is called an *in-line* thermal mass flowmeter. In-line meters are directly calibrated for the total gas mass flow rate flowing through the pipe. The several flow body sizes facilitate mass flow monitoring over the range of 10 mg s^{-1} to 10 kg s^{-1}.

Research Thermal Anemometers

Research thermal anemometers measure the point velocity and/or turbulence of clean gases and liquids in research, product development, and laboratory applications. Because of their more fragile nature, they are not used for industrial applications. Typically, the gas is ambient air. Constant-temperature, filtered, degasified water is the primary liquid application, but the technique has also been applied to clean

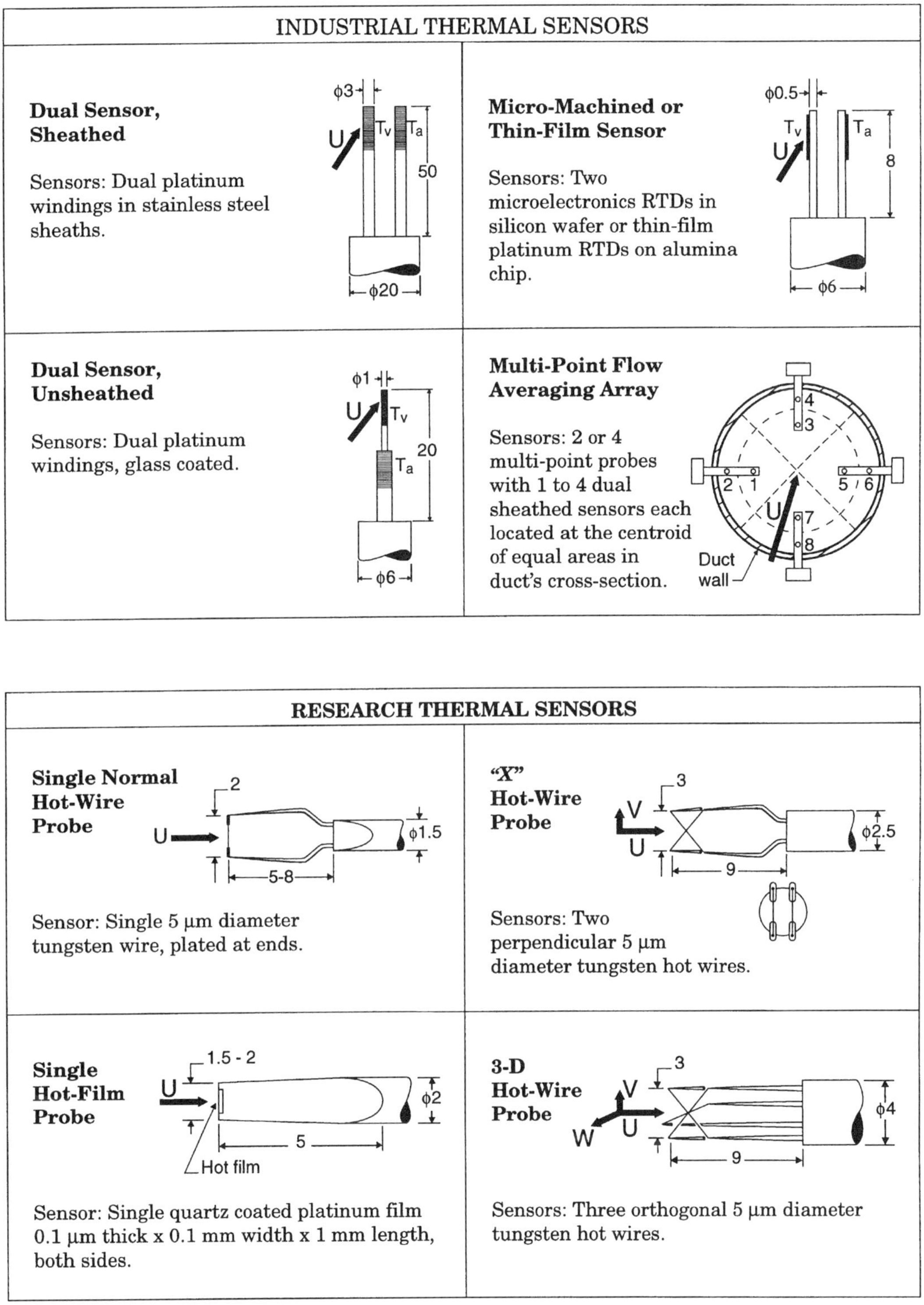

FIGURE 10.18 Typical industrial and research thermal anemometer sensors. All dimensions are in millimeters. T_v indicates the heated velocity sensor; T_a indicates the temperature sensor; U is the major velocity component in the x-direction; V is the transverse velocity component in the y-direction; and W is the transverse velocity component in the z-direction.

hydrocarbon liquids. As shown in Figure 10.18, the research anemometer's velocity sensor is either a hot wire or a hot film. Hot-wire sensors have a high frequency response and, therefore, are excellent for turbulence measurements in air and other gases. They are seldom used in liquid flows because they are susceptible to fouling and contamination. Hot-film sensors trade off lower frequency response for increased ruggedness and are used in gas and liquid flows. For liquid flows, hot-film sensors are designed to shed lint and other fouling or contaminating materials. Bruun [1] is an excellent reference source for the theory and applications of hot-wire and hot-film anemometers. Another comprehensive source is Fingerson and Freymuth [2].

Typical applications for hot-wire and hot-film anemometers include: one-, two-, and three-dimensional flow and turbulence studies; validation of computational fluid dynamics codes; environmental and micrometeorological measurements; turbomachinery; internal combustion engines; biological studies; heat-transfer research; boundary-layer measurements; supersonic flows; two-phase flows; and vorticity measurements. Freymuth [3] describes the 80-year history of research thermal anemometers. Today, hot-wire and hot-film anemometers have become the most widely used instruments for fluid mechanics research and development studies.

The typical hot-wire sensor is a fine tungsten wire welded at each end to miniature prongs designed to minimize their influence on the wire's flow field. The wire is usually gold or copper plated a short length at each end to define an active sensor length away from the two prongs. For work in water, the wire is quartz coated to prevent electrolysis or electrical shorting, but, in this case, cracking of the coating can occur. A typical tungsten wire has a diameter of 4 μm to 5 μm, an active length of 1 mm to 3 mm, and an electrical resistance of 2 Ω to 6 Ω. Because it oxidizes above 350°C in air, tungsten hot wires are usually operated at a temperature not exceeding 300°C. Platinum, 90% platinum + 10% rhodium, and 80% platinum + 20% iridium wires also are used. They can be soldered onto the prongs, but are weaker than tungsten. In cases where the fluid temperature T_a changes enough to cause measurement errors, a separate sensor is used to measure T_a and make temperature corrections. The temperature sensor is either a hot wire or a larger wire-wound RTD mounted either on a separate probe or integrally on the same probe stem as the velocity sensor. As shown in Figure 10.18, for two-dimensional or three-dimensional flow studies, probes with two perpendicular wires in an "X" pattern or three orthogonal wires are used, respectively. Special subminiature probes and probes with the prongs displaced from the probe stem are used for near-wall, boundary-layer work and small flow passages. Gibbings et al. [4, 5] describe hot-wire probes for use in near-wall, turbulent boundary-layer studies.

As shown in Figure 10.18, the typical hot-film sensor is a wedge-tipped or cone-tipped quartz rod with a thin 0.1 μm thick platinum film plated on its tip via cathode sputtering. The platinum film usually is coated with a 1 μm to 2 μm layer of quartz for protection and to avoid electrical shorting or electrolysis in water flows. Because hot-film sensors have a much larger mass than hot-wire sensors, their frequency response is not as flat as hot wires; hence, they are not quite as good for high frequency turbulence measurements. It also has been observed by Mikulla [6] that the shape of some hot-film sensors can suppress response to the turbulent velocity component normal to its surface. On the other hand, hot-film sensors have less breakage and a more stable geometry than hot-wire sensors. Other configurations of hot-film sensors include cylindrical quartz rods (approximately 25 μm to 150 μm in diameter); one or more split-film cylindrical sensors for multidimensional measurements; and flush-mounted sensors for wall heat-transfer measurements.

Principle of Operation

First Law of Thermodynamics

Figure 10.19 shows the first law of thermodynamics applied to a control volume consisting of the velocity sensor of either an industrial thermal anemometer, such as shown in Figure 10.17, or a research thermal anemometer. Application of the first law to thermal anemometer sensors provides the basis for determining point velocity. Applied to Figure 10.19, the first law states that the energy into the control volume

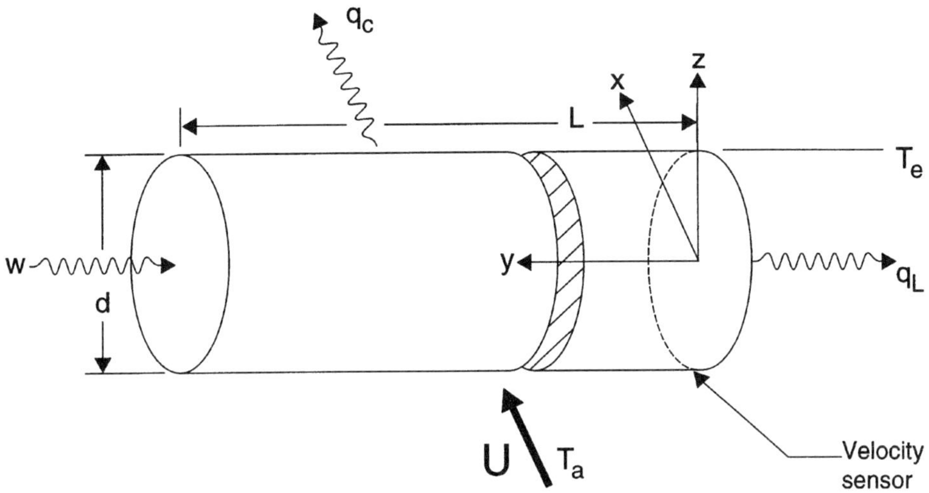

FIGURE 10.19 First law of thermodynamics applied to a thermal anemometer velocity sensor. The term w is the electric power (Watts) supplied to the sensor; q_c is the heat convected away from the sensor by the flowing fluid having a velocity U and temperature T_a; q_L is the conductive heat lost; T_e is the average surface temperature of the sensor over its length L; and d is the sensor's outside diameter.

equals the energy out plus the energy stored. Making the practical simplifying assumptions of steady-state operation (i.e., no energy stored) and no heat transfer via radiation, one obtains:

$$w = q_c + q_L \tag{10.28}$$

The heat transfer q_c due to natural and forced convection normally is expressed in terms of the heat transfer coefficient h as:

$$q_c = hA_v\left(T_e - T_a\right) \tag{10.29}$$

where $A_v = \pi dL$ is the external surface area of the velocity sensor. The electric power w usually is expressed as:

$$w = E_v^2 / R_v \tag{10.30}$$

where E_v is the voltage across the sensor, and R_v is its electric resistance.

For the industrial velocity sensor shown in Figure 10.17, q_L is the heat conducted from the end of the heated velocity sensor of length L to the remainder of the sensor's length. Most of this heat is convected away by the flowing fluid, and a small fraction is conducted to the probe stem. In the case of research hot-wire or cylindrical hot-film sensors, q_L is conducted to the two prongs, of which a major fraction is convected away and a minor fraction enters the probe stem. In well-designed velocity sensors, q_L is at most 10% to 15% of w, a fraction that decreases as velocity increases.

For research velocity sensors, the surface temperature T_e is identical to the wire or film temperature T_v. However, the surface temperature T_e of industrial velocity sensors with stainless-steel sheaths is slightly less than the temperature T_v of the platinum winding because a temperature drop is required to pass the heat q_c through the intervening "skin" — the cement layer and the stainless-steel tube. This is expressed as:

$$T_e = T_v - q_c R_s \tag{10.31}$$

where R_s is the thermal skin resistance in units of K W^{-1}. R_s is a constant for a given sensor and is the sum of the thermal resistances of the cement layer and the stainless-steel tube. For research velocity sensors, $R_s = 0$ and $T_e = T_v$ in Equation 10.31. In well-designed, sheathed, industrial velocity sensors, R_s is minimized and is approximately 1 K W^{-1}. As evidenced by Equation 10.31, the effect of skin resistance increases as velocity (i.e., q_c) increases. The effect is almost negligible at low velocity; but at high velocity, it is responsible for the characteristic droop in power vs. velocity flow-calibration curves.

Because the flow around cylinders in cross flow is confounded by boundary-layer separation and vortex shedding, it has defied analytical solution. Therefore, the film coefficient h in Equation 10.29 is found using empirical correlations. Correlations for h are expressed in terms of the following nondimensional parameters:

$$\mathrm{Nu} = \Im\left(\mathrm{Re},\ \mathrm{Pr},\ \mathrm{Gr},\ \mathrm{M},\ \mathrm{Kn}\right) \tag{10.32}$$

where Nu = hd/k = Nusselt number (the heat-transfer parameter)
Re = $\rho Vd/\mu$ = Reynolds number (the ratio of dynamic to viscous forces)
Pr = $\mu C_p/k$ = Prandtl number (the fluid properties parameter)
M = Mach number (the gas compressibility parameter)
Kn = Knudsen number (the ratio of the gas mean free path to d)

In the above, k is the fluid's thermal conductivity; μ is its viscosity; and C_p is its coefficient of specific heat at constant pressure. If one takes the practical case where: (1) natural convection is embodied in Re and Pr, (2) the velocity is less than one third the fluid's speed of sound (i.e., <100 m s^{-1} in ambient air), and (3) the flow is not in a high vacuum, then one can ignore the effects of Gr, M, and Kn, respectively. Thus,

$$\mathrm{Nu} = \Im\left(\mathrm{Re},\ \mathrm{Pr}\right) \tag{10.33}$$

Over the years, many attempts have been made to find universal correlations for the heat transfer from cylinders in cross flow. For an isothermal fluid at constant pressure, King [7] expresses Equation 10.33 as:

$$\mathrm{Nu} = A + B\mathrm{Re}^{0.5} \tag{10.34}$$

where A and B are empirical calibration constants that are different for each fluid and each temperature and pressure. Kramers [8] suggests the following correlation:

$$\mathrm{Nu} = 0.42\mathrm{Pr}^{0.2} + 0.57\mathrm{Pr}^{0.33}\mathrm{Re}^{0.50} \tag{10.35}$$

This correlation accounts for the variation in fluid properties (k, μ, and Pr) with temperature. Kramers [8] evaluates these properties at the so-called "film" temperature $(T_v + T_a)/2$, rather than at T_a itself. Another comprehensive correlation is given by Churchill and Bernstein [9]. Several other correlations are similar to Equation 10.35, but have exponents for the Reynolds number ranging from 0.4 to 0.6. Others have 0.36 and 0.38 for the exponent of the Prandtl number. Equations 10.34 and 10.35 are strictly valid only for hot-wire sensors with very high L/d ratios, in which case q_L and R_s are zero. The following universal correlation is suggested for real-world velocity sensors with variable fluid temperature and nonzero q_L and R_s:

$$\mathrm{Nu} = A + B\mathrm{Pr}^{0.33}\mathrm{Re}^{n} \tag{10.36}$$

where constants A, B, and n are determined via flow calibration. Equation 10.36 is applicable to most commercial industrial and research velocity sensors.

Combining Equations 10.28, 10.29, 10.30, and 10.36, and recognizing that $h = k\text{Nu}/d$, one obtains:

$$E_v^2 / R_v = \left(Ak + Bk\text{Pr}^{0.33}\text{Re}^n\right)\left(T_v - T_a\right) \tag{10.37}$$

where A and B are new constants. A, B, and n are determined via flow calibration and account for all nonidealities, including end conduction and skin resistance. Equation 10.37 is applicable to most commercial industrial and research velocity sensors. Manufacturers of industrial thermal anemometers can add other calibration constants to Equation 10.37 to enhance its correlation with flow-calibration data. The presence of end conduction means that the temperature of the velocity sensor varies with the axial coordinate y in Figure 10.19. The temperature T actually sensed by the velocity sensor is the *average* temperature over length L, or:

$$T_v = \left(1/L\right)\int_o^L T_v\left(y\right)\mathrm{d}y \tag{10.38}$$

Bruun [1] presents an analytical solution for T_v (y) for hot-wire sensors. Equation 10.38 is the correct expression for T_v in Equation 10.37 and is so defined hereafter.

For fluid temperatures less than 200°C, the electric resistance of the RTD velocity and temperature sensors is usually expressed as:

$$R_v = R_{v0}\left[1 + \alpha_v\left(T_v - T_0\right)\right] \tag{10.39}$$

$$R_T = R_{T0}\left[1 + \alpha_T\left(T_T - T_0\right)\right] \tag{10.40}$$

where R_{v0} and R_{T0} are, respectively, the electric resistances of the velocity sensor and the temperature sensor at temperature T_0 (usually 0°C or 20°C), and α_v and α_T are the temperature coefficients of resistivity at temperature T_0. Additional terms are added to Equations 10.39 and 10.40 when fluid temperatures exceed 200°C. When evaluated at the fluid temperature T_a, the resistance R_a of the velocity sensor is:

$$R_a = R_{v0}\left[1 + \alpha_v\left(T_a - T_0\right)\right] \tag{10.41}$$

For applications with wide excursions in fluid temperature, additional terms are added to Equations 10.39–10.41. At 20°C, α_v and α_T are approximately 0.0036°C^{-1} for tungsten wire; 0.0038°C^{-1} for pure platinum wire; 0.0016°C^{-1} for 90% platinum + 10% rhodium wire; 0.0024°C^{-1} for platinum film; and 0.0040°C^{-1} for tungsten film. R_v and R_a are called the "hot" and "cold" resistances of the velocity sensor, respectively. The ratio R_v/R_a is called the "overheat ratio." For gas flows, sheathed industrial velocity sensors are operated at overheat ratios from 1.1 to 1.4 ($T_v - T_a$ = 30°C to 100°C). For gas flows, the overheat ratio of tungsten hot-wire and hot-film sensors are usually set to approximately 1.8 ($T_v - T_a$ = 200°C to 300°C) and 1.4 ($T_v - T_a$ = 150°C to 200°C), respectively. For water flows, the overheat ratio of hot-film sensors is approximately 1.05 to 1.10 ($T_v - T_a$ = 10°C to 20°C). Mikulla [6] shows the importance of the effect of overheat ratio on frequency response.

Combining Equations 10.39 and 10.41, one obtains:

$$T_v - T_a = \frac{R_v - R_a}{\alpha_v R_{v0}} \tag{10.42}$$

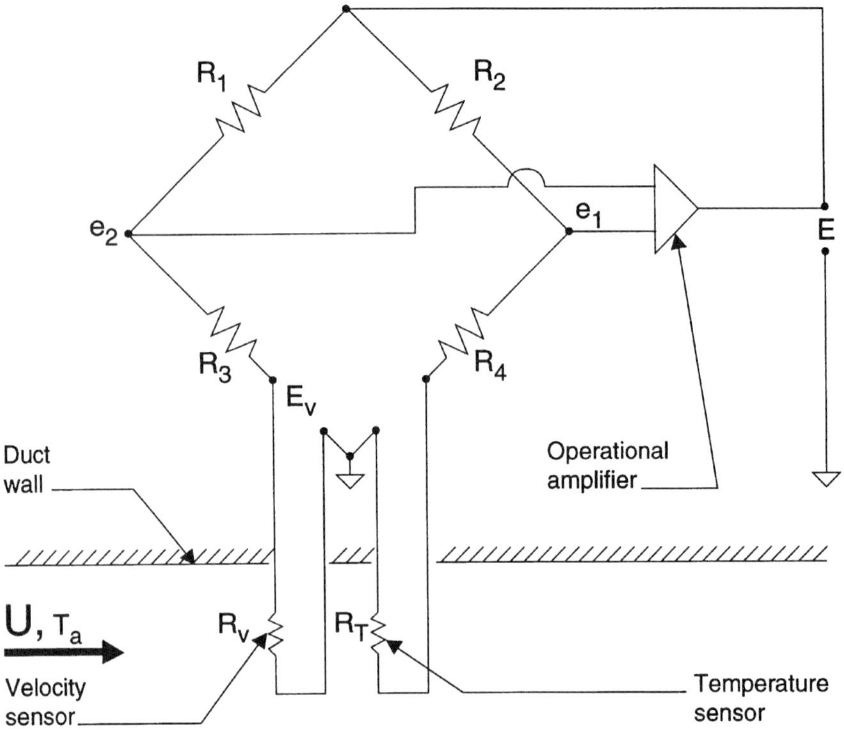

FIGURE 10.20 Constant-temperature thermal anemometer bridge circuit with automatic temperature compensation. R_1, R_2, and R_4 are fixed resistors selected to achieve temperature compensation; R_3 is the probe and cable resistance; R_v is the velocity sensor's resistance; R_T is the temperature sensor's resistance; and E is the bridge voltage output signal. For research anemometers operating in isothermal flows, the temperature sensor is eliminated and replaced with a variable bridge resistor. Some temperature compensation circuits have an additional resistor in parallel with R_T.

Inserting this into Equation 10.37 obtains:

$$\frac{E_v^2}{R_v\left(R_v - R_a\right)} = Ak + Bk\text{Pr}^{0.33}\text{Re}^n \tag{10.43}$$

where new constants A and B have absorbed the constants α_v and R_{v0}.

Figures 10.20 to 10.22 show three typical electronic drives for thermal anemometer sensors. Figure 10.20 shows the commonly used constant-temperature anemometer Wheatstone bridge circuit described by Takagi [10]. Figure 10.21 is similar, but is controlled and operated via a personal computer. In the constant-temperature mode, the hot resistance R_v, and hence the velocity sensor's temperature, remains virtually constant, independent of changes in velocity. With the addition of the temperature sensor shown in Figure 10.20, the bridge circuit also compensates for variations in fluid temperature T_a, as described later. Another common analog sensor drive is the constant-current anemometer. In this mode, a constant current is passed through the velocity sensor, and the sensor's temperature decreases as the velocity increases. Because the entire mass of the sensor must participate in this temperature change, the sensor is slower in responding to changes in velocity. Because the constant-temperature anemometer has a flatter frequency response, excellent signal-to-noise ratio [2], and is easier to use, it is favored over constant-current anemometers by most researchers and manufacturers for velocity and turbulence measurements. The constant-current anemometer with a very low overheat ratio is often used

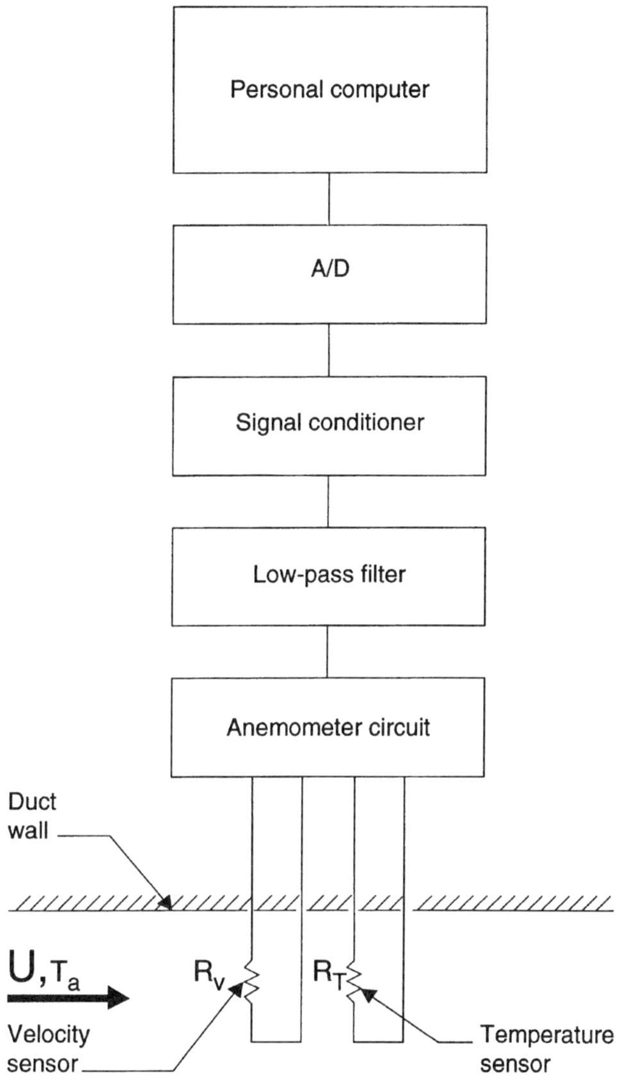

FIGURE 10.21 Personal computer-based digital thermal anemometer system. The signal conditioner matches the anemometer circuit's output to the ADC. For isothermal flows, the temperature sensor is eliminated.

as the temperature sensor. Subsequently, references made herein to sensor electronics will be based on the constant-temperature anemometer.

In the constant-temperature anemometer drive shown in Figure 10.20, the resistances R_1 and R_2 are chosen to: (1) maximize the current on the velocity-sensor side of the bridge so it becomes self-heated and (2) minimize the current on the temperature-sensor side of the bridge so it is not self-heated and is independent of velocity. Additionally, the temperature sensor must be sufficiently large in size to avoid self-heating. The ratio R_2/R_1 is called the "bridge ratio." A bridge ratio of 5:1 to 20:1 is normally used; but for optimum frequency response and compensation for long cable length, a bridge ratio of 1:1 can be used. In Figure 10.20, the operational amplifier, in a feedback control loop, senses the error voltage $(e_2 - e_1)$ and feeds the exact amount of current to the top of the bridge necessary to make $(e_2 - e_1)$ approach zero. In this condition, the bridge is balanced; that is,

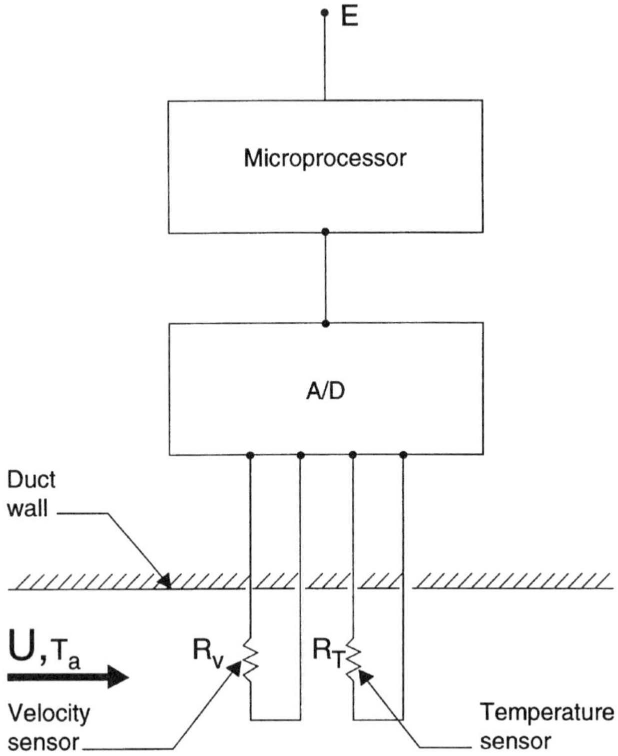

FIGURE 10.22 Microprocessor-based digital thermal anemometer. This system digitally maintains a constant temperature difference ($T_v - T_T$) and automatically corrects for the variation in fluid properties with temperature. The manufacturer provides a probe-mounted electronics package delivering an analog output signal E and/or a digital RS485 signal linearly proportional to gas mass velocity. (Reprinted with permission of Sierra Instruments, Inc.)

$$\frac{R_1}{R_v + R_3} = \frac{R_2}{R_T + R_4} \tag{10.44}$$

or

$$R_v = \frac{R_1}{R_2}\left(R_T + R_4\right) - R_3 \tag{10.45}$$

From Equation 10.45, one sees that R_v is a linear function of R_T. This relationship forms the basis for analog temperature compensation.

Expressing the voltage E_v across the velocity sensor in terms of the bridge voltage E, one obtains:

$$E_v = \frac{ER_v}{R_1 + R_3 + R_v} \tag{10.46}$$

Inserting this into Equation 10.43, one arrives at the generalized expression for the first law of thermodynamics for the thermal anemometer velocity sensor:

$$E^2 = G\left[Ak + Bk\left(\frac{\rho_s}{\mu}\right)^n \mathrm{Pr}^{0.33} U_s^n\right] \tag{10.47}$$

where $G = (R_1 + R_3 + R_v)^2 (R_v - R_a)/R_v$, and where A and B again are new constants. In Equation 10.47, one recognizes that conservation-of-mass considerations require that $\rho U = \rho_s U_s$, where ρ and U are referenced to the actual fluid temperature and pressure, and ρ_s and U_s are referenced to normal conditions of 0°C or 20°C temperature and 1 atm pressure. To write Equation 10.47 in terms of U, one simply replaces ρ_s by ρ and U_s by U.

Temperature Compensation

The objective of temperature compensation is to make the bridge voltage E in Equation 10.47 independent of changes in the fluid temperature T_a. This is accomplished if: (1) the term G in Equation 10.47 is independent of T_a and (2) compensation is made for the change in fluid properties (k, μ, and Pr) with T_a. Since these fluid properties have a weaker temperature dependence than G in Equation 10.47, for small temperature changes (less than ±10°C) in gas flows, only G requires compensation.

The two-temperature method is a typical procedure for compensating for both G and fluid properties. In this method, fixed-bridge resistors R_1, R_2, and R_4 in Figure 10.20 are selected so that E is identical at two different temperatures, but at the *same* mass flow rate. This procedure is accomplished during flow calibration and has variations among manufacturers.

The two-temperature method adequately compensates for temperature variations less than approximately ±50°C. In higher temperature gas flow applications, such as the flow of preheated combustion air and stack gas, temperature variations typically are higher. The microprocessor-based digital sensor drive in Figure 10.22 provides temperature compensation for temperature variations ranging from ±50°C to ±150°C. This sensor drive has no analog bridge. Instead, it has a virtual digital bridge that maintains $(T_v - T_a)$ constant within 0.1°C and has algorithms that automatically compensate for temperature variations in k, μ, and Pr. For this digital sensor drive, the first law of thermodynamics is found from Equation 10.37 as:

$$w = \left[Ak + Bk\left(\frac{\rho_s}{\mu}\right)^n \mathrm{Pr}^{0.33} U_s^n\right]\Delta T \tag{10.48}$$

where $\Delta T = (T_v - T_a)$ is now a known constant.

Flow Calibration

Figure 10.23 shows a typical flow calibration curve for the digital electronics drive shown in Figure 10.22. The curve is nonlinear of a logarithmic nature. The nonlinearity is disadvantageous because it requires linearization circuitry, but is advantageous because it provides rangeabilities up to 1000:1 for a single sensor. Additionally, the high-level output of several volts provides excellent repeatability and requires no amplification other than that for spanning. Since the critical dimensions of thermal anemometer sensors are so small, current manufacturing technology is incapable of maintaining sufficiently small tolerences to ensure sensor reproducibility. Therefore, each thermal anemometer must be flow calibrated, for example as in Figure 10.23, over its entire velocity range, either at the exact fluid temperature of its usage or over the range of temperatures it will encounter if it is to be temperature compensated. A 10 to 20 point velocity calibration is required to accurately determine the calibration constants A, B, and n in Equation 10.47. A least-squares curve-fitting procedure usually is applied. Proper flow calibration requires two critical elements: (1) a stable, reproducible, flow-generating facility and (2) an accurate velocity transfer standard. Bruun [1] and Gibbings et al. [4] provide more insight into curve fitting.

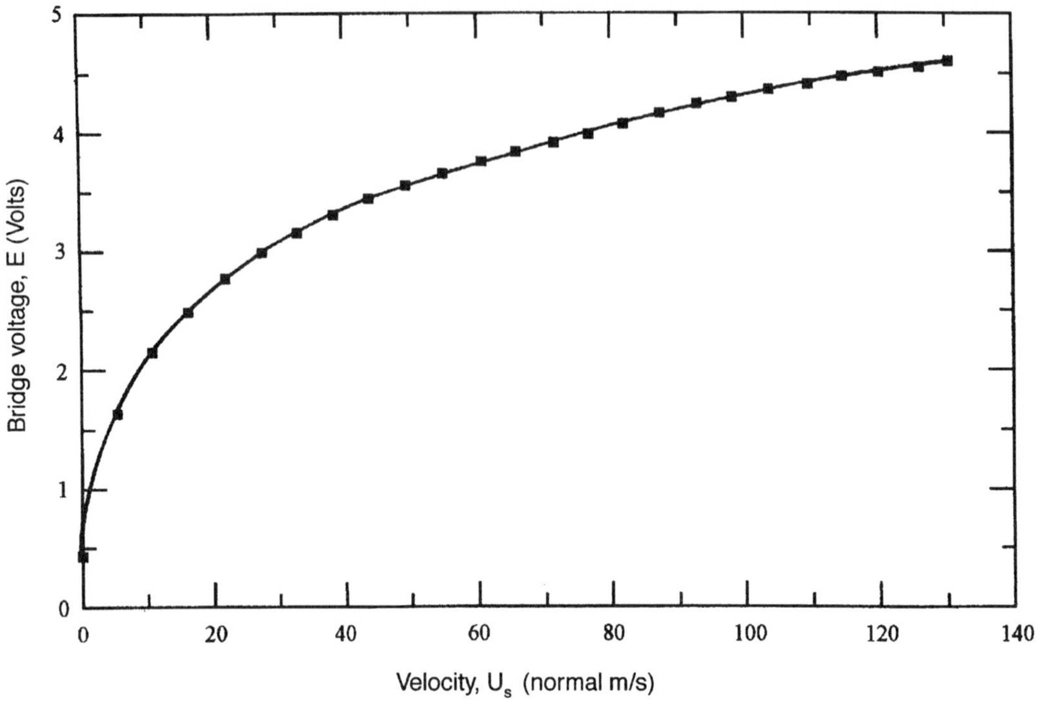

FIGURE 10.23 Typical flow calibration curve for an industrial thermal anemometer. The electronics drive is that shown in Figure 10.22. The constant temperature differential ($T_v - T_T$) is 50.0°C. The cold resistances R_{v0} and R_{T0} of the velocity and temperature sensors at 20°C are approximately 20 Ω and 200 Ω, respectively. (Reprinted with permission of Sierra Instruments, Inc.)

Flow-generating facilities are of two types — open loop and closed loop. An open-loop facility consists of: (1) a flow source such as a fan, pump, elevated tank, or compressed gas supply; (2) a flow-quieting section, such as a plenum with flow straighteners, screens, or other means to reduce swirling, turbulence, or other flow nonuniformities; (3) a nozzle to accelerate the flow and further flatten, or uniformize, the velocity profile; (4) a test section or free jet into which the thermal anemometer probe is inserted; and (5) a means for holding and sealing the thermal anemometer probe and velocity transfer standard. The test section or free jet must have: a velocity profile which is uniform within approximately 0.5% to 1.0% in its central portion; a turbulence intensity less than about 0.5%; and an area large enough so that the projected area of the velocity probe is less than 5% to 10% of the cross-sectional area. Manufacturers of small open-loop flow calibrators often determine the calibration flow velocity by measuring the pressure drop across the nozzle.

The closed-loop flow-generating facility, or wind tunnel, has the same components, but the exit of the test section is connected via ductwork to the inlet of the fan or pump so that the air mass inside the facility is conserved. Open-loop facilities are less expensive than closed-loop tunnels and are far more compact, making them suitable for flow calibrations in the field. But, a laboratory open-loop air-flow calibrator with a fan as the flow generator actually is *closed loop*, with the loop closing within the laboratory. For air velocities less than about 5 m s^{-1}, open-loop calibrators can experience shifts due to changing pressure, temperature, or other conditions in the laboratory. Properly designed closed-loop wind tunnels generate precise, reproducible air velocities from about 0.5 m s^{-1} to 150 m s^{-1}. When fitted with water chillers, they remove compression heating and provide a constant-temperature air flow within ±2°C. When fitted with an electric heater and proper thermal insulation, they provide air temperatures up to 300°C. Gibbings [4] describes a water box displacement rig for flow calibration at very low velocities in the range of 0.1 m s^{-1} to 4 m s^{-1}.

Pitot tubes and laser Doppler anemometers are the two most common velocity transfer standards used to calibrate thermal anemometers. Both have detailed descriptions earlier in this chapter. The Pitot tube usually has the classical "L" shape and an outside diameter of about 3 mm. Its tip is located in the same plane in the test section as the thermal anemometer probe but is no closer than approximately 3 cm. The focal volume of the laser Doppler anemometer is similarly located. The Pitot tube is far less expensive and easier to operate, but is difficult to use if air velocities are less than about 3 m s^{-1}. A proper Pitot-tube flow transfer standard should have its calibration recertified every 6 months by an accredited standards laboratory. On the other hand, the laser Doppler anemometer is a fundamental standard that accurately measures air velocity from approximately 0.5 m s^{-1} to 100 m s^{-1}. Since it provides noncontact anemometry, it is usable at high temperatures. Its primary disadvantages are high expense and complications associated with properly seeding the flow with particles.

Measurements

Point Velocity

Based on the first law of thermodynamics expressed by Equation 10.47, one now can solve for the desired quantity — either the actual point velocity U (m s^{-1}) or the point mass velocity U_s (normal m s^{-1}). Here, one assumes that the velocity vector is normal to the flow sensor. Two- and three-dimensional velocity measurements are discussed later. In the following, *A*, *B*, and *n* are constants, but are different for each case.

The simplest case is isothermal flow with a hot-wire sensor having a very high length-to-diameter ratio (L/d). In this case, the exponent n in Equation 10.47 is 0.5, as shown by Equation 10.34. The applicable first law and velocity expressions are:

$$E^2 = A + BU^{0.5} \tag{10.49}$$

and

$$U = \left[\frac{E^2 - A}{B}\right]^2 \tag{10.50}$$

In the case of a real-world sensor in an isothermal flow having either end loss only or both end loss and skin resistance, one obtains:

$$E^2 = A + BU^n \tag{10.51}$$

and

$$U = \left[\frac{E^2 - A}{B}\right]^{1/n} \tag{10.52}$$

Often, Equation 10.52 is replaced with a polynomial of the form $U = F(E^2)$, where the function *F*( ) is a fourth-order polynomial whose coefficients are determined from flow calibration data using least-squares curve-fitting software. For the same case as above, but with nonisothermal flow, the first law is expressed by Equation 10.47, and the velocity is expressed as:

$$U_s = \frac{\mu}{\rho_s}\left[\frac{E^2/G - Ak}{Bk\mathrm{Pr}^{0.33}}\right]^{1/n} \tag{10.53}$$

For the digital sensor drive of Figure 10.22, the first law is given by Equation 10.48, and the velocity by:

$$U_s = \frac{\mu}{\rho_s}\left[\frac{w/\Delta T - Ak}{Bk\mathrm{Pr}^{0.33}}\right]^{1/n} \tag{10.54}$$

Current commercial industrial thermal anemoneter systems have temperature-compensation and "linearization" electronics that automatically calculate U_s as a linear function of E or w, based on the foregoing relationships.

Turbulence

Turbulence measurements are the second most common application of research thermal anemometers. This measurement requires the high-freqency response of hot-wire and hot-film research anemometers operated in the constant-temperature mode. The vast majority of fluid flows are turbulent. Only flows with very low Reynolds numbers are nonturbulent, or laminar. Turbulent flows are time variant and usually are separated as follows into time-mean and fluctuating parts:

$$\begin{aligned} U(t) &= U + u \\ V(t) &= \overline{V} + v \\ W(t) &= \overline{W} + w \\ T_a(t) &= \overline{T}_a + \theta \\ E &= E + e \end{aligned} \tag{10.55}$$

where $U(t)$, $V(t)$, $W(t)$ are the orthogomal components in the x, y, and z directions, respectively, such as shown in Figure 10.18 for the 3-D hot-wire probe. $T_a(t)$ is the fluid temperature, and $E(t)$ is the bridge voltage. $\overline{U}$, $\overline{W}$, $\overline{V}$, $\overline{T}_a$, and $\overline{E}$ are the time-mean parts, and $u(t)$, $v(t)$, $w(t)$, $\theta(t)$, and $e(t)$ are the time-dependent fluctuating parts. The time-mean parts are averaged sufficiently long to become independent of turbulent fluctuations, yet respond to changes with time in the main flow. In the previous subsection, the expressions given were for the time-mean velocity. In the study of turbulence, one is primarily interested in the time average of the product of two fluctuating velocity components (turbulence correlations) because these terms appear in the time-averaged Navier–Stokes equation. Two important turbulence correlations are $\overline{u^2}$ and $\overline{uv}$. The correlation $(\sqrt{\overline{u^2}})/\overline{U}$ is called the *turbulence intensity*. Manufacturers of research anemometer systems provide electronics for automatically computing turbulence correlations.

For a fluid with changes in temperature sufficiently small that fluid properties are essentially constant, one can write Equation 10.47 in the following form:

$$E^2 = \left(A + BU^n\right)\left(T_v - T_a\right) \tag{10.56}$$

where A, B, and n are constant and where R_v is virtually constant because the anemometer is in the constant-temperature mode. Elsner [11] shows that the fluctuating voltage e is found by taking the total derivative of Equation 10.56, as follows:

$$e = S_u u + S_\theta \theta \tag{10.57}$$

where

$$S_u = \frac{\delta E}{\delta U} = \frac{nBU^{n-1}}{2}\left[\frac{\left(T_v - T_a\right)}{A + BU^n}\right]^{1/2} = \textit{Velocity sensitivity} \tag{10.58}$$

$$S_\theta = \frac{\delta E}{\delta T_a} = -\frac{1}{2}\left[\frac{\left(A + BU^n\right)}{\left(T_v - T_a\right)}\right]^{1/2} = \textit{Temperature sensitivity} \tag{10.59}$$

It is seen from Equations 10.58 and 10.59 that increasing ($T_v - T_a$), i.e., operating the sensor as hot as possible, maximizes the velocity sensitivity and minimizes the sensitivity to temperature fluctuations. This is why tungsten hot wires are operated at high temperatures (typically 200°C to 300°C).

The fluctuating components of velocity have a broad frequency spectrum, ranging from 10^{-2} Hz to 10^5 Hz, and sometimes even higher. Therefore, it is imperative that the frequency response of constant-temperature research anemometers have a flat frequency response, i.e., minimized attenuation and phase shift at higher frequencies. Blackwelder [12] and several other investigators have studied the frequency response of hot-wire anemometers. For turbulence measurements, Borgos [13] describes commercial research anemometer systems with features such as: low-pass filters to decrease electronics noise; a subcircuit for determining and setting overheat ratio; a square-wave generator for frequency response testing; and two or more controls to optimize the frequency response to fast fluctuations. Recent systems have electronics that compensate for frequency attenuation. When used with 5 μm diameter hot-wire sensors in air, commercial systems are capable of nearly flat frequency response and very small phase lag from 0 Hz to approximately 10^4 Hz. As reported by Nelson and Borgos [14], wedge and conical hot-film sensors in water have a relatively flat response from 0 Hz to 10 Hz for velocities above 0.3 m s^{-1}.

Two- and three-component velocity and turbulence measurements are made using hot-wire or hot-film research anemometers, such as shown in Figure 10.18. As described by Müller [15], hot-wire or cylindrical hot-film probes in the "X"-configuration are used to measure the U and V velocity components. In a three-sensor orthogonal array, they measure U, V, and W. Döbbeling, Lenze, and Leuckel [16] and other investigators have developed four-wire arrays for measurement of U, V, and W. Olin and Kiland [17] describe an orthogonal array of three cylindrical split hot-film sensors. Each of the three sensors in this array has two individually operated hot-films separated by two axial splits 180° apart along its entire length. The two split films take advantage of the nonuniform heat-transfer distribution around a cylinder in cross flow.

In multisensor arrays, the velocity vector is not necessarily normal to a cylindrical sensor. If the discussion is limited to isothermal flows, the first law expressed by Equation 10.47 becomes:

$$E^2 = A + BV_e^n \tag{10.60}$$

where V_e is the effective velocity sensed by a single cylindrical sensor in the array, and A, B, and n are constants. Jörgenson [18] describes V_e as follows:

$$V_e^2 = U_N^2 + a^2\, U_T^2 + b^2\, U_B^2 \tag{10.61}$$

where U_N = velocity component normal to the sensor
U_T = tangential component
U_B = component perpendicular to both U_N and U_T (i.e., binormal)

The constants a and b in Equation 10.61 are referred to as the sensor's yaw and pitch coefficients, respectively, and are determined via flow calibration. Typical values for a and b for a plated hot-wire sensor are 0.2 and 1.05, respectively. Inserting Equation 10.61 into Equation 10.60, we get the following expression for the output signal of a single sensor in the array:

$$E^2 = A + B\left(U_N^2 + a^2 U_T + b^2 U_B^2\right)^{n/2} \tag{10.62}$$

Expressions like this, or similar ones such as given by Lekakis, Adrian, and Jones [19], are written for all sensors in the array. These expressions and trigonometry are then used to solve for the components of velocity U, V, and W in the x, y, z spatially fixed reference frame.

Channel Flows

Based on the following relationship, a single-point industrial *insertion* thermal anemometer monitors the mass flow rate $\dot{m}$ (kg s^{-1}) in ducts, pipes, stacks, or other flow channels by measuring the velocity $U_{s,c}$ at the channel's centerline:

$$\dot{m} = \rho_s \gamma\, U_{s,c} A_c \tag{10.63}$$

where $U_{s,c}$ is the velocity component parallel to the channel's axis measured at the channel's centerline and referenced to *normal* conditions of 0°C or 20°C temperature and 1 atmosphere pressure; ρ_s, a constant, is the fluid's mass density at the same normal conditions; A_c, another constant, is the cross-sectional area of the channel; and γ is a constant defined as $\gamma = U_{s,ave}/U_{s,c}$, where $U_{s,ave}$ is the average velocity over area A_c. The velocity in channel flows is seldom uniform and therefore γ is not unity. If the flow channel has a length-to-diameter ratio of 40 to 60, then its flow profile becomes unchanging and is called "fully developed." In fully developed flows, the fluid's viscosity has retarded the velocity near the walls, and hence γ is always less than unity. If the channel's Reynolds number is less than 2000, the flow is laminar; the fully developed profile is a perfect parabola; and γ is 0.5. If the Reynolds number is larger than 4000, the flow is turbulent; the fully developed profile has a flattened parabolic shape; and for pipes with typical rough walls, γ is 0.79, 0.83, and 0.83 for Reynolds numbers of 10^4, 10^5, and 10^6, respectively. If the Reynolds number is between 2000 and 4000, the flow is transitioning between laminar and turbulent flows, and γ ranges between 0.5 and 0.8.

Unfortunately, in most large ducts and stacks, 40 to 60 diameters of straight run preceding the flow monitoring location does not exist. Instead, the flow profile usually is highly nonuniform, swirling, and, in air-preheater ducts and in stacks, is further confounded by temperature nonuniformities. In these cases, single-point monitoring is ill-advised. Fortunately, multipoint monitoring with industrial thermal anemometer flow-averaging arrays, such as shown in Figure 10.18, have proven successful in these applications. As described by Olin [20], this method consists of a total of N (usually, $N = 4$, 8, or 12) industrial thermal anemometer sensors, each similar to that shown in Figure 10.17, located at the centroid of an equal area A_c/N in the channel's cross-sectional area A_c. The individual mass flow rate $\dot{m}_i$ monitored by each sensor is $\rho_s\, U_{s,i}\, (A_c/N)$, where $U_{s,i}$ is the individual velocity monitored by the sensor at point i. The desired quantity, the total mass flow rate $\dot{m}$ through the channel, is the sum of the individual mass flow rates, or:

$$\dot{m} = \sum_{i=1}^{N} \dot{m}_i = \rho_s A_c U_{s,ave} \tag{10.64}$$

where $U_{s,ave}$ is the arithmetic average of the N individual velocities $U_{s,i}$. As described by Olin [21], industrial multipoint thermal anemometers are used as the flow monitor in stack continuous emissions monitoring systems required by governmental air-pollution regulatory agencies.

Table 10.3 Typical Commercial Thermal Anemometer Systems

Product Description	Average 1997 U.S. List Price
Industrial systems	
Insertion mass flow transducer	$1,900
50 mm (2 in.) NPT in-line mass flowmeter	$2,500
8-point smart industrial flow averaging array	$15,000
Research systems	
Single-channel hot-wire or hot-film anemometer system	$10,000
Three-component hot-wire anemometer system	$21,000
Portable air velocity meter	$1,000

Note: Prices listed are the average of the manufacturers listed in Table 10.4. Insertion probe is 25 cm in length. Insertion and in-line mass flowmeters have: probe-mounted FM/CENELEC approved, explosion-proof housing; ac line voltage input power; 5-0 V dc output signal; 316 SS construction; and ambient air calibration. In-line industrial mass flowmeter has built-in flow conditioning. Industrial flow averaging array has four 1 m long probes, 2 points per probe, 316 SS construction, line voltage input power, 0 to 5 V dc output signal, and smart electronics mounted on probe. Research anemometer systems have standard hot-wire probes, most versatile electronics, and include ambient air calibrations.

TABLE 10.4 Manufacturers of Thermal Anemometer Systems

Industrial Systems and Portable Air Velocity Meters	Research Systems and Portables
Sierra Instruments, Inc. 5 Harris Court Building L Monterey, CA 93940 Tel: (831) 373-0200 Fax: (831) 373-4402	TSI Inc. 500 Cardigan Road St. Paul, MN 55164 Tel: (612) 490-2811 Fax: (612) 490-3824
Fluid Components, Inc. 1755 La Costa Meadows Drive San Marcos, CA 92069 Tel: (619) 744-6950 Fax: (619) 736-6250	Dantec Measurement Technology, Inc. Denmark Tel: (45) 4492 3610 Fax: (45) 4284 6136
Kurz Instruments, Inc. 2411 Garden Road Monterey, CA 93940 Tel: (831) 646-5911 Fax: (831) 646-8901	

Instrumentation Systems

Table 10.3 lists examples of typical commercial thermal anemometer systems. Table 10.4 lists their major manufacturers. Thermal anemometer systems include three elements: sensors, probe, and electronics. Sensors and probes have been described in previous sections. The electronics of industrial systems are enclosed in an explosion-proof or other industrial-grade housing mounted either directly on the probe or remotely (usually within 30 m). The electronics is powered with a 24 V dc source or with 100, 115, or 230 V ac line voltage. The output signal typically is 0 to 5 V dc, 4 to 20 mA, RS232, or RS485 linearly proportional to gas mass velocity U_s over the range of 0.5 normal m s^{-1} to 150 normal m s^{-1}. In-line mass flowmeters have the same output-signal options and are calibrated directly in mass flow rate $\dot{m}$ (kg s^{-1}). In-line meters are now available with built-in flow conditioners that eliminate errors associated with upstream disturbances, such as elbows, valves, and pipe expansions. Systems are available either with lower cost analog electronics or with smart microprocessor-based electronics. The repeatability of these systems is ±0.2% of full scale. The typical accuracy of a smart industrial system is ±2% of reading over 10 to 100% of full scale and ±0.5% of full scale below 10% of full scale. Automatic temperature compensation facilitates temperature coefficients of ±0.04% of reading per °C within ±20°C of calibration temperature and ±0.08% of reading per °C within ±40°C. High-temperature applications have temperature compensation over a range of ±150°C. Pressure effects are negligible within ±300 kPa of calibration pressure.

Research thermal anemometer systems usually are coupled with a personal computer, as shown in Figure 10.21. The PC provides system set-up and control, as well as data display and analysis. Modern systems feature low-noise circuits, together with smart bridge optimization technology that eliminates tuning and automatically provides flat frequency response up to 300,000 Hz. Lower cost units provide flat response up to 10,000 Hz. A built-in thermocouple circuit simplifies temperature measurement. The PC's windows-based software provides near real-time displays of velocity, probability distribution, and turbulence intensity. Post-processing gives additional statistics, including: mean velocity; turbulence intensity; standard deviation; skewness; flatness; normal stress for one-, two-, and three-component probes; as well as shear stress, correlation coefficients and flow-direction angle for two- and three-dimensional probes. In addition, power spectrum, auto correlations, and cross correlations can be displayed. The software automatically handles flow calibration set-up and calculates calibration velocity. Systems are available in 1-, 2-, 8-, and 16-channel versions.

Commercial industrial and research thermal anemometer systems were first introduced in the early 1960s. At first, industrial thermal anemometers were not considered sufficiently durable for the rigors of industrial use. With the advent of stainless-steel sheathed sensors and microprocessor-based electronics, industrial thermal anemometers now enjoy the credibility formerly attributed to only traditional flowmeter approaches. Initial research systems required a high level of user knowledge and considerable involvement in operation. In contrast, current research systems have nearly flat frequency response, high accuracy, and easy-to-use controls providing the flexibility researchers require. Research systems based on personal computers have graphical user interfaces that enhance both performance and simplicity of operation.

References

1. H. H. Bruun, *Hot-Wire Anemometry: Principles and Signal Analysis,* Oxford: Oxford University Press, 1995.
2. L. M. Fingerson and P. Freymuth, Thermal anemometers, in R. J. Goldstein (ed.), *Fluid Mechanics Measurements,* Washington, D.C.: Hemisphere, 1983.
3. P. Freymuth, History of thermal anemometry, in N.P. Cheremisinoff and R. Gupta (ed.), *Handbook of Fluids in Motion,* Ann Arbor, MI: Ann Arbor Science Publishers, 1983.
4. J. C. Gibbings, J. Madadnia, and A.H. Yousif, The wall correction of the hot-wire anemometer, *Flow Meas. Instrum.,* 6(2), 127-136, 1995.
5. J. C. Gibbings, J. Madadnia, S. Riley, and A.H. Yousif, The proximity hot-wire probe for measuring surface shear in air flows, *Flow Meas. Instrum.,* 6(3), 201-206, 1995.
6. V. Mikulla, *The Measurement of Intensities and Stresses of Turbulence in Incompressible and Compressible Air Flow,* Ph.D. Thesis, University of Liverpool, 1972.
7. L. V. King, On the convection of heat from small cylinders in a stream of fluid: determination of the convection constants of small platinum wires with application to hot-wire anemometry, *Phil. Trans. Roy. Soc.,* A214, 373-432, 1914.
8. H. Kramers, Heat transfer from spheres to flowing media. *Physica,* 12, 61-80, 1946.

9. S. W. Churchill and M. Bernstein, A correlating equation for forced convection from gases and liquids to a circular cylinder in crossflow, *J. Heat Transfer*, 99, 300-306, 1997.
10. S. Takagi, A hot-wire anemometer compensated for ambient temperature variations, *J. Phys. E.: Sci. Instrum.*, 19, 739-743, 1986.
11. J. W. Elsner, An analysis of hot-wire sensitivity in non-isothermal flow, *Proc. Dynamics Flow Conf., Marseille*, 1972.
12. R. F. Blackwelder, Hot-wire and hot-film anemometers, in R.J. Emrich (ed.), *Methods of Experimental Physics: Fluid Dynamics*, New York: Academic Press, 18A, 259-314, 1981.
13. J. A. Borgos, A review of electrical testing of hot-wire and hot-film anemometers, *TSI Quart.*, VI(3), 3-9, 1980.
14. E.W. Nelson and J. A. Borgos, Dynamic response of conical and wedge type hot films: comparison of experimental and theoretical results, *TSI Quart.*, IX(1), 3-10, 1983.
15. U. R. Müller, Comparison of turbulence measurements with single, X and triple hot-wire probes, *Exp. in Fluids*, 13, 208-216, 1992.
16. K. Döbbeling, B. Lenze, and W. Leuckel, Four-sensor hot-wire probe measurements of the isothermal flow in a model combustion chamber with different levels of swirl, *Exp. Thermal and Fluid Sci.*, 5, 381-389, 1992.
17. J.G. Olin and R. B. Kiland, Split-film anemometer sensors for three-dimensional velocity-vector measurement, *Proc. Symp. on Aircraft Wake Turbulence*, Seattle, Washington, 1970, 57-79.
18. F. E. Jörgenson, Directional sensitivity of wire and fibre-film probes, *DISA Info.*, (11), 31-37, 1971.
19. I. C. Lekakis, R. J. Adrian, and B. G. Jones, Measurement of velocity vectors with orthogonal and non-orthogonal triple-sensor probes, *Experiments in Fluids*, 7, 228-240, 1989.
20. J. G. Olin, A thermal mass flow monitor for continuous emissions monitoring systems (CEMS), *Proc. ISA/93 Int. Conf. Exhibition & Training Program*, (93-404), 1993, 1637-1653.
21. J. G. Olin, Thermal flow monitors take on utility stack emissons, *Instrumentation and Control Systems*, 67(2), 71-73, 1994.

Further Information

J. A. Fay, *Introduction to Fluid Mechanics*, Cambridge, MA: MIT Press, 1994.
A. Bejan, *Convection Heat Transfer*, New York: John Wiley & Sons, 1995.

10.3 Laser Anemometry

Rajan. K. Menon

Laser anemometry, or *laser velocimetry*, refers to any technique that uses lasers to measure velocity. The most common approach uses the Doppler shift principle to measure the velocity of a flowing fluid at a point and is referred to as Laser Doppler Velocimetry (LDV) or Laser Doppler Anemometry (LDA). This technique (also known as dual beam, differential Doppler or fringe mode technique), incorporating intersecting (focused) laser beams, is also used to measure the motion of surfaces [1]. In some special flow situations, another approach using two *nonintersecting*, focused laser beams known as *dual focus* (also known as L2F) technique is used to measure flow velocity at a point [2]. More recently, laser illumination by light sheets is used to make global flow measurements and is referred to as *particle image velocimetry* (PIV) [3]. The strength of PIV (including particle tracking velocimetry) lies in its ability to capture turbulence structures within the flow and transient phenomena, and examine unsteady flows [4]. The development of this technique to obtain both spatial and temporal information about flow fields is making this a powerful diagnostic tool in fluid mechanics research [5, 6]. Other approaches to measure global flow velocities come under the category of molecular tagging velocimetry [7] or Doppler global velocimetry [8, 9].

The noninvasive nature of the LDV technique and its ability to make accurate velocity measurements with high spatial and temporal resolution, even in highly turbulent flows, have led to the widespread use of LDV for flow measurement. Flow velocities ranging from micrometers per second to hypersonic speeds have been measured using LDV systems. Measurements of highly turbulent flows [10], flows in rotating machinery [11], especially in the interblade region of rotors [12, 13], very high [14] or very low [15] velocity flows, flows at high temperatures [16] and in other hostile environments [17, 18], and flows in small spaces [19] have been performed using the LDV technique. The versatility and the widespread use of the LDV approach to measure flows accurately has resulted in referring to this technique as *laser velocimetry* or *laser anemometry.* Many details of the technique, including some of the early developments of the hardware, are provided in the book by Durst [20]. A bibliography of the landmark papers in LDV has been compiled by Adrian [21].

For the case of spherical scatterers, the technique has also been extended to measure size of these particles. In this case, the scattered light signal from a suitably placed receiver system is processed to obtain the diameter of the particle, using the phase Doppler technique [22].

The first reported fluid flow measurements using LDV principles was by Yeh and Cummins [23]. Although in this case an optical arrangement referred to as the reference beam system was used to measure the Doppler shift, in almost all measurement applications, what is referred to as the dual beam or differential Doppler arrangement [24] is used now. This arrangement, also referred to as the "fringe" mode of operation, uses two intersecting laser beams (Figure 10.24) to measure one velocity component.

The advantages of the LDV technique in measuring flows include (1) a small measuring region (i.e., point measurement), (2) high measurement accuracy, (3) the ability to measure any desired velocity component, (4) accurate measurement of high turbulence intensities, including flow reversals, (5) a large dynamic range, (6) no required velocity calibration, (7) no probe in the flow (does not disturb the flow; measures in hostile environments), and (8) good frequency response.

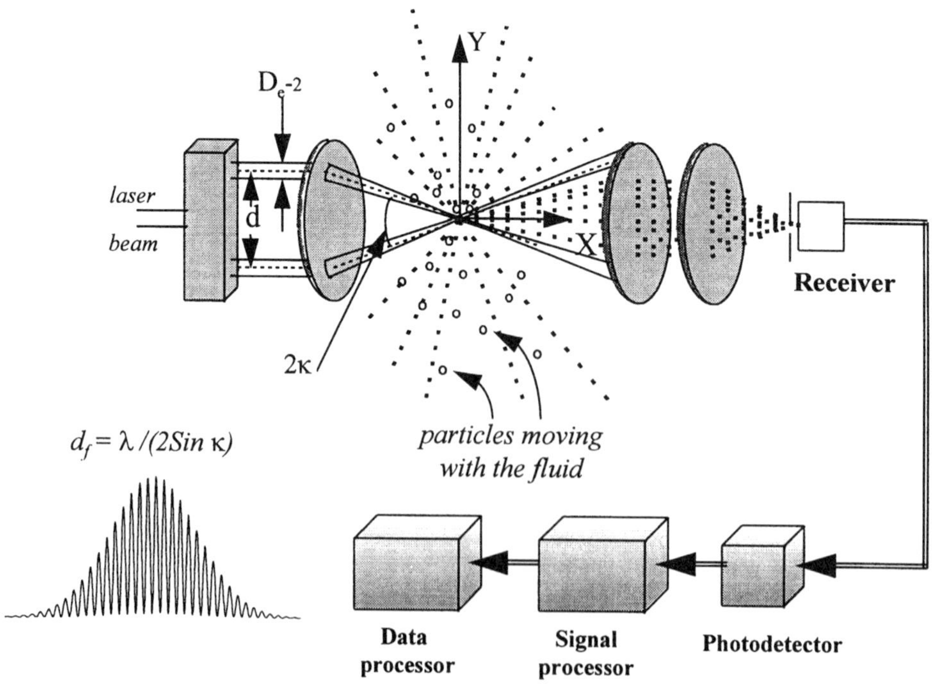

FIGURE 10.24 Schematic of a dual-beam system.

The LDV technique relies on the light scattered by scattering centers in the fluid to measure flow velocity. These scattering centers will also be referred to as *particles*, with the understanding that bubbles or anything else that has a refractive index different from that of the fluid could be the source of scattered light. The particles, whose velocities are measured, must be small enough (generally in the micron range) to follow the flow variations and large enough to provide signal strength adequate for the signal processor to give velocity measurements. It should be noted that the signal exists only when a "detectable" particle is in the measuring volume and, hence, is discontinuous. This, along with other properties of the signal, adds special requirements on the signal processing and the subsequent data analysis systems. The scattered light signal is processed to obtain the Doppler shift frequency and from that the velocity of the particle. Hence, the rate at which the velocity measurements are made depends on the rate of particle arrival. It is desirable to have a high particle concentration to obtain a nearly continuous update of velocity. In carefully controlled experiments, the LDV system can provide very high accuracy (0.1% or better) measurements in mean velocity. Thermal anemometer systems are generally able to measure lower turbulence levels compared to that by an LDV system [25]. While the direct measurement of the Doppler-shifted frequency from a single laser beam caused by a moving particle is possible [26], most LDV systems employ the heterodyne principle to obtain and process only the Doppler shift (difference) frequency.

Principle of Operation

Dual-Beam Approach

The dual-beam approach is the most common optical arrangement used for LDV systems for flow measurement applications. The schematic (Figure 10.24) shows the basic components of a complete LDV system to measure one component of velocity. The transmitting optics include an optical element to split the original laser beam into two parallel beams and a lens system to focus and cross the two beams. The intersection region of the two beams becomes the measuring region. The receiving optics (shown to be set up in the forward direction) collect a portion of the light scattered by the particles, in the fluid stream, passing through the beam-crossing region (measuring volume) and direct this light to a photodetector, which converts the scattered light intensity to an analog electrical signal. The frequency of this signal is proportional to the velocity of the particle. A signal processor extracts the frequency information from the photodetector output and provides this as a digital number corresponding to the instantaneous velocity of the particle. The data processing system obtains the detailed flow properties from these instantaneous velocity measurements. The idealized photodetector signal, for a particle passing through the center of the measuring volume, is shown in the lower left side of Figure 10.24. Actual signals will have noise superimposed on them; and the signal shape will vary, depending on the particle trajectory through the measuring volume [27].

Fringe Model Description.

While there are several ways to describe the features of a dual-beam system, the description based on a fringe model is, perhaps, the simplest. For simplicity, the diameter and the intensity of both the beams are assumed to be the same. After the beams pass through the transmitting lens, the diameter of each beam continuously decreases to a minimum value (beam waist) at the focal point of the lens, and then increases again. Thus, the beam waists cross where the two laser beams intersect (at the focal point of the lens), and the wavefronts in the beams interfere with each other, creating a fringe pattern [28]. In this pattern, assuming equal intensity beams and other needed qualities of the beams, the light intensity varies from zero (dark fringe) to a maximum (bright fringe), and the fringes are equally spaced. The particles in the flow passing through the intersection region (measuring region) scatter light in all directions. An optical system, including a receiving lens (to collimate the scattered light collected) and a focusing lens, is used to collect the scattered light and focus it onto the receiver. The aperture in front of the receiver is used to block out stray light and reflections and collect only the light scattered from the measuring region.

As a particle in the flow, with velocity u, moves across the fringes, the intensity pattern of the light scattered by the particle resembles that shown in the lower left of Figure 10.24. The velocity component,

u_y (perpendicular to the optical axis and in the plane of the incident beams) can be obtained from the ratio of the distance between fringes (or fringe spacing, d_f), and the time t (= $1/f_D$) for the particle to cross one pair of fringes, where f_D is the frequency of the signal. The amplitude variation of the signal reflects the Gaussian intensity distribution across the laser beam. Collection (receiving) optics for the dual-beam system can be placed at any angle, and the resulting signal from the receiving system will still give the same frequency. However, signal quality and intensity will vary greatly with the collection optics angle.

Doppler Shift Explanation.
The description of the dual-beam system using the Doppler shift principle is as follows. At the receiver, the frequencies of the Doppler-shifted light scattered by a particle from beam one and beam two are given by:

$$\nu_{D1} = \nu_{01} + \frac{\vec{u}}{\lambda}\left(\hat{r} - \hat{S}_1\right); \quad \nu_{D2} = \nu_{02} + \frac{\vec{u}}{\lambda}\left(\hat{r} - \hat{S}_2\right) \tag{10.65}$$

where ν_{01} and ν_{02} are the frequencies of laser beam 1 and laser beam 2; $\hat{r}$ is the unit vector directed from the measuring volume to the receiving optics; $\hat{S}_1$ and $\hat{S}_2$ are the unit vectors in the direction of incident beam 1 and incident beam 2; $\vec{u}$ is the velocity vector of the particle (scattering center); and λ is the wavelength of light. The frequency of the net (heterodyne) signal output from the photodetector system is given by the difference between ν_{D1} and ν_{D2}.

$$f_D = f_S + \frac{\vec{u}}{\lambda}\left(\hat{S}_2 - \hat{S}_1\right) \tag{10.66}$$

where $f_S = \nu_{01} - \nu_{02}$ is the difference in frequency between the two incident beams. This difference frequency is often intentionally imposed (see section on frequency shifting) to permit unambiguous measurement of flow direction and high-turbulence intensities. Assuming $f_S = 0$, the frequency detected by the photodetector is:

$$f_D = \frac{\dot{u}}{\lambda}\left(\hat{S}_2 - \hat{S}_1\right) = 2u_y \sin\kappa \tag{10.67}$$

Hence,

$$u_y = \frac{f_D \lambda}{2\sin\kappa} = f_D\, d_f \tag{10.68}$$

This is the equation for u_y and shows that the signal frequency f_D is directly proportional to the velocity u_y. The heterodyning of the scattered light from the two laser beams at the photodetector actually gives both the sum and difference frequency. However, the sum frequency is too high to be detected and so only the difference frequency ($\nu_{D1} - \nu_{D2}$) is output from the photodetector as an electrical signal. The frequency f_D is often referred to as the Doppler frequency of the output signal, and the output signal is referred to as the Doppler signal.

It can be seen from Equation 10.68 that the Doppler frequency is independent of the receiver location ($\hat{r}$). Hence, the receiver system location can be chosen based on considerations such as signal strength, ease of alignment, and clear access to the measuring region. The expressions for the other optical configurations can be reduced similarly [29], giving the identical equation for the Doppler shift frequency f_D. It should be noted that the fringe description does not involve a "Doppler shift" and is, in fact, not always appropriate. The fringe model is convenient and gives the correct expression for the frequency. However, it can be misleading when studying the details of the Doppler signal (e.g., signal-to-noise ratio) and other important parameters e.g., modulation depth or visibility ($\overline{V}$) of the signal [30].

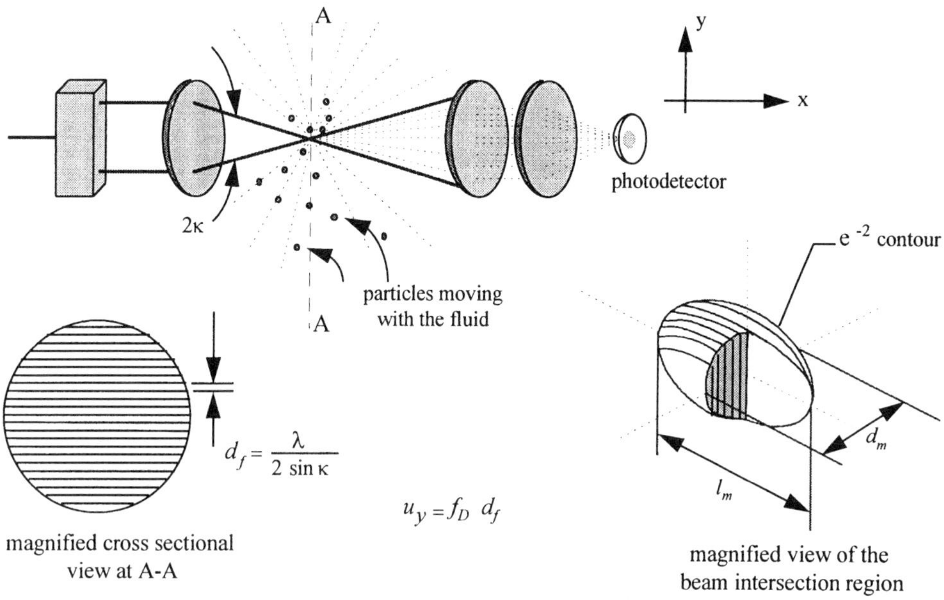

FIGURE 10.25 Details of the beam crossing.

The time taken by the particle to cross the measuring volume is referred to as *transit time, residence time,* or *total burst time,* τ_B, and corresponds to the duration of the scattered light signal. The number of cycles (N) in the signal (same as the number of fringes the particle crosses) is given by the product of the transit time (τ_B) and the frequency, f_D, of the signal.

It should also be noted that the fringe spacing (d_f), depends only on the wavelength of the laser light (λ) and the angle (2κ) between the two beams. It can be shown that the effect of the fluid refractive index on these two terms tends to cancel out and, hence, the value of fringe spacing is independent of the fluid medium [31]. The values of λ and κ are known for any dual-beam system and, hence, an actual velocity calibration is not needed. In some cases, an actual velocity calibration using the rim of a precisely controlled rotating wheel has been performed to overcome the errors in measuring accurately the angle between the beams.

The intensity distribution in a laser beam operating in the TEM_{00} mode is Gaussian [32]. Using wave theory and assuming diffraction-limited optics, the effective diameter of the laser beam and the size of the measurement region can be defined. The conventional approach to the definition of laser beam diameter and measuring volume dimensions is based on the locations where the light intensity is $1/e^2$ of the maximum intensity (at the center of the beam). This definition of the dimensions is analogous to that of the boundary layer thickness. The dimensions d_m and l_m of the ellipsoidal measuring volume (Figure 10.25) are based on the $1/e^2$ criterion and are given by:

$$d_m = 4f\lambda/\pi D_{e^{-2}}; \quad l_m = d_m/\tan\kappa; \quad N_{FR} = d_m/d_f \tag{10.69}$$

N_{FR} is the maximum number of fringes in the ellipsoidal measuring region. Note that as the value of $D_{e^{-2}}$ increases, the measuring volume becomes smaller. In flow measurement applications, this relationship is exploited to arrive at the desired size of the measuring volume.

The measuring volume parameters for the following sample situation are wavelength, λ = 514.5 nm (green line of argon-ion laser), $D_{e^{-2}}$ = 1.1 mm, d = 35 mm, and f = 250 mm. Then, κ = 4°, d_m = 149 μm, and l_m = 2.13 mm. The fringe spacing, d_f, is 3.67 μm and the maximum number (N_{FR}) of fringes (number of cycles in a signal burst for a particle going through the center of the measuring volume in the

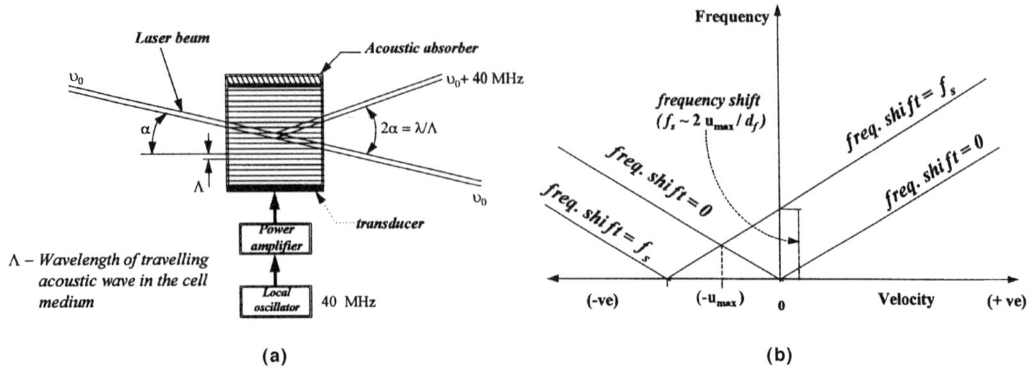

FIGURE 10.26 (*a*) Bragg cell arrangement; (*b*) velocity vs. frequency.

y-direction) in the measuring volume is 40. Consider a particle passing through the center of the measuring region with a velocity (normal to the fringes) of 15 m s^{-1}. This would generate a signal with a frequency of about 4.087 MHz. The transit time of the particle (same as duration of the signal) would be approximately 9.93 μs!

Frequency Shifting

The presence of high turbulence intensity and recirculating or oscillatory flow regions is common in most flow measuring situations. In the fringe model and the Doppler shift (with $f_3 = 0$) descriptions of the dual-beam system, the Doppler signal does not indicate the influence of the sign (positive or negative) of the velocity. Further, a particle passing through the measuring volume parallel to the fringes would not cross any fringes and, hence, not generate a signal having the cyclic pattern resulting in the inability to measure the zero normal (to the fringes) component of velocity. In addition, signal processing hardware used to extract the frequency information often requires the signals to have a minimum number of cycles. This, as well as the ability to measure flow reversals, is achieved by a method of frequency offsetting referred to as *frequency shifting*. Frequency shifting is also used to measure small velocity components perpendicular to the dominant flow direction and to increase the effective velocity measuring range of the signal processors [31].

By introducing a phase or frequency offset (f_s) to one of the two beams in a dual-beam system, the directional ambiguity can be resolved. From the fringe model standpoint, this situation corresponds to a moving (instead of a stationary) fringe system. A stationary particle in the measuring volume will provide a continuous signal at the photodetector output whose frequency is equal to the difference in frequency, f_s, between the two incident beams. In other words, as shown in Figure 10.26(*b*), the linear curve between velocity and frequency is offset along the positive frequency direction by an amount equal to the frequency shift, f_s. Motion of a particle in a direction opposite to fringe movement would provide an increase in signal frequency, while particle motion in the direction of fringe motion would provide a decrease in frequency. To create a signal with an adequate number of cycles even while measuring negative velocities (e.g., flow reversals, recalculating flows), a convenient "rule-of-thumb" approach for frequency shifting is often used. The approach is to select the frequency shift ($f_s \sim 2\, u_{max}/d_f$) to be approximately twice the frequency corresponding to the magnitude of the maximum negative velocity (u_{max}) expected in the flow. This provides approximately equal probability of measurement for all particle trajectories through the measuring volume [33, 34].

Frequency shifting is most commonly achieved by sending the laser beam through a Bragg cell (Figure 10.26(*a*)), driven by an external oscillator [35]. Typically, the propagation of the 40 MHz acoustic wave (created by a 40 MHz drive frequency) inside the cell affects the beam passing through the cell to yield a frequency shift of 40 MHz for that beam. By properly adjusting the angle the cell makes with the incoming beam and blocking off the unwanted beams, up to about 80% of input light intensity is recovered in the shifted beam. The Bragg cell approach will provide a 40 MHz frequency shift in the photodetector output signal. To improve the measurement resolution of the signal processor, the resulting photodetector signal is often "downmixed" to have a more appropriate frequency shift (based on the rule-of-thumb shift value) for the flow velocities being measured. Frequency shifting using two Bragg cells (one for each beam of a dual-beam system) operating at different frequencies is attractive to systems where the bandwidth of the photodetector is limited. However, the need to readjust the beam crossing with a change in frequency shift has not made this approach (double Bragg cell technique) attractive for applications where frequency shift needs to be varied [31].

More recently, Bragg cells have been used in a multifunctional mode to split the incoming laser beam into two equal intensity beams, with one of them having the 40 MHz frequency shift. This is accomplished by adjusting the Bragg cell angle differently. In addition to Bragg cells, rotating diffraction gratings and other mechanical approaches have been used for frequency shifting. However, limits on rotational speed and other mechanical aspects of these systems make them limited in frequency range [20]. Other frequency shifting techniques have been suggested for use with laser diodes [36, 37]. Because so many flow measurement applications involve recirculating regions and high turbulence intensities, frequency shifting is almost always a part of an LDV system used for flow measurement.

Signal Strength

Understanding the influence of various parameters of an LDV system on the signal-to-noise ratio (SNR) of the photodetector signal provides methods or approaches to enhance signal quality and hence improve the performance of the measuring system. The basic equation for the ratio of signal power to noise power (SNR) of the photodetector signal can be written as [38]:

$$\mathrm{SNR} = A_1 \frac{\eta_q P_0}{\Delta f} \left[\frac{D_a}{r_a} \frac{D_{e^{-2}}}{f} \right]^2 d_p^2 \overline{G} \overline{V}^2 \tag{10.70}$$

Equation 10.70 shows that higher laser power (P_0) provides better signal quality. The quantum efficiency of the photodetector, η_q depends on the type of photodetector used and is generally fixed. The SNR is inversely proportional to the bandwidth, Δf, of the Doppler signal. The term in brackets relates to the optical parameters of the system; the "*f*-number" of the receiving optics, D_a/r_a, and the transmitting optics, $D_{e^{-2}}/f$. The square dependence of SNR on these parameters makes them the prime choice for improving signal quality and, hence, measurement accuracy. The focal length of the transmitting (f) and receiving (r_a) lenses are generally decided by the size of the flow facility. Using the smallest possible values for these would increase the signal quality. The first ratio (D_a is the diameter of the receiving lens) determines the amount of the scattered light that is collected, and the second ratio determines the diameter of (and hence the light intensity in) the measuring volume. The last three terms are the diameter, d_p, of the scattering center and the two terms (scattering gain $\overline{G}$, visibility $\overline{V}$) relating to properties of the scattered light. These need to be evaluated using the Mie scattering equations [38] or the generalized Lorentz–Mie theory [39].

Measuring Multiple Components of Velocity

A pair of intersecting laser beams is needed to measure (Figure 10.24) one component of velocity. This concept is extended to measure two components of velocity (perpendicular to the optical axis) by having two pairs of beams that have an overlapping intersection region. In this case, the plane of each pair of beams is set to be orthogonal to that of the other. The most common approach to measure two components of velocity is to use a laser source that can generate multiwavelength beams so that the wavelength of one pair of beams is different from the other pair. The Doppler signals corresponding to the two components of velocity are separated by wavelength [31].

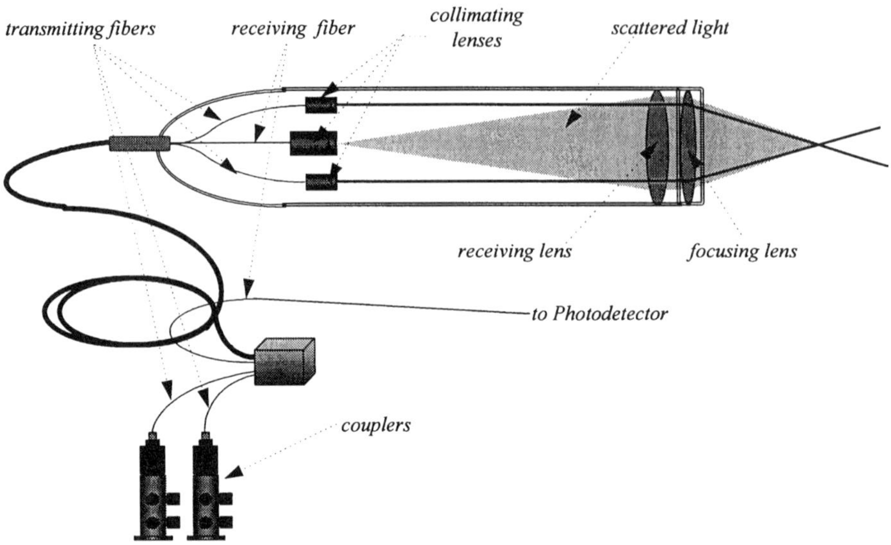

FIGURE 10.27 Schematic arrangement of a fiberoptic system.

Historically, LDV systems were assembled by putting together a variety of optical modules. These modules included beam splitters, color separators, polarization rotators, and scattered light collection systems. The size of such a modular system depended on the number of velocity components to be measured.

The use of optical fibers along with multifunctional optical elements has made the systems more compact, flexible, and easier to make measurements. The laser, optics to generate the necessary number of beams (typically, one pair per component of velocity to be measured), photodetectors, and electronics can be isolated from the measurement location [40]. The fibers carrying the laser beams thus generated are arranged in the probe to achieve the desired beam geometry for measuring the velocity components. Hence, flow field mapping is achieved by moving only the fiber-optic probes, while keeping the rest of the system stationary. To achieve maximum power transmission efficiency and beam quality, special single-mode, polarization-preserving optical fibers along with precision couplers are used. In most cases, these fiber probes also have a receiving system and a separate fiber (multimode) to collect (in back scatter) the scattered light and carry that back to the photodetector system.

A schematic arrangement of a fiber probe system to measure one component of velocity is shown in Figure 10.27. In flow measurement applications, LDV systems using these types of fiber-optic probes have largely replaced the earlier modular systems.

The best way to make three-component of velocity measurements is to use an arrangement using two probes [13]. In this case, the optical axis of the system to measure the third component of velocity (u_x) is perpendicular to that of the two-component system. Unfortunately, access and/or traversing difficulties often make this arrangement impractical or less attractive. In most practical situations, the angle between the two probes is selected to be less than 90°. Such an arrangement using two fiber-optic probes to measure three components of velocity simultaneously is shown in Figure 10.28.

Signal Processing

Nature of the Signal

Every time a particle passes through the measuring region, the scattered light signal level (Figure 10.29) suddenly increases ("burst"). The characteristics of the burst signal are (1) amplitude in the burst not constant, (2) lasts for only a short duration, (3) amplitude varies from burst to burst, (4) presence of noise, (5) high frequency, and (6) random arrival.

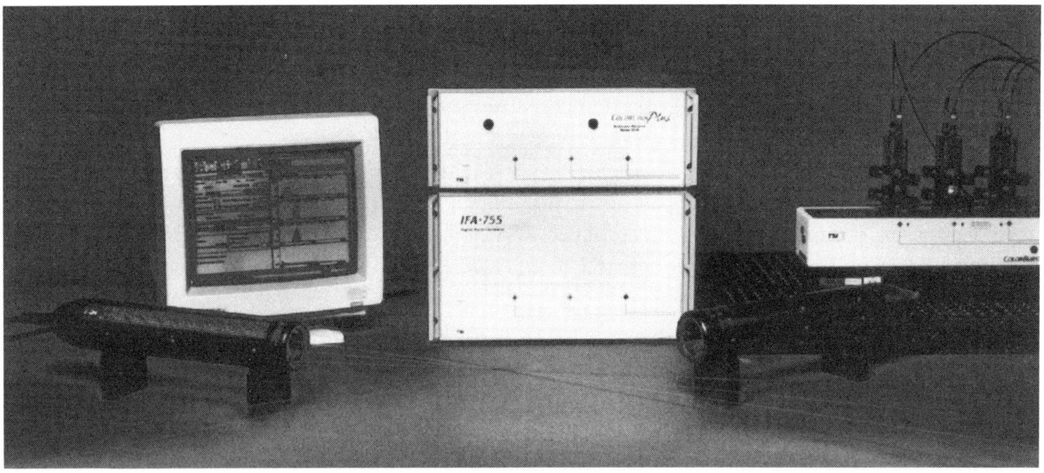

FIGURE 10.28 Three-component LDV system with fiber-optic probes.

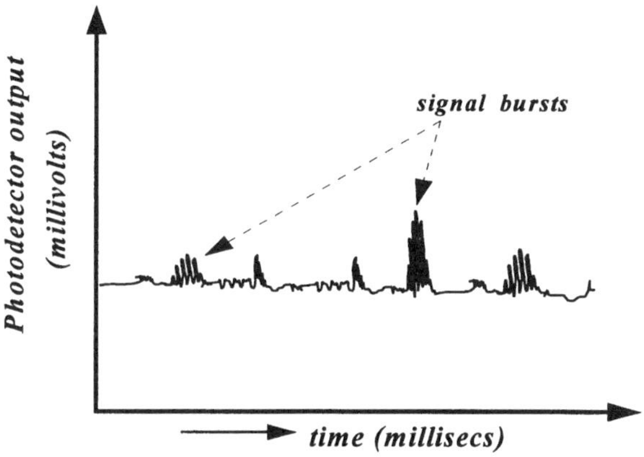

FIGURE 10.29 Time history of the photodetector signal.

The primary task of the signal processor is to extract the frequency information from the burst signal generated by a particle passing through the measuring volume, and provide an analog or digital output proportional to the frequency of the signal. The unique nature of the signals demands the use of a special signal processing system to extract the velocity information.

A variety of techniques has been used for processing Doppler signals. Signal processors have been based on spectrum analysis, frequency tracking, photon correlation, frequency counting, Fourier transform, and autocorrelation principles. The evolution of the signal processing techniques shows the improvement in their ability to handle more difficult measuring situations (generally implies noisier signals), give more accurate measurements, and have higher processing speed.

The traditional instrument to measure signal frequency is a spectrum analyzer. The need to measure individual particle velocities, and to obtain the time history and other properties of the flow, has eliminated the use of "standard" spectrum analyzers [29]. The "tracker" can be thought of as a fixed bandwidth filter that "tracked" the Doppler frequency as the fluid velocity changed. This technique of "tracking flow" worked quite well at modest velocities and where the concentration of scattering centers was high enough to provide an essentially continuous signal. However, too frequently these conditions could not be met in the flows of most interest [29].

When the scattered light level is very low, the photodetector output reveals the presence of the individual photon pulses. By correlating the actual photon pulses from a wide bandwidth photodetector, the photon correlator was designed to work in situations where the attainable signal intensity was very low (low SNR). However, as normally used, it could not provide the velocity of individual particles but only the averaged quantities, such as mean and turbulence intensities.

The "counter" type processor was developed next, and basically measured the time for a certain number (typically, eight) of cycles of the Doppler signal. Although it measured the velocity of individual particles, it depended on the careful setting of amplifier gain and, especially, threshold levels to discriminate between background noise and burst signals. Counters were the processors of choice for many years, and excellent measurements were obtained [42]. However, the reliance on user skill, the difficulty in handling low SNR signals, the possibility of getting erroneous measurements, the inclination to ignore signals from small particles, and the desire to make measurements close to surfaces and in complex flows led to the need for a better signal processor.

Digital Signal Processing

The latest development in signal processing is in the area of digital signal processors. Recent developments in high-speed digital signal processing now permit the use of these techniques to extract the frequency from individual Doppler bursts fast enough to actually follow the flow when the seeding concentration is adequate in a wide range of measurement situations. By digitizing the incoming signal and using the Fourier transform [43] or autocorrelation [44] algorithms, these new digital processors can work with lower SNR signals (than counters), while generally avoiding erroneous data outputs. While instruments using these techniques are certainly not new, standard instruments were not designed to make rapid individual measurements on the noisy, short-duration burst signals with varying amplitudes that are typical of Doppler bursts.

Because the flow velocity and hence the signal frequency varies from one burst to the next, the sampling rate needs to be varied accordingly. And because the signal frequency is not known *a priori*, the ability to optimally sample the signal has been one of the most important challenges in digital signal processing. In one of the digital signal processors, the question of deciding the sample rate is addressed by a burst detector that uses SNR to identify the presence of a signal [44]. In addition, the burst detector provides the duration and an approximate estimate of the frequency of each of the burst signals. This frequency estimate is used to select the output of the sampler (from the many samplers) that had sampled the burst signal at the optimum rate. Besides optimizing the sample rate for each burst, the burst detector information is also used to focus on and process the middle portion of the burst where the SNR is maximum. These optimization schemes, followed by digital signal processing, provide an accurate digital output that is proportional to the signal frequency, and hence the fluid velocity.

Seeding and Other Aspects

The performance of an LDV system can be significantly improved by optimizing the source of the signal, the scattering particle. The first reaction of many experimentalists is to rely on the particles naturally present in the flow. There are a few situations (e.g., LDV systems operating in forward scatter to measure water or liquid flows) where the particles naturally present in the flow are sufficient in number and size to provide good signal quality and hence good measurements. In most flow measurement situations, particles are added to the flow (generally referred to as seeding the flow) to obtain an adequate number of suitable scatterers. Use of a proper particle can result in orders of magnitude increase in signal quality (SNR), and hence can have greater impact on signal quality than the modification of any other component in the LDV system. Ideally, the seed particles should be naturally buoyant in the fluid, provide adequate scattered light intensity, have large enough number concentration, and have uniform properties from particle to particle. While this ideal is difficult to achieve, adequate particle sources and distribution systems have been developed [29, 45–47].

LDV measurements of internal flows such as in channels, pipes and combustion chambers result in the laser beams (as well as the scattered light) going through transparent walls or "windows." In many cases, the window is flat and, hence, the effect of light refraction can be a simple displacement of the measuring region. In the case of internal flows with curved walls, each beam can refract by different amounts and the location of the measuring region needs to be carefully estimated [48]. For internal flows in models with complex geometries, the beam access needs to be carefully selected so that the beams do cross inside. Further, to make measurements close to the wall in an internal flow, the refraction effect of the wall material on the beam path needs to be minimized. One of the approaches is to use a liquid [49] that has the same refractive index as that of the wall material.

Data Analysis

The flow velocity is "sampled" by the particle passing through the measuring volume, and the velocity measurement is obtained only when the Doppler signal, created by the particle, is processed and output as a data point by the signal processor. While averaging the measurements to get, for example, mean velocity would seem reasonable, this method gives the wrong answer. This arises from the fact that the number of particles going through the measuring region per unit time is higher at high velocities than at low velocities. In effect, there is a correlation between the measured quantity (velocity) and the sampling process (particle arrival). Hence, a simple average of the data points will bias the mean value (and other statistical parameters) toward the high-velocity end and is referred to as *velocity bias* [50]. The magnitude of the bias error depends on the magnitude of the velocity variations about the mean. If the variations in velocity are sufficiently small, the error might not be significant.

If the actual data rate is so high that the output data is essentially able to characterize the flow (time history), then the output can be sampled at uniform time increments. This is similar to the procedure normally used for sampling a continuous analog signal using an ADC. This will give the proper value for both the mean and the variance when the data rate is sufficiently high compared to the rates based on the Taylor microscale for the temporal variation of velocity. This is referred to as a high data density situation [29].

In many actual measurement situations, the data rate is not high enough (low data density) to actually characterize the flow. Here, sampling the output of the signal processor at uniform time increments will not work because the probability of getting an updated velocity (new data point) is higher at high velocity than at low velocity (velocity bias). The solution to the velocity bias problem is to weight the individual measurements with a factor inversely proportional to the probability of making the measurement.

$$\overline{U} = \frac{\Sigma u_j \tau_{Bj}}{\Sigma \tau_{Bj}} \tag{10.71}$$

where u_j = Velocity of particle j
τ_{Bj} = Transit time for particle j

Similar procedures can be used to obtain unbiased estimators for variance and other statistical properties of the flow [29]. Modern signal processors provide the residence time and the time between data points along with the velocity data. A comparison of some of the different approaches to do bias correction has been presented by Gould and Loseke [51]. Some of the other types of biases associated with LDV have been summarized by Edwards [52].

A variety of techniques to obtain spectral information of the flow velocity from the random data output of the signal processors have been tried. The goal of all these techniques has been to get accurate and unbiased spectral information to as high a frequency as possible. Direct spectral estimation of the digital output of the processors [53] exhibit the spectrum estimates at high frequency to be less reliable. The "slotting" technique [54, 55] of estimating the autocorrelation of the (random) velocity data followed by Fourier transform continues to be attractive from a computational standpoint. To obtain reliable spectrum estimates at high frequencies, a variety of methods aimed at interpolation of measured velocity values have been attempted. These are generally referred to as *data* or *signal reconstruction* techniques. A review article [37] emphasizes the need to correct for velocity bias in the spectrum estimates. It also covers some of the recent reconstruction algorithms and points out the difficulties in coming up with a general-purpose approach.

Extension to Particle Sizing

In LDV, the frequency of the scattered light signal provides the velocity of the scatterer. Processing the scattered light to get information about the scatterer other than velocity has always been a topic of great interest in flow and particle diagnostics. One of the most promising developments is the extension of the LDV technique to measure the surface curvature and, hence, the diameter of a spherical scatterer [22]. This approach (limited to spherical particles) uses the phase information of the scattered light signal to extract the size information. To obtain a unique and, preferably, monotonic relation between phase of the signal and the size of the particle, the orientation and the geometry (aperture) of the scattered light collection system needs to be carefully chosen. In the following, unless otherwise mentioned, the particles are assumed to be spherical.

The light scattered by a particle, generally, contains contributions from the different scattering mechanisms — reflection, refraction, diffraction, and internal reflection(s). It can be shown that, by selecting the position of the scattered light collection set-up, contributions from one scattering mechanism can be made dominant over the others. The aim in phase Doppler measurements is to have the orientation of the receiver system such that the scattered light collected is from one dominant scattering mechanism.

The popularity of the technique is evidenced by its widespread use for measuring particle diameter and velocity in a large number of applications, especially in the field of liquid sprays [56]. The technique has also been used in diagnosing flow fields associated with combustion, cavitation, manufacturing processes, and other two-phase flows.

Phase Doppler System: Principle

The phase Doppler approach, outlined as an extension to an LDV system, was first proposed by Durst and Zare [57] to measure velocity and size of spherical particles. The first practical phase Doppler systems using a single receiver were proposed by Bachalo and Houser [22].

A schematic arrangement of a phase Doppler system is shown in Figure 10.30(*a*). This shows a receiver system arrangement that collects, separates, and focuses the scattered light onto multiple photodetectors. In general, the receiving system aperture is divided into three parts and the scattered light collected through these are focused into three separate photodetectors. For simplicity, in Figure 10.30(*a*), the output of two detectors are shown. The different spatial locations of the detectors (receiving apertures) results in the signals received by each detector having a slightly different phase. In general, the difference in phase between the signals from the detectors is used to obtain the particle diameter whereas the signal frequency provides the velocity of the particle.

Fringe Model Explanation

The fringe model provides an easy and straightforward approach to arrive at the expressions for Doppler frequency and phase shift created by a particle going through the measuring volume. As the particle moves through the fringes in the measuring volume, it scatters the fringe pattern (Figure 10.30(*b*)). The phase shift in the signals can be examined by looking at the scattered fringe pattern. If the particle acts like a spherical mirror (dominant reflection) or a spherical lens (dominant refraction), it projects fringes from the measuring volume into space all around as diverging bands of bright and dark light, known as *scattered fringes.* Scattered fringes as seen on a screen placed in front of the receivers are shown in Figure 10.30(*b*). The spacing between the scattered fringes at the plane of the receiver is s_f. The receiver system shown in Figure 10.30(*b*) shows two apertures. The distance between (separation) the centroids

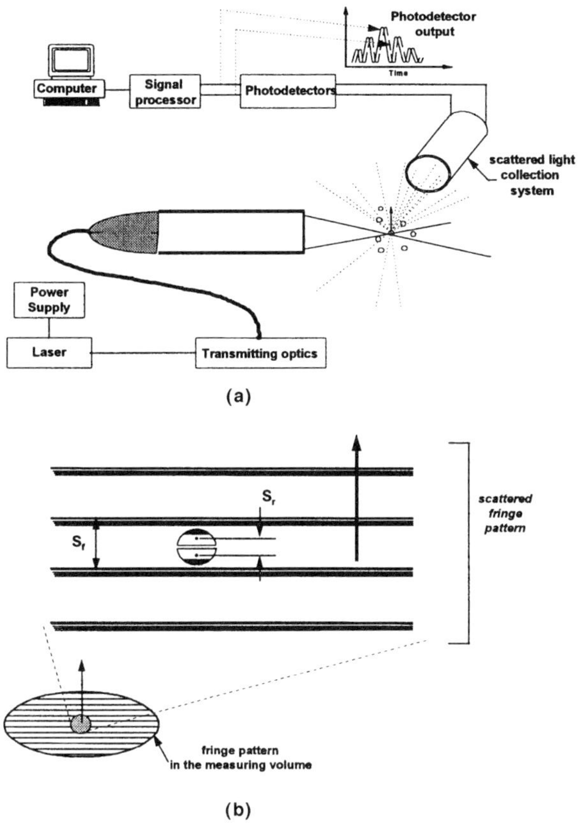

FIGURE 10.30 (*a*) Phase/Doppler system: schematic. (*b*) phase/Doppler System: fringe model.

of the two receiving apertures is s_r. Scattered fringes move across the receivers as the particle moves in the measuring volume, generating temporally fluctuating signals. The two photodetector output signals are shifted in phase by s_r/s_f times 360° [31]. Large particles create a scattered fringe pattern with a smaller fringe spacing (compared to that for small particles), i.e., particle diameter is inversely proportional to s_f, while s_f is inversely proportional to phase difference. Thus, the fringe model shows the particle diameter to be directly proportional to the phase difference. It can also be seen that the sensitivity (degrees of phase difference per micrometer) of the phase Doppler system can be increased by increasing the separation (s_r) between the detectors.

The phase Doppler system shown above measures the phase difference between two detectors in the receiver system to obtain particle diameter. This brings in the limitation that the maximum value of phase that could be measured is 2π. A three-detector arrangement in the receiver system is used to overcome this 2π ambiguity. Figure 10.31 shows the three-detector (aperture) arrangement. Scattered light collected through apertures 1, 2, and 3 are focused into detectors 1, 2, and 3. Φ_{13} is the phase difference between the detectors 1 and 3 and provides the higher phase sensitivity because of their greater separation compared to detectors 1 and 2. As Φ_{13} exceeds 2π, the value of Φ_{12} is below 360° and is used to keep track of Φ_{13}. It should be noted that the simplified approach in terms of geometrical scattering provides a linear relationship between the phase difference and diameter of the particle.

It has been pointed out that significant errors in measured size can occur due to trajectory-dependent scattering [58]. These errors could be minimized by choosing the appropriate optical configuration of the phase Doppler system [59]. An intensity-based validation technique has also been proposed to reduce the errors [60].

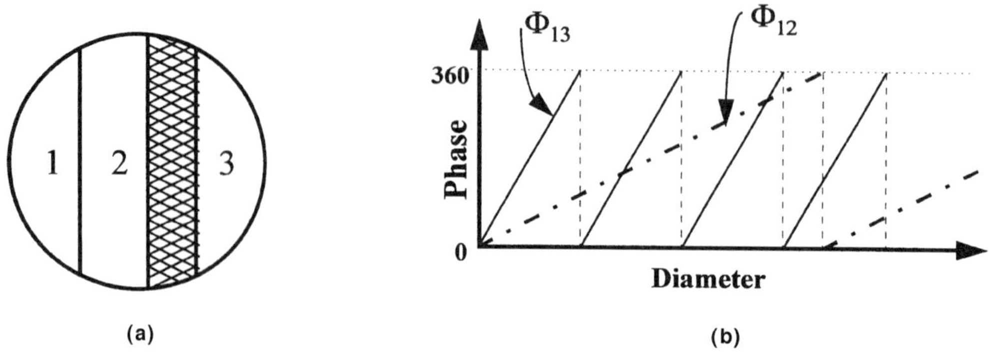

FIGURE 10.31 (*a*) Three-detector configuration; (*b*) phase–diameter relationship.

To explore the fundamental physical limits on applicability of the Phase Doppler technique, a rigorous model based on the electromagnetic theory of light has been developed. Computational results based on Mie scattering and comparison with and limitations of the geometric scattering approach have also been outlined by Naqwi and Durst [61]. These provided a systematic approach to develop innovative solutions to particle sizing problems. A new approach (PLANAR arrangement) to achieve high measurement resolution provided the ability to extend the measurement range to submicrometer particles. The Adaptive Phase Doppler Velocimeter (APV) system [59] that incorporates this layout uses a scattered light collection system that employs independent receivers. In the APV system, the separation between the detectors is selectable and is not dependent on the numerical aperture of the receiving system. Such a system was used for measuring submicrometer droplets in an electrospray [62]. By integrating a phase Doppler velocimeter system with a *rainbow refractometer* system, the velocity, size, and the refractive index of a droplet could be determined [63].

The velocity and diameter information is obtained by processing the photodetector output signals. The frequency of the photodetector output signal provides the velocity information. In general, the signal processing system for velocity measurements is expanded to measure the phase difference between two photodetector signals. The digital signal processing approaches described earlier have been complimented by the addition of accurate phase measurement techniques [64, 65].

Although the phase Doppler technique is limited to spherical particles, there has always been an interest in extending the technique to nonspherical particles. In the past, symmetry checks [66] and other similar techniques have been used to check on the sphericity of particles. An equivalent sphere approach has been used to describe these nonspherical particles. Sizing irregular particles is a more complex problem because the local radius of curvature concept is not meaningful in these cases. An innovative stochastic modeling approach has been used to study irregular particles using a phase Doppler system [67].

Conclusion

LDV has become the preferred technique for measuring flow velocity in a wide range of applications. The ability to measure noninvasively the velocity, without calibration, of any transparent flowing fluid has made it attractive for measuring almost any type of flow. Velocity measurement of moving surfaces by LDV is used to monitor and control industrial processes. Use of laser diodes, fiber optics, and advances in signal processing and data analysis are reducing both the cost and complexity of measuring systems. The extension of LDV to the phase Doppler technique provides an attractive, noncontact method for measuring size and velocity of spherical particles. Recent developments in the phase Doppler technique have generated a method to size submicrometer particles as well. These ideas have been extended to examine irregular particles also.

Acknowledgments

The input and comments from Dr. L. M. Fingerson and Dr. A. Naqwi of TSI Inc. have been extremely valuable in the preparation of this chapter section. The author is sincerely grateful to them for the help.

References

1. D. Niccum, A new tool for fiber spinning process control and diagnostics, *Int. Fiber J.*, 10(1), 48-57, 1995.
2. R. Schodl, On the extension of the range of applicability of LDA by means of a the laser-dual-focus (L-2-F) technique, *The Accuracy of Flow Measurements by Laser Doppler Methods*, Skovulunde, Denmark: Dantec Measurement Technology, 1976, 480-489.
3. R. J. Adrian, Particle imaging techniques for experimental fluid mechanics, *Annu. Rev. Fluid Mech.*, 23, 261-304, 1991.
4. I. Grant, *Selected Papers in Particle Image Velocimetry*, SPIE Milestone Series, MS 99, Bellingham, WA: SPIE Optical Engineering Press, 1994.
5. R. J. Adrian, *Bibliography of Particle Velocimetry Using Imaging Methods: 1917–1995*, TAM Report, University of Illinois Urbana-Champaign, Produced and distributed in cooperation with TSI Inc., March 1996. (Also available in electronic format.)
6. W. T. Lai, Particle image velocimetry: a new approach to experimental fluid research, in *Three Dimensional Velocity and Vorticity Measuring and Image Analysis Techniques*, Th. Dracos (ed.), Boston: Kluwer Academic, 1996, 61-92.
7. M. M. Koochesfahani, R. K. Cohn, C. P. Gendrich, and D. G. Nocera, Molecular tagging diagnostics for the study of kinematics and mixing in liquid phase flows, *8th Int. Symp. Appl. Laser Techniques Fluid Mechanics*, Lisbon, 1996.
8. H. Komine, S. J. Brosnan, A. B. Litton, and E. A. Stappaerts, Real time Doppler global velocimetry, *AIAA 29th Aerospace Sciences Meeting*, Paper No. AIAA-91-0337, January 1991.
9. R. L. McKenzie, Measurement capabilities of planar Doppler velocimetry using pulsed lasers, *Appl. Opt.*, 35, 948-964, 1996.
10. C. Berner, Supersonic base flow investigation over axisymmetric bodies, *Proc. 5th Inc. Conf. Laser Anemometry and Applications*, Netherlands, SPIE, 2052, 1993.
11. K. Jaffri, H. G. Hascher, M. Novak, K. Lee, H. Schock, M. Bonne, and P. Keller, Tumble and Swirl Quantification within a Four-valve SI Engine Cylinder Based on 3D LDV Measurements, SAE Paper No. 970792, Feb. 1997.
12. G. G. Podboy and M. J. Krupar, Laser Velocimeter Measurements of the Flow Field Generated by a Forward-Swept Propfan During Flutter, NASA Technical Memorandum 106195, 1993.
13. Y. O. Han, J. G. Leishman, and A. J. Coyne, Measurements of the velocity and turbulence structure of a rotor tip vortex, *AIAA J.*, 35, 477-485, 1997.
14. T. Mathur and J. C. Dutton, Velocity and turbulence measurements in a supersonic base flow with mass bleed, *AIAA J.*, 34, 1153-1159, 1996.
15. E. J. Johnson, P. V. Hyer, P. W. Culotta, and I. O. Clark, Laser velocimetry in nonisothermal CVD systems, *Proc. 4th Int. Conf. Laser Anemometry*, Cleveland, OH, August 1991.
16. R. W. Dibble, V. Hartmann, R. W. Schefer, and W. Kollmann, Conditional sampling of velocity and scalars in turbulent flames using simultaneous LDV-Raman scattering, *Exp. Fluids*, 5, 103-113, 1987.
17. D. V. Srikantiah and W. W. Wilson, Detection of a pulsed flow in an MHD environment by laser velocimetry, *Exp. Fluids*, 6, 500-503, 1988.
18. P. O. Witze, Velocity measurements in end-gas region during homogeneous-charge combustion in a spark ignition engine, *Laser Techniques and Applications in Fluid Mechanics*, Adrian, et al. (eds.), Lisbon: Ladoan, 1992, 518-534.

19. G. L. Morrison, M. C. Johnson, R. E. DeOtte, H. D. Thames, and B. J. Wiedner, An experimental technique for performing 3D LDA measurements inside whirling annular seals, *Flow Meas. Instrum.*, 5, 43-49, 1994.
20. F. Durst, A. Melling, and J. H. Whitelaw, *Principles and Practice of Laser Doppler Anemometry*, 2nd ed., New York: Academic Press, 1981.
21. R. J. Adrian (ed.), *Selected Papers on Laser Doppler Velocimetry*, SPIE Milestone Series, MS 78, Bellingham, WA: SPIE Optical Engineering Press, 1993.
22. W. D. Bachalo and M. J. Houser, Phase Doppler spray analyzer for simultaneous measurements of drop size and velocity distributions, *Opt. Eng.*, 23, 583-590, 1984.
23. Y. Yeh and H. Z. Cummins, Localized fluid flow measurements with an He-Ne laser spectrometer, *Appl. Phys. Lett.*, 4, 176-178, 1964.
24. C. M. Penney, Differential Doppler velocity measurements, *IEEE J. Quantum Electron.*, QE-5, 318, 1969.
25. L. M. Fingerson and P. Freymuth, Thermal anemometers, in *Fluid Mechanics Measurements*, R. J. Goldstein (ed.), New York: Hemisphere, 1983, 99-154.
26. G. Smeets and A. George, Michelson spectrometer for instantaneous Doppler velocity measurements, *J. Phys. E: Sci. Instrum.*, 14, 838-845, 1981.
27. D. Brayton, Small particle signal characteristics of a dual scatter laser velocimeter, *J. Appl. Opt.*, 13, 2346-2351, 1974.
28. F. Durst and W. H. Stevenson, Moiré patterns to visually model laser Doppler signals, *The Accuracy of Flow Measurements by Laser Doppler Methods*, Skovulunde, Denmark: Dantec Measurement Technology, 1976, 183-205.
29. R. J. Adrian, Laser Velocimetry, in *Fluid Mechanics Measurements*, R. J. Goldstein (ed.), New York: Hemisphere, 1983, 155-240.
30. R. J. Adrian and K. L. Orloff, Laser anemometer signal: visibility characteristics and application to particle sizing, *Appl. Opt.*, 16, 677-684, 1977.
31. L. M. Fingerson, R. J. Adrian, R. K. Menon, S. L. Kaufman, and A. Naqwi, Data Analysis, Laser Doppler Velocimetry and Particle Image Velocimetry, TSI Short Course Text, TSI Inc., St. Paul, MN, 1993.
32. H. Kogelnik and T. Li, Laser beams and resonators, *Appl. Opt.*, 5, 1550-1567, 1966.
33. M. C. Whiffen, Polar response of an LV measurement volume, *Minnesota Symp. Laser Anemometry*, University of Minnesota, 1975.
34. C. Tropea, A practical aid for choosing the shift frequency in LDA, *Exp. Fluids*, 4, 79-80, 1986.
35. M. K. Mazumder, Laser Doppler velocity measurement without directional ambiguity by using frequency shifted incident beams, *Appl. Phys. Lett.*, 16, 462-464,1970.
36. H. Muller, V. Tobben, V. Arndt, V. Strunck, H. Wang, R. Kramer, and D. Dopheide, New frequency shift techniques in laser anemometry using tunable semiconductor lasers and solid state lasers, *Proc. 2nd Int. Conf. Fluid Dynamic Measurement Applications*, Beijing, Oct. 1994, 3-19.
37. E. Muller, H. Nobach, and C. Tropea, LDA signal reconstruction: application to moment and spectral estimation, *Proc. 7th Int. Symp. Applications Laser Techniques Fluid Mechanics*, Lisbon, 1994b.
38. R. J. Adrian and W. L. Early, Evaluation of laser Doppler velocimeter performance using Mie scattering theory, *Proc. Minnesota Symp. Laser Anemometry*, University of Minnesota, 1975, 429-454.
39. G. Grehan, G. Gouesbet, A. Naqwi, and F. Durst, Trajectory ambiguities in phase Doppler systems: study of a new forward and a near-backward geometry, *Part. Part. Syst. Charact.*, 11, 133-144, 1994.
40. D. J. Fry, Model submarine wake survey using internal LDV probes, *Proc. ASME Fluids Engineering Meeting*, T. T. Huang, J. Turner, M. Kawahashi, and M. V. Otugen (eds.), FED- Vol. 229, August 1995, 159-170.
41. P. A. Chevrin, H. L. Petrie, and S. Deutsch, Accuracy of a three-component laser Doppler velocimeter system using a single lens approach, *J. Fluids Eng.*, 115, 142-147, 1993.

42. R. I. Karlsson and T. G. Johansson, LDV measurements of higher order moments of velocity fluctuations in a turbulent boundary layer, in *Laser Anemometry in Fluid Mechanics III*, Ladoan-Instituto Superior Technico, 1096 Lisbon Codex, Portugal, 1988, 273-289.
43. K. M. Ibrahim, G. D. Werthimer, and W. D. Bachalo, Signal processing considerations for laser Doppler and phase Doppler applications, *Proc. 5th Int. Symp. Applications Laser Techniques Fluid Mechanics*, Lisbon, 1990.
44. L. Jenson, LDV digital signal processor based on Autocorrelation, *Proc. 6th Int. Symp. Applications Laser Techniques Fluid Mechanics*, Lisbon, 1992.
45. W. W. Hunter and C. E. Nichols (compilers), Wind Tunnel Seeding Systems for Laser Velocimeters, NASA Conference Publication 2393, 1985.
46. A. Melling, Seeding gas flows for laser anemometry, AGARD CP-339, 1986, 8-1–8-11.
47. R. K. Menon and W. T. Lai, Key considerations in the selection of seed particles for LDV measurements, *Proc. 4th Int. Conf. Laser Anemometry*, Cleveland, OH, August 1991.
48. M. L. Lowe and P. H. Kutt, Refraction through cylindrical tubes, *Exp. Fluids*, 13, 315-320, 1992.
49. R. Budwig, Refractive index matching methods for liquid flow investigations, *Exp. Fluids*, 17, 350-355, 1994.
50. D. K. McLaughlin and W. G. Tiederman, Biasing correction for individual realization of laser anemometer measurements in turbulent flows, *Phys. Fluids*, 16, 2082-2088, 1973.
51. R. D. Gould and K. W. Loseke, A comparison of four velocity bias correction techniques in laser Doppler velocimetry, *J. Fluids Eng.*, 115, 508–514, 1993.
52. R. V. Edwards (ed.), Report on the special panel on statistical particle bias problems in laser anemometry, *J. Fluids Eng.*, 109, 89-93, 1987.
53. J. B. Roberts, J. Downie, and M. Gaster, Spectral analysis of signals from a laser Doppler anemometer operating in the burst mode, *J. Physics, E: Sci. Instrum.*, 13, 977-981, 1980.
54. W. T. Mayo, Spectrum measurements with laser velocimeters, *Proc. Dynamic Flow Conf. Dynamic Measurements in Unsteady Flows*, DISA Electronik A/S, Denmark, 1978, 851-868.
55. H. L. Petrie, Reduction of noise effects on power spectrum estimates using the discretized lag product method, *ASME Fluids Engineering Meeting*, FED-229, 139-144, 1995.
56. W. D. Bachalo, A. Brena de la Rosa, and S. V. Sankar, Diagnostics for fuel spray characterization, *Combustion Measurements*, N. Chigier (ed.), New York: Hemisphere, 1991, chap. 7.
57. F. Durst and M. Zare, Laser Doppler measurements in two-phase flows, *The Accuracy of Flow Measurements by Laser Doppler Methods*, Skovulunde, Denmark: Dantec Measurement Technology, 1976, 480-489.
58. M. Saffman, The use of polarized light for optical particle sizing, *Laser Anemometry in Fluid Mechanics III*, Adrian, et al. (eds.), Lisbon: Ladoan, 1988, 387-398.
59. A. Naqwi, Innovative phase Doppler systems and their applications, *Part. Part. Syst. Charact.*, 11, 7-21, 1994.
60. S. V. Sankar, D. A. Robart, and W. D. Bachalo, An adaptive intensity validation technique for minimizing trajectory dependent scattering errors in phase Doppler interferometry, *Proc. 4th Int. Congr. Optical Particle Sizing*, Nuremberg, Germany, March 1995.
61. A. Naqwi and F. Durst, Light scattering applied to LDA and PDA measurements. 2. Computational results and their discussion, *Part. Part. Syst. Charact.*, 9, 66-80, 1992.
62. A. Naqwi, *In-situ* measurement of submicron droplets in electrosprays using a planar phase Doppler system, *J. Aerosol Sci.*, 25, 1201-1211, 1994.
63. S. V. Sankar, D. H. Buermann, D. A. Robart, and W. D. Bachalo, An advanced rainbow signal processor for improved accuracy of droplet temperature measurements in spray flames, *Proc. 8th Int. Symp. Applications Laser Techniques Fluid Mechanics*, Lisbon, 1996.
64. J. Evenstad, A. Naqwi, and R. Menon, A device for phase shift measurement in an advanced phase Doppler velocimeter, *Proc. 8th Int. Symp. Applications Laser Techniques Fluid Mechanics*, Lisbon, 1996.

65. K. M. Ibrahim and W. D. Bachalo, A novel architecture for real-time phase measurement, *Proc. 8th Int. Symp. Applications of Laser Techniques to Fluid Mechanics,* Lisbon, 1996.
66. M. Saffman, P. Buchave, and H. Tanger, Simultaneous measurement of size, concentration and velocity of spherical particles by a laser Doppler method, in *Laser Anemometry in Fluid Mechanics II,* R. J. Adrian, et al. (eds.), Lisbon: Ladoan, 1986, 85-104.
67. A. Naqwi, Sizing of irregular particles using a phase Doppler system, *Proc. ASME Heat Transfer and Fluid Engineering Divisions,* FED-Vol. 233, 1995.

Further Information

C. A. Greated and T. S. Durrani, *Laser Systems and Flow Measurement,* New York: Plenum, 1977.

L. E. Drain, *The Laser Doppler Technique,* New York: John Wiley & Sons, 1980.

Proc. Int. Symp. (1 to 8) *on Applications of Laser Techniques to Fluid Mechanics,* Lisbon, Portugal, 1982, 1984, 1986, 1988, 1990, 1992, 1994, 1996.

P. Buchave, W. K. George, and J. L. Lumley, The measurement of turbulence with the laser Doppler anemometer, *Annu. Rev. Fluid Mech.,* 11, 443-504, 1979.

Proc. 5th Int. Conf. Laser Anemometry and Applications, Netherlands, SPIE, Vol. 2052, 1993.

Proc. ASME Fluids Engineering Meeting, T. T. Huang, J. Turner, M. Kawahashi, and M. V. Otugen, eds., FED- Vol. 229, August 1995.

L. H. Benedict and R. D. Gould, Experiences using the Kalman reconstruction for enhanced power spectrum estimates, *Proc. ASME Fluids Engineering Meeting,* T. T. Huang, J. Turner, M. Kawahashi, and M. V. Otugen (eds.), FED 229, 1-8, 1995.

D. Dopheide, M. Faber, G. Reim, and G. Taux, Laser and avalanche diodes for velocity measurement by laser Doppler anemometry, *Exp. Fluids,* 6, 289-297, 1988.

F. Durst, R. Muller, and A. Naqwi, Measurement accuracy of semiconductor LDA systems, *Exp. Fluids,* 10, 125-137, 1990.

A. Naqwi and F. Durst, Light scattering applied to LDA and PDA measurements. 1. Theory and numerical treatments, *Particle and Particle System Characterization,* 8, 245-258, 1991.

11

Viscosity Measurement

G.E. Leblanc
McMaster University

R.A. Secco
The University of Western Ontario

M. Kostic
Northern Illinois University

11.1 Shear Viscosity

An important mechanical property of fluids is *viscosity.* Physical systems and applications as diverse as fluid flow in pipes, the flow of blood, lubrication of engine parts, the dynamics of raindrops, volcanic eruptions, planetary and stellar magnetic field generation, to name just a few, all involve fluid flow and are controlled to some degree by fluid viscosity. *Viscosity* is defined as the internal friction of a fluid. The microscopic nature of internal friction in a fluid is analogous to the macroscopic concept of mechanical friction in the system of an object moving on a stationary planar surface. Energy must be supplied (1) to overcome the inertial state of the interlocked object and plane caused by surface roughness, and (2) to initiate and sustain motion of the object over the plane. In a fluid, energy must be supplied (1) to create viscous flow units by breaking bonds between atoms and molecules, and (2) to cause the flow units to move relative to one another. The resistance of a fluid to the creation and motion of flow units is due to the viscosity of the fluid, which only manifests itself when motion in the fluid is set up. Since viscosity involves the transport of mass with a certain velocity, the viscous response is called a *momentum transport process.* The velocity of flow units within the fluid will vary, depending on location. Consider a liquid between two closely spaced parallel plates as shown in Figure 11.1. A force, *F,* applied to the top plate causes the fluid adjacent to the upper plate to be dragged in the direction of *F.* The applied force is communicated to neighboring layers of fluid below, each coupled to the driving layer above, but with diminishing magnitude. This results in the progressive decrease in velocity of each fluid layer, as shown by the decreasing velocity vector in Figure 11.1, away from the upper plate. In this system, the applied force is called a *shear* (when applied over an area it is called a *shear stress*), and the resulting deformation rate of the fluid, as illustrated by the *velocity gradient* dU_x/dz, is called the *shear strain rate,* $\dot{\gamma}_{zx}$. The mathematical expression describing the viscous response of the system to the shear stress is simply:

$$\tau_{zx} = \frac{\eta dU_x}{dz} = \eta\dot{\gamma}_{zx} \tag{11.1}$$

where τ_{zx}, the shear stress, is the force per unit area exerted on the upper plate in the *x*-direction (and hence is equal to the force per unit area exerted by the fluid on the upper plate in the *x*-direction under the assumption of a no-slip boundary layer at the fluid–upper plate interface); dU_x/dz is the gradient of the *x*-velocity in the *z*-direction in the fluid; and η is the *coefficient of viscosity.* In this case, because one is concerned with a shear force that produces the fluid motion, η is more specifically called the *shear*

0-8493-0047-9/00/$0.00+$.50

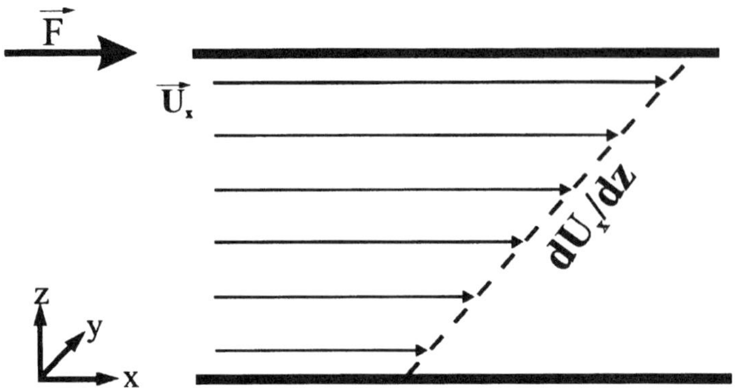

FIGURE 11.1 System for defining Newtonian viscosity. When the upper plate is subjected to a force, the fluid between the plates is dragged in the direction of the force with a velocity of each layer that diminishes away from the upper plate. The reducing velocity eventually reaches zero at the lower plate boundary.

dynamic viscosity. In fluid mechanics, diffusion of momentum is a more useful description of viscosity where the motion of a fluid is considered without reference to force. This type of viscosity is called the *kinematic viscosity,* ν, and is derived by dividing dynamic viscosity by ρ, the mass density:

$$\nu = \frac{\eta}{\rho} \tag{11.2}$$

The definition of viscosity by Equation 11.1 is valid only for *laminar* (i.e., layered or sheet-like) or streamline flow as depicted in Figure 11.1, and it refers to the molecular viscosity or *intrinsic viscosity.* The molecular viscosity is a property of the material that depends microscopically on bond strengths, and is characterized macroscopically as the fluid's resistance to flow. When the flow is turbulent, the diffusion of momentum is comprised of viscous contributions from the motion, sometimes called the *eddy viscosity,* in addition to the intrinsic viscosity. Viscosities of turbulent systems can be as high as 10^6 times greater than viscosities of laminar systems, depending on the Reynolds number.

Molecular viscosity is separated into *shear viscosity* and bulk or *volume viscosity,* η_v, depending on the type of strain involved. Shear viscosity is a measure of resistance to isochoric flow in a shear field, whereas volume viscosity is a measure of resistance to volumetric flow in a three-dimensional stress field. For most liquids, including hydrogen bonded, weakly associated or unassociated, and polymeric liquids as well as liquid metals, $\eta/\eta_v \approx 1$, suggesting that shear and structural viscous mechanisms are closely related [1].

The shear viscosity of most liquids decreases with temperature and increases with pressure, which is opposite to the corresponding responses for gases. An increase in temperature usually causes expansion and a corresponding reduction in liquid bond strength, which in turn reduces the internal friction. Pressure causes a decrease in volume and a corresponding increase in bond strength, which in turn enhances the internal friction. For most situations, including engineering applications, temperature effects dominate the antagonistic effects of pressure. However, in the context of planetary interiors where the effects of pressure cannot be ignored, pressure controls the viscosity to the extent that, depending on composition, it can cause fundamental changes in the molecular structure of the fluid that can result in an anomalous viscosity decrease with increasing pressure [2].

Newtonian and Non-Newtonian Fluids

Equation 11.1 is known as Newton's law of viscosity and it formulates Sir Isaac Newton's definition of the viscous behavior of a class of fluids now called Newtonian fluids.

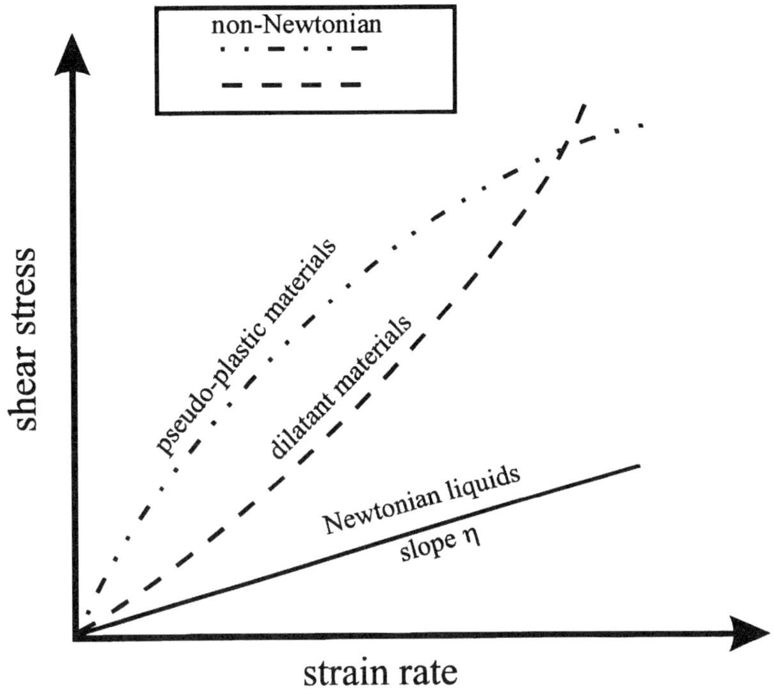

FIGURE 11.2 Flow curves illustrating Newtonian and non-Newtonian fluid behavior.

If the viscosity throughout the fluid is independent of strain rate, then the fluid is said to be a *Newtonian fluid.* The constant of proportionality is called the coefficient of viscosity, and a plot of stress vs. strain rate for Newtonian fluids yields a straight line with a slope of η, as shown by the solid line flow curve in Figure 11.2. Examples of Newtonian fluids are pure, single-phase, unassociated gases, liquids, and solutions of low molecular weight such as water. There is, however, a large group of fluids for which the viscosity is dependent on the strain rate. Such fluids are said to be non-Newtonian fluids and their study is called *rheology.* In differentiating between Newtonian and non-Newtonian behavior, it is helpful to consider the time scale (as well as the normal stress differences and phase shift in dynamic testing) involved in the process of a liquid responding to a shear perturbation. The velocity gradient, dU_x/dz, in the fluid is equal to the shear strain rate, $\dot{\gamma}$, and therefore the time scale related to the applied shear perturbation about the equilibrium state is t_s, where $t_s = \dot{\gamma}^{-1}$. A second time scale, t_r, called the *relaxation time,* characterizes the rate at which the relaxation of the strain in the fluid can be accomplished and is related to the time it takes for a typical flow unit to move a distance equivalent to its mean diameter. For Newtonian water, $t_r \sim 10^{-12}$ s and, because shear rates greater than 10^6 s^{-1} are rare in practice, the time required for adjustment of the shear perturbation in water is much less than the shear perturbation period (i.e., $t_r \ll t_s$). However, for non-Newtonian macromolecular liquids like polymeric liquids, for colloidal and fiber suspensions, and for pastes and emulsions, the long response times of large viscous flow units can easily make $t_r > t_s$. An example of a non-Newtonian fluid is liquid elemental sulfur, in which long chains (polymers) of up to 100,000 sulfur atoms form flow units that are easily entangled, which bind the liquid in a "rigid-like" network. Another example of a well-known non-Newtonian fluid is tomato ketchup.

With reference to Figure 11.2, the more general form of Equation 11.1 also accounts for the nonlinear response. In terms of an initial shear stress required for flow to start, $\tau_{xy}(0)$, an initial linear term in the Newtonian limit of a small range of strain rate, $\dot{\gamma}\partial\tau_{xy}(0)/\partial\gamma$, and a nonlinear term $O(\dot{\gamma}^2)$, the shear stress dependence on strain rate, $\tau_{xy}(\dot{\gamma})$ can be described as:

$$\tau_{xy}(\dot{\gamma}) = \tau_{xy}(0) + \frac{\dot{\gamma}\partial\tau_{xy}(0)}{\partial\dot{\gamma}} + O(\dot{\gamma}^2) \tag{11.3}$$

For a Newtonian fluid, the initial stress at zero shear rate is zero and the nonlinear function $O(\dot{\gamma}^2)$ is zero, so Equation 11.3 reduces to Equation 11.1, since $\partial\tau_{xy}(0)/\partial\dot{\gamma}$ then equals η. For a non-Newtonian fluid, $\tau_{xy}(0)$ may be zero but the nonlinear term $O(\dot{\gamma}^2)$ is nonzero. This characterizes fluids in which shear stress increases disproportionately with strain rate, as shown in the dashed-dotted flow curve in Figure 11.2, or decreases disproportionately with strain rate, as shown in the dashed flow curve in Figure 11.2. The former type of fluid behavior is known as *shear thickening* or dilatancy, and an example is a concentrated solution of sugar in water. The latter, much more common type of fluid behavior, is known as *shear thinning* or pseudo-plasticity; cream, blood, most polymers, and liquid cement are all examples. Both behaviors result from particle or molecular reorientations in the fluid that increase or decreases, respectively, the internal friction to shear. Non-Newtonian behavior can also arise in fluids whose viscosity changes with time of applied shear stress. The viscosity of corn starch and water increases with time duration of stress, and this is called *rheopectic behavior.* Conversely, liquids whose viscosity decreases with time, like nondrip paints, which behave like solids until the stress applied by the paint brush for a sufficiently long time causes them to flow freely, are called *thixotropic fluids.*

Fluid deformation that is not recoverable after removal of the stress is typical of the purely viscous response. The other extreme response to an external stress is purely elastic and is characterized by an equilibrium deformation that is fully recovered on removal of the stress. There are an infinite number of intermediate or combined viscous/elastic responses to external stress, which are grouped under the behavior known as *viscoelasticity.* Fluids that behave elastically in some stress range require a limiting or yield stress before they will flow as a viscous fluid. A simple, empirical, constitutive equation often used for this type of rheological behavior is of the form:

$$\tau_{yx} = \tau_y + \dot{\gamma}^n \eta_p \tag{11.4}$$

where τ_y is the *yield stress,* η_p is an *apparent viscosity* called the plastic viscosity, and the exponent n allows for a range of non-Newtonian responses: $n = 1$ is pseudo-Newtonian behavior and is called a *Bingham fluid*; $n < 1$ is shear thinning behavior; and $n > 1$ is shear thickening behavior. Interested readers should consult [3–9] for further information on applied rheology.

Dimensions and Units of Viscosity

From Equation 11.1, the dimensions of dynamic viscosity are M L^{-1} T^{-1} and the basic SI unit is the Pascal second (Pa·s), where 1 Pa·s = 1 N s m^{-2}. The c.g.s. unit of dyn s cm^{-2} is the poise (P). The dimensions of kinematic viscosity, from Equation 11.2, are L^2 T^{-1} and the SI unit is m^2 s^{-1}. For most practical situations, this is usually too large and so the c.g.s. unit of cm^2 s^{-1}, or the stoke (St), is preferred. Table 11.1 lists some common fluids and their shear dynamic viscosities at atmospheric pressure and 20°C.

TABLE 11.1 Shear Dynamic Viscosity of Some Common Fluids at 20°C and 1 atm

Fluid	Shear dynamic viscosity (Pa·s)
Air	1.8×10^{-4}
Water	1.0×10^{-3}
Mercury	1.6×10^{-3}
Automotive engine oil (SAE 10W30)	1.3×10^{-1}
Dish soap	4.0×10^{-1}
Corn syrup	6.0

TABLE 11.2 Viscometer Classification and Basic Characteristics

Drag Flow Types:
Flow set by motion of instrument boundary/surface using external or gravity force.

Type/Geometry	Basic characteristics/Comments
Rotating concentric cylinders (Couette)	Good for low viscosity, high shear rates; for $R_2/R_1 \cong 1$, see Figure 11.3; hard to clean thick fluids
Rotating cone and plate	Homogeneous shear, best for non-Newtonian fluids and normal stresses; need good alignment, problems with loading and evaporation
Rotating parallel disks	Similar to cone-and-plate, but inhomogeneous shear; shear varies with gap height, easy sample loading
Sliding parallel plates	Homogeneous shear, simple design, good for high viscosity; difficult loading and gap control
Falling body (ball, cylinder)	Very simple, good for high temperature and pressure; need density and special sensors for opaque fluids, not good for viscoelastic fluids
Rising bubble	Similar to falling body viscometer; for transparent fluids
Oscillating body	Needs instrument constant, good for low viscous liquid metals

Pressure Flow Types:
Fluid set in motion in fixed instrument geometry by external or gravity pressure

Type/Geometry	Basic characteristics/Comments
Long capillary (Poiseuille flow)	Simple, very high shears and range, but very inhomogeneous shear, bad for time dependency, and is time consuming
Orifice/Cup (short capillary)	Very simple, reliable, but not for absolute viscosity and non-Newtonian fluids
Slit (parallel plates) pressure flow	Similar to capillary, but difficult to clean
Axial annulus pressure flow	Similar to capillary, better shear uniformity, but more complex, eccentricity problem and difficult to clean

Others/Miscellaneous:

Type/geometry	Basic characteristics/Comments
Ultrasonic	Good for high viscosity fluids, small sample volume, gives shear and volume viscosity, and elastic property data; problems with surface finish and alignment, complicated data reduction

Adapted from C. W. Macosko, *Rheology: Principles, Measurements, and Applications,* New York: VCH, 1994.

Viscometer Types

The instruments for viscosity measurements are designed to determine "a fluid's resistance to flow," a fluid property defined above as viscosity. The fluid flow in a given instrument geometry defines the strain rates, and the corresponding stresses are the measure of resistance to flow. If strain rate or stress is set and controlled, then the other one will, everything else being the same, depend on the fluid viscosity. If the flow is simple (one dimensional, if possible) such that the strain rate and stress can be determined accurately from the measured quantities, the absolute dynamic viscosity can be determined; otherwise, the relative viscosity will be established. For example, the fluid flow can be set by dragging fluid with a sliding or rotating surface, falling body through the fluid, or by forcing the fluid (by external pressure or gravity) to flow through a fixed geometry, such as a capillary tube, annulus, a slit (between two parallel plates), or orifice. The corresponding resistance to flow is measured as the boundary force or torque, or pressure drop. The flow rate or efflux time represents the fluid flow for a set flow resistance, like pressure drop or gravity force. The viscometers are classified, depending on how the flow is initiated or maintained, as in Table 11.2.

The basic principle of all viscometers is to provide as simple flow kinematics as possible, preferably one-dimensional *(isometric) flow,* in order to determine the shear strain rate accurately, easily, and independent of fluid type. The resistance to such flow is measured, and thereby the shearing stress is

TABLE 11.3 Different Causes of Viscometers Errors

Error/Effect	Cause/Comment
End/edge effect	Energy losses at the fluid entrance and exit of main test geometry
Kinetic energy losses	Loss of pressure to kinetic energy
Secondary flow	Energy loss due to unwanted secondary flow, vortices, etc.; increases with Reynolds number
Nonideal geometry	Deviations from ideal shape, alignment, and finish
Shear rate non-uniformity	Important for non-Newtonian fluids
Temperature variation and viscous heating	Variation in temperature, in time and space, influences the measured viscosity
Turbulence	Partial and/or local turbulence often develops even at low Reynolds numbers
Surface tension	Difference in interfacial tensions
Elastic effects	Structural and fluid elastic effects
Miscellaneous effects	Depends on test specimen, melt fracture, thixotropy, rheopexy

TABLE 11.4 Viscometer Manufacturers

Manufacturers	Model	Description
Brookfield Eng. Labs Inc.	DV-I+	Concentric cylinder
Custom Scientific Inst. Inc.	CS245	Concentric cylinder
Reologica Inst.	Various	Falling sphere, capillary, rotational
Haake GmbH	Various	Falling sphere, rotational
Cannon Inst. Co.	Various	Extensive variety of capillary viscometers
Toyo Seikl Seisaku-Sho Ltd.	Capirograph	Capillary
Gottfert Werkstoff-Prufmaschinen GmbH	Various	Extensive variety of capillary viscometers
Cole-Palmer Inst. Co.	GV2100	Falling sphere
Paar Physica U.S.A. Inc.	Various	Concentric cylinder, falling sphere, capillary
Monsanto Inst. & Equipment	ODR 2000	Oscillating viscometer
Nametre Co.	Vibrational viscometer	Oscilating viscometer
Rheometric Scientific Inc.	RM180	Cone-and-plate, parallel plate
	RM265	Concentric cylinders
T.A. Instruments Inc.	Various	Concentric cylinder

Note: All the above manufacturers can be found via the Internet (World Wide Web), along with the most recent contact information, product description and in some cases, pricing.

determined. The shear viscosity is then easily found as the ratio between the shearing stress and the corresponding shear strain rate. Practically, it is never possible to achieve desired one-dimensional flow nor ideal geometry, and a number of errors, listed in Table 11.3, can occur and need to be accounted for [4–8]. A list of manufacturers/distributors of commercial viscometers/rheometers is given in Table 11.4.

Concentric Cylinders

The main advantage of the rotational as compared to many other viscometers is its ability to operate continuously at a given shear rate, so that other steady-state measurements can be conveniently performed. That way, time dependency, if any, can be detected and determined. Also, subsequent measurements can be made with the same instrument and sampled at different shear rates, temperature, etc. For these and other reasons, *rotational viscometers* are among the most widely used class of instruments for rheological measurements.

Concentric cylinder-type viscometers/rheometers are usually employed when absolute viscosity needs to be determined, which in turn, requires a knowledge of well-defined shear rate and shear stress data. Such instruments are available in different configurations and can be used for almost any fluid. There are models for low and high shear rates. More complete discussion on concentric cylinder viscometers/rheometers is given elsewhere [4–9]. In the *Couette-type viscometer,* the rotation of the outer cylinder, or cup, minimizes centrifugal forces, which cause Taylor vortices. The latter can be present in the *Searle-type viscometer* when the inner cylinder, or bob, rotates.

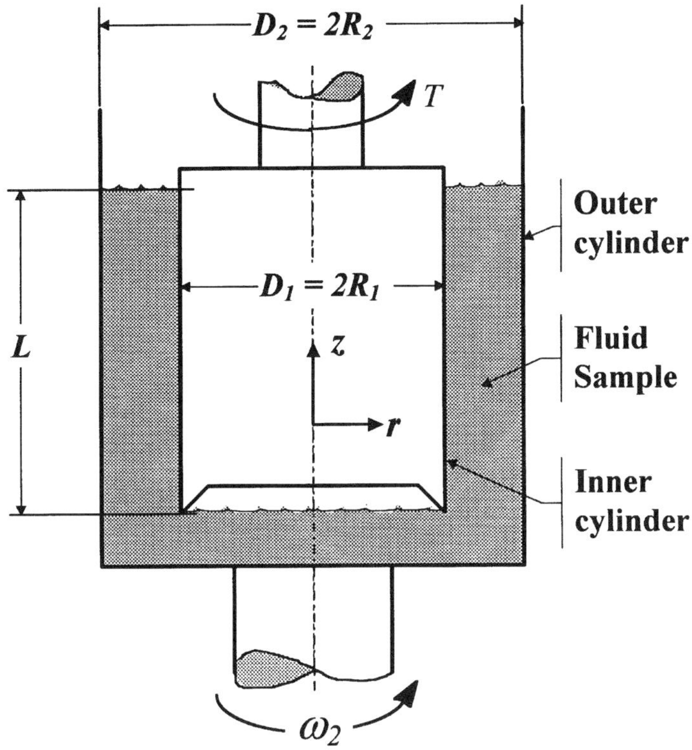

FIGURE 11.3 Concentric cylinders viscometer geometry.

Usually, the torque on the stationary cylinder and rotational velocity of the other cylinder are measured for determination of the shear stress and shear rate, which is needed for viscosity calculation. Once the torque, *T*, is measured, it is simple to describe the fluid shear stress at any point with radius *r* between the two cylinders, as shown in Figure 11.3.

$$\tau_{r\theta}(r) = \frac{T}{2\pi r^2 L_e} \tag{11.5}$$

In Equation 11.5, $L_e = (L + L_c)$, is the effective length of the cylinder at which the torque is measured. In addition to the cylinder's length *L*, it takes into account the end-effect correction L_c [4–8].

For a narrow gap between the cylinders ($\beta = R_2/R_1 \cong 1$), regardless of the fluid type, the velocity profile can be approximated as linear, and the shear rate within the gap will be uniform:

$$\gamma(r) \cong \frac{\Omega \bar{R}}{(R_2 - R_1)} \tag{11.6}$$

where $\Omega = (\omega_2 - \omega_1)$ is the relative rotational speed and $\bar{R} = (R_1 + R_2)/2$, is the mean radius of the inner (1) and outer (2) cylinders. Actually, the shear rate profile across the gap between the cylinders depends on the relative rotational speed, radii, and the unknown fluid properties, which seems an "open-ended" enigma. The solution of this complex problem is given elsewhere [4–8] in the form of an infinite series, and requires the slope of a logarithmic plot of *T* as a function of Ω in the range of interest. Note that for a stationary inner cylinder ($\omega_1 = 0$), which is the usual case in practice, Ω becomes equal to ω_2.

However, there is a simpler procedure [10] that has also been established by German standards [11]. For any fluid, including non-Newtonian fluids, there is a radius at which the shear rate is virtually independent of the fluid type for a given Ω. This radius, being a function of geometry only, is called the *representative radius*, R_R, and is determined as the location corresponding to the so-called representative shear stress, $\tau_R = (\tau_1 + \tau_2)/2$, the average of the stresses at the outer and inner cylinder interfaces with the fluid, that is:

$$R_R = R_1\left\{\frac{[2\beta^2]}{[1+\beta^2]}\right\}^{1/2} = R_2\left\{\frac{2}{[1+\beta^2]}\right\}^{1/2} \tag{11.7}$$

Since the shear rate at the representative radius is virtually independent on the fluid type (whether Newtonian or non-Newtonian), the representative shear rate is simply calculated for Newtonian fluid ($n = 1$) and $r = R_R$, according to [10]:

$$\dot{\gamma}_R = \dot{\gamma}_{r=R_R} = \omega_2\left\{\frac{[\beta^2+1]}{[\beta^2-1]}\right\} \tag{11.8}$$

The accuracy of the representative parameters depends on the geometry of the cylinders (β) and fluid type (n).

It is shown in [10] that, for an unrealistically wide range of fluid types ($0.35 < n < 3.5$) and cylinder geometries ($\beta = 1$ to 1.2), the maximum errors are less than 1%. Therefore, the error associated with the representative parameters concept is virtually negligible for practical measurements.

Finally, the (apparent) fluid viscosity is determined as the ratio between the shear stress and corresponding shear rate using Equations 11.5 to 11.8, as:

$$\eta = \eta_R = \frac{\tau_R}{\dot{\gamma}_R} = \left\{\frac{[\beta^2-1]}{[4\pi\beta^2 R_1^2 L_e]}\right\}\frac{T}{\omega_2} = \left\{\frac{[\beta^2-1]}{[4\pi R_2^2 L_e]}\right\}\frac{T}{\omega_2} \tag{11.9}$$

For a given cylinder geometry (β, R_2, and L_e), the viscosity can be determined from Equation 11.8 by measuring T and ω_2.

As already mentioned, in Couette-type viscometers, the Taylor vortices within the gap are virtually eliminated. However, vortices at the bottom can be present, and their influence becomes important when the Reynolds number reaches the value of unity [10, 11]. Furthermore, flow instability and turbulence will develop when the Reynolds number reaches values of 10^3 to 10^4. The Reynolds number, Re, for the flow between concentric cylinders is defined [11] as:

$$\text{Re} = \left\{\frac{[\rho\omega_2 R_1^2]}{2\eta}\right\}[\beta^2-1] \tag{11.10}$$

Cone-and-Plate Viscometers

The simple cone-and-plate viscometer geometry provides a uniform rate of shear and direct measurements of the first normal stress difference. It is the most popular instrument for measurement of non-Newtonian fluid properties. The working shear stress and shear strain rate equations can be easily derived in spherical coordinates, as indicated by the geometry in Figure 11.4, and are, respectively:

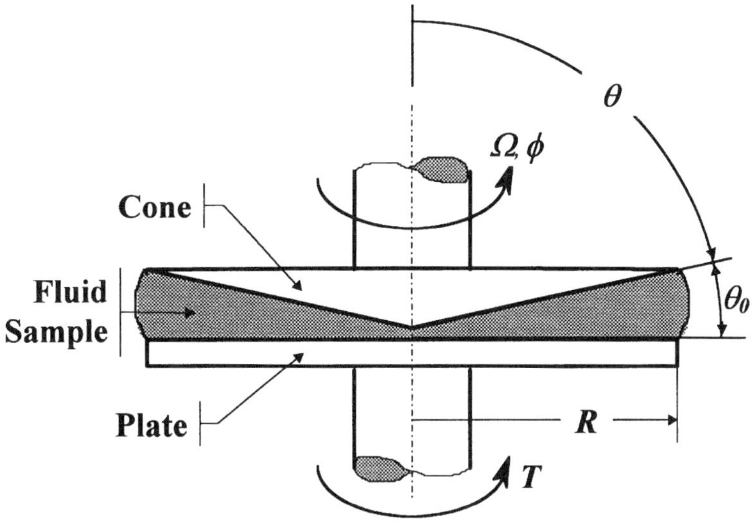

FIGURE 11.4 Cone-and-plate viscometer geometry.

$$\tau_{\theta\phi} = \frac{3T}{\left[2\pi R^3\right]} \tag{11.11}$$

and

$$\dot{\gamma} = \frac{\Omega}{\theta_0} \tag{11.12}$$

where R and $\theta_0 < 0.1$ rad (≈6°) are the cone radius and angle, respectively. The viscosity is then easily calculated as:

$$\eta = \frac{\tau_{\theta\phi}}{\dot{\gamma}} = \frac{\left[3T\theta_0\right]}{2\pi\Omega R^3} \tag{11.13}$$

Inertia and secondary flow increase while shear heating decreases the measured torque (T_m). For more details, see [4, 5]. The torque correction is given as:

$$\frac{T_m}{T} = 1 + 6 \cdot 10^{-4}\,\mathrm{Re}^2 \tag{11.14}$$

where

$$\mathrm{Re} = \frac{\left\{\rho\left[\Omega\theta_0 R\right]^2\right\}}{\eta} \tag{11.15}$$

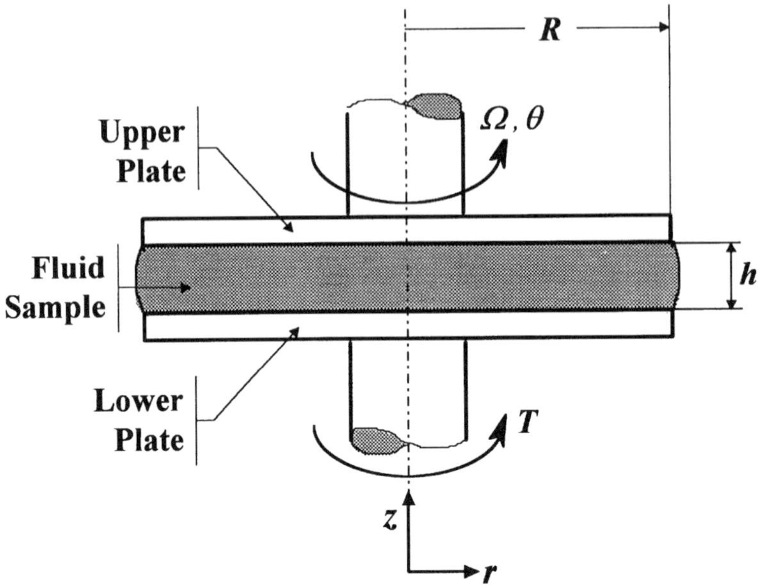

FIGURE 11.5 Parallel disks viscometer geometry.

Parallel Disks

This geometry (Figure 11.5), which consists of a disk rotating in a cylindrical cavity, is similar to the cone-and-plate geometry, and many instruments permit the use of either one. However, the shear rate is no longer uniform, but depends on radial distance from the axis of rotation and on the gap h, that is:

$$\dot{\gamma}(r) = \frac{r\Omega}{h} \tag{11.16}$$

For Newtonian fluids, after integration over the disk area, the torque can be expressed as a function of viscosity, so that the latter can be determined as:

$$\eta = \frac{2Th}{\left[\pi\Omega R^4\right]} \tag{11.17}$$

Capillary Viscometers

The *capillary viscometer* is based on the fully developed laminar tube flow theory (*Hagen–Poiseuille flow*) and is shown in Figure 11.6. The capillary tube length is many times larger than its small diameter, so that entrance flow is neglected or accounted for in more accurate measurement or for shorter tubes. The expression for the shear stress at the wall is:

$$\tau_w = \left[\frac{\Delta P}{L}\right] \cdot \left[\frac{D}{4}\right] \tag{11.18}$$

and

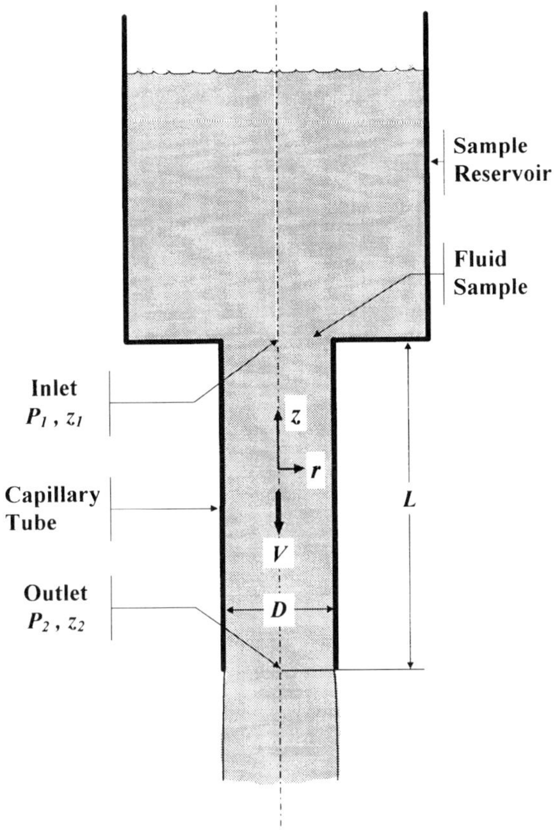

FIGURE 11.6 Capillary viscometer geometry.

$$\Delta P = \left(P_1 - P_2\right) + \left(z_1 - z_2\right) - \frac{\left[C\rho V^2\right]}{2} \tag{11.19}$$

where, $C \cong 1.1$, P, z, $V = 4Q/[\pi D^2]$, and Q are correction factor, pressure, elevation, the mean flow velocity, and the fluid volume-flow rate, respectively. The subscripts 1 and 2 refer to the inlet and outlet, respectively.

The expression for the shear rate at the wall is:

$$\dot{\gamma} = \left\{\frac{\left[3n+1\right]}{4n}\right\} \circ \left\{\frac{8V}{D}\right\} \tag{11.20}$$

where $n = \mathrm{d}[\log \tau_w]/\mathrm{d}[\log (8V/D)]$ is the slope of the measured $\log(\tau_w) - \log (8V/D)$ curve. Then, the viscosity is simply calculated as:

$$\eta = \frac{\tau_w}{\dot{\gamma}} = \left\{\frac{4n}{\left[3n+1\right]}\right\} \circ \left\{\frac{\Delta P D^2}{\left[32LV\right]}\right\} = \left\{\frac{4n}{\left[3n+1\right]}\right\} \circ \left\{\frac{\left[\Delta P D^4 \pi\right]}{\left[128QL\right]}\right\} \tag{11.21}$$

Note that $n = 1$ for a Newtonian fluid, so the first term, $[4n/(3n + 1)]$ becomes unity and disappears from the above equations. The advantages of capillary over rotational viscometers are low cost, high accuracy (particularly with longer tubes), and the ability to achieve very high shear rates, even with high-viscosity samples. The main disadvantages are high residence time and variation of shear across the flow, which can change the structure of complex test fluids, as well as shear heating with high-viscosity samples.

Glass Capillary Viscometers

Glass capillary viscometers are very simple and inexpensive. Their geometry resembles a U-tube with at least two reservoir bulbs connected to a capillary tube passage with inner diameter D. The fluid is drawn up into one bulb reservoir of known volume, V_0, between etched marks. The efflux time, Δt, is measured for that volume to flow through the capillary under gravity.

From Equation 11.21 and taking into account that $V_0 = (\Delta t) VD^2\pi/4$ and $\Delta P = \rho g(z_1 - z_2)$, the kinematic viscosity can be expressed as a function of the efflux time only, with the last term, $K/\Delta t$, added to account for error correction, where K is a constant [7]:

$$\nu = \frac{\eta}{\rho} = \left\{ \left[\frac{4n}{(3n+1)} \right] \cdot \frac{\left[\pi g (z_1 - z_2) D^4 \right]}{128 L V_0} \right\} (\Delta t - K \Delta t) \tag{11.22}$$

Note that for a given capillary viscometer and $n \cong 1$, the bracketed term is a constant. The last correction term is negligible for a large capillary tube ratio, L/D, where kinematic viscosity becomes linearly proportional to measured efflux time. Various kinds of commercial glass capillary viscometers, like Cannon-Fenske type or similar, can be purchased from scientific and/or supply stores. They are the modified original *Ostwald viscometer* design in order to minimize certain undesirable effects, to increase the viscosity range, or to meet specific requirements of the tested fluids, like opacity, etc. Glass capillary viscometers are often used for low-viscosity fluids.

Orifice/Cup, Short Capillary: Saybolt Viscometer

The principle of these viscometers is similar to glass capillary viscometers, except that the flow through a short capillary ($L/D \ll 10$) does not satisfy or even approximate the Hagen–Poiseuille, fully developed, pipe flow. The influences of entrance end-effect and changing hydrostatic heads are considerable. The efflux time reading, Δt, represents relative viscosity for comparison purposes and is expressed as "viscometer seconds," like the Saybolt seconds, or Engler seconds or degrees. Although the conversion formula, similar to glass capillary viscometers, is used, the constants k and K in Equation 11.23 are purely empirical and dependent on fluid types.

$$\nu = \frac{\eta}{\rho} = k\Delta t - \frac{K}{\Delta t} \tag{11.23}$$

where $k = 0.00226, 0.0216, 0.073$; and $K = 1.95, 0.60, 0.0631$; for Saybolt Universal ($\Delta t < 100$ s), Saybolt Furol ($\Delta t > 40$ s), and Engler viscometers, respectively [12, 13]. Due to their simplicity, reliability, and low cost, these viscometers are widely used for Newtonian fluids, like in oil and other industries, where the simple correlations between the relative properties and desired results are needed. However, these viscometers are not suitable for absolute viscosity measurement, nor for non-Newtonian fluids.

Falling Body Methods

Falling Sphere

The falling sphere viscometer is one of the earliest and least involved methods to determine the absolute shear viscosity of a Newtonian fluid. In this method, a sphere is allowed to fall freely a measured distance through a viscous liquid medium and its velocity is determined. The viscous drag of the falling sphere results in the creation of a restraining force, *F*, described by Stokes' law:

$$F = 6\pi\eta r_s U_t \tag{11.24}$$

where r_s is the radius of the sphere and U_t is the *terminal velocity* of the falling body. If a sphere of density ρ_2 is falling through a fluid of density ρ_1 in a container of infinite extent, then by balancing Equation 11.24 with the net force of gravity and buoyancy exerted on a solid sphere, the resulting equation of absolute viscosity is:

$$\eta = 2gr_s^2 \frac{(\rho_2 - \rho_1)}{9U_t} \tag{11.25}$$

Equation 11.25 shows the relation between the viscosity of a fluid and the terminal velocity of a sphere falling within it. Having a finite container volume necessitates the modification of Equation 11.25 to correct for effects on the velocity of the sphere due to its interaction with container walls (*W*) and ends (*E*). Considering a cylindrical container of radius *r* and height *H*, the corrected form of Equation 11.25 can be written as:

$$\eta = 2gr_s^2 \frac{(\rho_2 - \rho_1)W}{(9U_t E)} \tag{11.26}$$

where

$$W = 1 - 2.104\left(\frac{r_s}{r}\right) + 2.09\left(\frac{r_s}{r}\right)^3 - 0.95\left(\frac{r_s}{r}\right)^5 \tag{11.27}$$

$$E = 1 + 3.3\left(\frac{r_s}{H}\right) \tag{11.28}$$

The wall correction was empirically derived [15] and is valid for $0.16 \le r_s/r \le 0.32$. Beyond this range, the effects of container walls significantly impair the terminal velocity of the sphere, thus giving rise to a false high viscosity value.

Figure 11.7 is a schematic diagram of the falling sphere method and demonstrates the attraction of this method — its simplicity of design. The simplest and most cost-effective approach in applying this method to transparent liquids would be to use a sufficiently large graduated cylinder filled with the liquid. With a distance marked on the cylinder near the axial and radial center (the region least influenced by the container walls and ends), a sphere (such as a ball bearing or a material that is nonreactive with the liquid) with a known density and sized to within the bounds of the container correction, free falls the length of the cylinder. As the sphere passes through the marked region of length *d* at its terminal velocity, a measure of the time taken to traverse this distance allows the velocity of the sphere to be calculated. Having measured all the parameters of Equation 11.26, the shear viscosity of the liquid can be determined.

This method is useful for liquids with viscosities between 10^{-3} Pa·s and 10^5 Pa·s. Due to the simplicity of design, the falling sphere method is particularly well suited to high pressure–high temperature viscosity studies.

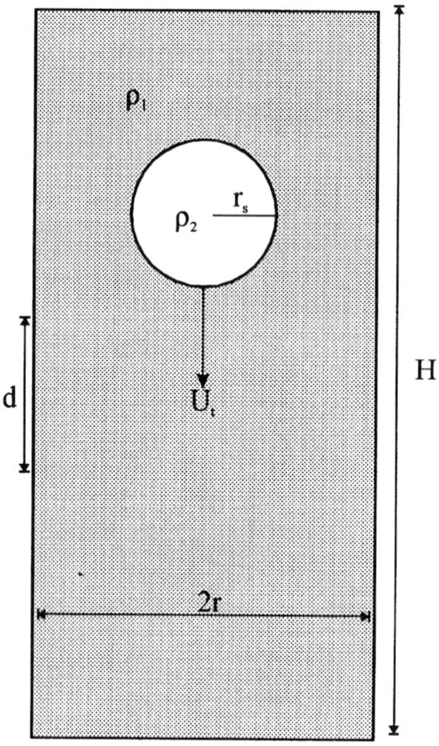

FIGURE 11.7 Schematic diagram of the falling sphere viscometer. Visual observations of the time taken for the sphere to traverse the distance *d*, is used to determine a velocity of the sphere. The calculated velocity is then used in Equation 11.24 to determine a shear viscosity.

Falling Cylinder

The *falling cylinder method* is similar in concept to the falling sphere method except that a flat-ended, solid circular cylinder freely falls vertically in the direction of its longitudinal axis through a liquid sample within a cylindrical container. A schematic diagram of the configuration is shown in Figure 11.8. Taking an infinitely long cylinder of density ρ_2 and radius r_c falling through a Newtonian fluid of density ρ_1 with infinite extent, the resulting shear viscosity of the fluid is given as:

$$\eta = g r_c^2 \frac{(\rho_2 - \rho_1)}{2U_t} \tag{11.29}$$

Just as with the falling sphere, a finite container volume necessitates modifying Equation 11.29 to account for the effects of container walls and ends. A correction for container wall effects can be analytically deduced by balancing the buoyancy and gravitational forces on the cylinder, of length *L*, with the shear force on the sides and the compressional force on the cylinder's leading end and the tensile force on the cylinder's trailing end. The resulting correction term, or geometrical factor, $G(k)$ (where $k = r_c/r$), depends on the cylinder radius and the container radius, *r*, and is given by:

$$G(k) = \frac{\left[k^2(1 - \ln k) - (1 + \ln k)\right]}{(1 + k^2)} \tag{11.30}$$

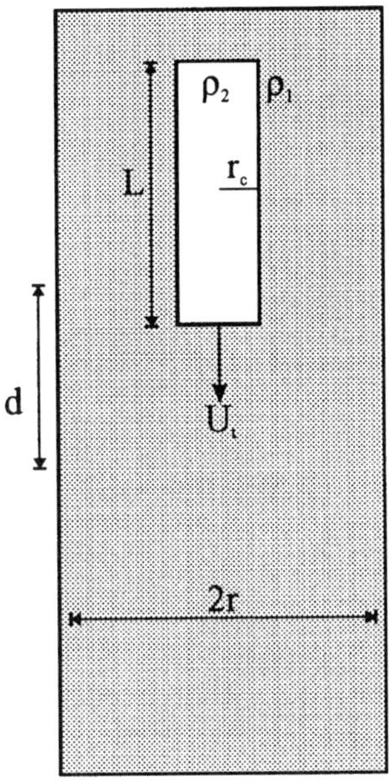

FIGURE 11.8 Schematic diagram of the falling cylinder viscometer. Using the same principle as the falling sphere, the velocity of the cylinder is obtained, which is needed to determine the shear viscosity of the fluid.

Unlike the fluid flow around a falling sphere, the fluid motion around a falling flat-ended cylinder is very complex. The effects of container ends are minimized by creating a small gap between the cylinder and the container wall. If a long cylinder (here, a cylinder is considered long if $\psi \geq 10$, where $\psi = L/r$) with a radius nearly as large as the radius of the container is used, then the effects of the walls would dominate, thereby reducing the end effects to a second-order effect. A major drawback with this approach is, however, if the cylinder and container are not concentric, the resulting inhomogeneous wall shear force would cause the downward motion of the cylinder to become eccentric. The potential for misalignment motivated the recently obtained analytical solution to the fluid flow about the cylinder ends [16]. An analytical expression for the end correction factor (ECF) was then deduced [17] and is given as:

$$\frac{1}{\text{ECF}} = 1 + \left(\frac{8k}{\pi C_w}\right)\left(\frac{G(k)}{\psi}\right) \tag{11.31}$$

where $C_w = 1.003852 - 1.961019k + 0.9570952k^2$. C_w was derived semi-empirically [17] as a disk wall correction factor. This is based on the idea that the drag force on the ends of the cylinder can be described by the drag force on a disk. Equation 11.31 is valid for $\psi \leq 30$ and agrees with the empirically derived correction [16] to within 0.6%.

With wall and end effects taken into consideration, the working formula to determine the shear viscosity of a Newtonian fluid from a falling cylinder viscometer is:

$$\eta = \frac{\left[g r_c^2 (\rho_2 - \rho_1)\, G(k)\right]}{\left(\frac{2U_t}{\text{ECF}}\right)} \tag{11.32}$$

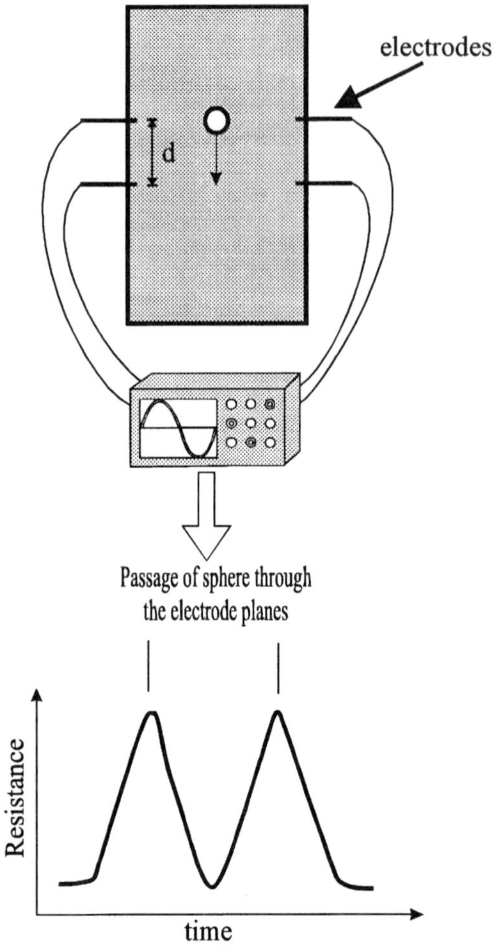

FIGURE 11.9 Diagram of one type of apparatus used to determine the viscosity of opaque liquids *in situ.* The electrical signal from the passage of the falling sphere indicates the time to traverse a known distance (d) between the two sensors.

In the past, this method was primarily used as a method to determine relative viscosities between transparent fluids. It has only been since the introduction of the ECF [16, 17] that this method could be rigorously used as an absolute viscosity method. With a properly designed container and cylinder, this method is now able to provide accurate absolute viscosities from 10^{-3} Pa·s to 10^{7} Pa·s.

Falling Methods in Opaque Liquids

The falling body methods described above have been extensively applied to transparent liquids where optical (often visual) observation of the falling body is possible. For *opaque liquids*, however, falling body methods require the use of some sensing technique to determine, often *in situ*, the position of the falling body with respect to time. Techniques have varied but they all have in common the ability to detect the body as it moves past the sensor. A recent study at high pressure [18] demonstrated that the contrast in electric conductivity between a sphere and opaque liquid could be exploited to dynamically sense the moving sphere if suitably placed electrodes penetrated the container walls as shown in the schematic diagram in Figure 11.9. References to other similar *in situ* techniques are given in [18].

Rising Bubble/Droplet

For many industrial processes, the rising bubble viscometer has been used as a method of comparing the relative viscosities of transparent liquids (such as varnish, lacquer, and beer) for decades. Although its use was widespread, the actual behavior of the bubble in a viscous liquid was not well understood until long after the method was introduced [19]. The rising bubble method has been thought of as a derivative of the falling sphere method; however, there are primary differences between the two. The major physical differences are (1) the density of the bubble is less that of the surrounding liquid, and (2) the bubble itself has some unique viscosity. Each of these differences can, and do, lead to significant and extremely complex rheological problems that have yet to be fully explained. If a bubble of gas or droplet of liquid with a radius, r_b, and density, ρ', is freely rising in some enclosing viscous liquid of density ρ, then the shear viscosity of the enclosing liquid is determined by:

$$\eta = \left(\frac{1}{\varepsilon}\right)\frac{\left[2gr_b^2\left(\rho-\rho'\right)\right]}{9U_t} \tag{11.33}$$

where

$$\varepsilon = \frac{\left(2\eta+3\eta'\right)}{3\left(\eta+\eta'\right)} \tag{11.34}$$

where η' is the viscosity of the bubble. It must be noted that when the value of η' is large (solid spheres), $\varepsilon = 1$, which reduces Equation 11.33 to Equation 11.25. For small values of η' (gas bubbles), ε becomes 2/3, and the viscosity calculated by Equation 11.33 is 1.5 times greater than the viscosity calculated by Equation 11.25. It is apparent from Equation 11.33 and 11.25 that if the density of the bubble is less than the density of the enclosing liquid, and the terminal velocity of the sphere is negative, which indicates upward motion since the downward direction is positive.

During the rise, great care must be taken to avoid contamination of the bubble and its surface with impurities in the surrounding liquid. Impurities can diffuse through the surface of the bubble and combine with the fluid inside. Because the bubble has a low viscosity, the upward motion in a viscous medium induces a drag on the bubble that is responsible for generating a circulatory motion within it. This motion can efficiently distribute impurities throughout the whole of the bubble, thereby changing its viscosity and density. Impurities left on the surface of the bubble can form a "skin" that can significantly affect the rise of the bubble, as the skin layer has its own density and viscosity that are not included in Equation 11.33. These surface impurities also make a significant contribution to the inhomogeneous distribution of interfacial tension forces. A balance of these forces is crucial for the formation of a spherical bubble. The simplest method to minimize the above effects is to employ minute bubbles by introducing a specific volume of fluid (gas or liquid), with a syringe or other similar device, at the lower end of the cylindrical container. Very small bubbles behave like solid spheres, which makes interfacial tension forces and internal fluid motion negligible.

In all rising bubble viscometers, the bubble is assumed to be spherical. Experimental studies of the shapes of freely rising gas bubbles in a container of finite extent [20] have shown that (to 1% accuracy) a bubble will form and retain a spherical shape if the ratio of the radius of the bubble to the radius of the confining cylindrical container is less than 0.2. These studies have also demonstrated that the effect of the wall on the terminal velocity of a rising spherical bubble is to cause a large decrease (up to 39%) in the observed velocity compared to the velocity measured within an unbounded medium. This implies that the walls of the container influence the velocity of the rising bubble sooner than its geometry. In this method, end effects are known to be large. However, a rigorous, analytically or empirically derived

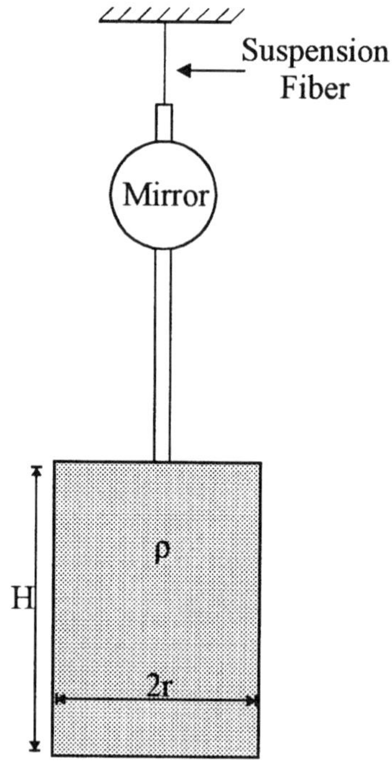

FIGURE 11.10 Schematic diagram of the oscillating cup viscometer. Measurement of the logarithmic damping of the amplitude and period of vessel oscillation are used to determine the absolute shear viscosity of the liquid.

correction factor has not yet appeared. To circumvent this, the ratio of container length to sphere diameter must be in the range of 10 to 100. As in other Stokian methods, this allows the bubble's velocity to be measured at locations that experience negligible end effects.

Considering all of the above complications, the use of minute bubbles is the best approach to ensure a viscosity measurement that is least affected by the liquid to be investigated and the container geometry.

Oscillating Method

If a liquid is contained within a vessel suspended by some torsional system that is set in oscillation about its vertical axis, then the motion of the vessel will experience a gradual damping. In an ideal situation, the damping of the motion of the vessel arises purely as a result of the *viscous coupling* of the liquid to the vessel and the viscous coupling between layers in the liquid. In any practical situation, there are also frictional losses within the system that aid in the damping effect and must be accounted for in the final analysis. From observations of the amplitudes and time periods of the resulting oscillations, a viscosity of the liquid can be calculated. A schematic diagram of the basic set-up of the method is shown in Figure 11.10. Following initial oscillatory excitation, a light source (such as a low-intensity laser) can be used to measure the amplitudes and periods of the resulting oscillations by reflection off the mirror attached to the suspension rod to give an accurate measure of the logarithmic decrement of the oscillations (δ) and the periods (T).

Various working formulae have been derived that associate the oscillatory motion of a vessel of radius r to the absolute viscosity of the liquid. The most reliable formula is the following equation for a cylindrical vessel [21]:

$$\eta = \left[\frac{I\delta}{\left(\pi r^3 HZ\right)} \right]^2 \left[\frac{1}{\pi \rho T} \right] \tag{11.35}$$

where

$$Z=\left(\frac{1+r}{4H}\right)a_0-\frac{\left(\frac{3}{2}+\frac{4r}{\pi H}\right)1}{p}+\frac{\left(\frac{3}{8}+\frac{9r}{4H}\right)a_2}{2p^2} \tag{11.36}$$

$$p=\left(\frac{\pi\rho}{\eta T}\right)^{1/2} r \tag{11.37}$$

$$a_0=1-\left(\frac{\delta}{4\pi}\right)-\left(\frac{3\delta^2}{32\pi^2}\right) \tag{11.38}$$

$$a_2=1+\left(\frac{\delta}{4\pi}\right)+\left(\frac{\delta^2}{32\pi^2}\right) \tag{11.39}$$

I is the mass moment of inertia of the suspended system and ρ is the density of the liquid.

A more practical expression of Equation 11.35 is obtained by introducing a number of simplifications. First, it is a reasonable assumption to consider δ to be small (on the order of 10^{-2} to 10^{-3}). This reduces a_0 and a_2 to values of 1 and –1, respectively. Second, the effects of friction from the suspension system and the surrounding atmosphere can be experimentally determined and contained within a single variable, δ_0. This must then be subtracted from the measured δ. A common method of obtaining δ_0 is to observe the logarithmic decrement of the system with an empty sample vessel and subtract that value from the measured value of δ. With these modifications, Equation 11.35 becomes:

$$\frac{(\delta-\delta_0)}{\rho}=\left[A\left(\frac{\eta}{\rho}\right)^{1/2}-B\left(\frac{\eta}{\rho}\right)+C\left(\frac{\eta}{\rho}\right)^{3/2}\right] \tag{11.40}$$

where

$$A=\left(\frac{\pi^{3/2}}{I}\right)\left[1+\left(\frac{r}{4H}\right)Hr^3T^{1/2}\right] \tag{11.41}$$

$$B=\left(\frac{\pi}{I}\right)\left[\left(\frac{3}{2}\right)+\frac{4r}{\pi H}\right]Hr^2T \tag{11.42}$$

$$C=\left(\frac{\pi^{1/2}}{2I}\right)\left[\left(\frac{3}{8}\right)+\frac{9r}{4H}\right]HrT^{3/2} \tag{11.43}$$

It has been noted [22] that the analytical form of Equation 11.40 needs an empirically derived, instrument-constant correction factor (ζ) in order to agree with experimentally measured values of η. The discrepancy between the analytical form and the measured value arises as a result of the above assumptions. However, these assumptions are required as there are great difficulties involved in solving the differential equations of motion of this system. The correction factor is dependent on the materials, dimensions, and densities of each individual system, but generally lies between the values of 1.0 and 1.08. The correction factor is obtained by comparing viscosity values of calibration materials determined by an individual system (with Equation 11.35) and viscosity values obtained by another reliable method such as the capillary method.

With the above considerations taken into account, the final working Roscoe's formula for the absolute shear viscosity is:

$$\frac{(\delta-\delta_0)}{\rho}=\zeta\left[A\left(\frac{\eta}{\rho}\right)^{1/2}-B\left(\frac{\eta}{\rho}\right)+C\left(\frac{\eta}{\rho}\right)^{3/2}\right] \tag{11.44}$$

The *oscillating cup method* has been used, and is best suited for use with low values of viscosity within the range of 10^{-5} Pa s to 10^{-2} Pa·s. Its simple closed design and use at high temperatures has made this method very popular when dealing with liquid metals.

Ultrasonic Methods

Viscosity plays an important role in the absorption of energy of an *acoustic wave* traveling through a liquid. By using ultrasonic waves (10^4 Hz $< f <$ 10^8 Hz), the elastic, viscoelastic, and viscous response of a liquid can be measured down to times as short as 10 ns. When the viscosity of the fluid is low, the resulting time scale for structural relaxation is shorter than the ultrasonic wave period and the fluid is probed in the relaxed state. High-viscosity fluids subjected to ultrasonic wave trains respond as a stiff fluid because structural equilibration due to the acoustic perturbation does not go to completion before the next wave cycle. Consequently, the fluid is said to be in an unrelaxed state that is characterized by dispersion (frequency-dependent wave velocity) and elastic moduli that reflect a much stiffer liquid. The frequency dependence of the viscosity relative to some reference viscosity (η_0) at low frequency, η/η_0, and of the absorption per wavelength, $\alpha\lambda$, where α is the absorption coefficient of the liquid and λ is the wavelength of the compressional wave, for a liquid with a single relaxation time, *t*, is shown in Figure 11.11. The maximum absorption per wavelength occurs at the *relaxation frequency* when $\omega\tau = 1$

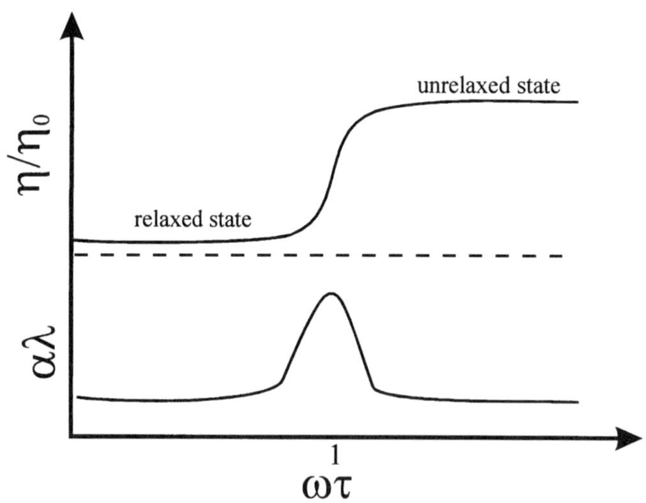

FIGURE 11.11 Effects of liquid relaxation (relaxation frequency corresponds to $\omega\tau = 1$ where $\omega = 2\pi f$) on relative viscosity (upper) and absorption per wavelength (lower) in the relaxed elastic ($\omega\tau < 1$) and unrelaxed viscoelastic ($\omega\tau > 1$) regimes.

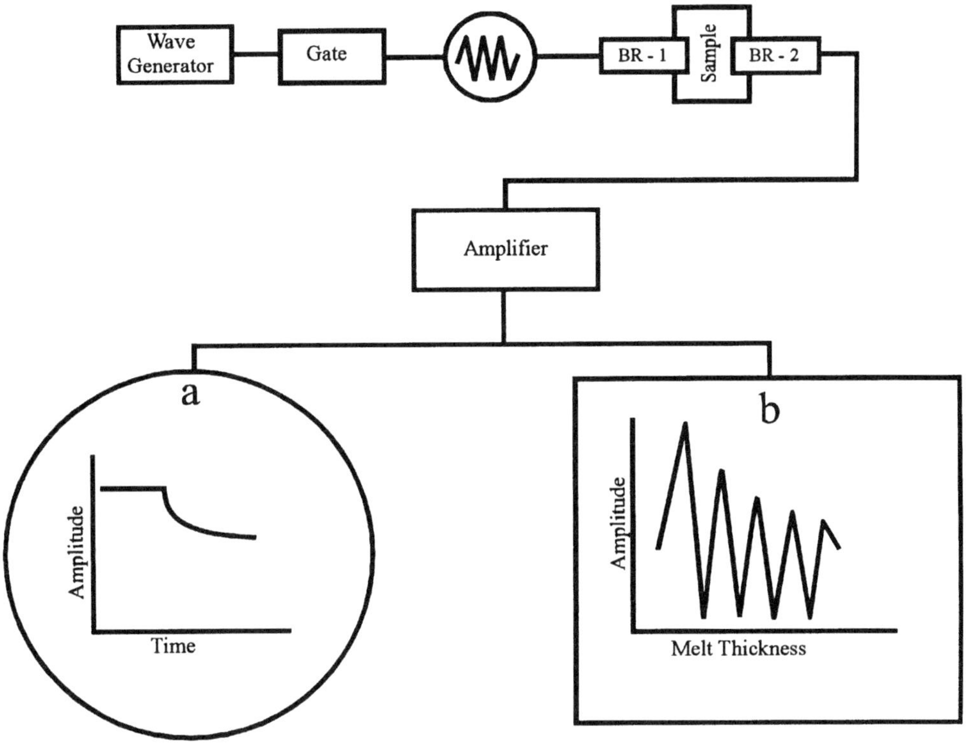

FIGURE 11.12 Schematic diagram of apparatus for liquid shear and volume viscosity determination by ultrasonic wave attenuation measurement showing the recieved signal amplitude through the exit buffer rod (BR-2) using (*a*) a fixed buffer rod (BR-2) using (*a*) a fixed buffer rod configuration, and (*b*) an interferometric technique with moveable buffer rod.

and is accompanied by a step in η/η_0, as well as in other properties such as velocity and compressibility. Depending on the application of the measured properties, it is important to determine if the liquid is in a relaxed or unrelaxed state.

A schematic diagram of a typical apparatus for measuring viscosity by the ultrasonic method is shown in Figure 11.12. Mechanical vibrations in a *piezoelectric* transducer travel down one of the buffer rods (BR-1 in Figure 11.12) and into the liquid sample and are received by a similar transducer mounted on the other buffer rod, BR-2. In the fixed buffer rod configuration, once steady-state conditions have been reached, the applied signal is turned off quickly. The decay rate of the received and amplified signal, displayed on an oscilloscope on an amplitude vs. time plot as shown in Figure 11.12(*a*), gives a measure of α. The received amplitude decays as:

$$A = A_0 e^{-(b+\alpha c)t'} \tag{11.45}$$

where A is the received decaying amplitude, A_0 is the input amplitude, b is an apparatus constant that depends on other losses in the system such as due to the transducer, container, etc. that can be evaluated by measuring the attenuation in a standard liquid, c is the compressional wave velocity of the liquid, and t' is time. At low frequencies, the absorption coefficient is expressed in terms of volume and shear viscosity as:

$$\left(\eta_v + \frac{4\eta}{3}\right) = \frac{\alpha\rho c^3}{2\pi^2 f^2} \tag{11.46}$$

One of the earliest ultrasonic methods of measuring attenuation in liquids is based on *acoustic interferometry* [23]. Apart from the instrumentation needed to move and determine the position of one of the buffer rods accurately, the experimental apparatus is essentially the same as for the fixed buffer rod configuration [24]. The measurement, however, depends on the continuous acoustic wave interference of transmitted and reflected waves within the sample melt as one of the buffer rods is moved away from the other rod. The attenuation is characterized by the decay of the maxima amplitude as a function of melt thickness as shown on the interferogram in Figure 11.12(*b*). Determining α from the observed amplitude decrement involves numerical solution to a system of equations characterizing complex wave propagation [25]. The ideal conditions represented in the theory do not account for such things as wave front curvature, buffer rod end nonparallelism, surface roughness, and misalignment. These problems can be addressed in the amplitude fitting stage but they can be difficult to overcome. The interested reader is referred to [25] for further details.

Ultrasonic methods have not been and are not likely to become the mainstay of fluid viscosity determination simply because they are more technically complicated than conventional viscometry techniques. And although ultrasonic viscometry supplies additional related elastic property data, its niche in viscometry is its capability of providing volume viscosity data. Since there is no other viscometer to measure η_v, ultrasonic absorption measurements play a unique role in the study of volume viscosity.

References

1. T. A. Litovitz and C. M. Davis, Structural and shear relaxation in liquids, in W. P. Mason (ed.), *Physical Acoustics: Principles and Methods, Vol. II. Part A, Properties of Gases, Liquids and Solutions*, New York: Academic Press, 1965, 281-349.
2. Y. Bottinga and P. Richet, Silicate melts: The "anomalous" pressure dependence of the viscosity, *Geochim. Cosmochim. Acta*, 59, 2725-2731, 1995.
3. J. Ferguson and Z. Kemblowski, *Applied Fluid Rheology*, New York: Elsevier, 1991.
4. R. W. Whorlow, *Rheological Techniques*, 2nd ed., New York: Ellis Horwood, 1992.
5. K. Walters, *Rheometry*, London: Chapman and Hall, 1975.
6. J. M. Dealy, *Rheometers for Molten Plastics*, New York: Van Nostrand Reinhold, 1982.
7. J. R. Van Wazer, J. W. Lyons, K. Y. Kim, and R. E. Colwell, *Viscosity and Flow Measurement*, New York: Interscience, 1963.
8. C. W. Macosko, *Rheology: Principles, Measurements, and Applications*, New York: VCH, 1994.
9. W. A. Wakeham, A. Nagashima, and J. V. Sengers (eds.), *Measurement of the Transport Properties of Fluids*, Oxford, UK: Blackwell Scientific, 1991.
10. J. A. Himenez and M. Kostic, A novel computerized viscometer/rheometer, *Rev. Sci. Instrum.*, 65(1), 229-241, 1994.
11. DIN 53018 (Part 1 and 2), 53019, German National Standards.
12. Marks' *Standard Handbooks for Mechanical Engineers*, New York: McGraw-Hill, 1978.
13. ASTM D445-71 standard.
14. W. D. Kingery, *Viscosity in Property Measurements at High Temperatures*, New York: John Wiley & Sons, 1959.
15. H. Faxen, Die Bewegung einer Starren Kugel Langs der Achsee eines mit Zaher Flussigkeit Gefullten Rohres: Arkiv for Matematik, *Astronomi och Fysik*, 27(17), 1-28, 1923.
16. F. Gui and T. F. Irvine Jr., Theoretical and experimental study of the falling cylinder viscometer, *Int. J. Heat and Mass Transfer*, 37(1), 41-50, 1994.
17. N. A. Park and T. F. Irvine Jr., Falling cylinder viscometer end correction factor, *Rev. Sci. Instrum.*, 66(7), 3982-3984, 1995.
18. G. E. LeBlanc and R. A. Secco, High pressure stokes' viscometry: a new *in-situ* technique for sphere velocity determination, *Rev. Sci. Instrum.*, 66(10), 5015-5018, 1995.
19. R. Clift, J. R. Grace, and M. E. Weber, *Bubbles, Drops, and Particles*, San Diego: Academic Press, 1978.

20. M. Coutanceau and P. Thizon, Wall effect on the bubble behavior in highly viscous liquids, *J. Fluid Mech.*, 107, 339-373, 1981.
21. R. Roscoe, Viscosity determination by the oscillating vessel method I: theoretical considerations, *Proc. Phys. Soc.*, 72, 576-584, 1958.
22. T. Iida and R. I. L. Guthrie, *The Physical Properties of Liquid Metals*, Oxford, UK: Clarendon Press, 1988.
23. H. J. McSkimin, Ultrasonic methods for measuring the mechanical properties of liquids and solids, in W.P. Mason (ed.), *Physical Acoustics: Principles and Methods, Vol. I Part A, Properties of Gases, Liquids and Solutions*, New York: Academic Press, 1964, 271-334.
24. P. Nasch, M. H. Manghnani, and R. A. Secco, A modified ultrasonic interferometer for sound velocity measurements in molten metals and alloys, *Rev. Sci. Instrum.*, 65, 682-688, 1994.
25. K. W. Katahara, C. S. Rai, M. H. Manghnani, and J. Balogh, An interferometric technique for measuring velocity and attenuation in molten rocks, *J. Geophys. Res.*, 86, 11779-11786, 1981.

Further Information

M. P. Ryan and J. Y. K. Blevins, *The Viscosity of Synthetic and Natural Silicate Melts and Glasses at High Temperatures and* 1 *Bar* (10^5 *Pascals) Pressure and at Higher Pressures*, U.S. Geological Survey Bulletin 1764, Denver, CO, 1987, 563, an extensive compilation of viscosity data in tabular and graphic format and the main techniques used to measure shear viscosity.

M. E. O'Neill and F. Chorlton, *Viscous and Compressible Fluid Dynamics*, Chichester: Ellis Horwood, 1989, mathematical methods and techniques and theoretical description of flows of Newtonian incompressible and ideal compressible fluids.

J. R. Van Wazer, J. W. Lyons, K. Y. Kim, and R. E. Colwell, *Viscosity and Flow Measurement: A Laboratory Handbook of Rheology*, New York: Interscience Publishers Div. of John Wiley & Sons, 1963, A comprehensive overview of viscometer types and simple laboratory measurements of viscosity for liquids.

12

Surface Tension Measurement

David B. Thiessen
California Institute of Technology

Kin F. Man
California Institute of Technology

The effect of surface tension is observed in many everyday situations. For example, a slowly leaking faucet drips because the force of surface tension allows the water to cling to it until a sufficient mass of water is accumulated to break free. Surface tension can cause a steel needle to "float" on the surface of water although its density is much higher than that of water. The surface of a liquid can be thought of as having a skin that is under tension. A liquid droplet is somewhat analogous to a balloon filled with air. The elastic skin of the balloon contains the air inside at a slightly higher pressure than the surrounding air. The surface of a liquid droplet likewise contains the liquid in the droplet at a pressure that is slightly higher than ambient. A clean liquid surface, however, is not elastic like a rubber skin. The tension in a piece of rubber increases as it is stretched and will eventually rupture. A clean liquid surface can be expanded indefinitely without changing the surface tension.

The mechanical model of the liquid surface is that of a skin under tension. Any given patch of the surface thus experiences an outward force tangential to the surface everywhere on the perimeter. The force per unit length of the perimeter acting perpendicular to the perimeter is defined as the *surface tension*, γ. Molecules in the interfacial region have a higher potential energy than molecules in the bulk phases because of an imbalance of intermolecular attractive forces. This results in an excess free energy per unit area associated with the surface that is numerically equivalent to the surface tension, as shown below. Consider a flat rectangular patch of fluid interface of width W and length L. In order to expand the length to $L + \Delta L$, an amount of work $\gamma W \Delta L$ must be done at the boundary. The product $W \Delta L$ is just the change in area ΔA of the surface. The work done to increase the area is thus $\Delta A \gamma$, which corresponds to the increase in surface free energy. Thus, the surface tension γ is seen to be equivalent to the surface free energy per unit area. Room-temperature organic liquids typically have surface tensions in the range of 20 mN m^{-1} to 40 mN m^{-1}, while pure water has a value of 72 mN m^{-1} at 25°C. The interface between two immiscible liquids, such as oil and water, also has a tension associated with it, which is generally referred to as the *interfacial tension.*

The surface energy concept is useful for understanding the shapes adopted by liquid surfaces. An isolated system in equilibrium is in a state of minimum free energy. Because the surface free energy contributes to the total free energy of a multiphase system, the surface free energy is minimized at

0-8493-0047-9/00/$0.00+$.50

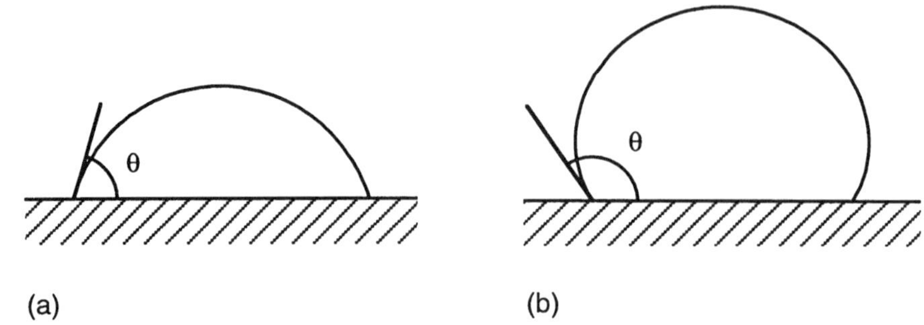

FIGURE 12.1 Illustration of contact angles and wetting. The liquid in (*a*) wets the solid better than that in (*b*).

equilibrium subject to certain constraints. Also, because the surface free energy is directly proportional to the surface area, surface area is also minimized. In the absence of gravity, a free-floating liquid droplet assumes a spherical shape because, for a given volume of liquid, the sphere has the least surface area. However, a droplet suspended from a needle tip on Earth does not form a perfect sphere because the minimum free energy configuration involves a trade-off between a reduction of the surface energy and a reduction of the gravitational potential. The droplet elongates to reduce its gravitational potential energy.

The surface energy concept is also useful for understanding the behavior of so-called surface active agents or *surfactants.* A two-component liquid mixture in thermodynamic equilibrium exhibits preferential adsorption of one component at the surface if the adsorption causes a decrease in the surface energy. The term surfactant is reserved for molecular species that strongly adsorb at the surface even when their concentration in the bulk liquid is very low. Surfactants are common in natural waters and are very important in many biological and industrial processes.

The interface between a solid and a fluid also has a surface free energy associated with it. Figure 12.1(*a*) shows a liquid droplet at rest on a solid surface surrounded by air. This system contains three different types of interfaces: solid–gas, solid–liquid, and liquid–gas, each with a characteristic surface free energy per unit area. The state of minimum free energy for the system then involves trade-offs in the surface area for the various interfaces. The region of contact between the gas, liquid, and solid is termed the *contact line.* The liquid–gas surface meets the solid surface with an angle θ measured through the liquid, which is known as the *contact angle.* The contact angle attains a value that minimizes the free energy of the system and is thus a characteristic of a particular solid–liquid–gas system. The system shown in Figure 12.1(*a*) has a smaller contact angle than that shown in Figure 12.1(*b*). The smaller the contact angle, the better the liquid is said to wet the solid surface. For $\theta = 0$, the liquid is said to be perfectly wetting.

Measurement of surface tension is important in many fields of science and engineering, as well as in medicine. A number of standard methods exist for its measurement. In many systems of interest, the surface tension changes with time, perhaps, for example, because of adsorption of surfactants. Several standard methods can be used to measure dynamic surface tension if it changes slowly with time. Special techniques have been developed to measure dynamic surface tensions for systems that evolve very rapidly.

12.1 Mechanics of Fluid Surfaces

Some methods of measuring surface tension depend on the mechanics at the line of contact between a solid, liquid, and gas. When the system is in static mechanical equilibrium, the contact line is motionless, meaning that the net force on the line is zero. Forces acting on the contact line arise from the surface tensions of the converging solid–gas, solid–liquid, and liquid–gas interfaces, denoted by γ_{SG}, γ_{SL}, and γ_{LG}, respectively (Figure 12.2). The condition of zero net force along the direction tangent to the solid surface gives the following relationship between the surface tensions and contact angle θ:

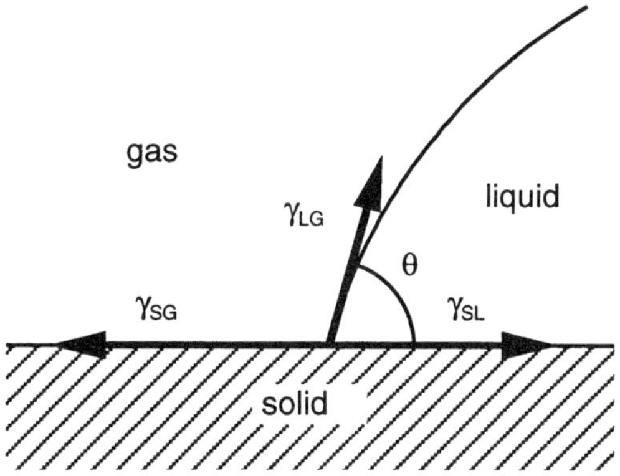

FIGURE 12.2 Surface tension forces acting on the contact line.

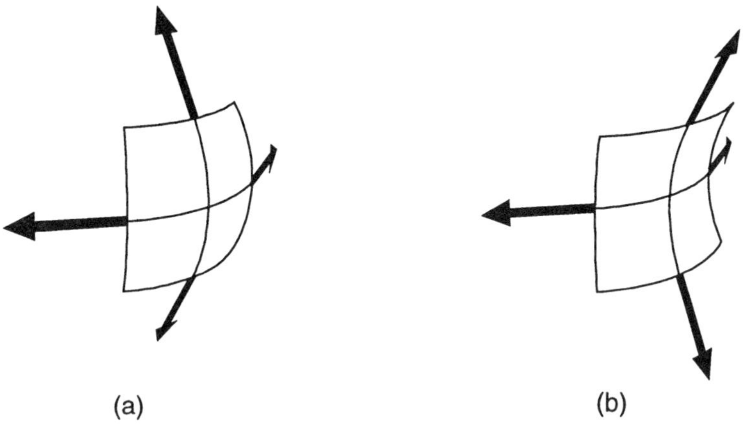

FIGURE 12.3 Mechanics of curved surfaces that have principal radii of curvature of: (*a*) the same sign, and (*b*) the opposite sign.

$$\gamma_{SG} = \gamma_{SL} + \gamma_{LG} \cos\theta \tag{12.1}$$

This is known as *Young's equation.* The contact angle is thus seen to be dependent on the surface tensions between the various phases present in the system, and is therefore an intrinsic property of the system.

As discussed in the introduction, the surface tension of a droplet causes an increase in pressure in the droplet. This can be understood by considering the forces acting on a curved section of surface as illustrated in Figure 12.3(*a*). Because of the curvature, the surface tension forces pull the surface toward the concave side of the surface. For mechanical equilibrium, the pressure must then be greater on the concave side of the surface. Figure 12.3(*b*) shows a saddle-shaped section of surface in which surface tension forces oppose each other, thus reducing or eliminating the required pressure difference across the surface. The mean curvature of a two-dimensional surface is specified in terms of the two principal radii of curvature, R_1 and R_2, which are measured in perpendicular directions. A detailed mechanical analysis of curved tensile surfaces shows that the pressure change across the surface is directly proportional to the surface tension and to the mean curvature of the surface:

$$P_A - P_B = \gamma\left(\frac{1}{R_1} + \frac{1}{R_2}\right) \tag{12.2}$$

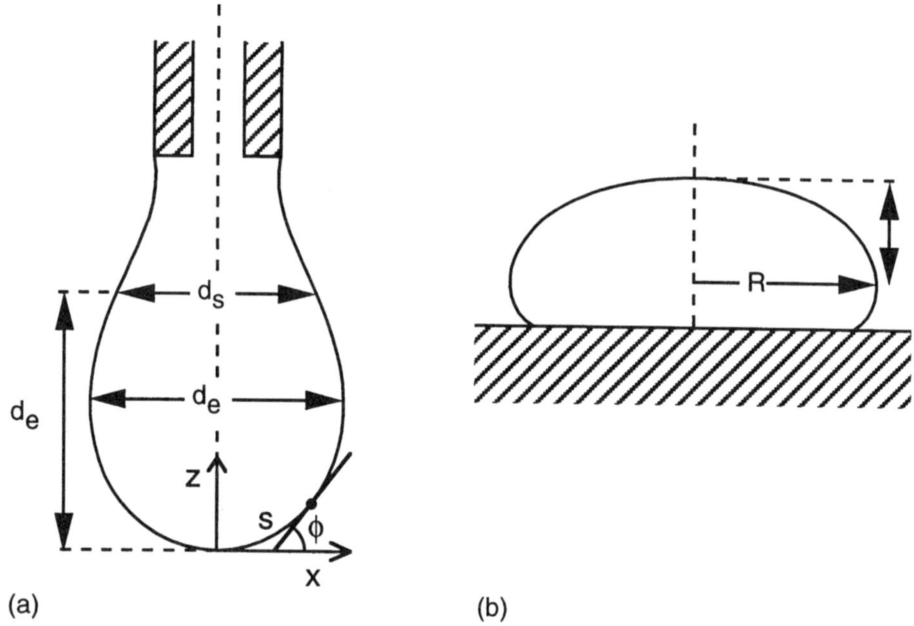

FIGURE 12.4 (*a*) A pendant drop showing the characteristic dimensions, d_e and d_s, and the coordinates used in the Young–Laplace equation. (*b*) A sessile drop showing the characteristic dimensions R and h.

where γ is the surface tension, and the quantity in brackets is twice the mean curvature. The sign of the radius of curvature is positive if its center of curvature lies in phase A and negative if it lies in phase B. Equation 12.2 is known as the *Young–Laplace equation*, and the pressure change across the interface is termed the *Laplace pressure*. Measurement of the Laplace pressure for a surface of known curvature then allows a determination of the surface tension.

Several methods of surface tension measurement are based on the measurement of the static shape of an axisymmetric drop or bubble or on the point of mechanical instability of such drops or bubbles. In a gravitational field a drop or bubble that is attached to a solid support assumes a nonspherical shape. Figure 12.4(*a*) shows the shape of a hanging droplet, also known as a pendant drop, and Figure 12.4(*b*) shows a so-called sessile drop. Axisymmetric air bubbles in water attain the same shapes as water drops in air, except that they are inverted. A bubble supported from below is thus called a hanging or pendant bubble, and a bubble supported from above is called a captive or sessile bubble. The reason for the deviation of the shape from that of a sphere can be understood from Equation 12.2. The hydrostatic pressure changes with depth more rapidly in a liquid than in a gas. The pressure difference across the surface of a pendant drop in air therefore increases from top to bottom, requiring an increase in the mean curvature of the surface according to Equation 12.2. The drop in Figure 12.4(*a*) has a neck at the top, which means that the two principal radii of curvature have opposite signs and cancel to some extent. At the bottom of the drop, the radii of curvature have the same sign, thus making the mean curvature larger. The Young–Laplace equation can be written as coupled first-order differential equations in terms of the coordinates of the interface for an axisymmetric surface in a gravitational field as:

$$\frac{dx}{ds} = \cos\phi$$
$$\frac{dz}{ds} = \sin\phi \tag{12.3}$$

$$\frac{d\phi}{ds} = \frac{2}{b} + \left(\frac{\Delta\rho g}{\gamma}\right) z - \frac{\sin\phi}{x}$$

$$x(0) = z(0) = \phi(0) = 0$$

where x and z are the horizontal and vertical coordinates, respectively, with the origin at the drop apex; s is the arc-length along the drop surface measured from the drop apex; and ϕ is the angle between the surface tangent and the horizontal (Figure 12.4(a)). The parameter b is the radius of curvature at the apex of the drop or bubble, $\Delta\rho$ is the density difference between the two fluid phases, and g is the acceleration of gravity. Numerical integration of Equation 12.3 allows one to compute the shape of an axisymmetric fluid interface. Comparison of computed shapes with experimentally measured shapes of drops or bubbles is a useful method of measuring surface tension. If all lengths in Equation 12.3 are made dimensionless by dividing them by b, the resulting equation contains only one parameter, $\beta = \Delta\rho g b^2/\gamma$, which is called the Bond number (or shape factor). The shape of an axisymmetric drop, bubble, or meniscus depends only on this one dimensionless parameter. The Bond number can also be written as $\beta = 2b^2/a^2$ where $a = \sqrt{2\gamma/\Delta\rho g}$ is known as the capillary constant and has units of length.

Several dynamic methods of measuring surface tension are based on capillary waves. Capillary waves result from oscillations of the liquid surface for which surface tension is the restoring force. The frequency of the surface oscillation is thus dependent on the surface tension and wavelength. Very low amplitude capillary waves with a broad range of frequencies are always present on liquid surfaces owing to thermal fluctuations. Larger amplitude capillary waves can be excited by purposely perturbing the surface.

12.2 Standard Methods and Instrumentation

A number of commonly used methods of measuring surface tension exist. The choice of a method depends on the system to be studied, the degree of accuracy required, and possibly on the ability to automate the measurements. In the discussion that follows, these methods are grouped according to the kind of instruments used in the measurements. Because the information presented for each method is necessarily brief, readers who are interested in constructing their own apparatus should consult the more detailed treatises in [1–4]. A list of commercially available instruments is given in Table 12.1, together with manufacturer names and approximate prices. Vendors can be contacted at the addresses given in Table 12.2.

TABLE 12.1 Commercially Available Instruments

Method	Instrument type	Manufacturer/Product name	Approximate price (range)
Capillary rise	Manual	Fisher	$79
Wilhelmy plate/du Noüy ring	Manual, mechanical balance	CSC, Fisher, Kahl	$2000–$4000
Wilhelmy plate/du Noüy ring	Manual, electrobalance	KSV, Lauda, NIMA	$4000–$11,000
Wilhelmy plate/du Noüy ring	Automatic, electrobalance	Cahn, Krüss, KSV, NIMA	$9000–$24,000
Maximum bubble pressure	Automatic	Krüss, Lauda, Sensa Dyne	$5000–$23,000
Pendant/sessile drop	Manual	Krüss, Rame-Hart	$7000–$10,000
Pendant/sessile drop	Automatic	ADSA, AST, FTA, Krüss, Rame-Hart, Temco	$10,000–$100,000
Drop weight/volume	Automatic	Krüss, Lauda	$16,000–$21,000
Spinning drop	Manual	Krüss	$20,100

Note: Price ranges reflect differences in degree of automation, the number of accessories included, or variation in price between manufacturers.

TABLE 12.2 Manufacturers and Suppliers of Instruments for Surface Tension Measurement

AST Products 9 Linnell Circle Billerica, MA 01821-3902 Tel: (508) 663-7652	Fisher Scientific 711 Forbes Ave. Pittsburgh, PA 15219-4785 Tel: (800) 766-7000
Applied Surface Thermodynamics Research Associates (ASTRA) (distributor of ADSA instrumentation) 15 Brendan Rd. Toronto, Ontario Canada, M4G 2W9 Tel: (416) 978-3601	Kahl Scientific Instrument Corp. P.O. Box 1166 El Cajon, CA 92022-1166 Tel: (619) 444-2158
Brinkmann Instruments, Inc. (distributor for Lauda tensiometers) One Catiaque Road P.O. Box 1019 Westbury, NY 11590-0207 Tel: (800) 645-3050	Krüss U.S.A. 9305-B Monroe Road Charlotte, NC 28270-1488 Tel: (704) 847-8933
Cahn Instruments 5225 Verona Rd., Bldg. 1 Madison, WI 53711 Tel: (800) 244-6305	KSV Instruments U.S.A. P.O. Box 192 Monroe, CT 06468 Tel: (800) 280-6216
CSC Scientific Company, Inc. 8315 Lee Highway Fairfax, VA 22031 Tel: (800) 458-2558	Rame-Hart, Inc. 8 Morris Ave. Mountain Lakes, NJ 07046 Tel: (201) 335-0560
CTC Technologies, Inc. (distributor for NIMA tensiometers) 7925-A North Oracle Road, Suite 364 Tucson, AZ 85704-6356 Tel: (800) 282-8325	Sensa Dyne Instrument Div. Chem-Dyne Research Corp. P.O. Box 30430 Mesa, AZ 85275-0430 Tel: (602) 924-1744
First Ten Angstroms (FTA) 465 Dinwiddie Street Portsmouth, VA 23704 Tel: (800) 949-4110	Temco, Inc. 4616 North Mingo Tulsa, OK 74117-5901 Tel: (918) 834-2337

Capillary Rise Method

If a glass capillary tube is brought into contact with a liquid surface, and if the liquid wets the glass with a contact angle of less than 90°, then the liquid is drawn up into the tube as shown in Figure 12.5(*a*). The surface tension is directly proportional to the height of rise, h, of the liquid in the tube relative to the flat liquid surface in the larger container. By applying Equation 12.2 to the meniscus in the capillary tube, the following relationship is obtained:

$$\Delta\rho g h = \frac{2\gamma}{b} \tag{12.4}$$

where b is the radius of curvature at the center of the meniscus and $\Delta\rho$ is the density difference between liquid and gas. For small capillary tubes, b is well approximated by the radius of the tube itself, assuming that the contact angle of the liquid on the tube is zero. For larger tubes or for increased accuracy, the

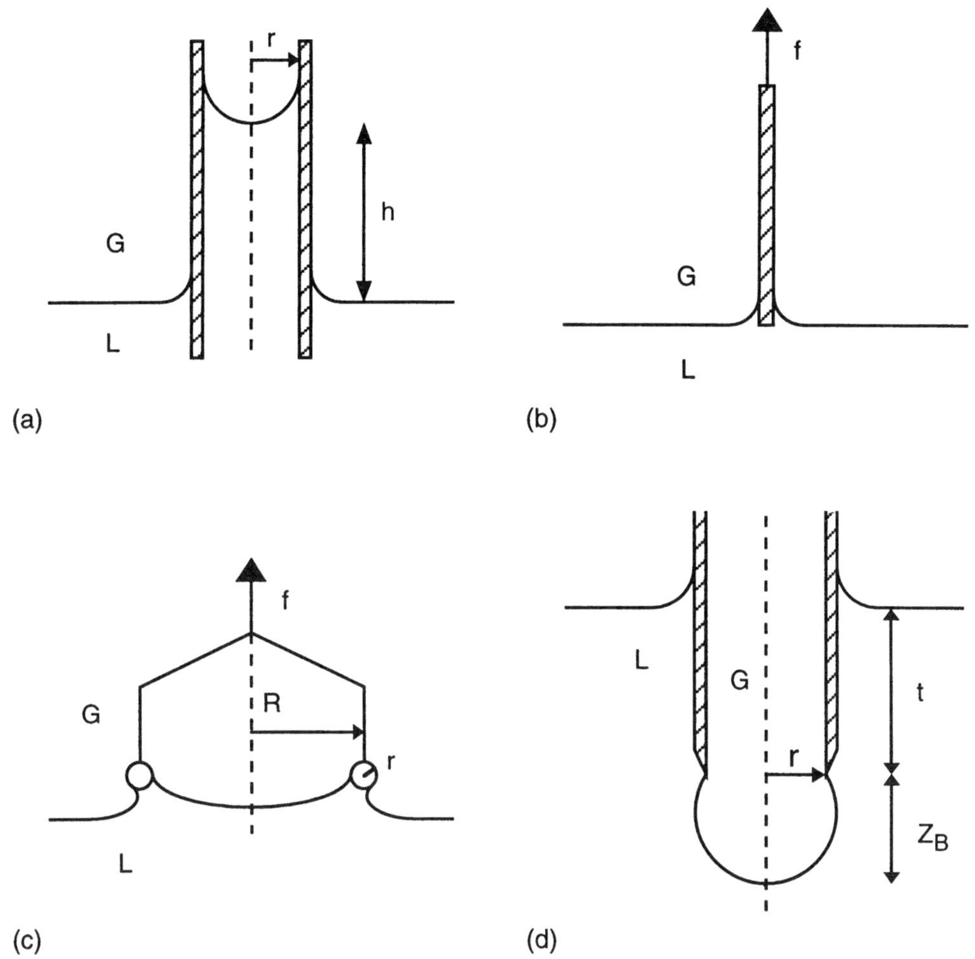

FIGURE 12.5 Geometries for: (*a*) capillary rise method, (*b*) Wilhelmy plate method, (*c*) du Noüy ring method, and (*d*) maximum bubble pressure method.

value of *h* must be corrected for gravitational deformation of the meniscus (p. 12 of [1]). Obtaining accurate results with the capillary rise method requires using a thoroughly clean glass capillary tube with a very uniform diameter of less than 1 mm. The container for the liquid should be at least 8 cm in diameter and the liquid must wet the capillary tube with a contact angle of zero. This method is primarily useful for pure liquids and is capable of high accuracy at relatively low cost.

Wilhelmy Plate and du Noüy Ring Methods

Measurement of the pull of a liquid surface directly on a solid object is the basis for two of the standard methods discussed here. In the Wilhelmy plate method, the solid object is a flat, thin plate that the test liquid should wet with a zero contact angle. The plate is suspended vertically from a delicate balance that is zeroed with the plate suspended in air. The test liquid is brought into contact with the bottom of the plate, causing the plate to be pulled down into the liquid by the surface tension force. The force applied to the plate from above is then increased to bring the bottom edge of the plate level with the flat surface of the liquid (Figure 12.5(*b*)). This avoids the necessity to make buoyancy corrections to the measurement. The surface tension is computed from the force measurement, *f*, using:

$$\gamma = \frac{f \cos\theta}{\left[2\left(l+t\right)\right]} \tag{12.5}$$

where l is the length of the plate and t is its thickness. The contact angle θ is often assumed to be zero for common liquids on clean glass or platinum plates, but one should be aware of the error caused by a non-zero contact angle. No other correction factors are necessary for this method and the fluid density does not need to be known.

The du Noüy ring method is known as a maximum pull method, of which there are several variations. The technique is to contact the liquid surface with a ring and then measure the force continuously as the surface is lowered until a maximum force, f_{max}, is recorded. The maximum force typically occurs just before the ring detaches from the surface. The surface tension is obtained from the formula:

$$\gamma = \left(\frac{f_{max}}{4\pi R}\right)\left[F\left(\frac{R^3}{V}, \frac{R}{r}\right)\right] \tag{12.6}$$

where R and r are the radii of the ring and wire, respectively, as indicated in Figure 12.5(c), V is the volume of liquid raised by the ring, and F is a correction factor (F is tabulated in Table 5, p. 132 of [4]). The du Noüy ring method requires knowledge of the liquid density, ρ_L, in order to determine V from $V = f_{max}/\rho_L$. This method requires the liquid to wet the ring with zero contact angle and is not suitable for solutions that attain surface equilibrium slowly.

A single instrument is normally capable of performing either Wilhelmy plate or du Noüy ring measurements. Some commercially available instruments can perform the complete measurement procedure automatically. Computer interfacing with a Wilhelmy plate instrument allows automatic data logging which can be used to follow changes in surface tension with time in surfactant solutions.

Maximum Bubble Pressure Method

The *maximum bubble pressure method* (MBPM) involves direct measurement of the pressure in a bubble to determine the surface tension. A tube is lowered to a depth t in the test liquid and gas is injected to form a bubble of height Z_B at the tip of the tube as shown in Figure 12.5(d). The increase in bubble pressure, P_b, over ambient pressure, P_a, arising from the interface is given by the sum of a hydrostatic pressure and Laplace pressure:

$$\delta p = P_b - P_a - \Delta\rho g t = \Delta\rho g Z_B + \frac{2\gamma}{b} \tag{12.7}$$

As a new bubble begins to form, Z_B increases while b, the radius of curvature at the bubble apex, decreases, resulting in an increase in pressure in the bubble. Ultimately, b increases as the bubble grows larger, thus reducing the pressure. The pressure in the bubble thus reaches a maximum when δp reaches a maximum, which in turn can be theoretically related to the surface tension. For $\delta p = \delta p_{max}$, Equation 12.7 can be rewritten in dimensionless form as follows:

$$\frac{r}{X} = \frac{r}{b} + \frac{r}{a}\frac{Z_B}{b}\left(\frac{\beta}{2}\right)^{1/2} \tag{12.8}$$

where r is the tube radius, X is a length defined as $X = 2\gamma/\delta p_{max}$, a is the capillary constant, and β is the Bond number. The dimensionless quantity r/X depends only on r/a, the relationship being determined by Equation 12.8 combined with numerical solutions of Equation 12.3. Tabulations of this relationship are used to calculate the surface tension by an iterative procedure (p. 18 of [1]). The standard MBPM requires a knowledge of the fluid densities, tube radius, and depth of immersion of the tube.

A differential MBPM uses two tubes of different diameters immersed to the same depth. The difference in the maximum bubble pressure for the two tubes, ΔP, is measured, eliminating the need to know the immersion depth and making the method less sensitive to errors in the knowledge of the liquid density. For the differential MBPM, surface tension is computed from (see [5]):

$$\gamma = A\Delta P\left[1+\left(\frac{0.69 r_2 \rho_L}{\Delta P}\right)\right] \tag{12.9}$$

where r_2 is the radius of the larger tube, ρ_L is the liquid density, and A is an apparatus-dependent constant that is determined by calibration with several standard liquids [6]. Automated MBPM units are commercially available (see Table 12.1). Sensa Dyne manufactures differential MBPM units that allow for online process measurements under conditions of varying temperature and pressure.

Pendant Drop and Sessile Drop Methods

The shape of an axisymmetric pendant or sessile drop (Figure 12.4) depends on only a single parameter, the Bond number, as discussed above. The Bond number is a measure of the relative importance of gravity to surface tension in determining the shape of the drop. For Bond numbers near zero, surface tension dominates and the drop is nearly spherical. For larger Bond numbers, the drop becomes significantly deformed by gravity. In principle, the method involves obtaining an image of the drop and comparing its shape and size to theoretical profiles obtained by integrating Equation 12.3 for various values of β and b. Once β and b have been determined from shape and size comparison, the surface tension is calculated from:

$$\gamma = \frac{\Delta\rho g b^2}{\beta} \tag{12.10}$$

In practice, the drop shape and size has traditionally been determined by the manual measurement of several characteristic dimensions (see Figure 12.4) of the drop from a photographic print. For pendant drops, the ratio d_s/d_e is correlated to a shape factor H from which surface tension is calculated (p. 27 of [3], [7]) according to:

$$\gamma = \frac{\Delta\rho g d_e^2}{H} \tag{12.11}$$

For sessile drops, various analytical formulae are available for computation of surface tension directly from the characteristic dimensions (p. 36 of [3]). Drop shape methods based on characteristic dimensions require very accurate measurement of the dimensions for good results. For more accurate results, methods that fit the entire shape of the edge of the drop to the Laplace equation are recommended.

In recent years, the entire procedure has been automated using digital imaging and computer image analysis [8, 9]. Typically, several hundred coordinates on the edge of the drop are located with subpixel resolution by computer analysis of the digital image. The size, shape, and horizontal and vertical offsets of the theoretical profile given by Equation 12.3 are varied by varying four parameters: b, β, and the pixel coordinates of the drop apex, x_0 and z_0. A best fit of the theoretical profile to the measured edge coordinates is obtained by minimizing an objective function. A digital image of a pendant drop can be analyzed for surface tension on a desktop computer in 1 or 2 s [10]. The speed of algorithms for pendant drop analysis on modern desktop computers has allowed this method to be used to track changes in surface tension for surfactant-covered surfaces by analyzing a sequence of images. The algorithms can simultaneously track the surface area and volume of the drop or bubble. Both soluble and insoluble surfactants have been studied using the pendant drop, sessile drop, pendant bubble, and captive bubble configurations [11–13]. Table 12.1 lists several manufacturers that can provide software for automated analysis of surface tension from drops or bubbles in pendant or sessile configurations. The increased accuracy and simplicity of the automated pendant drop procedure makes it a very flexible method that has been applied to measure ultralow interfacial tensions, pressure, temperature and time dependence of interfacial tension, relaxation of adsorption layers, measurement of line tensions, and film-balance measurements [14].

Drop Weight or Volume Method

A pendant drop will become unstable and detach from its support if it grows too large. The weight of the detached portion is related to the surface tension of the fluid by:

$$\gamma = \left(\frac{mg}{r}\right)\left[F\left(\frac{r}{V^{1/3}}\right)\right] \tag{12.12}$$

where mg is the weight of the detached drop, r is the radius of the tip from which the drop hangs, and V is the volume of the detached drop. An empirical correction factor, F, is tabulated as a function of $r/V^{1/3}$ (p. 50 of [3]). For Equation 12.12 to apply, drops must be formed slowly. Measurements typically involve weighing the accumulated liquid from a large number of drops to determine the average weight per drop. The density of the fluid must be known in order to determine the drop volume and then obtain the factor F. Another method involves measuring the volumetric flow rate of liquid to the tip while counting the drops. The density of the fluid must be known in order to determine the drop weight. The latter method allows for automation of measurements [15].

Spinning Drop Method

The spinning drop method is a shape-measurement method similar to the pendant and sessile drop methods. However, the deformation of the drop in this case is caused by radial pressure gradients in a rapidly spinning tube. This method is normally used for measuring interfacial tensions between immiscible liquids. A horizontal glass tube with sealed ends is filled with the more dense liquid through a filling port. The tube is then spun about its axis while a drop of the lower density liquid is injected into the tube. The pressure in the outer liquid increases from the center of the tube toward the walls as a result of the spinning motion. The pressure gradient forces the drop to move to the center of the tube and causes it to elongate, while surface tension opposes elongation. Measurement of the maximum drop diameter, $2r_{max}$, and length, $2h_{max}$, together with the angular velocity of rotation, Ω, allows for calculation of the surface tension according to:

$$\gamma = \frac{1}{2}\left(\frac{r_{max}}{r^*_{max}}\right)^3 \Delta\rho\Omega^2 \tag{12.13}$$

where r^*_{max} is correlated to the aspect ratio r_{max}/h_{max} [16]. The spinning drop method is particularly suited for measuring ultralow interfacial tensions (10^{-2} mN m^{-1} to 10^{-4} mN m^{-1}).

12.3 Specialized Methods

Dynamic Surface Tension

In an aqueous solution of soluble surfactant, the surface tension decreases following creation of new surface area because of adsorption of surfactant molecules. Surfactant adsorption kinetics can be studied by measuring the change in surface tension with time. For a dilute solution, the rate of change of surface tension is often slow enough that automated versions of static methods such as the Wilhelmy plate or pendant drop [17] methods can be used to follow the changes in surface tension. In concentrated solutions in which large changes in surface tension can occur within a fraction of a second following surface creation, a dynamic method must be used. A liquid jet emerging from an elliptical orifice has stationary waves on its surface, the wavelengths of which are related to the surface tension. The oscillating jet method has been used to measure surface tension for surface ages as low as 0.6 ms [18]. A dynamic version of the maximum bubble pressure method has been used to measure dynamic surface tension at surface ages down to 0.1 ms [19].

Surface Viscoelasticity

A liquid surface covered by a monolayer of surfactant exhibits viscoelastic behavior. In addition to surface tension, the surface rheology is characterized in terms of dilatational and shear elasticities as well as dilatational and shear viscosities. The dilatational properties in particular are important in a variety of situations from foam stability to the functioning of the human lung. The surface dilatational modulus is proportional to the change in surface tension for a given change in surface area. This modulus depends on the rate of change of surface area for both soluble and insoluble surfactant monolayers, which indicates that relaxation processes are active. These relaxation processes give rise to the surface dilatational viscosity. For the case of soluble surfactants, one of the relaxation processes is the adsorption or desorption of molecules at the surface. The equilibrium dilatational elasticity of an insoluble monolayer can be measured by slowly expanding or compressing the monolayer in a Langmuir trough while monitoring the surface tension with a Wilhelmy plate apparatus [20]. Studies of surface rheology at high rates of surface expansion or compression are of interest for both soluble and insoluble surfactants.

Surface tension relaxation following sudden expansion or compression of the surface for a soluble surfactant has been studied by the automated pendant drop method [21]. A method known as oscillating bubble tensiometry has been applied to measure the kinetics of adsorption and desorption for soluble surfactants [22, 23]. Other methods for studying dynamic dilatational viscoelastic properties are reviewed in [24], including transverse and longitudinal capillary wave methods, a modified maximum bubble pressure method, and an oscillating bubble method.

Measurements at Extremes of Temperature and Pressure

Several of the standard methods described in this chapter can be adapted to make surface or interfacial tension measurements at extreme temperatures and/or pressures. The most common methods used to measure the surface tension of high-temperature molten metals, alloys, and semiconductors are the maximum bubble pressure method [25] and the pendant or sessile drop method [26–28]. Measurement of the interfacial tension between oil and a second immiscible phase at high pressure and elevated temperature is of interest for understanding aspects of enhanced oil recovery. The pendant drop method has been applied under pressures of 82 MPa at 449 K [29], while a capillary wave method has been applied at 136 MPa and 473 K [30]. A pendant drop apparatus capable of measurements to 10,000 psi (69 MPa) and 350°F (450 K) is commercially available from Temco Inc. (Table 12.2).

Interfacial Tension

Measurement of the interfacial tension between two immiscible liquids can present special difficulties. Measurement by the capillary rise, du Noüy ring, or Wilhelmy plate method is problematic in that the contact angle is often nonzero. The pendant drop [7] and drop weight [31] methods can both be applied, provided the densities of the two liquids are sufficiently different. The pendant drop method, in particular, is widely used for interfacial tension measurement. Interfacial tension can be measured by a modified maximum bubble pressure method in which one measures the maximum pressure in a liquid drop injected into a second immiscible liquid [32]. The modified maximum bubble pressure method [32] and a liquid bridge method [33] have been used to measure interfacial tension between two liquids of equal density. Ultralow values of interfacial tension can be measured by the spinning drop [34], pendant drop [35], and capillary wave methods [34].

Defining Terms

Surface tension: A force per unit length that acts tangential to a liquid surface and perpendicular to any line that lies within the surface.

Surface energy: The excess free energy per unit area associated with a surface between two phases. For a liquid–fluid surface, the surface energy is numerically equivalent to the surface tension.

Acknowledgments

During the preparation of this chapter, one of us (DBT) was supported in part by the National Aeronautics and Space Administration (NASA) and by the Office of Naval Research. The work by one of us (KFM) was performed at the Jet Propulsion Laboratory, California Institute of Technology, under contract with NASA.

References

1. A. W. Adamson, *Physical Chemistry of Surfaces*, 5th ed., New York: John Wiley & Sons, 1990.
2. A. E. Alexander and J. B. Hayter, Determination of surface and interfacial tension, in A. Weissberger and B. W. Rossiter (eds.), *Physical Methods of Chemistry, Part V*, 4th ed., New York: John Wiley & Sons, 1971.
3. A. Couper, Surface tension and its measurement, in B. W. Rossiter and R. C. Baetzold (eds.), *Physical Methods of Chemistry, Vol. 9A*, 2nd ed., New York: John Wiley & Sons, 1993.
4. J. F. Padday, Surface tension. II. The measurement of surface tension, in E. Matijevic (ed.), *Surface and Colloid Science, Vol. 1*, New York: John Wiley & Sons, 1969.
5. S. Sugden, The determination of surface tension from the maximum pressure in bubbles. Part II, *J. Chem. Soc.*, 125, 27-31, 1924.
6. ASTM Standard D3825-90, Standard test method for dynamic surface tension by the fast-bubble technique, *1996 Annual Book of ASTM Standards, Vol. 05.02,* West Conshohocken, PA: ASTM, 1996, 575-579.
7. D. S. Ambwani and T. Fort, Jr., Pendant drop technique for measuring liquid boundary tensions, in R. J. Good and R. R. Stromberg (eds.), *Surface and Colloid Science, Vol. 11,* New York: Plenum Press, 1979.
8. Y. Rotenberg, L. Boruvka, and A. W. Neumann, Determination of surface tension and contact angle from the shapes of axisymmetric fluid interfaces, *J. Colloid Interface Sci.*, 93, 169-183, 1983.
9. P. Cheng, D. Li, L. Boruvka, Y. Rotenberg, and A. W. Neumann, Automation of axisymmetric drop shape analysis for measurements of interfacial tensions and contact angles, *Colloids Surf.*, 43, 151-167, 1990.
10. D. B. Thiessen, D. J. Chione, C. B. McCreary, and W. B. Krantz, Robust digital image analysis of pendant drop shapes, *J. Colloid Interface Sci.*, 177, 658-665, 1996.
11. S. Lin, K. McKeigue, and C. Maldarelli, Diffusion-controlled surfactant adsorption studied by pendant drop digitization, *AIChE J.*, 36, 1785-1795, 1990.
12. D. Y. Kwok, D. Vollhardt, R. Miller, D. Li, and A. W. Neumann, Axisymmetric drop shape analysis as a film balance, *Colloids Surf., A*, 88, 51-58, 1994.
13. W. M. Schoel, S. Schurch, and J. Goerke, The captive bubble method for the evaluation of pulmonary surfactant: surface tension, area, and volume calculations, *Biochim. Biophys. Acta*, 1200, 281-290, 1994.

14. S. Lahooti, O. I. Del Rio, P. Cheng, and A. W. Neumann, Axisymmetric drop shape analysis (ADSA), in A. W. Neumann and J. K. Spelt (eds.), *Applied Surface Thermodynamics*, New York: Marcel Dekker, 1996.
15. M. L. Alexander and M. J. Matteson, The automation of an interfacial tensiometer, *Colloids Surf.*, 27, 201-217, 1987.
16. J. C. Slattery and J. Chen, Alternative solution for spinning drop interfacial tensiometer, *J. Colloid Interface Sci.*, 64, 371-373, 1978.
17. D. Y. Kwok, M. A. Cabrerizo-Vilchez, Y. Gomez, S. S. Susnar, O. Del Rio, D. Vollhardt, R. Miller, and A. W. Neumann, Axisymmetric drop shape analysis as a method to study dynamic interfacial tensions, in V. Pillai and D. O. Shah (eds.), *Dynamic Properties of Interfaces and Association Structures*, Champaign, IL: AOCS Press, 1996.
18. W. D. E. Thomas and L. Potter, Solution/air interfaces. I. An oscillating jet relative method for determining dynamic surface tensions, *J. Colloid Interface Sci.*, 50, 397-412, 1975.
19. V. B. Fainerman and R. Miller, Dynamic surface tension measurements in the sub-millisecond range, *J. Colloid Interface Sci.*, 175, 118-121, 1995.
20. G. L. Gaines Jr., *Insoluble Monolayers at Liquid-Gas Interfaces*, New York: John Wiley & Sons, 1966, 44.
21. R. Miller, R. Sedev, K.-H. Schano, C. Ng, and A. W. Neumann, Relaxation of adsorption layers at solution/air interfaces using axisymmetric drop-shape analysis, *Colloids Surf.*, 69, 209-216, 1993.
22. D. O. Johnson and K. J. Stebe, Oscillating bubble tensiometry: a method for measuring the surfactant adsorptive-desorptive kinetics and the surface dilatational viscosity, *J. Colloid Interface Sci.*, 168, 21-31, 1994.
23. D. O. Johnson and K. J. Stebe, Experimental confirmation of the oscillating bubble technique with comparison to the pendant bubble method: the adsorption dynamics of 1-decanol, *J. Colloid Interface Sci.*, 182, 526-538, 1996.
24. D. A. Edwards, H. Brenner, and D. T. Wasan, *Interfacial Transport Processes and Rheology*, Boston: Butterworth-Heinemann, 1991.
25. C. Garcia-Cordovilla, E. Louis, and A. Pamies, The surface tension of liquid pure aluminium and aluminium-magnesium alloy, *J. Mater. Sci.*, 21, 2787-2792, 1986.
26. B. C. Allen, The surface tension of liquid transition metals at their melting points, *Trans. Metall. Soc. AIME*, 227, 1175-1183, 1963.
27. D. B. Thiessen and K. F. Man, A quasi-containerless pendant drop method for surface tension measurements on molten metals and alloys, *Int. J. Thermophys.*, 16, 245-255, 1995.
28. S. C. Hardy, The surface tension of liquid silicon, *J. Cryst. Growth*, 69, 456-460, 1984.
29. V. Schoettle and H. Y. Jennings, Jr., High-pressure high-temperature visual cell for interfacial tension measurement, *Rev. Sci. Instrum.*, 39, 386-388, 1968.
30. R. Simon and R. L. Schmidt, A system for determining fluid properties up to 136 MPa and 473K, *Fluid Phase Equilib.*, 10, 233-248, 1983.
31. K. Hool and B. Schuchardt, A new instrument for the measurement of liquid-liquid interfacial tension and the dynamics of interfacial tension reduction, *Meas. Sci. Technol.*, 3, 451-457, 1992.
32. A. Passerone, L. Liggieri, N. Rando, F. Ravera, and E. Ricci, A new experimental method for the measurement of the interfacial tension between immiscible fluids at zero Bond number, *J. Colloid Interface Sci.*, 146, 152-162, 1991.
33. G. Pétré and G. Wozniak, Measurement of the variation of interfacial tension with temperature between immiscible liquids of equal density, *Acta Astronaut.*, 13, 669-672, 1986.
34. D. Chatenay, D. Langevin, J. Meunier, D. Bourbon, P. Lalanne, and A. M. Bellocq, Measurement of low interfacial tension, comparison between a light scattering technique and the spinning drop technique, *J. Dispersion Sci. Technol.*, 3, 245-260, 1982.
35. D. Y. Kwok, P. Chiefalo, B. Khorshiddoust, S. Lahooti, M. A. Cabrerizo-Vilchez, O. Del Rio, and A. W. Neumann, Determination of ultralow interfacial tension by axisymmetric drop shape analysis, in R. Sharma (ed.), *Surfactant Adsorption and Surface Solubilization (ACS Symp. Ser. 615)*, Washington, D.C.: ACS, 1995, 374-386.

III

Mechanical Variables Measurement — Thermal

13

Temperature Measurement

Robert J. Stephenson
University of Cambridge

Armelle M. Moulin
University of Cambridge

Mark E. Welland
University of Cambridge

Jim Burns
Burns Engineering Inc.

Meyer Sapoff
MS Consultants

R. P. Reed
Proteun Services

Randy Frank
Motorola, Inc.

Jacob Fraden
Advanced Monitors Corporation

J.V. Nicholas
Industrial Research Limited

Franco Pavese
CNR Instituto di Metrologia "G. Colonnetti"

Jan Stasiek
Technical University of Golansk

Tolestyn Madaj
Technical University of Golansk

Jaroslaw Mikielewicz
Institute of Fluid Flow Machinery

Brian Culshaw
University of Strathclyde

0-8493-0047-9/00/$0.00+$.50

13.1 Bimaterials Thermometers

Robert J. Stephenson, Armelle M. Moulin, and Mark E. Welland

The first known use of differential thermal expansion of metals in a mechanical device was that of the English clockmaker John Harrison in 1735. Harrison used two dissimilar metals in a clock escapement to account for the changes in temperature on board a ship. This first marine chronometer used a gridiron of two metals that altered the flywheel period of the clock through a simple displacement. This mechanical actuation, resulting from the different thermal expansivities of two metals in contact, is the basis for all bimetallic actuators used today.

The bimetallic effect is now used in numerous applications ranging from domestic appliances to compensation in satellites. The effects can be used in two ways: either as an actuator or as a temperature measuring system. A bimetallic actuator essentially consists of two metal strips fixed together. If the two metals have different expansivities, then as the temperature of the actuator changes, one element will expand more than the other, causing the device to bend out of the plane. This mechanical bending can then be used to actuate an electromechanical switch or be part of an electrical circuit itself, so that contact of the bimetallic device to an electrode causes a circuit to be made. Although in its simplest form a bimetallic actuator can be constructed from two flat pieces in metal, in practical terms a whole range of shapes are used to provide maximum actuation or maximum force during thermal cycling.

As a temperature measuring device, the bimetallic element, similar in design to that of the actuator above, can be used to determine the ambient temperature if the degree of bending can be measured. The advantage of such a system is that the amount of bending can be mechanically amplified to produce a large and hence easily measurable displacement.

The basic principle of a bimetallic actuator is shown in Figure 13.1. Here, two metal strips of differing thermal expansion are bonded together. When the temperature of the assembly is changed, in the absence

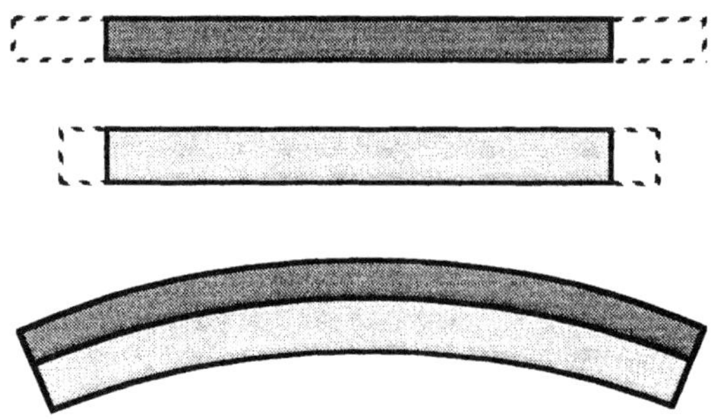

FIGURE 13.1 Linear bimetallic strip.

of external forces, the bimetallic actuator will take the shape of an arc. The total displacement of the actuator out of the plane of the metal strips is much greater than the individual expansions of the metallic elements. To maximize the bending of the actuator, metals or alloys with greatly differing coefficients of thermal expansion are normally selected. The metal having the largest thermal expansivity is known as the *active element*, while the metal having the smaller coefficient of expansion is known as the *passive element*. For maximum actuation, the passive element is often an iron–nickel alloy, Invar, having an almost zero thermal expansivity (actually between 0.1 and 1×10^{-6} K^{-1}, depending upon the composition). The active element is then chosen to have maximum thermal expansivity given the constraints of operating environment and costs.

In addition to maximizing the actuation of the bimetallic element, other constraints such as electrical and thermal conductivity can be made. In such cases, a third metallic layer is introduced, consisting of either copper or nickel sandwiched between the active and passive elements so as to increase both the electrical and thermal conductivity of the actuator. This is especially important where the actuator is part of an electrical circuit and needs to pass current in addition to being a temperature sensor.

Linear Bimaterial Strip

Basic Equations

The analysis of the stress distribution and the deflection of an ideal bimetallic strip was first deduced by Timoshenko [1], who produced a simple derivation from the theory of elasticity. Figure 13.2 shows the internal forces and moments that induce bending in a bimetallic strip followed by the ideal stress distribution in the beam. This theory is derived for bimetallic strips, but is equally applicable to bimaterial strips.

The general equation for the curvature radius of a bimetallic strip uniformly heated from T_0 to T in the absence of external forces is given by [1]:

$$\frac{1}{R} - \frac{1}{R_0} = \frac{6(1+m)^2(\alpha_2 - \alpha_1)(T - T_0)}{t\left[3(1+m)^2 + (1+mn)\left(m^2 + 1/mn\right)\right]} \tag{13.1}$$

where $1/R_0$ = Initial curvature of the strip at temperature T_0
α_1 and α_2 = Coefficients of expansion of the two elements: (1) low expansive material and (2) high expansive material
n = E_1/E_2, with E_1 and E_2 their respective Young's moduli
m = t_1/t_2, with t_1 and t_2 their respective thicknesses
t = $t_1 + t_2$ thickness of the strip

The width of the strip is taken as equal to unity.

Equation 13.1 applies for several strip configurations, including the simply supported strip and a strip clamped at one end (i.e., a cantilever). For a given configuration, the deflection of a strip can be determined by its relationship with curvature, $1/R$.

An example of a calculation of deflection is a bimetallic strip simply supported at its two ends. The initial curvature $1/R_0$ is assumed to be zero. Figure 13.3 shows the geometrical relationship between the radius R of the strip and the deflection d at its mid-point and is given by:

$$\left(R - t_2\right)^2 = \left(R - d - t_2\right)^2 + \left(\frac{L}{2}\right)^2 \tag{13.2}$$

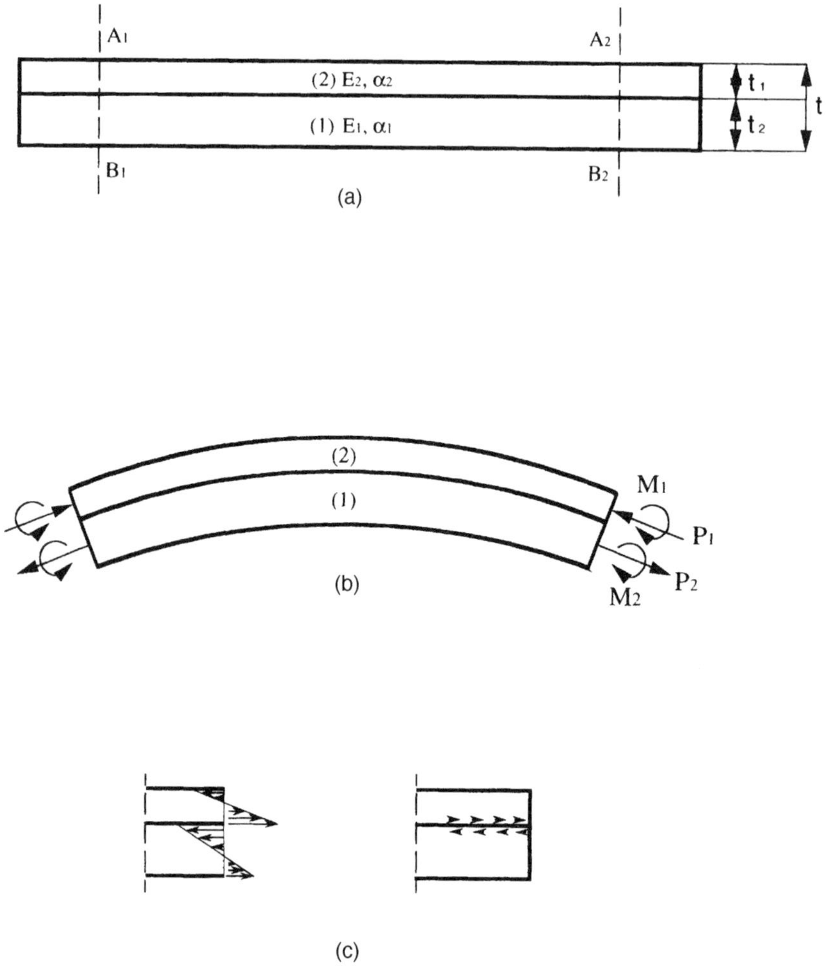

FIGURE 13.2 Bending of bimetallic strip uniformly heated with $\alpha_2 \geq \alpha_1$. (a) Bimetallic strip. A_1B_1–A_2B_2 is an element cut out from the strip. (*b*) Bending of the element A_1B_1–A_2B_2 when uniformly heated. Assuming $\alpha_2 > \alpha_1$, the deflection is convex up. The total force acting over the section of (1) is an axial tensile force P_1 and bending moment M_1, whereas over the section of (2) it is an axial compressive force P_2 and bending moment M_2. (*c*) Sketch of the internal resulting stress distribution. (Left): normal stresses over the cross section of the strip. The maximum stress during heating is produced at the interface between the two components of the strip. This stress is due to both axial force and bending. (Right): shearing stresses at the ends of the strip.

Hence,

$$\frac{1}{R} = \frac{8d}{L^2 + 4d^2 + 8dt_2} \tag{13.3}$$

Making the assumption that the deflection and the thickness are less than 10% of the length of the strip (which is true in most practical cases) means the terms $8dt_2$ and $4d^2$ are therefore negligible and the expression reduces to:

$$d = \frac{L^2}{8R} \tag{13.4}$$

or

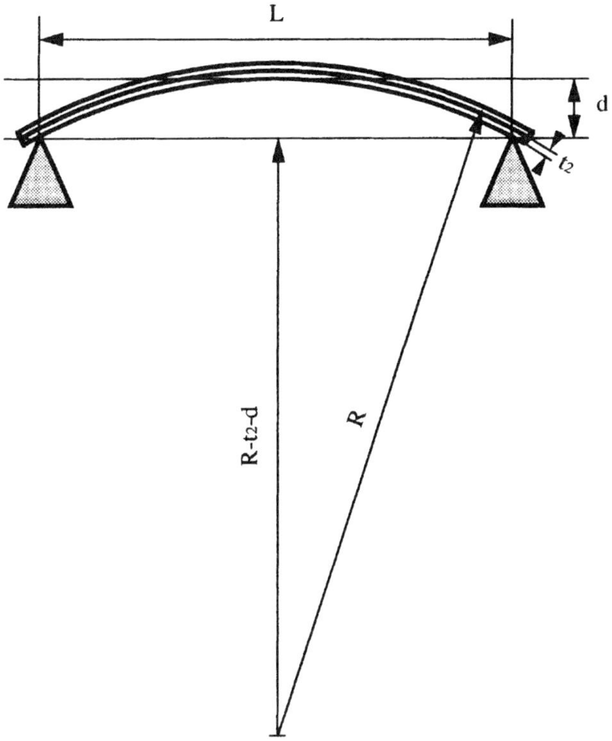

FIGURE 13.3 Bending of a strip simply supported at its ends.

$$d = L^2 \frac{3(1+m)^2}{4t\left[3(1+m)^2 + (1+mn)\left(m^2 + 1/mn\right)\right]}(\alpha_2 - \alpha_1)(T - T_0) \tag{13.5}$$

If a 100-mm strip is composed of two layers of the same thickness (0.5 mm) with the high-expansive layer being made of iron (from Table 13.1, $E_2 = 211$ GPa and $\alpha_2 = 12.1 \times 10^{-6}$ K^{-1}), the low-expansive layer made of Invar (from Table 13.1, $E_1 = 140$ GPa and $\alpha_1 = 1.7 \times 10^{-6}$ K^{-1}), and the temperature increases from 20°C to 120°C, then the theoretical bending at the middle of the strip will be 1.92 mm.

As a second example, consider the calculation of the deflection of the free end of a bimetallic cantilever strip as illustrated in Figure 13.4. In this case, the geometrical relation is:

$$(R + t_1)^2 = (R + t_1 - d)^2 + L^2 \tag{13.6a}$$

or

$$\frac{1}{R} = \frac{2d}{L^2 + d^2 - 2dt_1} \tag{13.6b}$$

Making the same assumptions as before, that is, $d^2 \ll L^2$ and $dt_1 \ll L^2$, then the deflection of the free end is given by:

$$d = \frac{L^2}{8R} \tag{13.7}$$

TABLE 13.1 Properties for Selected Materials Used in Bimaterial Elements

Material	Density (ρ) (kg m^{-3})	Young's Modulus (*E*) (GPa)	Heat capacity (*C*) (J kg^{-1} K^{-1})	Thermal expansion (10^{-6} K^{-1})	Thermal conductivity (W m^{-1} K^{-1})
Al	2700[c]	61–71[b]	896[a]	24[b]	237[c]
	2707[a]	70.6[c]	900[c]	23.5[c]	204[a]
Cu	8954[a]	129.8[c]	383.1[a]	17.0[c]	386[a]
	8960[c]		385[c]		401[c]
Cr	7100[c]	279[c]	518[c]	6.5[c]	94[c]
Au	19300[b,c]	78.5[b,c]	129[b,c]	14.1[b,c]	318[b,c]
Fe	7870[c]	211.4[c]	444[c]	12.1[c]	80.4[c]
Ni	8906[a]	199.5[c]	446[a]	13.3[c]	90[a]
	8900[c]		444[c]		90.9[c]
Ag	10524[a]	82.7[c]	234.0[a]	19.1[c]	419[a]
	10500[c]		237[c]		429[c]
Sn	7304[a]	49.9[c]	226.5[a]	23.5[c]	64[a]
	7280[c]		213[c]		66.8[c]
Ti	4500[c]	120.2[c]	523[c]	8.9[c]	21.9[c]
W	19350[a]	411[c]	134.4[a]	4.5[c]	163[a]
	19300[c]		133[c]		173[c]
Invar (Fe64/Ni36)	8000[c]	140–150[c]	—	1.7-2.0[c]	13[c]
Si	2340[c]	113[c]	703[c]	4.7-7.6[c]	80–150[c]
n-Si	2328[b]	130–190[b]	700[b]	2.6[b]	150[b]
p-Si	2300[b]	150–170[b]	770[b]	—	30[b]
Si_3N_4	3100[a]	304[b]	600–800[b]	3.0[b]	9–30[b]
SiO_2	2200[b]	57-85[b]	730[b]	0.50[b]	1.4[b]

[a] From Reference [13], Table A1 at 20°C.
[b] From Reference [13], Table A2 at 300K.
[c] From Goodfellow catalog 1995/1996 [14].

and combining this with Equation 13.1 yields:

$$d = L^2 \frac{3\left(1+m\right)^2}{t\left[3\left(1+m\right)^2 + \left(1+mn\right)\left(m^2 + 1/mn\right)\right]} \left(\alpha_2 - \alpha_1\right)\left(T - T_0\right) \tag{13.8}$$

If an aluminum and silicon nitride bimaterial microcantilever as used for sensor research [2] is considered, then $L = 200$ μm, $t_1 = 0.6$ μm, $t_2 = 0.05$ μm, $E_1 = 300$ GPa, $E_2 = 70$ GPa, $\alpha_1 = 3 \times 10^{-6}$ K^{-1}, $\alpha_2 = 24 \times 10^{-6}$ K^{-1} (see Table 13.1). In this situation, a temperature difference of 1°C gives a theoretical deflection of the cantilever of 0.103 μm.

Terminology and Simplifications

For industrial purposes, bimetallic thermostatic strips and sheets follow a standard specification — ASTM [3] in the U.S. and DIN [4] in Europe. Important parameters involved in this specification are derived directly from the previous equations, in which simplifications are made based on common applications.

It can be seen that the magnitude of the ratio $E_1/E_2 = n$ has no substantial effect on the curvature of the strip, and taking $n = 1$ implies an error less than 3%. Assuming again that the initial curvature is zero, Equation 13.1 can be simplified to:

$$\frac{1}{R} = \frac{6m}{t\left(m+1\right)^2} \left(\alpha_2 - \alpha_1\right)\left(T - T_0\right) \tag{13.9}$$

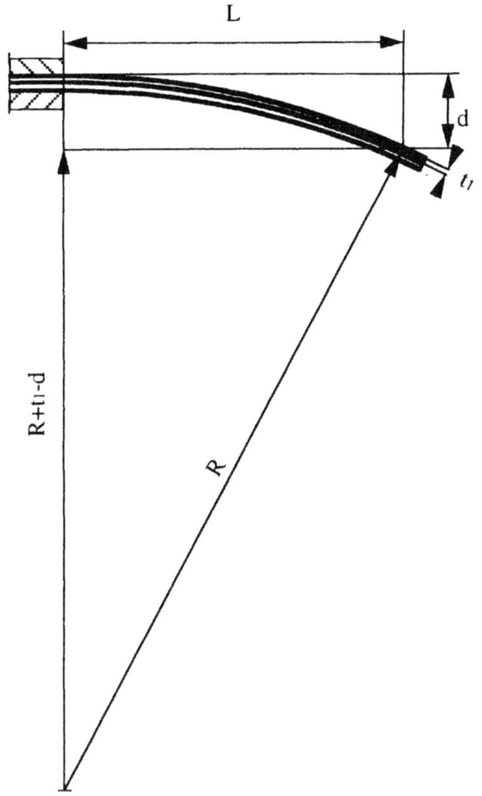

FIGURE 13.4 Bending of a strip fixed at one end.

In most industrial applications involving bimetallic elements, the thicknesses of the two component layers are taken to be equal ($m = 1$), thus Equation 13.6 becomes:

$$\frac{1}{R} = \frac{3}{2}\frac{(\alpha_2 - \alpha_1)(T - T_0)}{t} \tag{13.10}$$

The constant $\frac{3}{2}(\alpha_2 - \alpha_1)$ is known as flexivity in the U.S. and as specific curvature in Europe. Introducing the flexivity, k, and rearranging Equation 13.10 gives:

$$k = \frac{\frac{t}{R}}{T - T_0} \tag{13.11}$$

Flexivity can be defined as "the change of curvature of a bimetal strip per unit of temperature change times thickness" [5]. The experimental determination of the flexivity for each bimetallic strip has to follow the test specifications ASTM B388 [3] and DIN 1715 [4]. The method consists of measuring the deflection of the midpoint of the strip when it is simply supported at its ends. Using Equation 13.4 derived above and combining with Equation 13.11 gives:

$$k = \frac{8dt}{(T - T_0)L^2} \tag{13.12}$$

TABLE 13.2 Table of Selected Industrially Available ASTM Thermostatic Elements

Type (ASTM)	Flexivity 10^{-6} (°C^{-1})	Max. sensitivity temp. range (°C)	Max. operating temp. (°C)	Young's Modulus (GPa)
TM1	27.0 ± 5%[a] 26.3 ± 5%[b]	−18–149	538	17.2
TM2	38.7 ± 5%[a] 38.0 ± 5%[b]	−18–204	260	13.8
TM5	11.3 ± 6%[a] 11.5 ± 6%[b]	149–454	538	17.6
TM10	23.6 ± 6%[a] 22.9 ± 6%[b]	−18–149	482	17.9
TM15	26.6 ± 5.5%[a] 25.9 ± 5.5%[b]	−18–149	482	17.2
TM20	25.0 ± 5%[a] 25.0 ± 5%[b]	−18–149	482	17.2

[a] 10–93°C. From ASTM Designation B 388 [15].
[b] 38–149°C. From ASTM Designation B 388 [15].

Coming back to the second example of the calculation of the deflection (cantilever case), using Equation 13.10 and the same assumptions ($m = n = 1$), Equation 13.7 becomes:

$$d = \frac{k}{2}\frac{L^2}{t}\left(T - T_0\right) \tag{13.13}$$

In Europe, the constant $a = dt/(T - T_0)L^2$ (theoretically equal to $k/2$) is called specific deflection and is measured following the DIN test specification from the bending of a cantilever strip. It can be noted that the experimental value differs from the theoretical value as it takes into account the effect of the external forces suppressing the cross-curvature where the strip is fastened (i.e., the theory assumes that the curvature is equal along the strip; whereas in reality, the fact that the strip is fastened implies that the radius is infinite at its fixed end).

Tables 13.2 and Table 13.4 present a selection of bimetallic elements following ASTM and DIN standards, respectively. Flexivity (or specific curvature), linear temperature range, maximum operating temperature, and specific deflection (DIN only) are given. The details of the chemical composition of these elements are specified in Tables 13.3 and Table 13.5.

Industrial Applications

The mechanical thermostat finds a wide range of applications in temperature control in industrial processes and everyday life. This widespread use of thermostats is due to the discovery of Invar, a 36% nickel alloy that has a very low thermal expansion coefficient, and was so named because of its property of invariability [6].

There are two general classes of bimetallic elements based on their movement in response to temperature changes. Snap-action devices jump from one position to another at a specific temperature depending on their design and construction. There are numerous different shapes and sizes of snap-action elements and they are typically ON/OFF actuators. The other class of elements, creep elements, exhibit a gradual change in shape in response to a change in temperature and are employed in temperature gauges and other smooth movement applications. Continuous movement bimetals will be considered first. A linear configuration was covered previously, so the discussion will focus on coiled bimetallic elements.

Spiral and Helical Coil Configurations

For industrial or commercial measurements, a spiral or helical coil configuration is useful for actuating a pointer on a dial as the thermal deflection is linear within a given operating range. Linearity in this

TABLE 13.3 Composition of Selected Industrially Available ASTM Thermostatic Elements Given in Table 13.2

	Element	TM1	TM2	TM5	TM10	TM15	TM20
High-expansive material chemical composition (% weight)	Nickel	22	10	25	22	22	18
	Chromium	3	72	8.5	3	3	11.5
	Manganese	—	18	—	—	—	—
	Copper	—	—	—	—	—	—
	Iron	75	—	66.5	75	75	70.5
	Aluminum	—	—	—	—	—	—
	Carbon	—	—	—	—	—	—
Intermediate nickel layer		No	No	No	Yes	Yes	No
Low-expansive material chemical composition (% weight)	Nickel	36	36	50	36	36	36
	Iron	64	64	50	64	64	64
	Cobalt	—	—	—	—	—	—
Component ratio (% of thickness)	High	50	53	50	34	47	50
	Intermediate	—	—	—	32	6	—
	Low	50	47	50	34	47	50

From ASTM Designation B 388 [15].

TABLE 13.4 Table of Selected Industrially Available DIN Thermostatic Elements

Type (DIN)	Specific deflection (10^{-6} K^{-1})	Specific curvature (10^{-6} K^{-1}) ± 5%	Linear range (°C)	Max. operating temperature (°C)
TB0965	9.8	18.6	–20–425	450
TB1075	10.8	20.0	–20–200	550
TB1170A	11.7	22.0	–20–380	450
TB1577A	15.5	28.5	–20–200	450
TB20110	20.8	39.0	–20–200	350

Note: From DIN 1715 standard [4]. Specific deflection and curvature are for the range 20°C to 130°C.

TABLE 13.5 Composition of Selected Industrially Available DIN Thermostatic Elements Given in Table 13.4

	Element	TB0965	TB1075	TB1170A	TB1577A	TB20110
High-expansive chemical composition (% mass)	Nickel	20	16	20	20	10-16
	Chromium	—	11	—	—	—
	Manganese	6	—	6	6	Remainder
	Copper	—	—	—	—	18-10
	Iron	Remainder	Remainder	Remainder	Remainder	0.5
	Carbon	—	—	—	—	—
Low-expansive chemical composition (% mass)	Nickel	46	20	42	36	36
	Iron	Remainder	Remainder	Remainder	Remainder	Remainder
	Cobalt	—	26	—	—	—
	Chromium	—	8	—	—	—

From DIN 1715 Standard [4].

case means that the deflection does not vary by more than 5% of the deflection, as calculated from the flexivity [4]. The basic bimaterial ideas from the previous section still apply, with some additional equations relating the movement of a bimetal coil to a change in temperature. As in the previous section, standard methods for testing the deflection rate of spiral and helical coils exist and can be found in [7]. The following equations have been taken from the Kanthal Thermostatic Bimetal Handbook [8], with some change in notation. The angular rotation of a bimetal coil is given by (see Figure 13.5):

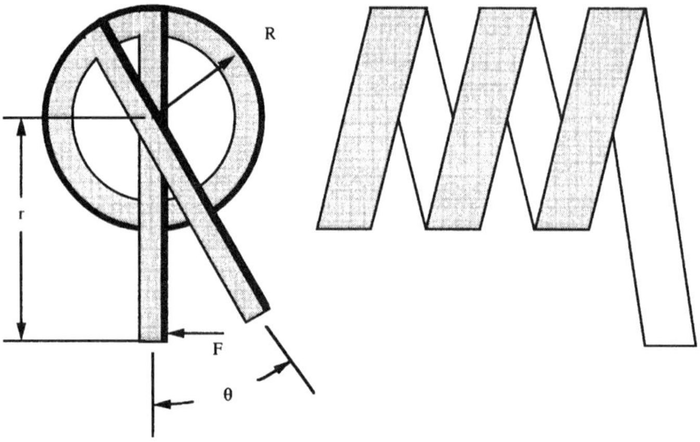

FIGURE 13.5 Helical coiled bimetal element.

$$\theta = \left(\frac{1}{R} - \frac{1}{R_0}\right)L \tag{13.14}$$

where L = length of the strip
R_0 and R = initial and final bending radii (assumed to be constant along the strip), respectively.

In terms of the specific deflection a, this can be written as:

$$\theta = \frac{2aL}{t}\left(T - T_0\right)\frac{360}{2\pi} \tag{13.15}$$

where t = thickness of the device
T_0 and T = initial and final temperatures, respectively.

An example would be a helical bimetal coil inside a steel tube with one end of the coil fixed to the end of the tube and the other connected to a pointer. The accuracy of a typical commercial product is 1% to 2% of full-scale deflection with an operating range of 0°C to 600°C [9].

If a change in temperature is required to both move a pointer and produce a driving force, then the angular rotation is reduced and is given by:

$$\theta = \left(\frac{2a\left(T - T_0\right)L}{t} - \frac{12\left(F - F_0\right)Lr}{wt^3E}\right)\frac{360}{2\pi} \tag{13.16}$$

where w = width of the element
r = distance from the center of the coil to the point of applied force, F.

This can be rewritten as:

$$T - T_0 = \frac{\theta t}{2aL}\frac{2\pi}{360} + \frac{6\left(F - F_0\right)r}{wt^2Ea} \tag{13.17}$$

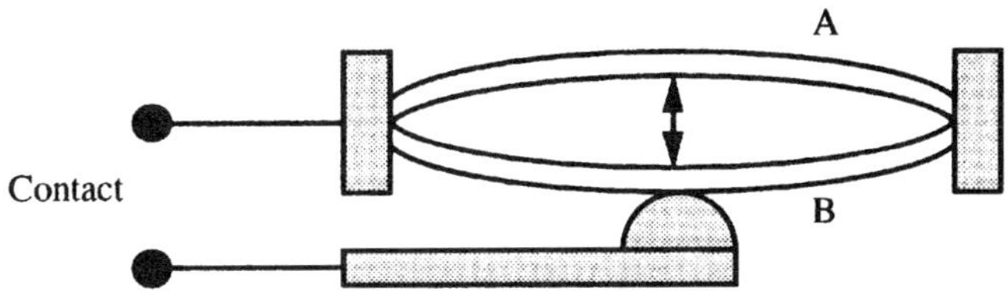

FIGURE 13.6 Snap-action bimetallic element.

where the first term represents the temperature associated with the angular rotation of the strip and the second term represents the temperature associated with the force generated by the strip. In general, the strip is designed so that the two are equal, as this leads to the minimum volume for the strip and consequently less weight and cost for the device.

Finally, if the coil is prevented from moving then the change in torque is given by

$$\left(F-F_0\right)r=\frac{1}{6}wt^2Ea\left(T-T_0\right) \tag{13.18}$$

Example: Consider a bimetal element that measures a temperature change from 20°C to 100°C and moves a lever 50 mm away with a force of 1 N. A dial reading range of 180° is required.

Choosing thermostatic bimetal TM2 gives the largest deflection per degree temperature change and TM2 meets the operating temperature requirements. Both a force and a movement are involved, so use Equation 13.7. Choosing each term equal to half the temperature change gives the minimum volume for the strip as:

$$\frac{1}{2}\left(T-T_0\right)=\frac{\theta t}{2aL}\frac{2\pi}{360}=\frac{6\left(F-F_0\right)r}{wt^2Ea} \tag{13.19}$$

Selecting a thickness of 1.0 mm and using a specific deflection of 19×10^{-6}°C^{-1} ($a=k/2$ from Table 13.2) gives a width of 29 mm. Similarly, the length of the bimetal strip is obtained from the second term in the equality, giving $L = 2.1$ m.

Snap-Action Configurations

Snap-action bimetal elements are used in applications where an action is required at a threshold temperature. As such, they are not temperature measuring devices, but rather temperature-activated devices. The typical temperature change to activate a snap-action device is several degrees and is determined by the geometry of the device (Figure 13.6). When the element activates, a connection is generally made or broken and in doing so, a gap between the two contacts exists for a period of time. For a mechanical system, there is no problem; however, for an electrical system, the gap can result in a spark that can lead to premature aging and corrosion of the device. The amount and duration of spark is reduced by having the switch activate quickly, hence the use of snap-action devices.

Snap-action elements also incorporate a certain amount of hysteresis into the system, which is useful in applications that would otherwise result in an oscillation about the set-point. It should be noted, however, that special design of creep action bimetals can also lead to different ON/OFF points, such as in the reverse lap-welded bimetal [8].

Sensitivity and Accuracy

Modern techniques are more useful where sensitivity and accuracy are concerned for making a temperature measurement; however, bimetals find application in commercial and industrial temperature control where an action is required without external connections. Evidently, geometry is important for bimetal systems as the sensitivity is determined by the design, and a mechanical advantage can be used to yield a large movement per degree temperature change. A demonstration of sensitivity using a helical coil was made by Huston [10] that gave 6 in. (15.2 cm) deflection per degree in their measurement system — yielding a sensitivity of 0.01°F per 1/16 in. (0.0035°C mm^{-1}). Huston also demonstrated a repeatable accuracy of 0.05°F (0.027°C) based on the use of a 0.1°F (0.056°C) accuracy calibration instrument.

The operating range for many bimetals is quite large; however, there is a range over which the sensitivity is a maximum. A bimetal element is generally chosen to operate in this range and specific details are provided by manufacturers in their product catalogs. Of particular note is that, despite extended thermal cycling, bimetal strips reliably return to the same position (i.e., show no hysteresis) at a given temperature and are very robust as long as they are not subjected to temperatures outside their specified operating range.

Advanced Applications

Thermostatic valves are a ready and robust means of measuring temperature and controlling heating and cooling in industrial settings. The basic designs have been around for many years and are the mainstay of many commercial temperature-control systems. New applications of bimaterials are being found in microactuators and microsensors.

Besides operating as temperature-measuring instruments, bimaterial devices can be used for a variety of applications where temperature is the controlling or triggering phenomenon, or indeed, other material properties are inferred from the temperature response. One such example is a nickel–silicon actuator developed by engineers at HP labs in Palo Alto, CA; the actuator operates by heating a thin nickel resistor on the silicon side of a bilayer device[11]. Both the silicon and nickel layers expand due to heating; however, the nickel layer expands more, thereby curling the device, which leads to the actuation of a tiny valve. The device can control gas flow rates from 0.1 to 600 standard cm^3 per minute with pressures ranging from 5 psi to 200 psi (34.5 kPa to 1379 kPa).

The use of micromachined thermal sensors compatible with modern silicon integrated circuit fabrication methods has recently received significant attention. One way of achieving highly sensitive thermal measurements is by using micromechanical bimetallic cantilevers and measuring the deflection as a result of thermal fluctuations. Rectangular or triangular cantilevers, typically 100 μm long made of silicon or silicon nitride, are coated with a thin high-expansive metallic layer (e.g., aluminum or gold). Precise measurement of the deflection of the end of the cantilever is achieved using an optical sensing arrangement commonly used in atomic force microscopes. The micromechanical nature of the cantilever-based sensor leads to significant advantages in the absolute sensitivity achievable. In this way, the device is capable of measuring temperature, heat, and power variations of 10^{-5} K, 150 fJ, or 100 pW, respectively [2]. In addition to their use as thermal sensors, the bimetallic cantilever systems have been used to investigate physical phenomena where heat is produced by the sample. Examples include photothermal spectroscopy studies of amorphous silicon [2] and the observation of oscillations in the catalyzed reaction of hydrogen and oxygen on platinum [12]. Thus, bimaterial sensors are becoming an increasingly important area of development.

Defining Terms

Linear coefficient of thermal expansion: The change in length of a material per degree change in temperature expressed as a fraction of the total length ($\Delta L/L$).

Flexivity: The change in radius of curvature of a bimaterial per degree change in temperature times the width of the element (See Equation 13.11).

Specific curvature: The European term for flexivity.

Specific deflection: Theoretically equal to half the flexivity. Specific deflection is measured by mounting the test element as a cantilever — supported at one end and free to move at the other.

References

1. S.P. Timoshenko, *The Collected Papers*, New York: McGraw-Hill, 1953.
2. J.R. Barnes, R.J. Stephenson, M.E. Welland, C. Gerber, and J.K. Gimzewski, Photothermal spectroscopy with femtojoule sensitivity using a micromechanical device. *Nature*, 372, 79-81, 1994.
3. ASTM Designation B 388.
4. DIN 1715. Part 1. Thermostat Metals. 1983.
5. ASTM Designation B 106.
6. M. Kutz, *Temperature Control*, New York: John Wiley & Sons, 1968.
7. ASTM Designation B 389.
8. *The Kanthal Thermostatic Bimetal Handbook*. Box 502. S-73401 Hallstammar, Sweden, 1987.
9. Bourdon Sedeme, F-41103 Vendome Cedex, France.
10. W.D. Huston, The accuracy and reliability of bimetallic temperature measuring elements, in C.M. Herzfeld and A.I. Dahl (eds.), *Temperature — Its Measurement and Control in Science and Industry*, New York: Reinhold, 1962.
11. L. O'Connor, A bimetallic silicon microvalve. *Mechanical Engineering*, 117(1), 1, 1995.
12. J.K. Gimzewski, C. Gerber, E. Meyer, and R. Schlittler, Observation of a chemical reaction using a micromechanical sensor. *Chem. Phys. Lett.* 217, 589-594, 1994.
13. G.C.M. Meijer and A.W. van Herwaarden, *Thermal Sensors*, Bristol, U.K.: Institute of Physics Publishing, 1994.
14. Goodfellow Cambridge Limited. Cambridge Science Park. U.K. CB4 4DJ.
15. American Society for Testing and Materials, *Annual Book of ASTM Standards*, Philadelphia, 1991.

Further Information

V.C. Miles, *Thermostatic Control — Principles and Practice*, Liverpool: C. Tinling and Co., 1965.

13.2 Resistive Thermometers

Jim Burns

Introduction to Resistance Temperature Detectors

One common way to measure temperature is by using Resistive Temperature Detectors (RTDs). These electrical temperature instruments provide highly accurate temperature readings: simple industrial RTDs used within a manufacturing process are accurate to ±0.1°C, while Standard Platinum Resistance Thermometers (SPRTs) are accurate to ±0.0001°C.

The electric resistance of certain metals changes in a known and predictable manner, depending on the rise or fall in temperature. As temperatures rise, the electric resistance of the metal increases. As temperatures drop, electric resistance decreases. RTDs use this characteristic as a basis for measuring temperature.

The sensitive portion of an RTD, called an element, is a coil of small-diameter, high-purity wire, usually constructed of platinum, copper, or nickel. This type of configuration is called a wire-wound element. With thin-film elements, a thin film of platinum is deposited onto a ceramic substrate.

Platinum is a common choice for RTD sensors because it is known for its long-term stability over time at high temperatures. Platinum is a better choice than copper or nickel because it is chemically inert, it withstands oxidation well, and works in a higher temperature range as well.

In operation, the measuring instrument applies a constant current through the RTD. As the temperature changes, the resistance changes and the corresponding change in voltage is measured. This measurement is then converted to thermal values by a computer. Curve-fitting equations are used to define this resistance vs. temperature relationship. The RTD can then be used to determine any temperature from its measured resistance.

A typical measurement technique for industrial thermometers involves sending a constant current through the sensor (0.8 mA to 1.0 mA), and then measuring the voltage generated across the sensor using digital voltmeter techniques. The technique is simple and few error-correcting techniques are applied.

In a laboratory where measurement accuracies of 10 ppm or better are required, specialized measurement equipment is used. High-accuracy bridges and digital voltmeters with special error-correcting functions are used. Accuracies of high-end measurement equipment can reach 0.1 ppm (parts per million). These instruments have functions to compensate for errors such as thermoelectric voltages and element self-heating.

In addition to temperature, strain on and impurities in the wire also affect the sensor's resistance vs. temperature characteristics. The Matthiessen rule states that the resistivity (ρ) of a metal conductor depends on temperature, impurities, and deformation. ρ is measured in (Ω cm):

$$\rho(\text{total}) = \rho(\text{temperature}) + \rho(\text{impurities}) + \rho(\text{deformation}) \qquad (13.20)$$

Proper design and careful material selection will minimize these effects so that resistivity will only vary with a change in temperature.

Resistance of Metals

Whether an RTD's element is constructed of platinum, copper, or nickel, each type of metal has a different sensitivity, accuracy, and temperature range. Sensitivity is defined as the amount of resistance change of the sensor per degree of temperature change. Figure 13.7 shows the sensitivity for the most common metals used to build RTDs.

Platinum, a noble metal, has the most stable resistance-to-temperature relationship over the largest temperature range –184.44°C (–300°F) to 648.88°C (1200°F). Nickel elements have a limited temperature range because the amount of change in resistance per degree of change in temperature becomes

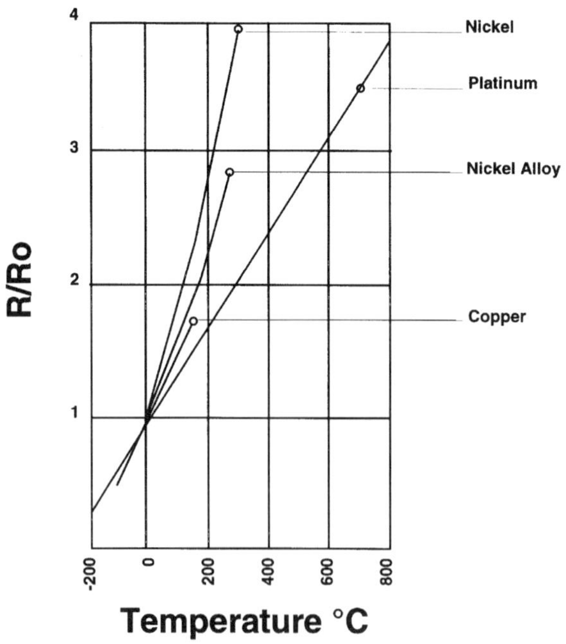

FIGURE 13.7 Of the common metals, nickel has the highest sensitivity.

TABLE 13.6

Probe	Basic application	Temperature	Cost	Probe style[a]	Handling
SPRT	Calibration of Secondary SPRT	–200 to 1000°C (–328 to 1832°F)	\$5000	I	Very fragile
Secondary SPRT	Lab use	–200 to 500°C (–328 to 932°F)	\$700	I, A	Fragile
Wirewound IPRT	Industrial field use	–200 to 648°C (–328 to 1200°F)	\$60–\$180	I, S, A	Rugged
Thin-film IPRT	Industrial field use	–50 to 260°C (–200 to 500°F)	\$40–\$140	I, S, A	Rugged

[a] I = immersion; A = air; S = surface.

very nonlinear at temperatures above 300°C (572°F). Copper has a very linear resistance-to-temperature relationship. However, copper oxidizes at moderate temperatures and cannot be used above 150°C (302°F).

Platinum is the best metal for RTD elements for three reasons. It follows a very linear resistance-to-temperature relationship; it follows its resistance-to-temperature relationship in a highly repeatable manner over its temperature range; and it has the widest temperature range among the metals used to make RTDs. Platinum is not the most sensitive metal; however, it is the metal that offers the best long-term stability.

The accuracy of an RTD is significantly better than that of a thermocouple within an RTD's normal temperature range of –184.44°C (–300°F) to 648.88°C (1200°F). RTDs are also known for high stability and repeatability. They can be removed from service and recalibrated for verifiable accuracy and checked for any possible drift.

Who Uses RTDs? Common Assemblies and Applications

Different applications require different types of RTDs. A direct-immersion Platinum Resistance Thermometer (PRT) and a connection head can be used for low-velocity pipelines, tanks, or air temperature measurement. A spring-loaded PRT, thermowell, and connection head are often used in pipelines or storage tanks. An averaging temperature element senses and measures temperatures along its entire sheath, which can range from 1 to 20 m in length. A heavy-duty underwater temperature sensor is designed for complete submersion under rivers, cooling ponds, or sewers. These are just a few examples of RTD configurations and applications.

Overview of Platinum RTDs

There are three main classes of Platinum Resistance Thermometers (PRTs): Standard Platinum Resistance Thermometers (SPRTs), Secondary Standard Platinum Resistance Thermometers (Secondary SPRTs), and Industrial Platinum Resistance Thermometers (IPRTs). Table 13.6 presents information about each.

Temperature Coefficient of Resistance

Each of the different metals used for sensing elements (platinum, nickel, copper) has a different amount of relative change in resistance per unit change in temperature. A measure of a resistance thermometer's sensitivity is its temperature coefficient of resistance. It is defined as the element's change in resistance per degree C change in temperature per ohm of sensor resistance over the range of 0°C to 100°C.

The alpha value is the average change in resistance per degree C per ohm resistance. The actual change in resistance per degree C per ohm is largest at –200°C and decreases steadily as the use temperatures increase.

The units for the coefficient are $\Omega/\Omega^{-1}/°C^{-1}$. This is called the alpha value and is commonly denoted by the Greek letter α. The larger the temperature coefficient, the greater the change in resistance for

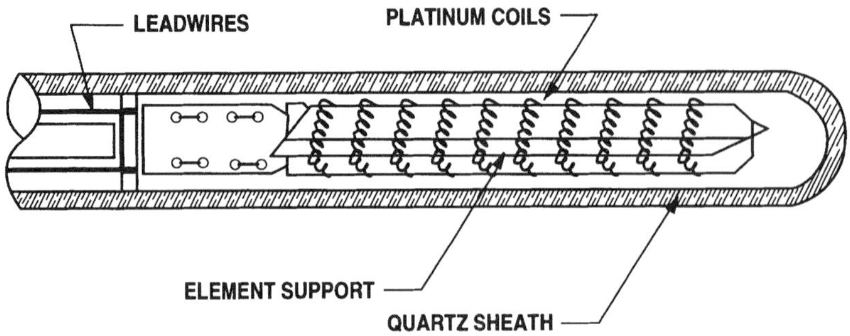

FIGURE 13.8 The Standard Platinum Resistance Thermometer is fragile and used only in laboratory environments.

a given change in temperature. Of the commonly used RTD metals, nickel has the highest temperature coefficient, 0.00672, while that of copper is 0.00427. The α value of the sensor is calculated using the equation:

$$\alpha = \frac{R_{100} - R_0}{100°\text{C} \times R_0} \tag{13.21}$$

where R_0 = the resistance of the sensor at 0°C
R_{100} = the resistance of the sensor at 100°C

Three primary temperature coefficients are specified for platinum:

1. ITS-90, the internationally accepted temperature scale, requires a minimum temperature coefficient of 0.003925 for SPRTs. This is achieved using high-purity wire (99.999% or better) wound in a strain-free configuration.
2. With reference-grade platinum wire used in industrial elements, the temperature coefficient is 0.003902.
3. IEC 751 [1] and ASTM 1137 [2] have standardized the temperature coefficient of 0.0038500 for platinum.

RTD Construction

Standard Platinum Resistance Thermometers (SPRTs), the highest-accuracy platinum thermometers, are fragile and used in laboratory environments only (Figure 13.8). Fragile materials do not provide enough strength and vibration resistance for industrial environments. SPRTs feature high repeatability and low drift, but they cost more because of their materials and expensive production techniques.

SPRT elements are wound from large-diameter, high-purity platinum wire. Internal leadwires are usually made from platinum and internal supports from quartz or fused silica. SPRTs are used over a very wide range, from –200°C (–328°F) to above 1000°C (1832°F). For SPRTs used to measure temperatures up to 660°C (1220°F), the ice point resistance is typically 25.5 Ω. For high-temperature thermometers, the ice point resistance is 2.5 Ω or 0.25 Ω. SPRT probes can be accurate to ±0.001°C (0.0018°F) if properly used.

Secondary Standard Platinum Resistance Thermometers (Secondary SPRTs) are also intended for laboratory environments (Figure 13.9). They are constructed like the SPRT, but the materials are less expensive, typically reference-grade, high-purity platinum wire, metal sheaths, and ceramic insulators. Internal leadwires are usually a nickel-based alloy. The secondary grade sensors are limited in temperature range — –200°C (–328°F) to 500°C (932°F) — and are accurate to ±0.03°C (±0.054°F) over their temperature range.

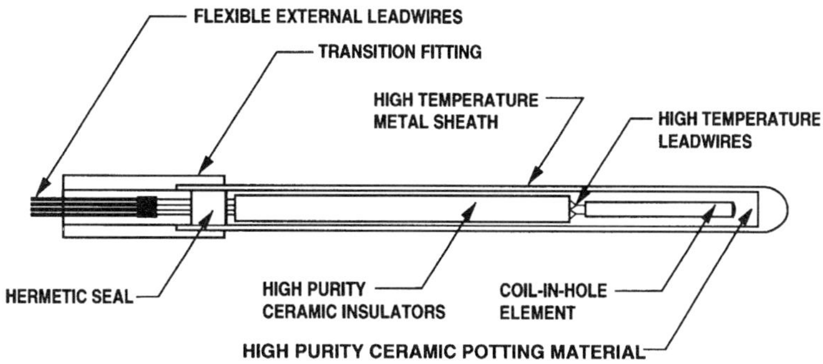

FIGURE 13.9 The Secondary Standard Platinum Resistance Thermometer is intended for laboratory environments.

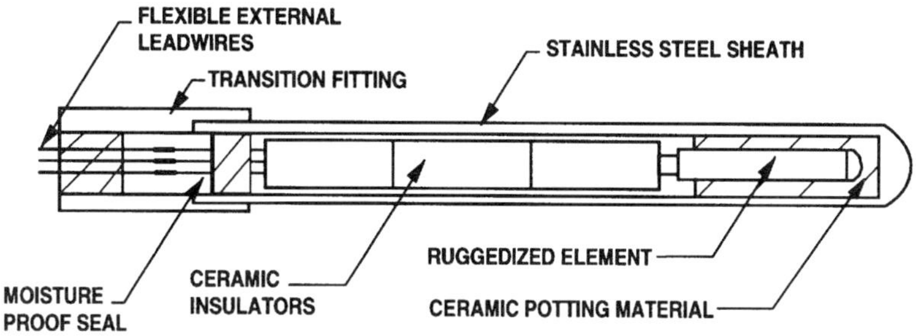

FIGURE 13.10 Industrial Platinum Resistance Thermometers are almost as durable as thermocouples.

Secondary standard thermometers can withstand some handling, although they are still quite strain-free. Rough handling, vibration, and shock will cause a shift in calibration. The nominal resistance of the ice point is most often 100 Ω. This simplifies calibration procedures when calibrating other 100-Ω RTDs. The temperature coefficient for secondary standards using reference-grade platinum wire is usually $0.00392\ \Omega\ \Omega^{-1}\ °C^{-1}$ or higher.

Industrial Platinum Resistance Thermometers (IPRTs) are designed to withstand industrial environments and are almost as durable as thermocouples (Figure 13.10). IEC 751 [1] and ASTM 1137 [2] standards cover the requirements for industrial platinum resistance thermometers. The most common temperature range is –200°C (–328°F) to 500°C (932°F). Standard models are interchangeable to an accuracy of ±0.25°C (±0.45°F) to ±2.5°C (±4.5°F) over their temperature range.

Several element designs are available for different applications. One common configuration is the wirewound element (Figure 13.11). This durable design was developed as a substitute for the fragile SPRT. The small platinum sensing wire (usually within 7 to 50 μm (0.0003 in. to 0.002 in. diameter) is noninductively wound around a cylindrical ceramic mandrel, and usually covered with a thin layer of material that provides electrical insulation and mechanical protection. Because the sensing element wire is firmly supported, it cannot expand and contract as freely as the SPRT's relatively unsupported platinum wire. This type of element offers higher durability than SPRTs and secondary standards, and very good accuracy for most industrial applications.

In another wirewound design, the coil suspension, a coil of fine platinum wire is assembled into small holes in a cylindrical ceramic mandrel (Figure 13.12). The coils are supported by ceramic powder or cement, and sealed at both ends. When ceramic powder is loosely packed in the bores of the mandrel, the element can expand and contract freely. This reduces the effects of strain on the resistance characteristics, resulting in very high accuracy and stability for use in secondary temperature standards and

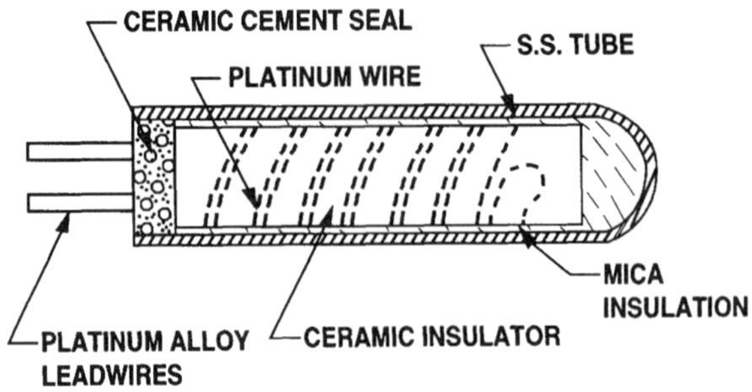

FIGURE 13.11 The wirewound RTD is noninductively wound around a cylindrical ceramic mandrel.

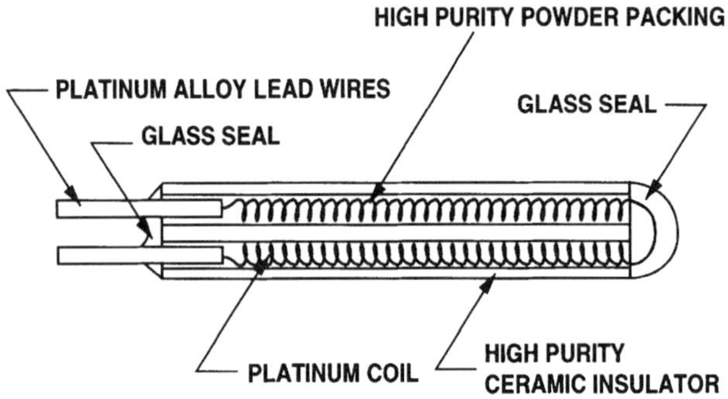

FIGURE 13.12 The coil suspension RTD has a coil of wire assembled into small holes.

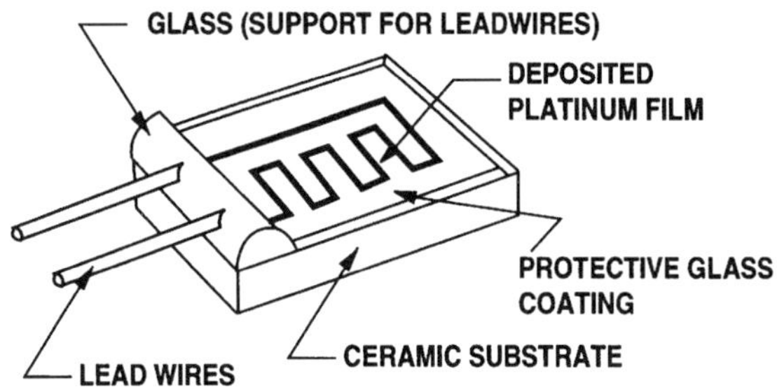

FIGURE 13.13 Thin-film elements have a thin film of platinum deposited onto a ceramic substrate.

docile industrial applications (with little or no vibration or shock). Recent improvements in ceramic materials give the sensing coil more stability— it will be capable of maintaining accuracies of 0.03°C after thousands of hours at temperatures of 500°C. These ceramic powders support the coils in the mandrel bores and hold them firmly in place with minimum strain.

Thin-film elements are extremely small, often less than 1.6 mm (1/16 in.) square (Figure 13.13). They are manufactured by similar techniques employed to make integrated circuits. First, a thin film of platinum is deposited onto a ceramic substrate. Some manufacturers use photolithography to etch the deposited platinum, leaving the element pattern on the ceramic substrate. Then, the element's surfaces are covered with glass material to protect the elements from humidity and contaminants.

The temperature range of thin film platinum elements is –50°C (–58°F) to 400°C (752°F); accuracy is from 0.5°C (0.9°F) to 2.0°C (3.6°F). The most common thin-film element has a 100-Ω ice point resistance and a temperature coefficient of 0.00385°C.

Thin-film RTDs can be extremely durable if the small-diameter leadwires and the thin element are properly protected. The accuracy and stability might not be as good as some wirewound elements due to hysteresis, long-term stability errors, and self-heating errors.

Self-heating Errors

The current that measures sensor resistance also heats the sensor. This is known as I^2R* heating or Joule heating. Because of this effect, the sensor's indicated temperature is somewhat higher than the actual temperature. This inconsistency is commonly called self-heating error. Self-heating errors, which are dependent on the application, can range from negligible values to 1°C. The greatest heating errors occur because of poor heat transfer between the sensing element and application, or excessive current used in measuring resistance.

The following are methods for reducing the self-heating error.

1. Minimize the power dissipation in the sensor. There is a tradeoff between the signal level and the self-heating of the sensor. Typically, 1 mA of current is used as the sensing current.
2. Use a sensor with a low thermal resistance. The lower the thermal resistance of the sensor, the better the sensor can dissipate the I^2R power and the lower the temperature rise in the sensor. Small time constants indicate a sensor with a low thermal resistance.
3. Maximize thermal contact between the sensor and the application.

Calibration

Testing programs are essential to verify the accuracy of PRTs. Some IPRTs are factory-calibrated to a temperature such as at the ice point, but PRT users might want to calibrate them at other temperatures, depending on their application.

The calibration results can be compared to prior calibrations from the same instrument. This will determine if it is necessary to repair or replace the instrument, or if calibration is required more often.

Since frequent repairs and recalibration are usually costly, purchasers and specifiers of PRTs might want to investigate various RTDs before installation by referring to ASTM Standard #E1137-95 [2], published by The American Society for Testing and Materials (ASTM).

Frequency. An RTD's stability depends on its working environment. High temperatures can cause drift and contamination of the platinum wire. The higher the temperature, the faster the drift occurs. Below 400°C, the high-temperature drift is not a significant problem, but temperatures reaching 500°C to 600°C are the most significant causes of drift — up to several degrees per year. Severe shock can damage a sensor instantly and cause failure. Shock, vibration, and rough handling will put strains in the platinum wire and change its characteristics, and ultimately damage the entire unit. If a sensor is not properly sealed, humidity can get into the sensor and cause some problems with the insulation resistance. Since the water in humidity is conductive, it will get between the lead wires and the sensing element, and basically shunt off the resistance of the element's wires. Under extreme operating conditions, a sensor should be calibrated on a monthly or bimonthly basis. If five or more calibrations are completed without a significant change, then the time between calibrations can be doubled; at least once a year is recommended, however.

Techniques. Two common calibration methods are the fixed point method and the comparison method.

Fixed point calibration, used for the highest accuracy calibrations, uses the triple point, freezing point or melting point of pure substances such as water, zinc, tin, and argon to generate a known and repeatable temperature (Figure 13.14). The fixed point cells are sealed to prevent contamination and to prevent atmospheric conditions from affecting the cell's temperature. These cells allow the user to reproduce actual conditions of the ITS-90 temperature scale.

Fixed point calibrations provide extremely accurate calibrations (within ±0.001°C), but the cells are time-consuming to use and can only accommodate one sensor at a time. For this reason, they are not widely used in calibrating industrial sensors. Each fixed point cell has a unique procedure for achieving the fixed point.

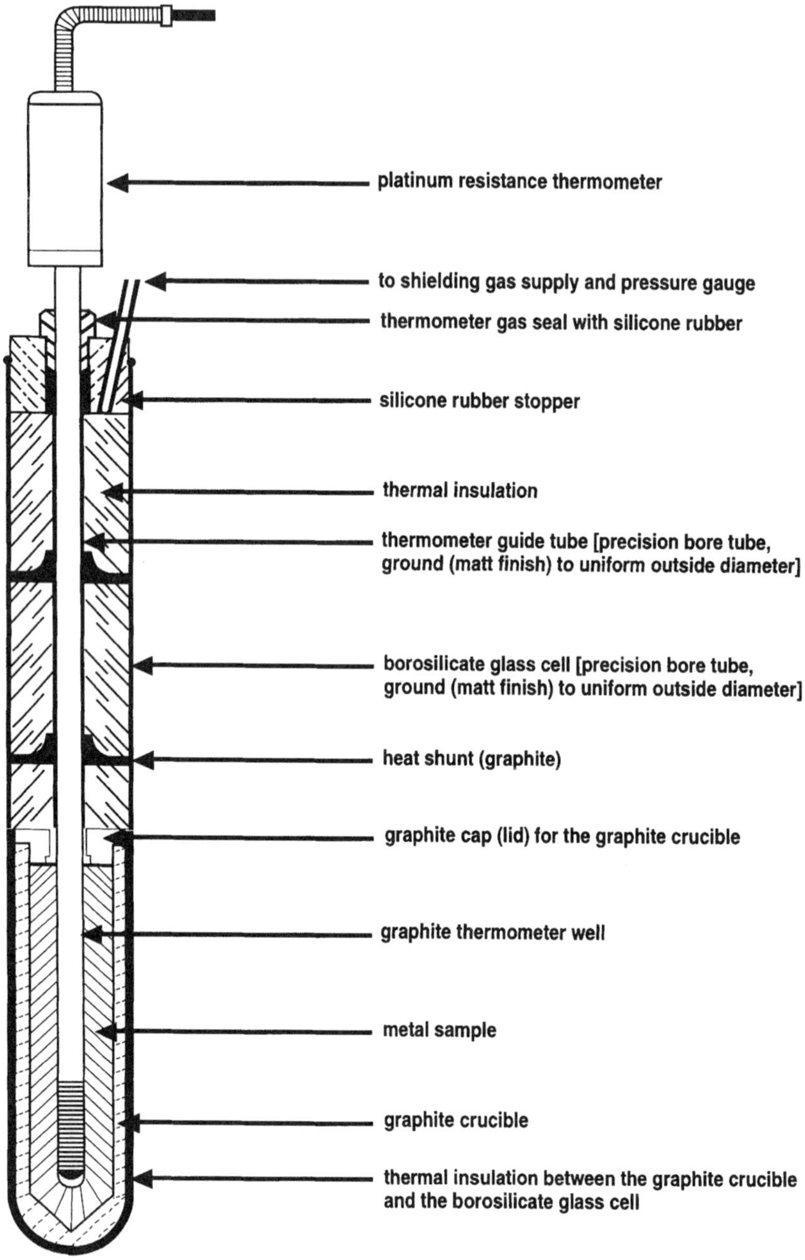

FIGURE 13.14 Fixed point calibration uses the triple point, freezing point, or melting point of water, zinc, tin, and argon.

A generalized procedure for fixed point calibration is as follows:

1. Prepare the cell. Different procedures exists for each fixed point.
2. Insert the thermometer to be calibrated.
3. Allow the system to stabilize. Stabilization times depend on thermometer/fixed point cells. Usually, 15 to 30 min is sufficient.
4. Measure the resistance of the thermometer. For the highest accuracy measurements, special resistance bridges are used. They have accuracies in the range of 10 ppm to 0.1 ppm.

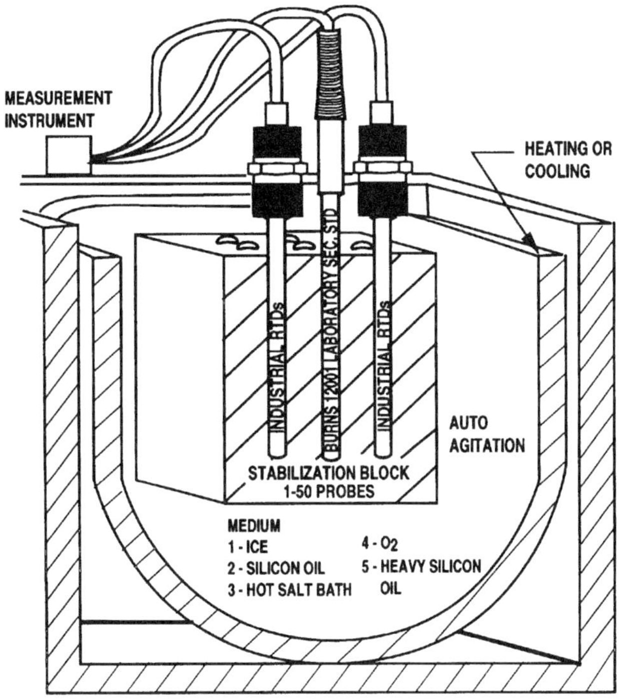

FIGURE 13.15 An isothermal bath permits calibration of industrial RTDs compared with a secondary standard.

A common fixed point calibration method for industrial-grade probes is the ice bath. The equipment is inexpensive, easy to use, and can accommodate several sensors at once. The ice point is designated as a secondary standard because its accuracy is ±0.005°C (±0.009°F), compared to ±0.001°C (±0.0018°F) for primary fixed points.

In *comparison calibrations*, commonly used with secondary SPRTs and industrial RTDs, the thermometers being calibrated are compared to calibrated thermometers by means of an isothermal bath whose temperature is uniformly stable (Figure 13.15). Unlike fixed point calibrations, comparisons can be made at any temperature between –100°C (–148°F) and 500°C (932°F). This method might be more cost-effective since several sensors can be calibrated simultaneously with automated equipment.

These isothermal baths, electrically heated and well-stirred, use silicone oils as the medium for temperatures ranging from –100°C (–148°F) to 200°C (392°F), and molten salts for temperatures above 200°C (392°F). At temperatures above 500°C (932°F), air furnaces or fluidized beds are used, but are significantly less uniformly stable.

The procedure for comparison calibration is as follows:

1. Insert the standard thermometer and thermometers being calibrated into the bath.
2. Allow the bath to stabilize.
3. Measure the resistance of the standard to determine the temperature of the bath.
4. Measure the resistance of each thermometer under calibration.

Deriving the resistance vs. temperature relationship of a PRT. After determining the PRT's resistance, the calibration coefficients can be determined. By plugging these values into an equation, temperature from any measured resistance can be derived. The two most common curve-fitting techniques are the ITS-90 and Callendar–Van Dusen equations.

On January 1, 1990, the International Temperature Scale of 1990 (ITS-90) became the official international temperature scale [3]. ITS-90 extends upward from 0.65 K (–272.5°C or –458.5°F) and defines temperatures of 0.65 K (0.65°C above absolute zero) and up by fixed points (see Table 13.7).

TABLE 13.7 Defining Fixed Points of the ITS-90

Material[a]	Equilibrium State[b]	Assigned value of temperature	
		T_{90} (K)	t_{90} (°C)
He	VP	3–5	–270.15 to –268.15
e-H_2	TP	13.8033	–259.3467
e-H_2 (or He)	VP (or GT)	≈17	≈–256.16
e-H_2 (or He)	VP (or GT)	≈20.3	≈–252.85
Ne	TP	24.5561	–248.5939
O_2	TP	54.3584	–218.7916
Ar	TP	83.8058	–189.3442
Hg	TP	234.3156	–38.8344
H_2O	TP	273.16	0.01
Ga	MP	302.9146	29.7646
In	FP	429.7485	156.5985
Sn	FP	505.078	231.928
Zn	FP	692.677	419.527
Al	FP	933.473	660.323
Ag	FP	1234.93	961.78
Au	FP	1337.33	1064.18
Cu	FP	1357.77	1084.62

[a] e-H_2 indicates equilibrium hydrogen; that is, hydrogen with the equilibrium distribution of its *ortho* and *para* forms at the corresponding temperatures. Normal hydrogen at room temperature contains 25% *para* and 75% *ortho* hydrogen. The isotopic composition of all materials is that naturally occurring.

[b] VP indicates vapor pressure point or equation; GT indicates gas thermometer point; TP indicates triple point; FP indicates freezing point; MP indicates melting point.

Two reference functions are used to define the temperature coefficient for an ideal SPRT: one for temperatures below 0°C and the other for temperatures above 0°C. When a PRT is calibrated on the ITS-90, the coefficients determined in the calibration are used to describe a deviation function that represents the difference between the resistance of the standard PRT and the reference function at all temperatures within the range. Using the calibration coefficients and the deviation functions, the SPRT can be used to determine any temperature from its measured resistance. Because ITS-90 equations are complex, computer software is necessary for accurate calculations.

ITS-90 affects:

- Standards and temperature calibration laboratories
- Users of standard and secondary SPRTs with traceability to standards laboratories
- Users of temperature measurement and control systems within companies concerned with verifiable total quality management

The National Institute for Standards and Technology (NIST) has published Technical Note 1265, *Guidelines for Realizing the International Temperature Scale of 1990* [4]. Not all PRT users need to follow the complex equations and computer programs associated with ITS-90. As a rule of thumb: if the minimum required uncertainty of measurement is less than 0.1°C, one probably will want to use ITS-90. For uncertainty of measurements greater than 0.1°C, the effect of the change in scales is relatively small and one will not be affected.

Callendar–Van Dusen equations are interpolation equations that describe the temperature vs. resistance relationship of industrial PRTs. These simple-to-use second- and fourth-order equations can be programmed easily into many electronic controllers. The equation for the temperature range of 0°C (32°F) to 850°C (1562°F) is:

$$R(t) = R_0\left(1 + At + Bt^2\right) \tag{13.22}$$

For the temperature range –200°C (–392°F) to 0°C (32°F):

$$R(t) = R_0\left[1 + At + Bt^2 + C(t-100)t^3\right] \tag{13.23}$$

where $R(t)$ = Resistance of the PRT at temperature t
t = Temperature in °C
R_0 = Nominal resistance of the PRT at 0°C
α, δ, β = Calibration coefficients

To determine the temperature from a measured resistance, a different set of equations and calibration coefficients is required. For temperatures greater than 0°C (measured resistances greater than the known ice point resistance of the PRT):

$$t(°\text{C}) = \left[(R_t - R_0)/(\alpha R_0)\right] + \delta\left[(t/100) - 1)(t/100)\right] \tag{13.24}$$

For temperatures less than 0°C (measured resistances less than the known ice point resistance of the PRT):

$$t(°\text{C}) = \left[(R_t - R_0)/(\alpha R_0)\right] + \delta\left[(t/100) - 1)(t/100)\right] + \beta\left[(t/100) - 1)(t/100)^3\right] \tag{13.25}$$

where t = Temperature to be calculated
$R(t)$ = Measured resistance at unknown temperature
R_0 = Resistance of the sensor at 0°C
α, δ, β = Coefficients

To correctly determine the temperature from a given resistance with these equations, one must iterate the equations a minimum of five times. After each calculation, the new value of temperature (t) is plugged back into the equations. The calculated temperature value will converge on its true value. After five iterations, the calculated temperature should be within ±0.001°C of the true value.

For industrial sensors, an alternative method would be to use nonlinear least squares curve fits to produce the temperature/resistance relationship. However, these methods should not be used for secondary and primary level thermometers as they cannot sufficiently match the defined ITS-90 scale. Curve fitting errors of up to 0.05°C are possible.

Usage of RTDs Today

Throughout the industry, usage of RTDs is increasing for many reasons. With the advent of the computer age, industries recognize the need for better temperature measurement, and an electrical signal to accompany advances in computerized process instrumentation. RTDs produce an electrical signal; and because of automatic control in industrial plants, it makes it simpler and easier to interface with process controllers. With the focus on reengineering, companies are searching for ways to improve processes. Improved temperature measurement and control is one good way to save energy, reduce material waste, reduce expenses and improve overall operating efficiencies.

Governmental regulations are another reason why RTDs are gaining popularity. In the pharmaceutical industry, the FDA [5] requires validation; among them the verification that temperature measurement is accurate. New regulations are currently being written for the food industry as well [6].

The growing worldwide acceptance of ISO 9000 standards has forced companies to calibrate their temperature measurement systems and instrumentation. In addition, the calibration must be documented, and must be traceable to a recognized national, legal standard.

RTDs are safer for the environment. With mercury thermometers, disposal of mercury is a problem. In many industries, the mere presence of mercury thermometers presents a risk.

Examples of Advanced Applications for Critical Temperature Measurement

RTD technology allows for custom design in a wide range of applications and industries. In many of these cases, the RTD becomes an integral part of an advanced application when temperature is critical.

Power plants use RTD sensors to monitor fuel and coolant temperatures entering and exiting heat exchangers. Accurate temperature measurement is also critical for nuclear power plants to perform pressure leak tests on the containment vessel surrounding the reactor core.

Microprocessor manufacturers require precise temperature control throughout their clean room areas. Air temperature is critical to production; many temperature measurement points need to be accurate ±0.028°C within (±0.05°F). To achieve this, an RTD is calibrated with a temperature transmitter. This matched pair ensures the highest level of system accuracy and eliminates the interchangeability error of the RTD.

The Future of RTD Technology

The future of RTD technology is driven by end-user needs and unsolved problems. For example, the need for high-temperature industrial RTDs exists for applications above 600°C (1200°F). In order to function at high temperatures, the RTD's platinum element must be protected from contamination. However, the sheath material can be a problem because, at high temperatures, it will react with the oxygen in the air and give off metal particles that can attach themselves to the platinum.

RTDs must be mechanically strong enough to survive higher temperatures as well. A high-temperature industrial RTD would require a thermally resistant sheath, perhaps Inconel 600. In addition, sensor components must be designed and manufactured to resist ultra-high temperatures. Some RTDs are specified to operate above 600°C (1200°F), but when tested, are not always reliable. Drift and nonrepeatability are problem areas that are in need of further attention.

Advances in RTD calibration provide significant improvements for temperature measurement and control. New measurement technologies combined with powerful computational techniques, have simplified the calibration process and made it more reliable.

Another current area of growth is in-house calibration and calibration baths. Governmental validation requirements and ISO-9000 standards are the driving forces in this area. Because third-party calibration services are expensive, companies want to be more productive and more cost-efficient. Wong [7] discusses the benefits of setting up an in-house calibration lab in addition to traceability concepts. Traceability refers to an unbroken chain of comparisons, linking the temperature measurement to a recognized national, legal standard. In the U.S., this national standard is maintained by the National Institute of Standards and Technology (NIST). All RTD manufacturers, laboratories, and calibration labs must adjust their standards to meet NIST standards.

Defining Terms

Accuracy: The degree of agreement between an actual measurement and its reference standard.

Alpha (α): The temperature coefficient of resistance of a PRT over the range 0°C to 100°C. For example, α for a standard platinum resistance thermometer (SPRT) is 0.003925.

Calibrate: To check, adjust or determine an RTD's accuracy by comparing it to a standard.

DIN (Deutsche Industrial Norm): A German organization that develops technical, scientific, and dimensional standards that are recognized worldwide.

Error: The difference between a correct value and the actual reading taken.

Primary Standard (or Standard PRT): A platinum resistance thermometer that meets the requirements for establishing calibrations according to the ITS-90. This highly accurate instrument is intended for laboratory environments and is accurate to 0.001°C.

Reliability: Used to designate precision for measurements made within a very restricted set of conditions.

Repeatability: The ability to give the same measurement under repeated, matching conditions.

Stability: The state of being resistant to change or deterioration.

Sensitivity: The amount of resistance change of the sensor per degree temperature change.

References

1. IEC, Industrial platinum resistance thermometer sensors, IEC International Standard 751.1995-07, Genéve, Suisse: Bureau Central de la Commission Electrotechnique Internationale, 1995.
2. ASTM Standards, Standard Specification for Industrial Platinum Resistance Thermometers, Standard E 1137-95, 1995.
3. H. Preston-Thomas, The International Temperature Scale of 1990 (ITS-90), *Metrologia*, 27(1), 3-10, 1990. For errata, see *ibid*, 27(2), 107, 1990.
4. NIST Technical Note 1265, Guidelines for Realizing the International Temperature Scale of 1990 (ITS-90), National Institute of Standards and Technology, 1990.
5. FDA validation for the pharmaceutical industry.
6. FDA validation for the food industry.
7. W. Wong, Traceability tops in-house calibration, *InTech*, 41(8), 27-29, 1994.

13.3 Thermistor Thermometers

Meyer Sapoff

A thermistor is a thermally sensitive resistor whose primary function is to exhibit a change in electric resistance with a change in body temperature. Unlike a wirewound or metal film resistance temperature detector (RTD), a thermistor is a ceramic semiconductor. An RTD exhibits a comparatively low temperature coefficient of resistance on the order of 0.4 to 0.5% °C^{-1}. Depending on the type of material system used, a thermistor can have either a large positive temperature coefficient of resistance (PTC device) or a large negative temperature coefficient of resistance (NTC device).

Two types of PTC thermistors are available. Silicon PTC thermistors rely on the bulk properties of doped silicon and exhibit resistance–temperature characteristics that are approximately linear. They have temperature coefficients of resistance of about 0.7 to 0.8% °C^{-1}. The most common application of silicon PTC thermistors is compensation of silicon semiconductor devices and circuits [1]. The materials used for switching-type PTC thermistors are compounds of barium, lead, and strontium titanates. Figure 13.16 shows the resistance–temperature characteristic of a typical switching-type PTC thermistor [2]. At low temperatures, from below 0°C to R_{min}, the resistance value is low, and R_T vs. T exhibits a small negative temperature coefficient of resistance on the order of –1% °C^{-1}. As the temperature increases, the temperature coefficient of resistance becomes positive and the resistance begins to rise. At a threshold or switching temperature, the rate of rise becomes very rapid and the PTC characteristic becomes very steep. Within its switching range, the temperature coefficient of resistance can be as high as 100% °C^{-1} and the device exhibits a high resistance value. At temperatures above the switching range, the resistance reaches a maximum value beyond which the temperature coefficient becomes negative again. The switching temperature can be varied between 80°C and 240°C by altering the chemical composition of the ceramic. Typical applications for switching-type PTC thermistors are over-temperature protection, current limiting, and self-regulated heating. The temperature coefficient of resistance of a unit used as a heating element typically is about 25% °C^{-1} at a switching temperature of 240°C [1, 2].

NTC thermistors consist of metal oxides such as the oxides of chromium, cobalt, copper, iron, manganese, nickel, and titanium. Such units exhibit a monotonic decrease in electric resistance with an increase in temperature. The resistance–temperature characteristics of NTC thermistors are nonlinear and approximate the characteristics exhibited by intrinsic semiconductors for which the temperature dependence of resistance is due to the excitation of carriers across a single energy gap. As such, the logarithm of resistance of an NTC thermistor is approximately a linear function of its inverse absolute temperature. Below room temperature, the slope of the function decreases and the thermistor behaves more like an extrinsic semiconductor. The actual conduction mechanism is comparable to the "hopping"

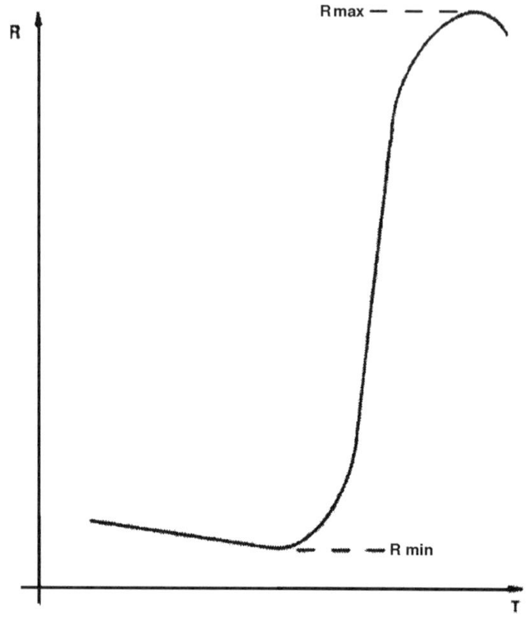

FIGURE 13.16 Resistance–temperature characteristic of a typical swtiching-type PTC thermistor. At low temperatures, from below 0°C to R_{min}, the resistance value is low and R_T vs. T exhibits a small negative temperature coefficient of resistance on the order of –1% °C^{-1}. As the temperature increases, the temperature coefficient of resistance becomes positive and the resistance begins to rise. At a threshold or switching temperature, the rate of rise becomes very rapid and the PTC characteristic becomes very steep. Within its switching range, the device exhibits a high resistance value. At temperatures above the switching range, the resistance reaches a maximum value beyond which the temperature coefficient becomes negative again.

mechanism observed in ferrites and manganites that have a spinel crystal structure. Conduction occurs when charge carriers hop from one ionic site in the spinel lattice to an adjacent site. Such hopping can occur when ions of the same element, with valences differing by 1, are present on equivalent lattice sites.

Because of its nonlinear resistance–temperature characteristic, the temperature coefficient of resistance of an NTC thermistor changes with temperature. Depending on the material system used, the temperature coefficient at 25°C typically is in the range of –3 to –5% °C^{-1}. At –60°C, it is in the range of –6.4 to –11.3% °C^{-1}; and at 100°C, it varies between –2.1 and –3.7% °C^{-1}. The corresponding resistance ratios with respect to 25°C, R_T/R_{25}, are 41 to 228 at –60°C and 0.13 to 0.03 at 100°C. The slope of the log R vs. $1/T$ characteristic β is relatively constant. The material systems for which the above data are presented exhibit β values of 2930 to 5135 K over the range of 25 to 125°C [1].

Although the resistance–temperature characteristic of an NTC thermistor is nonlinear, it is possible to achieve good linearity of the conductance–temperature and resistance–temperature characteristics using thermistor–resistor networks. The use of a resistor in series with a thermistor results in a linear conductance network, such as a voltage divider. The current obtained in response to a voltage applied to the network is linear with temperature. Consequently, the output of the voltage divider (voltage across the resistor) is linear with temperature. The parallel combination of a resistor and thermistor is a linear resistance network. Techniques exist for optimizing the linearity of single thermistor networks [3, 4]. Such networks exhibit S-shaped curves for their voltage–temperature or resistance–temperature characteristics. For temperature spans of 20°C to 30°C, the maximum deviation is small between any point on the curve and the best straight-line approximation of the curve. As the span increases beyond 30°C, the maximum deviation from linearity increases rapidly from about 0.15°C for a 30°C span to 0.7°C for a 50°C span. Using a two-thermistor network, a maximum linearity deviation of 0.22°C can be obtained over a range of 0°C to 100°C, as compared with a 2°C deviation obtained with a single-thermistor network [5]. Applications for NTC thermistors include temperature measurement, control, and compensation.

When current flows through a thermistor, there is a self-heating effect caused by the power dissipated in it. The thermistor temperature rises until the rate of heat loss from the thermistor to its surrounding environment equals the power dissipated. For this condition of thermal equilibrium, the temperature rise is directly proportional to the power dissipated in the thermistor. The constant of proportionality δ is the dissipation constant of the thermistor. By definition, δ is the ratio — at a specified ambient temperature, mounting condition, and environment — of the change in power dissipation in a thermistor to the resultant body temperature change. Factors that affect the dissipation constant are the thermistor surface area, the thermal conductivity and relative rate of motion of the medium surrounding the thermistor, the heat loss due to free convection in a still medium, the heat transfer between the thermistor and its mount through the thermistor lead wires, and heat loss due to radiation (significant for gases at low pressure). Because the thermal conductivities of fluids vary with temperature and free convection depends on the temperature difference between the thermistor and its ambient, δ is not a true constant. For gases in particular, the dissipation constant varies both with the thermistor body temperature and the amount of self-heating.

The thermal time constant τ is the 63.2% response to a step-function change in the thermistor temperature when the power dissipated in the thermistor is negligible. The thermal time constant only has meaning when there is a single exponential response. In such a case, an elapsed time of τ results in a 63.2% change between the initial and final steady-state temperatures, while an elapsed time of 5τ results in a 99.3% change. For a more complex structure, such as a thermistor encapsulated in a sensor housing, multiple exponentials can exist and one cannot predict that the 99% response will occur after an interval of five time constants. As with the dissipation constant, the thermal time constant is dependent on the rate of heat transfer between the thermistor and its environment. Consequently, all of the factors that increase the dissipation constant decrease the time constant.

The most common types of silicon PTC devices are the glass diode package having diameters of 1.8 mm to 2.5 mm and lengths of 3.8 mm to 7.5 mm, and the molded epoxy package with diameters of 3.6 mm to 6.0 mm and lengths of 10.4 mm to 15.0 mm. Such units have axial leads. Epoxy-coated chips and disks with radial leads also are available, with diameters of approximately 3 mm. Silicon PTC thermistors that comply with MIL-T-23648 have an operating temperature range of –55°C to 125°C. Commercial versions can be obtained with a range of –65°C to 150°C [6].

The most common configuration for switching-type PTC thermistors is the radial lead disk, with and without an insulating coating. Such units are available in diameters of 4 mm to 26 mm. The thickness dimension is in the range between 0.5 mm to 6.5 mm. They also are available as surface-mount devices, disks without leads, and in glass diode packages. Switching-type PTC disks have a storage temperature range of –25°C to 155°C and an operating range of 0°C to 55°C [1]

The emphasis of this text will be on NTC thermistors. Most thermistor applications use such devices. Two major categories of NTC thermistors exist. Bead-type thermistors have platinum alloy lead wires sintered into the body of the ceramic. Chip, disk, surface-mount, flake, and rod-type thermistors have metallized surface electrodes. The latter category includes glass diode packages in which dumet leads are compression-bonded to disks or chips.

Glass-coated beads include both adjacent and opposite lead configurations with diameters of 0.125 mm to 1.5 mm. Glass probes with diameters of 0.4 mm to 2.5 mm have lengths ranging from 1.5 mm to 6.35 mm. Probes with diameters of 1.5 mm to 2.5 mm have lengths of 3 mm to 12.7 mm, while larger probes with diameters of 2 mm to 2.5 mm have lengths of up to 50 mm. Glass rods typically are 6.3 mm long with diameters of 1.5 mm to 2.5 mm. A glass probe consists of a bare-bead-type thermistor sealed at the tip of a solid glass rod and radial dumet leads. A glass rod, by comparison, has its bead sealed in the center of a solid glass rod and has axial dumet leads. The temperature range specified in MIL-T-23648 is –55°C to 200°C for glass-coated beads and –55°C to 275°C for glass probes and glass rods. Commercial versions of such glass-enclosed devices typically have a range of –80°C to 300°C. Some units are rated for intermittent operation at 600°C, while special cryogenic devices are rated for operation in the range of –196°C to 25°C.

Chip thermistors with radial leads are available with cross-sections of 0.25 mm × 0.25 mm to 13 mm × 13 mm and a thickness range of 0.175 mm to 1.5 mm. Disks having diameters of 1 mm to 25 mm are available in both radial and axial lead configurations, with a thickness range of 0.25 mm to 6.35 mm. Chips and disks also are available in glass diode packages with axial dumet leads. Surface-mount chips are available in sizes ranging from 2 mm to 3.2 mm long × 1.1 mm to 1.6 mm wide × 0.36 mm to 1.3 mm thick. Flake thermistors with both adjacent and opposite leads are available in a thickness range of 0.025 mm to 0.125 mm, with cross-sections of 0.5 mm × 0.5 mm to 3 mm × 3 mm. The temperature range specified for chips, disks, rods, and glass diode packages in MIL-T-23648 is –55°C to 125°C. Commercial chips and disks are available with an operating temperature range of –80°C to 155°C [1]. Rods are available for use over the range of –60°C to 150°C [7], and glass diode packages are available with an operating range of –60°C to 300°C [8].

Thermal Properties of NTC Thermistors

The energy dissipated as heat in a thermistor connected to an electric circuit causes the thermistor body temperature to rise above the ambient temperature of its environment. At any instant, the applicable heat transfer equation is:

$$\frac{\mathrm{d}H}{\mathrm{d}t} = P = E_{\mathrm{T}} I_{\mathrm{T}} = \delta\left(T - T_{\mathrm{a}}\right) + cm\frac{\mathrm{d}T}{\mathrm{d}t} \qquad (13.26)$$

where $\mathrm{d}H/\mathrm{d}t = P = E_{\mathrm{T}}I_{\mathrm{T}}$ = Rate of thermal energy or heat supplied to the thermistor
δ = Dissipation constant
T_{a} = Ambient temperature
$\delta\,(T - T_{\mathrm{a}})$ = Rate of heat loss to the surrounding environment
c and m = Respectively, specific heat and mass of the thermistor
$cm(\mathrm{d}T/\mathrm{d}t)$ = Rate of heat absorbed by the thermistor

The solution of Equation 13.26 when P is constant is:

$$T = T_{\mathrm{a}} + \frac{P}{\delta}\left[1 - \exp\left(-\frac{\delta}{cm}t\right)\right] \qquad (13.27)$$

The transient solution of Equation 13.27 is the basis for the current–time characteristic of a thermistor. When, $t \gg cm/\delta$, then $\mathrm{d}T/\mathrm{d}t \to 0$, and the steady state solution of Equation 13.27 becomes:

$$P = E_{\mathrm{T}} I_{\mathrm{T}} = \delta\left(T - T_{\mathrm{a}}\right) \qquad (13.28)$$

where E_{T} and I_{T} are the steady-state thermistor voltage and current, respectively. Equation 13.28 is the basis for the voltage–current characteristic of a thermistor.

By reducing the thermistor power to a value that results in negligible self-heating, $P \to 0$, Equation 13.26 becomes:

$$\frac{\mathrm{d}T}{\mathrm{d}t} = -\frac{\delta}{cm}\left(T - T_{\mathrm{a}}\right) \qquad (13.29)$$

TABLE 13.8 Thermal Properties of Hermetically Sealed Beads and Probes

Style	Diameter (mm) Still air	Dissipation constant (mW °C^{-1}) Still water	Time constant (s) Still air	Water plunge	
Glass-coated bead	0.13	0.045	0.45	0.12	0.005
Glass-coated bead	0.25	0.09	0.9	0.5	0.010
Glass-coated bead	0.36	0.10	0.98	1.0	0.015
Ruggedized bead	0.41	0.12	1.1	1.2	0.016
Glass probe	0.63	0.19	1.75	1.9	0.020
Glass-coated bead	0.89	0.35	3.8	4.5	0.10
Glass-coated bead	1.1	0.40	4.0	5.5	0.14
Ruggedized bead	1.4	0.51	4.3	7.0	0.20
Glass probe	1.5	0.72	4.4	12.0	0.30
Glass probe	2.16	0.90	4.5	16.0	0.40
Glass probe	2.5	1.0	4.5	22.0	0.65

From Reference [1].

TABLE 13.9 Thermal Properties of Thermistors with Metallized Surface Electrodes

Style	Diameter (mm)	Dissipation constant (mW °C^{-1})	Time constant (s)
Chip or disk in glass diode package	2	2–3	7–8
Interchangeable epoxy coated chp or disk	2.4	1	10
Disk with radial or axial leads	2.5	3–4	8–15
Disk with radial or axial leads	5.1	4–8	13–50
Disk with radial or axial leads	7.6	7–9	35–85
Disk with radial or axial leads		10.28–11	28–150
Disk with radial or axial leads		12.75–16	50–175
Disk with radial or axial leads		19.615–20	90–300
Disk with radial or axial leads		25.424–40	110–230
Disk with radial or axial leads	1.3	25–3	16–20
Rod with radial or axial leads	1.8	4–10	35–90
Rod with radial or axial leads	4.4	8–24	80–125

From References [1, 6, 9, 10].

Equation 13.29 is a mathematical statement of Newton's law of cooling and has the solution:

$$T = T_a + \left(T_i - T_a\right)\exp\left(-\frac{t}{\tau}\right)$$

where T_i = Initial thermistor body temperature
T_a = Ambient temperature
$\tau = cm/\delta$ = Thermal time constant

Table 13.8 gives dissipation and time constant data for glass-coated beads, glass probes, and glass rods. Table 13.9 lists similar data for chip, disk, and rod-type thermistors with metallized surface contacts. Data for the dissipation and time constants are for the thermistor suspended by its leads at 25°C using the procedure specified in MIL-T-23648A. The temperature increment used for computing the dissipation constant results from self-heating the thermistors to 75°C from an ambient of 25°C. The time constant data result from allowing the thermistors to cool from 75°C to an ambient of 25°C. The water-plunge time constant data result from rapidly immersing the thermistors from room temperature air into still water. The transit time was negligible. The air temperature was approximately 25°C and the water temperature was about 5°C. The diameters specified for epoxy-coated interchangeable units are maximum diameters. To illustrate the effect of the environment on the thermal properties of these units, the

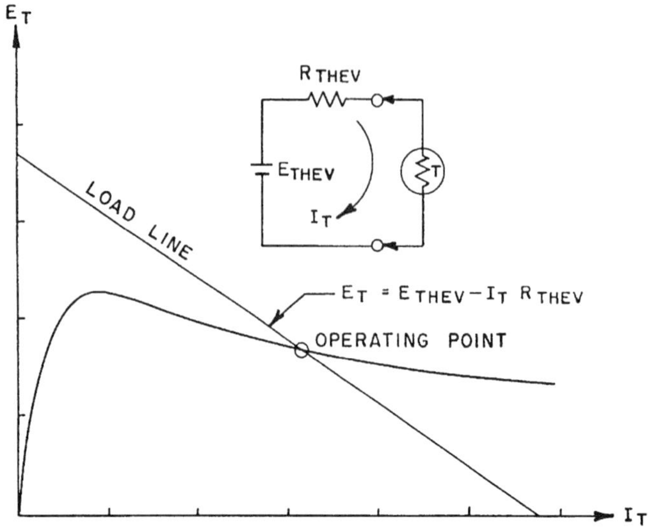

FIGURE 13.17 Typical voltage–current characteristic of an NTC thermistor. Due to self-heating of the thermistor, the slope of the E_T vs. I_T curve decreases with increasing current. This continues until a maximum value of E_T is reached for which the slope is equal to zero. Beyond this value, the slope continues to decrease and the thermistor exhibits a negative resistance characteristic. The Thevenin equivalent circuit with respect to the thermistor terminals provides a straight-line relationship between E_T and I_T that can be plotted as a load line. The intersection of the load line with the E_T vs. I_T curve is the operating point.

dissipation and time constants are 8 mW °C^{-1} and 1 s, respectively, when tested in stirred oil, as compared with 1 mW °C^{-1} and 10 s, respectively, when tested in still air. The dissipation and time constants for disks depend on the disk thickness, lead wire diameter, quantity of solder used for lead attachment, distance between the thermistor body and its mount, and the thermistor material. The dissipation and time constants for rods depend on the rod length as well as the variables specified above for disks.

Electrical Properties of NTC Thermistors

Thermistor applications depend on three fundamental electrical characteristics. These are the voltage–current characteristic, the current–time characteristic, and the resistance–temperature characteristic.

Voltage–Current Characteristics

Figure 13.17 shows a typical voltage–current curve. At low currents, the power dissipated is small compared with the dissipation constant and the self-heating effect is negligible. Under these conditions, the resistance is constant, independent of the current, and the voltage is proportional to the current. Consequently, at low currents, the characteristic curve is tangent to a constant resistance line equal to the zero-power resistance of the thermistor. As the current increases, the effects of self-heating become more evident; the thermistor temperature rises, and its resistance begins to decrease. For each subsequent incremental increase in current, there is a corresponding decrease in resistance. Hence, the slope of the voltage–current characteristic decreases with increasing current. This continues until the current reaches a value at which the slope becomes zero. Beyond this point, at which the voltage exhibits its maximum value, the slope continues to decrease and the thermistor exhibits a negative resistance characteristic. The temperature, voltage, and current corresponding to the peak of the curve are:

$$T_P = \frac{\beta - \sqrt{\beta^2 - 4\beta T_a}}{2} \tag{13.30}$$

TABLE 13.10 Applications Based on the Voltage–Current Characteristic of Thermistors

1. Variation in dissipation constant (fixed load line on a family of curves)
   - Vacuum manometers
   - Anemometers, flowmeters, fluid velocity
   - Thermal conductivity analysis, gas detectors, gas chromotography
   - Liquid level measurement, control, and alarm
2. Variation in circuit parameters (rotation and/or translation of load line on a fixed *E–I* curve)
   - Oscillator amplitude and/or frequency regulation
   - Gain or level stabilization and equalization
   - Volume limiters
   - Voltage compression and expansion
   - Switching devices
3. Variation in ambient temperature (fixed or variable load line on a family of curves)
   - Temperature control or alarm
4. Microwave power measurement
   - Bolometers

From References [1, 14].

$$E_P = \sqrt{R_P \delta (T_P - T_a)} \tag{13.31}$$

$$I_P = \sqrt{\delta (T_P - T_a)/R_P} \tag{13.32}$$

where T_P = Absolute temperature (in K) at which the peak occurs
T_a = Absolute ambient temperature (in K)
R_P = Thermistor resistance at T_P

Applications based on the voltage–current characteristics involve changes in the operating point on a single curve or family of curves. Such changes result from changes in the environmental conditions or variations in circuit parameters. Figure 13.17 shows a graphical solution for obtaining the operating point. The curve represents a nonlinear relationship between the thermistor voltage and current. The Thevenin equivalent circuit with respect to the thermistor terminals provides the relationship given by Equation 13.33.

$$E_T = E_{THEV} - I_T R_{THEV} \tag{13.33}$$

where E_T and I_T are the thermistor voltage and current, respectively. The straight-line relationship of Equation 13.33 is the load line of the self-heated thermistor. Its intersection with the voltage–current characteristic is the operating point. Table 13.10 categorizes the more familiar applications into four groups distinguished by the form of thermistor excitation [1, 14].

Current–Time Characteristics

The voltage–current characteristic discussed above deals with a self-heated thermistor operated under steady-state conditions. This condition, for which a decrease in thermistor resistance results from a current sufficiently high to cause self-heating, does not occur instantaneously. A transient condition exists in a thermistor circuit from the time at which power is first applied ($t = 0$) until the time equilibrium occurs ($t \gg \tau$). The relationship between the thermistor current and the time required to reach thermal equilibrium is the current–time characteristic. Generally, the excitation is a step function in voltage through a Thevenin equivalent source. Figure 13.18 shows the current–time characteristics for several

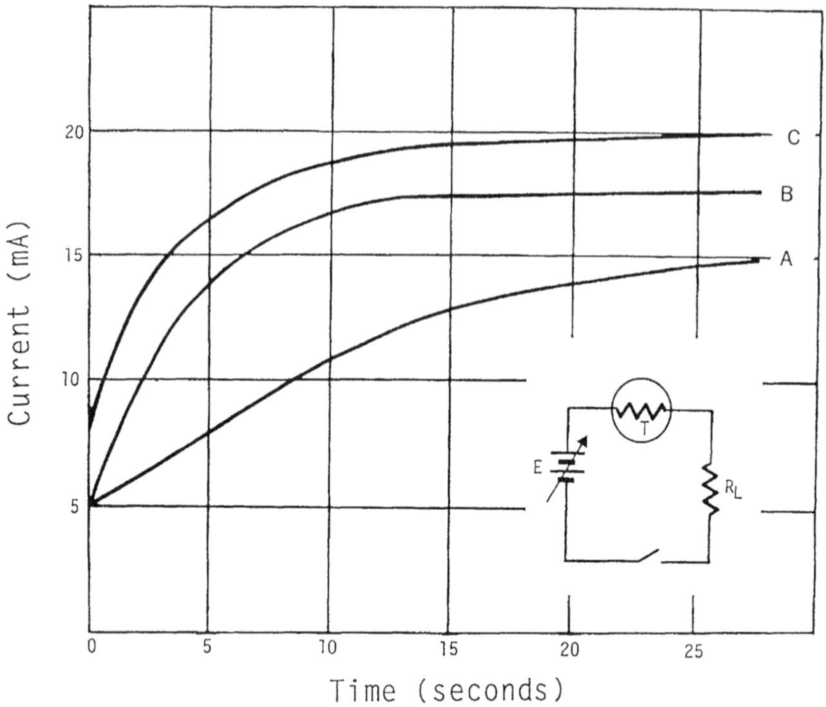

FIGURE 13.18 Current–time characteristics of Thermometrics, Inc. epoxy-coated chip thermistors. Curves A and C are for a DC95F402 having a zero-power resistance at 25°C of 4 kΩ. Curve B is for a DC9F802 having a zero-power resistance at 25°C of 8 kΩ. The excitation is a step function in voltage through a Thevenin equivalent source. The source voltage and resistance values for curve A are 24 V and 1 kΩ, respectively. For curves B and C, the source voltage is 48 V and source resistance is 2 kΩ.

thermistor disks and voltage sources. For any given characteristic, the source voltage, source resistance, and thermistor zero-power resistance determine the initial current. The source voltage, source resistance, and voltage–current characteristic determine the final equilibrium value. The curve between the initial and final values depends on the circuit design parameters, as well as the dissipation constant and heat capacity of the thermistor. The proper choice of thermistor and circuit design results in a transient time range of a few milliseconds to several minutes.

Applications based on the current–time characteristic are time delay, surge suppression, filament protection, overload protection, and sequential switching.

Resistance–Temperature Characteristics

The term zero-power resistance applies to thermistors operated with negligible self-heating. This characteristic describes the relationship between the zero-power resistance of a thermistor and its ambient temperature. Over small temperature spans, the approximately linear relationship between the logarithm of resistance and inverse absolute temperature is given by:

$$\ln R_{\mathrm{T}} \cong A + \frac{\beta}{T} \qquad (13.34)$$

where T = Absolute temperature (K)

β = Material constant of the thermistor

If one lets $R_T = R_{T_0}$ at a reference temperature $T = T_0$ and solves for R_T, one obtains:

$$R_T \cong R_{T_0} \exp\left[\frac{\beta\left(T_0 - T\right)}{TT_0}\right] \tag{13.35}$$

The temperature coefficient of resistance α is defined as:

$$\alpha \equiv \frac{1}{R_T}\frac{dR_T}{dT} \tag{13.36}$$

Solving Equation 13.34 for α yields:

$$\alpha = -\frac{\beta}{T^2} \tag{13.37}$$

In Equation 13.34, the deviation from linearity results in temperature errors of 0.01°C, 0.1°C, and 0.3°C for temperature spans of 10°C, 30°C, and 50°C, respectively, within the range of 0°C to 50°C. Using a polynomial for ln R_T vs. $1/T$ reduces the error considerably. The degree of the polynomial required depends on the temperature range and material system used. The use of a third-degree polynomial is adequate for most applications. Hence, more accurate expressions for the resistance–temperature characteristic are:

$$\ln R_T = A_0 + \frac{A_1}{T} + \frac{A_2}{T^2} + \frac{A_3}{T^3} \tag{13.38a}$$

$$\ln R_T = A_0 + \frac{A_1}{T} + \frac{A_2}{T^2} + \frac{A_3}{T^3} \tag{13.38b}$$

$$\frac{1}{T} = \alpha_0 + \alpha_1 \ln R_T + \alpha_2\left(\ln R_T\right)^2 + \alpha_3\left(\ln R_T\right)^3 \tag{13.39}$$

Steinhart and Hart proposed the use of Equation 13.39 for the oceanographic range of –2°C to 30°C [11]. Their analysis showed that no significant loss in accuracy occurred by eliminating the square term $a_2(\ln R_T)^2$. Consequently, they proposed the use of Equation 13.40.

$$\frac{1}{T} = b_0 + b_1 \ln R_T + b_3\left(\ln R_T\right)^3 \tag{13.40}$$

Mangum reported that the use of Equation 13.40 resulted in interpolation errors of approximately 0.001°C over the range of 0°C to 70°C [12].

The technical staff at Thermometrics, Inc. evaluated Equation 13.38 over the range of –80°C to 260°C using 17 different thermistor materials. The glass probes investigated encompassed a span of specific resistance values of 2 Ω cm to 300 kΩ cm and a resistance range at 25°C of 10 Ω to 2 MΩ. The results were presented at the *Sixth International Symposium on Temperature* and show that the interpolation errors do not exceed the total measurement uncertainties. For temperature spans of 100°C within the range of –80°C to 260°C, 150°C within the range of –60°C to 260°C, and 150°C to 200°C within the

TABLE 13.11 Interpolation Errors for $\ln R_T = C_0 + C_1/T + C_3/T^3$

Temperature range (°C)	Temperature span (°C)	Interpolation error (°C)
–80 to 0	50	0.002–0.01
0 to 200	50	0.001–0.003
–80 to 0	100	0.02–0.03
0 to 200	100	0.01
–60 to 90	150	0.1
0 to 150	150	0.045
50 to 200	150	0.015
0 to 200	200	0.08

range of 0°C to 260°C, the interpolation errors are 0.005°C to 0.01°C [13]. Lowering the temperature span to 50°C within the range of 0°C to 260°C reduces the interpolation error to 0.001°C to 0.003°C.

Sapoff and Siwek [14] evaluated the loss in accuracy introduced by eliminating the quadratic term in Equation 13.38. For this condition, Equation 13.38 reduces to Equation 13.41.

$$\ln R_T = C_0 + \frac{C_1}{T} + \frac{C_3}{T^3} \tag{13.41}$$

The interpolation errors introduced by Equations 13.40 and 13.41 depend on the material system, temperature range (nonlinearity increases at low temperatures), and the temperature span considered. Table 13.11 summarizes the errors associated with the use of Equation 13.41 for various temperature spans and ranges.

Equations 13.38 through 13.41 can be rewritten as:

$$R_T = \exp\left(A_0 + \frac{A_1}{T} + \frac{A_2}{T^2} + \frac{A_3}{T^3}\right) \tag{13.42}$$

$$T = \left[a_0 + a_1 \ln R_T + a_2 (\ln R_T)^2 + a_3 (\ln R_T)^3\right]^{-1} \tag{13.43}$$

$$T = \left[b_0 + b_1 \ln R_T + b_3 (\ln R_T)^3\right]^{-1} \tag{13.44}$$

$$R_T = \exp\left(C_0 + \frac{C_1}{T} + \frac{C_3}{T^3}\right) \tag{13.45}$$

The solutions for Equations 13.38, 13.39, 13.42, and 13.43 require four calibration points and the use of simultaneous equations. Similarly, the solutions for Equations 13.40, 13.41, 13.44, and 13.45 require three calibration points. The use of a polynomial regression analysis involving additional calibration points can minimize the effects of calibration uncertainties.

Applications that depend on the resistance–temperature characteristic are temperature measurement, control, and compensation. There also are applications for which the thermistor temperature depends on some other physical phenomenon. For example, an hypsometer is an instrument in which the temperature of a boiling liquid provides an indication of the liquid vapor pressure. Another example involves the use of thermistor-type cardiac catheters for thermodilution analysis. A saline or dextrose

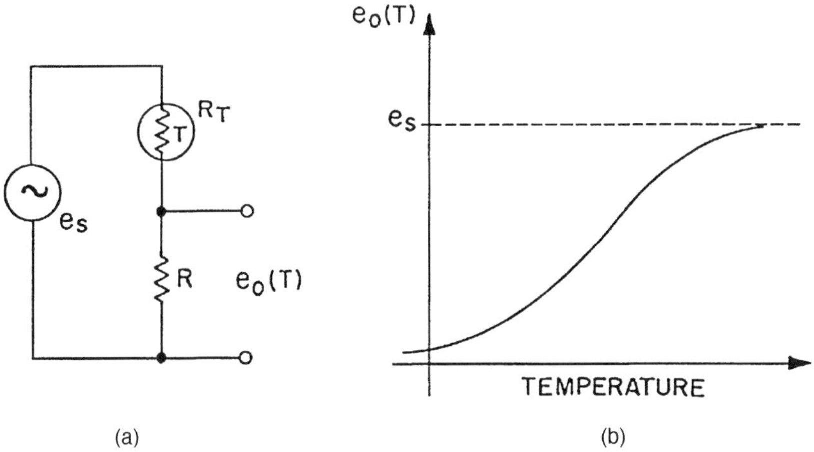

FIGURE 13.19 (a) Thermistor voltage divider. The resistor R represents the parallel combination of the load resistance and the fixed divider resistor. (b) The output, e_0 vs. T, is an S-shaped curve that is linear over a portion of the temperature range.

solution of known volume and temperature is injected into the bloodstream through one of the catheter lumens. The mixing of the solution with the blood dilutes the solution and decreases the temperature of the blood as it flows downstream past a thermistor located at the surface of another lumen in the catheter. The cardiac output computed from the temperature–time response data measured by the thermistor provides an indication of the heart pump efficiency.

Linearization and Signal Conditioning Techniques

The measured resistance and computed constants for Equations 13.43 and 13.44 can be used with a computer to determine temperature without the need for a linear network. This is useful for obtaining high accuracy over relatively wide temperature ranges. Most applications based on the resistance–temperature characteristics of thermistors, however, use some form of linearization or signal conditioning. The use of a constant current source and a linear resistance network provides a voltage output that is linear with temperature. By using the proper combination of current and resistance level, a digital voltmeter connected across the network provides a direct display of temperature. The use of a constant voltage source results in a linear conductance network with a current that is linear with temperature.

Linear Conductance Networks

Voltage dividers, ohmmeter circuits and Wheatstone bridge circuits are linear conductance networks. Consider the voltage divider circuit shown in Figure 13.19. Using the fixed resistor for the output eliminates the effect of the load resistance. For the purpose of analysis, the resistor denoted by R is the parallel combination of the load and fixed divider resistors. Another advantage is that the output voltage increases with temperature with this arrangement. The ratio of output voltage to input voltage is given by:

$$\frac{e_0}{e_s} = \frac{R}{R + R_T} = \frac{1}{1 + R_T/R} \tag{13.46}$$

where e_0 = Output voltage
e_s = Source voltage
R = Parallel combination of the load and fixed resistors
R_T = Thermistor resistance at a specified temperature T

If one normalizes the thermistor resistance with respect to its value at a specified reference temperature T_0, then:

$$r_T = \frac{R_T}{R_{T_0}} \tag{13.47}$$

where R_{T_0} = Thermistor resistance at T_0
r_T = Resistance ratio

Thermistor manufacturers typically supply resistance ratio–temperature characteristics in their catalogs.

Substituting Equation 13.47 in Equation 13.46 yields:

$$\frac{e_0}{e_s} = \frac{1}{1 + r_T R_{T_0} / R} \tag{13.48}$$

If the circuit constant $\sigma = R_{T_0}/R$, then Equation 13.48 becomes:

$$F(T) = \frac{e_0}{e_s} = \frac{1}{1 + r_T \sigma} \tag{13.49}$$

The output for the voltage divider of Figure 13.19*a* is the S-shaped curve shown in Figure 13.19*b*. The value of the circuit constant σ determines the temperature range for which good linearity exists between e_0 and T. References [3, 4] include families of S-shaped curves over the range $0.01 \le \sigma \le 20.0$ for the three basic material systems of MIL-T-23648. A good criterion for achieving optimum linearity is to equate the slopes of the function $F(T)$ at the end-points of the specified temperature range $T_L \le \sigma \le T_H$. Specifying that $dF(T_L)/dT = dF(T_H)/dT$ results in the following:

$$\sigma = \frac{X - Y}{Y r_{T_L} - X r_{T_H}} \tag{13.50}$$

where X and Y are determined by the end-point conditions and the equation constants for ln R_T vs. T. When Equation 13.41 is used, X and Y are given by:

$$X = T_H \sqrt{r_{T_L}\left(C_1 + 3\frac{C_3}{T_H^2}\right)} \tag{13.51}$$

$$Y = T_L \sqrt{r_{T_H}\left(C_1 + 3\frac{C_3}{T_L^2}\right)} \tag{13.52}$$

When Equation 13.38 is used, X and Y are given by:

$$X = T_H \sqrt{r_{T_L}\left(A_1 + 2\frac{A_2}{T_L} + 3\frac{A_3}{T_L^2}\right)} \tag{13.53}$$

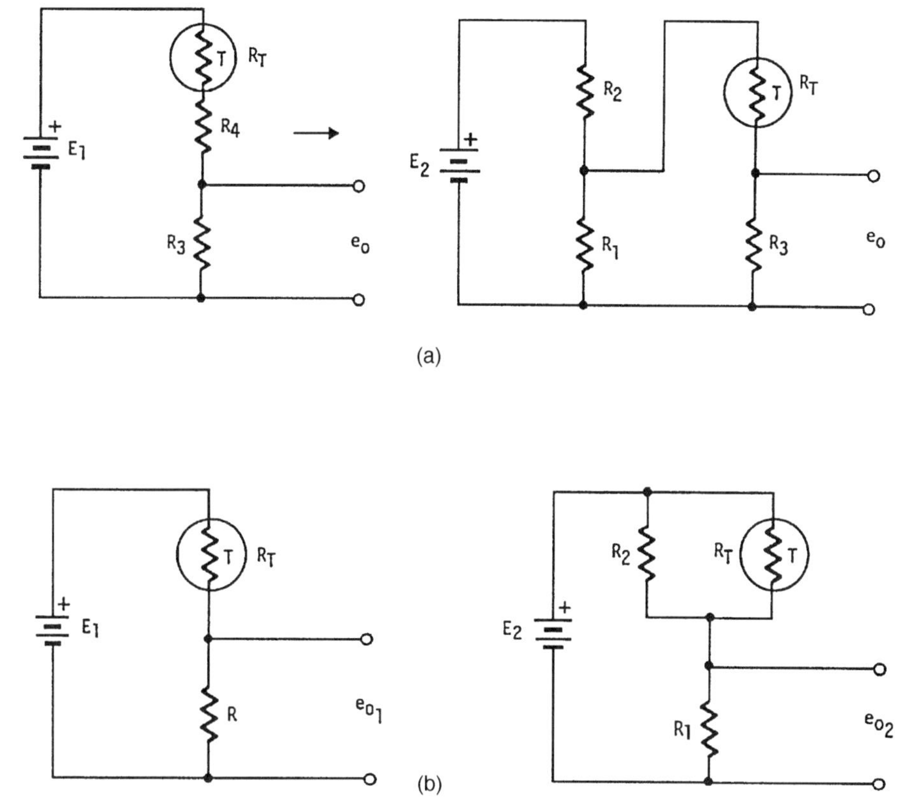

FIGURE 13.20 Equivalent thermistor voltage divider circuits for converting a network with a nonstandard voltage source to a network that uses a standard voltage source. Only a portion of the output is available in (*a*). Ohmmeter-type thermometers use such circuits. The conversion circuit of (*b*) provides the full available output voltage in a series with a bias voltage. (*b*) is used with bridge circuit-type thermometers.

$$Y = T_L \sqrt{r_{T_H}\left(A_1 + 2\frac{A_2}{T_H} + 3\frac{A_3}{T_H^2}\right)} \tag{13.54}$$

By allowing e_s and R to be the Thevenin equivalent voltage and resistance with respect to the thermistor terminals, respectively, the voltage divider analysis applies to any more complex circuit such as a Wheatstone bridge [3, 4].

Thermistor self-heating constraints typically result in nonstandard voltage sources for thermometer circuit designs. The modified divider circuits of Figure 13.20 provide equivalent circuits with standard voltage sources. Only a portion of the available output voltage appears across the output detector in Figure 13.20*a*. Ohmmeter-type thermometers use such circuits. The circuit of Figure 13.20*a* provides the full available output voltage. However, the conversion results in a bias voltage in series with the output. Bridge circuit-type thermometers use the circuit of Figure 13.20*b*. The conversion equations applicable to Figure 13.20*a* are:

$$K = \frac{R_1}{R_1 + R_2} = \frac{E_1}{E_2} = \frac{R_4}{R_2} \tag{13.55}$$

$$R = R_3 + R_4 = R_3 + \frac{R_1 R_2}{R_1 + R_2} \tag{13.56}$$

$$E_2 = \text{Desired source voltage} \tag{13.57}$$

$$R_2 = \frac{R_4}{K} \tag{13.58}$$

$$R_1 = \frac{R_2 R_4}{R_2 - R_4} \tag{13.59}$$

$$\left(\frac{R_3}{R_3 + R_4}\right) F(T) \tag{13.60}$$

The conversion equations applicable to Figure 13.20*b* are:

$$K = \frac{R_2}{R_1 + R_2} = \frac{E_1}{E_2} = \frac{R}{R_1} \tag{13.61}$$

$$R = \frac{R_1 R_2}{R_1 + R_2} \tag{13.62}$$

$$E_2 = \text{Desired source voltage} \tag{13.63}$$

$$R_1 = \frac{R}{K} \tag{13.64}$$

$$R_2 = \frac{RR_1}{R_1 - R} \tag{13.65}$$

$$e_{0_2} = (E_2 - E_1) + e_{0_1} \tag{13.66}$$

Reference [15] includes design examples that use these conversion equations.

Temperature Controllers.

Thermistor temperature controllers frequently use voltage divider and bridge circuits. The use of a thermistor in such a circuit results in much higher sensitivity than that obtainable with a thermocouple or RTD. The most sensitive standard thermocouples exhibit output voltage slopes of 50 to 55 μV °C^{-1} in the range of 0°C to 300°C. Thermistor voltage dividers typically exhibit output slopes of about 8 to 10 mV °C^{-1} per volt applied to the divider. Since the input to a thermistor voltage divider typically is in the range of 1 to 5 V, a thermistor provides about 200 to 1000 times the sensitivity of a thermocouple. The temperature coefficient of resistance of a thermistor is about 10 times that of an RTD. However, the temperature span about the control point of a thermistor controller is small compared with that available with a thermocouple or RTD.

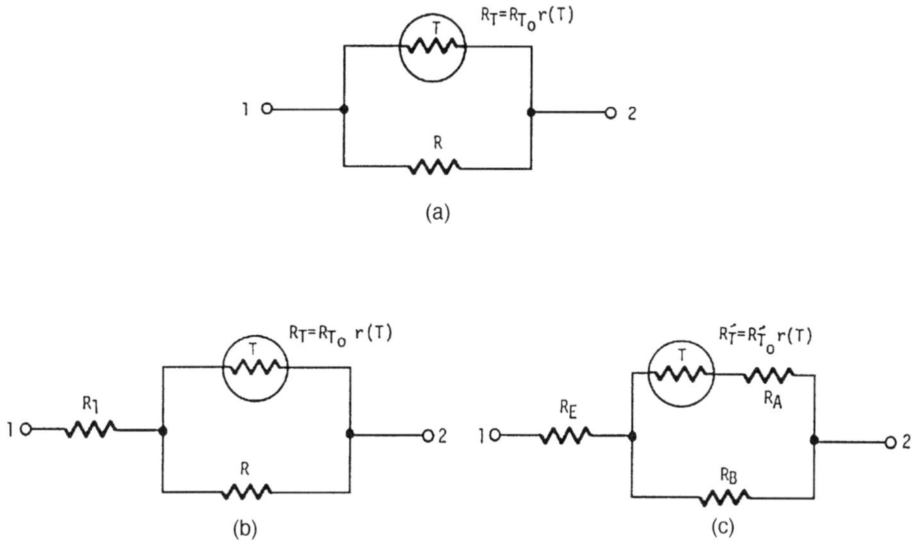

FIGURE 13.21 Equivalent linear thermistor networks for converting a network with a nonstandard R_{T_0} to a network with a standard catalog value for R'_{T_0}. A requirement is that both thermistors have the same resistance ratio–temperature characteristic.

Reference [1] includes some typical low-cost thermistor temperature controllers. Thermistor temperature controllers are available from Hart Scientific, Inc. that provide control stabilities of better than 0.001°C [16].

Linear Resistance Networks

A common technique for designing a thermometer circuit is to apply a constant-current source to a linear resistance network. This results in a voltage across the network that is linear with temperature. Temperature compensation for resistance changes that occur in coil windings, instruments, relays, motors, and generators also use such networks. For example, a copper coil has a positive temperature coefficient of resistance of approximately 0.39% $°C^{-1}$. The thermistor network has a negative temperature coefficient. The compensator is designed to provide a slope that is equal in magnitude, but opposite in sign to that of the copper coil. Hence, the Ω vs. °C slope of the compensator is equal to the Ω/°C change of the coil. This results in a current through the coil that is independent of temperature. Additional applications include compensation of drift in silicon strain gages, infrared detectors, and circuits that contain both passive and active components.

For the basic linear resistance network of Figure 13.21*a*:

$$R_{12} = R_C = \frac{RR_T}{R + R_T} = R\left(1 - \frac{1}{1 + R_T/R}\right) \tag{13.67}$$

Normalizing with respect to R and substituting $r_T = R_T/R_{T_0}$ and $R_{T_0}/R = \sigma$ in Equation 13.67 yields:

$$R_n = \frac{R_C}{R} = 1 - \frac{1}{1 + r_T\sigma} = 1 - F(T) \tag{13.68}$$

Consequently, the techniques used for optimizing the linearity of voltage dividers also apply to linear resistance networks. The use of the series resistor R_1 in Figure 13.21*b*, translates the curve to a higher resistance level and does not affect the Ω/°C output.

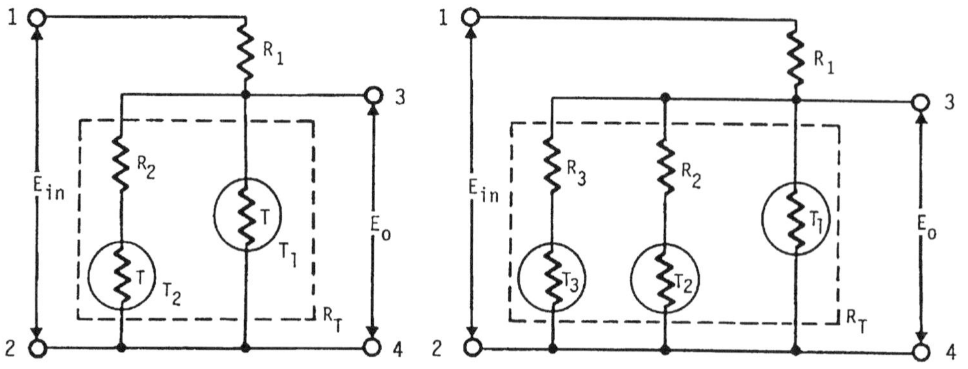

FIGURE 13.22 Two-thermistor and three-thermistor linear voltage dividers that provide improved linearity. Placing the resistors R_l across the output terminals 3–4 converts the networks to linear resistance networks.

Frequently, the design of a compensator that provides the desired Ω/°C output results in a thermistor that has a nonstandard R_{T_0} value. The conversion circuit of Figure 13.21*c* permits the use of a standard R_{T_0} value. The conversion equations are as follows:

$$K_1 = \sqrt{\frac{R'_{T_0}}{R_{T_0}}} \tag{13.69}$$

$$R_E = R_1 - R\left(K_1 - 1\right) \tag{13.70}$$

$$R_A = RK_1\left(K_1 - 1\right) \tag{13.71}$$

$$R_B = RK_1 \tag{13.72}$$

where R_{T_0} is the nonstandard value and R'_{T_0} is the standard value. A requirement is that both thermistors have the same resistance ratio–temperature characteristic. Setting $R_1 = R(K_1 - 1)$ in Figure 13.21*b* results in $R_E = 0$ for Figure 13.21*c*. Consequently, the conversion of Figure 13.21*a* to that of Figure 13.21*c* requires the use of the minimum insertion resistance $R_1 = R(K_1 - 1)$.

The network provides good linearity for small temperature spans of 10°C to 30°C. However, the error increases rapidly from about 0.15°C for a 30°C span to 0.7°C for a 50°C span (0°C to 50°C). The use of the two-thermistor network shown in Figure 13.22 results in a maximum linearity error of 0.22°C over the range of 0°C to 100°C, while the three-thermistor network shown reduces the error to 0.04°C for the same 0°C to 100°C range [5]. The networks shown in Figure 13.22 are linear voltage dividers. Placing the resistor R_l across the output terminals 3–4 converts these networks to linear resistance networks. Many manufacturers sell an interchangeable, three-wire, dual thermistor that is suitable for use in the two-thermistor circuit of Figure 13.2.

Interchangeable thermistors and the linearization techniques described above have been available for many years. In addition, the stability of NTC thermistors is better than that of thermocouples and frequently better than or equal to that of commercial RTDs [17]. However, the nonlinear resistance–temperature characteristics of thermistors continue to limit their use in industrial temperature measurement and control applications. The availability of low-cost microprocessors has eliminated this limitation. Such devices can use the equation constants of Equations 13.38 or 13.39 to compute and display temperature directly. They also can be used to compute lookup tables to provide interpolation uncertainties in the range of 0.001°C to 0.01°C. There are instruments described in the literature that utilize such microprocessor chips [18, 19]. Thermometrics, Inc. sells a commercial instrument that reads the equation constants from a chip in the connector of each probe supplied for use with the instrument [1]. The use of thermistors for industrial applications will continue to increase as the cost of microprocessor chips continues to fall.

References

1. Thermometrics, Inc., *Thermometrics NTC & PTC Thermistors,* Edison, NJ, 1993.
2. J. Fabien, Heating with PTC thermistors, *EDN Products Edition,* 41(12A), 10, 1996.
3. M. Sapoff and R. M. Oppenheim, The design of linear thermistor networks, *IEEE International Convention Record,* Part 8, 12, 1964.
4. M. Sapoff, Thermistors: Part 4, Optimum linearity techniques, *Measurements & Control,* 14(10), 1980.
5. C. D. Kimball and R. W. Harruff, Thermistor thermometry design for biological systems, *Temperature, Its Measurement and Control in Science and Industry,* Vol. 4, Pittsburgh, PA: Instrument Society of America, 1972, Part 2.
6. Ketema, Rodan Division, *Thermistor Product Guide,* Anaheim, CA, 1995.
7. Cesiwid Inc., *Alphalite® Bulk Ceramic NTC Thermistors,* Niagara Falls, NY, 1996.
8. Fenwal Electronics, Inc., *Standard Products Catalog,* Milford, MA, 1994.
9. Fenwal Electronics, Inc., *Thermistor Manual,* Milford, MA, 1974.
10. Victory Engineering Corporation, *Technical Corporation of Thermistors & Varistors,* Springfield, NJ, 1962.
11. J. S. Seinhart and S. R. Hart, Calibration curves for thermistors, *Deep Sea Research,* 15, 497, 1968.
12. B. W. Mangum, The triple point of succinonitrile and its use in the calibration of thermistor thermometers, *Rev. Sci. Instrum.,* 54(12), 1687, 1983.
13. M. Sapoff, W. R. Siwek, H. C. Johnson, J. Slepian, and S. Weber, The exactness of fit of resistance–temperature data of thermistors with third-degree polynomials, in J. F. Schooley (ed.), *Temperature, Its Measurement and Control in Science and Industry,* Vol. 5, New York, NY: American Institute of Physics, 1982, 875.
14. M. Sapoff and R. M. Oppenheim, Theory and application of self-heated thermistors, *Proc. IEEE,* 51, 1292, 1963.
15. M. Sapoff, Thermistors: Part 5, Applications, *Measurements & Control,* 14(12), 1980.
16. Hart Scientific, Inc., *Hart Scientific Calibration Equipment,* Pleasant Grove, UT, 1995.
17. W. R. Siwek, M. Sapoff, A. Goldberg, H. C. Johnson, M. Botting, R. Lonsdorf, and S. Weber, Stability of NTC thermistors, in J. F. Schooley (ed.), *Temperature, Its Measurement and Control in Science and Industry,* Vol. 6, New York, NY: American Institute of Physics, 1992, 497.
18. R. L. Berger, T. Clem, C. Gibson, W. Siwek, and M. Sapoff, A digitally linearized thermistor thermometer referenced to IPTS—26(13), 68, 1980.
19. W. R. Siwek, M. Sapoff, A. Goldberg, H. C. Johnson, M. Botting, R. Lonsdorf, and S. Weber, A precision temperature standard based on the exactness of fit of thermistor resistance–temperature data using third degree polynomials, in J. F. Schooley (ed.), *Temperature, Its Measurement and Control in Science and Industry,* Vol. 6, New York, NY: American Institute of Physics, 1992, 491.

13.4 Thermocouple Thermometers

R. P. Reed

The Simplest Thermocouple

Despite an increasing variety of temperature sensors, the self-generating thermocouple remains the most generally used sensor for thermometry because of its versatility, simplicity, and ease of use. Any pair of

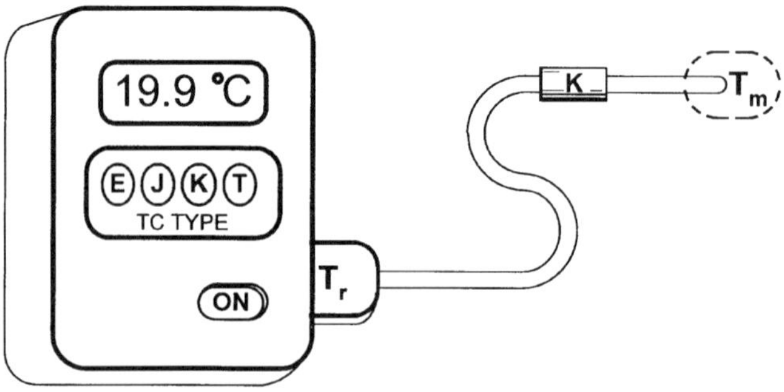

FIGURE 13.23 The simple modern digital thermocouple thermometer. Modern digital electronics has made casual thermometry very easy, but has obscured the continuing need to have an authentic understanding of thermoelectric principles for accurate thermometry with more complicated circuits and more important measurements.

electrically conducting and thermoelectrically dissimilar materials coupled at an interface is a *thermocouple* [1]. The legs are *thermoelements*. The *Seebeck effect* produces a voltage in all such thermoelements where they are not at a uniform temperature. Any electric interface between dissimilar electric conductors is a *real thermoelectric junction*. A free end of a thermoelement is a *terminus*, not a junction. Couplings between *identical* thermoelements are *splices* or *joins*, not junctions.

It is the thermoelements that determine thermocouple *sensitivity* and calibration; but, it is the temperatures of the end-points of thermoelements (i.e., junction temperatures) that determine the *net* emf observed in thermometry. The Seebeck effect, which converts temperature to voltage, is used for thermoelectric thermometry but is also a primary low-frequency noise source in all low-level electronic circuits [2].

Simple Thermocouple Thermometry

In the simplest applications, thermocouple thermometry now is as easy to use as is a multimeter to measure resistance. In fact, many present-day digital multimeters (in addition to voltage, current, and resistance) do provide a thermocouple temperature probe, and temperature measurement is just another button-selectable function. These thermocouple thermometers consist of an indicator that digitally displays the temperature of the tip of a plug-in thermocouple probe (Figure 13.23). The simpler of such versatile multimeters can be purchased for less than U.S.$60 [3]. Interchangeable thermocouple probes of different standard thermocouple material types and specialized sensing tips of widely varying designs are adapted to measurement from surface temperatures to internal temperatures in foodstuffs [4].

With some probes, temperatures up to 1370°C can be indicated merely by (1) ensuring that the selected thermocouple types of the indicator and probe correspond, (2) pressing a power-on button, and (3) applying the probe tip at the point where temperature is to be measured. Promptly, and without calibration, the present temperature of the probe tip is digitally displayed selectably in °C or °F, usually to a resolution of 0.1°C or 1°C. Some specifications claim "accuracy" of 0.3% (4°C at 1370°C) of reading. Some even offer certified *calibration traceability* to NIST or other national standards laboratory.

Modern digital electronics has created the illusion that very accurate thermocouple thermometry is no different than other routine electrical measurements. It has encouraged the perception that arbitrarily fine accuracy can be always be accomplished by calibration and guaranteed by certification. The remarkable simplification of instruments that is now commonplace and an abundance of misleading tutorials obscure the real need for a sound understanding of principles of thermocouple circuits for anything other than inconsequential thermometry.

Manufacturers can shelter users from many problems inherent in thermocouple application. Unfortunately, there are many pitfalls from which the manufacturer cannot isolate the user by design or construction. In less simple applications, it is necessary for the user to avoid problems by carefully learning and applying the true principles of thermoelectric circuits and thermocouple thermometry. Even the simplest indicators and probes can easily be misused and produce unrecognized substantial error.

Thermometry errors of only a few °C or even much less in energy, process, manufacturing, and research fields annually cost many millions of dollars in lost yield, fuel cost, performance bonuses, etc. Consequence of error can also be incurred as incorrect interpretation of data, failure of objective, equipment damage, personal injury, or even loss of life.

Unusual thermocouple circuits, installations, and special applications often produce *inconspicuous* error or else *peculiar* results that are very puzzling if an authentic model of thermoelectric circuits is not understood. This chapter section presents the factual principles of thermoelectric circuits that equip the user to easily apply thermocouples for reliable and critical thermometry, even in unusual circumstances with justified confidence. These principles are very simple, yet they justify study even by experienced thermocouple users as thermoelectric circuits are very often misrepresented or misunderstood in subtle ways that, although unrecognized, degrade measurement. The aim of the chapter section is to allow thermometry with all *practical* simplicity while avoiding possibly costly measurement errors.

Thermoelectric Effects

The three thermoelectric phenomena are the Seebeck, Peltier, and Thomson effects [1, 2, 5-7]. Of these, *only* the Seebeck effect converts thermal energy to electric energy and results in the thermocouple voltage used in thermometry. The current-dependent Peltier and Thomson effects are insignificant in practical thermometry. Neither produces a voltage, contrary to common misconceptions. The Peltier and Thomson effects only transport heat by electric current and redistribute it around a circuit. Thermocouple thermometry is properly conducted by *open-circuit* measurement. The Seebeck emf occurs even without current where Peltier and Thomson effects *necessarily* vanish. Related *thermo-magneto-electric* effects are significant only in the presence of large magnetic fields and infrequently degrade applied thermoelectric thermometry [2, 5, 6].

The Seebeck Effect

The Seebeck effect is the occurrence of a net source emf, the *absolute Seebeck emf*, between pairs of points in any individual electrically conducting material due to a difference of temperature between them [1, 2, 5, 7, 8]. The Seebeck emf occurs *without* dissimilar materials. It is *not* a junction phenomenon, nor is it related to Volta's contact potential.

Absolute Seebeck Properties

The *absolute Seebeck coefficient* expresses the measurement sensitivity (volts per unit of temperature) of the Seebeck effect. It is defined over any *thermoelectrically homogeneous* region of a slender individual conducting material by:

$$\sigma(T) = \lim_{\Delta T \to 0} \Delta E/\Delta T = dE/dT, \text{ or} \tag{13.73}$$

$$dE = \sigma(T)dT \tag{13.74}$$

The Seebeck coefficient is a transport property of *all* electrically conducting materials. Equation 13.74 will be acknowledged later as the functional law that governs thermoelectric emf. From Equation 13.74,

$$\Delta E = \int_{T_1}^{T_2} \sigma(T)dT = E(T_2) - E(T_1) \tag{13.75}$$

where ΔE is the increment of emf between a pair of points, separated by *any* distance, between which the temperature difference is $\Delta T = (T_2 - T_1)$.

From Equation 13.75, the net Seebeck effect for a particular material depends only on the temperatures at the two points and *not* on the values of temperature gradients between the two points. The Seebeck coefficient is a nonlinear function of temperature. It is not a constant. For accurate thermometry, the Seebeck coefficient must remain dependent on temperature alone. $\sigma(T)$ cannot vary along a thermoelement, nor can it vary significantly during the time interval of use. Although vulnerable to such environmental effects, for accurate thermometry it must not depend during measurement on such environmental variables as strain, pressure, or magnetic field.

The coefficient, $\sigma_M(T)$, is an *absolute Seebeck coefficient* for an individual material M [1, 2, 7, 8]. A corresponding *source* voltage within a single material is an *absolute Seebeck emf*, $E_M(T)$. The absolute Seebeck emf does physically exist but it is not simply observable. The absolute Seebeck coefficient can be determined indirectly by measuring the *Thomson coefficient*, τ, of the individual material and applying a Kelvin relationship,

$$\sigma = \int_0^{T_{abs}} \left(\tau / T_{abs}\right) dT \tag{13.76}$$

to deduce the thermodynamically related Seebeck coefficient [1, 2, 5, 7, 8].

Relative Seebeck Properties

The difference between the Seebeck emfs of two thermoelements, of materials A and B, of a thermocouple with their shared junction at temperature, T_m, and both their termini at a *physical reference temperature*, T_r, is their *relative Seebeck emf*, $E_{AB}(T)$, expressed as:

$$E_{AB}\left(T_m, T_r\right) = E_A\left(T_m, T_r\right) - E_B\left(T_m, T_r\right) \tag{13.77}$$

The corresponding relative Seebeck coefficient for the pair is:

$$\sigma_{AB}\left(T_m, T_r\right) = \sigma_A\left(T_m, T_r\right) - \sigma_B\left(T_m, T_r\right) \tag{13.78}$$

It is this relative Seebeck coefficient that has been called by the anachronistic and inept term "thermopower" [1]. It is these *relative* voltage or coefficient values that are directly observable and usually used in thermometry. These relative properties, defined for convenience in the *series* circuits of thermometry, have no general meaning for electrically paralleled thermoelements [2]. These are the relative values that are presented in the familiar tables of thermocouple emf vs. measuring junction temperature, T_m, referred to as a *designated reference temperature* $T_0 = T_r$ [1, 4, 9-12]. For convenience, T_0 is usually taken as 0°C, but the value is arbitrary.

Realistic Thermocouple Circuits

The thermocouple is often represented as only a pair of dissimilar thermoelements joined by two junctions in a closed circuit. One junction, at temperature T_m, is the *measuring junction*; the other, at temperature T_r, is the *reference junction*. The net Seebeck emf is proportional to the temperature difference between the two junctions and to a relative coefficient for the paired materials.

The Seebeck phenomenon has wrongly been characterized as the occurrence of current in the closed loop. The true nature of the Seebeck phenomenon is the occurrence of a *source emf* that, for accurate thermometry, must be measured in open-circuit mode that suppresses current. In practical thermometry, no realistic thermocouple circuit has only two dissimilar materials. Some have many and several of these can be expected to contribute some Seebeck emf. The most common thermometry circuits have two separate reference junctions, not one. Valid, but uncommon, circuits can simultaneously have more than one reference temperature [1].

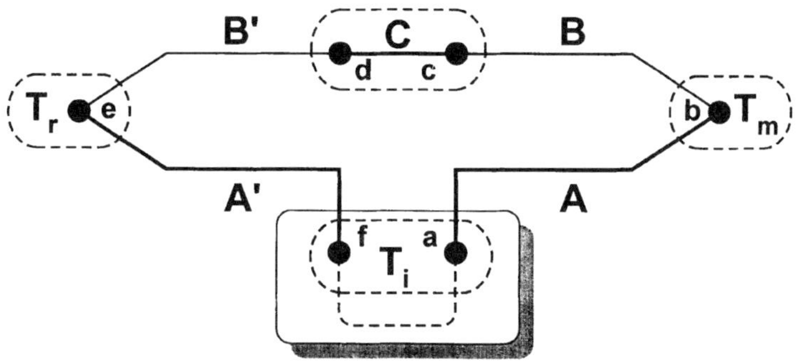

FIGURE 13.24 The basic thermocouple circuit with a *single* temperature reference junction, **e**. The Seebeck voltage measured in open-circuit mode at terminals **a** and **f** is proportional to the temperature difference between thermocouple *measuring junction* **b** and the necessary temperature *reference junction* **e**. For convenience, T_r is usually made to be 0°C. For thermometry, the zones at temperatures T_r and T_i must be isothermal.

Reference Temperature

The *physical* reference temperature, T_r, can be different from the *designated* reference temperature, T_0, of the characterizing relation. If T_r is not identical to T_0, then, to use standard scaling functions, the observed thermocouple emf must be corrected for the temperature difference by adding an emf equivalent to $E_{AB}(T_r) - E_{AB}(T_0)$ to the observed thermocouple emf [1, 7, 9-12]. This is often accomplished by separately monitoring T_r and applying a correction, either numerically or electrically, using a fixed $E_{AB}(T)$ relation that only *approximates* that of the actual thermocouple and the standardized characteristic over a limited range in the vicinity of T_0 and T_r. The error due to the slight discrepancy between the approximation and the actual $E_{AB}(T)$ is small if the two temperatures are similar.

Special thermocouple extension leads are used in most applied thermometry. Many industrial principal thermocouples are inflexibly metal sheathed [1, 4, 10-12]. Others have bare thermoelements separated by bead insulators. Often, these kinds of assemblies are housed in protective wells, have measuring junctions of complex construction, or are distant from the monitoring instrument. Short "pigtails" and extension leads can be of larger wire size, lower resistance, greater flexibility, and very different rugged cable construction than the principal thermocouple (Figure 13.23). All thermoelements must be very well electrically isolated except at measuring and reference junctions. Most modern thermometry is conducted with variants of two basic circuits.

Circuit with Single Reference Junction.

Figure 13.24 shows a thermometry circuit now used mostly in calibration laboratories. This form is convenient when a *fixed point temperature reference* such as an ice point bath or water triple point cell is used to impose the known *physical* reference temperature, T_r [1, 7, 9-12]. The circuit of the thermocouple indicator between **a** and **f** inconspicuously includes many incidental materials and complex circuitry within the instrument. It is necessary that $T_i = T_a = T_f$. Unless instrument temperatures are constant throughout measurement, any nonisothermal portion of the circuit within the instrument can contribute Seebeck emf as noise. This emf can be correctly offset *only* if it is constant.

Often, in circuits like Figure 13.24, the measuring junction, **b** of the principal thermocouple, and reference junction, **e**, are provided by *separately* acquired thermocouples **a-b-c** and **d-e-f.** The two might have significantly different calibrations although they are nominally of the same thermocouple material type. With T_m at 40°C and T_0 at 0°C the reference, thermocouple **d-e-f** contributes about half the Seebeck emf.

This commonplace circuit has at least four distinct thermoelement materials — A, A′, B, and B′ — each of which must be homogeneous. Also, they can be joined, as between **c** and **d**, by an intermediate uncalibrated linking material, *C* (unless $T_c = T_d$, material *C* contributes unwanted Seebeck emf). If B

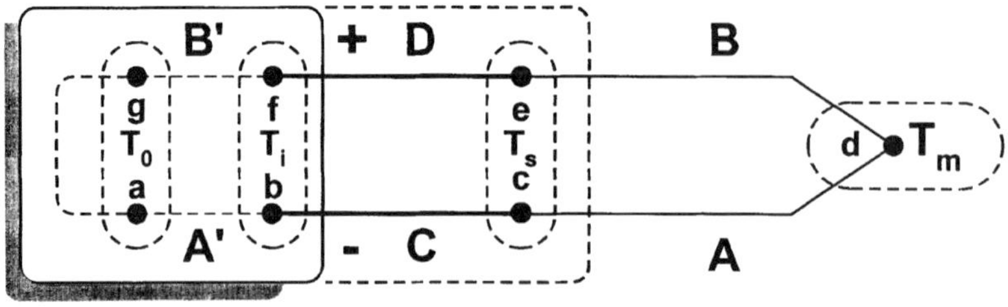

FIGURE 13.25 The basic thermocouple circuit with *dual* temperature reference junctions. The principal thermocouple is the AB pair. Pair CD is the extension lead. The Seebeck voltage measured in open-circuit mode between terminals **b** and **f** is proportional to the temperature difference between thermocouple *measuring junction* **d** and the necessary temperature *reference junctions.* Depending on the type of extension leads, the reference junctions might be either **c** and **e**, or else **b** and **f**. For thermometry, the zones at temperatures T_s and T_i must be isothermal.

and B′ are not *identical* in Seebeck characteristics, then **c** (and/or **d**) are *real* junctions. If so, it is also necessary that $T_i = T_c = T_d$. The termini **a** and **f**, when connected to a monitor, also become real junctions. They must be controlled so that both stay at the same temperature.

In Figure 13.24, for thermometry the physical reference temperature is intended to be $T_r = T_e$. However, monitoring instruments that internally compensate for reference junction temperature presume (*incorrectly for this circuit*) that $T_r = T_a = T_f$. Therefore, the *single-reference circuit of Figure 13.24 cannot be used directly with thermocouple instruments that automatically apply reference junction temperature compensation.*

Circuits with Dual Reference Junctions.
In the simple "black box" thermocouple thermometer (Figure 13.23), as well as in most applied thermoelectric thermometry, the most common circuit is that of Figure 13.25. This circuit is now the most commonly used in modern thermocouple thermometry. The thermocouple **c-d-e** with thermoelements joined at the measuring junction, **d**, is the *principal thermocouple.* This circuit might have only the principal thermocouple with a plug and jack at the indicator input. More often, as in Figure 13.23, the relatively inflexible principal thermocouple probe also has flexible *extension leads, pair C-D,* which can reside unseen within the indicator. Dashed pair A′-B′ schematically represents internal reference junction temperature compensation.

Circuits like Figure 13.25 have two separate *reference junctions* that must be held at the same *known* temperature. Net Seebeck terminal voltage is measured between **b** and **f**. When thermocouple leads *C* and *D* are connected to the monitor, the input terminals, **b-f**, might be intended to be reference junctions. If so, their reference temperature is accurately monitored and emf reference correction corresponding to the difference between T_r and T_0 is applied. Accurate thermocouple measurements cannot be made with ordinary voltmeters in which the temperatures of **b** and **f** are not deliberately controlled to be equal and known.

Extension Leads

Paired thermoelements *C* and *D* are thermoelectric *extension leads.* Extension leads are of three kinds: (1) *neutral,* (2) *matching,* and (3) *compensating.* All types are readily available; some are proprietary. Which of the junctions in Figure 13.25 must be reference junctions varies with the type of extension. Depending on the extension type, materials *A*, *B*, *C*, and *D* might, intentionally, all be of very different materials.

Neutral Extension Leads.
In the simplest application of Figure 13.25, legs *C* and *D* are thermoelectrically homogeneous and are carefully matched to have the same Seebeck coefficient ($\sigma_C = \sigma_D$). Such pairs are *neutral extension leads.* Provided that $T_c = T_e$ and $T_b = T_f$, they contribute no *net* Seebeck emf, so the like pair function effectively serve only as "passive" leads. Therefore, the leads could be made of any electrically conducting material. Such extension leads should be made of solid unplated copper, for which the Seebeck coefficient is small and uniform.

Neutral extension leads require that junctions **c** and **e** be the reference junctions at temperature $T_r = T_s$. *Thermocouple monitors that provide reference junction temperature compensation at input terminals **b** and **f** cannot be used with neutral extension leads.*

The reference temperature can be controlled on both junctions by a fixed point physical temperature reference such as an ice bath external to the voltage monitor [1, 9-12] Ice baths, very carefully made and maintained, can routinely approximate 0°C to within 0.1°C to 0.2°C [9, 11]. Carelessly applied and maintained, they can deviate from the "icepoint temperature" by up to 4°C [10]. Peltier thermoelectric refrigeration is also used. More often, in thermometry, that reference temperature is not physically imposed. Instead, the actual temperature of the reference junctions is accurately measured and electrical compensation is applied (schematically by A′-B′) for the difference between T_r and T_0. The particular value of the temperature at **b** and **f** does not matter if $T_b = T_f$ is compensated.

Complementary Extension Leads.

Two kinds of complementary extension leads are allowed to contribute to the circuit Seebeck emf: *matched* and *compensating* extensions. The legs *C* and *D* are thermoelectric extension leads that are usually uncalibrated, and possibly are of broader or even unknown tolerance. It is expected that T_s and T_i are similar so that the uncalibrated contribution is small and the error is negligible. For very accurate thermometry, this assumption might not be justified.

Complementary extension leads can be used directly with thermocouple indicators that internally compensate the reference junction temperature for the difference between the physical reference junction temperature, T_r, and the designated reference temperature, T_0. For both matched and compensating complementary extensions, the reference temperature is $T_r = T_i$ over the isothermal reference zone around the terminals **b** and **f**.

Matching Extension Leads.

Matching leads have $\sigma_C = \sigma_A$ and also $\sigma_D = \sigma_B$. It is essential that $T_b = T_e$, but it is not essential (although it is desirable) that $T_c = T_e$. Near room temperature, thermoelements are intended to have σ_{CD}, as a pair, be nearly the same as σ_{AB} for the temperature span near T_r. Error due to slightly mismatched and unproven calibration of the extension is minimized if $T_s \cong T_i$. Matched extensions are used with base metal thermocouples, not with refractory or precious metal thermocouples.

Compensating Extension Leads.

The third variation of the circuit in Figure 13.25 is often used for economy and convenience with expensive precious and refractory metal principal thermocouples that are intended for use at very high temperatures and in special environments [1, 7, 9-12]. Usually, only a portion of the principal thermocouple need be exposed to the adverse environment. Extension leads *C* and *D* need only survive a more benign environment. Therefore, they can be made of a less expensive or more conveniently handled material, use lower temperature insulation, be more flexible and of lower resistance, add bulky shielding and mechanical protection, and extend a great distance to a recording facility.

Compensating leads have σ_{CD} of the extension *C–D*, only as a pair, match σ_{AB} of the *A–B* pair. It is not necessary that material *A* be like *C*, nor that *C* and *D* be alike. A practical circuit can have four very dissimilar materials. However, to allow this, it is essential that $T_c = T_e$. Therefore, this circuit with compensating leads is more vulnerable to error from improperly matched temperatures of incidental junctions than with matching extension leads. The reference junction temperature is $T_r = T_b = T_f$, and only this temperature must be independently known. Reference junction compensation is the same as for the matching leads.

Modular Signal Conditioning Components

The myriad of diverse and capable thermocouple indicators and recorders now commercially available, ranging from simple and very inexpensive to sophisticated and versatile yet reasonably priced, makes it unnecessary for most applied thermometry to assemble special signal conditioning circuits. Some, seeking economy of hardware, have built custom systems. Too often, these have not achieved accurate measurement and have proved to be very costly because the critical distinctions between thermoelectric circuits and ordinary electric circuits were not appreciated. For special situations where thermocouples are incorporated into special measurement or operational components or systems, several manufacturers now offer modular and integrated circuit components that simplify special application and do protect the unaware from some pitfalls [13].

Thermocouple Signal Conditioning on a Chip.
As thermocouple signals are low level, it is sometimes desired to amplify them for improved resolution, recording, or control. It might be desirable to incorporate thermocouple sensing as a functional component in other instrumentation packages. There are now several miniature and inexpensive integrated circuit modules for thermocouples that provide reference junction compensation, linearization, isolation, open input indication, amplification, and set point control. Properly applied, these make the integration of thermocouple sensing into other devices, such as computer data acquisition boards, very simple [13].

When a user applies modular components such as these, it is particularly important to understand and to apply the authentic thermoelectric circuit model and principles introduced in the section "The Functional Model of Thermoelectric Circuits." Some precautions normally provided in commercial thermocouple monitors must be provided by the user.

Reference Temperature Sensors.
Thermocouples are self-generating. For casual approximate temperature and differential temperature measurement, they require neither excitation nor reference temperature. For accurate thermometry, unless a known reference temperature is physically imposed, an accurate measurement of the reference junction temperature by independent means is necessary. This sensing is usually by powered resistance temperature detectors, thermistors, transistors, or integrated circuit sensors [1, 4].

Reference Temperature Compensators.
Proper reference compensation requires (1) the establishment of an isothermal temperature zone that includes the terminals of the thermocouples, (2) sensing of the temperature of this zone, and (3) application of a complementary physical or numerical voltage to the thermocouple terminal voltage before scaling the total voltage to temperature.

An analog method includes the resistive zone temperature sensor in a bridge that nonlinearly modulates a supplied voltage according to an approximated nonlinear curve of the $E(T)$ characteristic of the thermocouple. This method is adapted to compensation of only a single thermocouple type. An alternate method numerically converts the sensed reference temperature to the appropriate compensating voltage value. The numerical approach allows applying a separate compensating voltage to each individual thermocouple and for different standard types. In principle, for accuracy, a numerical compensating voltage could be programmed by the user to conform to the specific calibration of the individual thermocouple, whether or not of standardized type.

Reference temperature compensators are internal to the more advanced thermocouple monitors. In advanced units, eight to ten thermocouple inputs are grouped on a separately removable isothermal terminal assembly so they can be removed for reference sensor calibration or for replacement. Like thermocouples, reference sensors occasionally drift, causing significant temperature error. For most thermocouple types, a 1° error in temperature reference produces a similar error in the measured temperature. A variety of external battery- or line-powered, single- or multichannel reference junction compensating units are also commercially available [4, 14].

Grounding, Shielding, and Noise

The technically well-founded principles of noise control for electric circuits apply also to thermoelectric circuits [15, 16]. There are, however, some additional considerations that are necessary for thermocouple circuits. Some noise control problems stem from the nature of thermoelectricity; others from commercial thermocouple design practices.

Noise Problems Peculiar to Thermocouples

Temperature control of all deliberate and incidental junctions and components, and the distributed nature of Seebeck source emf, cannot be ignored with impunity in thermocouple application. The unavoidable requirements of such control are best visualized using methods such as given later in "The Functional Model of Thermoelectric Circuits." Incidental circuit components such as balancing or swamping resistors, feedthrus, splices, and terminal strips, not intended to contribute Seebeck emf, must deliberately be maintained *isothermal.* Shields, unavoidably nonisothermal, made electrically common with the thermocouple must be explicitly recognized as latent sources of random dc noise which contributes Seebeck emf if not properly connected.

Electromagnetic (EM) Noise

In thermocouple probes of mineral-insulated metal-sheathed (MIMS) construction, ceramic-bead-insulated thermocouples, and in some paired insulated thermocouple wires, the paired thermoelements are not twisted and are well separated. Lengthy and larger diameter thermocouple probes present a significant circuit loop area to couple magnetic noise fields. Even if probes are entirely metal-sheathed, the standardized sheaths are very thin and of low magnetic permeability. They are scarcely effective for electromagnetic (EM) shielding. These features prevent the use of some classic techniques for the rejection of EM noise [15, 16]. To reduce troublesome EM noise, the principal thermocouple probe should be of the smallest practical diameter, and of minimum length. In instances where the EM source is localized and identified, the orientation of the probe relative to the source can be arranged to minimize EM coupling. For rejection of EM noise, extension leads should always be of twisted-pair construction, with a pitch small enough to reject high-frequency noise [15].

Electrostatic (ES) Noise

Electrostatic noise is more easily reduced. Shields of low resistance, though thin, can be effective if properly connected [15, 16]. Plated copper braid is commonly used and is effective for noise of moderate frequency. Continuous shields, such as the MIMS sheaths and aluminized mylar film, are effective for low-frequency ES noise and are more effective than braid shields for very-high-frequency noise. Electrostatic shields must be continuous (without gaps and holes as in braid) for maximum effectiveness. Optimum benefits from shielding require use with thermocouple monitors that provide three-wire input and multiplexers that, individually for each thermocouple, switch both the signal pair and their separate shield lead.

Grounding

Shielding for ES noise and pair twisting for EM noise are important for reduction of high-frequency noise. A secondary overall shield, isolated from inner shields and separately grounded, can improve the rejection of EM noise. Appropriate circuit grounding is also particularly important for both high-frequency EM and ES control, but it is even more significant for low-frequency and dc noise, which are often more consequential in thermocouple measurements. Appropriate grounding is complicated where intimate thermal contact of the measuring junction with an earth-grounded conducting test subject is needed for rapid transient thermometry or for accuracy in the presence of static temperature gradients.

Figure 13.27 illustrates appropriate grounding for several of many possible situations. Low-level thermocouple circuits should be grounded only at a single electric reference point. Because the measuring junction is not a localized site of source emf, the point of grounding must be carefully considered. Grounding should never be at any point of the thermocouple circuit other than measuring or reference junctions. Electric contact, and particularly shunting or shorting, on a thermoelement at any point between measuring and reference junctions usually introduces spurious Seebeck emf from incidental unrecognized thermoelements.

Allowable grounding might be dictated by the internal design of monitors and data loggers. Each instrument design addresses noise control in a distinctive way. Simple and inexpensive line-powered thermocouple thermometers might have only single-ended inputs with the negative lead internally

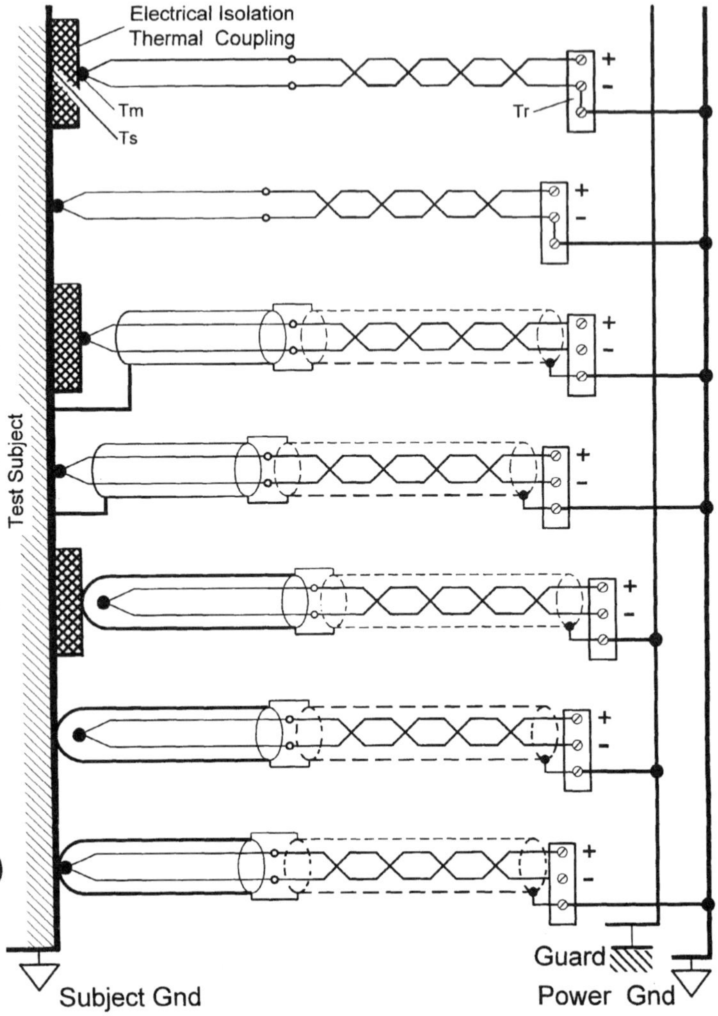

FIGURE 13.27 Preferred grounding and shielding for several thermometry situations. Electromagnetic (EM) and electrostatic (ES) noise must be controlled by different means. The design of the thermocouple monitoring instrument may dictate the grounding and shielding scheme that can be used.

connected to the power ground. This forces the sensor grounding to also refer to power ground. A grounded-junction arrangement might also be required for fast transient thermometry. In use, initially isolated thermocouples can short to the protective sheath or test subject. Although this does change the grounding configuration, resulting errors due to other effects usually are then predominant.

Balanced Thermocouple Circuits

The more sophisticated multichannel data acquisition systems specialized to thermocouple thermometry usually have high-resistance balanced inputs for high common-mode signal rejection. Some differential input amplifiers require significant input resistance for stability. Most have an internal chassis that provides a common signal reference guard surface that is well-isolated from power ground. In a few designs, the guard is electrically driven to a particular reference point. Because many variations are used, it is prudent to study and follow the specific grounding instructions recommended in the user's manual for the instrument. Most accommodate alternative grounding arrangements.

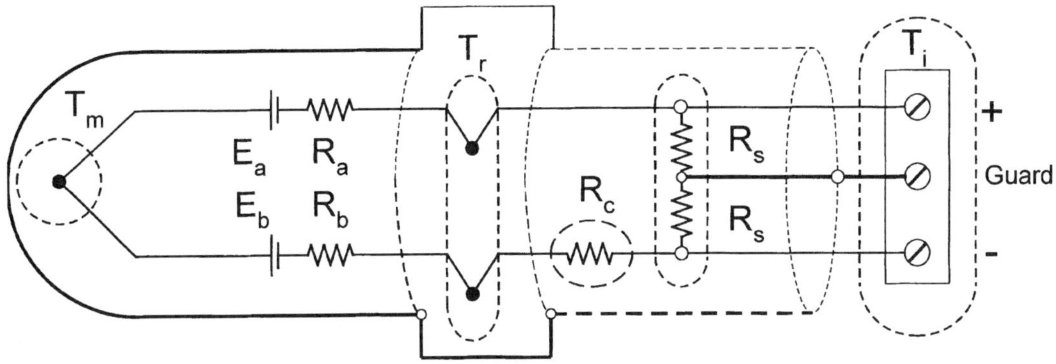

FIGURE 13.27 A circuit arrangement for balanced input recorders to improve common-mode noise rejection. The thermoelement balancing resistor, $R_c = |R_a - R_b|$, and the matched pair of high-resistance input resistors R_s must be held isothermal.

In sheathed thermocouple probes of standard MIMS dimensions, the thermoelements are of the same cross-section, parallel and well spaced, and located within the cylindrical sheath symmetrically [10]. Noise tends to be coupled equally to the paired thermoelements in common mode. However, lengthy or fine gage thermocouples have legs of high resistance compared to copper circuits. The ratio between the resistances of the dissimilar legs can be up to 25:1. Noise introduced in common mode adversely tends to convert to normal mode.

In addition, unlike most voltage sources, the Seebeck emf is not localized to the measuring junction (a material interface). The emf can arise anywhere along the thermoelements. In transient thermometry, it first occurs adjacent to, and then spreads away from the measuring junction. In the more commonplace *static* thermometry, the site of source emf is usually distant from the measuring junction.

Most specialized thermocouple recorders provide reference temperature compensation. They do not allow convenient user access to the circuit beyond the reference junctions. This restricts the noise control methods available to the user. For externally referenced thermocouples, some effective simple circuit modifications are possible. Figure 13.27 shows the addition of three resistors that, used with monitors of balanced input, improve noise rejection. Isothermal balancing resistor, $R_c = |R_a - R_b|$ is added in series with the thermoelement of lower resistance. Making the resistance of the two circuit branches equal reduces the undesired conversion of noise from common mode to normal mode. The matched pair of shunting resistors, R_s, presents a significant definite minimum source resistance to the amplifier input for stability and symmetrically couples the signals and sheath to the common reference point. The resistors must be large enough, $R_s \geq 1000\ (R_a + R_b)$, to avoid significant shunting resistive attenuation of the thermocouple source Seebeck emf. With great care to maintain these resistors strictly isothermal, the balancing and shunt resistors could be placed just at the input terminals *external* to the instrument-provided reference junction temperature compensation.

Filtering

For low-frequency thermometry, to complement the user-provided passive rejection of EM and ES noise by proper grounding and shielding, internal filtering provided by data acquisition monitor manufacturers can effectively reject high-frequency noise. Filtering before recording is preferred. This control can be performed by conventional filters; but for quasi-static thermocouple measurements, other very effective techniques such as "double-slope integration" are provided in some designs. These techniques effectively average the signal over several measured samples. The noise reduction and stability under real-life conditions is remarkable. Routine resolution and stability to 0.1 μV variation over a period of days is achieved by many modern thermocouple data loggers. However, this noise rejection method restricts the sampling speed so that samples can be limited to intervals no shorter than 10 ms to 20 ms that are unsuited to very fast transient thermometry.

Thermal Coupling

Accurate thermometry obviously requires that the temperature of thermocouple measuring junctions closely agree with the temperatures to be measured. In static measurements at thermal equilibrium, this is often easy to accomplish. In fast transient measurements, it can be very difficult.

Static Thermometry

Thermocouple thermometry of steady-state or slowly varying temperatures requires close thermal coupling of the measuring junctions to the point of measurement. This agreement of temperatures naturally occurs by equilibration if the point of measurement can be chosen to be well within a region that stays at uniform temperature during measurement. Even in such nearly static thermometry, the thermoelements should be as small as practical to reduce any conduction of heat along the thermocouple that could affect the temperature being measured.

In typical situations, however, the vicinity of the measuring point is not isothermal and, even at steady state, static temperature gradients can cause the measuring junction to be at a significantly different temperature from the subject [18]. This source of error is more severe for surface, liquid, and gas measurements than for internal solid measurements, and is aggravated for the measurement of subjects of low thermal mass. Efficient thermal coupling of the thermocouple to the subject, as in Figure 13.27, is essential.

Noise or safety might require electrical isolation of the thermocouple. Commonplace electrical insulating materials are poor thermal conductors. There are now special greases, coatings, and thin sheet materials available designed to provide high electrical isolation, yet good thermal contact [17]. Where needed, they can improve accuracy by efficient thermal coupling.

Transient Thermometry

Monitoring very rapid variations of temperature is most effectively accomplished by thermocouples because of the low sensing mass and small size of measuring junction that they allow. Fast thermal response requires the use of special thermoelement configurations in the vicinity of the measuring junction, special circuitry, special thermocouple monitoring instruments, and characterizing methods [16, 18]. Some suppliers offer very thin (10 μm thick) foil and fine wire thermocouples that are especially well adapted to such measurement [19, 20].

A thermocouple alone has no characteristic response time. The Seebeck effect occurs at electronic speed. The thermocouple source emf always promptly corresponds to the temperatures of all junctions in the circuit. However, the transient temperature response of a thermocouple installation is very much slower, and is primarily governed by the thermal interaction of the thermocouple components adjacent to the junctions and the adjacent mass of the test subject [18, 22, 23]. That slow conductive heat transfer is retarded many-fold by even a thin film of air, vacuum, or an insulating solid layer. In fast transient thermometry, the electrical parameters of the thermocouple and associated circuitry and dynamic response of the recording system can further degrade response time.

Intimate thermal coupling by special materials required for static measurement in a thermal gradient is even more critical to transient than to static thermometry [17]. Even the special coupling materials slow response. They should be as thin as feasible. For very fast thermometry, such coupling might not be tolerable. It may be necessary to place the bare measuring junction in direct electric contact with the subject (Figure 13.27).

Intrinsic Thermocouples.
Where the subject of thermometry is an electric conductor, it is possible to electrically fuse the two thermoelements separately to the test subject. This is an "intrinsic" thermocouple junction arrangement. Fine wires or thin foils spot-welded to the subject allow the fastest possible electrothermal response and reduce the influence of the thermocouple on the temperature being measured.

Observe two special precautions in the use of intrinsic junctions. First, the two thermoelements should be fused to the subject side-by-side, close together (not one over the other.) With intrinsic junctions, the bridging segment of the subject becomes a part of the series thermoelectric circuit. (There are two intrinsic junctions coupled by the intermediate subject material.) This produces the fastest response; but in very fast transients, the temperatures of the two junctions can very briefly be slightly different and thus introduce a momentary error, as the unknown Seebeck coefficient of the bridging subject material is different from that of one or both of the principal thermoelements [21].

Second, for transient thermometry, the fine thermoelements leading to the intrinsic junctions should be as short and fine as feasible to reduce their thermal loading; however, this results in a relatively high electric lead resistance. A strain-relieved transition from the fine filaments to more substantial low-resistance compensating lead wires should be made near the measuring junction to enhance electric response. The electrical effect of capacitive coupling of the circuit on electric transient response must also be minimized.

Numerical Correction.

All feasible physical techniques to achieve fast transient temperature response when applied might not be sufficient for a very fast measurement. An inadequate transient response can be further enhanced by numerical analysis [21]. Two additional steps must be taken.

First, an *authentic* experimental or *correctly* modeled transient response characterization of the overall thermal-electrical thermometry system must be made. The usual simplistic first-order representations of *in situ* thermocouple response might be inadequate [21–23].

Second, this authentic characterization must be applied by mathematical *deconvolution* to better estimate the true temperature history of data indicated by a system that had inadequate physical transient response [16, 19]. Critical and reliable improvements of effective response time by factors of four or more are often possible. If uncorrected, indicated transient peak temperatures can be in error by a much greater percentage than the error that results from static thermocouple calibration error. *No general uncertainty can be assigned to a transient measurement without such response characterization.*

Thermocouple Materials

Thermocouple Standardization

Many materials are in regular use as thermocouples for thermometry. Some pairs are *standardized;* some are not. The distinction is a matter of formal consensus group approval by balanced standards committees of experienced users, producers, and standards laboratory staff [10, 24]. The most extensive application data is available for standardized thermocouples. In the U.S., eight materials presently are letter-designated (Types *B, E, J, K, N, R, S,* and *T*) [9–12, 24]. Some properties of these are summarized in Table 13.12. The standardized $E(T)$ characteristics of these are now *defined exactly* by polynomial functions of high degree rather than by tabulated values [1, 9, 10]. Thermoelements should be used as selectively paired for thermocouples by the producer, as randomly paired thermoelements of the same type from the same or different manufacturers are not assured to conform to the standard pair values or tolerances [1, 9–12].

The properties of a few other popular pairs, such as precious metal and tungsten refractory alloys, have also been committee-proposed, but letter designations and color codes have not been formally assigned [10]. Infrequently, other popular materials are considered for standardization. Limited Seebeck and application information is available for a multitude of nonstandard materials [25, 26].

Low Temperature Thermometry

Most thermometry is at elevated temperatures. Thermocouple measurement of temperatures below the ice point requires special consideration. Cryogenic thermometry has been very loosely defined as measurement below 280 K (7°C) [1]. More restrictively, the defined cryogenic range has been limited to below 90 K (–183°C, the boiling point temperature of liquid oxygen at 1 atm) [1]. The former range broadly overlaps the measure of atmospheric temperatures (down to –50°C) below the ice point that, along with higher ambient temperatures, often are measured by a single thermocouple system. The latter definition favors characterizing as cryogenic only the extremely low-temperature regime over which thermometry involves distinctive severe problems and different techniques.

TABLE 13.12 Characteristics of U.S. Letter-Designated Thermocouples

Type	Common name	Color code	M.P. (°C)	Recommended range, (°C)[d]	emf at 400°C, (mV)	Uncertainty, +/– Special tolerance Normal tolerance	ρ (μΩ-cm)
B	—	Brown[a]	1810	870 to 1700	0.787	0.25%	34.4
BX	—	Gray[a]	—	—	—	0.50%	—
BP	Pt30Rh	Gray	1910	—	—	—	18.6
BN	Pt6Rh	Red	1810	—	—	—	15.8
E	—	Brown[a]	1270	–200 to 870	28.946	1.0°C or 0.40%	127
EX	—	Purple[a]	—	—	—	1.7°C or 0.50%	—
EP	Chromel[b]	Purple	1430	—	—	—	80
EN	Constantan	Red	1270	—	—		46
J	—	Brown[a]	1270	0 to 760	21.848	1.1°C or 0.40%	56
JX	—	White[a]	—	—	—	2.2°C or 0.75%	—
JP	Iron	White	1536	—	—	—	10
JN	Constantan	Red	1270	—	—	—	46
K	—	Brown[a]	1400	–200 to 1260	16.397	1.1°C or 0.40%	112
KX	—	Yellow[a]	—	—	—	2.2°C or 0.75%	—
KP	Chromel	Yellow	1430	—	—	—	80
KN	Alumel[b]	Red	1400	—	—	—	31
N	—	Brown[a]	—	0 to 1260	12.974	1.1°C or 0.40%	
NX	—	Orange[a]	—			2.2°C or 0.75%	—
NP	Nisil	Orange			—	—	—
NN	Nicrosil	Red					
R	—	Brown[a]	1769	0 to 1480	3.408	0.6°C or 0.10%	29
RX	—	Green[a]	—	—		1.5°C or 0.25%	—
RP	Pt13Rh	Green	1840	—	—		19
RN	Pt	Red	1769	—		—	10
S	—	Brown[a]	1769	0 to 1480	3.259	0.6°C or 0.10%	30
SX	—	Green[a]	—	—		1.5°C or 0.25%	—
SP	Pt10Rh	Green	1830	—	—		20
SN	Pt	Red	1769	—		—	10
T	—	Brown[a]	1083	–200 to 370	20.810	0.5°C or 0.40%	48
TX	—	Blue[a]	—	—		1.0°C or 0.75%	—
TP	Copper	Blue	1083	—		—	2
TN	Constantan	Red	1270	—		—	46

From References [1, 4, 9, 10]

[a] Overall jacket color.

[b] Chromel and Alumel are trademarks of Hoskins Mfg. Co.

[c] Initial tolerances are for material as manufactured and used within recommended temperature range, Table 13.15, protected in a benign environment.

[d] Recommended temperature range is a guideline for service in compatible environments and for short durations.

The Seebeck coefficient of all conductors is insignificant at 0 K and common materials progressively decrease in thermoelectric sensitivity below the ice point. A few natural superconductor metal elements experience an abrupt drop in Seebeck coefficient to zero at a characteristic superconducting threshold that is below 10 K for most unalloyed metal superconductors [27]. Special alloys have recently been developed to raise the superconducting threshold to about 120 K — well above the formal cryogenic range. Superconducting transitions complicate thermoelectric thermometry at the lowest cryogenic temperatures [7, 8].

Standard Seebeck characteristics are defined for Types *E, J, K, N,* and *T* down to –270°C. The characteristics for Types *R* and *S* extend only down to –50°C, and Type *B* is not characterized below 0°C. The standard polynomials that define the Seebeck properties of letter-designated thermocouples and the production tolerances for commercial materials are different for temperatures below and above the ice point. Materials manufactured for thermometry at elevated temperatures might conform less well to the standard cryogenic characteristics than alloys of the same type especially furnished for such use [1]. This quality issue should be discussed with the thermocouple supplier. Special alloys are available for cryogenic thermometry.

For the lower cryogenic range, of the letter-designated thermocouple materials, Type *E* is preferred for use down to –233°C (40 K) because of its higher relative Seebeck coefficient [1]. The less-sensitive Type *K* and Type *T* are also used in this range. Below 40 K, special alloy combinations, such as Type *KP* vs. Au/0.07 Fe, are recommended. Special thermoelectric relations apply to these materials in the cryogenic range [1].

As Peltier heating at junctions and Thomson heating along thermoelements are current-dependent thermoelectric effects, neither is a significant problem if thermometry is properly conducted by open-circuit measurement. There is no significant thermocouple self-heating as with resistance thermometers. Some thermoelement alloys experience grain growth and incur serious inhomogeneity under prolonged exposure to deep cryogenic temperatures. More sensitively at cryogenic temperatures than at elevated temperatures, the Seebeck coefficient of most thermocouple alloys is very strongly dependent on magnetic field. Strong magnetic fields are often involved in cryogenic experiments, so thermo-magneto-electric effects become significant in studies of superconductivity [2, 27].

Sensitivity

The need for large thermocouple output has been drastically reduced by the enhanced sensitivity, stability, and noise suppression of modern solid-state digital thermocouple indicators. These instruments can routinely indicate temperature to 0.1°C resolution and stability for all letter-designated types. The standard pairs differ significantly in their sensitivity (Table 13.12). The sensitivity of the Type *E* thermocouple is 10 times that of the Type *B* thermocouple at 1000°C. Because the Type *B* thermocouple has extremely small sensitivity around room temperature, it is intended for use only with the measuring junction above 870°C. The most sensitive standardized thermocouple, Type *E*, has a maximum Seebeck coefficient of 81 $\mu V\,°C^{-1}$ and, referenced to 0°C, has a maximum Seebeck emf of 76 mV. While initial tolerances for both normal- and special-grade material have been standardized, the difference between *commercial* and *premium* grades is small and, in use, special-grade materials can degrade to exceed the initial tolerances of the normal-grade material.

Letter Designations

The U.S. ANSI standard letter designations and color codes for eight particular thermocouple types (*B, E, J, K, N, R, S,* and *T*) were first established by the ISA in Standard MC-96.1 [28]. The same conventions are followed by Standards of the ASTM and ANSI [10, 24]. A first suffix to the type letter designator, *P* or *N*, as in types *KP* or *KN*, designates the positive or negative thermoelement of a thermocouple pair. A final "*X*" suffix designates an extension wire material, as in *KPX* for a positive type *K* extension thermoelement. For non-standardized material pairs, producers and vendors often apply their own unofficial letter designations, color codes, and trade names. These *commercial* identifiers of individual manufacturers have no universal meaning.

Color Codes

Intended to ease identification, the standards of many nations have assigned color codes to the different letter-designated thermocouples and to thermoelements and extensions. As the individual thermoelements determine both polarity and sensitivity, it is very important to properly identify each leg. Color codes now used in the U.S. are shown in Table 13.12.

TABLE 13.13 International Thermocouple Color Codes

Type	U.S.	IEC	England	China	France	Germany	Japan	Russia
B	Brown	—	—	—	—	—	—	— B
BX	Gray	—	—	—	—	Gray	Gray	— BX
BP	Gray	—	—	—	—	Red	Red	— BP
BN	Red	—	—	—	—	Gray	Gray	— BN
E	Brown	—	—	—	—	—	—	— E
EX	Purple	Purple	Brown	—	Purple	Black	Purple	— EX
EP	Purple	Purple	Brown	Red	Yellow	Red	Red	Purple or black EP
EN	Red	White	Blue	Brown	Purple	Black	White	Yellow or orange EN
J	Brown	—	Red	—	—	—	—	— J
JX	Black	Black	Black	—	Black	Blue	Yellow	— JX
JP	White	Black	Yellow	Red	Yellow	Red	Red	White JP
JN	Red	White	Blue	Purple	Black	Blue	White	Yellow or orange JN
K	Brown	—	—	—	—	—	—	— K
KX	Yellow	Green	Red	—	Yellow	Green	Blue	— KX
KP	Yellow	Green	Brown	Red	Yellow	Red	Red	Red KP
KN	Red	White	Blue	Blue	Purple	Green	White	Brown KN
N	Brown	—	—	—	—	—	—	— N
NX	Orange	—	—	—	—	—	—	— NX
NP	Orange	—	—	—	—	—	—	— NP
NN	Red	—	—	—	—	—	—	— NN
R	Brown	—	—	—	—	—	—	— R
RX	Black	Orange	Green	—	Green	White	Black	— RX
RP	Black	Orange	White	Red	Yellow	Red	Red	— RP
RN	Red	White	Blue	Green	Green	White	White	— RN
S	Brown	—	—	—	—	—	—	— S
SX	Black	Orange	Green	—	Green	White	Black	— SX
SP	Black	Orange	White	Red	Yellow	Red	Red	Red or pink SP
SN	Red	White	Blue	Green	Green	White	White	Green SN
T	Brown	—	—	—	—	—	—	— T
TX	Blue	Brown	Blue	—	Blue	Brown	Brown	— TX
TP	Blue	Brown	White	Red	Yellow	Red	Red	Red or pink TP
TN	Red	White	Blue	White	Blue	Brown	White	Brown TN
Std.	ANSI	IEC	BS	NMI	NFC4 2	DIN	JIS	
No.	MC96.1	584-3	1843		42-323	43714	1620	

From References [1, 4, 9, 10].

U.S.-standardized color codes have remained uniform for several decades; thus color code confusion of material types in the U.S. is mostly due to user carelessness. The present globally discordant color codes can cause costly misinterpretation in multinational use, particularly outside the U.S. where neighboring countries and trading partners have unlike or multiple color codes. Clearly, a single universal international color code accepted by all nations would be beneficial. Such an international color code is embodied in standard *IEC 584* that is being considered by several nations [28].

Unlike the standardized Seebeck characteristics that are fairly uniform worldwide, the uncoordinated color codes of different national standards are very inconsistent. The unfortunate Babel of national color codes that existed in 1998 is displayed in Table 13.13 [1, 4].

Unfortunately, and uniquely, in present U.S. thermocouple standards, the *negative* thermoelement is always *red,* contrary to customary U.S. electrical and instrument practices. This is also contrary to the historic national thermocouple color codes of China, Germany, and Japan, in which *red* designates the *positive* thermoelement. In English standards, the *negative leg* of all types is *blue.*

In France, the *positive* thermoelement is always coded *yellow*. However, a *yellow positive* leg in the U.S. standard designates Type *KP* material. The wire leads of some U.S. electric blasting caps use yellow insulated wire with a parallel red tracer that has been confused with Type *KX* thermocouple extension wire. In England, yellow denotes Type *JP* material. The *white positive* and *black negative* of the present U.S. ANSI color code for Type *J* are *transposed* in international standard *IEC* 584.

Despite the clear desirability for a uniform color code, there remains a huge quantity of material of different color codes in stock and in use worldwide. For any nation that switches to any new color standard, there would be, immediately and over an unavoidably lengthy transition period following acceptance, the new color-coded thermocouple material intermixed with the multitude of inconsistent legacy color codes. The immediate possibility for confusion of material type would greatly increase rather than decrease. Also, even the present color codes can become indistinct on long-installed material, colors can fade, and/or colors may have been incorrectly applied in manufacture.

Identifying Thermoelements

The color codes apply directly to extension lead wires and effectively to the principal thermocouple wire. Many principal thermoelements and thermocouples are not color coded. A user might not correctly recall the color code. The prudent user will, before use, *confirm* material type identification independent of the color code. *Definite* type identification must be by a combination of methods. Any single identification method can be indefinite, and no method is adaptable to all materials or circumstances.

Visual Identification.

TP thermoelements are of copper and thus are distinguished by their distinctive reddish color. *JP* thermoelements are iron and have a distinctive matte gray cast. Other base metal alloys and platinum and its alloys, if bare, all have a very similar bright silvery appearance unless, if bared from compressed mineral insulation, fabrication has effected a roughened gray matte surface appearance.

Magnetic Identification.

Type *JP* (iron) is strongly magnetic. Type *KN* (Alumel) is slightly magnetic. All other standard thermoelement materials are nonmagnetic. *JP* and *KN* thermoelements can be distinguished from the others by testing the attraction to them of a small magnet.

Resistive Identification.

Resistivities of thermoelements are given in Table 13.12. Although assembled thermocouples have a measuring junction that cannot generally be removed for testing, a bare junction can be directly accessible or it can be electrically accessible at a probe tip for resistance measurement of each leg if the junction is of the type made common at the measuring junction to a conductive sheath. Thermocouple assemblies or cable usually have paired thermoelements of the same length and cross-section. The resistance of each leg distinguishes the material if the wire size and length are known. In these instances, *with both ends of the cable at the same temperature*, the resistance of each thermoelement can be directly measured. The ratio between positive and negative leg resistances can aid type confirmation. For assembled thermocouples with inaccessible measuring junctions, only the loop resistance can be measured and compared with calculated loop resistances.

Thermoelectric Identification.

Less conveniently, a pair can be identified by the output for a known temperature of measuring junction and reference junction. *Complementary extension cables* can be thermally identified by temporarily forming a junction between a pair at one end. Identification can be definite using a less precise procedure than necessary for formal calibration. For identification, both reference and measuring junction temperatures must be independently known or measured. The approximate temperature of the reference junction can be determined by momentarily shorting the input directly at the indicator input terminals of an instrument that provides reference compensation. Because the thermocouple material might not correspond to the compensation applied, the temperature must be determined separately.

The temperature applied to the measuring junction for identification must be at least 200°C because uncertainty of the thermocouple calibration and of the imposed temperature makes unreliable the emf distinction of thermocouple pairs, such as *E* from *J*, and *K* from *N* or *T* using either ice or boiling water baths. Very similar Types *R* and *S* can be reliably distinguished only at much higher temperatures or by formal calibration.

The Functional Model of Thermoelectric Circuits

For simplicity, the relation between junction temperatures required for accurate measurement was merely asserted in the section on "Practical Thermocouple Circuits," without any physical explanation. Some subtle problems of realistic thermoelectric circuits are difficult to visualize without an explicit circuit model. A simple, practical, and general-purpose model of thermoelectric circuits now explains why those temperature structures are appropriate. More significantly, it makes clear the consequences of deviation from these temperatures. It illuminates the common problems of calibration and inhomogeneity. It explains why the commonplace misconception that the Seebeck emf is localized to junctions can lead to serious error in general thermocouple circuits.

Real thermocouple circuits involve several materials and incidental real junctions, often many more than in Figures 13.23 or 13.24. The incidental uncontrolled materials of feedthrus, terminals, splices, etc. might not be recognized as source elements, yet they can, unnoticed, contribute significant unwanted Seebeck emf to the measurement. Therefore, for practical thermocouple thermometry, it is essential to understand and use a descriptive circuit model that is simple to apply and that forces the attention to locations of potential error so that problems can be avoided.

One such authentic model, the *Functional Model of Thermoelectric Circuits* [1, 2, 5, 7, 29–31], combines (1) a basic thermoelectric circuit element, (2) a single fundamental law that describes the sensitivity of that element, (3) a set of practical corollaries from that law that illuminate its practical implications, and (4) a graphic tool for circuit visualization to simplify analysis. This very simple but nontraditional model is crafted for practical thermometry and is worth studying.

The Basic Thermoelectric Circuit Element

Any thermoelectrically homogeneous nonisothermal segment of arbitrary length of material M within any thermoelement is a *Seebeck cell.* Each such segment across which a net temperature difference exists (Figure 13.28) is a *non-ideal voltage source* with internal resistance $R(T)$. The Seebeck *source emf* must be observed in an "open-circuit" (null current) mode for the most accurate thermometry. Any iR voltage

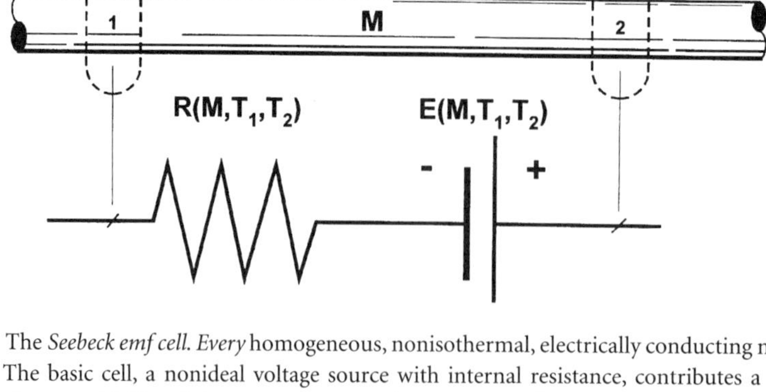

FIGURE 13.28 The *Seebeck emf cell. Every* homogeneous, nonisothermal, electrically conducting material is a source of Seebeck emf. The basic cell, a nonideal voltage source with internal resistance, contributes a Seebeck emf that depends only on the material M and the temperatures at the segment end-points 1 and 2.

drop due to current allowed by a low input resistance of the voltage monitor reduces the Seebeck source emf to a lower terminal Seebeck voltage. For thermometry, that voltage difference causes a temperature error unless corrected.

In *static* thermometry, as in calibration and in typical process measurement, the measuring junction and a substantial length of the adjacent thermoelements are immersed in a stationary and somewhat isothermal zone so that most of the Seebeck emf occurs well apart from the measuring junction in thermoelements where they cross remote temperature transition regions. In *transient* thermometry, the zone of principal temperature difference initially is adjacent to the measuring junction so that the emf arises across a region of spreading extent adjacent to, but not in, the measuring junction (a material interface).

The Law of Seebeck emf

Equation 13.74, $dE = \sigma(T)dT$, is the functional law that governs the emf of the Seebeck cell and, thus, the net voltage of the most complex thermoelectric circuits. It is the *Law of Seebeck emf.* Every thermoelectric aspect of circuit behavior follows from only this simple relation. Physical details of the process that leads to the Seebeck effect are very complex [5, 6]. Nevertheless, this one simple law is entirely consistent with all physical theory and is experimentally confirmed. If this simple relationship *does* apply to the basic Seebeck cell then accurate thermometry is possible. If it does not, then accurate and reliable thermoelectric thermometry is *not* possible.

The source emf of an individual cell of material M, from Equation 13.75, is:

$$\Delta E_{M}(T) = E_{M}(T_2) - E_{M}(T_1) \tag{13.79}$$

and, for *thermally paired* segments of materials A and B that happen *at any instant* to share a pair of end-point temperatures, *regardless of their residence or proximity in a circuit,*

$$\Delta E_{AB}(T) = E_{AB}(T_2) - E_{AB}(T_1) \tag{13.80}$$

They need not be directly joined at a junction (nor, indeed, be electrically joined). The values of $E_{AB}(T)$ are obtained directly from the standard thermocouple polynomial defining equations, simpler representations of those equations, tables, or graphs [1, 9–12]. Absolute Seebeck properties for many materials are also available, but are less commonly reported [1, 5, 7, 8]. As is evident from this model, absolute properties could always be used in thermoelectric analysis. They *must* be used in some realistic circumstances where the conventional relative properties are meaningless or where the relative properties are not known.

Either the absolute Seebeck coefficient, the temperature increment across the segment, or both of them, can be either positive or negative. Therefore, the polarity of a cell within a circuit depends both on the material and, unlike the electrochemical emf cell, on the *momentary* sense of temperature difference across the segment. *Polarity, and even function, changes with temperature distribution.*

Corollaries from the Seebeck Law

From the single Law of Seebeck emf (Equation 13.74), five particularly instructive practical corollaries that aid thermoelectric circuit analysis have been recognized. These are the corollaries of: (1) functional roles, (2) functional determinacy, (3) temperature determinacy, (4) emf determinacy, and (5) Seebeck emf [1, 2, 7, 29–31]. These are stated in Table 13.14, abbreviated from [29]. These revealing corollaries relate more directly to practical thermometry than do the three familiar thermocouple "laws" [32] that actually are only oblique alternative corollaries to the sole physical law (Equation 13.74).

The T/X Visualization Sketch

The practical significance of the fundamental law and its corollaries for realistic thermoelectric circuits is revealed by a simple graphic sketch (Figure 13.29). The *T/X sketch* is used for visualization and numerical analysis only. It is *not* drawn to scale. It is *not* used for graphic solution. It illuminates essential

TABLE 13.14 Corollaries from the Law of Seebeck emf

In any circuit of electrically conducting materials that have an absolute Seebeck coefficient $\sigma(T)$, that are each *thermoelectrically homogeneous*, and which follow the Seebeck Law, $dE(T) = \sigma(T)dT$:

1. **The *Corollary of Functional Roles***

   There are three thermoelectric functional roles: *junctions*, "*conductors*," and *Seebeck emf sources*:
   - *Real thermoelectric junctions* are interfaces that ohmically couple dissimilar materials,
   - "*Conductors*" are segments that, *in effect*, individually or in combination, contribute no *net* Seebeck emf, and
   - *All* other segments are sources of Seebeck emf.

2. **The *Corollary of Functional Determinacy***

   Instantaneous thermoelectric roles around a circuit *cannot* be predetermined by physical construction, material choice, or circuit arrangement alone; they are governed by temperature distribution.

3. **The *Corollary of Temperature Determinacy***

   In a circuit with multiple junctions, the temperature of a single junction can be determined from the net Seebeck emf only if the temperatures of all other real junctions are defined.

4. **The *Corollary of emf Determinacy***

   Seebeck emf is produced *only* by thermoelements, but the net Seebeck emf is governed by the temperatures only of *real junctions*.

5. **The *Corollary of Seebeck emf***

   The Seebeck emf of any segment of material M with end-point temperatures T_1 at segment endpoint X_1 and T_2 at segment endpoint X_2 is independent of *temperature distribution*, *temperature gradient*, or *cross-section* as it is determined by:

$$E_M(T) = \int_{T_1(X_1)}^{T_2(X_2)} \sigma_M(T)dT = E_M(T_2(X_2)) - E_M(T_1(X_1))$$

From References [1, 7, 29, 30].

facts that are not obvious from a conventional electrical schematic or $E(T)$ plot. The T/X sketch shows the temperatures of *all* real junctions in the sequence in which they occur around the circuit. Real junctions are indicated by closed circles. Junctions are joined by thermoelements. The sketch is not drawn to scale, so slopes do not represent temperature gradients.

The T/X sketch reveals that segments that span a temperature interval always occur in pairs — but significantly only in *series* circuits. It is this fact that allows the convenient use of relative Seebeck properties. This conventional simplification does *not* apply to circuits with paralleled branches [2]. The T/X sketch, applied to the thermocouple circuit of Figure 13.25, depicts the significant circuit elements (the junctions and thermoelements) in a way that focuses on their unavoidable thermoelectric *functions* (*Corollary 1*). Figure 13.29 shows a circuit temperature distribution with junction temperatures $T_b \neq T_e$, that were shown (intentionally improperly different) to illustrate a principle and the benefit of the sketch.

Temperatures of Incidental Junctions.
The *reference junction* temperatures must independently be accurately known for measurement. The unknown temperature of the *measuring junction* is to be deduced from the Seebeck voltage. The specific temperatures of all incidental real junctions of a circuit are rarely known accurately in practice. For use in the T/X sketch, *the actual values need not be accurately known.* Nevertheless, if the *relative* values of all incidental junctions are not properly controlled, as described in the section "Practical Thermocouple Circuits," and if some essential junction temperatures are not known well enough to draw the sketch, then *accurate* thermometry cannot be ensured! *The revealing T/X sketch requires no more information than is essential for the physical measurement.*

In the estimate of measurement consequences and to visualize how junction temperatures must be controlled in circuits of many materials, it is sufficient to assume, for qualitative analysis, plausible *relative*

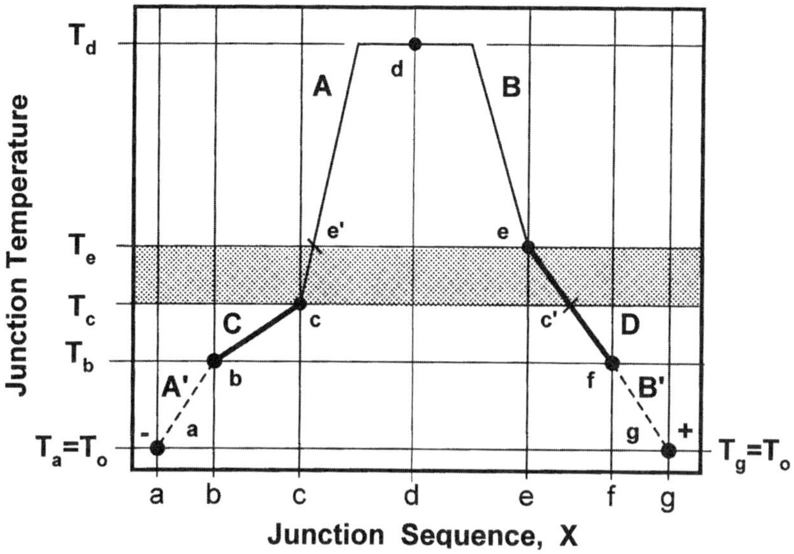

FIGURE 13.29 The *T*/*X* sketch for thermoelectric circuit visualization. The temperature of the reference junction(s) and the relation between (not necessarily the specific values of) the temperatures of *all* incidental real junctions in a circuit must be known for accurate thermometry, just as for the use of the *T*/*X* sketch. The simple sketch is an aid in recognizing temperatures of incidental junctions that must be controlled to the same temperature. It also makes clear which thermoelements are *thermally paired*, even if not directly joined in the series circuit.

temperature levels and their consequences. The sketch is most often used for visualizing consequences by inspection [1, 2, 29–31]. Nevertheless, it also can aid in quantitative analysis of error for *plausible* temperature distributions.

Virtual Junctions and Thermoelements.
On the *T*/*X* sketch, isotherms through *real* junctions intersect some thermoelements. It is useful for analysis to view these intercepts (e.g., **c′** and **e′**), each marked by a *tic*, as *virtual junctions.* For inspection and analysis, they conveniently delineate the arbitrary temperature end-points of segments. Also, on the diagram, virtual thermoelements, **b-c** and **f-g** are indicated by dashed lines. These segments represent the imaginary thermoelectric source of complementary emf that must be supplied to extend the physical reference temperature, T_r, to the designated reference temperature value, T_0. The virtual thermoelements represent the reference junction compensation.

The *T*/*X* sketch aids in assigning segment bounding temperatures and thus a polarity and an emf to each thermoelement in the circuit. The circuit is traversed on the sketch from one instrument terminal, conveniently the negative, to the other terminal. Then, if the absolute Seebeck coefficient of a thermoelement is positive, the emf contribution of the segment adds emf if the temperature increases in proceeding across the segment from the negative toward the positive instrument terminal, etc. For inspection and analysis, it is efficient to consider segments as thermally paired over a temperature zone bounded by isotherms.

Examples.
On the *T*/*X* sketch, consider the absolute Seebeck emfs supplied by the four thermoelements of Figure 13.25. As connected in this series circuit, proceeding from negative to positive terminals, the net Seebeck emf of the physical circuit between terminals *b* and *f* is:

$$E_{net} = E_C + E_A + E_B + E_D \tag{13.81}$$

where

$$\begin{aligned}
E_{\mathrm{C}} &= E_{\mathrm{C}}\Big|_{T_{\mathrm{b}}}^{T_{\mathrm{c}}} \\
E_{\mathrm{A}} &= E_{\mathrm{A}}\Big|_{T_{\mathrm{c}}}^{T_{\mathrm{d}}} \\
E_{\mathrm{B}} &= E_{\mathrm{B}}\Big|_{T_{\mathrm{d}}}^{T_{\mathrm{e}}} \\
E_{\mathrm{D}} &= E_{\mathrm{D}}\Big|_{T_{\mathrm{e}}}^{T_{\mathrm{f}}}
\end{aligned} \tag{13.82}$$

Recall that these individual Seebeck emfs *do* physically exist in the thermoelements whether or not they are connected by junctions as a circuit. It is the temperature dependence of both the magnitude and the momentary polarity of emf that distinguishes thermoelectric from ordinary electric circuit analysis (Corollary 5).

Note that emf from thermoelement *A* can be rewritten as:

$$E_{\mathrm{A}}\Big|_{T_{\mathrm{c}}}^{T_{\mathrm{d}}} = E_{\mathrm{A}}\Big|_{T_{\mathrm{c}}}^{T_{\mathrm{e}'}} + E_{\mathrm{A}}\Big|_{T_{\mathrm{e}'}}^{T_{\mathrm{d}}} \tag{13.83}$$

This arbitrarily breaks thermoelement *A* into two segments joined at *virtual* junction **e′**. Thermoelement *D* can be segmented as well. Now, across the shaded temperature zone between T_c and T_e, there are improperly thermally paired segments **c-e′** and **c′-e.** The net emf from these thermally paired segments is:

$$E_{\mathrm{net}}\Big|_{T_{\mathrm{c}}}^{T_{\mathrm{e}}} = E_{\mathrm{A}}\Big|_{T_{\mathrm{c}}}^{T_{\mathrm{e}'}} - E_{\mathrm{D}}\Big|_{T_{\mathrm{c}}}^{T_{\mathrm{e}}} \tag{13.84}$$

From Equation 13.75, recognize that this is the *relative* Seebeck emf for the *unintended A-D* pair over that arbitrary temperature zone even though they are not directly joined in the circuit. That improper pairing of segments clearly is avoided only if $T_c = T_e$, whatever the temperature. In the instance that Figure 13.25 represents a principal thermocouple *A-B* with matching extension leads *C-D*, materials *A* and *C* are alike and *B* and *D* are alike. If the legs of extension leads *C-D* each exactly match the corresponding legs of thermocouple *AB*, then

$$E_{\mathrm{CD}}\Big|_{T_{\mathrm{c}}}^{T_{\mathrm{e}}} = E_{\mathrm{AB}}\Big|_{T_{\mathrm{c}}}^{T_{\mathrm{e}}} \tag{13.85}$$

and (*for matching extension leads only*) the accidental pairing is benign.

Otherwise, with *compensating extension leads* ($\sigma_A = \sigma_C$ and $\sigma_B = \sigma_D$), error occurs even if σ_{CD} closely matches σ_{AB} as a pair but not individually. In the instance of *neutral extension leads* where *C* and *D* are alike, the net emf from the pair is *null* over the zone from T_b to T_c. In this instance, the physical reference junction is recognized as necessarily $T_r = T_c = T_e$, rather than $T_r = T_b = T_f$.

If T_c and T_e were interchanged, the error would be of different magnitude, not merely of opposite sign, simply because of the temperature distribution (Corollary 2). The unknown temperature of only one junction can be determined; the others, including incidental junctions, must be defined by value or, indirectly, as being isothermal (Corollary 4).

Note that such relative contributions (the null net contribution from two opposed like segments or the inappropriate thermal mispairing between *B* and *D*) are immediately evident simply *by inspection* of the informal *T*/*X* sketch without tedious algebra. These critical facts of thermoelectric circuits are not evident from the usual electrical schematic or from *E*(*T*) plots.

Most real circuits include several incidental uncalibrated materials such as connectors, terminals, splices, feedthrus, short pigtails, or extension leads that each have their own (usually indefinite) Seebeck properties. In some circuits, some thermoelements might accidentally be paralleled. It is important to include them in the sketch to recognize the unnoticed potential error that could be contributed by such circuit elements if they are not held isothermal. Also, multiple extension circuit elements might be improperly connected with crossed polarity. The specific voltage or temperature consequence of these situations is easily calculated for any plausible temperature distribution. The *T*/*X* plot was designed specifically to aid in circuit visualization to avoid these very common practical problems and to easily assess their possible impact.

Inhomogeneity

The Nature of Inhomogeneity

A slender thermoelement is inhomogeneous if σ_M (T) varies along its length. The environment during application can introduce irregular inhomogeneity in one or both thermoelements of a pair. The effect is as if one or more additional dissimilar materials had been added to the circuit.

The Significance of Inhomogeneity

Adequate thermoelectric homogeneity is the most critical assumption of thermocouple application. It usually is presumed; rarely is it confirmed. Inhomogeneity is a real but phantom problem. Inhomogeneity often remains undetected even while producing substantial error. It rarely is discovered by even the most careful conventional calibration [1, 7, 11, 29–31, 33, 34]. In physical thermometry, thermocouple *drift* is *invariably* a symptom of progressing *inhomogeneity.* Such change is progressive, often insidious, and usually is misinterpreted [34].

Rather than envisioned correctly as localized degradation of Seebeck coefficient, *drift* is often viewed improperly as a uniform "black-box" decalibration of the thermocouple rather than progressing inhomogeneity. For this reason, it is a far more commonplace problem than recognized by most experienced users. It can occur in use, in fabrication, or in calibration because of mechanical strain, thermal phase change, surface contamination, chemical interaction between materials, evaporation or migration of alloy constituents, transmutation under radiation, and a variety of other realistic causes.

Inhomogeneity error in thermocouples, as manufactured, is intended to be covered by standard tolerances [1, 9, 10, 28]. Troublesome inhomogeneity is most common in used and abused thermocouples, but can sometimes occur in new thermocouples that have been individually calibrated to high temperature. It can occur within and through apparently impervious protective metal sheaths or thermal wells and between bead insulators.

The example, Figure 13.29, illustrated the effect of improperly controlled junction temperatures that resulted in the subtle introduction of relative Seebeck emf from an unintended thermal pairing of segments of homogeneous thermoelements. The analogous *inhomogeneity* problem, best visualized with the *T*/*X* sketch, arises when a portion of one or both thermoelements locally changes in Seebeck coefficient over some nonisothermal span of the thermocouple. In effect, this introduces *phantom* dissimilar segments of indefinite graduated property and unrecognized location [35]. This most often occurs over a lengthy region near the measuring junction where the thermoelements are exposed to damaging environments in an oven or process.

The *maximum possible* error of inhomogeneity is determined by the location and magnitude of greatest deviation from normal of the Seebeck coefficient. The *actual* error depends on the momentary distribution of temperature during use. In thermometry and in calibration, the *likely* error introduced by inhomogeneity is moderated. While present only within an isothermal region, inhomogeneities introduce *no* error. Under unfavorable temperature distributions, inhomogeneity error can be extreme and results in peculiar puzzling responses [34]. Changes of relative Seebeck coefficient by more than 60% over 25-cm spans have been observed in individually calibrated, certified, premium-grade fine wire Type R MIMS thermocouples entirely enclosed within intact platinum sheaths and exposed to temperature within the tabulated temperature range [18].

Testing for Inhomogeneity

Many authentic sensitive and accurate tests for thermoelectric inhomogeneity have been developed, and their practical need has been demonstrated; however, they are rarely used [33]. All true inhomogeneity tests require moving an abrupt *step* of temperature progressively along thermoelements.

The commonplace application of a *very narrow symmetric* temperature pattern is the antithesis of an inhomogeneity test. It is *not* a test for inhomogeneity. Its popular use has misled many to discount inhomogeneity as a real and significant problem in accurate thermoelectric thermometry.

Valid tests range from simple methods of low resolution to advanced methods that can accurately resolve inhomogeneity with spatial resolution of a few millimeters [21, 33]. Regrettably, as inhomogeneity errors usually are not recognized, inhomogeneity is not authentically tested in commercial practice nor by calibration laboratories. Nevertheless, invalid though certified NIST-traceable certification to great *precision* is possible on a thermocouple that can be accurately measured to be *severely* inhomogeneous and of indefinite uncertainty [29–31, 33, 34].

Calibration

The measurement uncertainty of most kinds of sensors can be reduced by *individual* calibration. Initial calibration, periodic recalibration of unused thermocouples, and even of degraded *individual* thermocouples, although commonplace, is often less beneficial (or even harmful) and more costly. Such ill-advised thermocouple calibration has been mandated by some "quality assurance" programs.

Surveys reveal a trend for customers to demand progressively higher accuracy of thermometry [36]. In some applications, a 1°F (0.56°C) error now is deemed too large, and 0.1°F tolerance may be specified independent of temperature level. (Compare with tolerances in Table 13.12.) There are a few industrial applications that truly require such accuracy. More often, the specification merely presumes that such accuracy in thermocouple thermometry is routinely attainable merely by calibration. Calibration is an opiate of quality assurance. The illusion of accuracy provided by NIST-traceable certified calibration and purported *in situ* calibration is counterproductive if it is not authentic.

Consideration of the details of thermocouple calibration suggests why authentic accuracy at the 0.1°C level is unlikely in industrial thermoelectric thermometry. The approach to achieving authentic calibration of thermocouple system accuracy (*ignoring* thermal coupling and transient response errors) is illustrated in Figure 13.30. Figure 13.30(a) represents the thermocouple system to be calibrated. Calibration for thermoelectric thermometry must distinguish three system components: (1) the thermocouple circuit, (2) the reference junction temperature compensator, and (3) the monitoring instrument. Each separately affects temperature uncertainty.

The Principal Thermocouple

Calibration of principal thermocouples is performed by immersing the vicinity of the measuring junction in the isothermal region of a bath or oven [37]. Several compact dry-well calibrators available are convenient for calibration at the job site [38]. Fixed-point cells, liquid baths, and fluidized solid beds can be more accurate and are widely used in the calibration laboratory [39]. Sufficient depth of immersion into an isothermal calibration zone, usually at least 10 to 20 times the probe diameter, ensures that conduction of heat along the thermocouple from the environment does not affect the junction temperature.

The appropriate concern for the effect of longitudinal heat conduction on measuring junction temperature has led to the misperception that it is the *junction* that is being calibrated [30]. Clearly, thermocouple calibration is *not* of the measuring junction; it *is* of unidentified segments of thermoelements, remote from the measuring junction, only where they enter the isothermal calibration zone through a temperature transition.

Segments of service-induced inhomogeneity that seriously degrade measurement accuracy are often placed, during unsuccessful attempted recalibration, within the isothermal region where they contribute no Seebeck emf, so inhomogeneity is not discovered. The act of calibration at temperatures above 400°C can actually induce inhomogeneity and degrade accuracy [9, 10]. Not even costly "NIST-traceable"

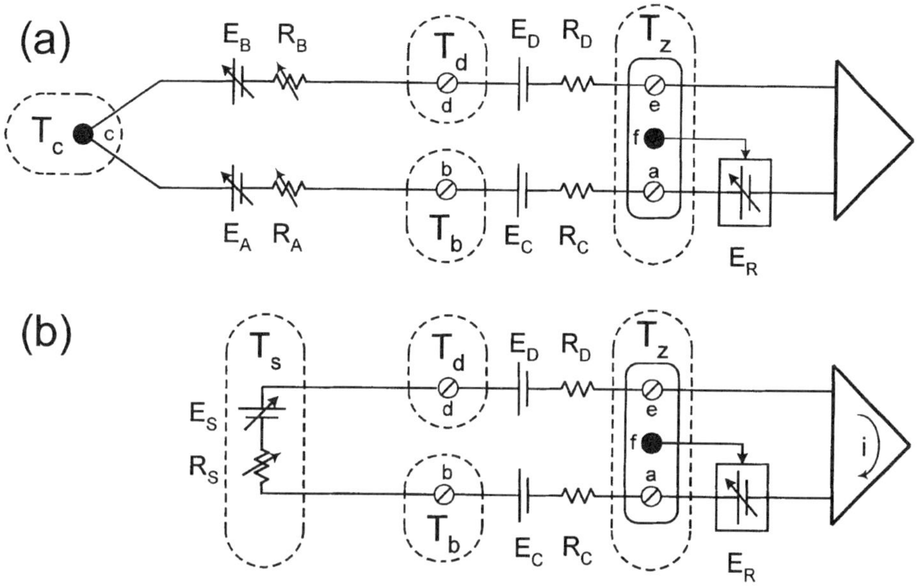

FIGURE 13.30 Contributions of Seebeck and compensating emf in system calibration: (a) the circuit to be calibrated or simulated, (b) a general calibration simulation of a system that includes internal reference junction temperature compensation.

calibration and certification by competent calibration laboratories, without specific assurance of homogeneity, are a guarantee of accuracy at *any* level.

Commercial Tolerances. Standardized tolerances for new thermocouple materials are established by the formal consensus judgment of many experienced producers, users, and calibration standards laboratories staff [10, 24, 28]. The tolerances include not only deviation from the overall Seebeck properties, but also cover inevitable uncertainty from inhomogeneity and irregular deviations from the smoothed standardized characteristic over small spans of temperature. They apply to material as delivered and not exposed to excessive temperature. The conservative tolerances for standard letter-designated thermocouple types as delivered are usually reliable until the thermocouples are exposed too long to excessive temperatures or adverse environments. Recalibration of used thermocouples without separate assurance of homogeneity can be misleading [31].

To address calibration problems presented by undiscovered inhomogeneity, two practices are common. For inexpensive base metal thermocouples, it is presumed that unused thermocouples are homogeneous and uniform within standardized tolerances. The typical Seebeck property of a production batch is characterized by the manufacturer or user by calibrating one or more expended surrogate samples. These are discarded after calibration. Their first-cycle calibration is taken as representative of other thermocouples of the same batch. Used principal thermocouples that have experienced drift are discarded.

A commonplace second approach is recalibration. This is usually in response to drift observed in service. Drift signals progressing inhomogeneity. Apart from an authentic inhomogeneity prescreening, recalibration is not recommended. Even widely promoted *in situ* recalibration of degraded thermocouples is ineffective if temperature distribution will vary in use [31].

Platinum-based thermocouple wire can be annealed full length by electrically heating in air [9–12]. Such annealed thermoelements can be presumed to be free of reversible strain-induced inhomogeneity and, thus, accurate recalibration might be justified. However, annealing cannot reverse decalibration from the migration, absorption, or evaporation of alloy constituents or other chemical contamination; therefore, recalibration, even of used precious metal thermocouples, should be performed only where the likelihood of homogeneity is factually based. Unlike base metal thermoelements, precious metal thermoelement materials often can be reconditioned. Also, precious metals have a significant material salvage value if recycled.

Extension Leads

Compensating extension leads, although Seebeck sources, usually are not calibrated. It is commonly assumed that they will be exposed to only a small fraction of the temperature span that is being measured and, if so, will contribute insignificantly to error. Extension insulations are rated for continuous use to maximum temperatures between 105°C and 540°C [43]. Seebeck characteristics might not be well approximated over this range [1, 10]. The possible contribution from extensions can be very large. The plausible error is easily estimated using the model in "The Functional Model of Thermoelectric Circuits." The more usual consequential errors due to extension leads result from failure to correctly control the temperatures of incidental splice junctions, as described in "Practical Thermocouple Circuits."

Reference Temperature Compensation

Modern thermocouple indicators usually provide internal reference junction temperature compensation. A few make compensation a selectable alternative. The accuracy of reference junction compensation, Figure 13.30(a), depends on the accuracy of the sensor(s), **f**, used to monitor reference junction temperature of isothermal terminations, but also on the conformity of the scaling algorithm or analog circuitry to the characteristic of the individual thermocouple. The zone sensor determines the compensating voltage, E_R. Some monitors have replaceable isothermal terminal blocks that include the zone temperature sensor. Most monitor designs unfortunately do not allow separate thermal calibration of the reference temperature sensor. Simply shorting the indicator input terminals, **b-d**, should produce a temperature indication near, and usually slightly above, ambient temperature. Specifications claim reference uncertainty on the order of 0.1°C to 0.5°C and error contributions usually *are* small but *infrequently* they have been *very* large and insidiously progressive up to *many* times the standard thermocouple tolerances [35].

Monitoring Instruments

The accuracy of conversion of the Seebeck voltage at terminals **b-d** to deduce the temperature T_c requires calibration of the monitoring instrument. Both the accuracy of voltage measurement and of linearization are sources of uncertainty. The scaling accuracy is based on an approximation of the defining $E(T)$, and *not* on the individual thermocouple characteristic.

Instrument designs are varied. The general principles of *authentic* calibration required to approach 0.1°C uncertainty for a thermocouple indicator that provides internal reference temperature compensation are illustrated with the circuit of Figure 13.30(b). A thermocouple simulator/calibrator supplies a well-known voltage E_S that corresponds, for the standard thermocouple type, to a desired temperature calibration point. The simulator must remain in thermal equilibrium during calibration, with irrelevant Seebeck voltage properly nulled .

If the input resistance of the indicator is very *large*, resistance matching of the simulator to the thermocouple is not required. If the indicator has a *low* input resistance or it internally produces current **i** when its terminals b–d are shorted, then a temperature error proportional to the indicator current and the loop resistance will be experienced. For accuracy with such indicators, the simulator R_S must also mimic the source resistance of the individual thermocouple ($R_A + R_B$). However, that thermocouple loop resistance may vary considerably in application as it depends on the temperature distribution around the thermocouple over a broad range of temperatures.

The input terminals of the indicator are **b** and **d**. For accurate calibration, it is essential that $T_b = T_d = T_z$. Some indicators and reference temperature compensators connect the isothermal zone block to the terminals with compensating thermocouple leads. For these, if $T_b \neq T_d \neq T_z$, an indefinite error results that cannot be overcome by calibration.

The simulator voltage must imitate the behavior of only the thermocouple, **b-c-d**. The voltage it must provide depends on whether or not the indicator supplies internal reference temperature compensation. If $T_b = T_d = T_z$, the proper total Seebeck voltage is $E_{AB}\Big|_{T_0=0°C}^{T_c} = E_{AB}\Big|_{T_z}^{T_c} + E_{AB}\Big|_{0°C}^{T_z}$. If the indicator *does not* apply reference temperature compensation, E_R, then the simulator must provide $E_S = E_{AB}\big|_{0°C}^{T_c}$. If the indicator *does* apply E_R, then the simulator must supply *only* $E_s = E_{AB}\big|_{T_z}^{T_c}$.

This requires that the internal T_z of the indicator be accurately known by the simulator. Some simulators allow setting a *presumed* T_z in calibration. Those that do not *cannot* be used directly without error for calibration of the commonplace reference temperature compensating indicators. Many convenient thermocouple instrument calibrators allow setting the desired calibration by temperature rather than by voltage for standardized thermocouple types. This convenience introduces an additional nonlinear scaling and only approximate conformity to the standard characteristic.

Thermocouple Failure and Validation

A thermocouple measurement has *failed* when its indications are beyond uncertainty limits required for a measurement. Thermocouples sometimes fail "open" as junctions separate or thermoelements corrode, yield, or melt. Explicit "open circuit" indication is a promoted feature of many modern thermocouple indicators. This popular indication is useful, but is not an adequate indicator of thermocouple integrity.

The more common but less apparent circuit failures are by inobvious shunting, shorting, or progressing inhomogeneity of thermoelements. These are not detected by the "open circuit" indication. The continual indication of a *plausible* temperature is not proof of *authentic* temperature measurement.

A deliberate or accidental electrical shunting or direct short between thermoelements at ambient temperature local to an electronic compensating reference junction should result in an indication near *ambient* temperature, not 0°C. A thermocouple failure resulting from direct shorting between thermoelements at some distance from the intended measuring or reference junctions can go undetected because it occurs in a temperature region that is very different from the reference temperature and where a valid or plausible temperature measurement can continue to be made *but at an irrelevant and unexpected and unrecognized location* [35]. In some situations, as in predictably hostile environments, such failure of some thermocouples in service is anticipated. Where the failure is because of thermocouple circuit damage, it may not be evident immediately (or ever) without special circuit diagnostics. For such critical situations, special thermoelectric circuits and monitoring methods have been developed to assess the continued circuit integrity, although not the accuracy, of thermometry [35].

Environmental Compatibility

A primary consideration in thermocouple selection is compatibility of the thermoelements with their protective enclosures and of both thermoelements with the environment of measurement. Thermoelements must be protected from corrosive environments and electrically conductive fluid or solid shunts. Plastic, ceramic, or fiber insulators, metallic sheaths, and thermowells are intended to serve this function but may fail in service [1, 9–12, 25]. The compressed granular insulation of mineral-insulated, metal-sheathed thermocouples, if exposed, rapidly absorbs moisture that can seriously degrade resistive isolation [1, 10–12]. When cut, ends of MIMS thermocouple sheaths must be quickly resealed to avoid moisture absorption. Fiber and bead insulators are easily contaminated and can degrade accuracy [1, 10].

Temperature Exposure

The *unavoidable* environmental variables in thermometry are *temperature* and *duration*. Although thermoelements may not be visibly affected by an environment, the Seebeck coefficient might be substantially degraded. Even if only a very thin surface layer of a thermoelement is modified, the Seebeck coefficient could be changed. The properties can be degraded by use for a long period of time, at extreme temperatures, and in adverse environments [1, 9–12]. For these reasons, application should be limited to within the suggested temperature limits (Table 13.15). Calibration should be performed quickly, allowing time only for equilibration, and only to the maximum temperature of intended use. Degradation, observed in use as *instability* or *drift*, is necessarily a symptom of progressing *inhomogeneity* [30, 31].

Sustained excessive temperature alone can quickly degrade a thermoelement. Melting points define the absolute upper temperature range of thermocouple use. However, below this definite value, there

TABLE 13.15 Temperature Upper Limits For Different Wire Diameters

Dia., mm	0.025	0.127	0.254	0.406	0.813	1.600	3.175	
Dia., in.	0.001	0.005	0.010	0.016	0.032	0.063	0.125	
Dia., AWG	50	35	30	26	20	14	8	
Type	Temperature limit,°C							Type
E	290	325	370	400	510	775	855	E
J	230	275	305	350	460	600	750	J
K, N	690	730	790	840	950	1095	1250	K, N
T	90	110	150	185	270	370	375	T

Note: Recommended limits are guidelines for continual use of bead-insulated thermocouples in closed-end protection tubes in compatible environments. Mineral insulated metal-sheathed thermocouples can have slightly higher limits and tolerate longer exposure.

From References [1, 9, 13].

TABLE 13.16 Environmental Tolerance of Letter Designated Thermocouples

	Environment							
Type	Oxygen rich	Oxygen poor	Reducing	Vacuum	Humid	Below 0°C	Sulfur traces	Neutron radiation
B	Good	Good	Poor	Fair	Good	Poor	Poor	Fair
E	Good	Poor	Poor	Poor	Good	Good	Poor	Poor
J	Fair	Good	Good	Good	Poor	Poor	Fair	Poor
K	Good	Poor	Poor	Poor	Good	Fair	Poor	Good
N	Good	Fair	Poor	Poor	Good	Good	Fair	Good
R,S	Good	Good	Poor	Poor	Good	Fair	Poor	Poor
T	Fair	Fair	Good	Good	Good	Good	Fair	Poor

From References [1, 4, 9, 10, 43].

are indefinite application limits at which stability might be substantially reduced [1, 9–12]. Standard thermocouple tables extend only to the greatest temperature of *recommended* use for benign protected environments, short durations, and for wire materials of 3 mm or greater diameter [1, 9, 10]. Significantly reduced temperature limits apply for materials of smaller cross-section, Table 13.14 [1]. The thermocouples of smaller cross-section degrade more quickly. The standard tolerances apply only to material as manufactured.

Chemical Environment

There are broad guidelines for environmental compatibility. These recommendations are conditional and critically depend on many specifics of exposure. A concise summary of environmental compatibility characteristics is given in Table 13.16. The references should be consulted for detailed compatibility information [1, 4, 9, 22, 23]. Many thermocouple catalogs include tables of *usually* tolerable chemical environments for thermoelements or thermowell materials [19, 20]. Service experience in working with customers in a wide variety of process environments sometimes enables manufacturers to advise users on special problems of compatibility. Often, suitability in a particular service can be assured only by trial.

Metallurgical Change

Alloyed thermoelements can locally change composition by evaporation of constituents when exposed to vacuum. Alloy constituents evaporated from a sheath or from one thermoelement can deposit and coat adjacent thermoelements even through insulators [1, 9–12, 24, 30, 31]. Alloy components can migrate from one thermoelement to the other through the measuring junction. Traces of contaminants can penetrate through pinholes and hairline cracks and can actually diffuse through intact protective sheaths and affect apparently isolated thermoelements. Minute trace impurities in sheaths or insulating materials can interact with thermoelements. Appropriate MIMS construction and material selection usually extends life and increases temperature limits, but damage can occur even within apparently fully sheathed assemblies [34]. Excessive strain can locally substantially modify the Seebeck coefficient of one or both thermoelements resulting in inhomogeneity. Very localized strain, as introduced by sharp bends, usually has little effect on practical thermometry.

Data Acquisition

Thermocouple Indicators

The instrument designer can shelter the user from many troublesome details that complicated measurement for thermocouple pioneers. Microvolt-level signal resolution, high input resistance, stability, noise reduction, nonlinear scaling, and reference temperature compensation are now routinely provided to allow the user to focus on the measurement rather than on details of its indication or recording.

Some thermocouple indicators are hand-held and accommodate only a single thermocouple and only of a single type. A few provide dual thermocouple inputs and allow direct differential temperature measurement for which reference junction compensation *must not* be directly applied. Other bench-top units accommodate several thermocouple inputs of the same or of intermixed thermocouple and other sensor types.

Claimed instrument accuracy could mislead the casual reader. Although many instruments are supplied complete with a thermocouple probe, some accuracy specifications describe only the accuracy of input terminal voltage measurement. The actual instrument accuracy should be determined by electrical calibration (see "Calibration" section). Occasionally, the error of the indicating instrument becomes substantially greater than specified due to drift of reference compensation, offset, or gain. The additional uncertainty of thermocouple calibration, Table 13.13, and the often larger discrepancy between measuring junction temperature and the actual temperature of the object being measured, cannot be included because the latter are governed by details of heat transfer and transient thermal response.

Thermocouple Transmitters

A less familiar form of thermocouple signal conditioning, the *thermocouple transmitter*, is commonplace in process industries [4, 39]. Most two-wire thermocouple transmitters (not an RF wireless transmitter) are single-channel devices that perform reference junction compensation, linearizing, and isolation, and transmit the conditioned information to a remote monitoring site [1]. The transmitter converts the temperature-scaled Seebeck emf to an analogous signal for wire transmission over long lines to a remote recording or monitoring location. Such inexpensive and compact single-sensor signal conditioners are available specialized to most process variables. Thermocouple transmitters are well suited for monitoring slowly changing temperatures at monitoring sites distant from the measuring junction. Many are electrically isolated for safety. Some are adapted for use with the DIN rail mounting system. A transmitter can be installed near each thermocouple. Some are small enough to install in the connection head of a thermowell.

The transmitted signal can take either current-modulating or voltage-modulating form following ISA standards. The current-style transmitter temperature modulates, in proportion to temperature, a dc current supplied from the monitoring site. The signal ranges between 4 mA and 20 mA. The current-style transmitter conveys information over great distances without the voltage loss attendant to long-line voltage transmission. Some of the voltage-style transmitters modulate a supplied voltage over the span from 1 Vdc to 5 Vdc. Offset and scaling for some are adjusted at the transmitter. Some transmitters now have a local digital readout, and the scaling and offset can be remotely adjusted [4, 39].

Recording

Some multichannel thermocouple data recorders are now battery powered for stand-alone use and small enough to be hand held [41]. Some simpler recorders now cost less than U.S.$200. More sophisticated units accommodate multiple thermocouples of the same or intermixed type. Larger digital data loggers with multiplexers that sequentially sample the output from individual thermocouples can accommodate up to several hundred inputs and can record a mix of both thermocouple and other signals at sample intervals programmable from a few milliseconds to hours or days. Desirable three-wire switching between thermocouples is commonplace.

Personal computers now accommodate inexpensive internal digital data acquisition boards that can convert the computer to a high-resolution temperature recording system. Laptop computers can accept special palm-sized PCMCIA plug-in cards for multiple thermocouple input. Digital computer based data acquisition systems with software specialized for data acquisition, analysis, and presentation, allow the user to apply special reference temperature compensation, customized linearization of thermocouple scaling functions for either standard or individual calibrations and of special thermocouple types. Computer-based thermocouple systems with suitable commercial data acquisition and other software, provide for experiment design, prediction, data acquisition, information management, analysis, reporting, archiving, and communication within a single compact fieldable laptop computer [42].

Signal Transmission

The length of the principal thermocouple should be as short as feasible. The thermoelements should be continuous between measuring junction and reference junctions for best accuracy, but lengthy extension cables can be used if necessary (see "Practical Thermocouple Circuits" section). In field experiments and in process monitoring, recording might be separated from the measuring junction by hundreds of meters. There are several satisfactory possibilities for signal transmission over great distances.

Extension Cable

Matching and compensating thermoelectric extension cables (see "Practical Thermocouple Circuits" section) are available in a wide variety of constructions and insulations [4, 43]. An extension cable, as a transmission line, can be a single twisted thermoelectric pair or a multipair cable designed for suppression of both electrostatic and electromagnetic noise and for environmental protection. While inexpensive unshielded insulated extension wire is available, in the common situation where electric noise is likely, each extension should be an insulated, individually shielded, and twisted pair. Where several extension pairs from a cluster of nearby thermocouples extend to a remote recorder, multipair thermoelectric cable is available that adds an overall electrostatic shield, mechanical strengthening reinforcement, and a robust environmental overjacket to protect from mechanical damage and chemical intrusion in harsh process environments [4, 42].

Such elaborate thermoelectric extension is more expensive than copper instrumentation cable; thus, alternative transmission systems should be considered. The proper use of twisted pair copper instrumentation cable as neutral extension leads, extending from external reference junction temperature compensation, can be sufficient and less costly than thermoelectric extensions.

Higher quality twisted-shielded-pair instrumentation cable can also improve signal fidelity in transient thermometry. For such neutral extension leads, insulated solid *unplated* copper instrumentation pair only should be used because nonuniform plating thickness can result in irregularly distributed inhomogeneity and cause Seebeck emf noise. Conventional *coaxial* instrumentation cable should never be used as lengthy neutral thermoelectric extension leads because the outer braid and center conductors have very different Seebeck coefficients. If not at uniform temperature, the cable can be a significant source of Seebeck noise emf.

Thermoelectric extensions and "pigtailed" assemblies should be checked full-length for hidden junctions between slightly mismatched spliced thermoelements and for deliberate, but hidden, pair splices incorrectly made in crossed polarity [10]. Pass a narrow heat source along the full length of the cable, while monitoring for local jumps of output local to such unintended junctions, using a thermocouple indicator. This test is very effective to pinpoint hidden junctions. It does not test for inhomogeneity [10, 30].

Optic fiber cable with electro-optic transmitters can be used for electric noise-free cable transmission over a long distance between signal conditioning that is local to the thermocouple and to a remote recorder or monitor.

Wireless Transmission

In several situations, copper wire transmission is less effective, more expensive, or less convenient than wireless methods. Several low-cost wireless (not thermocouple transmitter) data acquisition systems are now adapted to combine thermocouple data acquisition and remote signal transmission [44]. *Radio modems* communicate data from a measuring location to recording sites as much as 5 km away. Most provide spread spectrum transmission that does not require a communications license. Wireless data transmission can be particularly economical where the distance between the measuring junction and recorder is great, when many thermocouples must be recorded, when little setup time is available for field cabling, and where the measuring setup must be moved occasionally. Some systems combine data logging and radio communication in a single unit. Others require the use of a thermocouple transmitter, a data logger, and the radio modem to form a system.

Sources of Thermocouple Application Information

Technical Papers

The most concentrated and broadest sources of refereed technical papers addressing thermocouple thermometry are the serial proceedings of a decennial international symposium on temperature [25]. The six volumes to date are a rich and reliable source of application data. Proceedings of annual symposia of the ISA and NCSL also have a few papers describing current developments in thermometry. Reference [26] has thermoelectric data on the widest variety of common and nonstandard materials. Current papers devoted specifically, not incidentally, to thermocouples are infrequent and are distributed broadly across trade magazines and journals of professional societies. Essential details of thermocouple thermometry are more often (poorly) described only incidentally in reports of experimental studies. Some manufacturers publish subscription technical journals specializing in thermometry [45].

One measurement journal frequently publishes articles on applied thermometry and, annually in the June issue, publishes an extensive directory of manufacturers of thermocouple materials and related instrumentation [46]. The directory includes, for several dozen listed manufacturers, concise descriptions of selected products and current prices.

Books, Reports, and Standards

A very detailed and authoritative NIST monograph is the primary source for Seebeck properties and physical characteristics of letter-designated thermocouples [9]. Standards of the ASTM and ISA adopt the values and complement the NIST report [10, 24]. U.S. standards are reviewed and updated at intervals of 5 years or less. The ASTM also publishes interim tables of Seebeck emf for a few popular materials that have not been standardized [10]. Many standards are collected in a single, annually revised volume specializing in thermometry [10]. The hardcopy thermocouple tables from these volumes are now available as functional computer programs that calculate $E(T)$ and $T(E)$ of the NIST 175 document over ranges, at intervals, and in units selected by the user [46].

A comprehensive thermocouple application manual is published by the ASTM [1]. Many commercial publishers issue new reference and textbooks addressing measurement. Most of these include a brief obligatory section or chapter describing the bare elements of thermocouple thermometry. A few devote an entire volume to thermometry. There are many deeper scientific treatments of thermoelectricity, mostly in the historic literature [1, 11, 12].

Many national and international consensus standards organizations publish thermoelectric test methods and material characterizations [10, 21, 25]. Standards are instructive as well as being formalizations of procedures and materials. ASTM Standards and indexes to them are now available in computer CD-ROM file format. Some standards now are available on the World Wide Web or by FAX. Societies with committees that specialize in thermoelectric standards now have Internet Web sites (such as *http://www.isa.org* and *astm.org*) that provide information about available standards and solicit on-line technical questions from users.

Workshops

Annually, there are several excellent workshops of a few hours or days duration devoted to measurement, thermometry, and even thermocouples. Professional societies such as the ISA, in association with special symposia or annual meetings, present special tutorial workshops. Some are offered without fee by manufacturers. Other measurement courses are offered by independent measurement specialists. Announcements of these workshops are published in the technical and trade journals.

Trade Literature

Abundant, free, commercial catalog literature, handbooks, and Web site advertising describe thermocouple hardware and data acquisition products. Some manufacturers publish elaborate catalogs of thermocouple-related hardware, software, and books [4]. Most include tutorial material. A few include technical reprints, discussion of principles and practices, and present extensive tables of thermoelectric characteristics and physical properties of thermocouple materials. Demonstration programs and product information are available from manufacturers on CD-ROM and on the Internet. There are several trade journals that relate to measurement, explore current issues and developments, and are distributed free to qualified recipients.

Caveat

The reader is cautioned that the extensive current and historic literature of thermoelectric thermometry, tutorial articles (*even this one*), standards, specifications, and advice from "experts," all must be studied very critically with an authentic thermoelectric model in mind. At every level of sophistication, from the promotional to the most esteemed esoteric and sophisticated mathematical and physical thermoelectric theory, innocently propagated misconceptions concerning the thermoelectric effects are very commonplace.

Summary

The Seebeck effect can be used to measure temperature with finer spatial and time resolution, over a broader temperature range, in more diverse geometries, and at lower cost than most other electric temperature sensors. The unique versatility of the thermocouple ensures that it will continue as the thermometry means of choice for many applications despite competition from an increasing variety and abundance of alternative special-purpose thermometers. However, the misleading seeming simplicity of the idealized thermocouple and the convenience and apparent accuracy afforded by modern signal conditioning instruments can be misleading. Correct application is simple.

Properly used within its limitations, the thermocouple is capable of accurate reliable thermometry. Improperly understood, the thermocouple is subject to misapplication and inconspicuous, but very significant, error. This precautionary overview described pitfalls and their avoidance in thermocouple practice. This can provide the receptive reader with an authentic basis for the knowledgeable use of even the most complex series and generalized thermoelectric circuits in circuit design, diverse applications, particularly in thermometry. The simple Functional Model presented here (and perhaps requiring some study) was deliberately crafted to aid the user in clearly distinguishing the authentic information from the misleading, and to aid in evaluating the designs and specifications of manufacturers of thermoelectric products.

References

1. R. M. Park (ed.), *Manual on the Use of Thermocouples in Temperature Measurement*, MNL 12, 4th ed., Philadelphia, PA, American Society for Testing and Materials, 1993.
2. R. P. Reed, Thermal effects in industrial electronics circuits, in J. D. Irwin (ed.), *CRC Industrial Electronics Handbook*, Boca Raton, FL, CRC Press, 1996, 57–70.

3. *Catalog*, TRANSCAT/EIL, Rochester, NY, 1998.
4. *Catalog, The Temperature Handbook*, Issue 29, Omega Engineering, Inc., Stamford, CT, 1995.
5. D. D. Pollock, *Physics of Engineering Materials*, Englewood Cliffs, NJ, Prentice-Hall, 1990.
6. D. M. Rowe (ed.), *CRC Handbook of Thermoelectrics*, Boca Raton, FL, CRC Press, 1995.
7. R. P. Reed, Principles of thermoelectric thermometry, in R. M. Park (ed.), *Manual on the Use of Thermocouples in Temperature Measurement*, MNL 12, 4th ed., Philadelphia, PA, American Society for Testing and Materials, 1993, chap. 2, 4–42.
8. R. P. Reed, Absolute Seebeck thermoelectric characteristics — principles, significance, and applications, in J. F. Schooley (ed.), *Temperature, Its Measurement and Control in Science and Industry*, Vol. 6, Part 1, New York, American Institute of Physics, 1992, 503-508.
9. G. W. Burns, M. G. Scroger, G. F. Strouse, M. C. Croarkin, and W. F. Guthrie, *Temperature-Electromotive Force Reference Functions and Tables for the Letter-Designated Thermocouple Types Based on the ITS-90*, NIST, Monograph 175, Washington, D.C., Department of Commerce, 1993.
10. *Annual Book of ASTM Standards, Temperature Measurement*, 14.03, Philadelphia, PA, American Society for Testing and Materials, 1998.
11. J. V. Nicholas and D. R. White, *Traceable Temperatures — An Introduction to Temperature Measurement and Calibration*, New York, John Wiley & Sons, 1994.
12. T. W. Kerlin and R. L. Shepard, *Industrial Temperature Measurement*, Philadelphia, PA, Instrument Society of America, 1982.
13. *Catalog*, Analog Devices, Inc., Norwood, MA.
14. *Catalog, Equipment for Temperature Measurement and Control Systems.*, Hades Manufacturing Corp., Farmingdale, NY.
15. R. Morrison, *Noise and Other Interfering Signals*, New York, John Wiley & Sons, 1992.
16. H. W. Markenstein, Proper shielding reduces EMI, *Electron. Packaging & Production*, 37, 72-78, 1997.
17. *Catalog*, Thermagon, Inc., Cleveland, OH.
18. N. R. Keltner and J. V. Beck, Surface temperature measurement errors, *J. Heat Transfer*, 105, 312-318, 1983.
19. *Catalog*, RdF Corporation, Hudson, NH.
20. *Catalog, Temperature Measurement Handbook*, Vol. IX, NANMAC Corporation, Framingham, MA.
21. R. P. Reed, Convolution & deconvolution in measurement and control, professional course, *Measurements & Control*, (178-188), 1997, 1998.
22. H. M. Hashemian, K. M. Peterson, D. W. Mitchell, M. Hashemian, and D. D. Beverly, In situ response time testing of thermocouples, *ISA Trans.*, 29, 1986.
23. R. P. Reed, The transient response of embedded thin film temperature sensors, *Temperature, Its Measurement and Control in Science and Industry*, Vol. 4, Part 3, New York, Instrument Society of America, 1972.
24. *Standard, Temperature Measurement Thermocouples*, ISA/ANSI Standard MC96.1-1982, Research Triangle Park, NC, ISA International Society for Measurement and Control, 1982. (Standard withdrawn 1993.
25. *Temperature, Its Measurement and Control in Science and Industry*, Vols. 2–6 (Serial), New York, American Institute of Physics, 1942–1992.
26. P. A. Kinzie, *Thermocouple Temperature Measurement*, New York, Wiley-Interscience, 1973.
27. H. L. Anderson (ed.), *Physics Vade Mecum*, New York, American Institute of Physics, 1981.
28. *Standard, Norme Internationale/International Standard*, IEC 584, Parts 1–3, Thermocouples, International Electrotechnical Commission, Geneva, Switzerland, 1995.
29. R. P. Reed, Thermoelectric thermometry: A functional model, in J. F. Schooley (ed.), *Temperature, Its Measurement and Control in Science and Industry*, vol. 5, Part 2, New York, American Institute of Physics, 1982, 915-922.
30. R. P. Reed, Ya can't calibrate a thermocouple junction! *Measurements & Control*, Part 1. *Why not?*, 178, 137–145; Part 2. *So What?*, 179, 93–100, 1996.

31. R. P. Reed, Thermocouples: calibration, traceability, instability, and inhomogeneity, *Isotech J. Thermometry*, 7(2), 91-114, 1996.
32. W. F. Roeser, Thermoelectric thermometry, *J. Appl. Phys.*, 11, 213-232, 1940.
33. R. P. Reed, Thermoelectric inhomogeneity testing. Part I: Principles; Part II: Advanced methods, in J. F. Schooley (ed.), *Temperature, Its Measurement and Control in Science and Industry*, Vol. 6, Part 1, New York, American Institute of Physics, 1992, 519-530.
34. W. Rosch, A. Fripp, W. Debnum, S. Sorokach, and R. Simchick, Damage of fine diameter platinum sheathed Type R thermocouples at temperatures between 950 and 1100°C, in J. F. Schooley (ed.), *Temperature, Its Measurement and Control in Science and Industry*, Vol. 6, Part 1, New York, American Institute of Physics, 1992, 569-574.
35. R. P. Reed, Validation diagnostics for defective thermocouple circuits, in J. F. Schooley (ed.), *Temperature, Its Measurement and Control in Science and Industry*, 5, Part 2, New York, American Institute of Physics, 1982, 915-922.
36. *Survey, Measurement Needs Tracking Study — 1997*, Keithley Instruments, Cleveland, OH, 1997.
37. J. P. Tavener, Temperature calibration, *Measurements and Control*, Sept., 160-164, 1986.
38. T. B. Fisher, Selecting a dry well calibrator, *Measurements and Control*, (185), 105, 1997.
39. *Catalog, Reference Manual for Temperature Products and Services*, 1st ed., Isothermal Technology Ltd., Merseyside, England, 1997.
40. *Catalog*, Moore Industries, Sepulveda, CA, 1997.
41. *Catalog*, DCC Corporation, Pennsauken, NJ, 1997.
42. *Catalog, Instrumentation Reference and Catalogue*, National Instruments, Austin, TX, 1997.
43. *Catalog, Temperature Sensors, Wire and Cable*, Watlow Gordon, Richmond, IL, 1997.
44. *Catalog*, ENCOM Radio Services, Calgary, Alberta, Canada.
45. *Isotech Journal of Thermometry*, ISSN 0968-347X, Isothermal Technology, Ltd., Merseyside, England.
46. *Measurements and Control* magazine, ISSN 0148-0057, Measurements & Data Corporation, Pittsburgh, PA.
47. R. P. Reed, A comparison of programs that convert thermocouple properties to the 1990 international temperature and voltage scales, *Measurements and Control*, 30(177), 105-109, 1996.

* *Note:* Mention of *representative* products is to introduce the reader to the variety of available thermocouple-related hardware and is not an endorsement.

13.5 Semiconductor Junction Thermometers

Randy Frank

Temperature sensors can be easily produced with semiconductor processing technology by using the temperature characteristics of the *pn junction.* The batch processing and well-defined manufacturing processes associated with semiconductor technology can provide low cost and consistent quality temperature sensors. The temperature sensitivity of the *pn junction* is part of the transistor's defining equations and is quite predictable over the typical semiconductor operating range of –55°C to +150°C.

Most semiconductor junction temperature sensors use a diode-connected bipolar transistor (short-circuited collector-base junction) [1]. A constant current passed through the base-emitter junction produces a junction voltage between the base and emitter (V_{be}) that is a linear function of the absolute temperature (Figure 13.31). The overall forward voltage drop has a temperature coefficient of approximately 2 mV °C^{-1}.

When compared to a thermocouple or a resistive temperature device (RTD), the temperature coefficient of a semiconductor sensor is larger but still quite small. Also, the semiconductor sensor's forward voltage has an offset that varies significantly from unit to unit. However, the semiconductor junction voltage vs. temperature is much more linear than that of a thermocouple or RTD. In addition to the

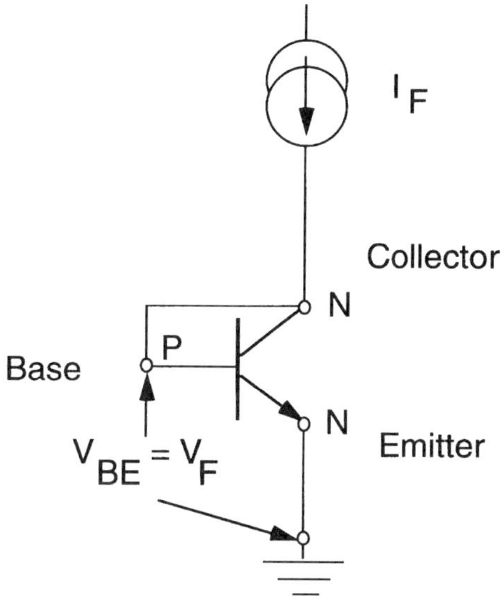

FIGURE 13.31 Bipolar transistor configured as a temperature sensor. The base of the transistor is shorted to the collector. A constant current flowing in the remaining *pn* (base to emitter) junction produces a forward voltage drop V_F proportional to temperature.

temperature-sensing element, circuitry is easily integrated to produce a monolithic temperature sensor with an output that can be easily interfaced to a microcontroller and to provide features that are useful in specific applications. For example, by using an *embedded temperature sensor* with additional circuitry, protection features can be added to integrated circuits (ICs). A temperature sensor becomes an embedded item in a semiconductor product when it has a secondary or supplemental purpose instead of the primary function.

The Transistor as a Temperature Sensor

A common semiconductor product for temperature sensing is a small-signal transistor such as a 2N2222 or a 2N3904. By selecting a narrow portion of the overall distribution of the V_{be} for these devices, a temperature sensor with a lower variation in characteristics can be obtained. The lower variation can provide a part-for-part replacement when a tolerance of only a few percent is acceptable. This device (formerly offered as an MTS102 but no longer in production) demonstrates the performance characteristics of the transistor used as a temperature sensor [2].

As shown in Figure 13.32 [2], a silicon temperature sensor has a nominal output of 730 mV at –40°C and an output of 300 mV at 150°C. The narrowly specified V_{be} ranges between 580 mV and 620 mV at 25°C. The linearity error, or variation from a straight line, of this device is shown in Figure 13.33 [2]. The total accuracy is within ±3.0 mV including nonlinearity which is typically within ±1°C in the range of –40°C to 150°C. These readings are made with a constant (collector) current of 0.1 mA, passing through the device to minimize the effect of self-heating of the junction. When the constant current applied is larger than 0.1 mA, the effect of self-heating in the device must be taken into account. The variation of the V_{be} with current is shown in Figure 13.34 [2].

Thermal Properties of Semiconductors: Defining Equations

A constant forward current supplied through an ideal silicon *pn* junction produces a forward voltage drop, V_F [3, 4]:

$$V_F = V_{be} = (kT/q)\ln(I_F/I_S) \tag{13.86}$$

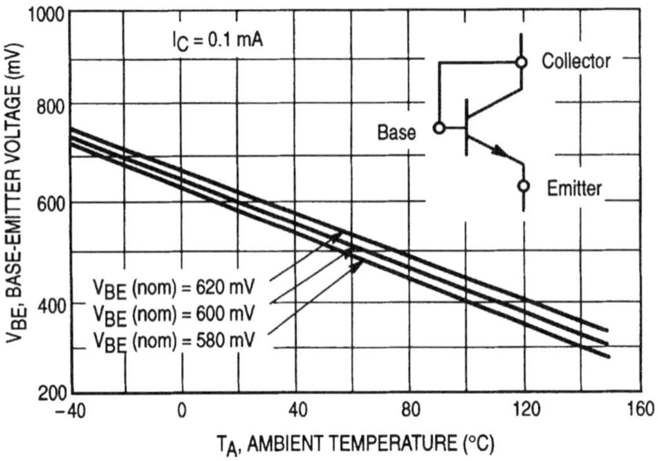

FIGURE 13.32 Base–emitter voltage vs. ambient temperature for a silicon temperature sensor.

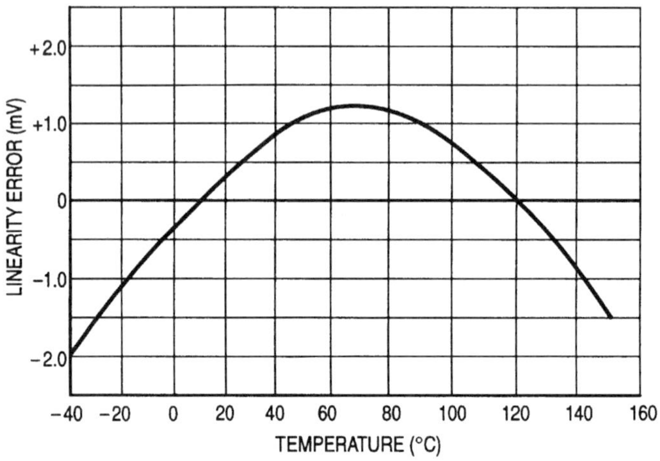

FIGURE 13.33 Linearity error (in mV) vs. temperature for a silicon temperature sensor.

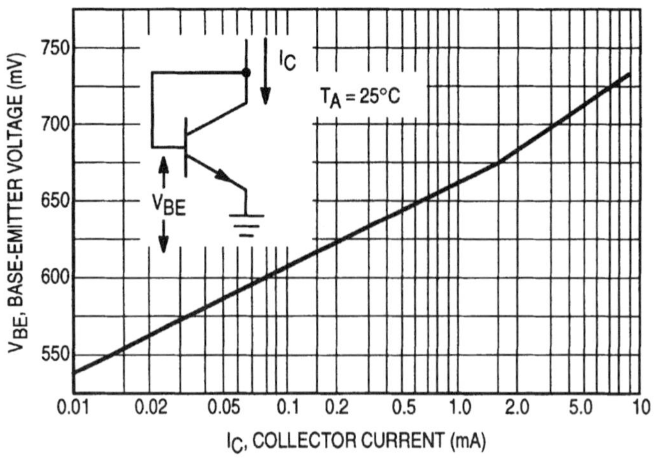

FIGURE 13.34 Base–emitter voltage vs. collector–emitter current.

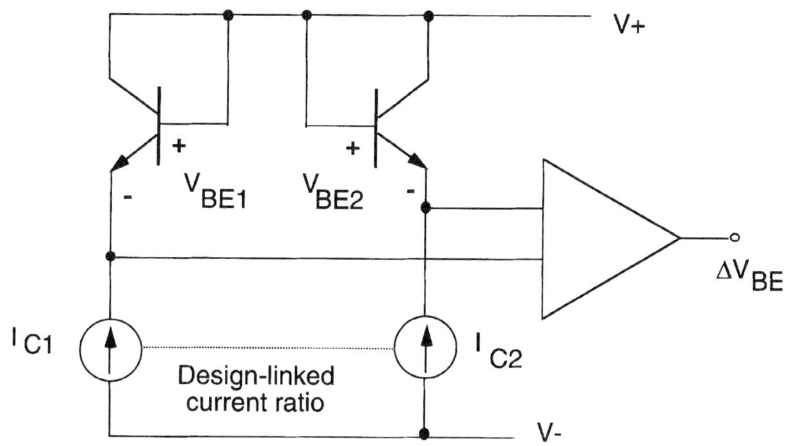

FIGURE 13.35 Differential pair formed by two *pn* junctions. The transistors are diode connected to form a temperature sensor independent of variations in source current.

where k = Boltzmann's constant (1.38×10^{-23} J K^{-1})
T = Temperature (K)
q = Charge of electron (1.6×10^{-19} C)
I_F = Forward current (A)
I_S = Junction's reverse saturation current (A)

For constant I_S, the junction voltage (V_{be}) would be directly proportional to absolute temperature. Unfortunately, I_S is temperature dependent and varies with the cube of absolute temperature. As a result, V_F has an overall temperature coefficient of approximately –2mV °C^{-1}.

To reduce the temperature variation, a *band gap reference* is formed based on two adjacent and essentially identical-behavior transistors with proportional emitter area designed in an integrated circuit process. The two base–emitter junctions are biased with different current densities (I/A), but the ratio of current densities is essentially constant over the operating temperature range (–55°C to +150°C). The following equation shows how the differential voltage (ΔV_{be}) is related to the current (I) and emitter area (A) of the respective transistors:

$$V_{be1} - V_{be2} = \Delta V_{be} = \left(kT/q\right)\ln\left(\left(I_1/A_1\right)/\left(I_2/A_2\right)\right) \tag{13.87}$$

The differential voltage appearing at the output can be amplified as shown in Figure 13.35 and used as a direct indication of absolute temperature. Additional circuitry can eliminate the offset voltage at 0°C and provide an output in degrees Celsius or degrees Fahrenheit.

The ability to obtain temperature sensing using semiconductor processing techniques has two significant consequences: (1) semiconductor processing and integrated circuit design can be used to improve the temperature sensor's performance for specific applications, and (2) temperature sensors can be integrated within other integrated circuits to obtain additional features. The next two sections explain these approaches.

Integrated Temperature Sensors

Once a temperature sensor can be manufactured using semiconductor processing techniques, a number of shortcomings of the sensor can be corrected by additional circuitry integrated into the sensor or by using circuit techniques external to the sensor. The linearity improvement, addition of precision voltage references, precision voltage amplifiers, and digital output for direct interface to a microcontroller (MCU)

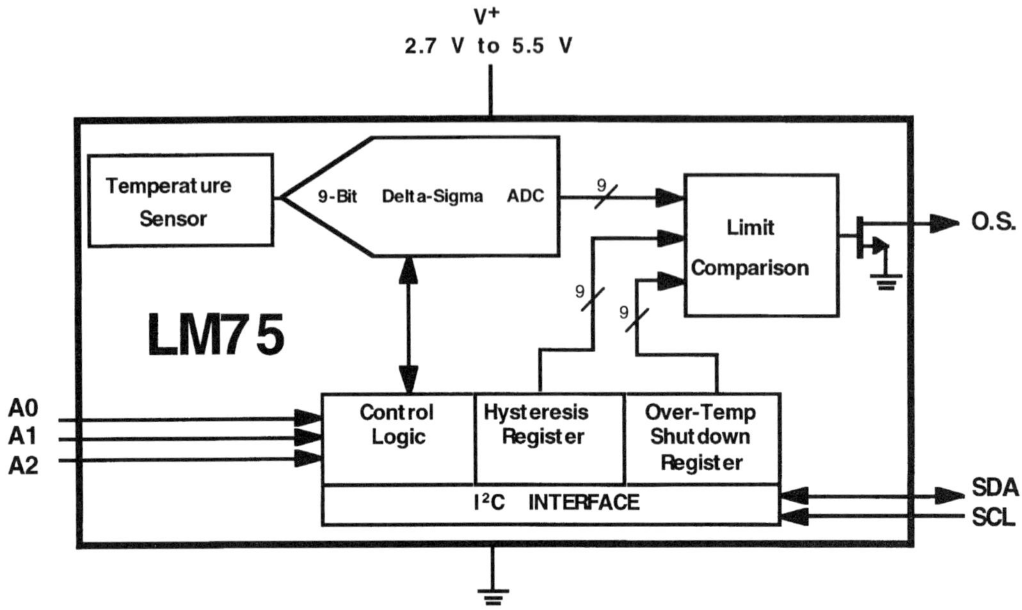

FIGURE 13.36 Block diagram of a monolithic digital-output temperature sensor. (Courtesy National Semiconductor Corp.)

are among the enhancements possible. Furthermore, resistance-measuring circuitry (i.e., RTD sensors) or cold junction compensation (i.e., thermocouple sensors) are not required. Three integrated circuits and one external circuit example are discussed.

Integrated Digital Temperature Sensor

A *monolithic* (one piece of silicon) semiconductor junction temperature sensor (LM75 from National Semiconductor) that incorporates several features, including a digital output, is shown in Figure 13.36 [3]. The analog signal of the temperature sensor is converted to digital format by an on-board sigma–delta converter [3]. Digital communication is provided directly to a host microcontroller through a serial two-wire interface. The sensor has a software-programmable setpoint that can be used to terminate the operation of the controller or implementing protection [5]. To avoid false triggering, a user-programmable number of comparisons (up to six successive over-temperature occurrences) can be implemented.

Eight different sensors can be operated on the bus. Resolution is ±1/2°C and the accuracy is ±2% from –25°C to +100°C. The sensor consumes 250 μA during operation and only 10 μA in sleep mode.

Analog Output Integrated Temperature Sensor

Another approach to integrated temperature sensing is shown in Figure 13.37 [4].The TMP-1 resistor-programmable temperature controller features a 5-mV °C^{-1} output high and low set points and over- and under-temperature output. A low-drift voltage reference is also included in the 70 mil × 78 mil (2.76 mm × 3.07 mm) design. Figure 13.38 shows a photomicrograph of the silicon die. The TMP-1 is specified for operation between –55°C and +125°C, with ±1°C accuracy over the entire range.

Digital Output Temperature Sensor

A direct-to-digital temperature sensor has been designed for multi-drop temperature sensing applications [6]. A unique serial number is etched onto each device. The 64-bit read-only memory (ROM) identifies the temperature of a particular sensor in a measurement system, with several sensors providing readings from different locations. The signal can be transmitted for distances up to 300 m.

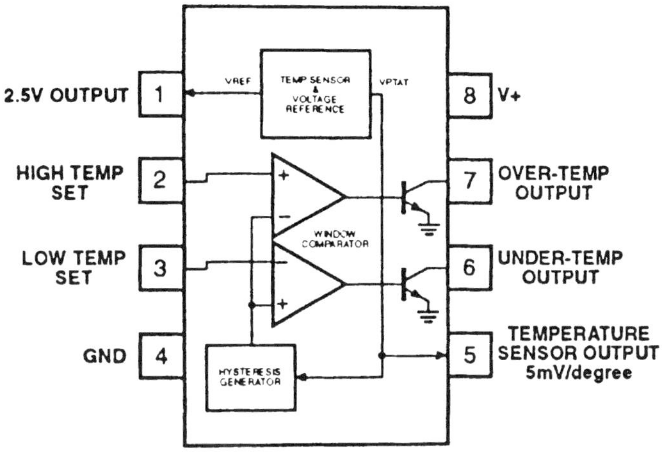

FIGURE 13.37 Block diagram and pinout of TMP-1 monolithic, programmable temperature controller. (Courtesy of Analog Devices, Inc.)

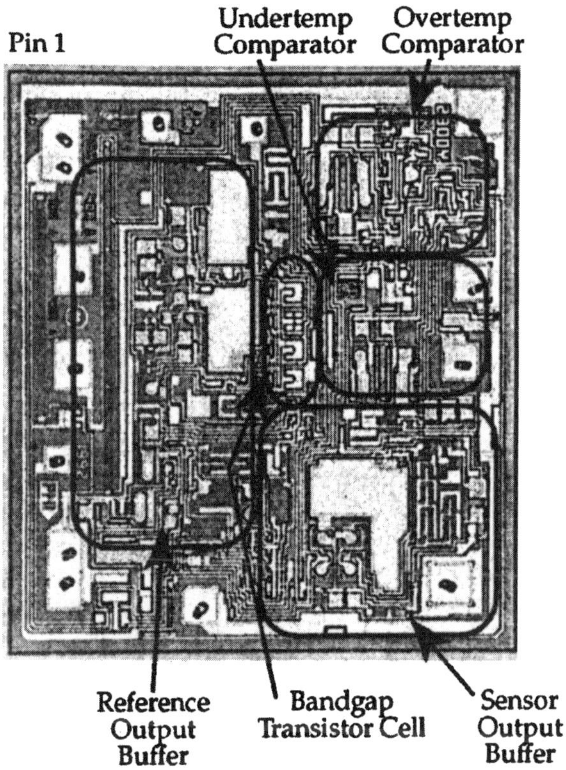

FIGURE 13.38 Detail of TMP-1 die. Note the relative size of the bandgap transistor cell to the other circuitry included in the monolithic device. (Courtesy of Analog Devices, Inc.)

The temperature sensor operates from –55°C to 125°C with the power supplied from the data line. The measurement is resolved in 0.5°C increments as a 9-bit digital value, with the conversion occurring within 200 ms. User-definable alarm settings are included in the device.

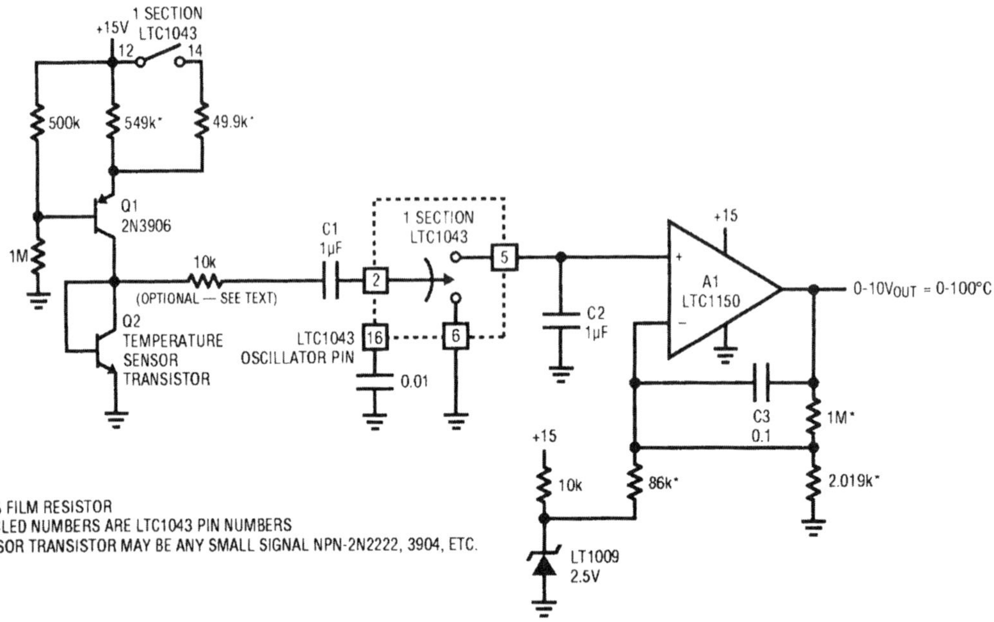

FIGURE 13.39 External circuitry provides ΔV_{be}-based thermometer that does not require calibration. (Courtesy of Linear Technologies Corporation.)

External Circuitry Eliminates Calibration

External circuitry can be designed to utilize the inherent sensing capability of low-cost transistors without incurring the additional cost of integrated circuitry or the need to calibrate each device. The circuit shown in Figure 13.39 uses an integrated circuit (LTC1043 from Linear Technology Corporation) to provide a 0 V to 10 V output from 0°C to 100°C, with an accuracy of ±1°C using any common small signal transistor as the temperature sensor [7]. The circuitry establishes a ΔV_{be} vs. current relationship that is constant regardless of the V_{be} diode's absolute value. Substituting different transistors from multiple sources showed a variation of less than 0.4°C.

Other Applications of Semiconductor Sensing Techniques

Several semiconductor parameters vary linearly over the operating temperature range. Power MOSFETs used to switch high levels of current (typically several amperes) at voltages that can exceed 500 V provide an example of these characteristics. As shown in Figure 13.40 [8], the gate threshold voltage of a power MOSFET changes from 1.17 to 0.65 times its 25°C value when the temperature increases from –40°C to 150°C. Also, the breakdown voltage of the power MOSFET varies from 0.9 to 1.18 times its value at 25°C over the same temperature range (Figure 13.41) [8]. These relationships are frequently used to determine the junction temperature of a semiconductor component in actual circuit operation during the design phase (see "Reliability Implications"). External package level temperature measurements can be many degrees lower than the junction temperature, especially during rapid, high-energy switching events. The actual junction temperature and the resulting effect on semiconductor parameters must be taken into account for the proper application of semiconductor devices.

Polysilicon diodes (and resistors) that are isolated from the power MOSFET can be produced as part of the semiconductor manufacturing process with minor process modifications. The diodes can be used as temperature sensing elements in an actual application [9]. The thermal sensing that is performed by the polysilicon elements is a significant improvement over power device temperature sensing that is performed by an external temperature sensing element. By sensing with polysilicon diodes, the sensor can be located close to the center of the power device near the source bond pads where the current

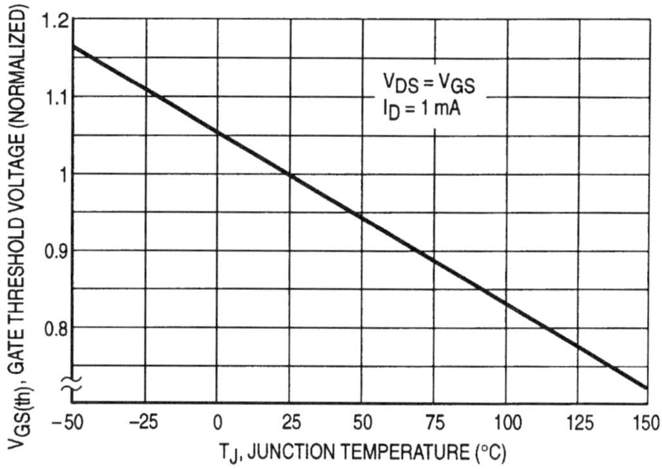

FIGURE 13.40 Power MOSFET's gate threshold variation vs. temperature.

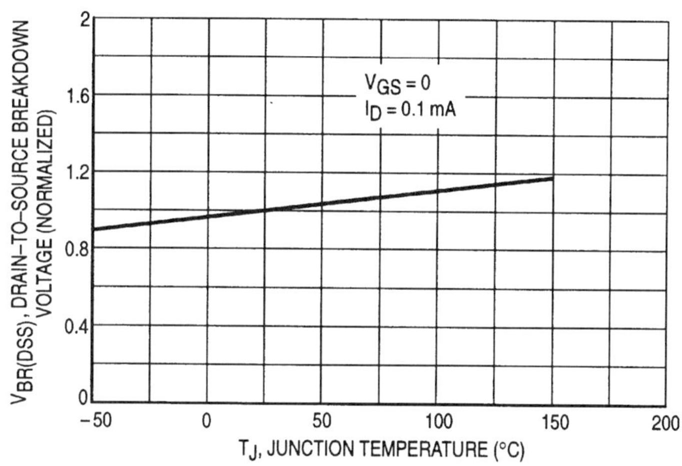

FIGURE 13.41 Power MOSFET's breakdown voltage variation vs. temperature.

density is the highest and, consequently, the highest die temperature occurs. The thermal conductivity of the oxide that separates the polysilicon diodes from the power device is 2 orders of magnitude less than that of silicon. However, because the layer is thin, the polysilicon element offers an accurate indication of the actual peak junction temperature.

A power FET that incorporates temperature sensing diodes is shown in Figure 13.42 [9]. By monitoring the output voltage when a constant current is passed through the integrated polysilicon diode(s), an accurate indication of the maximum die temperature is obtained. A number of diodes are actually provided in the design. A single diode in this design has a temperature coefficient of 1.90 mV °C^{-1}. Two or more can be placed in series if a larger output is desired. For greater accuracy, the diodes can be trimmed during wafer-level testing by blowing fusible links made from polysilicon. The response time of the diodes is less than 100 μs, which has allowed the device to withstand a direct connection across an automobile battery with external circuitry providing shutdown prior to device failure. The sensing capability also allows the output device to provide an indication (with additional external circuitry) if the heatsinking is not proper when the unit is installed in a module or if a change occurs in the application that would ultimately cause a failure.

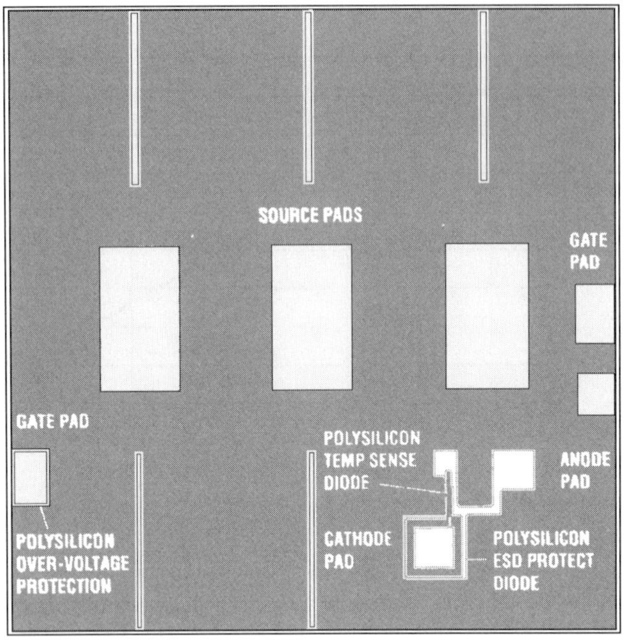

FIGURE 13.42 Photomicrograph of temperature sensor integrated in power MOSFET. Note the relative size of the temperature sensor compared to the total area of the power MOSFET and the source pads which allow attachment of 15 mil (0.60 mm) aluminum wire. (Courtesy of Motorola, Inc.)

Temperature Sensing in Power ICs for Fault Protection and Diagnostics

Sensing for fault conditions, such as a short-circuit, is an integral part of many smart power (or power) ICs. The ability to obtain temperature sensors in the semiconductor process provides protection and diagnostics as part of the features of these devices. The primary function of the power IC is to provide a microcontroller-to-load interface for a variety of loads. In multiple output devices, sensing the junction temperature of each device allows the status of each device to be provided to the microcontroller (MCU), and, if necessary, the MCU can shut down a particular unit that has a fault condition.

A *smart power IC* can have multiple power drivers integrated on a single monolithic piece of silicon [10]. Each of these drivers can have a temperature sensor integrated to determine the proper operating status and shut off only a specific driver if a fault occurs. Figure 13.43 shows an eight-output driver that independently shuts down the output of a particular driver if its temperature is excessive (i.e., between 155°C and 185°C) [10].

The octal serial switch (OSS) adds independent thermal sensing through over-temperature detection circuitry to the protection features. Faults can be detected for each output device, and individual shutdown can be implemented. In a multiple output power IC, it is highly desirable to shut down only the device that is experiencing a fault condition and not all of the devices that are integrated on the power IC. With outputs in various physical locations on the chip, it is difficult to predict the thermal gradients that could occur in a fault situation. Local temperature sensing at each output, instead of a single global temperature sensor, is required.

As shown in Figure 13.44, the eight outputs of the device with individual temperature sensors can be independently shut down when the thermal limit of 170°C is exceeded [10]. All of the outputs were connected to a 16-V supply at a room temperature ambient. A total current of almost 30 A initially flowed through the device. Note that each device turns off independently. The hottest device turns off first. Variations can result from differences in current level and thermal resistance. As each device turns off, the total power dissipation in the chip decreases and the devices that are still on can dissipate heat more effectively.

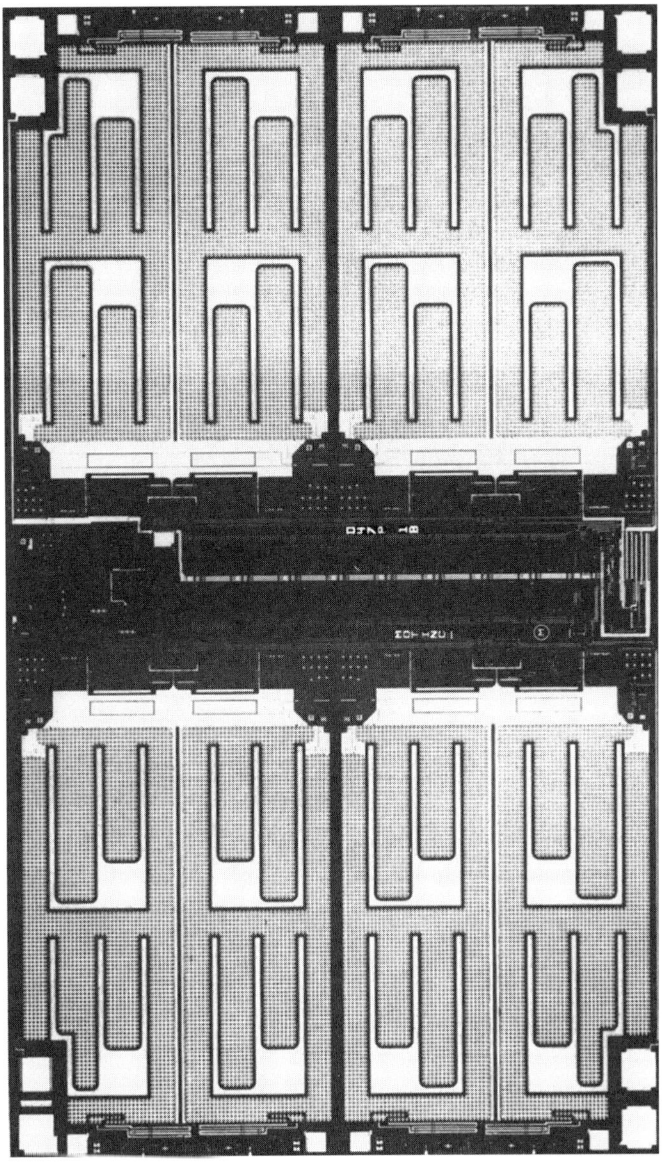

FIGURE 13.43 Photomicrograph of eight-output power IC. Note that the area of the eight output devices (two located at each corner of the die) are considerably larger than the circuitry in the center, and top and bottom that provides the temperature sensing, signal conditioning, and other control features. (Courtesy of Motorola, Inc.)

Connecting directly to the battery is a hard short that could have been detected by current limit circuitry. However, a soft short is below the current limit, but exceeds the power-dissipating capability of the chip, and can be an extremely difficult condition to detect. Soft shorts require over-temperature sensing to protect the IC from destructive temperature levels.

The over-temperature condition sensed by the power IC could mean that the device turns itself off to prevent failure in one case; and in another situation, a fault signal provides a warning to the MCU but no action is taken, depending on the fault circuit design. The remaining portion of the system is allowed to function normally. With the fault conditions supplied to the MCU, an orderly system shutdown can be implemented. Integrated temperature sensing is essential to provide this type of protection in a multiple-output power IC.

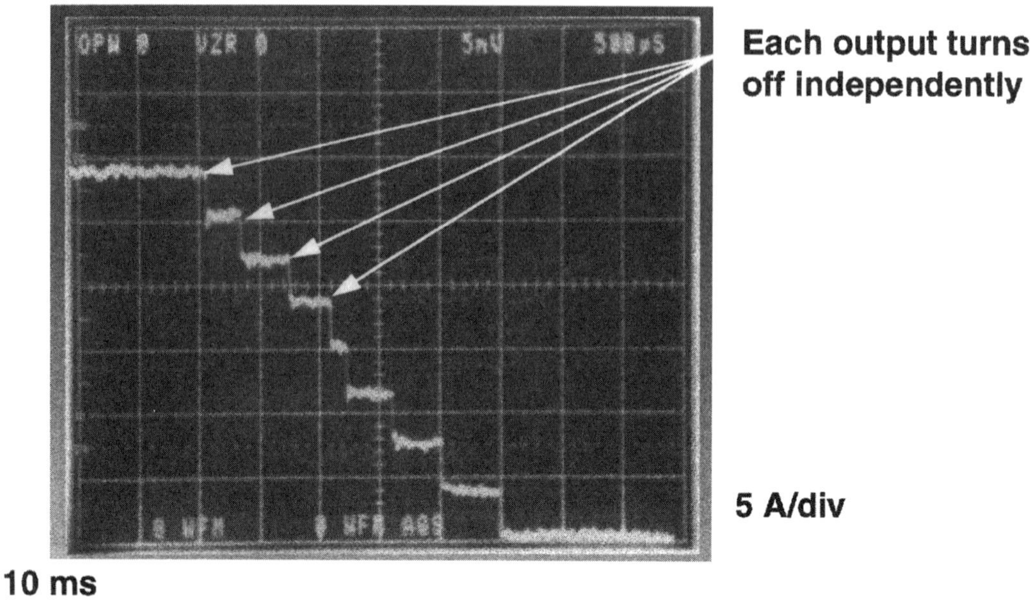

FIGURE 13.44 Independent thermal shutdown of an 8-output power IC. (Courtesy of Motorola, Inc.)

Reliability Implications of Temperature to Electronic Components

The effect of temperature on electronic components and their successful application in electronic systems is one of the issues that must be addressed during the design of the system. Temperature affects the performance and expected life of semiconductor components. Mechanical stress created by different coefficients of thermal expansion can cause failures in thermal cycling tests (air-to-air) or during thermal shock (water-to-water) transitions.

The typical failure rate for semiconductor component can be expressed by the Arrhenius equation [9]:

$$\lambda = Ae^{-\varnothing/KT} \tag{13.88}$$

where λ = Failure rate
A = Constant
$\varnothing$ = Activation energy (eV)
K = Boltzmann's constant (8.62×10^{-5} eV K^{-1})
T = Junction temperature (K)

The failure rate of semiconductor components is typically stated to double for every 10 to 15°C increase in operating (i.e., junction) temperature. However, increased testing and design improvements have minimized the failures due to specific failure mechanisms.

One of the temperature-related parameters that must be taken into account during the design phase of a power switch is the transient thermal response, which is designated as $r(t)\ R\varnothing_{JC}$, where $r(t)$ represents the normalized transient thermal resistance. The value of $r(t)$ is determined from the semiconductor manufacturer's data sheet using duty cycle and pulse duration used in the application. This reduced level of the thermal resistance (see "Junction Temperature Measurement"), based on the transistor operating in a switching mode and being off for a period of time, approaches the dc level within a second. Excessive temperatures can be generated quickly and must be detected within milliseconds to prevent failure.

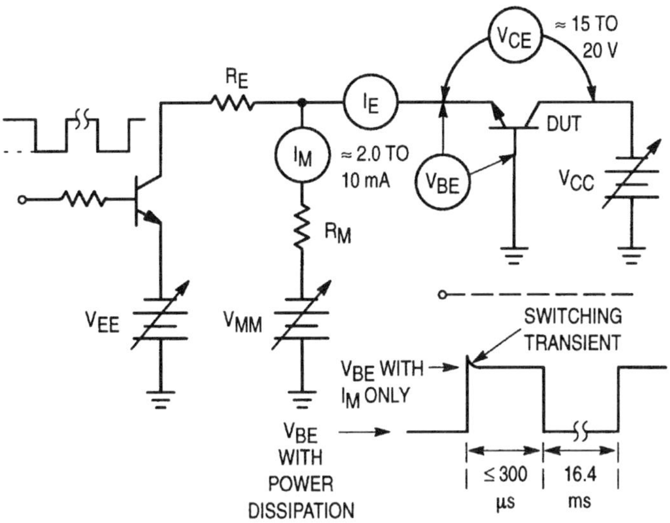

FIGURE 13.45 Example of steady-state thermal resistance test circuit for a bipolar power transistor. Power is applied for 16.4 ms and interrupted for ≤300 μs to measure the V_{BE}.

Junction Temperature Measurement

In a semiconductor, the change in temperature is directly related to the power dissipated through the thermal resistance. The steady-state dc thermal resistance junction-to-case, $R\o_{JC}$, is defined as the temperature rise per unit power above an external reference point (typically the case). The relationship is shown in Equation 13.89 [8].

$$R\O_{JC} = \Delta T / P_D \qquad (13.89)$$

where ΔT = Junction temperature minus the case temperature (°C)
P_D = Power dissipated in the junction (W)

The semiconductor device or silicon die is typically enclosed in a package that prevents a direct measurement of the junction temperature. The junction temperature is measured indirectly by measuring the case temperature, T_C; the heatsink temperature, T_S; for those higher-power applications that require a heatsink; the ambient temperature, T_A; and a temperature-sensitive electrical parameter of the device.

The first step of the process requires calibrating the temperature-sensitive parameter. Using a bipolar power transistor as an example, the base-emitter forward voltage is measured and recorded with a low calibration current (I_M) flowing through the device that is low enough to avoid self-heating (typically between 2 mA and 10 mA) and yet sufficiently high to be in the linear range of the forward voltage curve. The procedure is performed at room and elevated temperatures, typically 100°C.

After calibration, a power switching fixture (such as Figure 13.45) is used to alternately apply and interrupt the power to the device [8]. The on portion is long (typically several milliseconds) and the off portion is short (only a few 100 μs), so the temperature of the case is stabilized and junction cooling is minimal. The transistor is operated in its active region and the power dissipation is varied by adjusting the I_E and/or V_{CE} until the junction is at the calibration temperature. This point is known by measuring V_{BE} during the time that I_M is the only current flowing.

When the V_{BE} value equals the value on the calibration curve, the junction temperature is at the calibration temperature. Measurements of V_{BE}, T_C, and I_E allow the thermal resistance for the device to

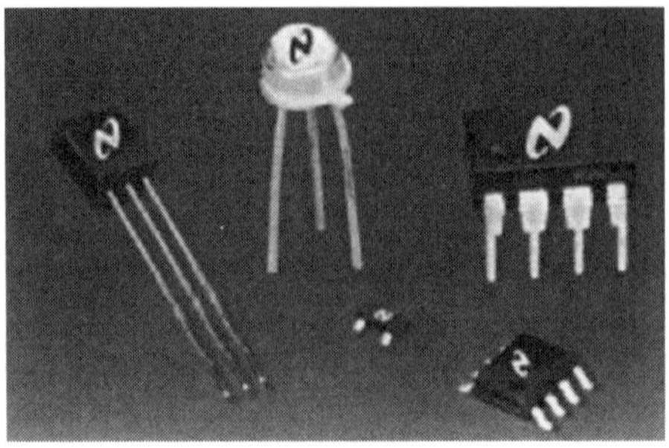

FIGURE 13.46 Plastic TO-92, TO-99 metal can, 8-lead DIP, 8-lead SOIC, and TinyPak™ SOT-23 plastic packages. (Courtesy National Semiconductor Corp.)

be calculated using Equation 13.89. Since $R\o_{JC}$ is a constant, subsequent measurements of T_C, V_{CE}, and I_E under different operating conditions can be used to calculate the junction temperature to keep the device within its safe operating range in the actual application. For devices with different electrical characteristics, such as the power MOSFETs discussed earlier, other parameters that have a linear relationship to temperature are used for calibration and measurement.

Semiconductor Temperature Sensor Packaging

Temperature sensors that are manufactured using semiconductor technology are typically packaged in packages common to the semiconductor industry. These include metal can (TO-99), ceramic, and more commonly available plastic (SOT-23, 8-lead DIP, TO- 92, 8-lead SOIC, etc.) packages. These packages are designed for circuit board solder attachment that can be either through-hole or surface-mount technology. As a result, package form factors can be considerably different from packages for temperature sensors manufactured using other technologies. Figure 13.46 shows examples of five available silicon temperature sensor packages.

Defining Terms

Bandgap reference: Forward-biased emitter junction characteristics of adjacent transistors used to provide an output voltage with zero temperature coefficient.

Die: An unpackaged semiconductor chip separated from the wafer.

Embedded sensor: A sensor included within an integrated circuit.

Integrated circuit: A multiplicity of transistors, as well as diodes, resistors, capacitors, etc., on the same silicon die.

Junction: The interface at which the conductivity type of a material changes from *p* type to *n* type.

Junction voltage: The voltage drop across a forward-biased *pn* interface in a transistor (V_{be}) or diode.

Junction temperature: The temperature of the *pn* interface in a transistor or diode.

Monolithic (integrated circuit): Constructed from a single piece of silicon.

Power IC or smart power IC: Hybrid or monolithic (semiconductor) device that is capable of being conduction-cooled, performs signal conditioning, and includes a power control function such as fault management and/or diagnostics.

Self-heating: Temperature rise within a (semiconductor) device caused by current flowing in the device.

Soft short: An excessive load condition that causes excessive temperature but is below the current limit of a device.

Thermal resistance: The steady-state dc thermal resistance junction-to-case, $R\o_{JC}$, is the temperature rise per unit power above an external reference point (typically the case).

TinyPak: A trademark of National Semiconductor Corp.

References

1. J. Carr, *Sensors and Circuits,* Englewood Cliffs, NJ, PTR Prentice-Hall, 1993.
2. *Pressure Sensor Device Data* DL200/D Rev. 1, Phoenix, AZ, Motorola, 1994.
3. K. Lacanette, "Silicon Temperature Sensors: Theory and Applications," *Measurements and Control,* pp. 120-126, April 1996.
4. R. Wegner and H. Hulsemann, "New Family of Monolithic Temperature Sensor and Controller Circuits Present Challenges In Maintaining Temperature Measurement Accuracy," *Proceedings of Sensors Expo West,* Anaheim, CA, Feb. 8-10, 1994.
5. W. Schweber, "Temperature sensors fill different needs," *EDN,* p. 20, 3/14/96.
6. "Digital Thermometer IC simplifies distributed sensing," *Electronic Products,* p. 56, 12/95.
7. J. Williams, "High Performance Signal Conditioning for Transducers," *Proceedings of Sensors Expo West,* San Jose, CA, March 2-4, 1993.
8. *TMOS Power MOSFET Transistor Data* DL135/D Rev. 4, Phoenix, AZ, Motorola Semiconductor.
9. R. K. Jurgen (ed.), *Automotive Electronics Handbook,* New York, McGraw-Hill, 1994.
10. R. Frank, *Understanding Smart Sensors,* Boston, MA, Artech House, 1995.

13.6 Infrared Thermometers

Jacob Fraden

Thermal Radiation: Physical Laws

In any material object, every atom and every molecule exist in perpetual motion. When an atom moves, it collides with other atoms and transfers to them part of its kinetic energy, thus losing some of its own energy in this perpetual bouncing. On the other hand, an atom having a smaller kinetic energy, after a collision gains some energy. Afterward, the material body consisting of such agitated atoms reaches the energetic equilibrium where all atoms, while not vibrating with exactly the same intensity, still can be described by an average kinetic energy. Such an average kinetic energy of agitated particles is represented by the *absolute temperature,* which is measured in degrees kelvin. In other words, what is commonly called temperature, is a measure of the atomic motion.

According to the laws of electrodynamics, a moving electric charge (all atoms are made of electric charges) is associated with a variable electric field. The field, in turn, produces an alternating magnetic field. And again, when the magnetic field changes, it results in a coupled with it variable electric field, and so on. Thus, a moving particle becomes a source of electromagnetic field that propagates outwardly with the speed of light and is called *thermal radiation.* This radiation is governed by the laws of optics — it can be reflected, filtered, focused, etc. Also, it can be used to measure the object's temperature.

Electromagnetic waves originating from mechanical movement of particles can be characterized by their intensities and wavelengths. Both of these characteristics relate to temperature; that is, the hotter the object, the shorter the wavelength. Very hot objects radiate electromagnetic energy in the visible portion of the spectrum — wavelengths between 0.4 μm (blue) and 0.7 μm (red). For example, a filament in an incandescent lamp is so hot that it radiates bright visible light. If such a lamp is controlled by a dimmer, the light intensity can be reduced by turning the knob and observing that the dimmed light becomes more yellowish, reddish, and finally disappears. Near the end of the dimmer control, the filament is still quite hot, yet one cannot see it because it emanates light in the invisible infrared spectral range — wavelengths greater than 0.8 μm. Cooler objects radiate light in the near-, mid-, and far-infrared spectral ranges, which one cannot see. For example, electromagnetic radiation emanating from human skin primarily is situated at wavelengths between 5 μm and 15 μm — in the mid- and far-infrared ranges and is not visible to human eyes; otherwise, we all would glow in the dark (sick people with fever would look even brighter). If one imagines that all atomic vibration stopped for some mysterious reason, no electromagnetic radiation would be emanated. Such an imaginable but impossible event is characterized by infinitely cold temperature, which is called *absolute zero.*

Because temperature is a measure of the average atomic kinetic energy, it is logical to assume that one can determine the object's temperature by measuring the intensity of the emanated electromagnetic radiation or its spectral characteristics. This presumption is the basis for noncontact temperature measurements that are known by various names, depending on the application: infrared thermometry, optical pyrometry, radiation thermometry, etc. *Pyrometry* is derived from the Greek word *pyr*, which means fire, and thus is more appropriate for measuring hot temperatures. For lower temperatures, *infrared thermometry* is used interchangeably with term *radiation thermometry.*

Planck's Law

A relationship between the magnitude of radiation at a particular wavelength λ and absolute temperature T is rather complex and is governed by Planck's law, which was discovered in 1901. It establishes radiant flux density W_λ as power of electromagnetic radiation per unit of wavelength:

$$W_\lambda = \frac{\varepsilon(\lambda) C_1}{\pi \lambda^5 \left(e^{C_2/\lambda T} - 1\right)} \tag{13.90}$$

where $\varepsilon(\lambda)$ = Emissivity of an object
C_1 = 3.74×10^{-12} Wcm^2 and C_2 = 1.44 cmK are Constants
e = Base of natural logarithms

Spectral densities for different temperatures are shown in Figure 13.47.

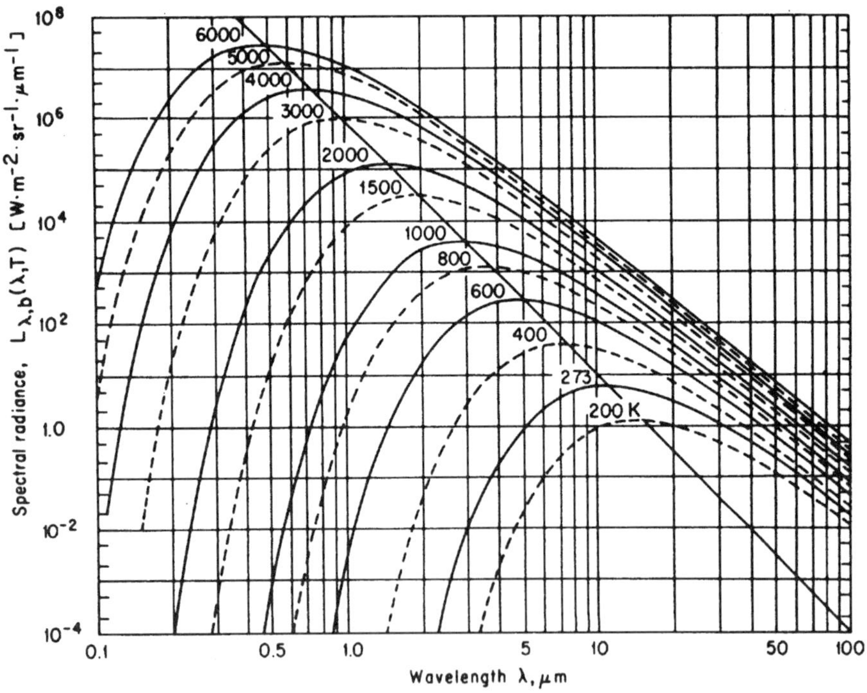

FIGURE 13.47 Spectral densities calculated within a solid angle of 1 steradian for blackbody source ($\varepsilon = 1$) radiation toward infinitely cold space (at absolute zero). (From J. C. Richmond and D.P. DeWitt (Eds). Application of radiation thermometry. ASTM PCN 04-895000-40. Used with permission.)

Wien's Law

Equation 13.90 does not lend itself to a simple mathematical analysis and thus is approximated by a simplified version, which is known as Wien's law:

$$W_\lambda = \frac{C_1}{\pi} \varepsilon(\lambda) \lambda^{-5} e^{\frac{C_2}{\lambda T}} \tag{13.91}$$

Because temperature is a statistical representation of an average kinetic energy, it determines the highest probability for the particles to vibrate with a specific frequency and to have a specific wavelength. This most probable wavelength follows from Wien's law by equating to zero a first derivative of Equation 13.91. The result of the calculation is a wavelength near which most of the radiant power is concentrated when light is emanated toward infinitely cold space at absolute zero:

$$\lambda_m = \frac{2898}{T} \tag{13.92}$$

where λ_m is in μm and T in kelvin. Wien's law states that the higher the temperature, the shorter the wavelength. This formula also defines the midpoint of spectral response of a pyrometer or infrared thermometer.

Stefan–Boltzmann Law

Theoretically, a thermal radiation bandwidth is infinitely wide. Yet, most of the emanated power is situated within quite a limited bandwidth. Also, one must account for the filtering properties of the real world windows used in instruments. In order to determine the total radiated power limited within a particular bandwidth, Equation 13.90 or 13.91 must be integrated within the limits from λ_1 to λ_2:

$$\Phi_{bo} = \frac{1}{\pi} \int_{\lambda_1}^{\lambda_2} \frac{\varepsilon(\lambda) C_1 \lambda^{-5}}{e^{C_2/\lambda T} - 1} \tag{13.93}$$

This integral can be resolved only numerically or by approximation. For a narrow bandwidth (λ_1 and λ_2 are close to one another), the solution can be approximated by:

$$\Phi_{bo} = kT^x \tag{13.94}$$

where k is constant and $x \approx (12/\lambda_2)(1200/T)$. For example, in the visible portion of spectrum at $\lambda_2 \approx$ 0.7 μm and for temperatures near 2000 K, the approximation is a 10th-order parabola. An approximation for a very broad bandwidth ($\lambda_2 \to \infty$ or practically, when the range between λ_1 and λ_2 embrace well over 50% of the total radiated power) is a 4th-order parabola, which is known as the *Stefan–Boltzmann law*:

$$\Phi_{bo} = A\varepsilon\sigma T^4 \tag{13.95}$$

where $\sigma = 5.67 \times 10^{-8}$ W/m^{2}K^4 (Stefan–Boltzmann constant), and ε is assumed to be wavelength independent. It is seen that with an increase in temperature, the intensity of electromagnetic radiation Φ_{bo} grows very fast due to the 4th power of T.

Kirchhoff's Law

While wavelengths of the radiated light are temperature dependent, the magnitude of radiation also is a function of the surface property. That property is called *emissivity*, ε. Emissivity is measured on a scale

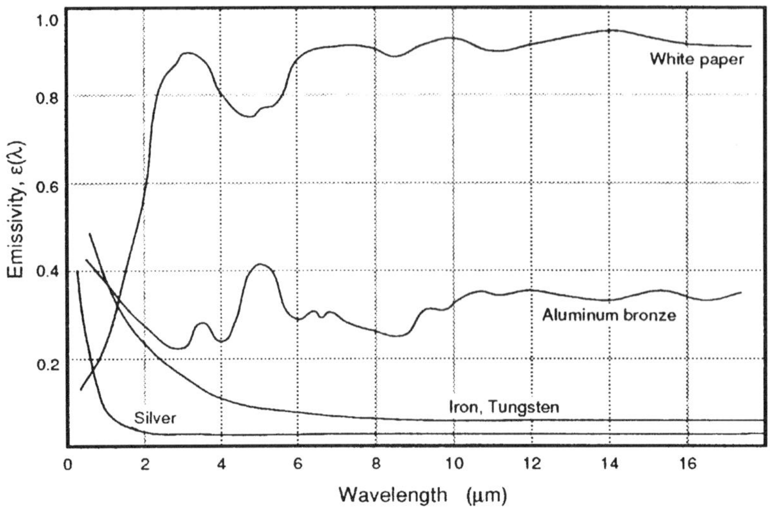

FIGURE 13.48 Wavelength dependence of emissivities. (From J. Fraden, *Handbook of Modern Sensors*, AIP Press, 1997. Used with permission.)

from 0 to 1. It is a ratio of electromagnetic flux that is emanated from a surface to the flux that would be emanated from the ideal emitter having the same temperature. *Reflectivity*, ρ, and *transparency*, γ, also on a scale from 0 to 1, show what portion of incident light is reflected and passed through, respectively. There is a fundamental equation that connects these three characteristics:

$$\varepsilon+\gamma+\rho=1 \tag{13.96}$$

Equation (13.96) indicates that any one of the three properties of the material can be changed only at the expense of the others. As a result, for an opaque object ($\gamma = 0$), reflectivity ρ and emissivity ε are connected by a simple relationship: $\rho = 1 - \varepsilon$, which, for example, makes a mirror a good reflector but a poor emitter.

Emissivity

The emissivity of a material is a function of its dielectric constant and, subsequently, refractive index *n*. It should be noted, however, that emissivity is generally wavelength dependent (Figure 13.48). For example, a white sheet of paper is very much reflective in the visible spectral range and emits no visible light. In the far-infrared spectral range, its reflectivity is low and emissivity is high (about 0.92), thus making paper a good emitter of thermal radiation. However, for many practical purposes in infrared thermometry, emissivity can be considered constant.

For nonpolarized far-infrared light in normal direction, emissivity can be expressed by the equation:

$$\varepsilon=\frac{4n}{\left(n+1\right)^{2}} \tag{13.97}$$

As a rule, emissivities of dielectrics are high and of metals are low. Due to the high emissivity of dielectrics, they lend themselves to easy and accurate noncontact temperature measurement. On the other hand, such measurements from nonoxidized metals are difficult, due to small amounts of emanated infrared flux. Table 13.17 gives typical emissivities of some opaque materials in a temperature range between 0°C and 100°C.

TABLE 13.17 Typical Emissivities of Different Materials (from 0 to 100°C)

Material	Emissivity	Material	Emissivity
Blackbody (ideal)	1.00	Green leaves	0.88
Cavity radiator	0.99–1.00	Ice	0.96
Aluminum (anodized)	0.70	Iron or steel (rusted)	0.70
Aluminum (oxidized)	0.11	Nickel (oxidized)	0.40
Aluminum (polished)	0.05	Nickel (unoxidized)	0.04
Aluminum (rough surface)	0.06–0.07	Nichrome (80Ni-20Cr) (oxidized)	0.97
Asbestos	0.96	Nichrome (80Ni-20Cr) (polished)	0.87
Brass (dull tarnished)	0.61	Oil	0.80
Brass (polished)	0.05	Silicon	0.64
Brick	0.90	Silicone rubber	0.94
Bronze (polished)	0.10	Silver (polished)	0.02
Carbon-filled latex paint	0.96	Skin (human)	0.93–0.96
Carbon lamp black	0.96	Snow	0.85
Chromium (polished)	0.10	Soil	0.90
Copper (oxidized)	0.6–0.7	Stainless steel (buffed)	0.20
Copper (polished)	0.02	Steel (flat rough surface)	0.95–0.98
Cotton cloth	0.80	Steel (ground)	0.56
Epoxy resin	0.95	Tin plate	0.10
Glass	0.95	Water	0.96
Gold	0.02	White paper	0.92
Gold-black	0.98–0.99	Wood	0.93
Graphite	0.7–0.8	Zinc (polished)	0.04

Source: J. Fraden, *Handbook of Modern Sensors,* AIP Press, 1997. Used with permission.

Unlike most solid bodies, gases in many cases are transparent to thermal radiation. When they absorb and emit radiation, they usually do so only in certain narrow spectral bands. Some gases, such as N_2, O_2, and others of nonpolar symmetrical molecular structure, are essentially transparent at low temperatures, while CO_2, H_2O, and various hydrocarbon gases radiate and absorb to an appreciable extent. When infrared light enters a layer of gas, its absorption has an exponential decay profile, governed by *Beer's law*:

$$\frac{\Phi_x}{\Phi_0} = e^{-\alpha_\lambda x} \tag{13.98}$$

where Φ_0 = Incident thermal flux
Φ_x = Flux at thickness x
α_λ = Spectral coefficient of absorption

The above ratio is called a monochromatic transmissivity γ_λ at a specific wavelength λ. If gas is nonreflecting, then its emissivity is defined as:

$$\varepsilon_\lambda = 1 - \gamma_\lambda = 1 - e^{-\alpha_\lambda x} \tag{13.99}$$

It should be emphasized that since gases absorb only in narrow bands, emissivity and transmissivity (transparency) must be specified separately for any particular wavelength. For example, water vapor is highly absorptive at wavelengths of 1.4, 1.8, and 2.7 µm, and is very transparent at 1.6, 2.2, and 4 µm.

All non-metals are very good diffusive emitters of thermal radiation with a remarkably constant emissivity defined by Equation 13.97 within a solid angle of about ±70°. Beyond that angle, emissivity begins to decrease rapidly to zero with the angle approaching 90°. Near 90°, emissivity is very low. A typical calculated graph of the directional emissivity of non-metals into air is shown in Figure 13.49A. It should be emphasized that the above considerations are applicable only to wavelengths in the far

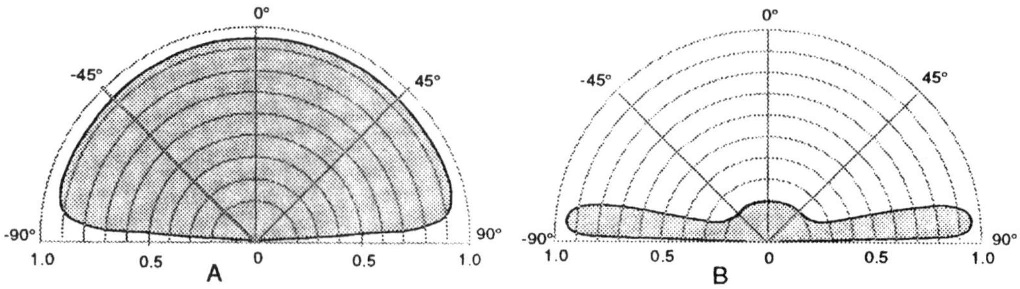

FIGURE 13.49 Spatial emissivities for non-metal (A) and a polished metal (B). (From J. Fraden, *Handbook of Modern Sensors,* AIP Press, 1997. Used with permission.)

infrared spectral range and are not true for visible light, since emissivity of thermal radiation is a result of electromagnetic effects that occur at an appreciable depth.

Metals behave quite differently. Their emissivities greatly depend on surface finish. Generally, polished metals are poor emitters within the solid angle of ±70°, while their emissivity increases at larger angles (Figure 13.49B). Oxidized metals start behaving more and more like dielectrics with increasing thickness of oxides.

Blackbody

By definition, the highest possible emissivity is unity. It is attributed to the so-called blackbody — an ideal emitter of electromagnetic radiation. If the object is opaque ($\gamma = 0$) and nonreflective ($\rho = 0$) according to Equation 13.96, it becomes an ideal emitter and absorber of electromagnetic radiation. The name blackbody implies its appearance at normal room temperatures — indeed, it does look black because it is not transparent and not reflective at any wavelength. In reality, a blackbody does not exist, and any object with a nonunity emissivity often is called a *graybody.* A practical blackbody (ε is about 0.99 or higher) is an essential tool for calibrating and verifying the accuracy of infrared thermometers.

Cavity Effect

To make a practical blackbody, a cavity effect is put to work. The effect appears when electromagnetic radiation is measured from a cavity of an object. For this purpose, a cavity means an opening in a concave void of a generally irregular shape whose inner wall temperature is uniform over an entire surface. The emissivity of a cavity opening dramatically increases, approaching unity at any wavelength, as compared with a flat surface. The cavity effect is especially pronounced when its inner walls have relatively high emissivity. Consider a non-metal cavity. All non-metals are diffuse emitters. Also, they are diffuse reflectors. It is assumed that the temperature and surface emissivity of the cavity are homogeneous over an entire area. The ideal emitter (blackbody) would emanate from area a, the infrared photon flux $\Phi_0 = a\sigma T_b^4$. However, the object has the actual emissivity ε_b and, as a result, the flux radiated from that area is smaller: $\Phi_r = \varepsilon_b \Phi_0$ (Figure 13.50). Flux emitted by other parts of the object toward area a is also equal to Φ_r (because the object is thermally homogeneous, one can disregard the spatial distribution of flux). A substantial portion of that incident flux Φ_r is absorbed by the surface of area a, while a smaller part is diffusely reflected:

$$\Phi_\rho = \rho\Phi_r = (1 - \varepsilon_b)\varepsilon_b \Phi_0 \tag{13.100}$$

and the combined radiated and reflected flux from area a is:

$$\Phi = \Phi_r + \Phi_\rho = \varepsilon_b \Phi_0 + (1 - \varepsilon_b)\varepsilon_b \Phi_0 = (2 - \varepsilon_b)\varepsilon_b \Phi_0. \tag{13.101}$$

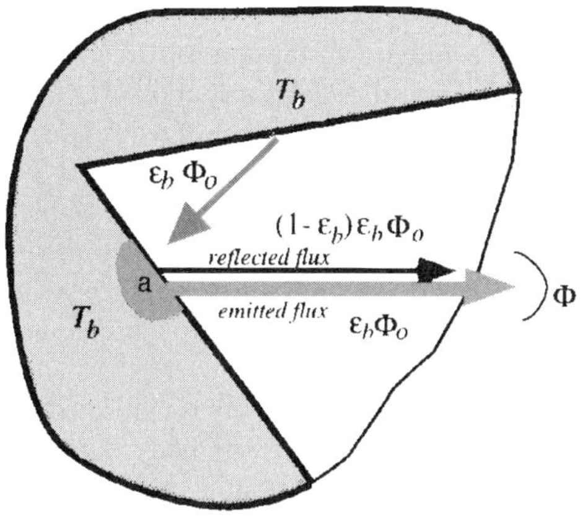

FIGURE 13.50 Cavity effect enhances emissivity. Note that $\varepsilon_e > \varepsilon_b$. (From J. Fraden, *Handbook of Modern Sensors,* AIP Press, 1997. Used with permission.)

As a result, for a single reflection, the *effective emissivity* can be expressed as:

$$\varepsilon_e = \frac{\Phi}{\Phi_0} = \left(2 - \varepsilon_b\right)\varepsilon_b \tag{13.102}$$

It follows from the above that due to a single reflection, a perceived (effective) emissivity of a cavity at its opening (aperture) is equal to the surface emissivity magnified by a factor of $(2 - \varepsilon_b)$. Of course, there may be more than one reflection of radiation before it exits the cavity. In other words, the incident on area *a* flux could already be a result of a combined effect from the reflectance and emittance at other parts of the cavity's surface. For a cavity effect to work, the effective emissivity must be attributed only to the cavity opening (aperture) from which radiation escapes. If a sensor is inserted into the cavity facing its wall directly, the cavity effect could disappear and the emissivity would be close to that of a wall surface.

Practical Blackbodies

A practical blackbody can be fabricated in several ways. Copper is the best choice for the cavity body material, thanks to its high thermal conductivity, which helps to equalize temperatures of the cavity walls. As an example, Figure 13.51 shows two practical blackbodies. One is a solid-state blackbody having

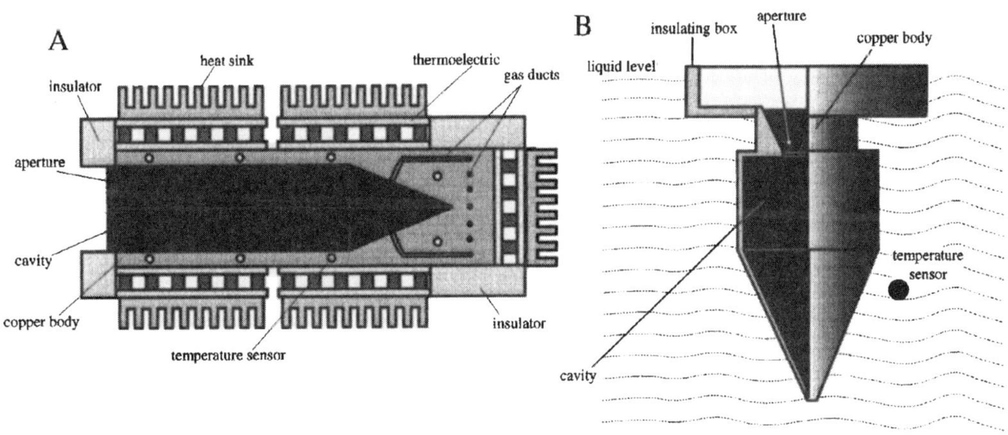

FIGURE 13.51 (A): Solid-state blackbody with thermoelectric elements. (B): Immersed blackbody.

thermoelectric elements (heat pumps) that provide either heating or cooling to the cavity body. The embedded temperature sensors are connected to the control circuit (not shown). The function of the multiple temperature sensors is to monitor thermal distribution over the length of the cavity. The inner shape of the cavity is partly conical to increase the number of reflections, and the entire surface is treated to provide it with as high emissivity as possible, typically over 0.9. This type of a blackbody has a relatively wide aperture that potentially can act as an entry for undesirable ambient air. The air can disturb the thermal uniformity inside the cavity, resulting in excessively high uncertainty of the radiated flux. To reduce this problem, the cavity is filled with dry air or nitrogen, which is continuously pumped in through the gas ducts. Before entering the cavity, the gas passes through the narrow channels inside the cavity and acquires the temperature of the blackbody.

Another example is an immersed blackbody. The cavity is fabricated of copper or aluminum and has relatively thin walls (a few millimeters). The entire cavity body is immersed into a stirred liquid bath, the temperature of which is precisely controlled by heating/cooling devices. The liquid assures uniform temperature distribution around the cavity with a typical thermal instability on the order of ±0.02°C. The inner surface of the cavity is coated with high-emissivity paint. The aperture of the cavity is relatively small. The ratio of the inner surface of the cavity to the aperture area should be at least 100, and preferably close to 1000.

Detectors for Thermal Radiation

Classification

Generally speaking, there are two types of sensors (detectors) known for their capabilities to respond to thermal radiation within the spectral range from the near-infrared to far-infrared; that is, from approximately 0.8 μm to 40 μm. The first type is quantum detectors; and the second type is thermal detectors. The latter, in turn, can be subdivided into passive (PIR) and active (AFIR) detectors.

Quantum Detectors

Quantum detectors (photovoltaic and photoconductive devices) relay on the interaction of individual photons with a crystalline lattice of semiconductor materials. Their operations are based on the photoeffect that was discovered by Einstein, and brought him the Nobel Prize. In 1905, he made a remarkable assumption about the nature of light: that at least under certain circumstances, its energy was concentrated into localized bundles, later named photons. The energy of a single photon is given by:

$$E = h\nu, \tag{13.103}$$

where ν = frequency of light
$h = 6.63 \times 10^{-34}$ J × s (or 4.13×10^{-15} eV s) = Planck's constant, derived on the basis of the wave theory of light

When a photon strikes the surface of a conductor, it can result in the generation of a free electron.

The periodic lattice of crystalline materials establishes allowed energy bands for electrons that exist within that solid. The energy of any electron within the pure material must be confined to one of these energy bands, which can be separated by gaps or ranges of forbidden energies.

In isolators and semiconductors, the electron must first cross the energy bandgap in order to reach the conduction band and the conductivity is therefore many orders of magnitude lower. For isolators, the bandgap is usually 5 eV or more; whereas, for semiconductors, the gap is considerably less.

When a photon of frequency ν_1 strikes a semiconductive crystal, its energy will be high enough to separate the electron from its site in the valence band and push it through the bandgap into a conduction band at a higher energy level. In that band, the electron is free to serve as a current carrier. The deficiency of an electron in the valence band creates a hole that also serves as a current carrier. This is manifested

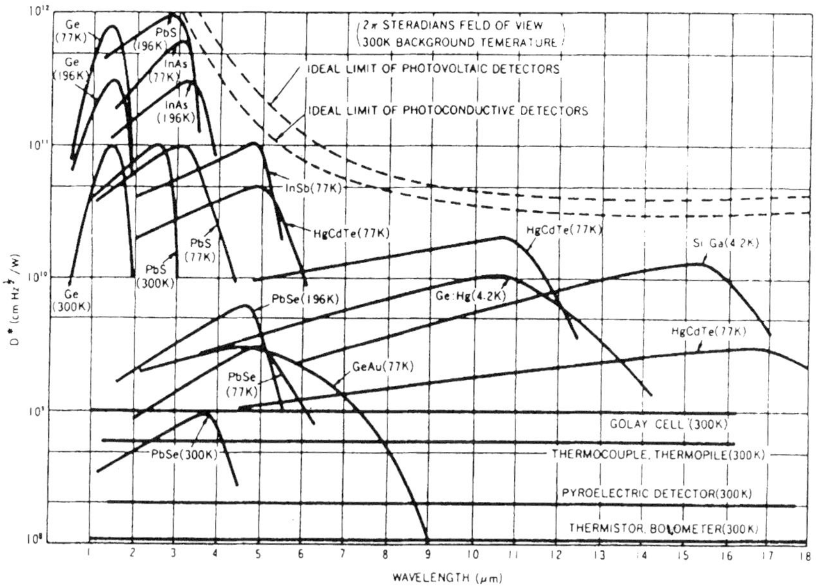

FIGURE 13.52 Operating ranges for some infrared detectors.

in the reduction of specific resistivity of the material. The energy gap serves as a photon energy threshold, below which the material is not light sensitive.

For measurements of objects emanating photons in the range of 2 eV or greater, quantum detectors having room temperature are generally used. For the smaller energies (longer wavelengths), narrower bandgap semiconductors are required. However, even if a quantum detector has a sufficiently small energy bandgap, at room temperature, its own intrinsic noise is much greater than a photoconductive signal. Noise level is temperature dependent; therefore, when detecting long-wavelength photons, a signal-to-noise ratio can become so small that accurate measurement becomes impossible. This is the reason why, for the operation in the near- and far-infrared spectral ranges, a detector not only should have a sufficiently narrow energy gap, but its temperature must be lowered to the level where intrinsic noise is reduced to an acceptable level. Depending on the required sensitivity and operating wavelength, the following crystals are typically used for the cryogenically cooled sensors (Figure 13.52): lead sulfide (PbS), indium arsenide (InAs), germanium (Ge), lead selenide (PbSe), and mercury-cadmium-telluride (HgCdTe).

The sensor cooling allows responses to longer wavelengths and increases sensitivity. However, response speeds of PbS and PbSe become slower with cooling. Methods of cooling include dewar cooling using dry ice, liquid nitrogen, liquid helium, or thermoelectric coolers operating on the Peltier effect.

Thermal Detectors

Another class of infrared radiation detectors is called *thermal detectors.* Contrary to quantum detectors that respond to individual photons, thermal detectors respond to heat resulting from absorption of thermal radiation by the surface of a sensing element. The heat raises the temperature of the surface, and this temperature increase becomes a measure of the net thermal radiation.

The Stefan–Boltzmann law (Equation 13.95) specifies radiant power (flux) which would emanate from a surface of temperature, *T,* toward an infinitely cold space (at absolute zero). When thermal radiation is detected by a thermal sensor, the opposite radiation from the sensor toward the object must also be taken in account. A thermal sensor is capable of responding only to a net thermal flux, i.e., flux from the object minus flux from itself. The surface of the sensor that faces the object has emissivity ε_s (and,

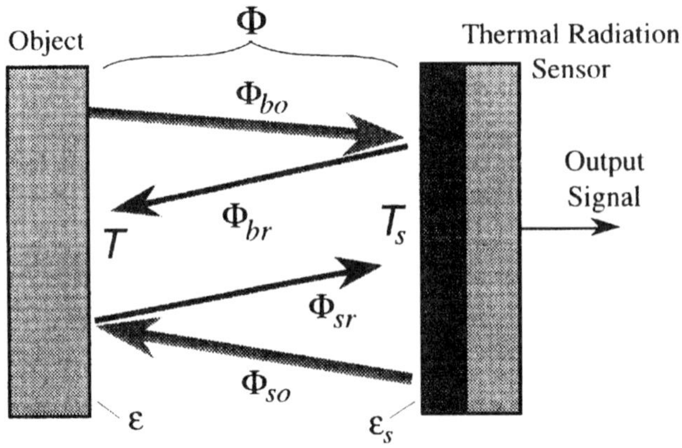

FIGURE 13.53 Heat exchange between the object and thermal radiation detector.

subsequently reflectivity $\rho_s = 1 - \varepsilon_s$). Because the sensor is only partly absorptive, the entire flux, Φ_{bo}, is not absorbed and utilized. A part of it, Φ_{ba}, is absorbed by the sensor, while another part, Φ_{br}, is reflected (Figure 13.53) back toward to object (here, it is assumed that there is 100% coupling between the object and the sensor and there are no other objects in the sensor's field of view). The reflected flux is proportional to the sensor's coefficient of reflectivity:

$$\Phi_{br} = -\rho_S \Phi_{bo} = -A\varepsilon\left(1-\varepsilon_S\right)\sigma T^4 \tag{13.104}$$

A negative sign indicates an opposite direction with respect to flux Φ_{bo}. As a result, the net flux originated from the object is:

$$\Phi_b = \Phi_{bo} + \Phi_{br} = A\varepsilon\varepsilon_S\sigma T^4 \tag{13.105}$$

Depending on its temperature T_s, the sensor's surface radiates its own net thermal flux toward the object in a similar way:

$$\Phi_s = -A\varepsilon\varepsilon_s\sigma T_s^4 \tag{13.106}$$

Two fluxes propagate in the opposite directions and are combined into a final net flux existing between two surfaces:

$$\Phi = \Phi_b + \Phi_s = A\varepsilon\varepsilon_s\sigma\left(T^4 - T_s^4\right) \tag{13.107}$$

This is a mathematical model of a net thermal flux that is converted by a thermal sensor into the output signal. It establishes a connection between thermal power, Φ, absorbed by the sensor, and the absolute temperatures of the object and the sensor. It should be noted that since the net radiation exists between the two bodies, the spectral density will have the maximum not described by Equation 13.92; but depending on the temperature gradient, it will be somewhat shifted toward the shorter wavelengths.

Dynamically, the temperature T_s of a thermal element in a sensor, in general terms, can be described by the first-order differential equation:

$$cm\frac{dT_s}{dt} = P - P_L - \Phi \tag{13.108}$$

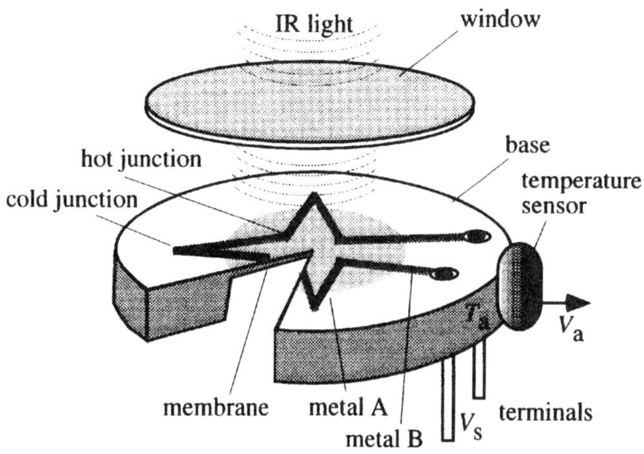

FIGURE 13.54 Thermopile sensor. "Hot" junctions are deposited on a membrane and "cold" junctions on the supporting ring.

where *P* is the power supplied to the element from a power supply or an excitation circuit (only in AFIR sensors; see below); P_L is a nonradiative thermal loss attributed to thermal conduction and convection; *m* and *c* are the sensor's mass and specific heat, respectively; and $\Phi = \Phi_\eta + \Phi_b$ is the net radiative thermal flux. We select a positive sign for power *P* when it is directed toward the element.

In the PIR detector (thermopiles, pyroelectric, and bolometers), no external power is supplied ($P = 0$); hence, the speed response depends only on the sensor's thermal capacity and heat loss, and is a first-order function that is characterized by a thermal time constant τ_T.

Thermopile Sensors.

Thermopiles belong to a class of PIR detectors. Their operating principle is the same as that of thermocouples. In effect, a thermopile can be defined as serially connected thermocouples. Originally, it was invented by Joule to increase the output signal of a thermoelectric sensor; he connected several thermocouples in series and thermally joined together their hot junctions. Presently, thermopiles have a different configuration. Their prime application is detection of thermal radiation.

A cut-out view of a thermopile sensor is shown in Figure 13.54. The sensor consists of a base having a relatively large thermal mass, which is the place where the "cold" junctions are positioned. The base can be thermally coupled with a reference temperature sensor or attached to a thermostat having a known temperature. The base supports a thin membrane whose thermal capacity and thermal conductivity are small. The membrane is the surface where the "hot" junctions are positioned.

The best performance of a thermopile is characterized by high sensitivity and low noise, which can be achieved by the junction materials having high thermoelectric coefficient, low thermal conductivity, and low volume resistivity. Besides, the junction pairs should have thermoelectric coefficients of opposite signs. This dictates the selection of materials. Unfortunately, most of metals having low resistivity (i.e., gold, copper, silver) have only very poor thermoelectric coefficients. The higher resistivity metals (especially bismuth and antimony) possess high thermoelectric coefficients and they are the prime selection for designing thermopiles. By doping these materials with Se and Te, the thermoelectric coefficient has been improved up to 230 $\mu V\ K^{-1}$ [1].

Methods of construction of metal junction thermopiles can differ to some extent, but all incorporate vacuum deposition techniques and evaporation masks to apply the thermoelectric materials, such as bismuth and antimony on thin substrates (membranes). The number of junctions varies from 20 to several hundreds. The "hot" junctions are often blackened (e.g., with goldblack or organic paint) to improve their absorptivity of the infrared radiation. A thermopile is a dc device with an output voltage that follows its "hot" junction temperature quite well. It can be modeled as a thermal flux-controlled voltage source that is connected in series with a fixed resistor. The output voltage V_s is nearly proportional to the incident radiation.

An efficient thermopile sensor can be designed using a semiconductor rather than double-metal junctions [2]. The thermoelectric coefficients for crystalline and polycrystalline silicon are very large and the volume resistivity is relatively low. The advantage of using silicon is in the possibility of employing standard IC processes. The resistivity and the thermoelectric coefficients can be adjusted by the doping concentration. However, the resistivity increases much faster, and the doping concentration must be carefully optimized for the high sensitivity–low noise ratios. Semiconductor thermopile sensors are produced with a micromachining technology by EG&G Heimann Optoelectronics GmbH (Wiesbaden, Germany) and Honeywell (Minneapolis, MN).

Pyroelectrics.

Pyroelectric sensors belong to a class of PIR detectors. A typical pyroelectric sensor is housed in a metal TO-5 or TO-39 for better shielding and is protected from the environment by a silicon or any other appropriate window. The inner space of the can is often filled with dry air or nitrogen. Usually, two sensing elements are oppositely, serially or in parallel, connected for better compensation of rapid thermal changes and mechanical stresses resulting from acoustic noise and vibrations [3].

Bolometers.

Bolometers are miniature RTDs or thermistors, which are mainly used for measuring rms values of electromagnetic signals over a very broad spectral range from microwaves to near-infrared. An external bias power is applied to convert resistance changes to voltage changes. For the infrared thermometers, the bolometers are often fabricated in the form of thin films having relatively large area. The operating principle of a bolometer is based on a fundamental relationship between the absorbed electromagnetic signal and dissipated power [3].

The sensitivity of the bolometer to the incoming electromagnetic radiation can be defined as [4]:

$$\beta = \frac{1}{2}\varepsilon\alpha_0 \sqrt{\frac{R_0 Z_T \Delta T}{\left(1+\alpha_0 \Delta T\right)\left[1+\left(\omega\tau\right)^2\right]}} \tag{13.109}$$

where $\alpha = (dR/dT)/R$ = TCR (temperature coefficient of resistance) of the bolometer
ε = Surface emissivity
Z_T = Bolometer thermal resistance, which depends on its design and the supporting structure
τ = Thermal time constant, which depends on Z_T and the bolometer's thermal capacity
ω = Angular frequency
ΔT = Bolometer's temperature increase

Bolometers are relatively slow sensors and are used primarily when no fast response is required. For thermal imaging, bolometers are available as two-dimensional arrays of about 80,000 sensors [5].

Active Far-Infrared Sensors.

In the active far-infrared (AFIR) sensor, a process of measuring thermal radiation flux is different from previously described passive (PIR) detectors. Contrary to a PIR sensing element — the temperature of which depends on both the ambient and the object's temperatures — the AFIR sensor's surface is actively controlled by a special circuit to have a defined temperature T_s that, in most applications, is maintained constant during an entire measurement process [6]. To control the sensor's surface temperature, electric power P is provided by a control (or excitation) circuit (Figure 13.55). To regulate T_s, the circuit measures the element's surface temperature and compares it with an internal reference.

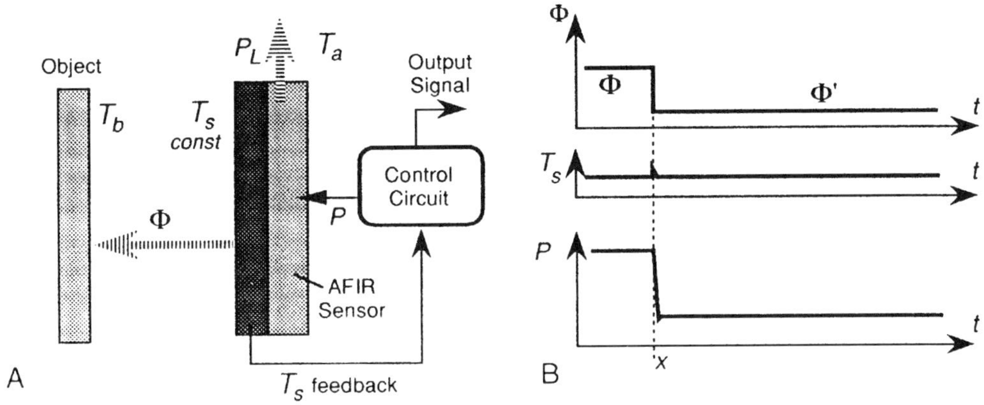

FIGURE 13.55 (A) AFIR element radiates thermal flux Φ_η toward its housing and absorbs flux Φ_b from the object. (B) Timing diagrams for radiative flux, surface temperature, and supplied power. (From J. Fraden, *Handbook of Modern Sensors,* AIP Press, 1997. Used with permission.)

Obviously, the incoming power maintains T_s higher than ambient, practically by just several tenths of a degree Celsius. Since the element's temperature is above ambient, the sensing element loses thermal energy toward its surroundings, rather than passively absorbing it, as in a PIR detector. Part of the heat loss is in the form of a thermal conduction; part is a thermal convection; and the other part is thermal radiation. The third part is the one that must be measured. Of course, the radiative flux is governed by the fundamental Stefan–Boltzmann law for two surfaces (Equation 13.99).

Some of the radiation power goes out of the element to the sensor's housing, while the other is coming from the object (or goes to the object). What is essential is that the net thermal flow (coductive + convective + radiative) must always come out of the sensor; that is, it must have a negative sign.

In the AFIR element, after a warm-up period, the control circuit forces the element's surface temperature T_s to stay constant; thus,

$$\frac{dT_s}{dt} = 0 \tag{13.110}$$

and Equation 13.108 becomes algebraic:

$$P = P_L + \Phi \tag{13.111}$$

It follows from the above that, under idealized conditions, its response does not depend on thermal mass and is not a function of time, meaning that practical AFIR sensors are quite fast. If the control circuit is highly efficient, since P_L is constant at given ambient conditions, the electronically supplied power P should track changes in the radiated flux Φ with high fidelity. Nonradiative loss P_L is a function of ambient temperature T_a and a loss factor α_s:

$$P_L = \alpha_s \left(T_s - T_a\right) \tag{13.112}$$

To generate heat in the AFIR sensor, it may be provided with a heating element having electrical resistance R. During the operation, electric power dissipated by the heating element is a function of voltage V across that resistance

$$P = V^2 / R \tag{13.113}$$

Substituting Equations 13.107, 13.112, and 13.113 into Equation 13.111, and assuming that $T = T_b$ and $T_s > T_a$, after simple manipulations, the object's temperature can be presented as function of voltage V across the heating element:

$$T_b = \sqrt{T_s^4 - \frac{1}{A\sigma\varepsilon_s\varepsilon_b}\left(\frac{V^2}{R} - \alpha_s \Delta T\right)} \tag{13.114}$$

where ΔT is the constant temperature increase above ambient. Coefficient α_s has a meaning of thermal conductivity from the AFIR detector to the environment (housing).

One way to fabricate an AFIR element is to use a discrete thermistor having a relatively large surface area (3 mm^2 to 10 mm^2) and operating in a self-heating mode. Electric current passing through the thermistor results in a self-heating effect that elevates the thermistor's temperature above ambient. In effect, the thermistor operates as both the heater and a temperature sensor.

Contrary to a PIR detector, an AFIR sensor is active and can generate a signal only when it works in orchestra with a control circuit. A control circuit must include the following essential components: a reference to preset a controlled temperature, an error amplifier, and a driver stage for the heater. In addition, it may include an *RC* network for correcting a loop response function and for stabilizing its operation; otherwise, an entire system could be prone to oscillations [7].

It can be noted that an AFIR sensor, along with its control circuit, is a direct converter of thermal radiative power into electric voltage and a quite efficient one. Its typical responsivity is in the range of 3000 V W^{-1}, which is much higher as compared with a thermopile, whose typical responsivity is in the range of 100 V W^{-1}. More detailed description of an AFIR sensor can be found in [3, 6].

Pyrometers

Disappearing Filament Pyrometer

Additional names for the disappearing filament pyrometer include: *optical pyrometer* and *monochromatic-brightness radiation thermometer.* This type of pyrometer is considered the most accurate radiation thermometer for temperatures over 700°C. This limitation is a result of human-eye sensitivity within a specific wavelength. The operating principle of this thermometer is based on Planck's law (Equation 13.90 and Figure 13.47) which states that intensity and color of the surface changes with temperature. The idea behind the design is to balance a radiation from an object having a known temperature against unknown temperature from a target. The pyrometer has a lens through which the operator views the target (Figure 13.56A). An image of a tungsten filament is superimposed on the image of the target. The filament

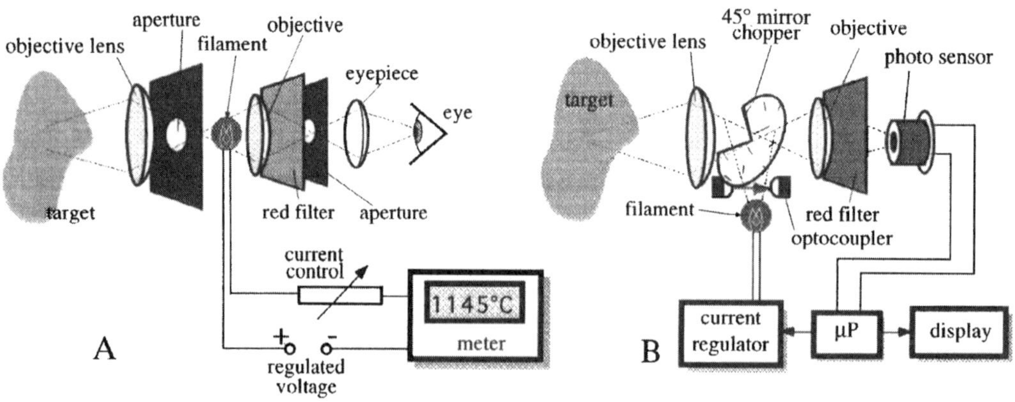

FIGURE 13.56 Disappearing filament optical pyrometer: (A): Manual, and (B): automatic versions.

is warmed up by electric current to glow. During the calibration, the relationship between the current and the filament temperature was established by measuring brightness of a blackbody of known temperature. The operator views the target through the eyepiece and manually adjusts the heating current to the level when an image of the glowing filament visible in the foreground disappears — that is, when both the target and the filament have the same brightness and color. A color component complicates the measurement somewhat; so to remove this difficulty, a narrow-band red filter ($\lambda_f = 0.65$ μm) is inserted in front of an eyepiece. Therefore, the operator has to balance only the brightness of an image visible in red color. Another advantage of the filter is that the emissivity ε_{bf} of the target needs to be known only at λ_f. The error in temperature measurement is given by [8]:

$$\frac{dT_b}{T_b} = -\frac{\lambda_f T_b}{C_2}\frac{d\varepsilon_{bf}}{\varepsilon_{bf}} \tag{13.115}$$

Thus, for a target at $T_b = 1000$ K, a 10% uncertainty in knowing emissivity of the target at the filter's wavelength results in only ±0.45% (±4.5 K) uncertainty in measured temperature.

The instrument can be further improved by removing an operator from the measurement loop. Figure 13.56B shows an automatic version of the pyrometer where a rotating mirror tilted by 45° has a removed sector that allows the light from a target to pass through to the photosensor. Such a mirror serves as a chopper, which alternately sends light to a photosensor, either from a target or from the filament. The microprocessor (μP) adjusts current through the filament to bring the optical contrast to zero. The optocoupler provides a synchronization between the chopper and the microprocessor.

Two-color Pyrometer

Since emissivities of many materials are not known, measurement of a surface temperature can become impractical, unless the emissivity is excluded from the calculation. This can be accomplished by use of a ratiometric technique: the so called "two-color radiation thermometer" or *ratio thermometer.* In such a thermometer, the radiation is detected at two separate wavelengths λ_x and λ_y for which emissivities of the surface can be considered nearly the same. The coefficients of transmission of the optical system at each wavelength respectively are γ_x and γ_y, then the ratio of two equations (2) calculated for two wavelengths is:

$$\phi = \frac{W_x}{W_y} = \frac{\gamma_x \varepsilon(\lambda_x)\frac{C_1}{\pi}\varepsilon(\lambda_x)\lambda_x^{-5} e^{-\frac{C_2}{\lambda_x T}}}{\gamma_y \varepsilon(\lambda_y)\frac{C_1}{\pi}\varepsilon(\lambda_y)\lambda_y^{-5} e^{-\frac{C_2}{\lambda_y T}}} \tag{13.116}$$

Because the emissivities $\varepsilon(\lambda_x) \approx \varepsilon(\lambda_y)$, after manipulations, Equation 13.116 can be rewritten for the displayed temperature T_c, where ϕ represents the ratio of the thermal radiation sensor outputs at two wavelengths:

$$T_c \approx C_2\left(\frac{1}{\lambda_y} - \frac{1}{\lambda_x}\right)\left(\ln\phi\frac{\gamma_y}{\gamma_x}\frac{\lambda_x^5}{\lambda_y^5}\right)^{-1} \tag{13.117}$$

It is seen that the equation for calculating temperature does not depend on the emissivity of the surface. Equation 13.117 is the basis for calculating temperature by taking the ratio of the sensor outputs at two different wavelengths. Figure 13.57 shows a block diagram of an IR thermometer where an optical modulator is designed in the form of a disk with two filters.

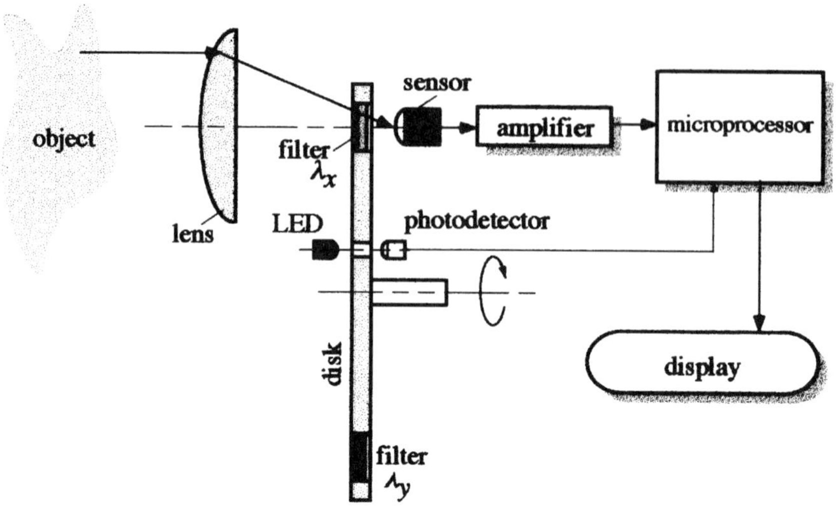

FIGURE 13.57 Two-color, non-contact thermometer. LED and photodetector are for synchronizing the disk rotation with the sensor response.

IR Thermometers

Operating Principle

Infrared (IR) Thermometer with PIR Sensors.
The operating principle of a noncontact infrared (IR) thermometer is based on the fundamental Stefan-Boltzmann law (Equation 13.118). For the purposes of calculating the object's temperature, when using the PIR sensor (thermopile, bolometer, or pyroelectric), the equation can be manipulated as:

$$T_c = \sqrt[4]{T_s^4 + \frac{\Phi}{A\sigma\varepsilon\varepsilon_s}} \tag{13.118}$$

where T_c is the calculated object's temperature in kelvin. Hence, to calculate the temperature of an object, one should first determine the magnitude of net thermal radiation flux Φ and the sensor's surface temperature T_s. The other parts of Equation 13.118 are considered as constants and must be determined during the calibration of the instrument from a blackbody. Emissivity of the object ε also must be known before the calculation. In practice, it is sometimes difficult to determine the exact temperature T_s of the sensor's surface, due to changing ambient conditions, drifts, handling of the instrument, etc. In such cases, the IR thermometer can be supplied with a reference target. Then, the calculation can still be done with the use of Equation 13.118; however, the value of Φ will have a meaning of a flux differential between the object and the reference target, and the value of T_s will represent the reference target temperature.

IR Thermometer with AFIR Sensor.
The object's temperature can be calculated from Equation 13.118; but first the sensing element surface temperature should be determined by measuring the temperature of the sensor's housing: $T_s = T_a + \Delta T$, where ΔT is the constant. In addition, the value of V must be measured. The AFIR sensor allows for continuous monitoring of temperature in a manner similar to thermopile sensors. Its prime advantage is simplicity and low cost.

Continuous Monitoring of Temperature

Depending on the type of thermal radiation sensor employed, a non-contact infrared thermometer for continuous monitoring can incorporate different components. Thus, if a sensor with a dc response

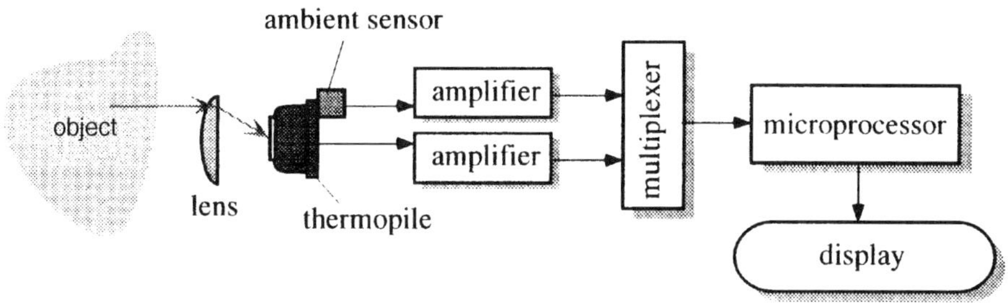

FIGURE 13.58 Infrared thermometer with a dc type of thermal radiation sensor.

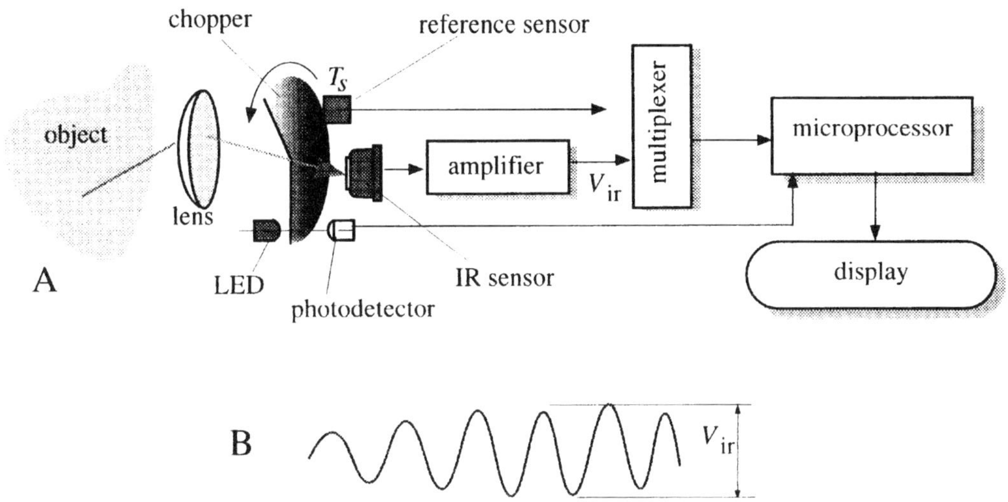

FIGURE 13.59 Infrared thermometer with a chopper.

is used, the circuit might look like the one shown in Figure 13.58. It has the following essential components: the optical system (the lens), the IR sensor (a thermopile in this example), the reference (ambient) sensor, and a conventional data acquisition circuit. The dc-type sensors are the thermopiles, bolometers, and AFIR. When the AFIR sensor is used, it should be supplied with an additional control circuit, as described above.

In the case when an ac-type of IR sensor is used (pyroelectric), the thermometer needs a chopper that breaks the IR radiation into series pulses, usually with a 50% duty cycle (Figure 13.59). The output of an amplifier is an alternate signal that has a magnitude dependent on both the chopping rate and the net IR radiation. The rotation of the chopper is synchronized with the data processing by the microprocessor by means of an optocoupler (LED and photodetector). It should be noted that the chopper not only converts the IR flux into an ac signal, but it also serves as a reference target, as described above. Use of a reference target can significantly improve IR thermometer accuracy; thus, the chopper is often employed, even with the dc-type sensors, although it is not essential for operation. Naturally, the chopper's surface emissivity will be high and known.

Intermittent Monitoring of Temperature

Obviously, any infrared thermometer capable of continuous monitoring of temperature can be made operational in the intermittent mode. In other words, the continuous temperature can be processed to extract a single value of temperature that might be of interest to the user. For example, it may be a maximum, a minimum, or an averaged over time temperature; yet it is possible to design a low-cost IR thermometer that takes only a "snapshot" of the temperature at any specific moment. Such a thermometer

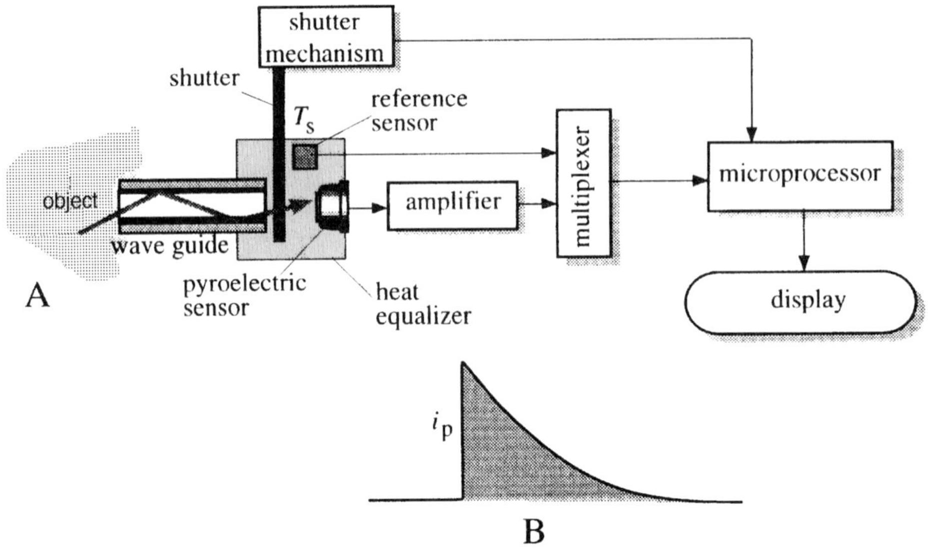

FIGURE 13.60 (A): Infrared thermometer with a pyroelectric sensor. Note that a waveguide is used as part of an optical system. (B) Output current from a pyroelectric sensor.

can be designed with a pyroelectric sensor that produces electric charge virtually instantaneously upon receiving the IR radiation. Figure 13.60A shows a block diagram of such a thermometer. A mechanical or electromechanical shutter is positioned in front of the pyroelectric sensor. The shutter surface serves as a reference target. When the shutter is closed, the sensor produces no output. Immediately upon shutter actuation, the pyroelectric current flows from the sensor into an amplifier that contains a charge-to-voltage converter. The sensor's current response has a shape close to the exponential function (Figure 13.60B). The magnitude of the spike is nearly proportional to the IR flux at the moment of the shutter opening. The pyroelectric thermometer is widely used for medical purposes and is dubbed the "instant thermometer."

Response Speed

All infrared thermometers are relatively fast; their typical response time is on the order of a second. Along with the non-contact way of taking temperature, this makes these devices very convenient whenever fast tracking of temperature is essential. However, an IR thermometer, while being very fast, still can require a relatively long warm-up time, and may not be accurate whenever it is moved from one environment to another without having enough time to adapt itself to new ambient temperature. The reason for this is that the reference and IR sensors (or the IR sensor and the shutter or chopper) must be in thermal equilibrium with one another; otherwise, the calculated temperature is erroneous.

Components of IR Thermometers

Optical Choppers and Shutters

The shutters or choppers must possess the following properties: (1) they should have high surface emissivity that does not change with time; (2) the opening and closing speed must be high; (3) the thermal conductivity between the front and back sides of the blade should be as small as possible; (4) the blade should be in good thermal contact with the reference sensor; and (5) while operating, the blade should not wear off or produce microscopic particles of dust that could contaminate the optical components of the IR thermometer.

For operation in the visible and near-infrared portions of the spectrum, when measured temperatures are over 800 K, the shutter can be fabricated as a solid-state device without the use of moving components

TABLE 13.18 Materials Useful for Infrared Windows and Lenses

Material	n	ρ	Wavelength (μm)	Note
AMTIR-1 ($Ge_{33}As_{12}Se_{55}$)	2.6	0.330	1	Amorphous glass
	2.5	0.310	10	
AMTIR-3 ($Ge_{28}Sb_{12}Se_{60}$)	2.6	0.330	10	Amorphous glass
As_2S_3	2.4	0.3290	8.0	Amorphous glass
CdTe	2.67	0.342	10.6	
Diamond	2.42	0.292	0.54	Best IR material
Fused silica (SiO_2)	1.46	0.067	3.5	Near-IR range
GaAs	3.13	0.420	10.6	
Germanium	4.00	0.529	12.0	Windows and lenses
Irtran 2 (ZnS)	2.25	0.258	4.3	
KRS-6	2.1	0.224	24	Toxic
Polyethylene	1.54	0.087	8.0	Low-cost IR windows; lenses
Quartz	1.54	0.088		Near-IR range
Sapphire (Al_2O_3)	1.59	0.100	5.58	Chemically resistant. Near- and mid-IR ranges
Silicon	3.42	0.462	5.0	Windows in IR sensors
ZnSe	2.4	0.290	10.6	IR windows; brittle

Note: n is the refractive index and ρ is the coefficient of reflection from two surfaces in air.

(e.g., employing liquid crystals). However, for longer wavelengths, only the mechanical blades are useful. Special attention should be paid with regard to the prevention of reflection by the shutter or chopper of any spurious thermal radiation that originates from within the IR thermometer housing.

Filters and Lenses

The IR filters and lenses serve two purposes: they selectively pass specific wavelengths to the sensing components, and they protect the interior of the instrument from undesirable contamination by outside pollutants. In addition, lenses — due to their refractive properties — divert light rays into specific direction [3]. In IR thermometry, the selection of filters and lenses is limited to a relatively short list. Table 13.18 lists some materials that are transparent in the infrared range. Note that most of these materials have a relatively high refractive index, which means that they have high surface reflectivity loss. For example, silicon (the cheapest material for the IR windows) reflects over 46% of incoming radiation which along with the absorptive loss, amounts to an even higher value of combined loss. To a certain degree, this problem can be solved by applying special antireflective (AR) coatings on both surfaces of the window or lens. These coatings are geared to specific wavelengths; thus, for a broad bandwidth, a multilayer coating may need to be deposited by a sputtering process.

Another problem with IR refractive materials is the relatively high absorption of light. This becomes a serious limitation for the lenses, as they need to be produced with appreciable thickness. The solution is to select materials with low absorption in the spectral range of interest. Examples are zinc selenide (ZnSe) and AMTIR. Another solution is the use of polyethylene Fresnel lenses, which are much thinner and less expensive [3]. Any window or lens that is absorptive will also emanate thermal radiation according to its own temperature. Hence, to minimize this effect on the overall accuracy of an IR thermometer, the refractive devices should be kept at the same temperature as the IR sensor, the shutter (chopper), and the reference sensor.

Waveguides

Waveguides are optical components intended for channeling IR radiation from one point to another. Usually, they are used when an IR sensor cannot be positioned close to the source of thermal radiation, yet must have a wide field of view. These components employ light reflection and are not focusing, even if they are designed with refractive materials. If focusing is required, the waveguides can be combined with lenses and curved or tilted mirrors. A waveguide has relatively wide entry and exit angles. A typical application is in a medical IR thermometer, where the waveguide is inserted into an ear canal (Figure 13.60).

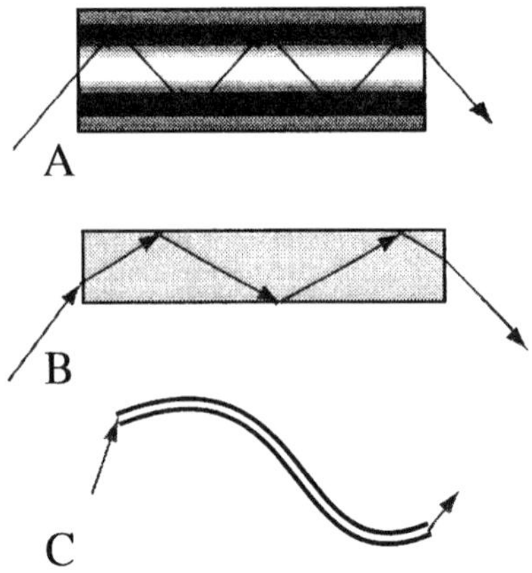

FIGURE 13.61 Light inside the barrel (A), rod (B) and fiber-optic (C) propagates on a zigzag path.

There are three types of waveguides: hollow tubes (barrels), optical fibers, and rods [3, 9, 10]. The latter two are made of IR-transparent materials, such as ZnSe or AMTIR, and use the angle of total internal reflection to propagate light inside in a zigzag pattern (Figure 13.61). The barrels are hollow tubes, polished and gold-plated on the inner surface.

Error Sources in IR Thermometry

Any error in detection of either radiated flux (Φ) or reference temperature (T_a) will result in inaccuracy in the displayed temperature. According to Equation 13.118, the emissivities of both the sensor (ε_s) and the object (ε) must be known for the accurate detection of thermal flux. Emissivity of the sensor usually remains constant and is taken care of during the calibration. However, uncertainty in the value of emissivity of the object can result in significant uncertainty in temperatures measured by non-contact infrared thermometers (Figure 13.62).

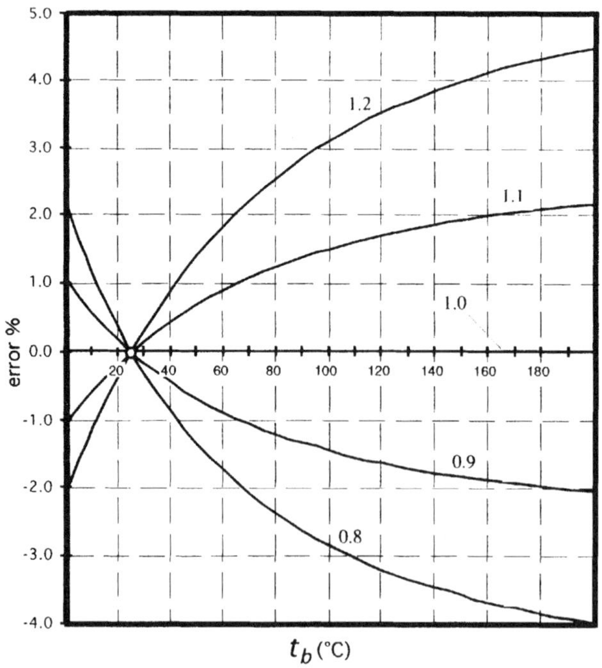

FIGURE 13.62 Error in temperature measurement resulted from uncertainty in the value of object's emissivity.

Another source of an error is spurious heat sources, which can emanate their thermal radiation either directly into the optical system of an IR thermometer, or by means of reflection from the measured surface [11]. Since no surface is an ideal emitter (absorber), its reflectivity may have significant value. For example, an opaque object with emissivity 0.9 has reflectivity of 0.1 (Equation 13.54); thus, about 10% of the radiation from a hot object in the vicinity of a measured surface is reflected and can greatly disturb the measurement.

Since emissivities of metals are quite small and one never can be sure about their actual values, it is advisable, whenever practical, to coat a portion of a metal surface with a dielectric material, such as metal oxide or organic paint having known emissivity. Alternatively, temperatures should not be taken from a flat surface, but rather from a hole or cavity that inside has a nonmetallic surface.

Another source of error is the thermal instability of the thermal sensing element. For example, in a thermopile or pyroelectric sensor, upon exposure to thermal radiation, the element's temperature might increase above ambient by just a few millidegrees. Hence, if for some reason, the element's temperature varies by the same amount, the output signal becomes indistinguishable from that produced by the thermal flux. To minimize these errors, the infrared sensors are encapsulated into metal bodies having high thermal capacity and poor coupling to the environment (high thermal resistance), which helps to stabilize the temperature of the sensor. Another powerful technique is to fabricate the IR sensor in a dual form; that is, with two identical sensing elements. One element is subjected to an IR signal, while the other is shielded from it. The signals from the elements are subtracted, thus canceling the drifts that affect both elements [3].

Some Special Applications

Semiconductor Materials

Temperature measurement of semiconductor material during processing, such as growth of epitaxial films in vacuum chambers, always has been a difficult problem. Various process controls require accurate temperature monitoring, often without physical contact with the substrates. As a rule, heating of a substrate is provided by resistive heaters. The substrates are often loaded into the chambers through various locks and rotated during processing. Therefore, it is very difficult to attach a contact sensor, such as a thermocouple, to the wafer. And, even if one does so, the thermal gradient between the sensor and the substrate may be so large that one should never trust the result. The attractive alternative is optical pyrometry; however, this presents another set of difficulties. The major problem is that semiconductors are substantially transparent in the spectral region where thermal radiation would be emanated. In other words, the emissivity of a semiconductor is negligibly small and the amount of thermal radiation from a semiconductor is also not only small, but due to wafer transparency, the radiation from the underlying devices (e.g., heater) will go through to the IR thermometer.

One relatively simple method is to coat a small portion of the semiconductor with a material having high emissivity in the IR spectral range. Then, the thermal radiation can be measured from that patch. An example of such a material is nichrome (see Table 13.17).

An attractive method of temperature monitoring, when no emissive patch can be deposited, is the use of temperature dependence of bandgaps of common semiconductors. The bandgap is determined from the threshold wavelength at which the radiation from the heaters behind the substrate is transmitted through the substrate [12]. Another method is based on the temperature dependence of diffuse reflection of a semiconductor. In effect, this method is similar to the former; however, it relies on reflection, rather than on transmission of the semiconductor. An external lamp is used for the measurement of a threshold wavelength from a front polished and backside textured substrate [13]. The temperature measurement arrangement is shown in Figure 13.63 where the diffused light is detected by cryogenically cooled quantum detector. The monochromator has resolution of 3 nm and scans through the threshold area at a rate of 100 nm/s.

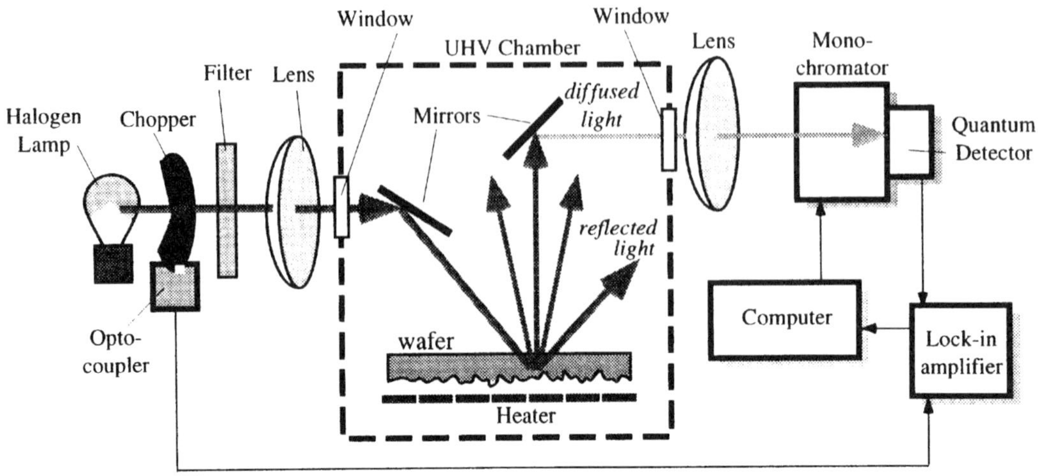

FIGURE 13.63 Diffused-light thermometer for measuring temperature of GaAs wafers. (After [12].)

Medical Applications

Medical infrared thermometry has two distinct types of measurements: *body* temperature measurement and *skin* surface temperature measurement. Skin temperature measurements have specific applications in determining surface temperature of a human body. That temperature greatly depends on both the skin blood perfusion and environmental conditions. Therefore, skin temperature cannot be independently correlated with the internal body temperature.

Now, it is customary to measure the internal body temperature by placing the probe of an IR thermometer into the ear canal aiming it at a tympanic membrane [14]. The tympanic membrane has temperature close to that of blood. As a rule, the probe is designed with a gold-plated waveguide (Figure 13.60) which is protected either by a semiconductor or polymer window. Because a medical IR thermometer is used in direct contact with patient tissues, it is imperative to protect the probe from becoming a carrier of soiling compounds and a transmitter of infection from one patient to another (cross-contamination) or even from re-infecting the same patient (recontamination). Thus, a probe of a medical IR thermometer is protected by a special probe cover made of a polymer film (polyethylene, polypropylene, etc.) which is transparent in the spectrum range of thermal radiation. In effect, the probe cover becomes a part of the optical system. This demands that covers be produced with very tight tolerances so that they will not significantly affect transmission of IR signal.

References

1. A. Völklein, A. Wiegand, and V. Baier, *Sensors and Actuators A*, 29, pp: 87-91, 1991.
2. J. Schieferdecker, R. Quad, E. Holzenkämpfer, and M. Schulze, Infrared thermopile sensors with high sensitivity and very low temperature coefficient. *Sensors and Actuators* A, 46-47, 422-427,1995.
3. J. Fraden, *Handbook of Modern Sensors.* 2nd ed., Woodbury, NY: AIP Press, 1996.
4. J-S. Shie and P.K. Weng, Fabrication of micro-bolometer on silicon substrate by anizotropic etching technique. *Transducers'91, Int. Conf. Solid-state Sensors and Actuators.* 627-630, 1991.
5. R.A. Wood, Uncooled thermal imaging with silicon focal plane arrays. *Proc. SPIE* 2020, *Infrared Technology XIX*, pp: 329-36, 1993.
6. J. Fraden, Active far infrared detectors. In *Temperature, Its Measurement and Control in Science and Industry.* Vol. 6, Part 2, New York: American Institute of Physics, 1992, 831-836.

7. C.J. Mastrangelo and R.S. Muller, Design and performance of constant-temperature circuits for microbridge-sensor applications. *Transducers'91. Int. Conf. Solid-state Sensors and Actuators.* 471-474, 1991.
8. E. O. Doebelin, *Measurement Systems. Application and Design,* 4th ed., New York: McGraw-Hill Co., 1990.
9. A. R. Seacord and G. E. Plambeck. *Fiber optic ear thermometer.* U.S. Patent No. 5,167,235.
10. J. Fraden, *Optical system for an infrared thermometer.* U.S. Patent No. 5,368,038.
11. D.R. White and J.V. Nicholas, Emissivity and reflection error sources in radiation thermometry, in *Temperature, Its Measurement and Control in Science and Industry.* Vol. 6, Part 2, New York: American Institute of Physics, 1992, 917-922.
12. E.S. Hellman et al., *J. Crystal Growth,* 81, 38, 1986.
13. M.K. Weilmeier et al., A new optical temperature measurement technique for semiconductor substrates in molecular beam epitaxy. *Can. J. Phys.,* 69, 422-426, 1991.
14. J. Fraden, Medical infrared thermometry (review of modern techniques), in *Temperature, Its Measurement and Control in Science and Industry.* Vol. 6, Part 2, New York: American Institute of Physics, 1992, 825-830.

13.7 Pyroelectric Thermometers

Jacob Fraden

Pyroelectric Effect

The pyroelectric materials are crystalline substances capable of generating an electric charge in response to heat flow [1]. The pyroelectric effect is very closely related to the piezoelectric effect. The materials belong to a class of ferroelectrics. The name was given in association with ferromagnetics and is rather misleading because most such materials have nothing to do with iron.

A crystal is considered to be pyroelectric if it exhibits a spontaneous temperature-dependent polarization. Of the 32 crystal classes, 21 are noncentrosymmetric and 10 of these exhibit pyroelectric properties. Besides pyroelectric properties, all these materials exhibit some degree of piezoelectric properties as well — they generate an electric charge in response to mechanical stress.

Pyroelectricity was observed for the first time in tourmaline crystals in the 18th century (some claim that the Greeks noticed it 23 centuries ago). Later, in the 19th century, Rochelle salt was used to make pyroelectric sensors. A large variety of materials became available after 1915: KDP (KH_2PO_4), ADP ($NH_4H_2PO_4$), $BaTiO_3$, and a composite of $PbTiO_3$ and $PbZrO_3$ known as PZT. Presently, more than 1000 materials with reversible polarization are known. They are called ferroelectric crystals. Most important among them are triglycine sulfate (TGS) and lithium tantalate ($LiTaO_3$).

A pyroelectric material can be considered a composition of a large number of minute crystallites, where each behaves as a small electric dipole. All these dipoles are randomly oriented, however, along a preferred direction. Above a certain temperature, known as the Curie point, the crystallites have no dipole moment.

When temperature of a pyroelectric material changes, the material becomes polarized. In other words, an electric charge appears on its surface. It should be clearly understood that the polarization occurs not as a function of temperature, but only as function of a *change in temperature* of the material. There are several mechanisms by which changes in temperature will result in pyroelectricity. Temperature changes can cause shortening or elongation of individual dipoles. It can also affect the randomness of the dipole orientations due to thermal agitation. These phenomena are called *primary pyroelectricity.* There is also *secondary pyroelectricity,* which, in a simplified way, can be described as a result of the piezoelectric effect; that is, a development of strain in the material due to thermal expansion.

The dipole moment, M, of the bulk pyroelectric sensor is:

$$M = \mu A h \tag{13.119}$$

TABLE 13.19 Physical Properties of Pyroelectric Materials

Material	Curie temperature °C	Thermal conductivity W mK^{-1}	Relative permittivity ε_r	Pyroelectric charge coeff. C $(m^2K)^{-1}$	Pyroelectric voltage coeff. V $(mK)^{-1}$	Coupling k_p^2 (%)
Single Crystals						
TGS	49	0.4	30	3.5×10^{-4}	1.3×10^6	7.5
$LiTa0_3$	618	4.2	45	2.0×10^{-4}	0.5×10^6	1.0
Ceramics						
$BaTiO_3$	120	3.0	1000	4.0×10^{-4}	0.05×10^6	0.2
PZT	340	1.2	1600	4.2×10^{-4}	0.03×10^6	0.14
Polymers						
PVDF polycrystalline layers	205	0.13	12	0.4×10^{-4}	0.40×10^6	0.2
$PbTiO_3$	470	2 (monocrystal)	200	2.3×10^{-4}	0.13×10^6	0.39

Note: The above figures may vary depending on manufacturing technologies.
From Reference [2].

where μ = Dipole moment per unit volume
A = Sensor's area
h = Thickness

The charge, Q_a, which can be picked up by the electrodes, develops the dipole moment across the material:

$$M_0 = Q_a h \tag{13.120}$$

M must be equal to M_0, so that:

$$Q_a = \mu A \tag{13.121}$$

As the temperature varies, the dipole moment also changes, resulting in an induced charge.

Thermal absorption can be related to a dipole change, so that μ must be considered a function of both temperature, T_a, and an incremental thermal energy, ΔW, absorbed by the material:

$$\Delta Q_a = A\mu\left(T_a, \Delta W\right) \tag{13.122}$$

The above equation shows the magnitude of electric charge resulting from absorption of thermal energy. To pick up the charge, the pyroelectric materials are fabricated in the shapes of a flat capacitor with two electrodes on opposite sides and the pyroelectric material serving as a dielectric.

Pyroelectric Materials

To select the most appropriate pyroelectric material, energy conversion efficiency should be considered. It is, indeed, the function of the pyroelectric sensor to convert thermal energy into electrical. "How effective is the sensor?" — is a key question in the design practice. A measure of efficiency is: k_p^2 which is called the pyroelectric coupling coefficient [2, 3]. It shows the factor by which the pyroelectric efficiency is lower than the Carnot limiting value $\Delta T/T_a$. Numerical values for k_p^2 are shown in Table 13.19.

Table 13.19 shows that triglycine sulfate (TGS) crystals are the most efficient pyroelectric converters. However, for a long time they were quite impractical for use in sensors because of a low Curie temperature. If the sensor's temperature is elevated above that level, it permanently loses its polarization. In fact, TGS sensors proved to be unstable even below the Curie temperature, with the signal being lost quite spontaneously [4]. It was discovered that doping of TGS crystals with L-alanine (LATGS process patented by

Philips) during its growth stabilizes the material below the Curie temperature. The Curie temperature was raised to 60°C, which allows its use at the upper operating temperature of 55°C, which is sufficient for many applications.

Other materials, such as lithium tantalate and pyroelectric ceramics, are also used to produce pyroelectric sensors. Polymer films (KYNAR from AMP Inc.) have become increasingly popular for a variety of applications. During recent years, deposition of pyroelectric thin films has been intensively researched. Especially promising is the use of lead titanate ($PbTiO_3$), which is a ferroelectric ceramic having both a high pyroelectric coefficient and a high Curie temperature of about 490°C. This material can be easily deposited on silicon substrates by the so called sol-gel spin casting deposition method [5].

In 1969, Kawai discovered strong piezoelectricity in the plastic materials, polyvinyl fluoride (PVF) and polyvinylidene fluoride (PVDF) [6]. These materials also possess substantial pyroelectric properties. PVDF is a semicrystalline polymer with an approximate degree of crystallinity of 50% [7]. Like other semicrystalline polymers, PVDF consists of a lamellar structure mixed with amorphous regions. The chemical structure contains the repeat unit of doubly fluorinated ethene CF_2-CH_2:

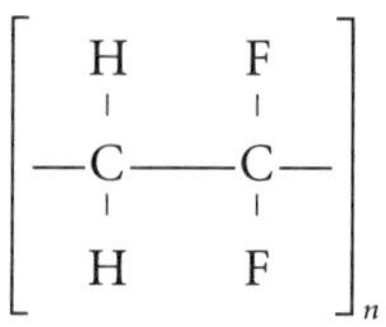

The molecular weight of PVDF is about 10^5, which corresponds to about 2000 repeat units. The film is quite transparent in the visible and near-IR regions, and is absorptive in the far-infrared portion of the electromagnetic spectrum. The polymer melts at about 170°C. Its density is about 1780 kg m^{-3}. PVDF is mechanically durable and flexible. In piezoelectric applications, it is usually drawn, uniaxially or biaxially, to several times its length. Elastic constants, for example, Young modulus, depend on this draw ratio. Thus, if the PVDF film was drawn at 140°C to the ratio of 4:1, the modulus value is 2.1 GPa; while for the draw ratio of 6.8:1, it was 4.1 GPa. Resistivity of the film also depends on the stretch ratio. For example, at low stretch, it is about 6.3×10^{15} Ω cm, while for the stretch ratio 7:1 it is 2×10^{16} Ω cm.

Since silicon does not possess pyroelectric properties, such properties can be added on by depositing crystalline layers of pyroelectric materials. The three most popular materials are zinc oxide (ZnO), aluminum nitride (AlN), and the so-called solid solution system of lead-zirconite-titanium oxides $Pb(Zr,Ti)O_3$ known as PZT ceramic, which is basically the same material used for fabrication of discrete piezoelectric and pyroelectric sensors. One of the advantages of using zinc oxide is the ease of chemical etching. The zinc oxide thin films are usually deposited on silicon by employing sputtering technology. Note, however, that silicon has a large coefficient of thermal conductivity. That is, its thermal time constant is very small (see below), so the pyroelectric sensors made with silicon substrates possess low sensitivity yet are capable of fast response.

Manufacturing Process

Manufacturing of ceramic PZT elements begins with high-purity metal oxides (lead oxide, zirconium oxide, titanium oxide, etc.) in the form of fine powders having various colors. The powders are milled to a specific fineness, and mixed thoroughly in chemically correct proportions. In a process called "calcining," the mixtures are then exposed to an elevated temperature, allowing the ingredients to react to form a powder, each grain of which has a chemical composition close to the desired final composition. At this stage, however, the grain does not yet have the desired crystalline structure.

The next step is to mix the calcined powder with solid and/or liquid organic binders (intended to burn out during firing) and mechanically form the mixture into a "cake" that closely approximates a shape of the final sensing element. To form the "cakes" of desired shapes, several methods can be used. Among them are pressing (under force of a hydraulic-powered piston), casting (pouring viscous liquid

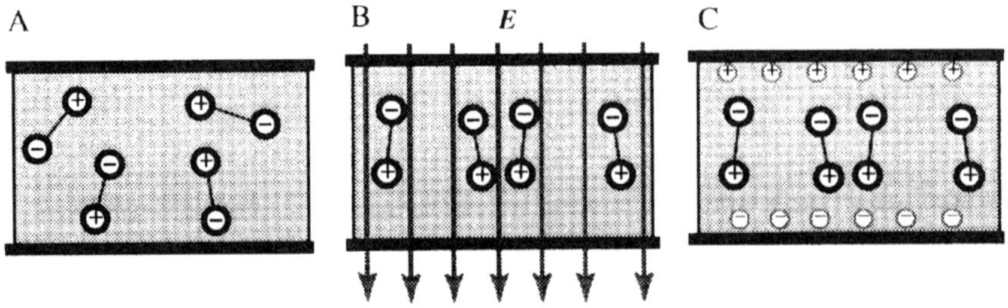

FIGURE 13.64 Poling of a pyroelectric crystal in a strong electric field. The sensor must be stored and operated below the Curie temperature.

into molds and allowing to dry), extrusion (pressing the mixture through a die, or a pair of rolls to form thin sheets), and tape casting (pulling viscous liquid onto a smooth moving belt).

After the "cakes" have been formed, they are placed into a kiln and exposed to a very carefully controlled temperature profile. After burning out of organic binders, the material shrinks by about 15%. The "cakes" are heated to a red glow and maintained at that state for some time, which is called the "soak time," during which the final chemical reaction occurs. The crystalline structure is formed when the material is cooled down. Depending on the material, the entire firing may take 24 h.

When the material is cold, contact electrodes are applied to its surface. This can be done by several methods. The most common are: fired-on silver (a silk-screening of silver-glass mixture and refiring), electroless plating (a chemical deposition in a special bath), and sputtering (an exposure to metal vapor in partial vacuum).

Crystallities (crystal cells) in the material can be considered electric dipoles. In some materials, like quartz, these cells are naturally oriented along the crystal axes, thus giving the material sensitivity to stress. In other materials, the dipoles are randomly oriented and the materials need to be "poled" to possess piezoelectric properties. To give a crystalline material pyroelectric properties, several poling techniques can be used. The most popular poling process is thermal poling, which includes the following steps:

1. A crystalline material (ceramic or polymer film) that has randomly oriented dipoles (Figure 13.64A) is warmed to slightly below its Curie temperature. In some cases (for a PVDF film), the material is stressed. High temperature results in stronger agitation of dipoles and permits one to more easily orient them in a desirable direction.
2. The material is placed in strong electric field, **E**, (Figure 13.64B) where dipoles align along the field lines. The alignment is not total. Many dipoles deviate from the filed direction quite strongly; however, statistically predominant orientation of the dipoles is maintained.
3. The material is cooled down while the electric field across its thickness is maintained.
4. The electric field is removed and the poling process is complete. As long as the poled material is maintained below the Curie temperature, its polarization remains permanent. The dipoles stay "frozen" in the direction that was given to them by the electric field at high temperature (Figure 13.64C). The above method is used to manufacture ceramic and plastic pyroelectric materials.

Another method, called a corona discharge poling, can be used to produce polymer piezo/pyroelectric films. The film is subjected to a corona discharge from an electrode at several million volts per centimeter of film thickness for 40 s to 50 s [8, 9]. Corona polarization is uncomplicated to perform and can be easily applied before electric breakdown occurs, making this process useful at room temperature.

The final operation in preparing the sensing element is shaping and finishing. This includes cutting, machining, and grinding. After the piezo (pyro) element is prepared, it is installed into a sensor's housing, where its electrodes are bonded to electrical terminals and other electronic components.

After poling, the crystal remains permanently polarized; however, it remains electrically charged for a relatively short time. There is a sufficient amount of free carriers that move in the electric field setup

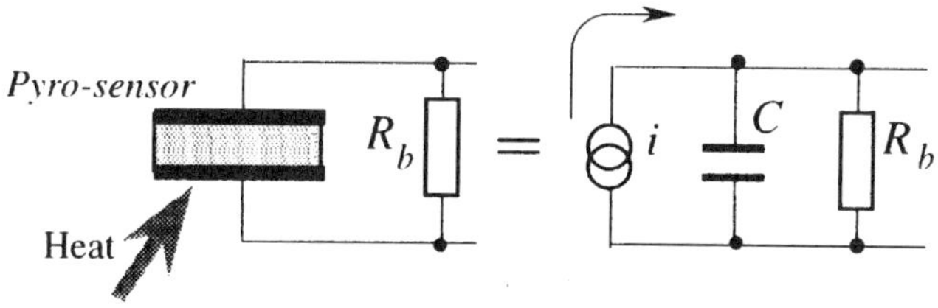

FIGURE 13.65 Pyroelectric sensor and its equivalent circuit.

inside the bulk material and there are plenty of charged ions in the surrounding air. The charge carriers move toward the poled dipoles and neutralize their charges (Figure 13.64C). Hence, after a while, the poled piezoelectric material becomes electrically discharged as long as it remains under steady-state conditions. When temperature changes and thermally induced stress develops, the balanced state is degraded and the pyroelectric material develops an electric charge. If the stress is maintained for a while, the charges again will be neutralized by the internal leakage. Thus, a pyroelectric material is responsive only to a changing temperature rather than to a steady level of it. In other words, a pyroelectric sensor is an ac device, rather than a dc device. Sometimes, it is called a *dynamic* sensor, which reflects the nature of its response.

Pyroelectric Sensors

To make sensors, the pyroelectric materials are used in the form of thin slices or films with electrodes deposited on the opposite sides to collect the thermally induced charges (Figure 13.65). The pyroelectric detector is essentially a capacitor that can be charged by an influx of heat. The detector does not require any external electrical bias (excitation signal). It needs only an appropriate electronic interface circuit to measure the charge. Contrary to thermoelectrics (thermocouples), which produce a steady voltage when two dissimilar metal junctions are held at steady but different temperatures, pyroelectrics generate charge in response to a change in temperature. Since a change in temperature essentially requires propagation of heat, a pyroelectric device is a heat flow detector rather than a heat detector. Figure 13.65 shows a pyroelectric detector (pyro-sensor) connected to a resistor R_b that represents either the internal leakage resistance or a combined input resistance of the interface circuit connected to the sensor. The equivalent electrical circuit of the sensor is shown on the right. It consists of three components: (1) the current source generating a heat induced current, *i*, (remember that a current is a movement of electric charges), (2) the sensor's capacitance, *C*, and (3) the leakage resistance, R_b. Since the leakage resistance is very high and often unpredictable, an additional bias resistor is often connected in parallel with the pyroelectric material. The value of that resistor is much smaller than the leakage resistance, yet its typical value is still on the order of 10^{10} Ω (10 GΩ).

The output signal from the pyroelectric sensor can be taken in the form of either charge (current) or voltage, depending on the application. Being a capacitor, the pyroelectric device is discharged when connected to a resistor, R_b (Figure 13.65). Electric current through the resistor and voltage across the resistor represent the heat flow-induced charge. It can be characterized by two pyroelectric coefficients [2]:

$$P_Q = \frac{dP_S}{dT} \quad \text{Pyroelectric charge coefficient}$$

$$P_V = \frac{dE}{dT} \quad \text{Pyroelectric voltage coefficient} \qquad (13.123)$$

where P_s = Spontaneous polarization (which is the other way to say "*electric charge*")
E = Electric field strength
T = Temperature in K

Both coefficients are related by way of the electric permittivity, ε_r, and dielectric constant, ε_0:

$$\frac{P_Q}{P_V} = \frac{dP_s}{dE} = \varepsilon_r \varepsilon_0 \tag{13.124}$$

The polarization is temperature dependent and, as a result, both pyroelectric coefficients in Equation 13.123 are also functions of temperature.

If a pyroelectric material is exposed to a heat source, its temperature rises by ΔT, and the corresponding charge and voltage changes can be described by the following equations.

$$\Delta Q = P_Q A \Delta T \tag{13.125}$$

$$\Delta V = P_V h \Delta T \tag{13.126}$$

Remembering that the sensor's capacitance can be defined as:

$$C_e = \frac{\Delta Q}{\Delta V} = \varepsilon_r \varepsilon_0 \frac{A}{h} \tag{13.127}$$

then, from Equations 13.124, 13.126, and 13.127, it follows that:

$$\Delta V = P_Q \frac{A}{C_e} \Delta T = P_Q \frac{\varepsilon_r \varepsilon_0}{h} \Delta T \tag{13.128}$$

It is seen that the peak output voltage is proportional to the sensor's temperature rise and pyroelectric charge coefficient and inversely proportional to its thickness.

Figure 13.66 shows a pyroelectric sensor whose temperature, T_0, is homogeneous over its volume. Being electrically polarized, the dipoles are oriented (poled) in such a manner as to make one side of the material positive and the opposite side negative. However, under steady-state conditions, free charge carriers (electrons and holes) neutralize the polarized charge and the capacitance between the electrodes does not appear to be charged. That is, the sensor generates zero charge. Now, assume that heat is applied to the bottom side of the sensor. Heat can enter the sensor in a form of thermal radiation that is absorbed by the bottom electrode and propagates toward the pyroelectric material via the mechanism of thermal conduction. The bottom electrode can be given a heat-absorbing coating, such as goldblack or organic paint. As a result of heat absorption, the bottom side becomes warmer (the new temperature is T_1), which causes the bottom side of the material to expand. The expansion leads to flexing of the sensor, which, in turn, produces stress and a change in dipole orientation. Being piezoelectric, stressed material generates electric charges of opposite polarities across the electrodes. Hence, one can regard secondary pyroelectricity as a sequence of events: a thermal radiation — a heat absorption — a thermally induced stress — an electric charge.

The temperature of the sensor T_s is a function of time. That function is dependent on the sensing element: its density, specific heat, and thickness. If the input thermal flux has the shape of a step function of time, for the sensor freely mounted in air, the output current can be approximated by an exponential function, so that:

$$i = i_0 e^{-t/\tau_T} \tag{13.129}$$

where i_0 = Peak current

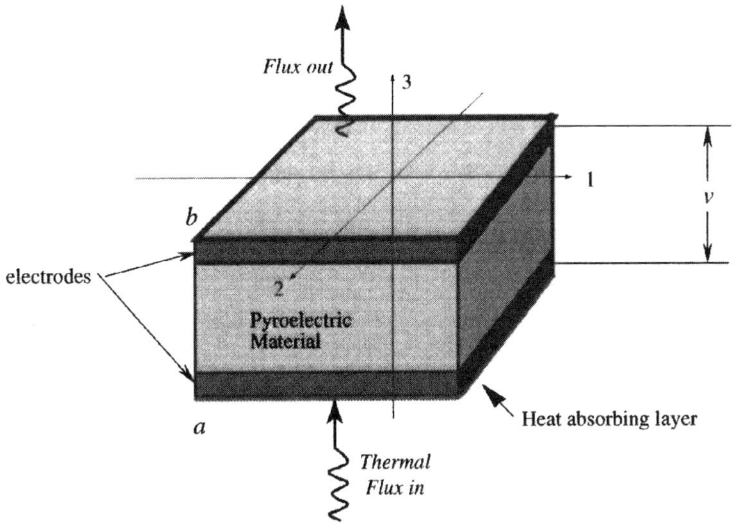

FIGURE 13.66 Pyroelectric sensor has two electrodes at the opposite sides of the crystalline material. Thermal radiation is applied along axis 3.

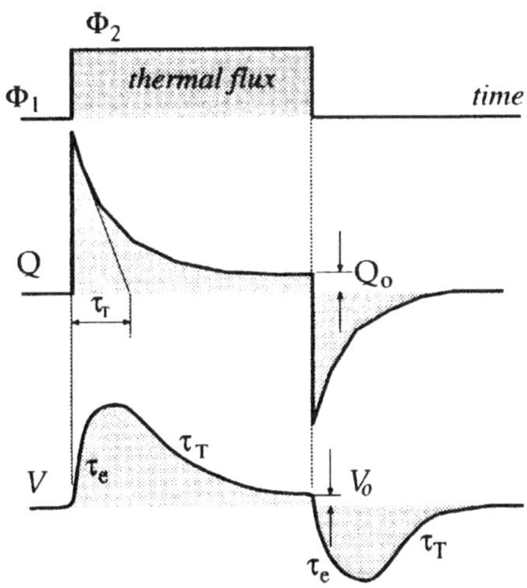

FIGURE 13.67 Response of a pyroelectric sensor to a thermal step function. The magnitudes of charge Q_0 and voltage v_0 are exaggerated for clarity.

Figure 13.67 shows timing diagrams for a pyroelectric sensor when it is exposed to a step function of heat. It is seen that the electric charge reaches its peak value almost instantaneously, and then decays with a *thermal time constant*, τ_T. This time constant is a product of the sensor's thermal capacitance, C, and thermal resistance, r, which defines a thermal loss from the sensing element to its surroundings:

$$\tau_T = Cr = cAhr \tag{13.130}$$

where c = Specific heat of the sensing element.

The thermal resistance r is a function of all thermal losses to the surroundings through convection, conduction, and thermal radiation. For low-frequency applications, it is desirable to use sensors with τ_T as large as practical; while for the high-speed applications (e.g., to measure the power of laser pulses), a

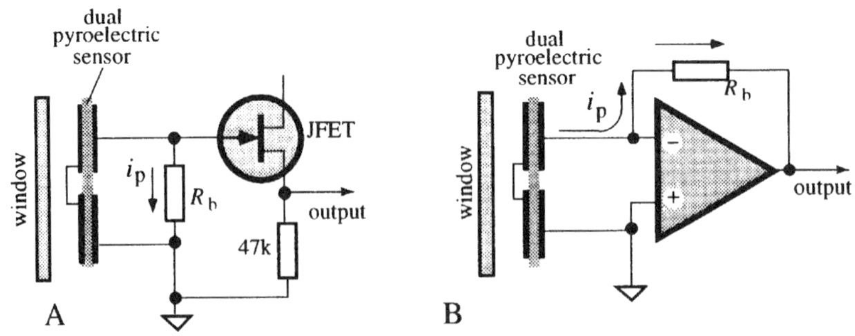

FIGURE 13.68 Interface circuits for pyroelectric sensors operating in voltage (A) and current (B) modes.

thermal time constant should be dramatically reduced. For that purpose, the pyroelectric material can be laminated with a heat sink: a piece of aluminum or copper.

When a heat flow exists inside the pyroelectric crystal, there is an outflow of heat from the opposite side of the crystal, as shown in Figure 13.66. Thermal energy enters the pyroelectric material from side *a*. Since the other side *b* of the sensor faces a cooler environment, part of the thermal energy leaves the sensor and is lost to its surroundings. Because the sides *a* and *b* face objects of different temperatures (one is the temperature of a target and the other is the temperature of the environment), a continuous heat flow exists through the pyroelectric material. As a result, in Figure 13.67, charge Q and voltage V do not completely return to zero, no matter how much time has elapsed. Electric current generated by the pyroelectric sensor has the same shape as the thermal current through its material. An accurate measurement can demonstrate that as long as the heat continues to flow, the pyroelectric sensor will generate a constant voltage V_0 whose magnitude is proportional to the heat flow.

Applications

The pyroelectric sensors are useful whenever changing thermal radiation or heat flow need to be measured. Examples are motion detectors for the security systems and light control switches [1], instant medical infrared thermometers, and laser power meters. Depending on the application, a pyroelectric sensor can be used either in *current* or in *voltage* mode. The voltage mode (Figure 13.68A) uses a voltage follower with a very high input resistance. Hence, JFET and CMOS input stages are essential. As a rule, in the voltage mode sensor, the follower is incorporated inside the same package along with the element and bias resistor. Advantages of the voltage mode are simplicity and lower noise. The disadvantages are slower speed response due to high capacitance of the sensor (typically on the order of 30 pF) and other influences of the sensor capacitance on the quality of output voltage. The output voltage of the follower is shown in Figure 13.67 (V). It is seen that it rises slowly with electric time constant τ_e and decays with thermal time constant τ_T.

The current mode uses an electronic circuit having a "virtual ground" as its input (Figure 13.68B). An advantage of this circuit is that the output signal is independent of the sensor's capacitance and, as a result, is much faster. The signal reaches its peak value relatively fast and decays with thermal time constant τ_T. The output voltage V_0 has the same shape of charge Q in Figure 13.67. The disadvantages of the circuit are higher noise (due to wider bandwidth) and higher cost.

Note that Figure 13.68 shows dual pyroelectric sensors, where two sensing elements are formed on the same crystalline plate by depositing two pairs of electrodes. The electrodes are connected in a serial-opposite manner. If both sensors are exposed to the same magnitude of far-infrared radiation, they will produce nearly identical polarizations and, due to the opposite connection, the voltage applied to the input of the transistor or the current passing through resistor R_b will be nullified. This feature allows for cancellation of undesirable common-mode input signals in order to improve stability and reduce noise. Signals that arrive only to one of the elements will not be canceled.

References

1. J. Fraden, *Handbook of Modern Sensors*, 2nd ed., Woodbury, NY: AIP Press, 1997.
2. H. Meixner, G. Mader, and P. Kleinschmidt, Infrared sensors based on the pyroelectric polymer polyvinylidene fluoride (PVDF), *Siemens Forsch. Entwickl. Ber.*, 15(3), 105-114, 1986.
3. P. Kleinschmidt, Piezo- und pyroelektrische Effekte. Heywang, W., ed., in *Sensorik*, Kap. 6: New York: Springer, 1984.
4. Semiconductor Sensors, *Data Handbook*, Philips Export B.V, 1988.
5. C. Ye, T. Tamagawa, and D.L. Polla, Pyroelectric $PbTiO_3$ thin films for microsensor applications, *Transducers'91. Int. Conf. Solid-State Sensors and Actuators*. 904-907, 1991.
6. H. Kawai, The piezoelectricity of poly(vinylidene fluoride), *Jap. J. Appl. Phys.*, 8, 975-976, 1969.
7. A. Okada, Some electrical and optical properties of ferroelectric lead-zirconite-lead-titanate thin films, *J. Appl. Phys.*, N. 48, 2905, 1977.
8. P.F. Radice, *Corona Discharge poling process*, U.S. Patent No. 4,365,283; 1982.
9. P.D. Southgate, *Appl. Phys. Lett.*, 28, 250, 1976.

13.8 Liquid-in-Glass Thermometers

J.V. Nicholas

The earliest form of thermometer, known as a thermoscope, was a liquid-in-glass thermometer that was open to the atmosphere and thus responded to pressure. By sealing a thermoscope so that it responded only to temperature, the modern form of a liquid-in-glass thermometer resulted. As a temperature sensor, its use dominated temperature measurement for at least 200 years. Liquid-in-glass thermometers had a profound effect on the development of thermometry and, in popular opinion, they are the only "real" thermometers! Liquid-in-glass thermometer sensors were developed in variety to fill nearly every niche in temperature measurement from –190°C to +600°C, including the measurement of temperature differences to a millikelvin. In spite of the fragile nature of glass, the popularity of these thermometers continues because of the chemical inertness of the glass, as well as the self-contained nature of the thermometer.

Measurement sensor designers are unlikely to develop their own liquid-in-glass thermometers, but many will use them to check the performance of a new temperature sensor. The emphasis in this chapter section will therefore be on the use of mercury-in-glass thermometers — the most successful liquid-in-glass thermometer — as calibration references. Mercury-in-glass thermometers provide a stable temperature reference to an accuracy of 0.1°C, provided they are chosen and used with care. The extra requirements to achieve higher accuracy are indicated, but are beyond the scope of this section.

The trend is, however, to move away from mercury-in-glass thermometers for specialized uses (in particular, where the risk from glass or mercury contamination is not acceptable; for example, in the food or aviation industries). Other forms of temperature sensors are more suitable.

General Description

A common form of mercury-in-glass is a solid-stem glass thermometer illustrated in Figure 13.69. The other major form of liquid-in-glass thermometer is the enclosed scale thermometer, for which the general principles discussed will also apply.

There are four main parts to the liquid-in-glass thermometer:

- The bulb is a thin glass container holding the bulk of the thermometric liquid. The glass must be of suitable type and properly annealed. The thinness is essential for good thermal contact with the medium being measured, but it can result in instabilities due to applied stress or sudden shock. Some lower-accuracy, general-purpose thermometers are made with a thicker glass bulb to lower the risk of breakage.

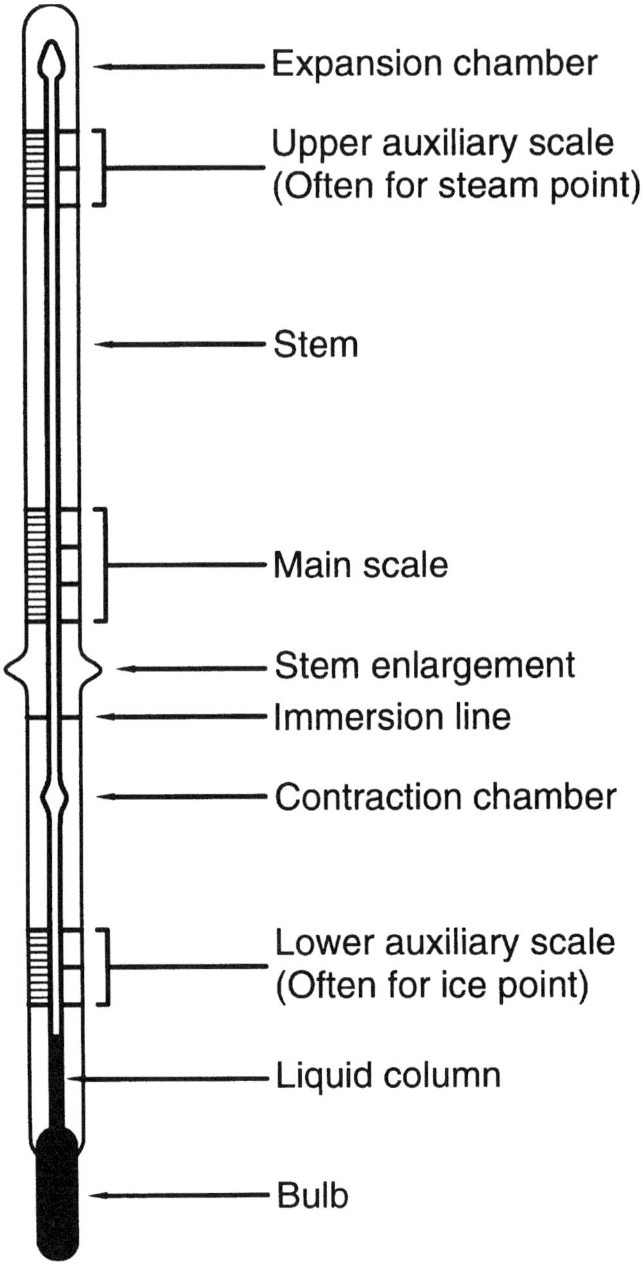

FIGURE 13.69 The main features of a solid-stem glass thermometer. The thermometer can have an enlargement in the stem or an attachment at the end of the stem to assist in the positioning of the thermometer. Thermometers will display several of these features, but seldom all of them.

- The stem is a glass capillary. Again, a suitable glass is necessary and may differ from that of the bulb. The bore can be gas-filled or vacuous. The volume of the bore must be somewhat smaller than the volume of the bulb for good sensitivity. In addition, the bore must be smooth, with a uniform cross section.
- The liquid is usually mercury for best precision, or an organic liquid for lower temperature ranges.

- The markings are usually etched or printed onto the stem. The markings include the scale, to allow direct reading of the temperature, as well as other important information.

Figure 13.69 illustrates the main parts of a mercury-in-glass thermometer, along with a nomenclature for other features commonly found. Not all of the features will be found on all thermometers.

The operation of liquid-in-glass thermometers is based on the expansion of the liquid with temperature; that is, the liquid acts as a transducer to convert thermal energy into a mechanical form. As the liquid in the bulb becomes hotter, it expands and the liquid is forced up the capillary stem. The temperature of the bulb is indicated by the position of the top of the mercury column with respect to the marked scale. The flattest part of the liquid meniscus is used as the indicator: for mercury, this is the top of the curve; for organic liquids, the bottom.

The thermometers appear to have a simplicity about them, but this is lost when accurate measurements are required. By accuracy, we mean any reading where the temperature needs to be known to within 1°C or better. The chief cause of inaccuracy is that not all of the liquid is at the required temperature due to its necessary presence in the stem. Thus, the thermometer is also sensitive to the stem temperature. The main advantage of a liquid-in-glass thermometer is that it is self-contained; but this means that the thermometer stem has to be seen to read the scale. Even in a well-designed apparatus, a good part of the stem will not be at the temperature of the bulb. For example, with a bulb immersed in boiling water and the entire stem outside, an error of 1°C results from the cooler stem. Correction for this error can be incorporated in the scale of a partial immersion thermometer, or the error can be corrected with a chart of stem corrections.

The next most significant cause of error comes from the glass, a substance with complex mechanical properties. Like mercury, it expands rapidly on heating but does not contract immediately on cooling. This produces a hysteresis which, for a good glass, is about 0.1% of the temperature change. A good, well-annealed thermometer glass will relax back over a period of days. An ordinary glass might never recover its shape. Besides this hysteresis, the glass bulb undergoes a secular contraction over time; that is, the thermometer reading increases with age, but fortunately the effect is slow and calibration checks at the ice point, 0°C, can correct for it.

A bewildering number of types of liquid-in-glass thermometers are available, with many of the variations being designed with different dimensions and temperature ranges to suit specific applications.

For best performance, solid-stem mercury-in-glass thermometers should be restricted to operation over the maximum range –38 °C to 250°C. The purchase should be guided by a specification as published by a recognized standards body. Such bodies include the International Standards Organisation (ISO) [1]; the American Society for Testing and Materials (ASTM) [2]; the British Standards Institute (BS) [3]; or the Institute for Petroleum (IP). Be aware that some type numbers are the same, yet may refer to different thermometers, such as in the IP and ASTM ranges. Make sure the specification body is referred to; for example, an order for a 16C thermometer could result in either an ASTM 10C, the equivalent of an IP 16C, or an IP 61C, the equivalent of ASTM 16C.

One's choice of thermometer will most probably be a compromise between the best range, scale division, and length for the purpose. If good precision is required, then the thermometer range will be constrained to avoid extremely long and unwieldy thermometers. Table 13.20 gives the specification for precision thermometers based on the compromise as seen by the ASTM [2]. The cost of these thermometers depends on the range and varies from $50 to $180 at 1996 prices. The best precision for the ASTM liquid-in-glass thermometers is around 0.1°C, with the thermometers supplied being accurate to one scale division. Consult the references at the end of this chapter section if higher-resolution thermometers are required. As a rule, choose thermometers subdivided to the accuracy desired, and do not rely on visual interpolation to increase the accuracy. If relying heavily on interpolation, then a better choice of thermometer should be made.

Table 13.20 has thermometers with an ice point either in the main scale or as an auxiliary scale. The ice point is a very convenient way to check on the on-going performance of a thermometer, and without it, more expensive time-consuming procedures may be needed.

TABLE 13.20 Summary of Requirements for ASTM Precision Thermometers

ASTM Thermometer Number	Range (°C)	Maximum length (mm)	Graduation at each (°C)	Maximum error (°C)
62C	–38 to +2	384	0.1	0.1
63C	–8 to +32	384	0.1	0.1
64C	–0.5 to +0.5 and 25 to 55	384	0.1	0.1
65C	–0.5 to +0.5 and 50 to 80	384	0.1	0.1
66C	–0.5 to +0.5 and 75 to 105	384	0.1	0.1
67C	–0.5 to +0.5 and 95 to 155	384	0.2	0.2
68C	–0.5 to +0.5 and 145 to 205	384	0.2	0.2
69C	–0.5 to +0.5 and 195 to 305	384	0.5	0.5
70C	–0.5 to +0.5 and 295 to 405	384	0.5	0.5

TABLE 13.21 Working Range of Some Thermometric Liquids and Their Apparent Thermal Expansion Coefficient in Thermometer Glasses Around Room Temperature

Liquid	Typical apparent expansion coefficient (°C^{-1})	Possible temperature range (°C)
Mercury	0.00016	–35 to 510
Ethanol	0.00104	–80 to 60
Pentane	0.00145	–200 to 30
Toluene	0.00103	–80 to 100

Liquid Expansion

The equation that best describes the expansion of the mercury volume is:

$$V = V_0\left(1+\alpha t+\beta t^2\right) \tag{13.131}$$

where V_o = Volume of mercury at 0°C
α and β = Coefficients of thermal expansion of mercury, with

$$\alpha = 1.8 \times 10^{-4}\,°\mathrm{C}^{-1}$$

$$\beta = 5 \times 10^{-8}\,°\mathrm{C}^{-2}$$

See Table 13.21 for the expansion coefficients of other liquids and their range of use.

Equation 13.131 is the ideal equation for a mercury-in-glass thermometer. In practice, several factors modify the ideal behavior because of the way in which the thermometers are constructed.

Because the glass of a mercury-in-glass thermometer also expands, it is the apparent expansion coefficient due to the differential expansion of the mercury with respect to the glass that is of interest. Glass used in a typical thermometer has a value of $\alpha = 2 \times 10^{-5}\,°\mathrm{C}^{-1}$, about 10% that of mercury. Hence, both the glass and the mercury act as temperature transducers and thus justify the description "mercury-in-glass."

The mercury also serves as the temperature indicator in the stem and consequently might not be at the same temperature as the mercury in the bulb. While this effect is small for mercury, where the bulb volume is 6250 times the volume of the mercury in a 1°C length of the capillary stem, thermometers used in partial immersion often need correcting.

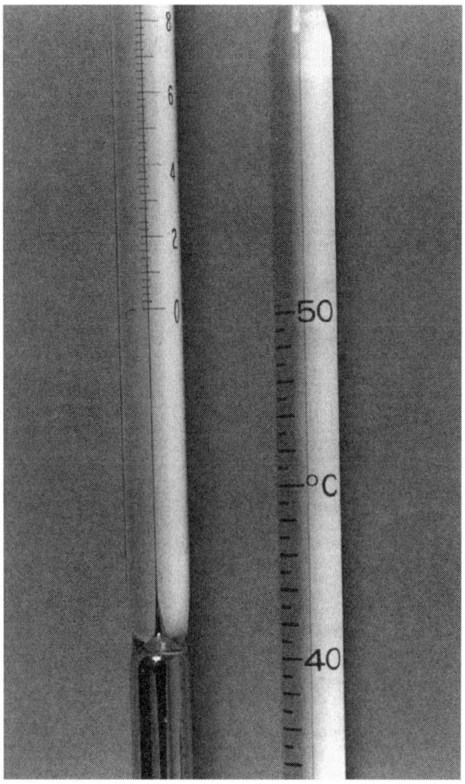

FIGURE 13.70 Calibration marks are usually scratched on at both ends of a thermometer's scale to locate the ruling of the scale. Left: A good quality thermometer. The calibration mark is immediately alongside the 0°C mark. Right: A general-purpose thermometer. Here, the calibration mark is about ¼ scale division above the 50°C mark. Since it is a cheaper thermometer, the manufacturer is content to locate the scale within the ¼ scale division, and this would vary from thermometer to thermometer in the same batch. Readings could be expected to be accurate to about one scale division, 0.5°C in this instance.

The bore in the stem needs to be smooth and uniform. An allowed departure from uniformity is a contraction chamber that, by taking up a volume of the expanding mercury, allows the overall length of the thermometer to be kept a reasonable size. The chamber shape must be very smooth to prevent bubbles of gas being trapped. An auxiliary scale is usually added for the ice point if a contraction chamber is used.

The marked scale allows the user to read the column length as a temperature. For a well-made thermometer, the change in length is proportional to the change in volume and hence to the temperature, as per Equation 13.131. In order to make the scale, the manufacturer first places "calibration" marks on the thermometer stem, as shown in Figure 13.70. Depending on the range and accuracy, more than two calibration marks can be used and, thus, the thermometer stem is divided into segments. A ruling engine is then used to rule a scale between each pair of marks, with careful alignments between the adjacent segments if they occur. The scale rulings will be spaced to approximate Equation 13.131 to the accuracy expected for the thermometer type. A good indicator of the quality of a thermometer is how close these calibration scratches are to the scale markings.

Because of the segmented ruling, it pays to check that the scale markings are uniform in appearance with no obvious glitches. Quite marked discontinuities in the scale are sometimes found. The markings are individual to each thermometer and the total scale length can vary from thermometer to thermometer. This can be an inconvenience if a thermometer has to be replaced; but fortunately, most quality thermometer specifications restrict the amount of variation permissible.

TABLE 13.22 Time Constants for a Mercury-In-Glass Thermometer with a 5-mm Diameter Bulb

Medium	Still (s)	0.5 m s^{-1} flow (s)	Infinite flow velocity (s)
Water	10	2.4	2.2
Oil	40	4.8	2.2
Air	190	71	2.2

Time-Constant Effects

The time constant is determined almost entirely by the diameter of the bulb because heat must be conducted from the outside to the center of the bulb. A typical bulb of diameter 5 mm has a relatively short time constant. The length of the bulb is then determined by the sensitivity required of the thermometer, given that there is a minimum useful diameter for the capillary bore.

The choice of bore diameter is a compromise involving several error effects. A large-diameter bore requires a larger volume bulb to achieve a given resolution, thus increasing the thermal capacity. A small-diameter bore not only becomes difficult to read but also suffers from stiction — the mercury moving in fits and starts due to the surface tension between the mercury and the bore wall. Stiction should be kept less than 1/5 of a scale division.

Table 13.22 gives the $1/e$ time constants in various media for a 5-mm diameter bulb. Time constants for other diameters can be estimated by scaling the time in proportion to the diameter. The table clearly indicates that the thermometer is best used with flowing (or stirred) fluids.

Thermal Capacity Effects

Glass thermometers are bulky and have considerable mass, especially high-precision thermometers that have a long bulb. The high thermal mass or heat capacity can upset temperature measurements, making high precision difficult. Inappropriate use of liquid-in-glass thermometers occurs when the thermometer is too massive to achieve the precision required. Preheating the thermometer can alleviate the worst of the problem. For higher precision and low mass, choose a platinum resistance thermometer or thermistor instead.

Simple estimates of the heat requirements are made by measuring the volume of thermometer immersed, and assuming that 2 J are required to raise 1 cm^3 of the thermometer volume (glass or mercury) by 1°C.

Separated Columns

A common problem is for a part of the thermometric liquid in the stem to become separated from the main volume. While this will show as an ice point shift, it is still important to make a simple visual check when using the thermometer.

With organic liquids, the problem might be more difficult to identify because liquid adheres to the surface of the capillary and may not be visible. Spirit thermometers need to be held vertically to allow the thermometric liquid to drain down. Warm the top of the thermometer to prevent condensation of any vapor. Allow time for drainage of the liquid in the thermometer if the temperature is lowered quickly (approximately 3 min per centimeter). Cool the bulb first in order to keep the viscosity of the liquid in the stem low for better drainage.

For mercury, the separation is usually visible. Two causes can be identified: boil-off and mechanical separation (Figure 13.71).

To help retard the boil-off of mercury vapor at high temperatures (e.g., above 150°C), the capillary tube is filled with an inert gas when manufactured. Usually, dry nitrogen is used under pressure to prevent

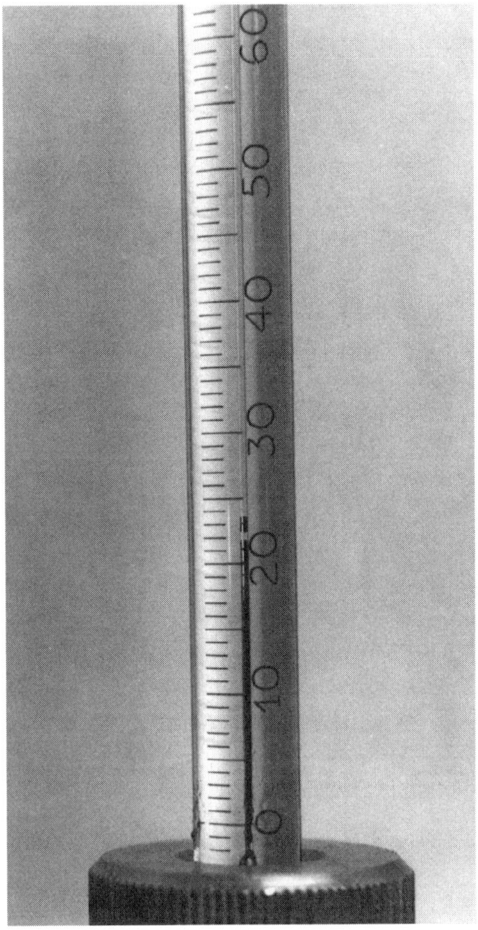

FIGURE 13.71 A typical break in the mercury column of a thermometer.

oxidation of the mercury. The expansion chamber must be kept cooler than the bulb to prevent a high pressure build-up. The high pressure can permanently distort the bulb even if rupture does not occur.

Mechanical separation of the liquid column is, unfortunately, a common occurrence, particularly after shipment. A gas fill will help prevent this separation but, conversely, the gas makes it more difficult to rejoin. There is also a risk of trapped gas bubbles in the bulb or expansion chambers and careful inspection is needed to locate them. A vacuum in the capillary tube will give rise to more breaks, but they are easily rejoined.

With care, it is often possible to rejoin the column and still have a viable thermometer. However, it must be realized that attempts to join a broken column could also result in the thermometer being discarded if the procedure is not successful. Column breaks that occur only when the thermometer is heated often require that the thermometer be discarded.

Various procedures for joining a broken mercury column can be tried. The procedures below are given in order of preference.

- Lightly tap the thermometer while holding it vertically. This may work for a vacuous thermometer.
- Apply centrifugal force, but avoid a flicking action, and be careful to avoid striking anything. This can be best done by holding the bulb alongside the thumb, protecting it with the fingers, and with the stem along the arm. Raise the arm above the head and bring it down quickly to alongside the leg.
- If both the above are unsuccessful, a cooling method can be attempted. This method relies on sufficient cooling of the bulb for all the mercury to contract into the bulb, leaving none in the

stem. The column should be rejoined when it has warmed to room temperature. Carry out the warming slowly so that all the mercury is at the same temperature. More than one attempt might be needed. The first two methods might also need to be applied to assist movement of the mercury.

Three cooling mediums readily available are:

1. Salt, ice, and water (to –18°C)
2. Dry ice, i.e., solid CO_2 (–78°C)
3. Liquid nitrogen (–196°C)

The last two refrigerants require more care as they could freeze the mercury. An excessive cooling rate could stress the glass. **Cold burns to the user could also occur.**

If the broken column has been successfully rejoined, then an ice point (or other reference point) check should be made. If the reading is the same as obtained previously, then the join can be considered completely successful and the thermometer ready for use. (It is essential to keep written records here.) However, a significant ice-point shift indicates that the join was not successful and that the thermometer should be discarded. If the ice-point shift is within that specified for the thermometer type, then treat the thermometer with suspicion until there is evidence of long-term stability, i.e., no significant ice-point changes after use.

Immersion Errors

It was previously mentioned that problems are expected if not all the liquid in a liquid-in-glass thermometer is at the same temperature as the bulb. Because the scale must be read visually, liquid-in-glass thermometers are used at various immersion depths, which results in different parts of the thermometric liquid being at different temperatures. In addition, the clutter around an apparatus often necessitates the thermometer being placed in a nonideal position.

Three distinct immersion conditions are recognized for a liquid-in-glass thermometer, and each requires a different error treatment. Figure 13.72 illustrates the three conditions.

- *Complete immersion*: By definition, if the complete bulb and stem are immersed at the same temperature, the thermometer is completely immersed. This condition is not common, except at room temperature, and should be avoided at higher temperatures. High pressure build-up in the thermometer can cause it to rupture, spreading deadly mercury vapor throughout the laboratory. Laboratories where there is a danger of mercury exposure to high temperatures should be kept well ventilated. In other words, DO NOT put a mercury thermometer completely inside an oven to measure the temperature. Specialized applications that use complete immersion, take into account pressure effects on the thermometers.
- *Total immersion*: Total immersion applies to the situation where all the thermometric liquid, i.e., all the mercury in the bulb, the contraction chamber, and the stem, is at the temperature of interest. The remaining portion of the stem will have a temperature gradient to room temperature (approximately). Especially at high temperatures, the expansion chamber should be maintained close to room temperature to avoid pressure build-up. A very small part of the mercury column can be outside the region of interest, to allow visual readings to be made. The error introduced by this can be estimated by the procedures given below for partial-immersion thermometers. Obviously, the thermometer will have to be moved to allow a range of temperatures to be measured. Total-immersion thermometers are generally calibrated at total immersion and therefore do not need additional corrections.
- *Partial immersion*: One way around the problem of scale visibility and the need to move the thermometer is to immerse the thermometer to some fixed depth so that most, but not all, of the mercury is at the temperature of interest. The part of the mercury column not immersed is referred to as the emergent column. Corrections will be needed to compensate for the error arising from the emergent column not being at the same temperature as the bulb. Many thermometers are

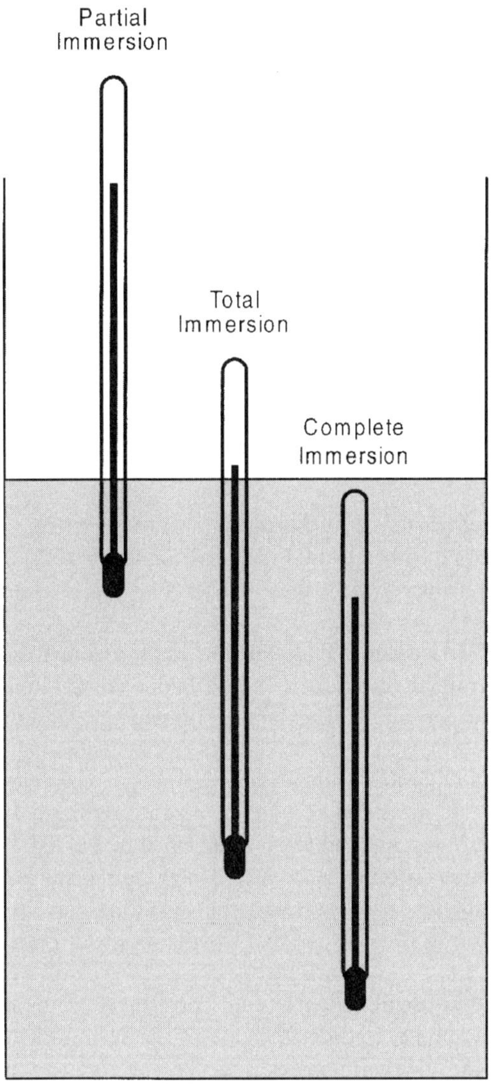

FIGURE 13.72 Three types of immersion conditions used for liquid-in-glass thermometers. The preferred immersion condition is usually marked as a line or distance on the stem for partial-immersion thermometers or is given by the thermometer specification.

designed and calibrated for partial immersion and are marked accordingly on the stem with an immersion depth or an immersion line (see Figure 13.69).

A partial-immersion thermometer is properly defined when the temperature profile of the emergent column is also specified. Usually, an average stem temperature is quoted to represent the temperature profile of the emergent column. Thermometer specifications can define the expected stem temperature for a set of test temperatures, but they do not usually define stem temperatures for all possible readings.

A measure of the stem temperature is required if the accuracy of the thermometer reading is to be assessed. The traditional way to measure the stem temperature is with a Faden thermometer. These are mercury-in-glass thermometers with a very long bulb, and various bulb lengths available. The bulb is mounted alongside the part of the stem containing the emergent column with the bottom of the bulb in the fluid. An average stem temperature is obtained as indicated by the Faden thermometer. Other ways of measuring the temperature profile are to use thermocouples along the length of the thermometer, or even several small mercury-in-glass thermometers. The stem temperature can be calculated as a simple average; but strictly speaking, it should be a length-weighted average.

Because the measured stem temperature might not be the same as that given on the calibration certificate, it is necessary to make corrections for the difference. For partial immersion thermometers, the true temperature reading t is given by:

$$t = t_i + N \times \left(t_2 - t_1\right) \times k \tag{13.132}$$

where t_i = Indicated temperature
N = Length of emergent column expressed in degrees, as determined by the thermometer scale
t_2 = Mean temperature of the emergent column when calibrated (i.e., the stem temperature on a certificate for partial immersion or the thermometer reading for a total-immersion certificate)
t_1 = Mean temperature of the emergent column in use
k = Coefficient of apparent expansion of the thermometric liquid used in the glass of which the thermometer stem is made

See Table 13.21 for suitable values to use for normal temperature ranges.

The use of Equation 13.132 with typical k values from Table 13.21 is estimated to give a 10% accuracy for the correction. Consequently, the correction is a major source of uncertainty for large temperature differences.

Figure 13.73 gives a chart derived from Equation 13.132 for mercury thermometers that enables the stem correction to be determined graphically. One should become familiar enough with it to make quick estimates in order to determine whether the immersion condition error is significant and therefore needs correction.

Thermometers are usually calibrated at their stated immersion conditions and the actual stem temperatures during calibration are measured and quoted on the certificate. In many applications, a thermometer is used for a standard test method (such as specified by the ASTM or IP). For these instances, the expected stem temperature is specified and there is no requirement for the stem temperature to be measured. The user will, however, need to adjust the certificate correction terms to the immersion conditions of the specification in order to see that the thermometer corrections meet the appropriate quality criteria.

The chart of Figure 13.73 is useful, either to find corrections or to show faults with a particular measurement method. For example, consider the case of measuring boiling water in a beaker with a total-immersion thermometer. The thermometer is too long to immerse and the water level is around the 20°C mark. The emergent column is therefore 80°C long and one assumes that the stem is close to room temperature of 20°C, resulting in an 80°C temperature difference from calibration conditions. On the chart, one finds that the intersection of the 80°C emergent line and the 80°C difference line gives a correction value of just over 1°C. If this value is unacceptable, then clearly a redesign of the measurement method is warranted. More detailed examples can be found in the text of Nicholas and White [5].

Organic Liquids

Thermometers with organic liquids have three possible uses:

- To measure temperatures below –38°C
- In situations where mercury is to be avoided
- For inexpensive thermometers

The utility of spirit thermometers is limited because of the lower achievable accuracy, the high nonlinearities, and the volatile nature of the liquids. Organic-liquid thermometers are also difficult to read because of the very clear liquid and concave meniscus. However, the use of a suitable dye and wide bore

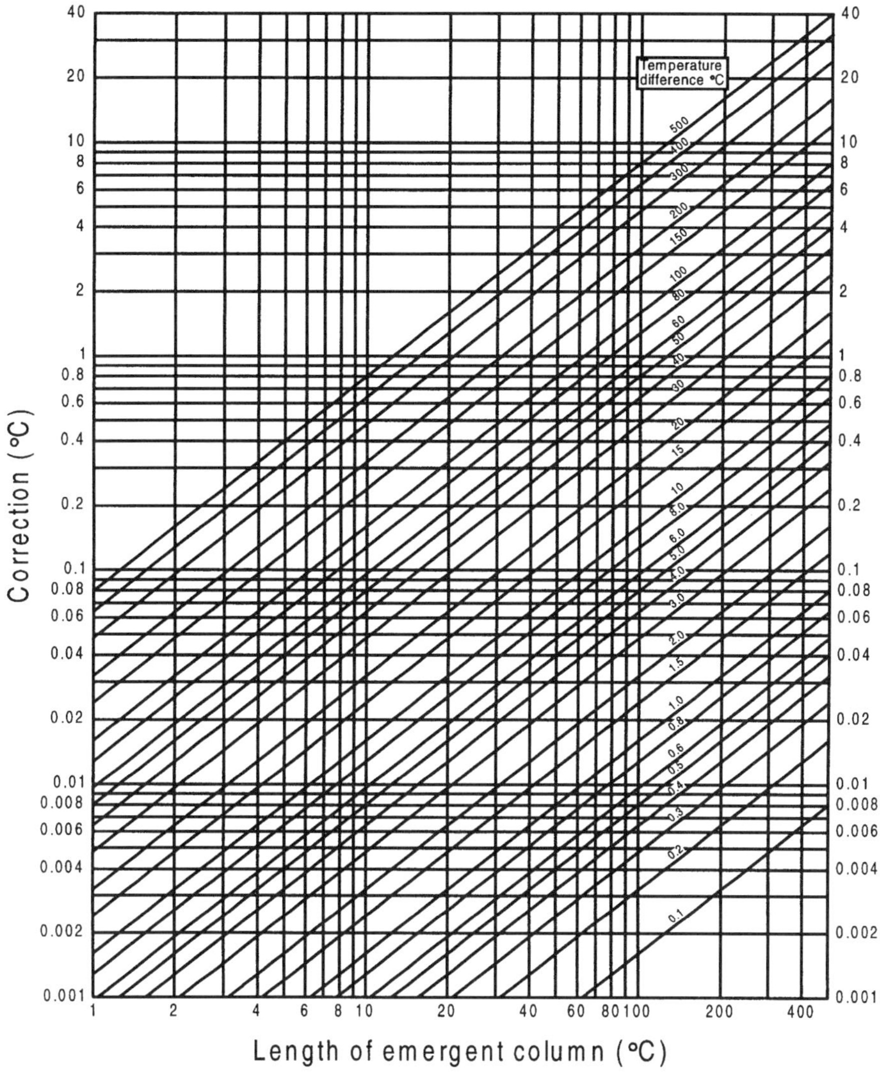

FIGURE 13.73 Chart of stem exposure corrections for mercury-in-glass thermometers with $k = 0.00016°C^{-1}$.

can give them as good a readability as mercury. Follow the recommendations of the section on Separated Columns and the section on Storage to get the best result from organic-liquid thermometers.

Storage

Most mercury-in-glass thermometers can be stored horizontally on trays in cabinets, care being taken to avoid any weight or pressure on the bulbs (one reason for the horizontal position). Avoid vibration. Corrugated cardboard, or similar material, can be used as a liner for a tray to prevent the thermometers from rolling.

Thermometers whose main range is below 0°C are better stored vertically, bulb down, in a cool place, but do not rest the thermometer on its bulb. This particularly applies to organic-liquid thermometers, which also should be shielded from light sources, as ultraviolet radiation can often degrade the liquid. If the top of the bore of a spirit thermometer is kept at a slightly higher temperature than the rest of the thermometer, then the volatile liquid will not condense in the expansion chamber.

High Accuracy

If a higher accuracy than 0.1°C is sought, the user will need to not only consult the references [4, 5], but also consult their calibration laboratory. Not all authorities give the same advice on how to achieve higher accuracy. It is important to apply very consistent procedures in line with the calibration. Below are the more important factors that will need further consideration.

Control of the *hysteresis effect* is important. The thermometers should not be used for 3 days after being exposed to elevated temperatures. Residual effects may be detectable for many weeks. That is, temperature measurements must always be made under rising temperature conditions. Three days is needed for the glass to relax, and in some cases, prolonged conditioning at a fixed temperature is required.

Avoidance of *parallax reading errors* is important because interpolation of the finer scale is essential. Optical aids are used, but they increase the parallax error. Good mechanical alignment is therefore required to keep parallax to a minimum.

The *pressure on the bulb* due to the length of mercury in the stem becomes important. Thermometers will give different readings in the horizontal and vertical positions. *External pressure variations* should also be considered.

In general, total immersion use of the thermometer is required to achieve the higher accuracy.

Defining Terms

See Figure 13.69 for an illustration of the terms used to describe the parts of a liquid-in-glass thermometer.

Emergent column: The length of thermometric fluid in the capillary that is not at the temperature of interest.
Ice point: The equilibrium between melting ice and air-saturated water.
Thermometric liquid: The liquid used to fill the thermometer.

References

1. ISO issue documentary standards related to the liquid-in-glass thermometer:
   ISO 386-1977 *Liquid-in-glass laboratory thermometer — Principles of design, construction and use.*
   ISO 651-1975 *Solid-stem calorimeter thermometers.*
   ISO 653-1980 *Long solid-stem thermometers for precision use.*
   ISO 654-1980 *Short solid-stem thermometers for precision use.*
   ISO 1770-1981 *Solid-stem general purpose thermometer.*
2. ASTM in their standards on Temperature Measurement, Vol. 14.03 include two standards related to liquid-in-glass thermometers:
   E1-95 *Specification for ASTM Thermometers.*
   E77-92 *Test Method for Inspection and Verification of Liquid-in-glass Thermometers.*
3. BSI publish a series of documentary specifications for thermometers, including:
   BS 593:1989 *Laboratory Thermometers.*
   BS 791:1990 *Thermometers for Bomb Calorimeters.*
   BS 1704:1985 *General Thermometers.*
   BS 1900:1976 *Secondary Reference Thermometers.*
4. J. A. Wise, *Liquid-in-glass Thermometer Calibration Service,* Natl. Inst. Stand. Technol. Spec. Publ., 250-23, 1988. A good treatment of calibration practice for liquid-in-glass thermometers, with a wider coverage than given here.
5. J. V. Nicholas and D. R. White, *Traceable Temperatures,* Chichester: John Wiley & Sons, 1990. The present chapter section was extracted and adapted from this text. The text explains how to make traceable calibrations of various temperature sensors to meet international requirements.

13.9 Manometric Thermometers

Franco Pavese

Manometric thermometers are defined in this Handbook as those thermometers that make use of the pressure of a *gaseous* medium as the physical quantity to obtain temperature. Very seldom are they available from commercial sources; for example, the temperature control of a home freezer is often of this kind. Consequently, instead of simply buying one, every user must build his own if this kind of thermometer is needed. They can be a quite useful choice since, in the era of electronic devices and sensors, it is still possible to make a totally nonelectronic thermometer, which in addition keeps its calibration indefinitely, as long as the quantity of substance sealed in it remains unchanged. The range of temperatures that can be covered by this kind of thermometer depends on the principle and on the substance used. When the thermodynamic equilibrium between the *condensed* phase of a substance (either liquid or solid) and its vapor is used, one has a "vapor-pressure thermometer" and the temperature range spanned by each substance is generally narrow. In addition, only substances that are gaseous at room temperature (i.e., condensed at temperatures lower than 0°C) are normally used, confining the working range to below room temperature; however, some substances that are liquid at room temperature and have a high vapor pressure (i.e., which easily evaporate) have bee used, but do not result in a sizable extension of the working range much above room temperature. A special case of vapor pressure being used at high temperature is the device called a "heat pipe," which is not used as a thermometer, but instead as an accurate thermostat [1]; using sodium, the working range is pushed up to ~1100°C. When a pure substance is used only in its gaseous state, one has a "gas thermometer," whose temperature range can be very wide, especially for moderate accuracy, depending mainly on the manometer; on the other hand, its fabrication is somewhat more complex and its use less straightforward.

Both thermometers can be built to satisfy the state-of-the-art accuracy of national standards (uncertainty better than ±0.001 K) or for lower accuracies, down to an uncertainty of ±1% or higher. Both thermometers require the measurement of pressure, in the range from less than 1 Pa up to 100 bar. Directions about the choice of the manometer can be found in Chapter 5.1 of this Handbook. A complete and specialized treatment on both vapor-pressure and gas thermometers up to room temperature and on pressure measurement instruments and techniques for gaseous media can be found in [2]. Gas thermometry above room temperature is treated in [3, 4].

In consideration of the fact that these kinds of thermometers typically must be built by the users, the following will concentrate on the basic guidelines for their design and fabrication.

Vapor Pressure

Figures 13.74 and 13.75 show the pressure values and the sensitivities in the temperature range allowed for each of the most common gaseous substances, considering also, in addition to the liquid phase, the use of the solid phase (where vapor pressure is lower) below the triple point. The lower end of the range is determined by the manometer uncertainty (for a given accuracy), the upper end by the full-scale pressure of the manometer (or by the full evaporation of the condensed phase).

Table 13.23 reports "certified" vapor pressure equations, linking the measured pressure p to the unknown temperature T. The reader might prefer them to the plethora of equations found in the literature, since they have been checked by official bodies and T is expressed in the ITS-90, the International Temperature Scale [5, text in 2]. More checked equations can be found in [6].

Figure 13.76 shows the general layout of a vapor-pressure thermometer. The fabrication of a vapor-pressure thermometer is not exceedingly difficult if a few guidelines are followed. Table 13.24 summarizes the most critical ones — design criteria and filling information — in a compact form [2]. Much more constructional details can be found in [1, 3]. In most cases, the manometer is located at room temperature, and the bulb is connected to it via a small-bore tube (the "capillary tube") without critical drawbacks. The accuracy of these thermometers ranges from ± 0.0001 K using very pure substances in calorimeters and precision mercury manometers, to ≈±1% using dial manometers.

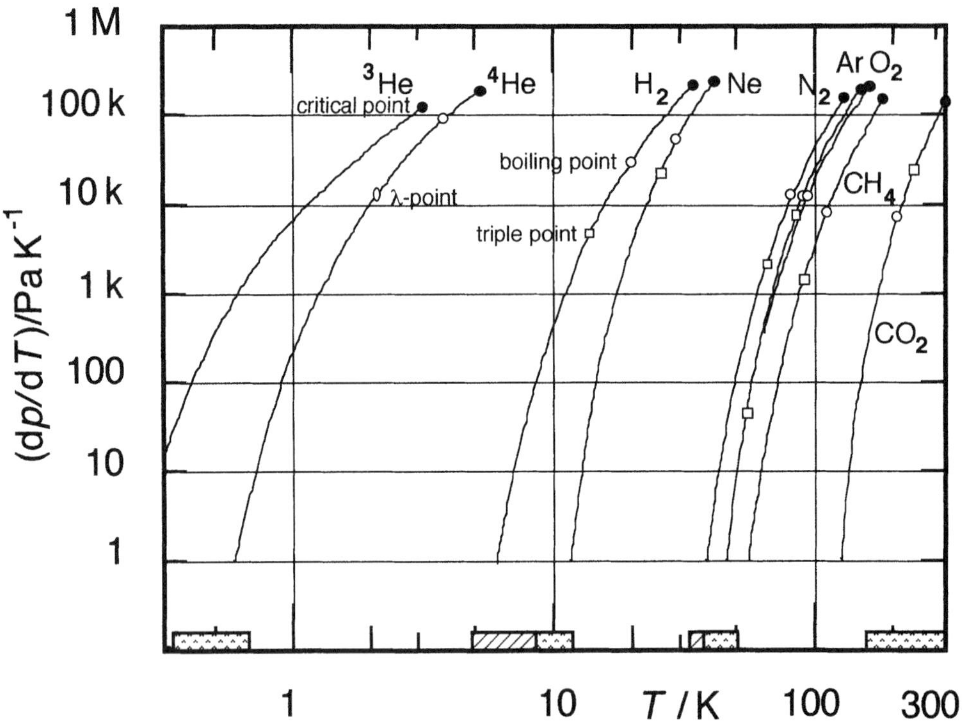

FIGURE 13.74 Range for vapor-pressure thermometry of various gases. The shaded parts indicate regions where it is less common or less accurate. [hatched], not available; [dotted], lower accuracy; ●, critical point; ○, triple point; 0, lambda-point.

Gas Thermometry

The layout of Figure 13.76 also applies to the design of a *constant-volume* gas thermometer (more common than the constant-pressure type), with the differences indicated in the relevant caption. The lower temperature end of the working range of a gas thermometer is stated, well before condensation of the substance takes place, by the increase of the uncertainty due to the increase in the nonideality of the gas — i.e., deviation from linearity of the relationship $p(T)$ — which takes place when approaching the condensed state or for increasing gas densities, or due to excessive absorption of gas on the bulb surface, thereby changing the quantity of the thermometric substance. All these conditions act at the lower end of the range. The upper end is stated by technological reasons or by the manometer full-scale capability. The best substances are, as listed, helium (either isotopes), hydrogen, and nitrogen.

From a design point of view, Table 13.25 summarizes the most critical issues. The major problem, apart from gas purity and ideality, is meeting the constant-volume requirement. Being that the manometer is generally at room temperature, the fraction of gas filling the connecting "capillary" tube is subtracted from the total amount of thermometric substance amount filling the system, and since this fraction is not constant, but depends on temperature and on technical conditions, it tends to increase the measurement uncertainty, which is contrary to the case of the vapor-pressure thermometer; this error is called the *dead-volume error.* Also, the bulb volume itself changes with temperature, due to thermal expansion and, to a much smaller extent, to the change in internal pressure. Design and fabrication criteria and measuring procedures are given in great detail in [2]. The case where the gas thermometer is *calibrated* at a number of fixed points is also described, with a discussion of the simplification in the use of the gas thermometer introduced with this instrument (called the *interpolating gas thermometer,* defined in the ITS-90 for use between 3 K and 26 K).

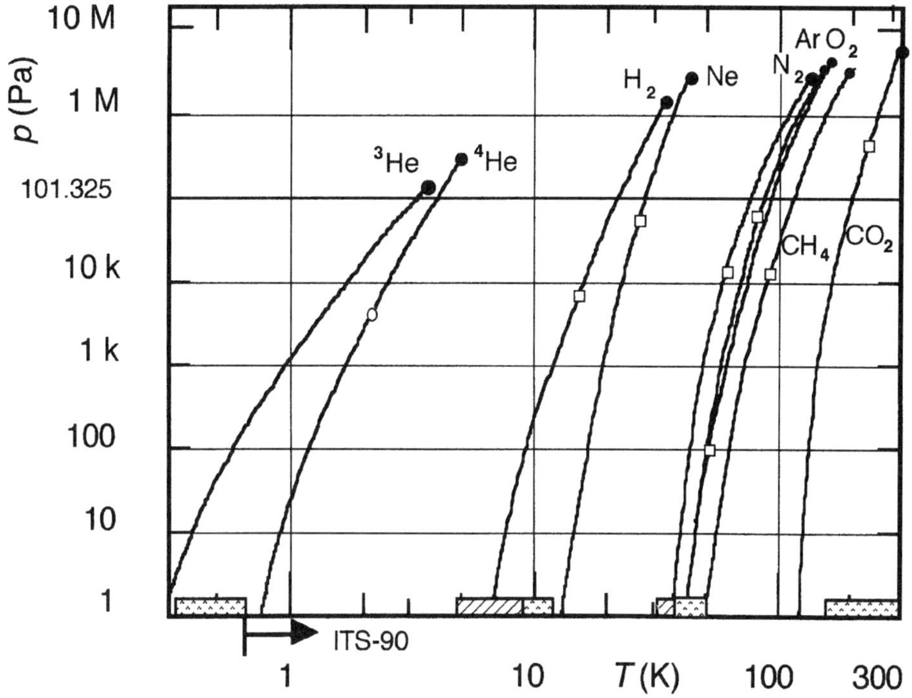

FIGURE 13.75 Sensitivity d*p*/d*T* of vapor-pressure thermometry for selected gases. The shaded parts indicate regions where it is less common or less accurate. (hatched), not available; (dotted), lower accuracy; ●, critical point; ○, triple point; 0, lambda-point.

TABLE 13.23 Vapor Pressure Equations

Equilibrium state	T_{90} (K)	Uncertainty $\pm\delta T$ (mK)	Purity of material[1] (vol%)
Liquid-vapor phases of helium-4	1.25–2.1768	0.1	99.9999
	$T_{90}/K = A_0 + \sum_{i=1}^{9} A_i\left[\left(\ln\left(p/\text{Pa}\right) - B\right)/C\right]^i$		
	$A_0 = 1.392408$	$A_1 = 0.527153$	$A_2 = 0.166756$
	$A_3 = 0.050988$	$A_4 = 0.026514$	$A_5 = 0.001975$
	$A_6 = -0.017976$	$A_7 = 0.005409$	$A_8 = 0.013259$
	$B = 5.6$	$C = 2.9$	
	2.1768–5.0	0.1	99.9999
	$A_0 = 3.146631$	$A_1 = 1.357655$	$A_2 = 0.413923$
	$A_3 = 0.091159$	$A_4 = 0.016349$	$A_5 = 0.001826$
	$A_6 = -0.004325$	$A_7 = -0.004973$	$B = 10.3$
	$C = 1.9$		
Liquid-vapor phases of equilibrium hydrogen	13.8–20.3	1[b]	99.99
	$p/\text{Pa} = \left(p_0/\text{Pa}\right)\exp\left[A + \frac{B}{T_{90}/K} + C\,T_{90}/K\right] + \sum_{i=0}^{5} b_i\left(T_{90}/K\right)^i$		
	$A = 4.037592968$	$B = -101.2775246$	
	$C = 0.0478333313$		
	$b_0 = 1902.885683$	$b_1 = -331.2282212$	$b_2 = 32.25341774$
	$b_3 = -2.106674684$	$b_4 = 0.060293573$	$b_5 = -0.000645154$

TABLE 13.23 (continued) Vapor Pressure Equations

Equilibrium state	T_{90} (K)	Uncertainty $\pm\delta T$ (mK)	Purity of material[1] (vol%)
Liquid-vapor phases of natural neon[c]	24.6–40	2	99.99
	$\log\left(\frac{p}{p_0}\right) = A + \frac{B}{T_{90}/K} + C\left(T_{90}/K\right) + D\left(T_{90}/K\right)^2$		
	$A = 4.61948943$	$B = -106.478268$	
	$C = -0.0369937132$	$D = 0.00004256101$	
Solid-vapor phases of nitrogen	56.0–63.1	2	99.999
	$\log\left(\frac{p}{p_0}\right) = A + \frac{B}{T_{90}/K} + C\left(T_{90}/K\right)$		
	$A = 12.07856655$	$B = -858.0046109$	$C = -0.009224098$
Liquid-vapor phases of nitrogen	63.2–125	5	99.999
	$\ln\left(\frac{p}{p_c}\right) = \frac{T_c}{T_{90}}\left[A\tau + B\tau^{0.5} + C\tau^3 + D\tau^6\right];\ \tau = 1 - \frac{T_{90}}{T_c}$		
	$A = -6.10273365$	$B = 1.153844492$	$C = -1.087106903$
	$D = -1.759094154$	$T_c = 126.2124$ K	$p_c = 3.39997$ MPa
Liquid-vapor phases of oxygen	54–154	2	99.999
	$\ln\left(\frac{p}{p_c}\right) = \frac{T_c}{T_{90}}\left[A\tau + B\tau^{1.5} + C\tau^3 + D\tau^7 + D\tau^9\right]$		
	$\tau = 1 - T_{90}/T_c$	$A = -6.044437278$	
	$B = 1.176127337$	$C = -0.994073392$	$D = -3.449554987$
	$E = 3.343141113$	$T_c = 154.5947$ K	$p_c = 5.0430$ MPa
Liquid-vapor phases of argon	83.8–150	5	99.999
	$\ln\left(\frac{p}{p_c}\right) = \frac{T_c}{T_{90}}\left[A\tau + B\tau^{1.5} + C\tau^3 + D\tau^6\right];\ \tau = 1 - \frac{T_{90}}{T_c}$		
	$A = -5.906852299$	$B = 1.132416723$	$C = -0.7720072001$
	$D = -1.671235815$	$T_c = 150.7037$ K	$p_c = 4.8653$ MPa
Liquid-vapor phases of methane	90.7–190	5[d]	99.99
	$\ln\left(\frac{p}{p_c}\right) = \frac{T_c}{T_{90}}\left[At + B\tau^{1.5} + C\tau^{2.5} + D\tau^5\right];\ \tau = 1 - \frac{T_{90}}{T_c}$		
	$A = -6.047641425$	$B = 1.346053934$	$C = -0.660194779$
	$D = -1.304583684$	$T_c = 190.568$ K	$p_c = 4.595$ MPa
Liquid-vapor phases of carbon dioxide	216.6–304	15	99.99
	$\ln\left(\frac{p}{p_c}\right) = A_0\left(1 - \frac{T_{90}}{T_c}\right)^{1.935} + \sum_{i=1}^{4} A_i\left(\frac{T_c}{T_{90}} - 1\right)^i$		
	$p_c = 7.3825$ MPa	$T_c = 304.2022$ K	$A_0 = 11.37453929$
	$A_1 = -6.886475614$	$A_2 = -9.589976746$	$A_3 = 13.6748941$
	$A_4 = -8.601763027$		

Note: For the relevant references and more gases, see [6]. $p_0 = 101325$ Pa, except when otherwise indicated.

[a] Minimum purity of the material to which the listed values of temperature and uncertainty apply.

[b] The summation term in the equation adds to the value of p a pressure amounting to the equivalent of 1 mK maximum.

[c] These values are for neon with an isotopic composition close to that specified in the ITS-90 definition.

[d] Above 100 K. It increases to 15 mK at 91 K, and to 10 mK near the critical point.

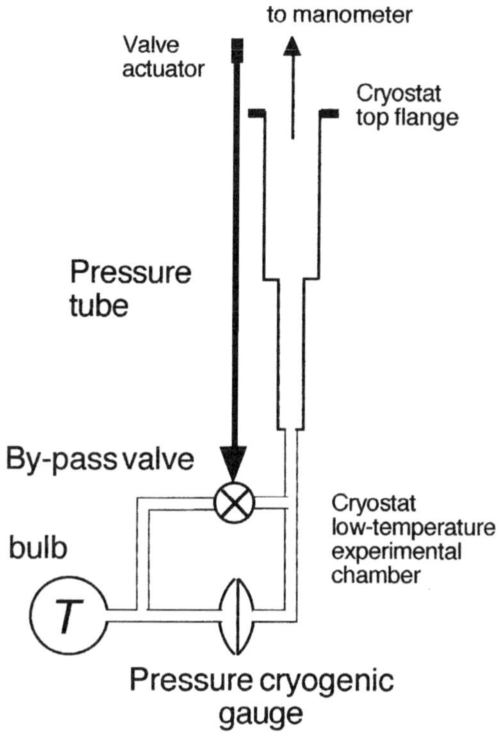

FIGURE 13.76 The general layout of a manometric thermometer. It is shown with a cryogenic diaphragm pressure transducer; when the transducer is placed instead at room temperature, the bypass valve is also placed at room temperature. *Vapor-pressure thermometer*: the diameter of the pressure tube increases in steps when pressures lower than ≈10 Pa must be measured, in order to decrease the thermomolecular pressure drop. *Gas thermometer* (constant-volume): the diameter of the pipes connecting the bulb to the cryogenic pressure transducer must be small in order to reduce the so-called "dead-volume." This requirement is much more stringent when the pressure transducer is moved up to room temperature. In this case, in order to reduce the error due to the "dead-volume," the bulb volume must be increased significantly.

TABLE 13.24 Summary of Design Criteria for Vapor-Pressure Thermometers

	Example
1. Choice of working substance:	$T_{max} = p_c$
• Temperature range: Each substance spans only a narrow temperature interval. T_{max}/T_{min} <2–3	T_{min} = 100 Pa K^{-1}
(including solid-vapor range), except helium. The limit:	T_{max}/T_{min} =
– T_{max} set by maximum manometer pressure.	^{3}He ≈ 10
– T_{min} set by manometer sensitivity.	^{4}He ≈ 9
• Accuracy:	H_2 ≈ 3
– Manometer: No single manometer spans whole range from ≈ 1 Pa (dp/dT ≈ 100 Pa K^{-1}) and	Ne ≈ 3
critical point ($p_c > 10^6$ Pa, except helium) with high or constant accuracy, or with sufficient	N_2 ≈ 2.5
sensitivity.	O_2 ≈ 2.5
– Substance: Not all substances allow for maximum accuracy, due to purity or to thermal	Ar ≈ 2.5
problems related to a low thermal diffusivity value.	CO_2 ≈ 2
2. Choice of pressure measuring system:	(solid ≈ 1.5)
Sensitivity and accuracy must be matched to the range of dp/dT and of p, i.e., T, to be measured.	
• *Without* separating diaphragm: Can be used only for low to medium accuracy, as thermometric gas also fills the entire manometric apparatus, with problems of contamination and increases in vapor volume.	
– Dial manometers: Used only for accuracy > ±1%.	
– Metal diaphragm or bellows (electronic) manometers: Can achieve a ±0.1–0.03% accuracy.	

TABLE 13.24 (continued) Summary of Design Criteria for Vapor-Pressure Thermometers

– Quartz bourdon gages: can approach a ±0.01% accuracy, but helium leaks through quartz.
– Cryogenic pressure transducers: None commercially available with accuracy better than ±0.1% (after cryogenic calibration). Eliminate need of the connecting tube in sealed thermometers, but transducer must withstand high room-temperature pressure.

- *With* separation diaphragm: Mandatory for high or top accuracy. Only zero reproducibility and a moderate linearity near zero are important.
  – Capacitive diaphragms: Several commercial models, when properly used, allow zero sensitivity and reproducibility better than ±0.1 Pa.
  – Cryogenic diaphragms: Only laboratory-made diaphragms available, some with high zero reproducibility. Allow to confine thermometric gas at low temperatures, but the tube connecting the diaphragm to room-temperature manometer is still necessary.

Room-temperature manometers: When a cryogenic diaphragm is used, only manometers allowing helium as manometric gas can be used.

3. Choice of sealed vs. "open" thermometer:
   - Sealed: Low-accuracy only (e.g., dial) thermometers.
     – Medium-accuracy sealed thermometers still very simple when using cryogenic manometer and reducing vapor volume, but room-temperature pressure can be higher than 10 MPa. Therefore, only low-sensitivity manometers can be used and thermometer measures only upper part of vapor-pressure scale.
     – High-accuracy sealed thermometers can be made, using ballast room-temperature volume and precision room-temperature diaphragm.
   - "Open": Vapor-pressure thermometers using gases are open only since working substance does not stay permanently in working bulb, but (new) samples are condensed in it only during measurements. Requires permanent use of a gas-handling system.
4. Gas purity, isotopic composition and spin equilibrium:
   - Purity: Must be known, and possibly checked, e.g., by performing a triple-point temperature measurement. Dew-boiling point difference measurement must also routinely be performed, before sealing in the case of sealed devices.
   - Isotopic composition: Some gases show irreproducibility in results due to sample-to-sample changes in isotopic composition. It is impossible to obtain top accuracy with these substances, unless pure isotopes are used.
   - Spin equilibrium: With some gases, showing different spin species, equilibrium must be ensured with use of a suitable catalyst.
5. Thermometer filling:
   - Amount of substance n_{max} at $T_{min} \rightarrow V^L \approx V_b$:

$$n_{max} \leq \frac{p_{min}}{R\,T_r}\left[\frac{2V_c T_r}{T_r + T_{min}} + V_r\right] + \frac{V_b}{M}\rho_{min}$$

   - Amount of substance n_{min} at $T_{max} \rightarrow V^L = V_\varepsilon^L \approx 0$:

$$n_{min} \geq \frac{V_\varepsilon^L \rho_{max}}{M} + \frac{p_{max}}{R\,T_r}\left[\frac{2V_c T_r}{T_r + T_{min}} + V_r + \frac{T_r}{T_{max}}\left(V_{max} - V_\varepsilon^L\right)\right]$$

   - Bulb volume V_b:

$$V_b\left[\frac{\rho_{min}}{M} - \frac{p_{max}}{R\,T_{max}}\right] \leq \frac{V_r}{R\,T_r}\left[p_{max} - p_{min}\right] + V^L\left[\frac{\boldsymbol{\rho}_{\mathbf{max}}}{\mathbf{M}} - \frac{\mathbf{p}_{\mathbf{max}}}{\mathbf{R\,T}_{\mathbf{max}}}\right]$$

$$+\frac{2V_c}{R}\left[\frac{\mathbf{p}_{\mathbf{max}}}{T_r + \mathbf{T}_{\mathbf{max}}} - \frac{\mathbf{p}_{\mathbf{min}}}{T_r + V_{\mathbf{min}}}\right]$$

to a first approximation the terms in **bold** can be omitted.

Example

(for item 4) Problems only for high accuracy
Kr, Xe
H_2, D_2

(for item 5) ^{4}He thermometer
$T_{min} = 2.2$ K
$p_{min} = 5.263$ kPa
$\rho_{min} = 146$ kg m^{-3}
$T_{max} = 5.2$ K
$p_{max} = 227.5$ kPa
$\rho_{max} = 67.5$ kg m^{-3}
$T_f = 4.2$ K
$p_f = 100$ kPa
$T_r = 300$ K
$p_r = 200$ kPa
$V_r = 220$ cm^3
$V_c = 16$ cm^3
$V_T = 500$ cm^3
$M = 4$ g mol^{-1}

It follows:
$V_b \geq \approx 2$ cm^3
Taking the minimum volume
$0.074 \geq n$
$n \geq 0.034$

TABLE 13.24 (continued) Summary of Design Criteria for Vapor-Pressure Thermometers

	Example
• Calculation of the amount of substance n to condense in the thermometer: the gas is stored at p_r in the room-temperature reservoir of volume V_T. When the substance is condensed in the thermometer bulb at a temperature T_f, a residual $n_o = p_f V_T / R\, T_r$ remains in V_T. Therefore, in order to condense a quantity N, one must have in the system: $$n' = \frac{p_r V_T}{R\, T_r}\left[1 - \frac{p_f}{p_r}\right] + \left(V_b + V_c + V_r\right)$$	In order to seal-in 0.05 mol, the filling system must contain 0.069 mol

Symbol caption: V^L = volume of the liquid phase; ρ = density; p = pressure; V = volume; subscript r = room temperature, c = capillary, b = bulb.

TABLE 13.25 Summary of Design Criteria for an Absolute Constant-Volume Gas Thermometer (CVGT) in the Low-Temperature Range ($T < 273.16$ K)

1. Choice of temperature range and of span $T_{min} \leftrightarrow T_{max}$:
   This choice is preliminary to the choice of most of the design parameters.
   - Below 273.16 K, ^{4}He gas thermometry is limited down to 2.5 K. With ^{3}He, accurate virial corrections available down to 1.5 K.
   - Only CVGTs of special design can be used in full span. Being that $p \propto T$, the 2.5 K to 273.16 K range corresponds to $p_{max}/p_{min} > 100$. For top accuracy, $\delta p/p < 0.01\%$, corresponding at p_{min} to $\delta p < 10^{-6}$. p_{max}, generally not achievable.
   - Being that $p \propto n/V$, molar density must generally be changed over the range to optimize accuracy, but n/V must be limited to avoid third virial correction, especially below ≈ 2 K.
   - In general, a CVGT is designed for work only below or only above a temperature between 25 K and 100 K.
2. Choice of reference temperature T_0:
   - Truly absolute thermometer: only one choice possible — 273.16 K.
     - Two-bulb CVGT: Avoids necessity to bring up to T_0 the bulb measuring $T_{min} > T < T_{max}$. Useful with thermometers designed for use at $T \ll T_0$.
     - Single-bulb CVGT: Same bulb spans the entire range up to T_0.
   - Low-temperature reference temperature T_0^* (≈ from 25 K to 90 K):
     - Single-bulb CVGT commonly used. T_0^* value assigned by an independent experiment, and, therefore, not exact by definition. However, the additional uncertainty is a minor inconvenience with respect to the advantage of limiting bulb temperature within the span $T_{min} \leftrightarrow T_{max}$.
3. Choice of thermometric gas and filling density:
   - Thermometric gas:
     - Nitrogen: Low-medium accuracy.
     - e-Hydrogen: Not used for over 50 years, but still suitable for low-medium accuracy and temperature range above ≈20 K.
     - Helium-4: Commonly employed in recent gas thermometers. Use limited to above 2.5 K.
     - Helium-3: Considered more in modern gas thermometry. Use presently limited to above 1.5 K; potential for use down to <1 K.
   - Filling density: $p \propto n/V$ and $dp/dT \propto n/V$ (1 kPa K^{-1} ≙ 121 mol m^{-3}). Always advantageous increasing n/V, up to an upper boundary set by need of third virial correction. As a rule, $n/V < 250$ mol m^{-3} above ≈2.5 K, $n/V < 160$ mol m^{-3}, down to 1.2 K and $n/V < 30$ mol m^{-3} at 0.8 K.
4. Choice of the pressure measuring system:
   - See Table 13.24.
5. CVGT parameter design:
   A. Room-temperature pressure transducer
   - Bulb: Top accuracy, 1 L volume typical; low accuracy, as low as 50 cm^3.
   - Dead-volume: Top accuracy, <10 cm^3; low accuracy: up to 10% of bulb volume.

   B. Cryogenic pressure transducer
   No difference with respect to a vapor-pressure thermometer.
6. Bulb design:
   - Volume may not be constant, because of:
     - Compression modulus: Walls must be thick to limit deflection due to pressure, or bulb must be enclosed in a guard chamber kept at bulb pressure. Stress in bulb material must be relieved by annealing after machining.

TABLE 13.25 (continued) Summary of Design Criteria for an Absolute Constant-Volume Gas Thermometer (CVGT) in the Low-Temperature Range ($T < 273.16$ K)

- Thermal expansion: Nothing can be done to suppress this effect, except using glass; must be corrected for. Small effect below ≈30 K.
- Amount of "active" gas might not be constant, because of:
  - Gas adsorption: Physicochemical interaction of bulb walls with the gas determines the amount adsorbed. Copper often gold-plated to limit adsorption: this prevents heating the bulb above 50–70°C.
  - Impurity molecules on walls and leaks: Clean machining used for metal bulbs, followed by physicochemical cleaning. The bulb sealing gaskets must be stable in shape and leak-proof at working temperatures.

7. Dead-volume design:
   Dead-volume effect comes from combination of geometrical volume, working pressure, and gas density distribution, i.e., from the amount of substance contained in it.
   - Room-temperature dead-volume: Consists of all volumes of the gas measuring system at room temperature. Must be kept at uniform temperature (except diaphragm, often thermostated at ≈40°C), to be measured within 0.1–1°C.
   - Low-temperature dead-volume: (Part of) capillary tube between room and bulb temperature. Temperature and density change from one end to the other. Tube diameter is a tradeoff between geometrical volume and thermomolecular pressure effect: typical values between 0.5 mm and 3 mm. Advantageous keeping the parts of tube where temperature variations occur as short as possible. For medium-high accuracy, temperature distribution must be known accurately.
8. Gas handling and measuring system (for non-sealed CVGTs):
   - Handling system: Must ensure purity, checked on-line with a mass spectrometer for the highest accuracy, and include gas recovery with cryogenic pumps and clean storage (or purification).
   - Measuring system (case A): Separating diaphragm requires valve system for zero check, including constant-value valves and provisions to avoid (or to restore) thermometric gas losses and contamination from the manometric gas. For this purpose, a second diaphragm separator can be used.

References

1. R. E. Bedford, G. Bonnier, H. Maas, and F. Pavese, *Techniques for approximating the ITS-90,* Monograph 90/1 of the Bureau International des Poids et Mesures, Sèvres: BIPM, 1990.
2. F. Pavese and G. F. Molinar, *Modern gas-based temperature and pressure measurements,* International Monograph Series on Cryogenic Engineering, New York: Plenum Publishing, 1992, and references therein.
3. J. F. Schooley, *Thermometry,* Boca Raton, FL: CRC Press, 1986.
4. T. J. Quinn, *Temperature,* London: Academic Press, 1983.
5. R. E. Bedford, G. Bonnier, H. Maas, and F. Pavese, Recommended values of temperature on the ITS-90 for a selected set of secondary reference points, *Metrologia,* 33, 133-154, 1996.
6. F. Pavese, Recalculation on ITS-90 of accurate vapour-pressure equations for e-H_2, Ne, N_2 O_2, Ar, CH_4 and CO_2, *J. Chem. Thermodynam.*, 25, 1351-1361, 1993.

13.10 Temperature Indicators

Jan Stasiek, Tolestyn Madaj, and Jaroslaw Mikielewicz

Temperature indicators serve for approximate determination of bodies' temperatures and are used to control a variety of temperature treatment processes. The temperatures are determined based on knowledge of characteristic rated temperatures, which are mean critical temperatures of the indicator. However, it should be stressed that the accuracy of these measurements is satisfactory only if the measurement conditions are similar to the standard conditions for which the temperature indicators were calibrated. Otherwise, the critical temperatures of the indicators can be different from their rated temperatures listed in the standards and the measurements can have considerable errors.

The temperature indicators can be classified into two groups, each group using different physical properties for the determination of the temperature. The indicators belonging to the first group melt at certain temperatures. For some of these indicators, such as pyrometric cones, thermoscope bars and rings, the process of melting manifests itself as a shape/size deformation for which the temperature is determined by measuring the degree/rate of deformation of the indicator. For others, such as melting pellets, liquids, crayons, and monitors, the melting means turning entirely into a liquid smear. This can also be accompanied by color changing. The second group consists of color-change indicators containing pigments that at different temperatures, show different colors by selectively reflecting incident white light. Among this group are reversible and irreversible paints, color-change crayons, and liquid-crystal indicators.

Melting and Shape/Size Changing Temperature Indicators

The latest British Standard BS 1041, Part 7, 1988 [1] lists the following temperature indicators: Seger cones, thermoscope bars, and Bullers rings. Some temperature indicators used in the past — such as Watkin cylinders and Holcdorft bars — are no longer used and are of historical value only. In the U.S., melting pellets, crayons, liquids and monitors are available on the market and widely used.

The rated temperature for Seger cones (pyrometric cones) and thermoscope bars is defined by appropriate shape deformation resulting from the transformation of a certain amount of the indicator substance from the solid to liquid state. For chemically pure elements and compounds at a constant pressure, the temperature during the entire process of phase change remains constant. If the pressure changes within the range of changes for atmospheric pressure, then the temperature changes are insignificant and can be neglected even during precise measurements. The pyrometric cones and thermoscope bars are prepared from complex mixes of frits, fluxes, clays, calcium and magnesium compounds, silica, etc. The melting temperatures of the indicators, also referred to as the critical or rated temperatures, vary with the proportions of the above compounds. Therefore, a set of indicators differing in the proportions of the compounds is capable of covering a required range of rated temperatures. The melting temperatures can also change, to a degree, with the proportion of phases. For mixes that constitute pyrometric cones and thermoscope bars, the temperature difference between the beginning and the end of the melting process can be as large as 25 to 40°C. At the rated temperature, one can assume that the temperature is either at the beginning or at the end of the melting process. In practice, an intermediate value is assumed, referring to an expected shape deformation of the temperature indicator.

For Bullers rings, the rated temperature is determined by a shape deformation that can be described as a temperature shrinkage. An indication of the required temperature is a proper contraction of the outer diameter of the ring made of special clay (a mix of appropriate materials) that contracts uniformly with the increase in temperature throughout the operating range.

For melting pellets, crayons, liquids, and monitors, the rated temperature is that of the beginning of the melting process when the indicator turns entirely into a liquid smear. Usually, on cooling, the liquid mark solidifies and becomes glossy-transparent or translucent in appearance. The entire process can be accompanied by a change in color — mostly because the color of the workpiece surface or the back of an adhesive label, which enables the contact of the indicator with the surface, will show up from under the transparent mark. However, the moment of melting — not a color change — is the temperature signal.

Seger Cones

The pyrometric cones are typically slender, truncated, trihedral pyramids, about 25 mm to 60 mm in height. The base of the pyramid is a regular triangle of side 7 mm to 16 mm. One edge of the pyramid is vertical or slightly leaned outward (see Figure 13.77a). The recommended height of the standard cones is 60 mm; the laboratory cones are 30 mm high.

The pyramids are manufactured by pressing a powder mixture of a number of minerals mixed in different proportions throughout the required range of rated temperatures. The main components are silicon oxide (SiO_2), aluminium oxide (Al_2O_3) with additives in the form of oxides (MgO, K_2O, Na_2O, CaO, B_2O_3, PbO), and an organic binder. The following equation is an example of chemical constitution of the pyrometric cone for the temperature range of 600°C to 900°C:

$$X\left(2SiO_2 + Al_2O_3\right) + \left(1 - X\right)\left(0.5Na_2O + 0.5PbO + B_2O_3 + 2SiO_2\right)$$

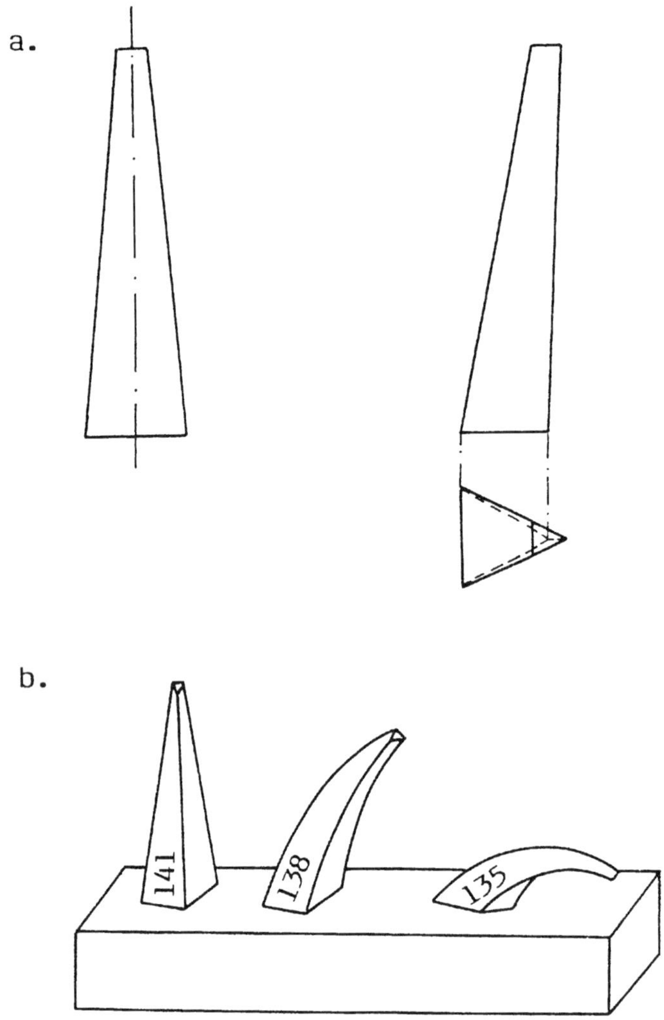

FIGURE 13.77 Pyrometric cones (a) in cross-sections; (b) on a plaque during firing.

where X is a mass unit. The pyramids designed for higher rated temperatures are prepared based on similar equations.

Touch-down Temperatures for Seger Cones.
As the heating progresses, the cone used for the measurement begins to soften and bends until its tip touches down on the surface on which it was placed. The rated temperature referring to this deformation is called the *touch-down temperature.* The touch-down temperatures for the Seger cones are determined in an electric kiln with clean atmospheric air at a heating rate of 60°C h^{-1}.

According to the earlier German standard DIN 51063 [2], the range of touch-down temperatures from 600°C to 2000°C at 10 to 50°C steps is covered by a series of Seger cones denoted traditionally by numbers from 022 to 42 (see Table 13.26a).

According to the latest British Standard BS 1041, Part 7, 1988, the touch-down temperatures within the range of 600°C to 1535°C at temperature intervals of 15 to 35°C are realized by a series of Seger cones numbered from 022 to 20 (see Table 13.26b).

The precision of determination of the touch-down temperatures for the industrial cones should be better than ±15°C; for the laboratory cones, better than ±10°C. If the heating rate undergoes change

TABLE 13.26a Approximate Touch-Down Temperatures of Pyrometric Cones (DIN 51063)

Cone no.	Temperature (°C)	Cone no.	Temperature (°C)	Cone no.	Temperature (°C)	Cone no.	Temperature (°C)
022	600	07a	960	9	1280	29	1650
021	650	06a	980	10	1300	30	1670
020	670	05a	1000	11	1320	31	1690
019	690	04a	1020	12	1350	32	1710
018	710	03a	1040	13	1380	33	1730
017	730	02a	1060	14	1410	34	1750
016	750	01a	1080	15	1435	35	1770
015a	790	1a	1100	16	1460	36	1790
014a	815	2a	1120	17	1480	37	1825
013a	835	3a	1140	18	1500	38	1850
012a	855	4a	1160	19	1520	39	1880
011a	880	5a	1180	20	1530	40	1920
010a	900	6a	1200	26	1580	41	1960
09a	920	7	1230	27	1610	42	2000
08a	940	8	1250	28	1630		

TABLE 13.26b Approximate Touch-Down Temperatures of Pyrometric Cones (BS 1041)

Cone no.	Temperature (°C)	Cone no.	Temperature (°C)	Cone no.	Temperature (°C)	Cone no.	Temperature (°C)
022	600	011	880	1	1135	11	1310
021	615	010	900	2	1150	12	1330
020	630	09	925	3	1165	13	1350
019	665	08	950	4	1180	14	1380
018	700	07	975	5	1195	15	1410
017	730	06	1000	6	1210	16	1435
016	760	05	1030	7	1230	17	1460
015	790	04	1060	8	1250	18	1485
014	810	03	1085	9	1270	19	1510
013	830	02	1105	10	1290	20	1535
012	860	01	1120				

Note: 1. Each temperature given in the table is that at which the tip of a cone will bend sufficiently to touch the base in an electric kiln with a heating rate of 60°C h^{-1}. 2. The touch-down temperature depends on the rate of heating: reports on firing behavior should quote the cone number, not the temperature taken from the above table. 3. Intermediate degrees of bending can be referred to the hands of a clock, e.g., 3 o'clock would represent a cone bent halfway to the stand.

within the range of 20 to 150°C h^{-1}, then the rated temperatures can change by –40°C for the above lower limiting value of the heating rate up to 60°C for the upper value. A 0.35% content of SO_2 in the atmosphere increases the rated temperatures by about 35°C. Also, the presence of soot in the atmosphere slightly raises the rated temperatures.

How to Use the Materials.

While single cones are sometimes used, usually three or four consecutively numbered cones, including a cone whose rated temperature is equal to the required temperature of the heat treatment and two cones of neighboring numbers (one less and one more) are employed for the temperature determination (see Figure 13.77b). They are installed into specially unfired plaques with tapered holes and protrusions that hold the cones firmly. The plaques are mounted to a workpiece surface to allow observations. A cone can be set up in other ways, such as inserting its base into refractory clay. However, it is necessary to assure a correct angle and firm hold of the cones during the firing cycle. Failure in these respects will

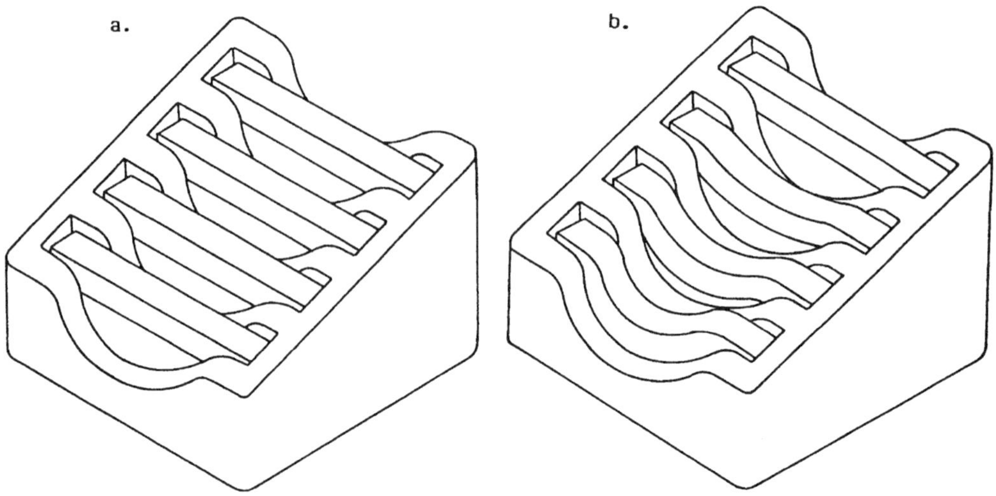

FIGURE 13.78 Thermoscope bars on a stand before and after firing.

cause the cone to bend unpredictably and give incorrect assessment of the heat treatment. If the process of heating takes place with a standard heating rate, then the rated temperature is reached when the tip of the central cone touches the base of the plaque. With further prolongation of the firing cycle, the cone will melt completely to form a blob on the plaque. The process of reaching the rated temperature is signaled in advance by the cone of one-less number. The cone of one-more number is there to prove that the required temperature value is not exceeded. Placing a series of cones with lower numbers (lower rated temperatures) provides the opportunity to carry out the process of heating at a required rate.

Typical Application.
Seger cones are used for the control of firing processes in the ceramics industry and artistry.

Thermoscope Bars

These indicators have the shape of bars of rectangular cross-sections. The typical dimensions of the bars are: length, 57 mm; width, 8 mm; and height, 6 mm. Bars of consecutive numbers (rated temperatures) are placed horizontally on a refractory stand as in Figure 13.78. The set of thermoscope bars is a more convenient and slightly modified form of Holcdorft bars. The bars are made of the same composites (mineral mixes and organic binder) as the pyrometric cones. The mixed powders are pressed and can be hardened by prefiring at relatively low temperatures, below those at which bending should occur.

Bending Temperatures of Thermoscope Bars.
The rated temperatures of thermoscope bars, referred to as the bending temperatures, are found during the calibration in an electric kiln with a heating rate of 60°C h^{-1} when the bars start to exhibit deformation (i.e., begin to bend).

According to the British Standard BS 1041, the range of rated temperatures from 590°C to 1525°C at temperature intervals of 15 to 35°C is covered by 42 thermoscope bars (see Table 13.27).

The precision of determination of the bending is about ±15°C; the other properties of the thermoscope bars referring to changes of the standard conditions are the same as for the pyrometric cones.

How to Use the Materials.
Four thermoscope bars of consecutive numbers — the first two having lower bending temperatures, the third one having the bending temperature equal or close to the required temperature, the fourth one having a higher bending temperature — are placed in sequence on a special refractory stand as in Figure 13.78a. The set is mounted to the workpiece surface where observations take place. If the process of heating takes place with a standard heating rate, then the beginning of deformation (bending) of the

TABLE 13.27 Approximate Bending Temperatures of Thermoscope Bars (BS 1041)

Bar no.	Temperature (°C)	Cone no.	Temperature (°C)	Cone no.	Temperature (°C)	Cone no.	Temperature (°C)
1	590	12	870	23	1130	33	1300
2	610	13	890	24	1145	34	1320
3	625	14	915	25	1160	35	1340
4	650	15	940	26	1175	36	1365
5	685	16	965	27	1190	37	1395
6	715	17	990	28	1205	38	1425
7	745	18	1015	29	1220	39	1450
8	775	19	1045	30	1240	40	1475
9	800	20	1075	31	1260	41	1500
10	820	21	1095	32	1280	42	1525
11	845	22	1115				

Note: 1. Each temperature given in the table is that at which the bar starts to bend in an electric kiln with a heating rate of 60°C h^{-1}. 2. The bending temperature depends on the rate of heating: reports on firing behavior should quote the bar number, not the temperature taken from the above table. 3. The bar can be expected to bend sufficiently to touch the stand at a temperature of 10°C to 30°C higher than the values given in the table, depending on the composition of the bar.

third bar indicates that the required temperature is reached. The process of reaching the rated temperature for the third bar is signaled in advance by the deformation of the proceeding bars whose behavior allows the evaluation of the heating rate. The unbent fourth bar testifies that the required temperature is not exceeded (see Figure 13.78b).

Typical Application.
The application of the thermoscope bars is identical to that of the Seger cones.

Bullers Rings

These temperature indicators in the form of rings have the following dimensions: outer diameter, 63 mm; inner diameter, 22 mm; and width, 8 mm. The appropriate measuring unit consists of a Bullers ring and a specially prescaled device for measurement of the temperature shrinkage of the ring. This contraction gage measures the outer diameter of the heated ring, based on which the heating work is assessed. The full measuring range of rated temperatures from 960°C to 1440°C is covered by four types of rings manufactured by pressing powders of ceramics mixes, with a binder, and without prefiring.

1. Rings denoted as 55 of brown color, suitable for temperatures from 960°C to 1100°C are used in the firing of glost ware and common building bricks where the finishing temperatures are relatively low.
2. Rings numbered 27/84, colored green, suitable for temperatures from 960°C to 1250°C are used for firing earthenware at the medium finishing temperatures, as well as tiles and bricks refractory with respect to low temperatures.
3. Rings numbered 75/84, colored natural, recommended for firing temperatures from 960°C to 1320°C, allow for higher finishing temperatures and are used for firing electrical porcelain, china, grinding wheels, and bricks refractory with respect to higher temperatures.
4. Rings numbered 73, colored yellow, recommended for temperatures from 1280°C to 1440°C for slow firing conditions as used in the manufacture of high-temperature ceramics and heavy refractories.

Approximate rated temperatures for the Bullers rings and corresponding readings of the contraction gage according to the British Standard BS 1041 are presented in Table 13.28.

How to Use the Materials.
One or more rings of the same type are placed vertically in a prefired stand and mounted to a workpiece surface. To determine the temperature as the firing progresses, the heated rings are withdrawn from their

TABLE 13.28 Approximate Rated Temperatures for Bullers Rings

Temperature (°C)	Gage readings			
	Ring no. 55	Ring no. 27/84	Ring no. 75/84	Ring no. 73
960	3	0	0	
970	7	1	1	
980	11	2.5	2	
990	15	4	3	
1000	18	5.5	4	
1010	21	7	5	
1020	24	8.5	6	
1030	27	10	7	
1040	30	11.5	8.5	
1050	32	13	10	
1060	34	14	11	
1070	36	15.5	12.5	
1080	37	17	14	
1090	38	18.5	15.5	
1100	39	20	17	
1110		21.5	18	
1120		23	20	
1130		24.5	21	
1140		26	22	
1150		27	23	
1160		28.5	24.5	
1170		30	26	
1180		31.5	27	
1190		33	28	
1200		34.5	29	
1210		36	30	
1220		37.5	31	
1230		38.5	32	
1240		40	33	
1250		41.5	34.5	
1260			36.5	
1270			38.5	
1280			40	29.5
1290			42	30
1300			44	31
1320			46	34
1340				37
1360				40.5
1380				44
1400				48
1420				51
1440				54

Note: These values should be used with caution because they are dependent on the firing cycle to which the rings are subjected.

stands, cooled to the ambient temperature, and then measured for contraction. This measurement is carried out on a gage consisting of a brass base plate on which a radial arm with a pointer moving over a scale and two steel dowel pins, against which the ring is pressed by the movable arm, are mounted (see Figure 13.79). A contraction of the ring gives rise to an amplified movement of the pointer over the scale. The divisions on the scale are numbered from –5 to 60. More heavily fired rings contract more and give higher readings on the gage. The divisions below 0 indicate expansion of the ring; above 0 indicates contraction. Rings should be measured across several diameters by turning them around in the gage so as to find the mean value to which the temperature can be assigned with the help of Table 13.28. Placing several rings in the stand in a manner that allows their easy withdrawal gives the possibility of measuring

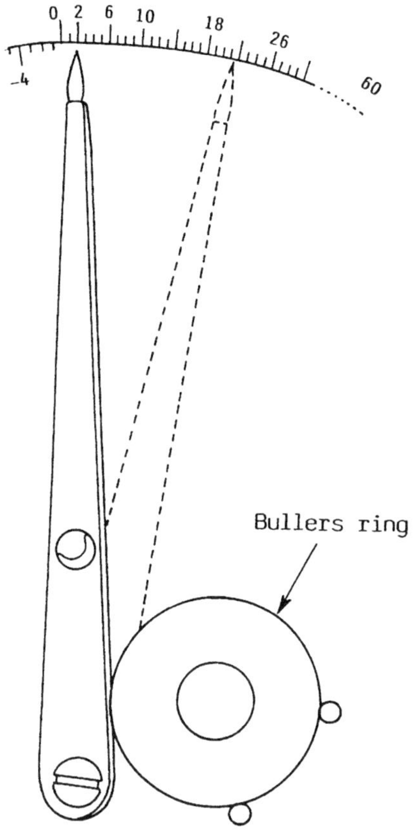

FIGURE 13.79 Contraction gage.

the heating rate. In a similar way, distribution of a number of rings throughout the furnace enables the determination of the temperature field in the furnace.

Usually, one or more test pieces from a series of rings are picked out for the sake of calibration so as to compare the obtained readings with the standard values enclosed in Table 13.28. Intermediate measurements are also carried out to evaluate the effect of the heating rate on the temperature shrinkage of the rings. The accuracy of the temperature determination for the standard heating conditions is ±0.5 of a single division of the scale.

Typical Application.

The application of the Bullers rings is similar to the Seger cones and thermoscope bars. An inconvenience is the fact that the Bullers rings require gage measurements and the temperature cannot be solely determined based on naked-eye observations.

Temperature-Indicating Pellets, Liquids, Crayons, and Monitors

Temperature-Indicating Pellets.

Temperature-indicating pellets are manufactured by pressing powders of mineral mixes of certain melting temperatures and an indifferent binder. Melting pellets are recommended as standard tablets ϕ7/16 × 1/8 and miniature tablets ϕ1/8 × 1/8 (see Figure 13.80). There are 112 different pellets which cover the temperature range from 40°C (100°F) to 1650°C (3000°F); see Table 13.29. The accuracy of the temperature determination is ±1% of the rated temperature. There are also available pellets for temperature control in strongly reducing atmospheres.

How to Use the Materials.

A pellet of the rated temperature equal to the required temperature of heat treatment is placed on the investigated surface before the heating starts. When the heating progresses, the beginning of melting

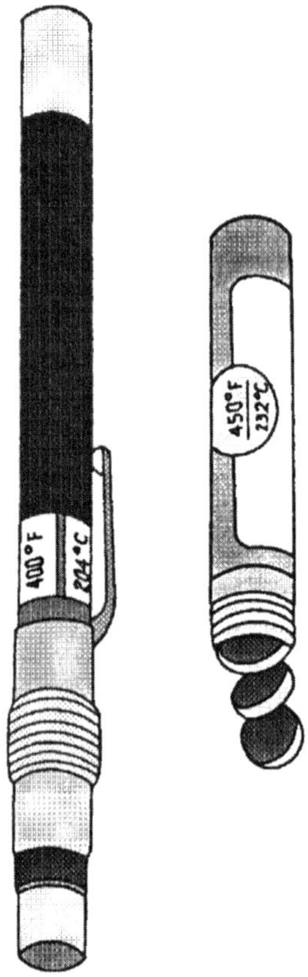

FIGURE 13.80 Temperature indicating crayon and pellets.

signals that the rated temperature is reached. Placing more pellets with the rated temperatures lower and higher than the required temperature enables more precise control of the heating process.

Typical Application.
Typical applications are checking furnace temperatures, control of heat treating of large units, as well as other applications involving long-duration heating.

Temperature-Indicating Liquids.
Temperature-indicating liquids are solutions of powdered mineral mixes in indifferent highly volatile solvents. They are available for use by brushing or spraying. There are over 100 different liquids with the rated temperatures from 40°C (100°F) to 1371°C (2500°F); see Table 13.29. The accuracy of the temperature determination is ±1% of the rated temperature.

How to Use the Materials.
A thin coating of the liquid is put on the clean and dry surface by brushing or spraying before the heating starts. It dries almost instantly to a dull opaque mark. When the required temperature is reached, this mark liquefies. The melted coating does not revert to its original opaque appearance but remains glossy-transparent on cooling. It should be noted that color changes do not signal the required temperature. The melting, not the color change, is the temperature signal.

Typical Application.
They are recommended for temperature control on fabrics, rubber, plastics, on smooth surfaces such as glass or polished metals, as well as for monitoring critical temperatures in electronic fields.

TABLE 13.29 Rated Temperatures for Temperature-Indicating Pellets, Liquids, and Crayons

°F	°C	°F	°C	°F	°C
100	38	325	163	1200[a]	649
103	39	331	166	1250[a]	677
106	41	338	170	1300[a]	704
109	43	344	173	1350[a]	732
113	45	350	177	1400[a]	760
119	48	363	184	1425	774
125	52	375	191	1450[a]	788
131	55	388	198	1480	804
138	59	400	204	1500[a]	816
144	62	413	212	1550	843
150	66	425	218	1600	871
156	69	438	226	1650	899
163	73	450	232	1700	927
169	76	463	239	1750[a]	954
175	79	475	246	1800	982
182	83	488	253	1850	1010
188	87	500	260	1900[a]	1032
194	90	525	274	1950	1066
200	93	550	288	2000	1093
206	97	575	302	2050	1121
213	101	600	316	2100	1149
219	104	625	329	2150[a]	1177
225	107	650[a]	343	2200[a]	1204
231	111	675	357	2250[a]	1232
238	114	700	371	2300[a]	1260
244	118	725	385	2350[a]	1288
250	121	750[a]	399	2400	1316
256	124	800[a]	427	2450	1343
263	128	850[a]	454	2500[a]	1371
269	132	900	482	2550[b]	1390
275	135	932	500	2600[b]	1427
282	139	950	510	2650[b]	1454
288	142	977	525	2700[b]	1482
294	146	1000	538	2800[b]	1538
300	149	1022	550	2900[b]	1593
306	152	1050[a]	566	3000[b]	1649
313	156	1100	593		
319	159	1150	621		

[a] Series "R" pellets for use in strongly reducing atmospheres.
[b] Available in pellets only.

Temperature-Indicating Crayons.
Temperature-indicating crayons are sticks manufactured from powders of mineral mixes of certain melting temperatures and an indifferent binder. The crayons are put in specially adjustable metal holders with labels saying their rated temperatures; see Figure 13.80. Similar to the temperature-indicating liquids, there are over 100 different crayons that cover the temperature range from 40°C (100°F) to 1371°C (2500°F); see Table 13.29. The accuracy of the temperature determination is also ±1% of the rated temperature.

How to Use the Materials.
During heating, the workpiece should be struck repeatedly by the crayon. Below its rated temperature, the crayon leaves a dry opaque mark. When the rated temperature is reached, the crayon leaves a liquid

smear. On cooling, the liquid mark will solidify with a transparent or translucent appearance. For temperatures below 700°F, the mark can be put on the workpiece surface before the heating process. The mark will liquefy when the rated temperature is reached. It should be remembered that the moment of melting, not any change in color, is the temperature signal.

Typical Application.
The crayons can be applied in welding, forging, heat treating and fabrication of metals, molding of rubber and plastics, wherever the workpiece is accessible during the heating process. Very smooth surfaces are excluded.

Temperature Monitors (Labels).
These temperature indicators are adhesive-backed labels with one or more heat-sensitive indicators under transparent circular windows. The indicators turn black (show black paper backing) when the rated temperature is reached. The rated temperatures are from 40°C (100°F) to 320°C (600°F). They are available as single temperature or multi-temperature indicators with 10°, 25°, or 50° steps. The tolerance of the temperature determination is ±1°C (±1.8°F) below 100°C, and ±1% of the rated temperature above 100°C. Exemplary rated temperatures for a series of 4-temperature (4-dot) indicators are presented in Table 13.30. A 4-dot temperature monitor is displayed in Figure 13.81.

TABLE 13.30 Rated Temperatures for 4-Temperature (4-dot) Labels

Model no.	°F	°C	°F	°C	°F	°C	°F	°C
4A-100	100	38	110	43	120	49	130	54
4A-110	110	43	120	49	130	54	140	60
4A-120	120	49	130	54	140	60	150	66
4A-130	130	54	140	60	150	66	160	71
4A-140	140	60	150	66	160	71	170	77
4A-150	150	66	160	71	170	77	180	82
4A-160	160	71	170	77	180	82	190	88
4A-170	170	77	180	82	190	88	200	93
4A-180	180	82	190	88	200	93	210	99
4A-190	190	88	200	93	210	99	220	104
4A-200	200	93	210	99	220	104	230	110
4A-210	210	99	220	104	230	110	240	116
4A-220	220	104	230	110	240	116	250	121
4A-230	230	110	240	116	250	121	260	127
4A-240	240	116	250	121	260	127	270	132
4A-250	250	121	260	127	270	132	280	138
4A-260	260	127	270	132	280	138	290	143
4A-270	270	132	280	138	290	143	300	149

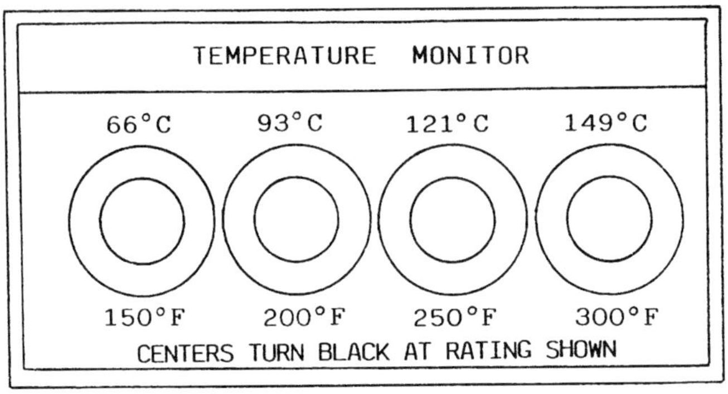

FIGURE 13.81 Four-dot label.

How to Use the Materials.
After removing the backing, the label is pressed firmly to the dry and clean workpiece surface. A change in color to black is the temperature signal. Application of multitemperature-indicating labels allows more precise temperature determination.

Typical Application.
They are especially applied for monitoring the safe operating temperature of equipment and processes, safeguarding temperature-sensitive materials during storage and transport.

The melting temperature indicators are described in the catalogs of their manufacturers [3, 4]. As the melting indicators are widely used in the U.S., temperatures in Fahrenheit are also given.

Color-Change Temperature Indicators

Color-change indicators comprise temperature-indicating paints, crayons, as well as liquid crystal indicators.

Temperature-Indicating Paints and Crayons

Temperature-indicating paints are basically acrylic lacquers containing finely dispersed temperature-sensitive inorganic pigments. The principle of operation of these indicators draws on the change in color of incident light reflected from the surface of the paint due to chemical reactions which the dispersed pigments undergo and creation of new compounds at specific transition temperatures. The color-change temperatures are also determined by the heating time. According to the British Standard, it is assumed that the rated temperatures of the paints correspond to the change in color at a heating interval of 10 min. To make the characteristics of temperature-indicating paints complete, the manufacturers also provide, together with the paints, graphs of trigger temperature vs. heating time relationships.

Temperature-indicating paints and crayons can be divided into two groups:

- Irreversible indicators: where the change of color becomes permanent
- Reversible indicators: after cooling and some time, they revert to previous colors.

Irreversible color-change indicators are complex compounds containing various metals, including cobalt, chromium, molybdenum, nickel, copper, vanadium, or uranium. However, they are lead- and sulphur-free. They are available on the market in the form of paints and crayons.

Irreversible Color-Change Paints.
The irreversible paints can change color once or several times during the heating process. Thus, another division can be made on single-change and multichange paints. The range of rated temperatures is from 40°C to 1350°C at 10 to 200°C steps. At the standard conditions, the tolerance of measurements is ±5°C for lower temperature values and ±1% for higher temperatures. Exemplary single-change paints with two critical temperatures — the initial trigger temperature for which the paint changes color after 10 min heating, and the cut-off temperature being the lowest temperature for which the color change is achieved for long-duration heating — are collected in Table 13.31. Color changes and critical temperatures for some multichange paints (changing color 2, 3, 3 or 6 times throughout the heating cycle) are presented in Table 13.32.

How to Use the Materials.
A thin layer of the paint is applied to a workpiece surface by brushing or spraying like an ordinary paint and allowed to dry before the heating starts. During the heating, when a point of the surface reaches or exceeds the critical temperature, a color change will take place. To determine the distribution of temperature, a multichange paint can be applied. With a nonuniform temperature rise, a number of colored bands separated by isothermal lines will appear on the workpiece surface, allowing the thermal record to be made of the temperature gradient across the surface.

Typical Application.
The temperature-indicating paints are widely used in industrial applications for observing heat patterns, detecting high and low temperature points on surfaces of heat engines, pipelines, and refrigeration fins.

TABLE 13.31 Single-Change Paints

Original color	Signal color	Initial trigger temperature[a] (°C)	Cut-off temperature (°C)
Pink	Blue	48	30
Pink	Blue	135	110
Mauve pink	Blue	148	120
Blue	Dark green	155	46
Yellow	Red	235	180
Blue	Fawn	275	150
Mauve red	Grey	350	220
Mauve	White	386	290
Green	Salmon pink	447	312
Green	White	458	312
Orange	Yellow	555	482
Red	White	630	450

[a] Color-change temperature for 10-min heating.

TABLE 13.32 Multichange Paints

Original color	Signal color	Initial trigger temperature[a] °C	Cut-off temperature °C
Light tan	Bronze green	160	150
Bronze green	Pale indian red	230	210
Reddish orange	Dark gray	242	193
Dark gray	Medium gray	255	211
Medium gray	Dirty white	338	228
Purple	Pink	395	355
Pink	Fawn	500	386
Fawn	Blue	580	408
Red	Dusty gray	420	310
Dusty gray	Yellow	555	328
Yellow	Orange	610	450
Orange	Green	690	535
Green	Brown	820	621
Brown	Green/gray	1050	945

[a] Color-change temperature for 10-min heating.

They can be also used for controlling temperatures of powered elements and surfaces that are inaccessible or revolve at high speeds.

Color-Change Crayons.

Color-change crayons, available in more than 10 distinct colors, similar in shape to regular crayons for drawing, cover the temperature range from 65°C to 670°C at 10 to 100°C temperature intervals. Exemplary single-change crayons are presented in Table 13.33. The accuracy of the temperature determination is the same as for the temperature-sensitive paints. They can be used for evaluating the temperature on already heated surfaces. They change color 2 min after reaching the rated temperature. Easy to use and inexpensive, they are invaluable for occasional temperature control in auto repairs, soldering, welding, electrical wiring, enameling, and for any operation involving boiling, baking, and other forms of heating.

TABLE 13.33 Single-Change Crayons

Original color	Signal color	Initial trigger temperature (°C)
Ivory	Light green	65
Yellow & green	Light green	75
Light pink	Blue	100
Gray & white	Light blue	120
Light ivory	Pink	150
Light blue	Black	200
Green	Black	280
Light green	Gray & brown	300
Blue	White	320
Brown	Red orange	350
White	Yellow	410
Light pink	Black	450
Ochre	Black	500
Blue	White	600
Green	White	670

Reversible Color-Change Indicators.
Reversible color-change indicators are available on the market as paints and in label form. The thermal pigments of these temperature indicators are mercury-based complexes. Therefore, they cannot be applied directly to metal surfaces as this causes decomposition. They also tend to decompose after long exposure to heat, but the decomposition can be retarded by using a clear over-lacquer. The pigments find their most successful application when encapsulated into labels.

The rated temperatures for the reversible color-change paints do not exceed 170°C. For temperatures up to 70°C under standard conditions, the tolerance of measurements is ±1°C; for 70 to 150°C, ±2°C; and for 150 to 170°C, ±3°C.

How to Use the Materials.
A thin layer of a reversible paint is applied to a workpiece surface by brushing or spraying, or a label is pressed to the surface. During the heating, a color change will take place when the temperature of the surface reaches or exceeds the critical temperature.

Typical Application.
The reversible color-change paints are widely used in the electrical industry, especially on busbars, live conductors, and connectors in high-current switches and in electronic fault-finding. They also find application as warning and indicating devices of domestic appliances. They are invaluable for controling lower temperatures when it is necessary to detect undesirable temperature excursions, correct faults, and revert to normal conditions.

Thermochromic Liquid Crystals

Liquid crystals constitute a class of matter unique in exhibiting mechanical properties of liquids (fluidity and surface tension) and optical properties of solids (anisotropy to light, birefringence). Certain liquid crystals are thermochromic and react to changes in temperature by changing color. They can be painted on a surface or suspended in a fluid and used to make the distribution of temperature visible. Normally clear, or slightly milky in appearance, liquid crystals change in appearance over a narrow range of temperatures called the color-play bandwidth (the temperature interval between first red and last blue), centered around the nominal event temperature (midgreen temperature). The displayed color is red at the low temperature margin of the color-play interval and blue at the high end. Within the color-play interval, the colors range smoothly from red to blue as a function of temperature; see Figure 13.82. Liquid

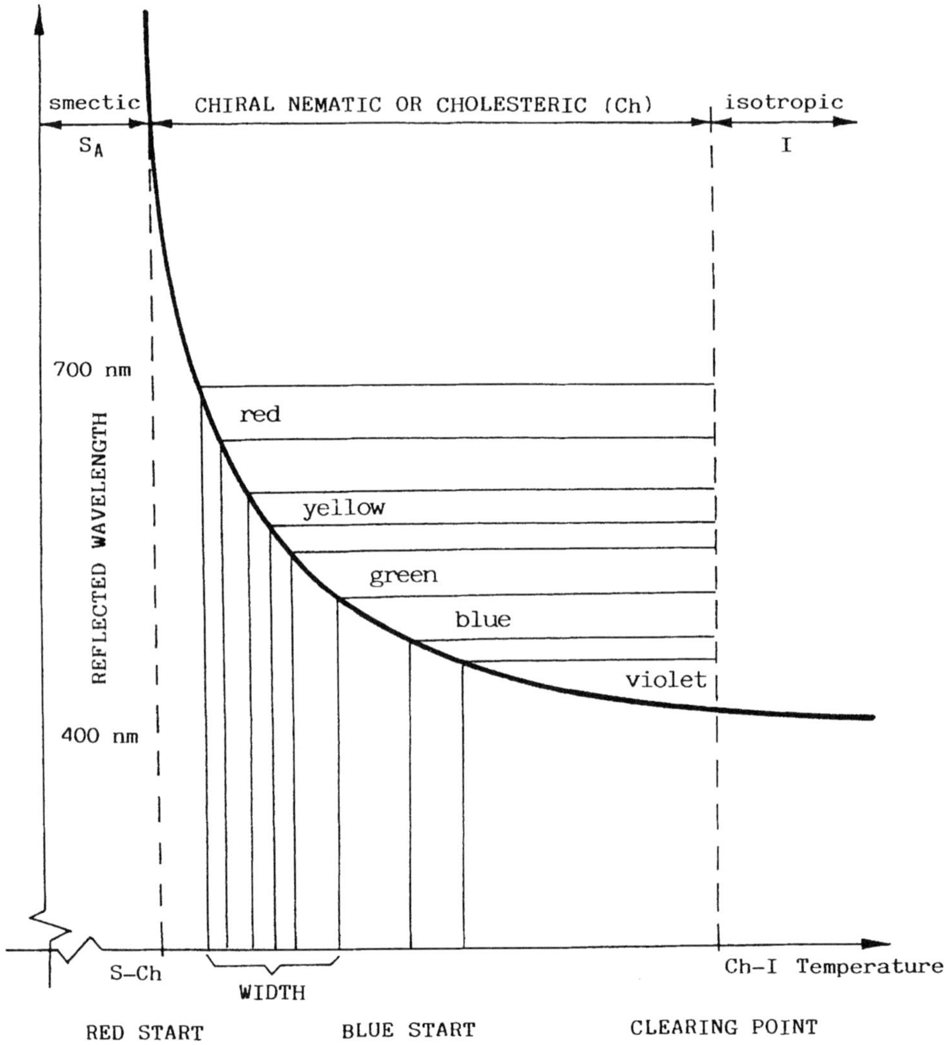

FIGURE 13.82 Typical pitch vs. temperature response of thermochromic liquid crystals.

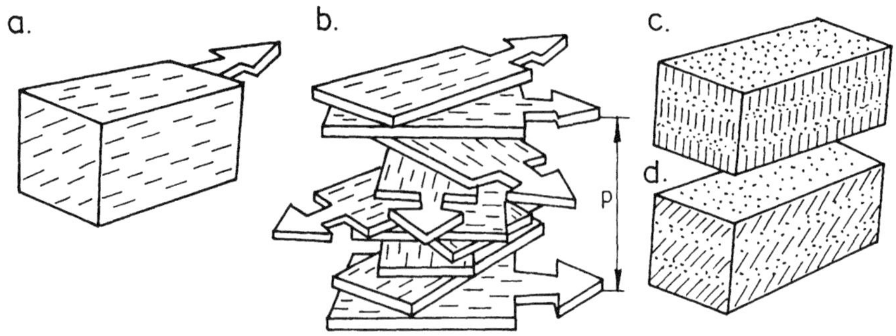

FIGURE 13.83 Structures of liquid crystals (a) nematic; (b) choleteric; (c) smectic A; (d) smectic B.

crystals or mesophases have been classified as smectic, chiral nematic, cholesteric, and blue. The structure of liquid crystals is shown schematically in Figure 13.83.

Temperature-Sensitive and Shear-Sensitive Formulations.
Temperature-sensitive liquid crystals show colors by selectively reflecting incident white light. Conventional temperature-sensitive mixtures turn from colorless (or black against a black background) to red at a given temperature and, as the temperature is increased, pass smoothly through the other colors of the visible spectrum in sequence (orange, yellow, green, blue, violet) before turning colorless (or black) again in the ultraviolet at a higher temperature. The color changes are reversible and on cooling the color change sequence is reversed.

Temperature-insensitive (sometimes called shear-sensitive) formulations can also be made. These mixtures show just a single color below a given transition temperature (called the clearing point) and change to colorless (black) above it. The working temperature range is thus below the clearing point. Both reversible and hysteretic (memory) formulations can be made. All liquid crystal mixtures should be viewed against nonreflecting backgrounds (ideally black, totally absorbing) for best visualization of the colors.

Color-Play Properties and Resolution.
Temperature-sensitive thermochromic mixtures have a characteristic red start or midgreen temperature and color-play bandwidth. The bandwidth is defined as the blue start temperature minus the red start temperature. The color play is defined by specifying either the red start or midgreen temperature and the bandwidth. For example, R35C1W describes a liquid crystal with a red start at 35°C and a bandwidth of 1°C, i.e., a blue start 1°C higher, at 36°C; G100F2W describes a liquid crystal with a midgreen temperature at 100°F and a bandwidth of 2°F.

Both the color-play bandwidth and the event temperature of a liquid crystal can be selected by its proper chemical composition. The event temperatures of liquid crystals range from –30°C to 115°C with color-play bands from 0.5°C to 20°C, although not all combinations of event temperature and color-play bandwidth are available. Liquid crystals with color-play bandwidths of 1°C or less are called narrow-band materials, while those whose bandwidth exceeds 5°C are referred to as wide-band. The type of material to be specified for temperature indicating should depend very much on the type of available image interpretation technique — human observers, intensity-based image processing, or true-color image processing systems (see [7]). The uncertainty associated with direct visual inspection is about 1/3 the color-play bandwidth, given an observer with normal color vision — about ±0.2°C to 0.5°C. The uncertainty of true-color image processing interpreters using wide-band liquid crystals is of the same order as the uncertainty assigned to human observers using narrow-band materials, and depends on the pixel-to-pixel uniformity of the applied paint and the size of the area averaged by the interpreter (about ±0.05°C can be achieved). Using a multiply filtered, intensity-based system, the resolution is better than ±0.1°C.

How to Use the Materials.
Liquid-crystal indicators can be used in a number of different forms: as unsealed liquids (also in solutions), in the microencapsulated form (as aqueous slurries or coating formulations), and as coated (printed) sheets. The different forms of the materials have selective advantages and suit different temperatures and flow visualization applications. Individual products are described in more detail in relevant booklets issued by the manufacturers of liquid crystals [5, 6].

Typical Application.
Liquid-crystal indicators are ideal for monitoring temperatures of electronic parts, transformers, relays, and motors. They are invaluable for a fast visual indication of temperatures.

References

1. BS 1041: Part 7. Temperature Measurement.
2. DIN 51063: Part 1. Testing of Ceramic Raw and Finished Materials, Pyrometric cone of Seger. Part 2. Testing of Ceramic Materials.
3. OMEGA International Corp. P.O. Box 2721, Stanford, CT 06906 (The Temperature Handbook).
4. TEMPIL Division, Big Three Industries, Inc. South Plainfield, NJ 07080 (Catalog GC-75).
5. HALLCREST Products Inc. 1820 Pickwick Lane, Glenview, IL 60025.
6. MERC Industrial Chemicals, Merc House, Poole Dorset, BH15 1TD, U.K.
7. Moffat, R.J., Experimental heat transfer, *Proc. 9th Int. Heat Transfer Conf.*, Jerusalem, Israel, 1990.

13.11 Fiber-Optic Thermometers

Brian Culshaw

Optical fiber sensing is a remarkably versatile approach to measurement. A fiber sensor guides light to and from a measurement zone where the light is modulated by the measurand of interest and returned along the same or a different optical fiber to a detector at which the optical signal is interpreted. The measurement zone can be intrinsic within the fiber that transports the optical signal, or can be extrinsic to the optical waveguide. The versatility of the fiber sensing medium arises in part because of the range of optical parameters that can be modulated and in part because of the diversity of physical phenomena that involve environmentally sensitive interactions with light.

For example, highly coherent light from a laser source can be introduced into a fiber and its phase modulated by a parameter of interest. The resulting phase changes can then be detected interferometrically. The phase change is simply a modification to the optical path length within the fiber, and can be modulated by shifts in temperature, strain, external pressure field or inertial rotation. A well-designed interferometer can detect 10^{-7} radians — equivalent to 10^{-14} m!

The laser light could also be Doppler shifted through reflection from a moving object. Its state of polarization can be changed. Its throughput intensity can be modified or the light can be used to stimulate some secondary emissions, which in turn can be monitored to produce the relevant optical signal. If the light is incoherent, then its wavelength distribution (color) can be modified, in addition, of course, to the possibilities for polarization changes and intensity changes.

The physical phenomena capable of imposing this modulation are again many and varied. They include, for example, periodically bending an optical fiber to introduce a localized loss that depends on the sharpness of the bend (usually referred to as microbend loss); changing the relative refractive indices of the core and the cladding of the optical fiber and thereby changing the guiding properties and again introducing a loss; modifying an optical phase delay by introducing a change in refractive index or a change in physical length; examining changes in birefringence introduced through modifications to physical stress and/or temperature; using external indicators to color modulate a broadband source and relate the color distribution to temperature, chemical activity, etc. These are all linear effects where the input optical frequency is the same as the output optical frequency (regarding Doppler shift as a rate of change of phase of an optical carrier) and where, for a given system setting, the output at all frequencies is directly proportional to the input.

Nonlinear effects are also widely exploited. Of these, the most important are fluorescence, observed usually in fluorophores external to the optical fiber, and Raman and Brillouin scattering, usually observed within the fiber itself. In all these phenomena, the light is absorbed within a material and re-emitted as a different optical wavelength from the one that was observed. The difference in optical wavelengths depends on the material and usually on strain and temperature fields to which the material is subjected. These major features of optical fiber sensors are encapsulated in Figure 13.84.

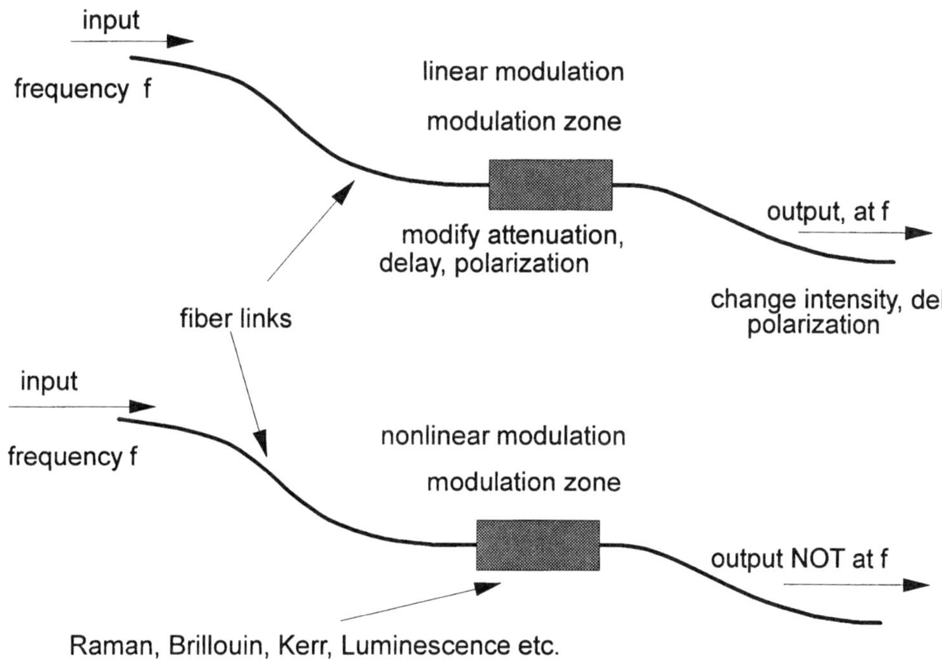

FIGURE 13.84 Linear and nonlinear optical processes for measurement using optical fiber sensors.

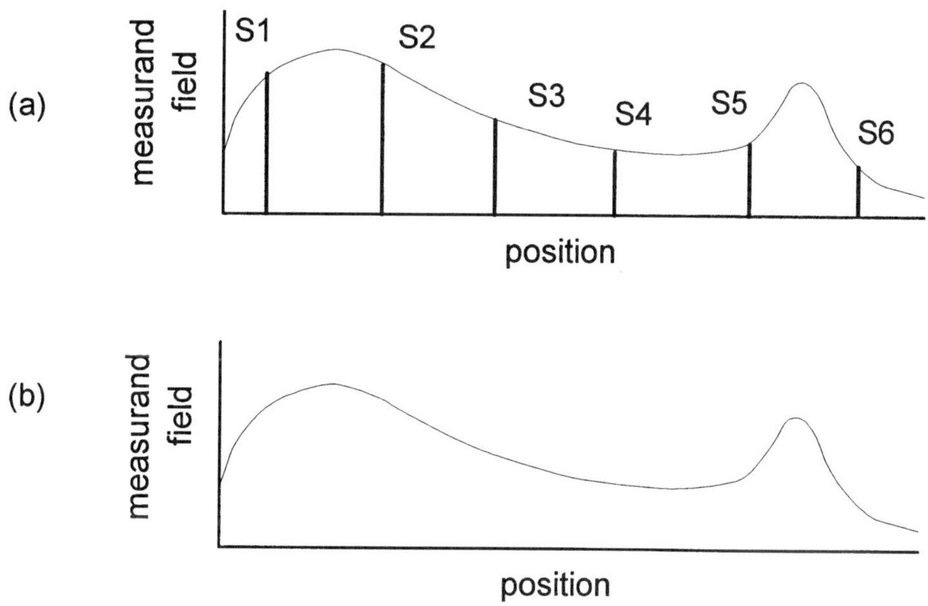

FIGURE 13.85 Sensor system outputs for (a) point array and (b) distributed sensor systems.

Optical fiber sensors have an additional feature that is unique to the medium — namely, the abilities for intrinsic networking in either distributed, quasi-distributed/multiplexed, or discrete (point) configurations. The essential features of these achitectures are sketched in Figure 13.85. For intrinsic sensors, the fiber responds to the measurand throughout its length, and the output in transmission is an integral of this linear response. However, using an interrogation scheme in reflection that incorporates a delay proportional to distance along the fiber enables the system to retrieve the measurand as a function of position. These are distributed sensors. The quasi-distributed architecture examines separately identified individual (usually adjacent) lengths of fiber and extracts the integral of the measurand along each of these individual lengths. Distributed and quasi-distributed sensors effectively convolve the measurand field along the interrogation fiber with a window determined by either the time resolution of the interrogating electronics (distributed architectures) or the defined lengths of the individual fiber sections (quasi-distributed systems).

Point and multiplexed systems address the measurement as essentially a point sample located at a specific distance along the interrogating fiber. All these architectures can realized in all optical fiber form and have been demonstrated to address a very wide range of measurements, often within a single network. The availability of distributed sensing is unique to optical fiber technology, as indeed are optical fiber-addressed passive arrays.

In summary, the optical fiber approach to measurement has the demonstrated capability to address a wide range of physical, chemical, and biological parameters. It must take its place along side other competing technologies against which its merits must be assessed. The principal benefits of using fiber optics include:

- Immunity to electromagnetic interference within the sensor system and within the optical feed and return leads
- The capacity for intrinsic distributed measurements
- Chemical passivity within the sensor system itself and inherent immunity to corrosion
- Small size, providing a physically, chemically, and electrically noninvasive measurement system
- Mechanical ruggedness and flexibility: optical fibers are exceptionally strong and elastic — they can withstand strains of several percent
- High temperature capability — silica melts at over 1500°C

There remain cost and user acceptability deterrents within the exploitation of optical fiber sensor technology. Consequently, the majority of field experience in optical fiber sensors is targeted at addressing the specialized problems where these aforementioned benefits are paramount. Many of these lie in the area of temperature measurement.

Fiber Optic Temperature Sensors

The important phenomena that have been exploited in the optical techniques for temperature measurement include:

- Collection and detection of blackbody radiation
- Changes in refractive index of external media with temperature
- Changes in fluorescence spectra and/or fluorescence rise times with temperature
- Changes in Raman or Brillouin scatter with temperature
- Phase transitions in carefully selected materials imposing mechanical modulation on optical fiber transmission properties
- Changes within an optical path length with temperature, either within the fiber or an external interferometer element

Within these phenomena, Brillouin and Raman scatter and mechanical phase transitions have been primarily used in distributed measurement systems. Some distributed measurement/quasi distributed measurement systems based on modulated to phase delay have also been evaluated, although they have yet to reach commercial reality. The remaining phenomena are almost exclusively used in point sensor systems.

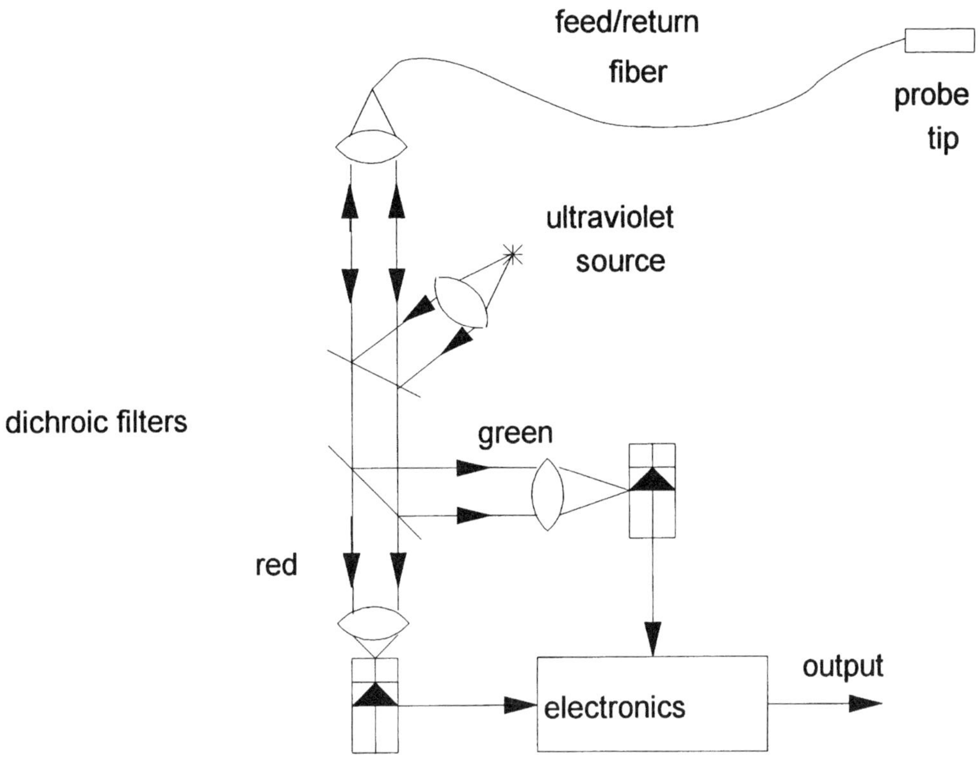

FIGURE 13.86 Optical fiber fluorescent thermometer.

Fiber Optic Point Temperature Measurement Systems

One of the first commercial optical fiber sensors was a fluorescence-based temperature probe introduced in the early 1980s by the Luxtron Corporation of Mountain View, CA. Successors to these early sensors are still commercially available and are a very effective, but expensive, approach to solving specific measurement problems. These include monitoring temperature profiles in both domestic and industrial microwave ovens, examining temperatures in power transformer oils, motor/generator windings, and similar areas where (primarily) the issue is the operation of a reasonably precise temperature probe within very high electromagnetic fields. In such circumstances, a metallic probe either distorts the electromagnetic field significantly (e.g., in microwave ovens) or is subjected to very high levels of interference, producing spurious readings. Other applications sectors exploit the small size or chemical passivity of the device, including operation within corrosive solvents or examination of extremely localized phenomena such as laser heating or in determining the selectivity of radiation and diathermy treatments.

The principles of the probe are quite simple and are shown in Figure 13.86. The rare earth phosphor is excited by an ultraviolet light source (which limits the length of the silica-based feed fiber to a few tens of meters) and the return spectrum is divided into "red" and "green" components, the intensity ratios of which are a simple single-valued function of phosphor temperature. For precision measurement, the detectors and feed fiber require calibration and, especially for the detectors, the calibration is a function of ambient temperature. However, this can be resolved through curve fitting and interrogation of a thermal reference. The instrument, which has now gone through several generations to improve upon the basic concept, is capable of accuracies of about ±0.1°C within subsecond integration times over a temperature range extending from approximately –50°C to ±200°C. Since its introduction, this particular

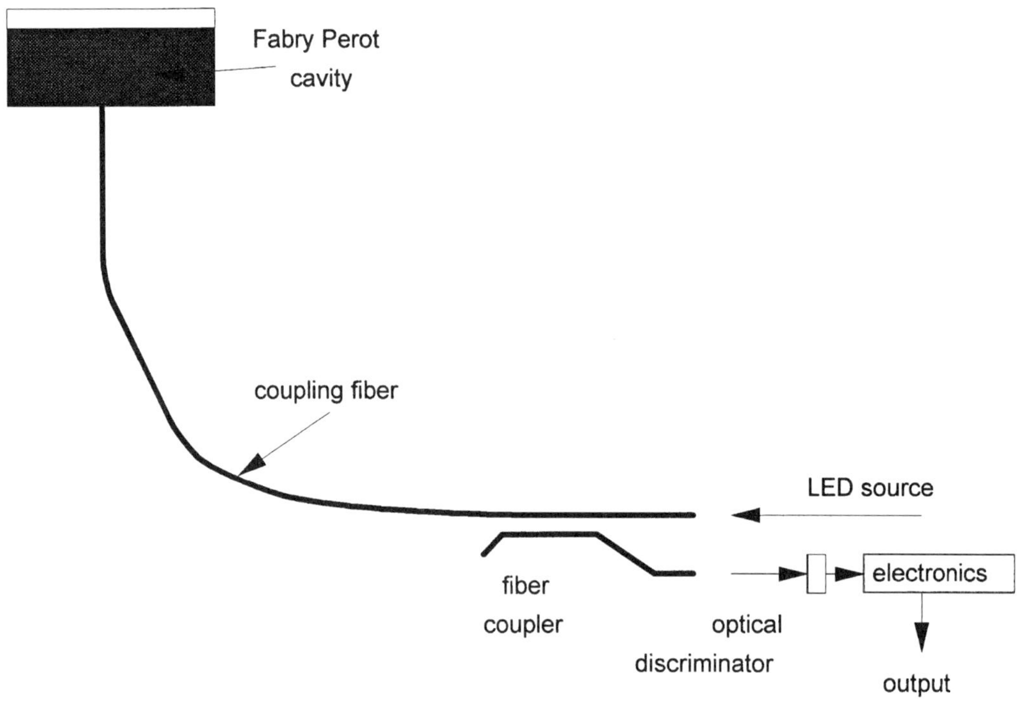

FIGURE 13.87 Optical fiber thermometry using short temperature-sensitive Fabry Perot cavity.

probe has accumulated extensive field experience in a wide variety of applications and remains among the most widely exploited fiber optic sensor concepts.

A number of temperature probes based on fluorescence decay time measurements have also been demonstrated. The level of commercial activity exploiting these concepts has, to date, been very modest, partly because the accurate measurement of decay times can be problematic.

Measuring the temperature response of dyes and other thermally sensitive color-selective materials can afford a very simple approach to temperature measurements. Among the most successful of these has been the temperature probe examining the bandedge of gallium arsenide introduced by ASEA (now ABB), again in the early 1980s and now transferred to Takaoka. The bandedge can either be monitored through examining the spectra of induced fluorescence or through interrogating the absorption characteristics of the material when subjected to a constant spectrum excitation. The accuracy and temperature range of this probe are comparable with those of the Luxtron system, and this particular version of the bandedge probe has the additional benefit of operating primarily in the near-infrared range of the spectrum, thereby accessing the best transmission characteristics of the optical fiber medium. The probe was originally conceived to address ASEA's internal needs in monitoring electrical power system components. Similar bandedge probes have also been demonstrated based on absorption edge detection in materials such as ruby.

Refractometry and interferometry are potential extremely sensitive thermal probes. Several have been demonstrated, some of which achieve microkelvin resolution. Interferometric detection or exploitation of sensitive mode coupling phenomena is the source of this very high sensitivity, although rarely is such high sensitivity required in practice. The relatively simple Fabry Perot probe shown in Figure 13.87 has been introduced commercially with simplified spectral analysis and a semiconductor source, although as yet its market penetration has been relatively modest.

Optical pyrometry is a well-established approach to measuring temperatures in the hundreds to thousands of degrees Centigrade. The disappearing filament pyrometer has been used in this fashion for over half a century. The optical fiber equivalent has also found a few niche applications. The general

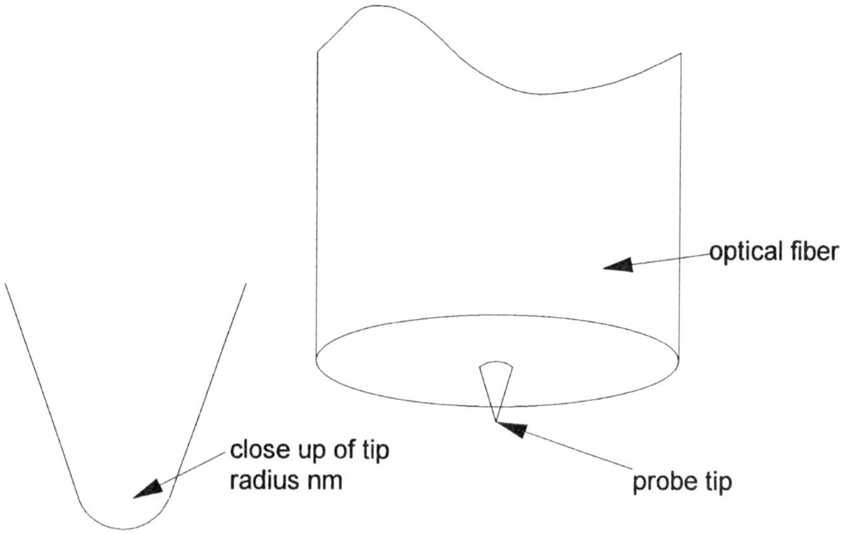

FIGURE 13.88 Probe for photon tunneling microscope and nano optrode.

form of such a sensor is to place the black body at the end of the fiber and place it with the fiber into the hot zone. The consequent radiation within the transmission spectrum of the fiber is then monitored using a semiconductor photodetector that can be based on III-V materials or silicon. The received radiation is then primarily within the red and near-infrared from about 600 nm to, depending on the detector, 1.8 μm. Blackbodies radiate significantly in this range at temperatures in the hundreds of degrees Centigrade and above. The most significant success of this approach has been in the fabrication of the reference standard temperature probe at NIST for temperatures above 1200°C. This uses a sapphire rather than silica collection fiber because of its superior optical and thermal properties within the temperature and the wavelength ranges of interest. It defines these high temperatures with subdegree precision.

Optical fibers also have the capacity to make unique nanoprobes — the opposite end of the scale by orders of ten from the distributed sensors discussed below. These (Figure 13.88) are tapered optical fibers with the end reduced in diameter to tens of nanometers. The tapered region is coated with a metal, often aluminium or silver, to confine the optical field. This produces an intense spot of light at the fiber tip which irradiates an area nanometers in dimensions. The tip can be coated with the dye or the fluorescent thermally sensitive material and used to monitor temperature over extremely small areas. This enables thermal profiles within cellular dimensions to be assessed. In other formats, the same probe can also be used to address chemical activity and chemical composition.

Fiber optic temperature sensing can be realized using a wide variety of techniques primarily, but not exclusively, based on the variation of optical properties of materials with temperature. An example of the exceptions is the optical excited vibrating element probe shown in Figure 13.89. This probe has been primarily used for pressure measurement and is now available commercially for pressure assessment down-hole in oil wells. It can also be configured to exhibit extremely high temperature sensitivity with accuracies and resolutions in the millikelvin region. It uses the beneficial features of mechanical resonators and the consequential frequency read-out in parallel with optomechanical excitation and direct optical interrogation to produce a probe that can be reliably exploited over interrogation lengths of tens of kilometers.

Fiber optic point sensors for temperature measurement are now a relatively mature technology. Most of the devices mentioned above were first introduced and characterized 10 or more years ago and have since been refined to address specific applications sectors. They remain expensive, especially when compared to the ubiquitous thermocouple, but their unique capability for noninvasive electrically passive interference immune measurement give them a very specific market address that cannot be accessed using alternative technologies. Within these market areas, the probes have been extremely successful.

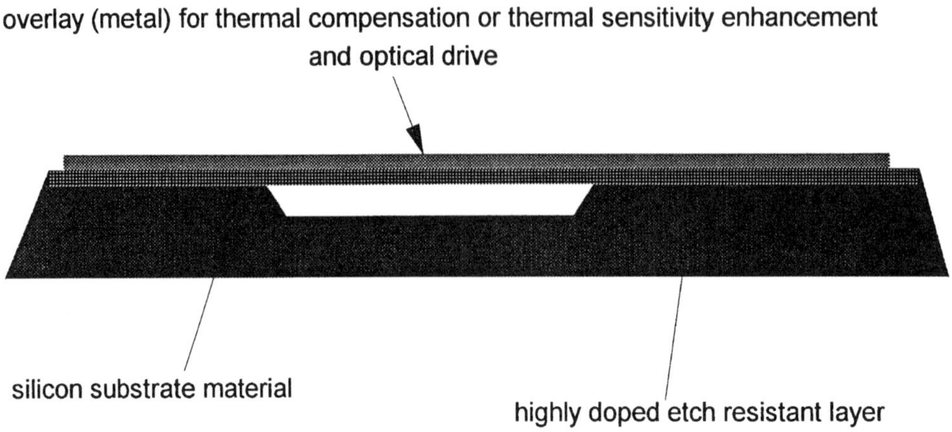

FIGURE 13.89 Longitudinal section of silicon optically excited microresonator.

Distributed and Quasi-distributed Optical Fiber Temperature Measurement Systems

These systems all exploit the unique capability for optical fibers to measure and resolve environmental parameters as a function of position along the fiber length. This generic technology is unique to optical fiber systems and, while there are a few commercial distributed temperature sensor systems available, the research in this sector continues.

The stimulated Raman scatter (SRS) distributed temperature probe is the most well established of these and, in common with many of the point sensors, was originally introduced commercially in the late 1980s. The principle (Figure 13.90) is quite simple. Within the Raman backscatter process (and also within the spontaneous Brillouin backscatter process), the amplitudes of the Stokes and anti-Stokes lines are related to the energy gap between these lines by a simple $\exp(-\Delta E/kT)$ relationship. Therefore, measuring this ratio immediately produces the temperature. Furthermore, this ratio is uniquely related to temperature and cannot be interfered with by the influence of other external parameters. The system block diagram is shown in Figure 13.91. The currently available performance from such systems enables resolutions of around 1 K in integration times of the order of 1 min, with resolution lengths of one to a few meters over total interrogation lengths of kilometers. The interrogation can extend to tens of kilometers if either the interrogation times are increased or the temperature and/or spatial resolutions are relaxed. The system is available from both European and Japanese manufacturers. The applications are very specific, as indeed they must be to accommodate an instrument price that is typically in tens of thousands of dollars. The instruments have been used in a variety of highly specific areas, ranging from monitoring temperature profiles in long process ovens to observing the thermal characteristics within large volumes of concrete during the curing process.

Distributed temperature alarms triggering on and locating the presence of either hot or cold spots along the fiber can be realized at significantly lower costs and have been modestly successful as commercial systems. The first of these — and probably the simplest — was originally conceived in the 1970s. This uses a simple step index fiber in which the refractive index of the core material has a different temperature coefficient than that of the cladding material. The temperature coefficient is designed such that at a particular threshold temperature, the two indices become equal and thereafter that of the cladding exceeds that of the core. Within this section, light is no longer guided. Simple intensity transmittance measurement is then very sensitive to the occurrence of this threshold temperature at a particular point along the fiber. If used with an optical time domain reflectometer, the position at which this first event occurs can be located. This system is now in use as a temperature alarm on liquefied natural gas storage tanks. Here, the core and cladding indices for a plastic-clad silica fiber cross at a temperature in the region of

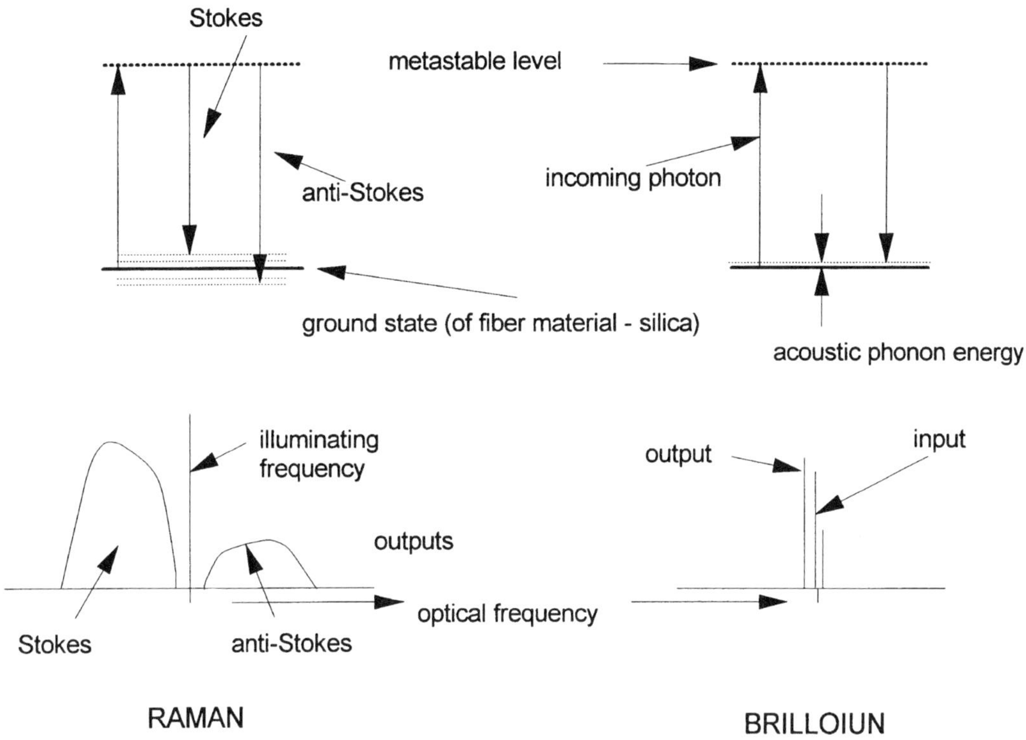

FIGURE 13.90 Thermally sensitive nonlinear scattering processes in optical fibers.

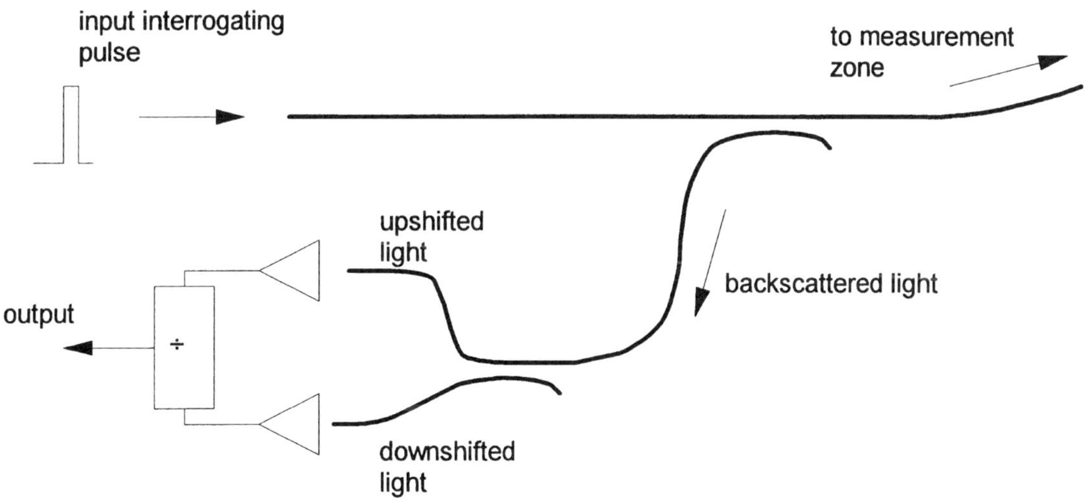

FIGURE 13.91 Raman distributed temperature probe: basic schematic.

50°C. Such temperatures can only be achieved when a leak occurs. Further, the system has the obvious benefit of intrinsic safety and total compatibility with use within potentially explosive atmospheres.

A heat — as opposed to cold — alarm system that has also been introduced commercially is shown in Figure 13.92. In this system, the central tube is filled with a wax that expands by typically 20% when passing through its melting point. This expansion in turn forces the optical fiber against the helical

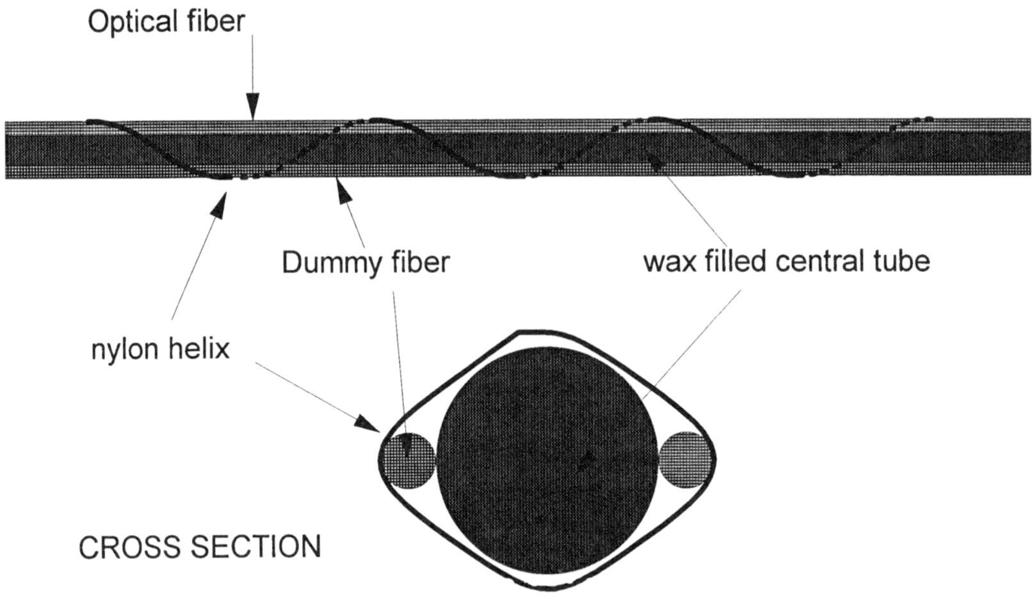

FIGURE 13.92 Fiber optic distributed heat (fire) alarm.

binding, introducing a periodic microbend and thereby increasing the local loss within the fiber. The wax transition temperatures can be defined over a relatively wide range (say 30 to 70°C) and a low-cost OTDR system enables location of the hot spot to within a few meters. This system presents a very cost-effective over-heat or fire alarm when such systems are required and are in intrinsically safe areas or in regions of very high electromagnetic interference. Again, it is the unique properties of the optical fiber sensing medium — especially intrinsic safety and electromagnetic immunity — which provide this system with its market address.

Brillouin scatter is very similar in character to Raman scatter except that in Brillouin scatter the interaction is with an acoustic phonon rather than an optical phonon. The frequency shifts are then correspondingly significantly smaller (typically 10 to 15 GHz). Additionally in Brillouin scatter, the frequency of the scattered light depends on the acoustic velocities in the medium within which the light is scattering. Consequently, the Brillouin scatter spectrum is a function of both temperature (through variations of modulus and density) and strain applied to the optical fiber. Usually this is exploited as a complex but very effective means for measuring strain distributions along an optical fiber. The Brillouin scatter cross-section is much higher than that for Raman scatter so that distributed strain distributions can be measured over distances well in excess of 100 km. This measurement is particularly effective when exploiting stimulated Brillouin scatter that guides the scattered light back toward the source. However, since the apparent value of strain depends on temperature through the variation of acoustic velocity with temperature, temperature correction is required in most practical situations and, in principle, this correction can be implemented by measuring spontaneous Brillouin scatter and specifically the intensity ratio of this in the upper and lower sidebands. This particular correction technique is currently in its infancy, and accuracies in the degree kelvin range are the current state of the art. The difficulty in temperature measurement is that the energy gap between the two sidebands is very small so that the ratio of the amplitudes is close to unity but must be measured very accurately in order to invert the exponential.

The optical Kerr effect manifests itself as an intensity-dependent refractive index. Consequently, this nonlinearity gives rise to either second harmonic generation or sum and difference frequencies. It has been investigated for distributed temperature sensing using pump:probe configurations and birefringent fiber from which the beat length is a function of temperature and strain. This beat length in turn

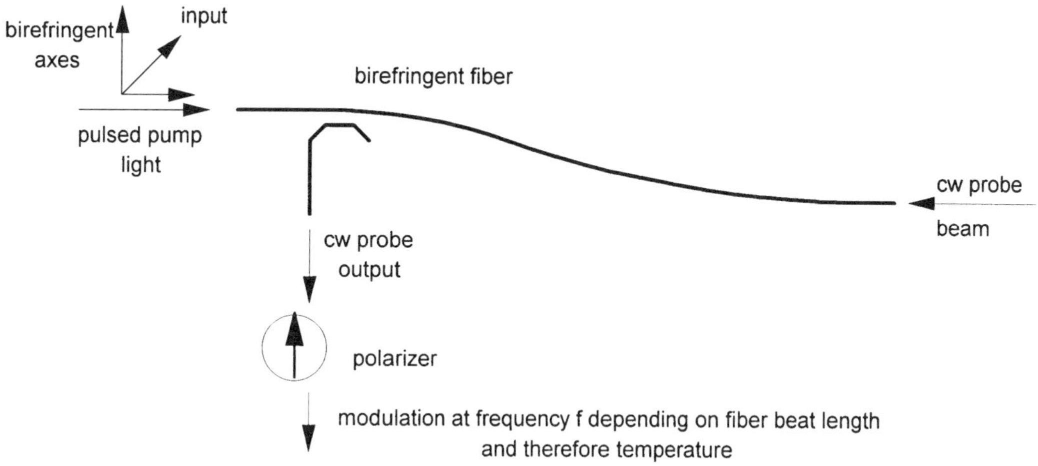

FIGURE 13.93 Distributed Kerr effect probe for temperature or strain field measurements.

determines the frequency offset through phase matching conditions of the mixed pump and probe signal (Figure 13.93). The overall situation is conceptually similar — this offset frequency depends on temperature and strain although in principle, dual measurements and adequate calibration can retrieve both, or alternatively the probe can measure a temperature field in the absence of strain. Again, the actual experimental results that have been achieved remain in the laboratory and the accuracies and resolutions are modest.

In quasi-distributed sensing and point multiplexed systems, temperature probes have, as yet, been but little exploited. There are many variations on the basic theme of a marked optical fiber within which the optical interrogation system measures the optical path length between the marks. These marks can be introduced using partially reflective gratings, mode coupling Bragg gratings, partially reflective splices or connectors, low reflectivity directional couplers, or a multitude of other arrangements. Similarly, the optical delay between the markings can be measured directly as an optical or subcarrier phase or indirectly through monitoring dispersion between adjacent modes typically in low moded or birefringent fibers. Yet again, the different delays depend on both temperature and strain so that for temperature measurement, a strain-free mounting is ideal. The context within which most, if not all, of this class of system has been evaluated is that of smart structures and here the technique does offer some promise as a means for deconvolving strain, mechanical, and thermal effects, and assessing structural integrity. It could also function as a temperature measurement probe, but to date has been minimally addressed in this application.

In multiplexed systems, the current fashion, again primarily for combined strain and temperature measurement, is to incorporate arrays of Bragg gratings used here as wavelength filters rather than as broadband reflectors. In this configuration, the Bragg grating presents a combined temperature/strain field at its location encoded within the reflection wavelength. Multiple addressing schemes can deconvolve temperature and strain sensitivities. There have also been a few demonstrations of discrete temperature-sensitive elements inserted at strategic points along an optical fiber. Of these the use of bandedge shifting in ruby crystals interrogated using a pulsed source observed in reflection has probably been the most successful. In this arrangement, the reflectors are replaced by the crystals and sample the temperature field at these points. The receiver then determines the return to spectrum as a function of time.

Applications for Optical Fiber Temperature Probes

Instrumentation is a very applications-specific discipline and, in particular for sensors, a particular technology is usually only relevant in a limited number of application sectors. As the technology becomes more and more specialized and expensive, these applications niches become much more tightly defined.

For optical fiber sensors and their use in temperature probes, the more conventional approaches (thermocouples, thermisters, platinum resistance thermometers, etc.) are always easier and simpler. The fiber optic technology must exploit one or more of electromagnetic immunity, small size, noninvasiveness, chemical immunity, or the capacity for distributed measurement.

Point optical sensors are therefore primarily used as measurement probes in regions of very high electromagnetic fields, in zone zero intrinsically safe areas, and as *in vivo* medical probes.

The distributed capability of fiber sensors is especially relevant in structural monitoring and in other specialized areas such as measuring the temperature distribution along underground power lines, tunnels or similar structures or in experimental circumstances such as the measurement of curing processes in large volumes of concrete.

Fiber optics then is exactly similar to all other sensing techniques — it is an inappropriate temperature probe for the majority of applications; but for those for which it is appropriate, it offers a unique and effective solution to frequently otherwise intractable measurement challenges. As a technology evolves and becomes both more widely accepted and readily available, the applications address will no doubt expand.

Further Information

Additional information on optical fiber temperature measurements can be obtained from the following.

B. Culshaw and J. P. Dakin, *Optical Fiber Sensors, Vol. I–IV,* Boston: Artech House, Vol. 1, 1988; Vol. II, 1989; Vols. III and IV, 1997.

B. Culshaw, *Optical Fiber Sensing and Signal Processing,* Stevenage, UK: IEE/Peter Perigrinus, 1983.

International Conferences on Optical Fiber Sensors (OFS) are regarded as the principal forum for the dissemination of research results OFS(1), London 1983 to OFS(12) Williamsburg, 1997, various publishers.

Proceedings of series of *Distributed and Multiplexed Optical Fiber Sensors and of Laser and Fiber Sensors Conf.* available from SPIE, Bellingham, Washington.

E. Udd (ed.), *Optical Fiber Sensors,* New York: John Wiley & Sons, 1991.

14

Thermal Conductivity Measurement

William A. Wakeham
Imperial College

Marc J. Assael
Aristotle University of Thessaloniki

There are three mechanisms whereby energy can be transported from one region of space to another under the influence of a temperature difference. One is by transmission in the form of electromagnetic waves (radiation); the second is the process of convection, in which a bulk or local motion of the material effects the transport; and the final process is that of thermal conduction, when energy is transported through a medium. In most practical situations, energy transport is accomplished by all three processes to some extent, but the relative importance of each contribution varies markedly. For example, within an evacuated region, radiation is the sole mechanism of transport; whereas, in an opaque solid, conduction is the only mechanism possible.

These processes of heat transfer are often very important in a wide variety of scientific and industrial applications. In the cooling of cast or crystalline materials (e.g., metals, semiconductors, or polymers) from a molten state to a solid state, the heat transfer within the material can have a profound effect on the final properties of the solid. Equally, the heat transfer in a foodstuff is a determinant of its cooking, freezing, or processing time, while the size of equipment needed to heat or cool the liquid or gas stream in a chemical plant depends sensitively on the heat transfer within and between one stream and another.

For these reasons, there has been great interest in the understanding and description of all three heat transfer processes. Among the three, that of thermal conduction is the simplest to describe in principle, since the empirical law of Fourier simply states that the heat transported by conduction per unit area in a particular direction is proportional to the gradient of the temperature in that direction. The coefficient of proportionality in this law is known as the thermal conductivity and denoted here by the symbol λ. Many important materials, whether made of pure chemical components or mixtures, are of uniform composition throughout and for them the thermal conductivity is a true physical property of the material, depending often only on the temperature, pressure, and composition of the sample. However, particularly in the solid state, the thermal conductivity can depend on the direction of the heat flow, for example, in the case of a molecular crystal.

It is also conventional to speak of the thermal conductivity of various types of composite materials such as bricks, glass-fiber insulation, carbon-fiber composites, or polymer blends. In this case, the thermal conductivity is taken to be the empirical constant of proportionality in the linear relationship between a measured heat transport per unit area and the temperature difference over a prescribed distance in the material. The thermal conductivity is not then, strictly, a property of the material, since it can often

0-8493-0047-9/00/$0.00+$.50

depend on a large number of parameters, including the history of the material, its method of manufacture, and even the character of its surface. However, this distinction between homogeneous and inhomogeneous materials is often ignored and leads to more than a little confusion, especially where intercomparisons among measurements are concerned.

The fact that in most practical situations all three heat transfer mechanisms are present greatly complicates the process of measurement of the thermal conductivity. Thus, much early work in the field is substantially in error, and it has been really quite difficult to devise methods of measurement that unequivocally determine the thermal conductivity. For that reason, the instruments to be described in the following sections often seem to be rather far removed from the apparent simplicity implied by Fourier's Law.

14.1 Fundamental Equations

The essential constitutive equation for thermal conduction relates the heat flux in a material to the temperature gradient by the equation:

$$Q = -\lambda \nabla T \tag{14.1}$$

It is not possible to measure local heat fluxes and gradients; thus, all experimental techniques must make use of an integrated form of the equation, subject to certain conditions at the boundaries of the sample. All experiments are designed so that the mathematical problem of the ideal model is reduced to an integral of the one-dimensional version of Equation 14.1, which yields, in general:

$$Q_a = G\lambda\,\Delta T \tag{14.2}$$

in which G is constant for a given apparatus and depends on the geometric arrangement of the boundaries of the test sample. Typical arrangements of the apparatus, which have been employed in conjunction with Equation 14.2, are two flat, parallel plates on either side of a sample, concentric cylinders with the sample in the annulus and concentric spheres.

Techniques that make use of Equation 14.2 are known as steady-state techniques and they have found wide application, some of which are discussed below. They are operated usually by measuring the temperature difference ΔT that is generated by the application of a measured heat input Q_a at one of the boundaries. The absolute determination of the thermal conductivity, λ, of the sample contained between the boundaries then requires a knowledge of the geometry of the cell contained in the quantity G. In practice, it is impossible to arrange an exactly one-dimensional heat flow in any finite sample so that great efforts have to be devoted to approaching these circumstances and then there must always be corrections to Equation 14.2 to account for departures from the ideal situation.

If the application of heat to one region of the test sample is made in some kind of time-dependent fashion, then the temporal response of the temperature in any region of the sample can be used to determine the thermal conductivity of the fluid. In these transient techniques, the fundamental differential equation that is important for the conduction process is:

$$\rho C_p \frac{\partial T}{\partial t} = \nabla \cdot \left(\lambda \nabla T\right) \tag{14.3}$$

which arises from an elementary energy balance in the absence of any other processes and in which ρ is the density of the material and C_p its isobaric heat capacity. In most, but not all, circumstances, it is acceptable to ignore the temperature dependence of the thermal conductivity in this equation and to write:

$$\frac{\partial T}{\partial t} = \frac{\lambda}{\rho C_P} \nabla^2 T = a \, \nabla^2 T \qquad (14.4)$$

in which a is known as the thermal diffusivity.

Experimental techniques for the measurement of the thermal conductivity based on Equation 14.4 generally take the form of the application of heat at one surface of the sample in a known time-dependent manner, followed by detection of the temperature change in the material at the same or a different location. In most applications, every effort is made to ensure that the heat conduction is unidirectional so that the integration of Equation 14.4 is straightforward. This is never accomplished in practice, so some corrections to the integrated form of Equation 14.4 are necessary. The techniques differ among each other by virtue of the method of generating the heating, of measuring the transient temperature rise, and of the geometric configuration. Interestingly, in one geometric configuration only, is it possible to determine the thermal conductivity essentially independently of a knowledge of ρ and C_P, which has evident advantages. More usually, it is the thermal diffusivity, a, that is the quantity measured directly, so that the evaluation of the thermal conductivity requires further, independent measurements.

In the following sections, brief descriptions of the specific applications of these general principles are given. The examples chosen for study are intended to cover the full spectrum of materials and thermodynamic states and, in each case, attention is concentrated on a method that has proved most accurate and is widely used. The steady-state and transient techniques are considered separately.

14.2 Measurement Techniques

Steady-State Methods

The steady-state methods employed for the measurement of the thermal conductivity of fluids and solids have most often employed the geometry of parallel plates so that it is that configuration described here in two variants. Coaxial cylinder equipment has largely been used within the preserve of the research laboratory, with the apparatus of Tufeu and Le Neindre [1] an excellent example of the genre.

Parallel-Plate Instrument

A schematic diagram of a guarded parallel-plate instrument is shown in Figure 14.1. The sample is contained in the gap between two plates (upper and lower) maintained a distance d apart by spacers. A small amount of heat, Q_a, is generated electrically in the upper plate and is transported through the sample to the lower plate. Around the upper plate, and very close to it, is placed a guard plate. This plate is, in many instruments, maintained automatically at the same temperature as the upper plate so as to reduce heat losses from the upper surface of the upper plate and to most nearly secure a one-dimensional heat flow at the edges of the sample.

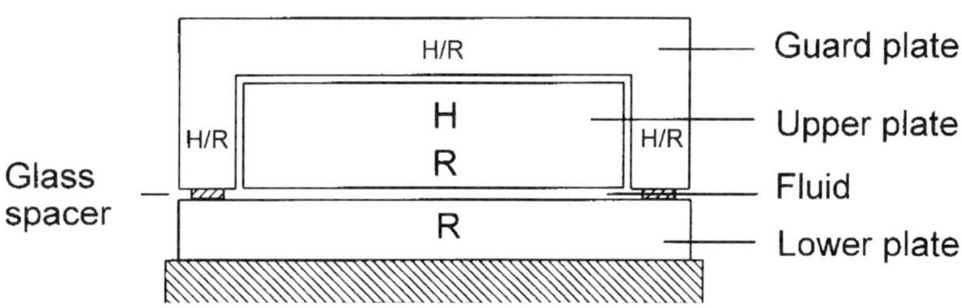

FIGURE 14.1 Schematic diagram of a guarded parallel-plate instrument. (H = heater, R = resistance thermometer).

The temperatures at the surfaces of the upper and lower plates are measured very precisely, as is the electric input of energy, so that the thermal conductivity can, in principle, be evaluated from the equation:

$$Q_a = \frac{A\lambda\Delta T}{d} \qquad (14.5)$$

where ΔT is the measured temperature difference, and A is the area of the upper plate.

Whether the sample is a fluid or a solid, the electric energy generated in the upper plate is not all conducted to the lower plate. Thus, it is necessary in all cases to account for spurious heat losses and for all except opaque materials for the radiative transfer between the two surfaces. When the sample is transparent to radiation, this correction is straightforward and can be reduced by means of surface coating the plates to reduce their emissivity; but when the material adsorbs radiation, the problem is much more complicated and has been the subject of some controversy in the past, which has since been resolved (p.147 of [2]). In many cases, the effort of performing absolute measurements cannot be justified so that the ratio A/d is determined by calibration with a material of known thermal conductivity.

Measurement on Fluids

Measurements with parallel-plate instruments on fluids have been performed for a considerable period of time. The technique has particular advantages in some special regions of thermodynamic space but requires great attention to detail if accurate results are to be obtained. In the most accurate instruments for fluids, the gap between upper and lower plates is kept as small as possible (perhaps as small as 0.2 mm). This has the benefit of reducing the effect of heat flows that are not normal to the heat surfaces, but more importantly the small gap contributes to the reduction of the heat transferred by bulk convective motion of the fluid. Indeed, if very small temperature differences are employed by heating the top plate (to have a stable density gradient) *and* considerable care is taken to align the parallel plates normal to the earth's gravitational acceleration, then the effects of convective heat transfer can be rendered negligible (p.154 of [2]). It is a fact of history that the necessary care with this instrument has been taken by only a few workers so that despite the fact that the instrument has been used in the temperature range from 4 K to 800 K, and for pressures up to 250 MPa, only some of the measurements are reliable.

An example of what can be achieved with a parallel-plate instrument is provided by the work of Mostert [3] and Sakonidou [4] at the van der Waals laboratory in Amsterdam. They have used the technique near the critical point of a pure fluid or mixture where the extreme values of the compressibility make the fluid exceedingly prone to convection. In such circumstances, the small vertical extent of the fluid layer required for this technique avoids large density variations in the test layer and, combined with the fact that very small temperature differences (0.3 mK) can be employed, has enabled measurements of the thermal conductivity to be conducted to within 100 mK of the critical temperature at the critical density. Under these circumstances, the thermal conductivity of a pure fluid reveals an enhancement that is, in principle, infinite at the critical point itself.

The unique characteristics of the parallel-plate instrument mean that it is the method of choice for work near the critical region of a material. The arguments above pertaining to the care required for reliable measurements militate against the production of commercial instruments.

Measurements on Solids

A parallel-plate instrument of exactly the same type as has been described for fluids has been employed for solids. However, the spacing of the two plates is normally significantly greater for solids, owing to the difficulty of preparing very thin samples of solids. One essential difficulty with solid samples in this configuration is the contact between the two heater surfaces and those of the sample. Unlike the case for fluids, the contact cannot be made uniform at the molecular level. There is therefore always the possibility of an unaccounted interfacial heat transfer resistance. It seems likely that these considerations contribute to the wide differences between values reported for the same sample by different authors.

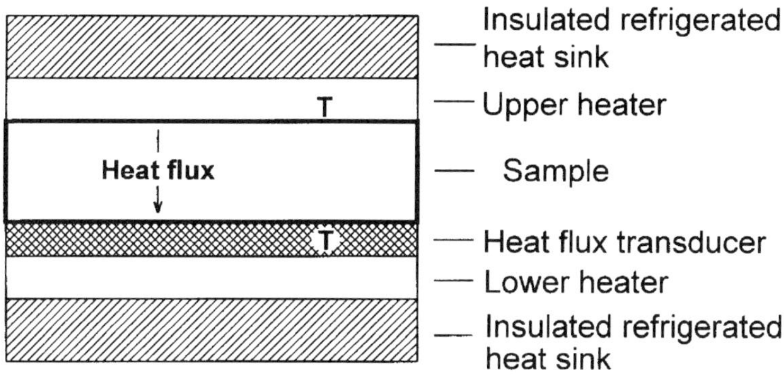

FIGURE 14.2 Schematic diagram of a heat flow-meter instrument. (T = thermocouple).

A more popular implementation of the parallel-plate configuration for solids is the so-called heat flow-meter instrument. When applied to materials such as building insulation, the dimensions of this type of instrument can be very large.

In such instruments (see Figure 14.2), the upper heater plate is set at a higher temperature than the lower one. The hot and cold surface temperatures of the sample are measured with the two thermocouples permanently installed on the adjacent surface plates, while a precalibrated heat flow transducer on the lower plate measures the magnitude of the heat flux through the sample. The thermal conductivity is calculated directly from Equation 14.5. In some cases, contact resistances (i.e., insulation) can be characterized and controlled by employing a pressurized gas in the sample chamber. Commercial instruments of this type are included in the listing of Table 14.1.

Transient Methods

There are rather more transient techniques that have achieved popularity than steady-state instruments. This is because transient techniques generally require much less precise alignment and dimensional knowledge and stability. Furthermore, some of the techniques have distinct advantages that arise from the speed of the measurement. Here we have space to describe only one technique in detail which has, in a variety of ways, far greater applicability.

TABLE 14.1 Companies That Make Thermal Conductivity Instruments

Type of instrument	Temperature range	Supplier	Approximate price (U.S.$)
Transient hot disk			
Thermal Analyser TPS	290–1,000 K	K-analys AB	$20,000
Guarded parallel plate			
Thermatest GHP-300	290–1,000 K	Holometrix Inc.	Variable
TCT 416	290–340 K	NETZSCH	$30,000
Heat flow meter			
Rapid-k RK-70	290–500 K	Holometrix Inc.	Variable
Unitherm 2021	290–500 K	ANTER	$18,000
Radial heat-flow			
Orton D.C.A.	290–1400 K	Orton	$50,000
Laser flash			
Thermaflash 1100	290–1300 K	Holometrix Inc.	Variable

Note: Prices are only indicative.

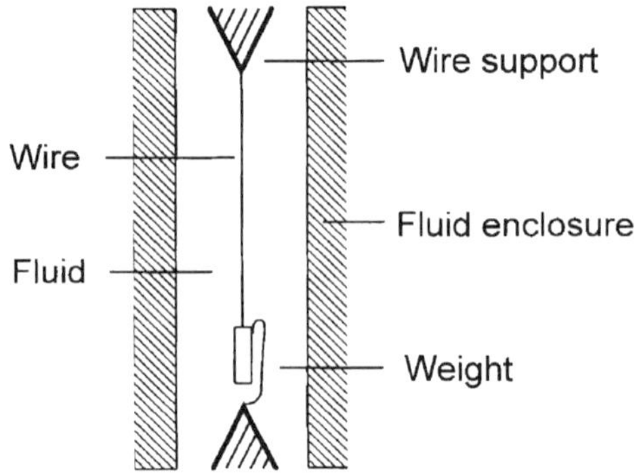

FIGURE 14.3 Schematic diagram of a transient hot-wire instrument for fluids. (Note that the hot wire is kept under constant tension by a weight.)

Transient Hot-Wire Technique

In this technique, the thermal conductivity of a material is determined by observing the temporal evolution of the temperature of a very thin metallic wire (see Figure 14.3) after a step change in voltage has been applied to it. The voltage applied results in the creation of a line source of nearly constant heat flux in the fluid. As the wire is surrounded by the sample material, this produces a temperature field in the material that increases with time. The wire itself acts as the temperature sensor and, from its resistance change, its temperature change is evaluated and this is related to the thermal conductivity of the surrounding material.

According to the ideal model of this instrument, an infinitely long, line source of heat possessing zero heat capacity and infinite thermal conductivity is immersed in an infinite isotropic material, with physical properties independent of temperature and in thermodynamic equilibrium with the line source at time $t = 0$ at a temperature T_0. The heat transferred from the line source to the sample when a stepwise heat flux, q, per unit length is applied, is assumed to be entirely conductive. Then the temperature rise of the material at a radial distance, r_0, which it transpires, is the same as the temperature rise at the surface of a wire of radius r_0, is $\Delta T_i(r_0, t)$ is given by:

$$\Delta T_i\left(r_0,t\right) = T\left(r_0,t\right) - T_0 = \frac{q}{4\pi\lambda}\left[\ln\left(\frac{4at}{r_0^2 C}\right) + \frac{r_0^2}{4at} + \cdots\right] \tag{14.6}$$

In the above equation, C is a known constant [5]. The equation suggests that, provided the radius of the wire is chosen small enough so that the second term on the right-hand side of Equation 14.6 is negligible, the thermal conductivity of the fluid can be obtained from the slope of the line ΔT_i vs. $\ln t$. Any practical implementation of this method of measurement inevitably deviates from this ideal model. The success of the technique, however, rests on the fact that by proper design, it is possible to construct an instrument that can operate very closely to the ideal model, while at the same time small departures can be treated by a first-order analysis [5].

The transient hot-wire technique was first developed in the 1930s to measure the effective thermal conductivity of powders. However, its application to other materials was somewhat slower until in the late 1960s the new technology associated with electronics made it possible to measure small, transient resistance changes with high accuracy in a period of less than 1 s. This development, pioneered by Haarman [6], made it possible to complete the transient heating process so quickly that, despite the inevitability of convective motion in the fluid from time zero, the inertia of the fluid ensures that the fluid velocity is sufficiently small that there is no significant contribution to heat transfer. This fact prompted a rapid development of the measurement technique for fluids — first in gases and then in liquids — that was then followed by further developments in solids. The differences in the technique between solids and fluids are rather small; thus, some aspects of the instrumentation for liquids are briefly discussed and the differences for solids are just outlined.

In the case of fluids, the instrumentation generally involves a wire some 7 μm to 25 μm in diameter (in order to reduce the correction owing to its heat capacity) and some 150 mm long. The wire is mounted vertically in a cylindrical cell containing the test sample. Often, a second wire differing only in length is employed to compensate automatically for effects at the ends of the wires via the electrical measurement system, but this can also be accomplished with potential taps [5]. Whenever possible, platinum is used for the wire material because its resistance/temperature characteristics are well known and it can be readily obtained in the form of wires with a diameter as small as 5 μm. When the material under test is electrically conducting, it is necessary to insulate electrically the wire from the fluid. A variety of techniques have been employed for this purpose that enjoy different degrees of success depending on the range of conditions to be studied. Near ambient temperature over a range of pressures, it has been found adequate to use a tantalum wire as the sensor that is electrolytically anodized to cover the wire with an insulating layer of tantalum pentoxide 100 nm thick [7]. Under more aggressive conditions, it has been necessary to employ ion-plating of the wire with a ceramic to secure the isolation [8]. In either case, the theory has been modified by the addition of a small correction.

In the case of solids, the need for the wire to be straight and vertical is removed by virtue of the rigidity of the material. Thus, Bäckström and colleagues [9] were able to employ a wire embedded as an arc within the compressed solid matrix of the material under study, particularly at very high pressures (up to 4 GPa).

The transient hot-wire technique has a unique advantage among transient methods that the thermal conductivity of the test material can be evaluated directly from the slope of the line relating the temperature rise of the wire to the logarithm of time. The heat capacity and density of the test material are required only to evaluate small corrections. Furthermore, the exact dimensions of the heating element and the cell are also unimportant so that the method avoids the intricate alignment problems of the parallel-plate technique while securing absolute measurements of the property. Despite these advantages and its wide application to measurements in gases, solids, and liquids, there has been no commercial development of an instrument of this kind, presumably because of the delicacy of the long, thin wire in the case of devices for fluids, and the difficulty of sample preparation for solids.

Hot-Disk Instrument

A transient technique for which there is a commercial version suitable for solid materials is the transient hot-disk instrument shown in Figure 14.4.

The sensor in this case comprises a thin metal strip, often of nickel, wound in the form of a double-spiral in a plane. It is printed on, and embedded within, a thin sandwich formed by two layers of a material that is a poor electric conductor but a good thermal conductor. This disk heater is then, in turn, placed either between two halves of a disk-shaped sample of solid material or affixed to the outside of the sample.

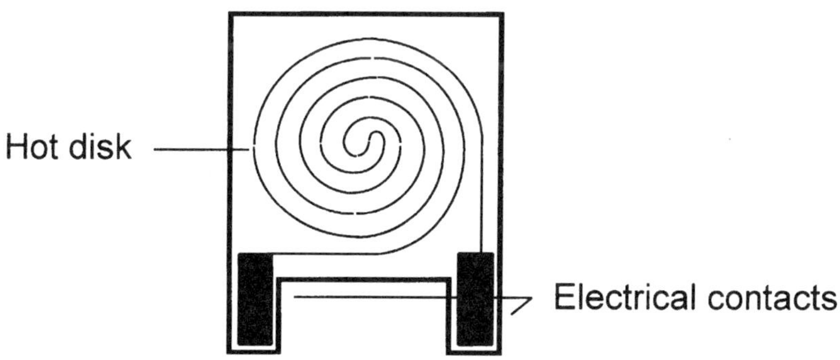

FIGURE 14.4 Schematic diagram of a transient hot-disk instrument.

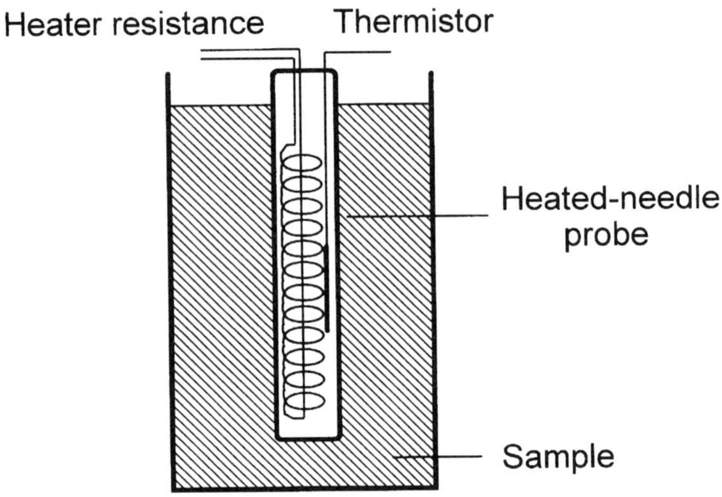

FIGURE 14.5 Schematic diagram of a transient heated-needle probe.

However it is configured, the essential measurement performed is the same as that for the hot-wire technique, and the temperature history of the sensor when subject to known electrical dissipation is inferred from its resistance charge. In the most recent version of the instrument, developed by Gustafsson [10], and also available commercially, the interpretation of the data is accomplished via a numerical solution of the differential equation rather than by some analytical approximation to it. The technique is used frequently for studies of polymer composites, glasses, superconductors, and insulating materials.

Heated-Needle Probe

A further commercial device exists for the measurement of the thermal conductivity of granular materials such as powders and soils, natural materials such as rock and concrete and, indeed, of food. The probe is shown schematically in Figure 14.5 where it is seen that it consists of a thin, hollow, metallic needle (diameter 3 mm) containing an electric heater and a separate thermistor as a probe of the temperature history of the needle following initiation of a heat pulse [11]. The temperature history of the probe is generally interpreted with the aid of the equation appropriate to a transient hot-wire instrument but in a relative manner whereby its response is calibrated against known standards. This rather simplistic approach to the analysis of a somewhat complex cell inevitably restricts the accuracy that can be achieved, but does provide a measurement capability where no other technique is viable. It is often employed for measurements in inhomogeneous samples such as rocks or soils where it is simply the effective thermal conductivity that is required.

Laser-flash Instrument

A final transient technique is that which has become known as the laser-flash technique developed originally for measurements in solids but occasionally used on liquids, particularly at high temperatures.

Figure 14.6 contains a schematic diagram of the instrument as it is available today in a commercial form. The sample is illuminated on one face with a laser pulse of very short duration and high intensity. The absorption of the laser energy on the front face of the sample causes the generation of heat at that front surface, which is subsequently transmitted throughout the sample to the back face of the sample where the temperature rise is detected with an infrared remote sensor. The interpretation of measurements is based on a one-dimensional solution of Equation 14.4 subject to an initial condition of an instantaneous heat pulse at one location.

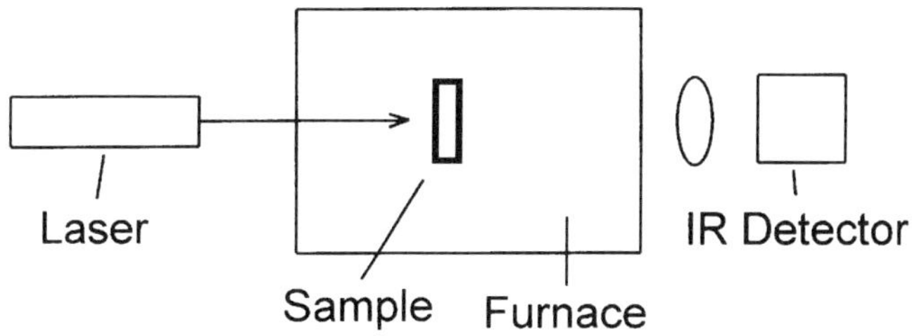

FIGURE 14.6 Schematic diagram of a laser-flash instrument.

The temperature rise at the back face of a sample of thickness l and radius r, is therefore given by [12]:

$$\Delta T(l,t) = \frac{Q}{\rho C_p l \pi r^2}\left[1 + 2\sum_{n=1}^{\infty}(-1)^n \exp\left(-\frac{n^2\pi^2 at}{l^2}\right)\right] \tag{14.7}$$

where Q is the energy absorbed at the front surface at time zero. The thermal diffusivity of the sample, a, is then often deduced from the measurement of the time taken for the back face of the sample to reach one half of its maximum value. The technique has the very distinct advantage that it does not require physical contact between the test sample and the heater or detector. For this reason, it is a particularly appropriate technique for use at high temperatures or in aggressive environments.

However, there are a number of precautions that must be taken to ensure accurate results. First, the theory should be modified to account for non-unidirectional heat flow. Secondly, care must be taken to ensure that no radiation incident on the front face penetrates to the back face for transparent samples. Due care must also be taken to match the laser power to the system being studied so that there is neither fusion nor ablation at the front face that can distort the results. Finally, when the fluid state is studied, due care should be taken to eliminate convective heat transport. Seldom are all of these precautions adopted in routine work, so that some results obtained with the technique are of dubious validity. Nevertheless, the method has seen widespread application to a wide range of materials, including composites, polymers, glasses, metals, refracting materials, insulating solids, and coatings.

Finally, the radial heat-flow method should also be mentioned. This is a transient technique in which the sample is heated and cooled continuously [13]. From the recording of the temperature gradient via thermocouples, the thermal diffusivity is obtained and thus the thermal conductivity is calculated. The advantage of the radial heat-flow method is that the measurements are fast and only small temperature gradients are necessary in the sample. A commercially available instrument operating according to this technique is listed in Table 14.1.

14.3 Instrumentation

Table 14.1 lists various instruments for the measurement of the thermal conductivity of solids, while the addresses of the suppliers are shown in Table 14.2. As already mentioned, to our knowledge, there is no company that produces instruments specifically for the measurement of the thermal conductivity of fluids.

14.4 Appraisal

Naturally, the technique to be employed for the measurement of thermal conductivity depends on the type of sample to be studied and the range of conditions to be employed. For fluid samples under most conditions, a variant of the transient hot-wire method must be the preferred technique. Under favorable

TABLE 14.2 Addresses of Companies That Make Thermal Conductivity Instruments

ANTER Corporation 1700 Universal Road Pittsburgh, PA 15235-3998 Tel: (412) 795-6410 Fax: (412) 795-8225	NETZSCH Gerätebau GmbH Wittelsbacherstraβe 42 D-95100 Selb/Bayern, Germany Tel: (9287) 88136 Fax: (9287) 88144
HOLOMETRIX Inc. 25 Wiggins Avenue Bedford, MA 0170-2323 Tel: (617) 275-3300 Fax: (617) 275-3705	ORTON The Edward Orton JR. Ceramic Foundation P.O. Box 460 Westerville, OH 43081 Tel: (614) 895-2663 Fax: (614) 895-5610
K-ANALYS AB Seminariegatan 33 H S-752 28 Uppsala, Sweden Tel: (46) 18 50 01 66 Fax: (46) 18 54 36 38	

conditions, an accuracy of ±0.3% can be achieved and a level of ±1% is possible under all but the most aggressive circumstances. Near the critical region of fluids, a parallel-plate instrument is essentially the only viable method. For molten materials at high temperature, while a variant of the hot-wire system has advantages, the laser-flash technique is very attractive but an accuracy of no better than 10% is then to be expected.

For solids, the hot-disk or laser-flash techniques have many features that recommend them when the sample is amenable to appropriate preparation. In those cases, an accuracy of a few percent should be possible but is rarely attained. For samples such as rocks, the heated needle-probe is undoubtedly the only viable technique.

It should be emphasized again here that when a sample is inhomogeneous, the quantity determined is not the thermal conductivity of any element of it, but rather an effective value suitable for engineering purposes. It is not an intrinsic thermophysical property of the material.

References

1. B. Le Neindre, *Contribution à l' étude expérimentale de la conductivité thermique de quelques fluides a haute température et à haute pression*, Ph.D. thesis, Université de Paris, 1969.
2. M. Sirota, Steady-State Measurements for Thermal Conductivity, in A. Nagashima, G. V. Sengers, and W. A. Wakeham (eds.), *Experimental Thermodynamics. Vol. III. Measurement of the Transport Properties of Fluids*, London: Blackwell Scientific Publications, 1991, chap. 6.
3. R. Mostert, H. R. van der Berg, and P. S. van der Gulik, The thermal conductivity of ethane in the critical region, *J. Chem. Phys.*, 92, 5454-5462, 1990.
4. E. Sakonidou, *The thermal conductivity of methane and ethane mixtures around the critical point*, Ph.D. thesis, University of Amsterdam, 1996.
5. M. J. Assael, C. A. Nieto de Castro, H. M. Roder, and W. A. Wakeham, Transient Methods for Thermal Conductivity, in A. Nagashima, J. V. Sengers, and W. A. Wakeham (eds.), *Experimental Thermodynamics. Vol. III. Measurement of the Transport Properties of Fluids*, London: Blackwell Scientific Publications, 1991, chap. 7.
6. J. W. Haarman, *Thermal conductivity measurements with a transient hot-wire method*, Ph.D. thesis, Technische Hogeschool Delft, 1969.
7. A. Alloush, W. B. Gosney, and W. A. Wakeham, A transient hot-wire instrument for thermal conductivity measurements in electrically conducting liquids at elevated temperatures, *Int. J. Thermophys.*, 3, 225-234, 1982.

8. Y. Nagasaka and A. Nagashima, Absolute measurements of the thermal conductivity of electrically conducting liquids by the transient-hot wire method, *J. Phys., E,* 14, 1435, 1981.
9. P. Andersson and G. Bäckström, Measurement of the thermal conductivity under high pressures, *Rev. Sci. Instrum.*, 47, 205, 1976.
10. S. E. Gustafsson, E. Karawacki, and M.N. Khan, Transient hot-strip method for simultaneously measuring thermal conductivity and thermal diffusivity of solids and fluids, *J. Phys. D: Appl. Phys.*, 12, 1411, 1979.
11. J. Nicolas, Ph. André, J. F. Rivez, and V. Debaut, Thermal conductivity measurements in soil using an instrument based on the cylindrical probe method, *Rev. Sci. Instrum.*, 64, 774-780, 1993.
12. W. J. Parker, R. J. Jenkins, C. P. Butler, and G. L. Abbott, Flash method for determining thermal diffusivity, heat capacity and thermal conductivity, *J. Appl. Phys.*, 32, 1679-1684, 1960.
13. G. S. Sheffield and J. R. Schorr, Comparison of thermal diffusivity and thermal conductivity methods, *Ceram. Bull.*, 70, 102, 1991.

15

Heat Flux

Thomas E. Diller
Virginia Polytechnic Institute

Thermal management of materials and processes is becoming a sophisticated science in modern society. It has become accepted that living spaces should be heated and cooled for maximum comfort of the occupants. Many industrial manufacturing processes require tight temperature control of material throughout processing to establish the desired properties and quality control. Examples include control of thermal stresses in ceramics and thin films, plasma deposition, annealing of glass and metals, heat treatment of many materials, fiber spinning of plastics, film drying, growth of electronic films and crystals, and laser surface processing.

Temperature control of materials requires that energy be transferred to or from solids and fluids in a known and controlled manner. Consequently, the proper design of equipment such as dryers, heat exchangers, boilers, condensers, and heat pipes becomes crucial. The constant drive toward higher power densities in electronic, propulsion, and electric generation equipment continually pushes the limits of the associated cooling systems.

Although the measurement of temperature is common and well accepted, the measurement of heat flux is often given little consideration. Temperature is one of the fundamental properties of a substance. Moreover, because temperature can be determined by human senses, most people are familiar with its meaning. Conversely, heat flux is a derived quantity that is not easily sensed. It is not enough, however, to only measure the temperature in most thermal systems. How and where the thermal energy goes is often equally or more important than the temperature. For example, the temperature of human skin can indicate the comfort level of the person, but has little connection with the energy being dissipated to the surroundings, particularly if evaporation is occurring simultaneously. Wind chill factor is another common example of the importance of convection heat transfer in addition to air temperature.

Maximizing or minimizing the thermal energy transfer in many systems is crucial to their optimum performance. Consequently, sensors that can be used to directly sense heat flux can be extremely important. The subsequent material in this chapter is intended to help individuals understand and implement heat flux measurements that are best suited for the required applications.

0-8493-0047-9/00/$0.00+$.50

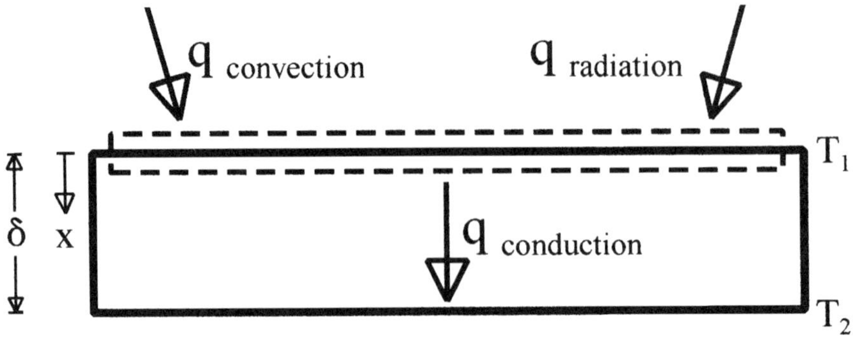

FIGURE 15.1 Illustration of energy balance.

15.1 Heat Transfer Fundamentals

The movement of thermal energy is known as "heat" and the rate of this transfer is commonly called "heat transfer." It is given the symbol q and has the units of watts. The heat transfer per unit area is termed the "heat flux" and is given the symbol q'' with the units of W m^{-2}. Although in some cases only the overall heat transfer from a system is required, often the spatial and temporal variation of the heat flux is important to performance enhancement. Methods for measuring the spatial *or* temporal distribution of heat flux are identified and discussed in this chapter. Detailed simultaneous measurements of *both* spatial and temporal distributions of heat flux, however, are generally not feasible at this time.

One of the most important principles concerning heat transfer is the first law of thermodynamics, which states that the overall energy transfer to and from a system is conserved. It includes all types of energy transfer across the system boundary, including the three modes of heat transfer — conduction, convection, and radiation. For the simple example shown in Figure 15.1, the transient energy balance on the control volume marked can be expressed as:

$$mC\frac{\partial T}{\partial t} = q_{\text{convection}} + q_{\text{radiation}} - q_{\text{conduction}} \tag{15.1}$$

where m is the mass of the system and C is the corresponding specific heat. The effect of the thermal capacitance (mC) of the material causes a time lag in the temperature response of the material to a change in heat transfer. A short summary of important heat transfer principles follows, with many engineering textbooks available in the field that give additional details.

Conduction

Conduction encompasses heat transfer through stationary materials by electrons and phonons. It is related to the temperature distribution by Fourier's law, which states that the heat flux vector is proportional to and in the opposite direction of the temperature gradient:

$$\vec{q}'' = -k\vec{\nabla}T \tag{15.2}$$

The constant k is the thermal conductivity of the material. Measuring this temperature gradient is one of the basic methods for determining heat flux.

For a homogeneous material in Cartesian coordinates Equation 15.1 becomes:

$$\frac{\partial T}{\partial t} = a\left(\frac{\partial^2 T}{\partial x^2} + \frac{\partial^2 T}{\partial y^2} + \frac{\partial^2 T}{\partial z^2}\right) \tag{15.3}$$

where a is the thermal diffusivity of the material with a density of ρ, $a = k/(\rho C)$. Measuring the temperature response of the system according to this equation is the second major method for determining the heat transfer. Because of the complexity of solutions to Equation 15.3, this method can be complicated if multidimensional effects are present.

If steady-state one-dimensional heat transfer can be assumed throughout the planar solid in Figure 15.1, the temperature distribution in the direction of heat flux q'' is linear. Equation 15.2 becomes simply:

$$q'' = k\frac{T_1 - T_2}{\delta} \tag{15.4}$$

where the temperatures are specified on either side of the material of thickness δ.

As illustrated in Figure 15.1, convection and radiation are the other modes of heat transfer typically present at the surface of a solid. These are usually the quantities of interest to measure with a heat flux sensor. Both are present at least to some extent in virtually all cases, although the effects of one or the other are often purposely minimized to isolate the effects of the other.

Convection

Although heat transfer by convection occurs by the same physical mechanisms as conduction, the fluid is free or forced to move relative to the surface. The fluid motion greatly complicates the analysis by coupling the heat transfer problem with the fluid mechanics. Particularly when the flow is turbulent, the fluid equations are generally impossible to solve exactly. Consequently, the heat transfer and fluid mechanics are commonly isolated by introduction of a heat transfer coefficient, which encompasses all of the fluid flow effects.

$$q'' = h_T\left(T_r - T_s\right) \tag{15.5}$$

The temperature of the fluid is represented by T_r, which for low-speed flows is simply the fluid temperature away from the surface. The recovery temperature is used for high-speed flows because it includes the effect of frictional heating in the fluid. The subscript T on the heat transfer coefficient, h_T, implies that the boundary condition on the surface is a constant temperature, T_s. Although other surface temperature conditions can be encountered, it is then important to carefully document the surface temperature distribution because it can have a profound effect on the values of h and q'' [1, 2].

Radiation

Heat transfer by radiation occurs by the electromagnetic emission and adsorption of photons. Because this does not rely on a medium for transmission of the energy, radiation is very different from conduction and convection. Radiation has a spectrum of wavelengths dependent on the temperature and characteristics of the emitting surface material. Moreover, the surface properties are often dependent on the wavelength and angular direction of the radiation. One classic example is material that looks black to the visible spectrum, but is transparent to the longer wavelengths of the infrared spectrum. Consequently, special coatings are sometimes applied to surfaces to control the absorption characteristics. For example, the surface of a radiation sensor is often coated with a high absorptivity paint or graphite.

Because the power emitted from a surface is proportional to the fourth power of the absolute temperature, radiation detectors are usually cooled sufficiently for the power emitted from the detector itself to be negligible. In this case, the temperature distribution of the sensor is not important. This is a big advantage over convection measurements where the temperature distribution on the surface has a big influence on the measurement.

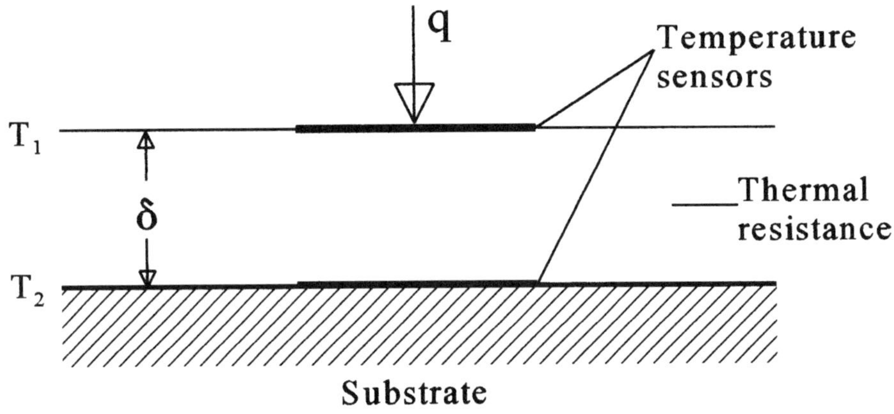

FIGURE 15.2 One-dimensional planar sensor concept.

15.2 Heat Flux Measurement

Most of the many methods for measuring heat flux are based on temperature measurements on the surface or close to the surface of a solid material. Usually this involves insertion of a device either onto or into the surface, which has the potential to cause both a physical disruption and a thermal disruption of the surface. As with any good sensor design, the goal for good measurements must be to minimize the disruption caused by the presence of the sensor. It is particularly important to understand the thermal disruption caused by the sensor because it cannot be readily visualized and because all heat flux sensors have a temperature change associated with the measurement. Consequently, wise selection of the sensor type and operating range is important for good heat flux measurements. The following sections emphasize important factors in using the currently available heat flux sensors, followed by short summaries of sensors used in research and possible future developments. They are grouped by the general type of sensor action.

15.3 Sensors Based on Spatial Temperature Gradient

The heat flux at the material surface can be found at a location if the temperature gradient can be determined at that position, as indicated in Equation 15.2. Because it is difficult to position temperature sensors with the requisite accuracy inside the material, sensors to measure heat flux are either applied on the surface or mounted in a hole in the material. In the following sections, the different types of commercially available sensors are discussed first and listed in a table by manufacturer. Shorter sections briefly describing sensors used in research labs or currently being developed follow.

One-Dimensional Planar Sensors

The simplest heat flux sensor in concept is illustrated in Figure 15.2. The one-dimensional heat flux perpendicular to the surface is found directly from Equation 15.4 for steady-state conditions:

$$q'' = \frac{k}{\delta}\left(T_1 - T_2\right) \tag{15.6}$$

The thickness of the sensor δ and thermal conductivity k are not known with sufficient accuracy for any particular sensor to preclude direct calibrations of each sensor. An adhesive layer may also be required between the sensor and surface to securely attach the sensor, which adds an additional thermal resistance

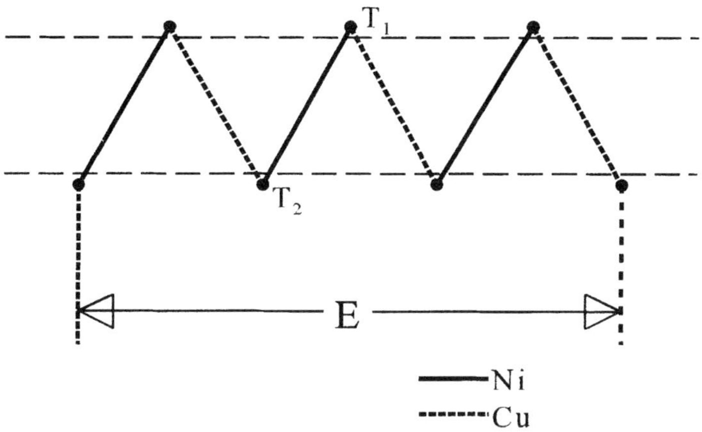

FIGURE 15.3 Thermopile for differential temperature measurement [2].

and increases the thermal disruption. Temperature measurements on the sensor and on the surrounding undisturbed material are recommended to quantify this disruption.

Although the temperature difference can be measured by any number of methods, the most commonly used are thermocouples. Thermocouples have the advantage that they generate their own voltage output corresponding to the temperature difference between two junctions. Consequently, they can be connected in series to form a thermopile that amplifies the output from a given temperature difference. An illustration of a thermopile for measuring a temperature difference is given in Figure 15.3. Most any pair of conductors (e.g., copper–constantan) can be used for the legs of the thermopile, but the output leads should be of the same material so that additional thermocouple junctions are not created. The voltage output, E, is simply:

$$E = N S_T \left(T_1 - T_2\right) \tag{15.7}$$

where N represents the number of thermocouple junction pairs, and S_T is the Seebeck coefficient or thermoelectric sensitivity of the materials, expressed in volts per degree Centigrade. The corresponding sensitivity of the heat flux sensor is:

$$S = \frac{E}{q''} = \frac{N S_T \delta}{k} \tag{15.8}$$

Although the sensitivity is determined in practice from a direct calibration, the last part of the equation can be used to determine the effects of different parameters for design purposes.

One successful design using a thermopile was described by Ortolano and Hines [3] and is currently manufactured by RdF Corp., as listed in Table 15.1. Thin pieces of two types of metal foil are alternately wrapped around a thin plastic (Kapton) sheet and butt-welded on either side to form thermocouple junctions, as illustrated in Figure 15.4. A separate thermocouple is included to provide a measure of the sensor temperature. The flexible, micro-foil sensors are 75 μm to 400 μm thick and can be glued to a variety of surface shapes, but are limited to temperatures below (250°C) and heat fluxes less than 100 kW m^{-2}. This covers many general-purpose industrial and research applications. The time response can be as fast as 20 ms, but transient signals can be attenuated unless the frequency of the disturbance is less than a few hertz. First-order systems, such as these sensors, give 70% response to a sinusoidal input with a period six times the exponential time constant.

TABLE 15.1 Heat Flux Instrumentation

Manufacturer	Sensor	Description	Approximate price (U.S.$)
RdF	Micro-foil	Foil thermopile	$100
Vatell	HFM	Microsensor thermopile	$900
Vatell	Episensor	Thermopile	$100–250
Concept	Heat flow sensor	Wire-wound thermopile	$100–300
Thermonetics	Heat flux transducer	Wire-wound thermopile	$50–900
ITI	Thermal flux meter	Thermopile	$150–350
Vatell	Gardon gage	Circular foil design	$250–500
Medtherm	Gardon gage	Circular foil design	$400–800
Medtherm	Schmidt-Boelter	Wire-wound thermopile	$500–800
Medtherm	Coaxial thermocouple	Transient temperature	$250–450
Medtherm	Null-point calorimeter	Transient temperature	$650–800
Hallcrest	Liquid crystals	Temperature measurement kit	$200
Image Therm Eng.	TempVIEW	Liquid crystal thermal system	$30k–50k

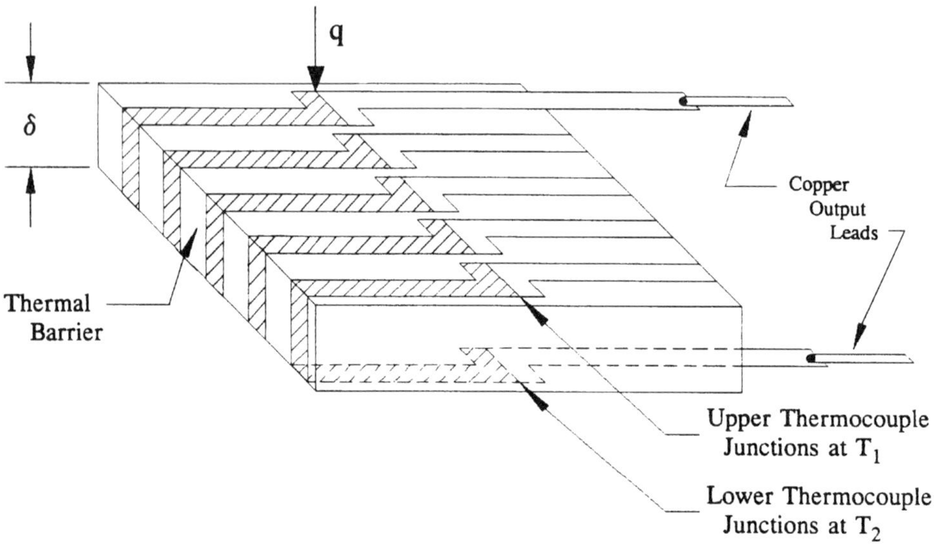

FIGURE 15.4 Thermopile heat flux sensor [3].

A similar design uses welded wire to form the thermopile across a sensor about 1 mm thick. This gives a higher sensitivity to heat flux, but also a larger thermal resistance. Time constants are on the order of 1 s and the upper temperature limit is 300°C. These are manufactured in a range of sizes by International Thermal Instrument Co., as listed in Table 15.1. Applications include heat transfer in buildings and physiology. Sensors with higher sensitivity are made with semiconductor thermocouple materials for geothermal applications. Lower sensitivity sensors are made for operating temperatures up to 1250°C.

A much thinner thermopile sensor called the Heat Flux Microsensor (HFM) was described by Hager et al. [4] and is manufactured by Vatell Corp., as listed in Table 15.1. Because it is made with thin-film sputtering techniques, the entire sensor is less than 2-μm thick. The thermal resistance layer of silicon monoxide is also sputtered directly onto the surface. The resulting physical and thermal disruption of the surface due to the presence of the sensor is extremely small. Use of high-temperature thermocouple materials allows sensor operating temperatures to exceed 800°C for the high-temperature models. They are best suited for heat flux values above 1 kW m^{-2}, with no practical upper limit. Because the sensor is so thin, the thermal response time is less than 10 μs [5], giving a good frequency response well above 1 kHz. A temperature measurement that is integrated into the sensor is very useful for checking the heat flux calibration [6] and determining the heat transfer coefficient. The high temperature and fast time response capabilities are useful for aerodynamic applications, combusting flows in engines and propulsion systems, and capturing high-speed events such as shock passage.

Terrell [7] describes a similar sensor design made with screen printing techniques of conductive inks. A copper/nickel thermocouple pair was used with a dielectric ink for the thermal resistance layer. The inks were printed onto anodized aluminum shim stock for the substrate. Although the entire package is 350-μm thick, the thermal resistance is low because of the high thermal conductivity of all of the materials. These are currently offered commercially by Vatell Corp., as listed in Table 15.1. Because of the large number of thermocouple pairs (up to 10,000), sensitivities are sufficient to measure heat fluxes as low as 0.1 W m^{-2}. The thermal time constant is about 1 s, and the upper temperature limit is approximately 150°C. The aluminum base allows some limited conformance to a surface. Applications include the low heat fluxes typical of building structures, biomedicine, and fire detection.

Another technique for measuring the temperature difference across the thermal resistance layer is to wrap wire and then plate one side of it with a different metal. A common combination is constantan wire with copper plating. The resulting wire-wound sensor looks similar to the sensor shown in Figure 15.4. The difference is that the constantan wire is continuous all around the sensor, so it does not form discrete thermocouple junctions. A summary of the theory is given by Hauser [8] and a general review is given by van der Graaf [9]. Concept Engineering offers a range of these type of sensors at moderate cost. Because of the hundreds of windings on these sensors around 2-mm thick plastic strips, the sensitivity to heat flux is high. The corresponding thermal resistance is also large and time constants are around 1 s. Temperatures are limited to about 150°C. Thermonetics also makes a plated wire-wound heat flux sensor. Thicknesses range from 0.5 mm to 3 mm, with time constants greater than 20 s. Some of the units are flexible and can be wrapped around objects. The normal temperature limit is 200°C, but ceramic units are available for operation above 1000°C. The main use for these sensors is to measure heat flux levels less than 1 kW m^{-2}, with applications including building structures, insulation, geothermal, and medicine.

One popular version of plated wire sensors uses a small anodized piece of aluminum potted into a circular housing, commonly known as a Schmidt-Boelter gage. Kidd [10] has performed extensive analyses on these gages to determine the effect of the potting on the measured heat flux. Neumann [11] discusses applications in aerodynamic testing. The sensors are commercially available from Medtherm Corp. in sizes as small as 1.5-mm diameter. There is also some ability to contour the surface of the sensor to match a curved model surface for complex test article shapes.

Circular Foil Gages

The circular foil or Gardon gage consists of a hollow cylinder of one thermocouple material with a thin foil of a second thermocouple material attached to one end. A wire of the first material is attached to the center of the foil to complete a differential thermocouple pair between the center and edge of the foil as illustrated in Figure 15.5. The body and wire are commonly made of copper with the foil made of constantan. Heat flux to the gage causes a radial temperature distribution along the foil as illustrated in Figure 15.5. The circular foil gage was originated by Robert Gardon [12] to measure radiation heat transfer. For a uniform heat flux typical of incoming radiation the center to edge temperature difference is proportional to the heat flux (neglecting heat losses down the center wire):

$$T_o - T_s = \frac{q'' R^2}{4k\delta} \tag{15.9}$$

The thickness of the foil is δ and the active radius of the foil is R. The temperature difference produced from center to edge of the foil is measured by a single thermocouple pair, typically copper-constantan. The output voltage is proportional to the product of the temperature difference in Equation 15.9 and the thermoelectric sensitivity of the differential thermocouple.

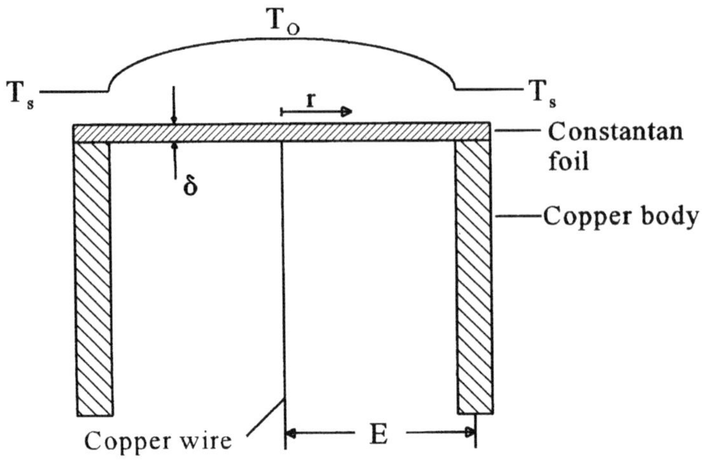

FIGURE 15.5 Circular foil heat flux gage.

These sensors are reasonably rugged and simple. They are manufactured by two companies at moderate cost (Medtherm and Vatell), and are often used as secondary standards for measurement of radiation. One important application is the measurement of the heat flux occurring during fire tests to check flammability of materials. The biggest problems with the circular foil gages arise when they are used with any type of convection heat transfer. It has been shown analytically and experimentally that the output is incorrect for convective heat transfer because of the distortion of the temperature profile in the foil from the assumed radially symmetric, parabolic profile of radiation [13]. Because the amount of error is a function of the gage geometry, the fluid flow, and the heat transfer coefficient, it is difficult to reliably correct. The errors become particularly large when the sensor is used in a flow that has a shear flow component [14], which encompasses most convection situations. Consequently, great care must be used to keep the temperature difference across the gage $T_o - T_s$ small if Gardon gages are used to measure convective heat transfer.

When the gages are used in high heat flux situations, such as combustors, water cooling is usually supplied through the body of the sensor to keep the temperature from exceeding material limits. Because of the resulting temperature mismatch of the gage and surrounding material in which it is mounted, a water-cooled gage is not recommended for convection heat transfer measurements. It is also important for a water-cooled gage to ensure that condensation does not occur on the sensor face.

Although most heat flux sensors are designed to measure the total heat flux, sensors have been developed to separate convection from radiation. The most common method is to put a transparent window over the sensor to eliminate convection to the sensor face. Because the resulting sensor only measures radiation, it is termed a radiometer. The field of view is limited, however, in these radiometers and must be included in the interpretation of results. In a dirty environment where the transmission of the window could be degraded, air is blown across the face of the window to keep the particles away from the sensor. Both manufacturers of the circular foil gages (Medtherm and Vatell) make these radiometer versions. Applications include use in high-ash boilers and gas turbine combustors.

Research Sensors

Although not commercially available, heat flux sensors using RTDs (resistance temperature devices) have been developed to measure the required temperature difference across a heat flux sensor. They are not as convenient for measuring small temperature differences as thermocouples, however, because RTDs require two individual temperature readings to be subtracted. Conversely, a thermocouple pair reads the temperature difference between the two junctions directly and allows the formation of thermopiles with many pairs of junctions to amplify the signal.

Researchers at MIT have developed a sensor like the Micro-Foil thermopile sensor using a nickel resistance element on each side of the plastic sheet [15]. One advantage of knowing the individual temperatures rather than the temperature difference is that the time response of the sensor can be analytically enhanced up to 100 kHz with appropriate modeling. Hayashi et al. [16] used vacuum deposition to create thin-film heat flux sensors like the HFM sputtered thermopile sensors except using a nickel RTD on either side of the silicon monoxide thermal resistance layer. The frequency response was estimated from shock-tunnel experiments to be 600 Hz.

Future Developments

The most exciting recent advances in the field of heat flux measurement have been provided by thin-film technology. Continued improvements in size, sensitivity, price, and time response are anticipated. As the size of sensors continues to decrease, the deposition of thin-film heat flux sensors directly on parts as they are manufactured should become a reality.

15.4 Sensors Based on Temperature Change with Time

Equations 15.1 to 15.3 give the form of the relationship between the unsteady response of temperature and surface heat transfer. If the thermal properties of the wall material are known along with sufficient detail about the temperature history and distribution, the heat transfer as a function of time can, in principle, be determined. Although temperature sensors are available from manufacturers, the necessary data manipulation must be done by the user to obtain heat flux. There are two types of solutions used to reduce the temperature history to heat flux. These are discussed separately as the semi-infinite solution and calorimeter methods. In addition, a variety of methods for measuring the required temperature history are discussed.

Semi-Infinite Surface Temperature Methods

An important technique for short-duration heat flux tests is to measure the surface temperature history on a test object with a fast-response temperature sensor. For short enough times and sufficiently thick material, it can be assumed that the transfer is one-dimensional and that the thermal effects do not reach the back surface of the material. Equation 15.3 reduces to a one-dimensional, semi-infinite solution, which is simple to implement for this case. For example, the surface temperature for a step change of heat flux at time zero is

$$T_s - T_i = \frac{2q_o''\sqrt{t}}{\sqrt{\pi k\rho C}} \tag{15.10}$$

The substrate properties are the thermal conductivity, density, and specific heat, represented by the product $k\rho C$. T_i is the uniform initial temperature of the substrate and T_s is the surface temperature as a function of time.

A good criteria for the time limit on the test is the time before 1% error occurs in the heat flux [17]:

$$t = 0.3\frac{L^2}{a} \tag{15.11}$$

Here, a represents the thermal diffusivity of the substrate material and L is the substrate thickness. For a typical ceramic substrate (MACOR), the corresponding minimum thickness for 1 s of test time is 1.6 mm. High conductivity materials, such as metals, have much larger required thicknesses.

Data analysis of the measured temperature record can be performed by several methods. The simplest is to use the analytical solution with each sampled data point to recreate the heat flux signal.

The most popular equation for this conversion is that attributed to Cook and Felderman [18] for uniformly sampled data:

$$q''\left(t_{\mathrm{n}}\right)=\frac{2\sqrt{k\rho C}}{\sqrt{\pi\Delta t}}\sum_{j=1}^{n}\frac{T_{\mathrm{j}}-T_{\mathrm{j\text{-}1}}}{\sqrt{n-j}+\sqrt{n+1-j}} \tag{15.12}$$

This can easily be implemented for digital data with a short computer program to perform the summation for the measured points. Modifications are also available to provide more solution stability [17]. More complex techniques include the use of parameter estimation techniques [19] and numerical solutions to account for changes in property values with the changing temperature [20]. Because of the noise amplification inherent in the conversion from temperature to heat flux, analog methods have been developed to convert the temperature signal electronically before digitizing the signal [21].

There are many methods for measuring surface temperature that can be used to determine heat flux. Two broad categories are point measurements using thermocouples or RTDs and optical methods that allow for simultaneous measurement of temperatures over the entire surface. They all require substantial effort from the user to initiate the test procedure and reduce the data to find heat flux. Places for additional information and a few sources for temperature sensors are given.

Point temperature measurements for determining convective heat flux are often made with thin-film RTDs. A metallic resistance layer is sputtered, painted and baked, or plated onto the surface. Because the resulting thickness of the sensor is less than 0.1 μm, the response time is a fraction of 1 μs and there should be no physical or thermal disruption of the measured temperature due to the sensor. Most researchers develop techniques for instrumenting models themselves. Transient flow facilities provide an easy method for quickly initiating the flow and heat transfer, as required by this transient method. However, the model can also be injected into the flow or the flow can be quickly diverted to provide the fast initiation of heat flux. The method is used for basic measurements applicable to gas turbine engines, rockets, internal combustion engines, and high-speed aerodynamics [2].

A special type of thermocouple is made for surface temperature measurement, called coaxial thermocouples [22]. It has one thermocouple wire inside the second thermocouple material with an insulating layer in between. One end is mounted into a metal sheath for press fitting into the surface material for testing. A thin thermocouple is formed by combining the two metals right at the end of the assembly. Response times are typically 1 ms or less, which although slower than the thin-film sensors is sufficient for most applications [11]. The cost per sensor is moderate and they are available from Medtherm Corp., as listed in Table 15.1.

A new approach to measure the transient surface temperature at a point is being developed using a fiber optic probe embedded in the surface [23]. A Fabry-Perot interferometer is the basis for the technique, which has the advantages of high spatial resolution and no electrical connections.

Null-point calorimeters [24] are a further extension of the semi-infinite surface temperature method. They are designed for measurement of extremely high levels of heat flux (over 1000 kW m^{-2}). To protect the thermocouple and wires, they are mounted in a cavity behind the surface. The geometry of the null-point calorimeter is designed, however, for the thermocouple to measure a temperature that would match the surface temperature of a semi-infinite material so that Equation 15.14 can be used for data reduction [24]. Medtherm is the current supplier of null-point calorimeters, as indicated in Table 15.1.

Optical methods give the opportunity to measure the entire temperature field over a section of the surface. Consequently, much data can be collected over each test, but interpretation to obtain quantitative heat flux values is more challenging than measurements with point sensors. The visual display of the temperature over the surface can be very qualitatively informative, however.

The most popular optical method for measuring temperature is to record the color change using liquid crystals. These are specially prepared molecules that reversibly change their color reflection through several distinct colors as a function of temperature, typically in the range of 25°C to 40°C. The best for transient measurements are the chiral-nematic form that have been microencapsulated to stabilize their properties. A variety of types can be obtained from Hallcrest (in Table 15.1), which can be used over the temperature range from 5°C to 150°C. They can easily be spray-painted onto a blackened surface for testing. Setting the lighting for reproducible color, temperature calibration, image acquisition, and accurately establishing the starting temperature are crucial steps. Detailed procedures for accurate measurements have been established by several groups [25–28]. The basic materials are cheap, but the associated equipment is expensive. As listed in Table 15.1, Image Therm Engineering offers a complete system for temperature measurement, including a high-quality video camera, lighting system, calibration system, computer hardware and software for image processing, and a liquid crystal kit.

As previously discussed for radiation heat transfer, all surfaces emit radiation with an intensity and wavelength distribution that can be related to the surface temperature. The advent of high-speed infrared scanning radiometers has made it feasible to record the transient temperature field for determination of the heat flux distribution [29]. A coating is usually applied to establish a known, high absorptivity surface. To convert the measured radiation emission to surface temperature, the radiation field of the surroundings is also required. The camera and associated equipment are quite expensive.

Thermographic phosphors emit radiation in the visible spectrum when illuminated with ultraviolet light. The intensity of emission at specific wavelengths can be related to the temperature over a wide range of surface temperatures. The potential high-temperature applications are particularly appealing [30]. A CCD camera is required to record the transient optical images and calibration is challenging.

Calorimeter Methods

A *calorimeter* is a device for measuring the amount of absorbed thermal energy. The slug calorimeter [31] assumes that the temperature throughout the sensor is uniform while it changes with time. When exposed to a fluid at a temperature T_r and heat transfer coefficient h over an active area of A, the solution for the temperature change is simply an exponential:

$$\frac{T - T_r}{T_i - T_r} = e^{-t/\tau} \tag{15.13}$$

where T_i is the initial temperature and the time constant is:

$$\tau = \frac{mC}{hA} \tag{15.14}$$

with the active surface area represented by A. The thermal capacitance is the product of the mass of the sensor and the specific heat. The time constant can be found from the temperature response of the system, which can then be used in Equation 15.14 to quantify the heat transfer coefficient, h. Although these calorimeters are simple in principle, it is often difficult to obtain reliable results because of heat losses and nonuniform temperatures.

A more useful device, called the plug-type heat flux gage, was developed by Liebert [32] at NASA Lewis. An annulus is created on the backside of the surface by electrical discharge machining. Four thermocouples are attached along the remaining plug to estimate both the temperature gradient and the change in thermal energy content in the plug. This gives a better estimate of the heat flux than the slug calorimeter. An additional advantage is that the measurement surface is physically undisturbed.

Another calorimeter technique, called the *thin skin method*, uses the entire test article as the sensor. Models are constructed of thin metals and instrumented with thermocouples on the back surface. The temperature is assumed constant throughout the material at any location, but varies with time and position around the model. The main errors to be avoided are: (1) lateral conduction along the surface material, (2) heat loss by conduction down the thermocouple wires, and (3) heat loss from the back surface, which is usually considered adiabatic. Because of the recent advances in thin-film and optical surface temperature measurement, the thin-skin method is considered outdated for most modern aerodynamic testing [11].

15.5 Measurements Based on Surface Heating

For research on convective heat transfer, electric heating provides an easy method of controlling and measuring the heat flux to the surface. A combination of guard heaters and proper insulation allows control of the heat losses to give an accurate knowledge of the heat flux leaving the surface based on the total electrical power supplied once steady-state conditions have been established. The temperature of the surface and fluid are used in Equation 15.5 to give the heat transfer coefficient for the surface:

$$h = \frac{q''}{T_r - T_s} \tag{15.15}$$

As with the transient temperature methods, it requires considerable experimental design and expertise by the user to begin making measurements.

15.6 Calibration and Problems to Avoid

Calibration of heat flux is a complicated issue because some heat flux sensors respond to different modes and conditions of heat flux differently. For example, a sensor calibrated by radiation can have a substantially different response to the same amount of heat flux in convection.

For the low heat fluxes seen in building applications, the guarded hot-plate method has been well established [33]. The National Institute of Standards and Technology (NIST) maintains calibration devices for this range of conduction heat transfer. Calibrated insulation samples are readily available to check other guarded hot-plate calibrators. Calibration of sensors at elevated temperatures has demonstrated that there is a dependence of the heat flux sensitivity on sensor temperature [34].

There have been several important advances in heat flux calibration for more general industrial applications within the past few years. NIST is completing three heat flux calibration facilities. A blackbody radiation facility operating to 100 kW m^{-2}, a laminar flow convection facility, and a helium conduction facility are currently being completed and tested [35–37]. This combination of facilities will allow comparison of sensor response under the different modes of heat transfer. In addition, Arnold Engineering Development Center (AEDC) has recently acquired a radiation calibration facility for elevated sensor temperatures (up to 800°C). The temperature dependence of the heat flux sensitivity is thought to be substantial for many sensors. Because in the past most all calibrations have only been performed with the sensors at room temperature, this is an important new facility.

As with many other measurements, the major problem with heat flux measurement is the error caused in the heat flux by the disruption of the sensor itself. For sensors based on the spatial temperature gradient methods, a larger signal implies a larger temperature difference and a larger temperature disruption. For the second type of sensors based on the transient temperature change, the surface temperature is changing while the measurement occurs. The larger the temperature change, the easier the determination of the sensor heat flux, but the larger the error from the sensor temperature disruption.

The error caused by the thermal disruption of the sensor can be estimated for conduction [38]. In convection, however, the effect of the surface temperature disruption on the developing thermal boundary layer is much more difficult to estimate and the effect on the heat flux can be much larger than the percentage change of the temperature [2]. Therefore, it is imperative in convection measurements to keep the thermal disruption of the sensor to a minimum.

15.7 Summary

There are a large number of off-the-shelf heat flux sensors available. Those commercially available have been listed in Table 15.1, and the information for contacting the manufacturers is given in Table 15.2.

TABLE 15.2 Companies That Make Sensors for Heat Flux Measurement

Concept Engineering 43 Ragged Rock Road Old Saybrook, CT 06475 (860) 388-5566	RdF Corporation P.O. Box 490 Hudson, NH 03051-9981 (603) 882-5195
Hallcrest Liquid Crystal Division 1820 Pickwick Lane Glenview, IL 60025 (312) 998-8580	Thermonetics Corp. Box 9112 San Diego, CA 92109 (619) 488-2242
International Thermal Instrument Co. P.O. Box 309 Del Mar, CA 92014 (619) 755-4436	Vatell Corporation P.O. Box 66 Christiansburg, VA 24073 (540) 961-3576
Medtherm Corporation P.O. Box 412 Huntsville, AL 35804 (205) 837-2000	Image Therm Eng., Inc. 159 Summer St. Waltham, MA 02154 (781) 893-7793

The differential temperature devices provide a direct readout of the heat flux over the surface of the sensor. With the proper choice of sensor for the application and care in measurement method, the results are easily interpreted and used. Alternatively, the transient temperature methods can provide more surface details, but the output is a surface temperature history that must be analyzed to obtain the corresponding heat flux. Although certain components of these systems are off-the-shelf, more work is required of the user to interpret the results. Issues of calibration and errors have been addressed briefly here. More detail on all aspects of heat flux measurement can be obtained from the manufacturers and the references listed.

References

1. R. J. Moffat, Experimental heat transfer, in G. Hetsroni (ed.), *Heat Transfer 1990,* Vol. 1, New York: Hemisphere, 1990, 187-205.
2. T. E. Diller, Advances in heat flux measurement, in J. P. Hartnett et al. (eds.), *Advances in Heat Transfer,* Vol. 23, Boston: Academic Press, 1993, 279-368.
3. D. J. Ortolano and F. F. Hines, A simplified approach to heat flow measurement. *Advances in Instrumentation,* Vol. 38, Part II, Research Triangle Park: ISA, 1983, 1449-1456.
4. J. M. Hager, S. Onishi, L. W. Langley, and T. E. Diller, High temperature heat flux measurements, *AIAA J. Thermophysics Heat Transfer,* 7, 531-534, 1993.
5. D. G. Holmberg and T. E. Diller, High-frequency heat flux sensor calibration and modeling, *ASME J. Fluids Eng.,* 117, 659-664, 1995.
6. J. M. Hager, J. P. Terrell, E. Sivertson, and T. E. Diller, *In-situ* calibration of a heat flux microsensor using surface temperature measurements, *Proc. 40th Int. Instrum. Symp.,* Research Triangle Park, NC: ISA, 1994, 261-270.
7. J. P. Terrell, New high sensitivity, low thermal resistance surface mounted heat flux transducer, *Proc. 42nd Int. Instrum. Symp.,* Research Triangle Park, NC: ISA, 1996, 235-249.
8. R. L. Hauser, Construction and performance of *in situ* heat flux transducers, in E. Bales et al. (eds.), *Building Applications of Heat Flux Transducers,* ASTM STP 885, 1985, 172-183.
9. F. Van der Graaf, Heat flux sensors, in W. Gopel et al. (eds.), *Sensors,* Vol. 4, New York: VCH, 1989, 295-322.
10. C. T. Kidd and C. G. Nelson, How the Schmidt-Boelter gage really works, *Proc. 41st Int. Instrum. Symp.,* Research Triangle Park, NC: ISA, 1995, 347-368.

11. D. Neumann, Aerothermodynamic instrumentation, AGARD Report No. 761, 1989.
12. R. Gardon, An instrument for the direct measurement of intense thermal radiation, *Rev. Sci. Instrum.*, 24, 366-370, 1953.
13. C. H. Kuo and A. K. Kulkarni, Analysis of heat flux measurement by circular foil gages in a mixed convection/radiation environment, *ASME J. Heat Transfer*, 113, 1037-1040, 1991.
14. N. R. Keltner, Heat flux measurements: theory and applications, Ch. 8, in K. Azar (ed.) *Thermal Measurements in Electronics Cooling*, Boca Raton, FL: CRC Press, 1997, 273-320.
15. A. H. Epstein, G. R. Guenette, R. J. G. Norton, and Y. Cao, High-frequency response heat-flux gauge, *Rev. Sci. Instrum.*, 57, 639-649, 1986.
16. M. Hayashi, S. Aso, and A. Tan, Fluctuation of heat transfer in shock wave/turbulent boundary-layer interaction, *AIAA J.*, 27, 399-404, 1989.
17. T. E. Diller and C. T. Kidd, Evaluation of Numerical Methods for Determining Heat Flux With a Null Point Calorimeter, in *Proc. 43rd Int. Instrum. Symp.*, Research Triangle Park, NC: ISA, 357-369, 1997.
18. W. J. Cook and E. M. Felderman, Reduction of data from thin film heat-transfer gages: a concise numerical technique, *AIAA J.*, 4, 561-562, 1966.
19. D. G. Walker and E. P. Scott, One-dimensional heat flux history estimation from discrete temperature measurements, in R. J. Cochran et al. (eds.), *Proc. ASME Heat Transfer Division*, Vol. 317-1, New York: ASME, 1995, 175-181.
20. W. K. George, W. J. Rae, P. J. Seymour, and J. R. Sonnenmeier, An evaluation of analog and numerical techniques for unsteady heat transfer measurement with thin film gages in transient facilities, *Exp. Thermal. Fluid Sci.*, 4, 333-342, 1991.
21. D. L. Schultz and T. V. Jones, Heat transfer measurements in short duration hypersonic facilities, AGARDograph 165, 1973.
22. C. T. Kidd, C. G. Nelson, and W. T. Scott, Extraneous thermoelectric EMF effects resulting from the press-fit installation of coaxial thermocouples in metal models, *Proc. 40th Int. Instrum. Symp.*, Research Triangle Park, NC: ISA, 1994, 317-335.
23. S. R. Kidd, P. G. Sinha, J. S. Barton, and J. D. C. Jones, Wind tunnel evaluation of novel interferometric optical fiber heat transfer gages, *Meas. Sci. Technol.*, 4, 362-368, 1993.
24. ASTM Standard E598-77, Standard method for measuring extreme heat-transfer rates from high-energy environments using a transient null-point calorimeter, *Annu. Book of ASTM Standards*, 15.03, 381-387, 1988.
25. D. J. Farina, J. M. Hacker, R. J. Moffat, and J. K. Eaton, Illuminant invariant calibration of thermochromic liquid crystals, *Exp. Thermal Fluid Sci.*, 9, 1-12, 1994.
26. J. W. Baughn, Liquid crystal methods for studying turbulent heat transfer, *Int. J. Heat Fluid Flow*, 16, 365-375, 1995.
27. Z. Wang, P. T. Ireland, and T. V. Jones, An advanced method of processing liquid crystal video signals from transient heat transfer experiments, *ASME J. Turbomachinery*, 117, 184-189, 1995.
28. C. Camci, K. Kim, S. A. Hippensteele, and P. E. Poinsatte, Evaluation of a hue capturing based transient liquid crystal method for high-resolution mapping of convective heat transfer on curved surfaces, *ASME J. Heat Transfer*, 115, 311-318, 1993.
29. G. Simeonides, J. P. Vermeulen, and H. L. Boerrigter, Quantitative heat transfer measurements in hypersonic wind tunnels by means of infrared thermography, *IEEE Trans. Aerosp. Electron. Syst.*, 29, 878-893, 1993.
30. D. J. Bizzak and M. K. Chyu, Use of laser-induced fluorescence thermal imaging system for local jet impingement heat transfer measurement, *Int. J. Heat Mass Transfer*, 38, 267-274, 1995.
31. ASTM Standard E457-72, Standard method for measuring heat-transfer rate using a thermal capacitance (slug) calorimeter, *Annual Book of ASTM Standards*, 15.03, 299-303, 1988.
32. C. H. Liebert, Miniature convection cooled plug-type heat flux gages, *Proc. 40th Int. Instrum. Symp.*, Research Triangle Park, NC: ISA, 1994, 289-302.

33. M. Bomberg, A workshop on measurement errors and methods of calibration of a heat flow meter apparatus, *J. Thermal Insulation and Building Environments*, 18, 100-114, 1994.
34. M. A. Albers, Calibration of heat flow meters in vacuum, cryogenic, and high temperature conditions, *J. Thermal Insulation and Building Environments*, 18, 399-410, 1995.
35. W. Grosshandler and D. Blackburn, Development of a high flux conduction calibration apparatus, in M. E. Ulucakli et al. (eds.), *Proc. ASME Heat Transfer Division*, Vol. 3, New York: ASME, 1997, 153-158.
36. A. V. Murthy, B. Tsai, and R. Saunders, Facility for calibrating heat flux sensors at NIST: an overview, in M. E. Ulucakli et al. (eds.), *Proc. ASME Heat Transfer Division*, Vol. 3, New York: ASME, 1997, 159-164.
37. D. Holmberg, K. Steckler, C. Womeldorf, and W. Grosshandler, Facility for calibrating heat flux sensors in a convective environment, in M. E. Ulucakli et al. (eds.), *Proc. ASME Heat Transfer Division*, Vol. 3, New York: ASME, 1997, 165-171.
38. S. N. Flanders, Heat flux transducers measure *in-situ* building thermal performance, *J. Thermal Insulation and Building Environments*, 18, 28-52, 1994.

16
Thermal Imaging

Herbert M. Runciman
Pilkington Optronics

All objects at temperatures above absolute zero emit electromagnetic radiation. Radiation thermometry makes use of this fact to estimate the temperatures of objects by measuring the radiated energy from selected regions. Thermal imaging takes the process one stage further and uses the emitted radiation to generate a picture of the object and its surroundings, usually on a TV display or computer monitor, in such a way that the desired temperature information is easily interpreted by the user.

Thermal imagers require no form of illumination to operate, and the military significance of this, together with their ability to penetrate most forms of smoke, has been largely responsible for driving thermal imager development. Although thermal imagers intended for military or security applications can be used for temperature measurement, they are not optimized for this purpose since the aim is to detect, recognize, or identify targets at long ranges by their shape; thus, resolution and sensitivity are favored over radiometric accuracy.

16.1 Essential Components

All thermal imagers must have a detector or array of detectors sensitive to radiation in the required waveband, and optics to form an image of the object on the detector. In modern thermal imagers, the detector array might have a sufficient number of sensitive elements to cover the focal plane completely (a "staring array"), in the same way as a CCD television camera. Some of the most recent staring arrays can give good performance without cooling. In other imagers, the detector might take the form of a single row or column of elements, in which case a scanning mechanism is required to sweep the image across the detector array. If a single-element detector, or a very small detector array, is used, a means

0-8493-0047-9/00/$0.00+$.50

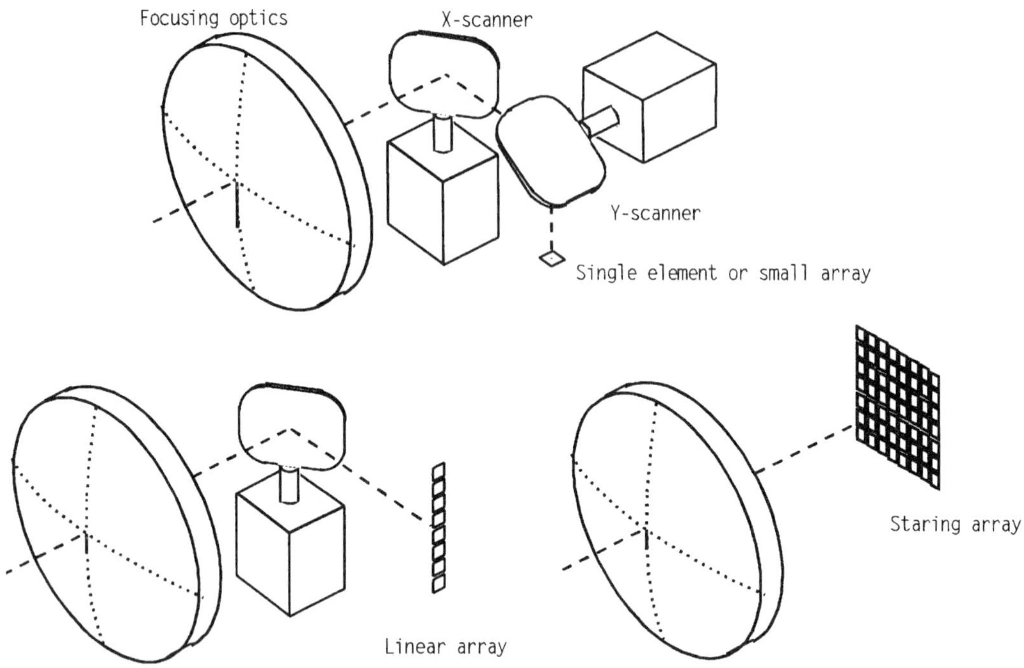

FIGURE 16.1 Thermal imaging options: (a) 2-D scanning for small detector array or single element; (b) 1-D scanning with linear detector array; (c) staring array without scanning.

of providing a two-dimensional scan is required. (Figure 16.1 shows these options schematically.) For scanning imagers, it is necessary to cool the detectors (usually to about 80 K to 120 K) to achieve adequate performance.

Although in principle it would be possible to deduce target temperature from the absolute value of the detector signal, it is necessary in practice to estimate temperature by comparison with one or more reference bodies of known temperature. The temperature references are usually internal to the equipment, and accessed by mechanical movement of the reference (which may take the form of a rotating chopper) or by deflecting the optical path using a mirror.

16.2 Thermal Imaging Wavebands

The optimum waveband for thermal imaging is determined partly by the wavelength distribution of the emitted radiation, partly by the transmission of the atmosphere, and partly by the chosen detector technology.

The power radiated from a given area of an object depends only on its temperature and the nature of its surface. If the surface absorbs radiation of all wavelengths completely, it is referred to as a "blackbody." It then also emits the maximum amount possible, which can be calculated using the Planck equation (given later). Figure 16.2(a) shows the way in which blackbody emission varies with wavelength for several temperatures. It will be seen that for objects near normal ambient temperature, maximum output occurs at a wavelength of about 10 μm, or about 20 times the wavelength of visible light. At wavelengths below about 3 μm, there is generally insufficient energy emitted to allow thermal imaging of room-temperature objects. The emissivity at any wavelength is defined as the ratio of the energy emitted at that wavelength to the energy that would be emitted by a blackbody at the same wavelength.

It is important that the atmosphere should have sufficient transparency to permit the target to be observed. There are two important "atmospheric windows" — one between 3 μm and 5 μm (with a notch at 4.2 μm due to carbon dioxide absorption) and one between 7.5 μm and 14 μm. These are commonly

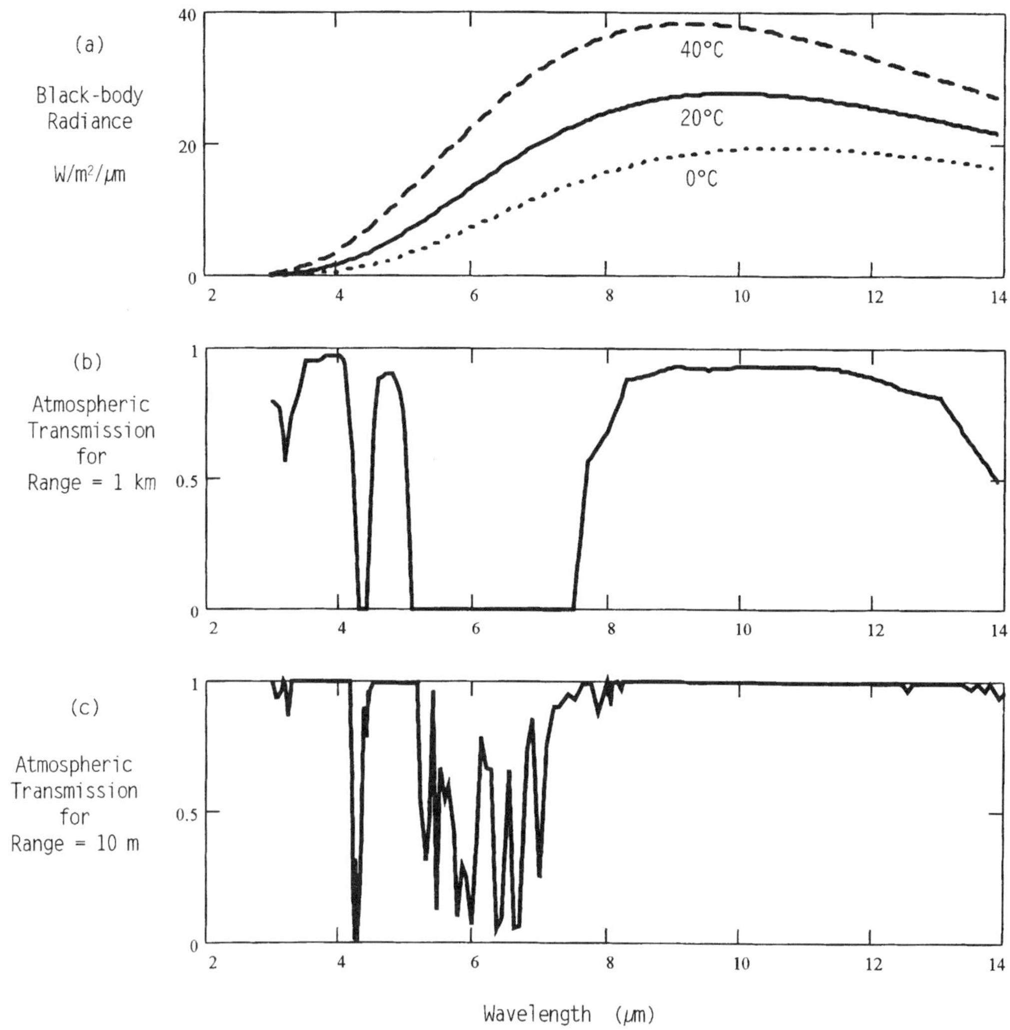

FIGURE 16.2 Factors determining thermal imaging wavebands. Imager must operate in regions where radiance is sufficiently high (a) and atmospheric transmission is good (b) and (c).

referred to as the medium-wave infrared (MWIR) and long-wave infrared (LWIR) windows, respectively. For thermal measurement over short ranges in the laboratory, it is possible to work outside these bands, but most instruments are optimized for either the MWIR or LWIR. Typical transmissions through 1 km and 10 m of a clear U.S. Standard Atmosphere are shown in Figures 16.2(b) and 16.2(c).

Emissivity for most naturally occurring objects and organic paints is high (>0.8) in the LWIR, but is lower and more variable in the MWIR. Metallic surfaces have low emissivity in both bands. Solar radiation in the MWIR is significant, and can cause errors in measurements made outdoors. These considerations favor the LWIR for quantitative imaging, but the band chosen can also be influenced by the chosen detector technology, the latter frequently being determined by cost. Scanning imagers can be used in either band, but are more sensitive for a given detector architecture in the LWIR. Cooled staring arrays give similar sensitivity in either band, but are currently more readily available in the MWIR. Uncooled staring arrays work well only in the LWIR band.

For temperature measurement, the electronics can be used to encode signal level as false color, a color scale derived from the thermal references being injected into the display to allow the user to identify the temperature of the object under examination. For general surveillance, a conventional gray-scale image is usually preferred. Imagers for thermography can also include emissivity compensation. If accurate results are required for an object of low emissivity, it is important to ensure that the temperature of anything that might be reflected by the object is known and that the emissivity is accurately known. If the object is accessible, another object placed beside it with the same surface characteristics but known temperature can be used for calibration.

16.3 Emission from Source

The spectral radiance $W(\lambda,T)$ of a blackbody at temperature T and wavelength λ is given by the Planck equation [1]. For temperature differences between the target and the reference of a few degrees, it is frequently sufficiently accurate to assume a linear dependence of radiance on temperature difference, making the temperature derivative of the blackbody equation, $dW(\lambda,T)/dT$, more relevant. In the case of photon detectors, the detector output is proportional to the photon flux, which can be derived from the radiance using the fact that photon energy $E(\lambda) = hc/\lambda$, where h is the Planck constant. The total photon flux $N(\lambda)$ and its derivative with respect to temperature are thus relevant in this case. The equations are as follows:

$$W(\lambda,T)=\frac{c_1}{\lambda^5\left(e^{\frac{c_2}{\lambda T}}-1\right)},\quad \text{W m}^{-2}\ \mu^{-1} \tag{16.1}$$

$$N(\lambda,T)=\frac{c_3}{\lambda^4\left(e^{\frac{c_2}{\lambda T}}-1\right)},\quad \text{photons s}^{-1}\ \text{m}^{-2}\ \mu\text{m}^{-1} \tag{16.2}$$

$$\frac{dW(\lambda,T)}{dT}=\frac{c_1 c_2 e^{\frac{c_2}{\lambda T}}}{\lambda^6 T^2\left(e^{\frac{c_2}{\lambda T}}-1\right)},\quad \text{W m}^{-2}\ \mu\text{m}^{-1}\ \text{K}^{-1} \tag{16.3}$$

$$\frac{dN(\lambda,T)}{dT}=\frac{c_3 c_2 e^{\frac{c_2}{\lambda T}}}{\lambda^5 T^2\left(e^{\frac{c_2}{\lambda T}}-1\right)},\quad \text{photons s}^{-1}\ \text{m}^{-2}\ \mu\text{m}^{-1}\ \text{K}^{-1} \tag{16.4}$$

where: Numerical values of the constants are:

$c_1 = 3.742 \times 10^8$

$c_2 = 1.439 \times 10^4$

$c_3 = 1.884 \times 10^{27}$

The unit of wavelength is chosen for convenience to be the micrometer (μm).

The above values are for radiation into a hemisphere. The intensities (watts per steradian, photons per steradian, etc.) are obtained by dividing the above values by π. The actual radiances for real targets are obtained by multiplying by the spectral emissivity $\varepsilon(\lambda)$; but since target reflectivity $\rho(\lambda) = 1 - \varepsilon(\lambda)$, some caution is required. For example, a target at temperature T surrounded by a background of temperature T_b will appear to emit $W(\lambda,T)\varepsilon(\lambda) + W(\lambda,T_b)\rho(\lambda) = [W(\lambda,T) - W(\lambda,T_b)]\varepsilon(\lambda) + W(\lambda,T_b)$.

Provided that the background surrounds the target and that the target is reasonably small, the background acts as an isothermal enclosure, which can be shown [2] to behave as an ideal blackbody (i.e., $\varepsilon(\lambda) = 1$). The differential spectral radiance against background $\Delta W(\lambda)$ is thus $[W(\lambda, T) - W(\lambda, T_b)]\varepsilon(\lambda)$, which for a small temperature difference ΔT is simply:

$$\Delta W(\lambda) = \varepsilon(\lambda) \frac{\mathrm{d}W(\lambda, T)}{\mathrm{d}T} \Delta T \qquad (16.5)$$

The spectral emissivity of a wide variety of natural and man-made objects is also given in [2].

A major difference between thermal imaging and visual imaging is the very low contrast. In the MWIR, the contrast calculated from Equation 16.1 due to 1 K at the target is about 4%, falling to about 2% in the LWIR.

16.4 Atmospheric Transmission

Provided that the absorption bands shown in Figures 16.2(b) and 16.2(c) are avoided, atmospheric transmission can frequently be ignored in the laboratory or industrial context. For longer ranges, an atmospheric transmission model must be used or calibrating sources must be placed at the target range. The standard atmospheric transmission model is LOWTRAN [3], currently at version 7. The atmospheric transmission $T_a(\lambda)$ reduces the differential signal from the target proportionately, but has no effect on the background flux if the atmosphere is at background temperature. Atmospheric transmission in the LWIR is severely affected by high humidity, making the MWIR the band of choice for long-range operation in the Tropics. (Many gases and vapors such as methane or ammonia have very strong absorptions in the infrared, making thermal imaging a possible means of leak detection and location.)

16.5 Detectors

Photon Detectors

In photon detectors, the response is caused by photons of radiation that generate free carriers in a semiconductor, which in turn increase the conductivity (for photoconductive detectors) or generate a potential difference across a junction (for photovoltaic detectors). Photovoltaic devices have the advantage of not requiring a bias current (important to reduce the heat load on the cooling system), and they have 40% lower noise because the electric field at the junction separates the carriers, thereby eliminating recombination noise. Whether or not the lower noise is achieved in practice depends on the read-out electronics. The photon energy in the LWIR is only about 1/20th of that of a photon in the visible region of the spectrum, so a semiconductor with a much smaller bandgap than silicon must be used. The most widely used material is a compound of mercury, cadmium, and tellurium (MCT or CMT) since not only is the quantum efficiency excellent (70% or more), but the bandgap can be tuned to the desired wavelength (in either waveband) by altering the composition. Cooling of the detector to about 80 K is desirable for the LWIR, but about 120 K is acceptable for the MWIR. For the MWIR, indium antimonide (InSb) is also an excellent material; and since it is a true stoichiometric compound, it is easier to achieve good uniformity of response, but cooling to 80 K is required.

Detectors for use in scanning systems are frequently arranged so that several elements are scanned over the same part of the scene in rapid succession, the output of each element being delayed and added to the previous one to enhance the signal-to-noise ratio (SNR). This approach (Figure 16.3) is termed

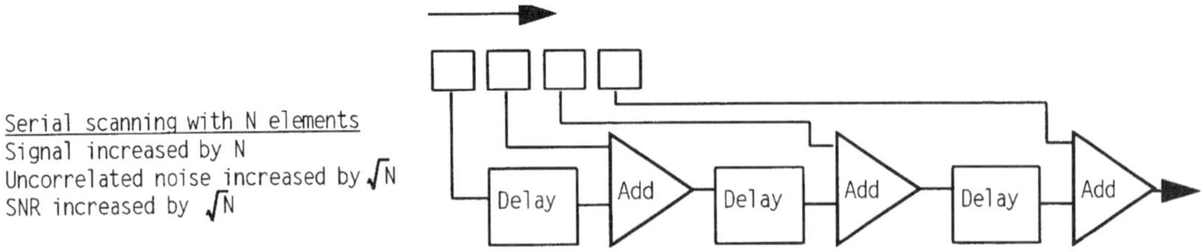

FIGURE 16.3 Use of serial scanning to enhance signal-to-noise ratio. The delay times are chosen to match the speed at which the image is swept along the detector array. Serial scanning is usually combined with parallel scanning using a detector matrix.

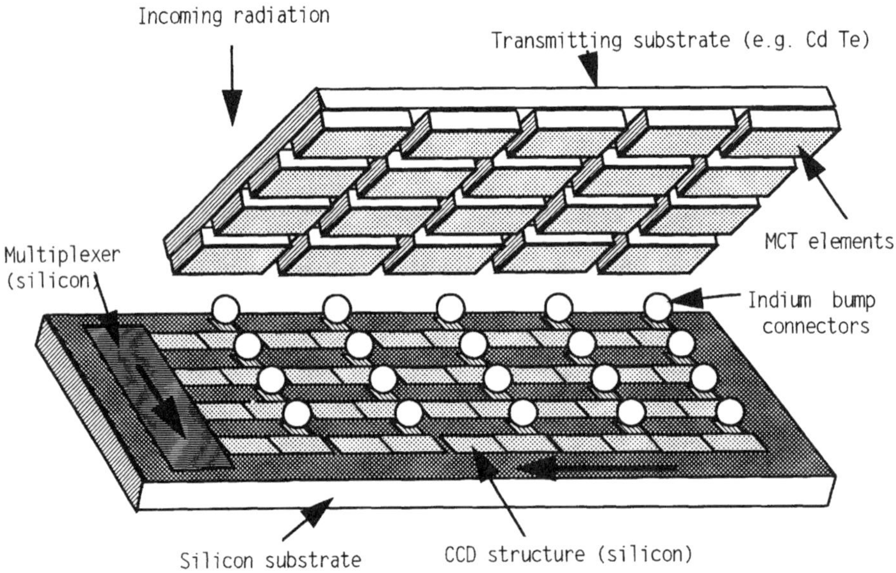

FIGURE 16.4 Typical hybrid detector construction. A typical element size is 30 μm. A large array of this type to match U.S. TV standard would have 640 × 480 elements.

serial scanning or "time-delay and integrate" (TDI) mode, and gives a theoretical gain in the SNR equal to the square root of the number of elements in TDI. It is also possible to perform TDI in the detector material itself. In the SPRITE detector (Signal Processing In The Element), the sensitive element is an elongated strip of CMT. Photons incident on the device generate carriers that drift toward a read-out electrode near one end. If the image is scanned along the detector at the same speed as the carrier drift, the signal builds up along the length of the device. The useful length is limited by carrier recombination, while diffusion of the carriers limits spatial resolution.

Large arrays of photon detectors are generally of hybrid construction, the sensitive elements being bonded to a silicon CCD or CMOS addressing circuit using indium "bumps" (Figure 16.4). An exception is the Schottky barrier detector (e.g., platinum silicide), which can be manufactured by a monolithic process, and thus tends to be lower in cost, but quantum efficiency is much lower and operation is usually limited to the MWIR band. Detector arrays and read-out architectures are discussed in depth in [2] Vol. 3, p. 246-341 and [4].

The detector assembly is encapsulated in a Dewar as shown in Figure 16.5. In front of the detector is a "cold shield" that limits the acceptance angle of the radiation to match that of the optics.

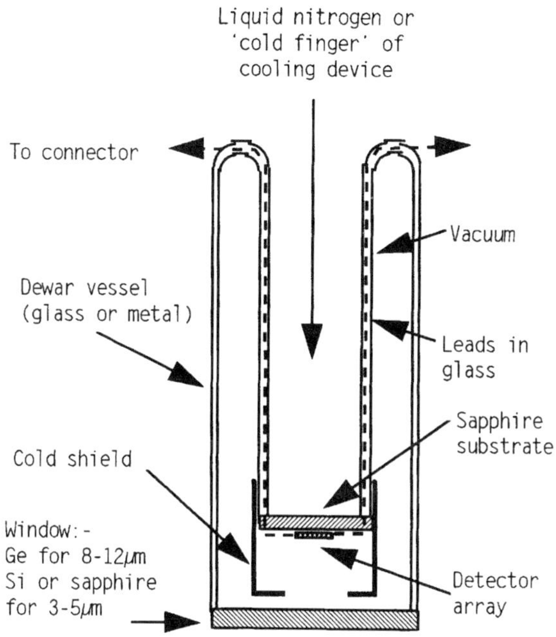

FIGURE 16.5 Construction of typical cooled detector. Cooling can be by liquid nitrogen, Joule-Thomson expansion of compressed gas, or a cooling engine.

Thermal Detectors

Thermal detectors rely on the heating effect of the incoming radiation, the change in temperature causing a change in resistance, capacitance, or electrical polarization that might be detected electrically. They are generally fairly slow in response (several milliseconds) but have the advantage that cooling is not essential (although it can be of considerable benefit with some types). The detectivity of uncooled thermal detectors is typically 1/100 that of cooled photon detectors, so real-time imaging requires the use of staring arrays. The essentials of a pyroelectric array are shown in Figure 16.6. Incoming radiation is absorbed by the blackened electrode, and the heat generated is transferred to the pyroelectric layer, which comprises a dielectric material that has been polarized by means of a high electric field during manufacture. The change in electrical polarization with temperature gives the electric signal. One of the most important parameters is the thermal isolation of the sensitive elements, so some kind of insulating support structure is necessary; and for good performance, the device must be evacuated to prevent convection. In a variant of this approach, the dielectric bolometer, the rapid variation of the dielectric constant at temperatures near the Curie point, causes the capacitance of the sensitive element to change, and hence the voltage for a constant charge. A detailed description of this approach is given in [5]. In both techniques, the detector responds only to change in temperature, so it is necessary to modulate the incoming radiation using a chopper. In the technique used initially by Wood [6] (now licensed to several manufacturers), the sensitive elements are vanadium dioxide coatings that undergo a large change in resistivity for a small temperature change. The elements are supported by silicon strips that are micromachined from the substrate and give excellent thermal insulation of the element. Changes in resistivity are read out by circuitry on the substrate, and no chopping is required, but the array must be maintained at a precise and uniform temperature.

Detector Performance Measures

The wavelength-dependent power responsivity of a detector $R(\lambda)$ is defined as the output potential or current that would result from 1 W of radiation at wavelength λ, assuming that linearity was maintained at such a high flux level. The units are $V\,W^{-1}$ or $A\,W^{-1}$. Photon responsivities in V photon^{-1} s^{-1} and A photon^{-1} s^{-1} are similarly defined.

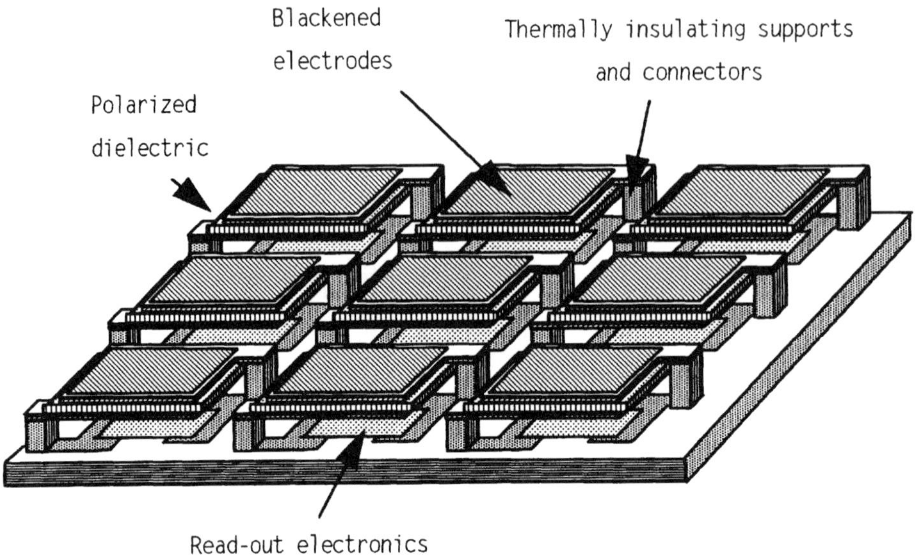

FIGURE 16.6 Essentials of an uncooled thermal detector array. The device is shown as pyroelectric or dielectric bolometer, but the essentials for a resistive bolometer are similar if the capacitors are replaced by resistors.

A thermal detector has a power responsivity that is essentially independent of wavelength, the limits of response being determined by the transparency of the window and the absorption spectrum of the element or the material used to blacken it.

In an ideal photon detector, the quantum efficiency η (defined as the number of carriers generated per photon) would be constant at all wavelengths for which the photon energy is greater than the bandgap, i.e., the photon responsivity is independent of wavelength up to the chosen cut-off wavelength. Since a given number of Watts corresponds to a number of photons proportional to the wavelength, the power responsivity ($V\,W^{-1}$ or $A\,W^{-1}$) would increase linearly with wavelength until the cut-off. In practice, the cut-off is spread over about 0.5 μm and shortwave performance is modified by window transmission and antireflection coatings.

The sensitivity of a detector is limited by noise that may be due to the detector itself or due to the background radiation (as is discussed later). Noise-equivalent power NEP(λ) is defined as the power incident on the detector at wavelength λ, which gives a signal equal to the rms noise when the measurement is made with a 1-Hz bandwidth. The NEP depends also on the modulation frequency, the latter effect being large for thermal detectors, but frequently negligible for quantum detectors. For many types of detector, the noise is proportional to the square root of the sensitive area A_d and the electrical bandwidth B, so that $\sqrt{(A_d B)}$/NEP is constant. A performance figure that is proportional to sensitivity can then be defined as specific detectivity $D^*(\lambda) = \sqrt{(A_d B)}/\mathrm{NEP}(\lambda)$. For historical reasons, specific detectivity is usually given in units of $\mathrm{cm}\sqrt{\mathrm{Hz}}\,\mathrm{W}^{-1}$, so it is important to remember to convert this to SI units or to measure detector area in square centimeters. Since noise is an electrical quantity particular to the detector under the conditions for which D^* is defined and is independent of wavelength, NEP(λ) is proportional to $1/R(\lambda)$; so if the value of detectivity D^*_p at the wavelength of peak responsivity R_p is known, for other wavelengths $D^*(\lambda) = D^*_p(R(\lambda)/R_p)$. Sometimes, "blackbody $D^*(T)$" figures are quoted rather than D^*_p. If the blackbody temperature is T, the value of D^*_p is given by:

$$D_p^* = \frac{D^*(T) R_p \int_0^\infty W(\lambda, T)\,d\lambda}{\int_0^\infty W(\lambda, T) R(\lambda)\,d\lambda} \tag{16.6}$$

The rms noise voltage V_n is simply the NEP multiplied by the responsivity. Since the ratio $D^*(\lambda)/R(\lambda)$ is independent of wavelength, one obtains:

$$V_n = \sqrt{A_d B}\,\frac{R_p}{D_p^*} \tag{16.7}$$

The detector also affects the modulation transfer function (MTF) of the imager, defined as the ratio of the modulation depth of the signal due to a target with sinusoidally varying brightness of spatial frequency f cycles per milliradian to the modulation depth from a similar target at very low frequency. If there are no limitations due to time constant, the MTF of a detector is due to its instantaneous field of view (IFOV), which is given by IFOV = x/F where F is the focal length of the optics. For example, a lens of 500-mm focal length used with a 50-μm square detector would give an IFOV of 0.1 mrad. Then,

$$\mathrm{MTF_d} = \frac{\sin\left(\pi \times f \times \mathrm{IFOV}\right)}{\pi \times f \times \mathrm{IFOV}} \tag{16.8}$$

For staring arrays, frequencies above 1/2 cycle per element pitch (the Nyquist frequency) give aliasing, and it is undesirable to rely on performance above this frequency.

Detector Cooling

The method with the lowest cost is to use liquid nitrogen poured directly into the detector Dewar, and many detector manufacturers supply detectors in Dewars with sufficient capacity for many hours of use per filling. Outside the laboratory, this technique is seldom practical, and a cooling engine (usually based on the Stirling thermodynamic cycle) is commonly used. The cost of such an engine has now fallen to a level where it no longer dominates the cost of the instrument, and power consumption is only a few watts to give cooling to 80 K. Dimensions vary widely, but a typical low-power cooling engine is about 40 mm × 40 mm × 60 mm excluding the length of the cold finger that lies inside the detector Dewar. A disadvantage is the relatively slow cool-down time (several minutes typically) and, where this is critical (as in military or security applications), Joule-Thomson cooling can be used. This method operates by expansion of air or nitrogen compressed to about 15 MPa through a nozzle, and can give cool-down times of a few seconds. Thermoelectric cooling (using the Peltier effect) can be used for temperatures down to about 200 K. Some detectors (e.g., lead selenide) have been designed to operate at this temperature in the 3 μm to 5 μm band, but thermoelectric cooling is inadequate for most types of photon detectors. A survey of cooling methods and devices is given in [2], Vol. 3, p. 345–431.

16.6 Electronics

The electronics architecture depends on the type of detector and the application, but typically electronics are required to provide bias and clocking signals to the detector, to amplify the low-level signals from the detector, to equalize the responses of the outputs from different detector elements, to provide scan conversion to a form suitable for display, and to provide image processing suitable to the application.

In scanning imagers with a small number of detector elements, the output of each detector element can be amplified continuously, the amplifier outputs then being multiplexed to give the required display. The bandwidth (important for estimating sensitivity) must be at least sufficient to accommodate the data rate, but too wide a bandwidth gives excess high-frequency noise. An approach frequently used for the measurement of sensitivity (e.g., [1], p. 167) is to make the electronics response equivalent to that of a single-pole filter with the 3-dB point placed at a frequency corresponding to 1/(2 × dwell time), in which case the bandwidth is:

$$B = \frac{\pi \times FOV_h \times FOV_v \times f_f}{4E \times N \times IFOV^2} \tag{16.9}$$

where FOV_h and FOV_v = Horizontal and vertical fields of view, respectively
f_f = Field rate
E = Ratio of the active scan period to the total period
N = Number of parallel detector channels

Often high-frequency boost filters are used to compensate for optics and detector MTFs, in which case the noise bandwidth can be much increased. A commonly used criterion for the electronic filtering is to make the perceived noise independent of frequency up to the cut-off frequency of the detector.

To maintain good spatial resolution, it is desirable that the detector output be sampled at least twice during the dwell time so that the Nyquist frequency is not a major limitation [7]. Frequencies above the Nyquist frequency are changed by the sampling process into lower frequencies, a process known as aliasing. Thus, the fidelity of the image is affected so that noise that might be expected to be of sufficiently high frequency to be filtered out can appear within the passband. To avoid aliasing, a steep-cut filter is used to eliminate frequencies above the Nyquist frequency before sampling. In staring systems, the Nyquist sampling frequency is 1/2 cycle per element pitch, so a 256 × 256 element array would be limited to a resolution of 128 cycles per line. This can be overcome by using "microscan" (or μscan), in which the image is collected over a number of fields with an image shift performed optically between each field, the commonest patterns being diagonal (low implementation cost) and 2 × 2 with four fields per frame.

For large detector arrays, charge is usually accumulated on a capacitor associated with each pixel for an integration time τ (which can often be controlled independently of the frame rate or dwell time to prevent saturation), the capacitor being discharged when the pixel is read out. The effective bandwidth of such an integrator is:

$$B = \frac{1}{2\tau} \tag{16.10}$$

With state-of-the-art amplifiers, it is generally possible to make amplifier and read-out noise less than the detector noise, the exceptions being systems of small aperture that are "photon starved" and uncooled systems.

In multichannel systems, a crucial role of the electronics is to provide channel equalization. The importance of this is due to the very low contrast of the target against the background. If the full sensitivity of an imager with a typical NETD of 0.1 K is to be realized, the difference in response between adjacent detector channels must be less than 0.4% in the MWIR or 0.2% in the LWIR. In a real-time imager where the eye integrates over several frames, the requirement is 2 to 3 times more stringent, since nonuniformity, unlike signal-to-noise ratio, is not improved by eye integration.

16.7 Optics and Scanning

The materials commonly used for visual optics are opaque in the thermal wavebands. In the LWIR, germanium is by far the most widely used material. It has a refractive index of 4, and chromatic dispersion is sufficiently low that it is frequently unnecessary to use a second material for color correction. These properties allow high performance to be achieved with few optical components, largely offsetting the relatively high cost of the material. In the MWIR, germanium has fairly high dispersion, but silicon/germanium doublets give highly achromatic performance. Zinc selenide and zinc sulfide are commonly used for color correction, some forms of the latter being transparent also in the visible band. The high refractive indices make anti-reflection coating essential to reduce surface losses — the transmission of a thin piece

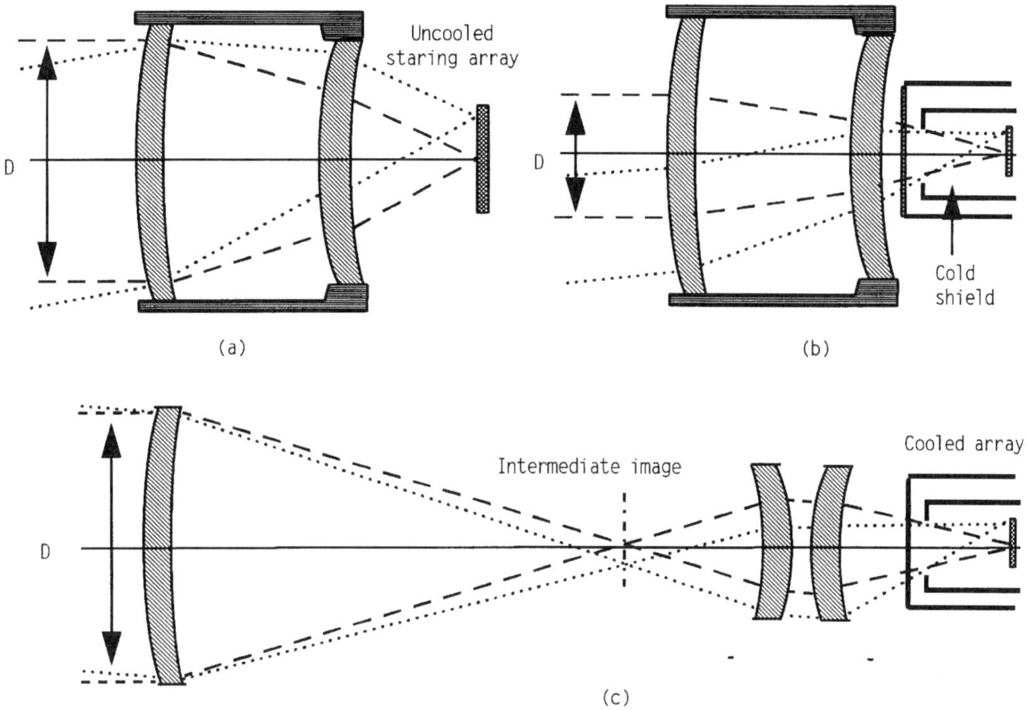

FIGURE 16.7 Optics for staring arrays. The same lens as is used for uncooled operation (a) can be used as in (b) for a cooled device, but re-imaging optics (c) are needed for a cooled imager to use the full aperture. In practice, the optics in (c) would usually have more components than shown.

of uncoated germanium is only 40%, rising to better than 99% when coated. Front-surface mirrors coated with aluminum or gold perform well in the thermal bands, and optics based on parabolic mirrors can be used if the detector array is small. For most applications, it is necessary to seal the imager to prevent ingress of dust or moisture, in which case a window is required, making reflecting optics less attractive than it first appears, since the cost of a mirror and window can be a little less than that of a lens. Plastic materials in general have poor transmission, although a thin polythene or "cling-wrap" window might be acceptable for laboratory use.

For uncooled staring arrays, the sole function of the optics is to focus an image of the scene on the detector. Good performance in the 8 μm to 12 μm band can be achieved with a two-element Petzval lens [8] with aspheric surfaces (Figure 16.7(a)). With a cooled array, it is desirable that the cold shield inside the detector should form the aperture stop of the system, since any radiation from the interior of the instrument will add to the system noise and might give shading effects. If the field of view is reasonably small, it is sometimes possible to use the same type of lens in the manner shown in Figure 16.7(b), but it can be seen that the beam diameter that can be accepted is now much smaller than the lens diameter; thus, for long-range applications requiring large beam diameters, the lens becomes very costly. The solution is to use re-imaging optics as shown in simplified form in Figure 16.7(c). The relay stage not only re-images the scene on the detector, but images the cold shield on the objective lens so that the latter need be no larger than the input beam. The intermediate image can be useful to allow insertion of temperature references or microscanning.

Scanning, when required, is most commonly performed by moving reflective surfaces since electro-optic and acousto-optic techniques either provide insufficient deflection angle or are highly wavelength dependent, causing smearing over the thermal wavebands. Rotating refractive polygons are now less used than previously. For fast line scanners, rotating reflective polygons (sometimes with curved facets) are

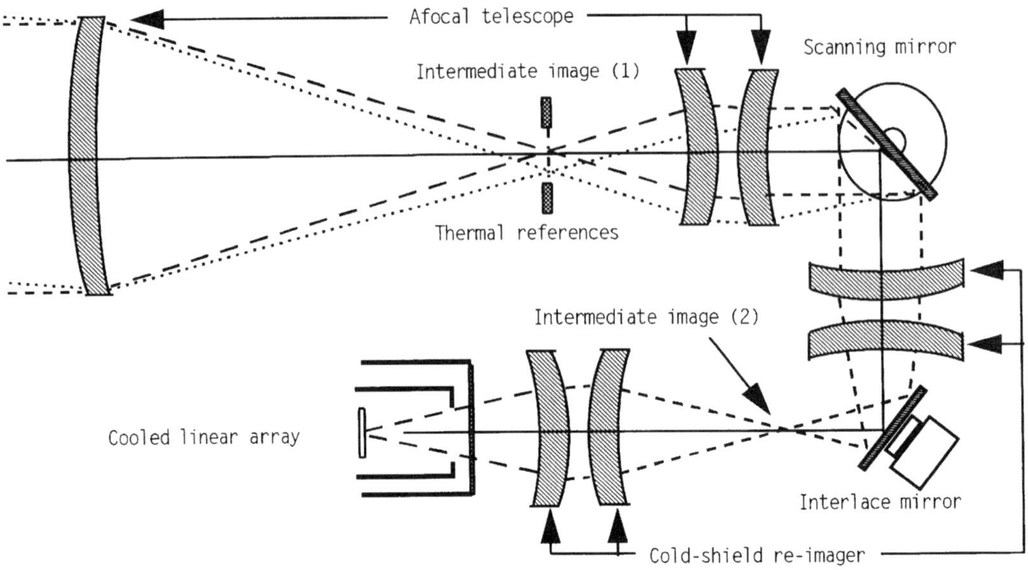

FIGURE 16.8 Optics for a linear array imager. The cold shield is imaged on the scanning mirror, and the scanning mirror is placed at the exit pupil of the telescope.

used, in some cases using gas-bearing motors operating in helium to reduce windage. For frame scanning, a plane mirror driven by a powerful galvanometer is customary. This can give very linear sweeps at 60 Hz with scan efficiencies of about 80% when operated in a closed loop. For microscan or interlace, a small image movement can be achieved by tilting a mirror using piezoelectric actuators or by tilting a refractive plate using a galvanometer. For scanning systems, it is generally necessary to re-image the detector cold shield at each scan mechanism, since otherwise the optics must be enlarged to accommodate pupil movements. Figure 16.8 shows how this is done in a typical imager using a linear array. The movement of the interlace mirror is sufficiently small that pupil re-imaging at this mirror is not required.

Figure 16.9 shows a typical arrangement for a 2-D scanner, a high-speed polygon rotor being used to generate the line scan.

The main optical parameters to be specified are transmission T_o, focal length F, and *f*/number $F_{\#}$ or numerical aperture (NA). Since IFOV = x/F, the focal length determines the spatial resolution for a given detector. The NA is defined as the sine of the semi-cone angle of the output beam from the optics. It can be shown that if a diffuse (Lambertian) source emits W W m^{-2} into a hemisphere, the radiance due to the source in the focal plane is $WT_o(\text{NA})^2$. $F_{\#}$ is defined as the ratio of focal length to diameter, and for a distant object NA = $1/(2\,F_{\#})$, so the ratio of the irradiance in the focal plane to the source radiance is:

$$\frac{\text{Irradiance in focal plane}}{\text{Radiance from extended object}} = \frac{T_o(\lambda)T_a(\lambda)}{4F_{\#}^2} \tag{16.11}$$

Transmission of thermal imaging optics is typically 60% to 90%, depending on complexity. The transmission of any optics between the temperature reference and the target must be known for quantitative measurement.

The imaging performance of a lens can be indicated by the size of the spot generated in the focal plane by a distant point object, and accurate results can be obtained if the intensity as a function of the angle α from the center of the image (point spread function, PSF) is known. A more usual approach is to use

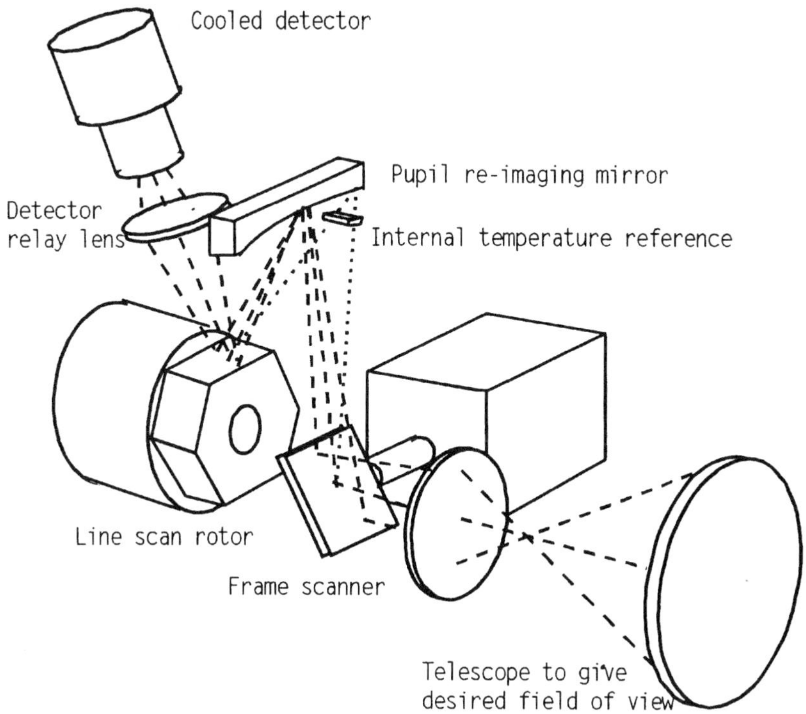

FIGURE 16.9 Simplified layout of imager using 2-D scanning (based on Pilkington Optronics HDTI). The line scan uses a high-speed polygon, and the concave strip mirror images the line scan pupil on the frame scanner.

the modulation transfer function (MTF), defined as the ratio of the contrast of the image of a target with a sinusoidal spatial variation in intensity to the contrast of the target itself. The MTF depends on the spatial frequency *f* of the target (expressed in cycles per radian) and on the wavelength of the radiation. For perfect optics of diameter *D*, diffraction gives the following values for PSF and MTF, which are also plotted in Figure 16.10:

$$\mathrm{PSF}(a) = 4\left[\left(\frac{\lambda}{\pi D a}\right) J_1\left(\frac{\pi D a}{\lambda}\right)\right]^2 \quad (16.12)$$

where J_1 is the Bessel function.

$$\mathrm{MTF}_0(f) = \frac{2}{\pi}\left\{a\cos\left(\frac{\lambda f}{D}\right) - \frac{\lambda f}{D}\sqrt{1-\left(\frac{\lambda f}{D}\right)^2}\right\} \quad (16.13)$$

In practice, optics are frequently far from the diffraction limit, so manufacturers' figures must be used. If a thermal imager is to be used for measurement, correction for MTF will be required unless the IFOV and the PSF are both smaller than the region over which the temperature is to be measured. The signal level for a subresolution point source can be obtained by integrating the PSF over the detector area. For diffraction-limited optics, a rule of thumb for the LWIR band is that the diffraction spot diameter in mrad is the reciprocal of the lens diameter in inches (or 25/*D* when *D* is in millimeters).

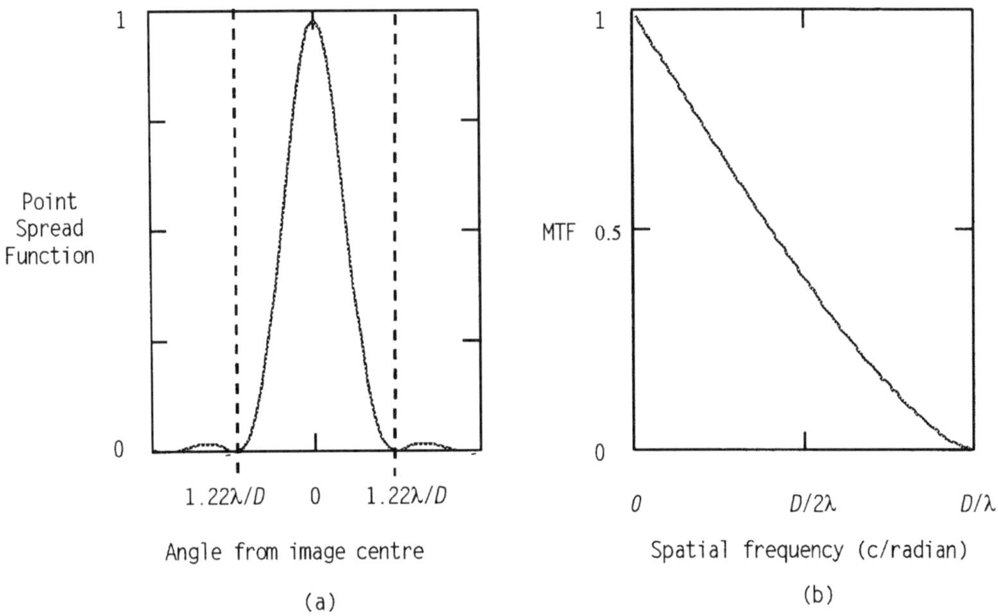

FIGURE 16.10 Point Spread Function (a) and MTF (b) for ideal optics of diameter D at wavelength λ. The first dark ring of the PSF has angular diameter $2.44\lambda/D$, and encircles 86% of the total energy. The MTF is zero for all frequencies above D/λ.

16.8 Temperature References

If a thermal imager is to be used for measurement, it is generally necessary to compare the signal from the target with that from one or more bodies at known temperature. The most precise results are obtained if two bodies at different known temperatures and having the same emissivity as the target (e.g., painted with the same paint type) are placed adjacent to the target and the target temperature is deduced by interpolation. There is then no dependence on emissivity, atmospheric transmission, and optical transmission. Outside the laboratory this is seldom practical, however, and it is necessary to use thermal references within the imager itself. The fewer optical components there are between the reference and the scene, the better will be the accuracy. If the interior of a cavity of uniform temperature is viewed through a small hole, the emission from the hole follows the Planck equation accurately, irrespective of the emissivity of the interior surface of the cavity. Although very accurate, such a blackbody cavity is usually inconveniently large, and it is more usual to use a blackened surface with deep grooves or pits to give high emissivity.

Unless the detector has only one element, thermal references are also desirable if not essential to allow the outputs of the different detector elements to be equalized.

16.9 Imager Performance

SNR and NETD

If the target is larger than the IFOV and the PSF, the signal level due to a small temperature difference is found by multiplying the source differential output, the atmospheric and optical transmissions, the geometric attenuation due to the *f*/number, the detector area, and the responsivity as derived or defined in the previous sections, and integrating over the imager passband to give:

$$V = \frac{A_d \Delta T}{4F_\#^2} \int R(\lambda) T_t(\lambda) T_a(\lambda) \varepsilon(\lambda) \frac{dW(\lambda,T)}{dT} d\lambda \tag{16.14}$$

The noise voltage V_n is given by Equation 16.7, so the signal-to-noise ratio is obtained. If the number of serial detector elements is N_s, the signal level is increased by this factor; but since the noise is uncorrelated between the elements, it increases only as $\sqrt{N_S}$, so SNR improves as $\sqrt{N_S}$. The SNR also improves as the square root of the number of parallel channels, since the dwell time is increased, giving a reduction in bandwidth B.

The sensitivity of a thermal imager for large targets is normally defined in terms of its noise equivalent temperature difference (NETD) — the temperature difference between a large blackbody at zero range and its background — which gives a signal equal to the rms noise. This is given by V_n/V when $\Delta T = 1$, $T_a = 1$ and $\varepsilon(\lambda) = 1$, and is found to be:

$$\text{NETD} = \frac{4F_\#^2}{\int_0^\infty D^*(\lambda) T_t(\lambda) \frac{dW(\lambda,T)}{dT} d\lambda} \sqrt{\frac{B}{A_d N_s}} \tag{16.15}$$

A good indication of signal-to-noise ratio is given if it is assumed that T_a and ε are constant within the imager passband, in which case SNR = $\varepsilon\ T_a\ \Delta T/\text{NETD}$.

Minimum Resolvable Temperature Difference

The performance of a thermal imager is frequently defined by its minimum resolvable temperature difference (MRTD). MRTD(f) is the temperature difference between a four-bar square test pattern of frequency f c mrad^{-1} and its background required for an observer to count the imaged bars. The test is subjective, but has the advantage of characterizing the complete system and display, including any effects of nonuniformity. MRTD is proportional to NETD, and inversely proportional to the MTF and the square root of the number of frames presented within the integration time of the eye τ_e. The constant of proportionality depends on the degree of overlap between the scan lines. Discussion of the full MRTD model is beyond the scope of this chapter, but a simple expression (based on [1] p.167) — which gives an indication of performance for an imager with square detector elements without overlap and at least two samples per IFOV, and in which the electronics bandwidth is the same as that used for NETD calculation — is:

$$\text{MRTD}(f) = \frac{3 \cdot \text{NETD} \cdot f \cdot \text{IFOV}}{\text{MTF}_o \text{MTF}_d \sqrt{\tau_e f_f}} \tag{16.16}$$

Calculation of MRTD is best performed using standard models, the most widely adopted being FLIR 92 [9] .

16.10 Available Imagers

Recent advances in detector technology are only now beginning to be incorporated in commercially available systems. The result is that product ranges are currently changing rapidly, and prices are very unstable. The prices for uncooled systems in particular are likely to drop significantly in the near future as the market size increases. An indication of cost at present is:

Military high-performance imagers $100,000–300,000
Medium-performance imagers for measurement $30,000–100,000
Uncooled imagers $10,000–30,000 (but falling rapidly)

TABLE 16.1 Typical Commercially Available Thermal Imagers

Manufacturer	Model	Data	Description
AGEMA	880 LWB	LWIR, CMT NETD = 0.07	Thermal measurement system 175 pixels (50% MTF)
Amber	Radiance1	MWIR, InSb NETD = 0.025	Compact imager for measurement 256 × 256 pixels, InSb
Amber	Sentinel	LWIR NETD = 0.07K	Uncooled compact imager 320 × 240 pixels
Cincinnati	IRRIS-160ST	MWIR, InSb NETD = 0.025	Compact imager 160 pixels/line
FLIR Systems	2000F	LWIR, CMT NETD = 0.1K	Surveillance imager >350 pixels/line
GEC Sensors	Sentry	LWIR	Uncooled low-cost imager 100 × 100 pixels
Hamamatsu	THEMOS 50	MWIR NETD = 0.2 K	Microscope, 4-μm resolution 256 × 256 pixels
Inframetrics	ThermaCAM	MWIR, PtSi NETD<0.1K	Thermal measurement system 256 × 256 pixels
Mitsubishi	IR-M600	MWIR, PtSi NETD = 0.08K	High-definition imager 512 × 512 pixels
Nikon	LAIRD-3	MWIR, PtSi NETD = 0.1K	High-definition imager 768 × 576 pixels
Pilkington Optronics	LITE	LWIR NETD = 0.2K	Hand held surveillance imager 350 × 175 pixels
Quest	TAM200	MWIR NETD = 0.05K	Microscope with probe facility Bench system with 12.5-μm resolution

Prices reflect not only performance, but ruggedness, environmental survivability, and image processing software. Military imagers in many countries are based on "common modules" that are sometimes multisourced, and that must be configured for specific applications. Table 16.1 lists some typical commercial imagers, which were selected to emphasize the wide variety of imager types, and the list must not be taken as a comprehensive survey. Some compact imagers weigh less than 2 kg, while some of the bench systems weigh over 100 kg. Accuracy of temperature measurement is not generally specified, but ±2 K or ±2% is a good figure for a calibrated imager. The information is based on published brochures, and there is every likelihood that improved models will be available by the time this book is published. It is strongly recommended that prospective purchasers should contact as many manufacturers as possible to obtain specifications and prices.

16.11 Performance Trade-offs

The expression for NETD presented earlier is appropriate for evaluation of existing systems where D^* is known. To predict performance of future systems based on photon detectors, and to carry out design trade-offs, it is important to appreciate that D^* is very dependent on conditions of use and on waveband. One reason is that the photons from an object of given temperature, although having a well-defined average flux, are emitted at random times. Statistical theory shows that if on average N photons are collected within a given time interval, the standard deviation is $\sqrt{N}$ if N is large enough to make the distribution Gaussian (as is almost always the case in the infrared due to the high background flux). This "photon noise" frequently predominates, in which case the detector is referred to as BLIP (background-limited photodetector). The D^* is then determined more by the conditions of use than by the detector itself. If the efficiency of the cold shield η_c is defined as the ratio of the effective *f*/number for the background flux to that of the signal, the number of background electrons generated within an integration time τ is given by:

$$N_e = \frac{\tau A_d}{4F_\#^2 \eta_c^2} \int_0^\infty N(\lambda, T)\eta(\lambda)\mathrm{d}\lambda \tag{16.17}$$

For the reasons discussed above, the rms electron noise for a BLIP device within time τ is simply the square root of the above figure. If the detector also has read-out noise of N_n electrons rms, this could be regarded as the noise that would be caused by a background flux that caused the generation of N_n^2 photoelectrons, so the noise can be written as:

$$\text{Noise} = \sqrt{N_e + N_n^2} \text{ electrons rms} \tag{16.18}$$

The signal due to a temperature difference of 1 K between target and background expressed in electrons within the integration time is:

$$\text{Signal} = \frac{\tau A_d}{4F_\#^2} \int_0^\infty T_t(\lambda) T_a(\lambda) \eta(\lambda) \frac{\mathrm{d}N(\lambda, T)}{\mathrm{d}T} \mathrm{d}\lambda \tag{16.19}$$

Thus, the signal-to-noise ratio for $\Delta T = 1$ K is found, NETD being simply 1/SNR when $T_a = 1$.

For staring systems, and some scanning systems, the integration time is limited by saturation. If the maximum number of electrons that can be stored is N_m, we find that the maximum integration time is:

$$\tau_m = \frac{4F_\#^2 \eta_c^2 N_m}{A_d \int_0^\infty N(\lambda)\eta(\lambda)\mathrm{d}\lambda} \tag{16.20}$$

When the integration time is controlled to restrict the number of electrons generated within the sampling interval to τ_m, the noise is given by letting $N_e = N_m$ in Equation 16.18, so is independent of *f*/number. If one also substitutes τ_m for τ in Equation 16.19, and divides Equation 16.19 by Equation 16.18, to obtain SNR, one obtains:

$$\text{SNR} = \frac{\Delta T \eta_c^2 N_m \int T_o(\lambda) T_a(\lambda) \eta(\lambda) \frac{\mathrm{d}N(\lambda, T)}{\mathrm{d}T} \mathrm{d}\lambda}{\sqrt{N_m + N_n^2} \int N(\lambda)\eta(\lambda)\mathrm{d}\lambda} \tag{16.21}$$

The ratio of integrals is effectively the contrast of the scene behind the optics, which is further reduced by the inefficiency of the cold shield; thus, performance depends mainly on image contrast at the detector and the number of electrons stored. (Quantum efficiency essentially cancels out; and for a saturated device, $\sqrt{N_m}$ is usually greater than N_n.)

The above formulae are sufficiently general to allow trade-offs for photon detectors to be carried out against aperture, waveband, range, and frame rate, the main uncertainty being the read-out noise. For staring array detectors with processing on the focal plane, a typical figure is $N_n = 1000$, but much better (and worse) values are possible. The SNR and NETD values obtained in this way are for a single pixel. If the samples overlap spatially, the SNR is improved by the square root of the overlap factor. In practice, nonuniformity of the detector array will introduce spatial noise that will affect MRTD; and even after

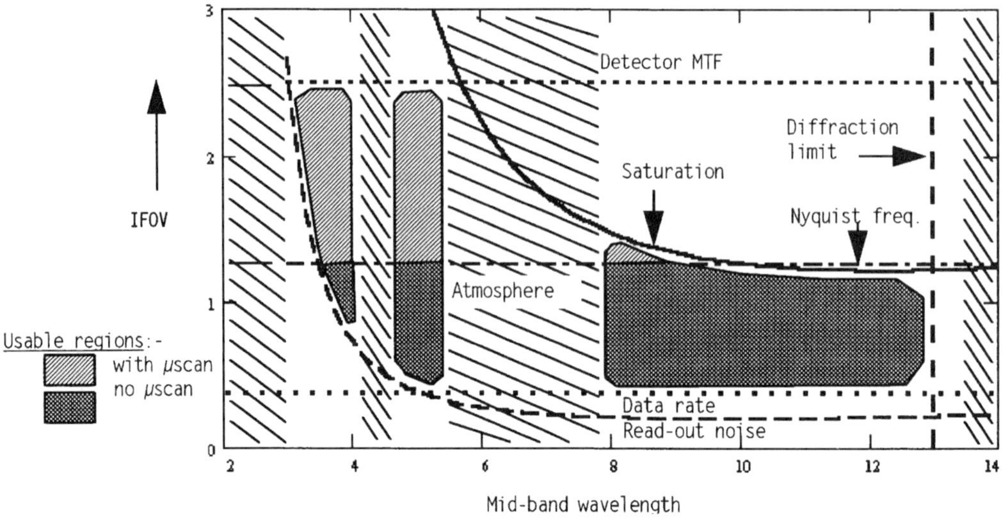

FIGURE 16.11 Physical limitations to thermal imaging when a specified spatial frequency must be resolved. The shaded regions indicate the combinations of wavelength and IFOV for which the task can be achieved. All the lines on the graph depend on system parameters such as aperture, quantum efficiency, read-out noise, and electron storage, but can be calculated from the equations given in the text.

electronic correction, this is frequently the main limitation on performance. Nonuniformity effects are discussed in [10] and [11].

The resulting trade-offs are discussed at some length in [12]. It is observed that if the integration time is controlled to prevent saturation, NETD is independent of *f*/number and quantum efficiency, but depends strongly on image contrast. This tends to favor the MWIR band, and makes high cold-shield efficiency crucial. The problem can be overcome in the LWIR by operating at a high frame rate, which has no effect on NETD but improves MRTD due to eye integration. For long-range applications, it is frequently necessary to use long focal lengths, giving large *f*/numbers if the optics diameter is to be kept within reasonable bounds. Under these conditions, saturation is less likely, and the high photon flux in the LWIR is necessary to overwhelm the read-out noise. If high spatial resolution rather than sensitivity is required, the MWIR band has a much better diffraction limit. Figure 16.11 indicates the ways in which the various limitations combine to define the combination of IFOV and waveband that gives optimum performance when it is necessary to resolve a given spatial frequency; the positions of the lines on this chart are of course peculiar to the system under investigation.

16.12 Future Trends in Thermal Imaging

At the high-performance end of the market, the developments are most likely to be driven initially by military requirements. Many countries already have thermal imaging common module programs, and many developments will be aimed at productionizing these modules to reduce cost and enhance performance. Multispectral instruments to aid in camouflage penetration and to broaden the conditions of operability through the atmosphere can be taken out of the laboratory into service. Currently, these are mainly scanning devices using adjacent long linear arrays of LWIR and MWIR detectors. For spectral agility, detector arrays based on multiquantum-well (MQW) technology can be used, since they can be tuned to some extent by varying the electrical bias. Multispectral refracting optics are complex and have poor transmission, so reflecting optics are generally used. Multispectral systems will allow more accurate compensation for emissivity, but their high cost might limit their use except for high-value applications such as earth resources surveys in aircraft or satellites. Techniques employed include frame-sequential filtering and imaging Fourier transform spectroscopy.

TABLE 16.2 Companies That Make Commercial Thermal Imagers

AGEMA Infrared Systems Inc. 550 County Avenue Secaucus, NJ 07094 Tel: (201) 867-5390	Mitsubishi Electronics America Inc. 5665 Plaza Drive, P.O. Box 6007 Cypress, CA 90630-0007 Tel: (800) 843-2515
Amber Engineering Inc. 57566 Thornwood Drive Goleta, CA 93117-3802 Tel: (800) 232-6237, Fax: (805) 964-2185	Nikos Corporation (Nikon) 1502 West Campo Bello Drive Phoenix, AZ 85023 Tel: (602) 863-6182
Cincinnati Electronics Corporation Detector and Microcircuit Devices Laboratories 7500 Innovation Way Mason, OH 45040-9699 Tel: (513) 573-6275, Fax: (513) 573-6290	Pilkington Optronics Inc 7550 Chapman Ave. Garden Grove, CA 92841 Tel: (714) 373-6061, Fax: (714) 373-6074
Hamamatsu Photonic Systems 360 Foothill Rd. Bridgewater, NJ 08807-0910 Tel: (908) 231-1116, Fax: (908) 231-0852	Quest Integrated Inc. 21414-68th Avenue South Kent, WA 98032 Tel: (206) 872-9500, Fax: (206) 872-8967
Inframetrics Imaging Radiometer Group 12 Oak Park Drive Bedford, MA 01730 Tel: (617) 275-8990, TWX: (710) 326-0659	

For commercial thermal imaging, the main thrust is likely to be cost reduction via the use of uncooled imagers, and in packaging to provide increased ease of use and functionality using in-built software. Optics cost can be reduced or performance improved by the use of hybrid aspheric components, which are now practical due to improvements in diamond turning. A diffractive surface is generated on one surface of a refracting lens. Because the power of a diffractive component is proportional to wavelength, a very low diffractive power can compensate for the chromatic aberration of the lens. This allows the use of materials such as zinc sulfide, which would otherwise be unacceptable due to chromatic aberration, but which have other desirable properties such as low cost or a low thermal coefficient of refractive index.

References

1. J.M. Lloyd, *Thermal Imaging Systems*, New York, Plenum Press, 1975, 20-21.
2. J.S. Acetta and D.L. Shumaker (eds.), *The Infrared & Electro-Optical Systems Handbook*, Bellingham, WA: SPIE Optical Engineering Press, 1993, Vol. 1, 52, 251-254.
3. F.X. Kneizys, E.P. Shettle, L.W. Abreu, G.P. Anderson, J.H. Chetwynd, W.O. Gallery, J.E.A. Selby, and S.A. Clough, *Users' Guide to LOWTRAN 7*, Report no. AFGL-TR-88-0177 Hanscom, Air Force Geophysics Laboratory, 1988.
4. M. Bass (ed.), *Handbook of Optics*, New York, McGraw-Hill, 1995, Vol. 1, chap. 23.
5. C.M. Hanson, Uncooled ferroelectric thermal imaging, in *Proc. SPIE 2020, Infrared Technology XIX*, 1993.
6. R.A. Wood, Uncooled thermal imaging with monolithic silicon focal plane arrays, *Proc. SPIE 2020, Infrared Technology XIX*, 1993, 329.
7. G.C. Holst, *Testing and Evaluation of Infrared Imaging Systems*, Winter Park, FL, JCD Publishing, 1993, 36-42.
8. M.J. Riedl, *Optical Design Fundamentals for Infrared Systems*, Bellingham, WA, SPIE Optical Engineering Press, 1995, 55.

9. L. Scott and J. D'Agostino, NVEOD FLIR92 thermal imaging systems performance model, *Proc. SPIE, Vol. 1689, Infrared Imaging,* 1992, 194-203.
10. A.F. Milton, F.R. Barone, and M.R. Kruer, Influence of nonuniformity on infrared focal plane performance, *Opt. Eng.*, 24, 855-862, 1985.
11. H.M. Runciman, Impact of FLIR parameters on waveband selection, *Infrared Phys. Technol.*, 37, 581-593, 1996.

Further Information

G. Gaussorgues, *Infrared Thermography,* London, Chapman and Hall, 1994.

F. Grum and R.J. Becherer (eds.), *Optical Radiation Measurements,* New York, Academic Press, 1979, Vol. 1.

G.C. Holst, *Electro-optical Imaging System Performance,* Winter Park, FL, JCD Publishing, 1993, 36-42.

C.L. Wyatt, *Radiometric Calibration: Theory and Methods,* New York, Academic Press, 1978.

17

Calorimetry Measurement

Sander van Herwaarden
Xensor Integration

17.1 Heat and Other Thermal Signals

Calorimetry is the science of measuring heat or thermal signals. This chapter begins by explaining thermal and other signals. For a correct understanding of calorimetry, knowledge of thermodynamics is necessary; however, space permits an introduction only. An overview of the most important types of calorimeters, followed by some of the many applications, will then be presented. For those who want to do measurements without having to make their own devices, an overview of the most important calorimeters, and their vendors and approximate prices will be given. The chapter concludes with some

0-8493-0047-9/00/$0.00+$.50

hints on present and future developments, and on further reading, since there is a large amount of literature on calorimeters.

Signal Domains

In measurement science, physical quantities can be distinguished by six different so-called signal domains [1]. All signals in calorimeters are either thermal signals or other signals that are transduced into thermal signals. The thermal signal (heat, for example) is then transduced into an electrical signal. In addition to the domains of thermal and electric signals, there is also the domains of the chemical, the mechanical, the magnetic, and the radiant signals. Calorimetric measurement of signals occurs from all domains [2]. Calorimeters usually measure the thermal effects of (bio)chemical or mechanical processes, or they measure the thermal effect of temperature changes on matter.

Heat and Temperature

One of the forms in which energy can be present in a system is the random, internal kinetic energy of the particles (molecules or atoms) of a system, which can intuitively be called "thermal energy." This is to be distinguished from the average, external movement of a system of particles as a whole, which can be called the "mechanical energy" of the system. For gases, thermal energy is closely related to the random velocity of the molecules; and in the case of multi-atom molecules, the rotations and vibrations of the atoms within the molecules. The zeroeth law of thermodynamics states that, if two systems are each in equilibrium with a third system, they will be in equilibrium with each other. They are said to have the same temperature. For gases, statistical mechanics shows the direct relation between the thermal energy (or heat) stored in the system and temperature [3]. This law, however, also applies to liquids and solids, although the quantitative relation between thermal energy (agitation of the particles) and temperature is less straightforward and not so easy to calculate as for gases. Temperature and thermal energy distribution can be viewed as the result of statistical processes, such as diffusion. Diffusion ensures that if ever there is a surplus of thermal energy (i.e., of fast molecules or electrons, or a higher density of phonons) in some area, some of it will flow toward areas with a lower thermal energy density until thermal equilibrium has been established. This flow of energy P (in W = J s^{-1}) is called heat flow. Heat flow is therefore the transfer of (thermal) energy from one (part of a) system to another. Note that heat is not conserved, because a change in thermal energy of a system may be achieved by heat exchange with the environment, but also by mechanical interaction, and heat can be used to do mechanical work (motor!) just as well as heating up a system. Heat can be viewed as "thermal energy on the move," but whether it ends up as thermal, mechanical or another form of energy depends on the circumstances.

Work and Enthalpy

The first law of thermodynamics is also of importance for calorimetry. In a simple form, in the absence of other forms of energy exchange, it states that the change in internal energy ΔU of a system is equal to the heat Q supplied by the ambient to the system, minus the work (mechanical energy) $p\Delta V$ done by the system on the ambient:

$$\Delta U = Q - p\Delta V \tag{17.1}$$

This is of importance when measuring the specific heat capacity c (in J kg^{-1} K^{-1}) of matter. This measurement can be performed under two conditions: at constant volume V and at constant pressure p. At constant volume, no work will be done by the system, since $p\Delta V = 0$. So, the specific heat capacity at constant volume c_v is simply the change in internal energy. The specific heat capacity of a system at constant pressure c_p is (usually) higher, since additional energy (heat) is needed to perform work on the ambient: $p\Delta V$. The quantity combining these contributions to the energy at constant pressure is called

TABLE 17.1 Overview of the Classification Criteria for Calorimeters

Relation to surroundings	Heat measurement
Isothermal	Phase-transition compensation of heat
	Thermoelectric compensation of heat
Adiabatic	Measurement of temporal temperature difference
Isoperibol	Measurement of spatial temperature difference

enthalpy H, and the change in enthalpy vs. temperature (in J K^{-1}) at constant pressure is the specific heat capacity C_p of a sample at constant pressure: $C_p = (dH/dT)_p$.

17.2 Calorimeters Differ in How They Relate to Their Surroundings

In essence, a calorimeter performs three functions: it encloses a chamber in which a thermal experiment is carried out; it measures the heat exchange between the sample under test and the calorimeter (and often other quantities are being measured as well, such as temperature and amount of substance); and it thermally separates the experimental chamber from its surroundings. There exist many types of calorimeters. They all have an experimental chamber. Apart from this, four essential criteria can be used to classify calorimeters [4, 5]. The first criterion is, what does the calorimeter do with the heat that is generated (or absorbed) by the experiment? The second criterion is, how does the calorimeter relate to its surroundings? The third criterion is, does it measure a single experimental sample, or is it a twin design with a reference compensating for common mode errors? The fourth criterion is, does the calorimeter function at a fixed temperature, or can it scan a temperature range? Table 17.1 lists the various possibilities for the relation to the surroundings, and for the heat measurement. In principle, almost every combination is possible. In practice, some combinations naturally go together because they compensate for their strengths and weaknesses. For example, calorimeters in which the experimental temperature is scanned often use a twin configuration to eliminate the common mode errors arising from the continually changing temperature.

Isothermal Calorimeters

In the isothermal calorimeter, the experiment is always kept at a fixed temperature. This is attained by instantly removing (or supplying) any heat that the experiment releases (or absorbs). The isothermal calorimeter was the first to be developed. In 1780, Lavoisier and Laplace made the "ice calorimeter" in which the heat generated by the experiment is used to melt ice into water. If enough ice is available, the calorimeter will remain at 0°C, regardless of the progress of the experiment. The experimentally generated heat Q (J) is calculated by weighing the mass m (kg) of the melt water and multiplying by the heat of the ice-water transition q_{fus} (J kg^{-1}):

$$Q = m\, q_{fus} \qquad (17.2)$$

The experiment with the melting ice is enveloped within a thermostat, which consists of a double-walled vessel with melting ice between the walls, which is always at 0°C as well. Thus, the isothermal calorimeter has a perfect thermal isolation between experiment and surroundings in the form of a second guard-vessel that buffers all the heat coming from the surroundings.

Phase-Transition or Thermoelectric Compensation of the Heat

In the isothermal calorimeter, the heat generated by the experiment is immediately absorbed by the calorimeter. This can be accomplished by phase-transition compensation of the heat, e.g., by melting of solids or evaporation of liquids. But nowadays, compensation by electrical means is preferred because it can be measured so easily. Thus, heat Q (J) absorbed by the experiment is replaced using Joule heating (dissipation of heat by a current I (A) through an electric heater with resistance R_h (Ω)), while heat generated by the experiment is absorbed by Peltier coolers:

$$Q = \int I^2 R_h \mathrm{d}t \tag{17.3}$$

17.3 Adiabatic Calorimeters Often Measure Time-Dependent Temperature Differences

In the adiabatic calorimeter, no heat exchange with the surroundings is allowed, and all the heat generated by the experiment is used to increase the temperature of the calorimeter. The amount of heat Q (J) generated follows from the temperature increase ΔT (K), divided by the heat capacity of the calorimeter C_c (K J^{-1}):

$$Q = \Delta T / C_c \tag{17.4}$$

The absence of heat exchange with the surroundings of the calorimeter is obtained by immersing the experimental chamber of the adiabatic calorimeter in an outer vessel; see Figure 17.1. The temperature of the outer vessel is kept at the same (increasing) temperature as the experimental chamber by means of electronic feedback, heating the outer vessel to maintain a practically zero temperature difference. In the adiabatic calorimeter, one must wait a few minutes after the experiment has finished to allow the heat to spread itself uniformly over the chamber and to obtain the final temperature. In Figure 17.1, the experimental chamber is a so-called "calorimetric bomb" in which fuel is fully burned to measure its heat of reaction. The bomb is immersed in a vessel filled with water — the inner vessel. A stirrer assures fast heat exchange between the bomb and the inner vessel, thermometers measure the temperature of the inner vessel and the outer guard vessel. The outer vessel is regulated to the inner-vessel temperature by means of electrical heaters and refrigerator coolers.

Isoperibol and Heat Flux Calorimeters

The term *isoperibol* was devised to indicate a calorimeter having "uniform surroundings." In this calorimeter, the outer shell provides a reference temperature, and customarily, the experiment starts at the same temperature. The experimental chamber is connected to the outer shell by a well-defined thermal conductance. Any heat generated by the experiment will cause a well-defined temperature difference ΔT (K) across the thermal conductance G_{th} (W K^{-1}), and this temperature difference is subsequently measured as a "local temperature difference." Calorimeters utilizing this way of measuring the heat are referred to as "heat flux calorimeters," and often measure the power P (J s^{-1} or W) generated by the experiment, rather than the energy:

$$P = \Delta T\, G_{th} \tag{17.5}$$

Scanning Calorimetry Sweeps the Experiment Temperature

While calorimeters are over 200 years old, a recent innovation is that of the *scanning calorimeter*. In the scanning calorimeter, the experimental chamber is not kept at (approximately) one temperature, but it is swept over a temperature range. The temperature is increased at given rate (e.g., 10 K min^{-1}) by electric heating, or decreased by, for example, cooling with liquid nitrogen. To achieve the temperature sweep, the experiment is placed inside a computer-controlled oven. To compensate for common-mode errors

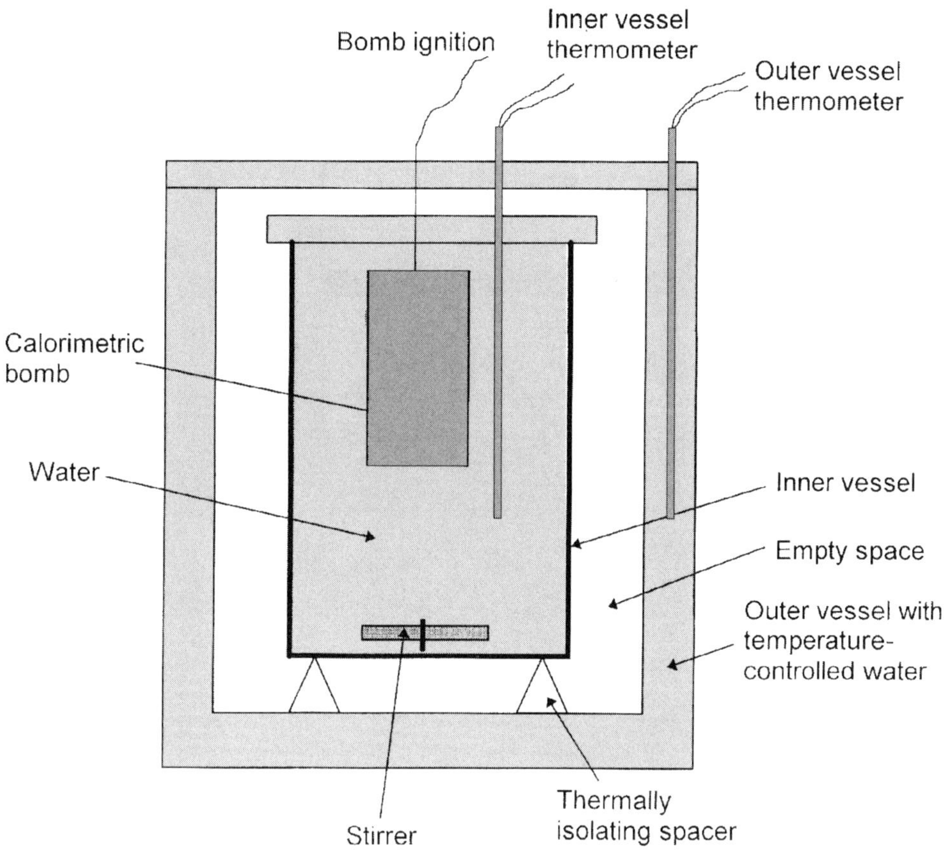

FIGURE 17.1 Adiabatic calorimeter, with inner experimental vessel being heated by a chemical reaction in the calorimetric bomb, and outer guard vessel electronically adjusted to the temperature of the inner vessel to prevent heat loss of the inner vessel. The temperature increase of the inner vessel is a measure of the heat of burning of fuel in the bomb.

(i.e., back-logging or deviation of the sample temperature from the oven temperature, and imperfections in the oven temperature profile with respect to time and location), scanning calorimeters are often built with twin experimental sites for which the difference is measured. One site is for the sample under test; the other site serves as reference, which is either left empty or contains material resembling the sample under test as much as possible, except for the phenomena to be measured. Such calorimeters are called *differential scanning calorimeters*, DSCs (see "Further Information" for a general book on DSC). Three different types of DSCs are common. The first type is the DSC based on heat flux measurement; see Figure 17.2. In this, the reference and the sample are both heated by the oven through a thermal resistance (gaseous or a solid circular disk). In Figure 17.2, the heating block is the oven, while the (constantan) disk is the path from oven to sample and reference, assuring both heat conduction from oven to sample/reference. The disk also forms a well-defined heat resistance from sample to oven, to measure the heat generation and absorption in the sample using thermocouples that measure the resulting temperature differences. Because of the small size of the experimental site and sample (typically less than 1 cm diameter and 0.1 mL volume), the DSC is usually very fast, and the temperature curve is an accurate representation of the momentaneous heat production in the sample. In the second type of DSC — the power compensation DSC — the temperature of the sample and the reference are both measured with platinum resistors. Sample and reference each have an individual heating source, which is electronically

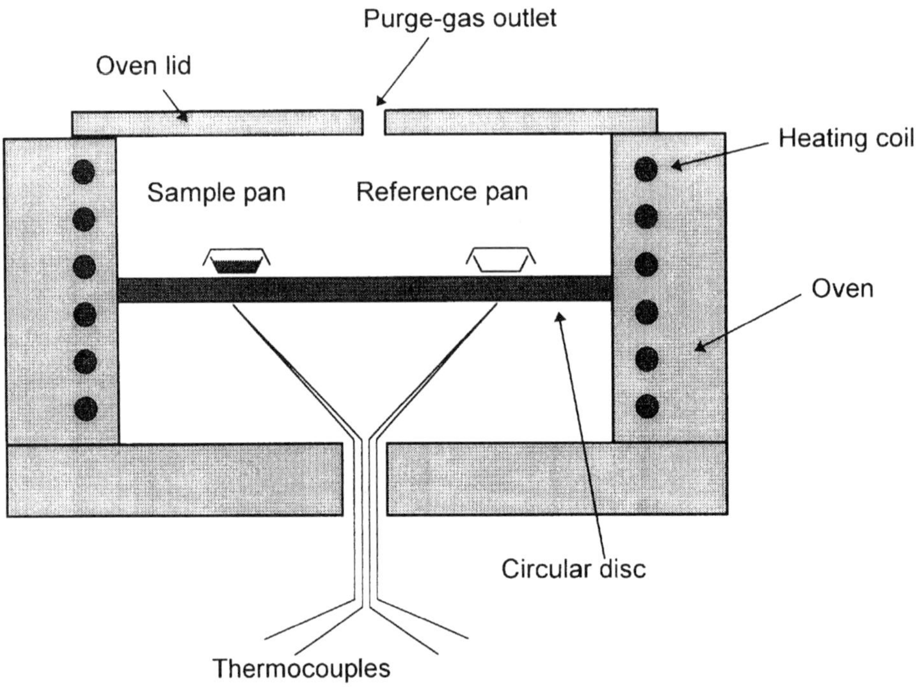

FIGURE 17.2 The heat flux differential scanning calorimeter (DSC) consists of two experimental sites on a heat-conducting disk. Heat generation in the sample pan results in a temperature increase of the sample with respect to the reference, measured by thermocouple (the disk being one part of the couple).

controlled to maintain sample and reference on the required temperature course. The difference between reference and sample heating power immediately gives the experimental heat. In the third (less accurate) type, only the temperature at which phenomena occur is registered. This type can be used for bigger samples (2 to 3 mL against 0.1 mL for the heat-flux DSC), or at very high temperatures up to 2000°C.

Converting the Measured Curve to the Actual Progress of the Experiment

Due to the thermal resistance and the heat capacitance inside the experimental chamber of a calorimeter, there is a time constant associated with distributing the experimental heat over the chamber, and reaching the final temperature. A heat pulse does not create a temperature pulse, but rather a smeared-out curve. In a DSC, this smearing-out is usually not significant because of the small time constant (1 to 3 s). For some other instruments, so-called *curve desmearing* has to be carried out using convolution integrals to extract all the information from the measurement curve. With commercially available instruments, software is often supplied that does the job for you, but the accuracy of this software is not satisfactory in all cases, so one might need to do some further study on desmearing [4].

Calibration of Calorimeters

The inaccuracy of the calorimeter can be reduced by *calibration.* Heat production can easily be simulated electrically. Unfortunately, the thermal leakage of the electric leads reduces the accuracy of this method. More common, therefore, is the use of reference materials with a known heat of transition or reaction for calibration. For bomb calorimeters, the response can be calibrated using the heat of reaction of benzoic acid, which has been carefully determined to be 26,457 J g^{-1} [6]. For a DSC, a range of materials can be used for calibration. This is due to the wide range of applications, and also to the wide range of temperatures used. Generally, one or more metals are used for calibration of both the power and the temperature scale; for example, indium, which has a melting point of 156.6°C, and an enthalpy of melting of 28.6 J g^{-1}. The uncertainty of the value for the heat of fusion (in this case, of melting) is about 0.5%. This is the limit when calibrating in this manner. Basically, DSC inaccuracy is around a few percent. For up-to-date details on calibration and the calibration materials, please consult [7, 8], the references given there, and the most recent data published, since calibration procedures for the DSC are still improving. Specific heat capacity measurements are often performed in three steps. First, the baseline-offset of the DSC is measured with empty reference and empty sample sites (one measures the difference between sample and reference baseline, i.e., the systematic asymmetry of the instrument). Then, a measurement of a reference material with accurately determined specific heat capacity is made; for this, sapphire (crystalline Al_2O_3) is often used. Finally, a measurement of the unknown sample is carried out. By correcting for the baseline-offset, and division of the measurements of the known and the unknown samples, the specific heat capacity of the unknown sample can be determined. Currently, these corrections are all made by the computer controlling the DSC [9].

17.4 Typical Applications of Calorimeters

Specific Heat, Transition Heat and Temperatures

Specific heat capacity can be measured by accurate point-wise measurements, but measurement with a DSC is much more efficient. With the DSC, one also obtains the temperature and heat of phase transitions by enthalpy measurement at the transition temperature. Many materials also exhibit changes in crystallization at some given temperature (e.g., glass transition points). Also, these are transition points with their specific heat and temperature, and these can be measured with DSC as well.

Analysis of Chemical and Physical Reactions

Calorimetry is very well suited for analysis of chemical reactions. In particular, the heat of exothermic and endothermic reactions can be determined. An important example is the calorific value of fuels. Households and factories buying fuels are primarily interested in the calorimetric value of their purchases. The so-called oxygen bomb calorimeter is indispensable for primary calibration in this application. The oxygen bomb is usually an adiabatic calorimeter, in which a sample of the fuel (such as solid coal, liquid oil, or gaseous methane) is brought together with an excessive amount of oxygen (e.g., at a pressure of 30 bar). Then, the mixture is ignited, and the heat of burning will spread over the bomb, until equilibrium is reached. The temperature increase of the bomb, divided by the heat capacity of the device, gives the heat of reaction. In general, corrections have to be made for volume and pressure changes (work!) as liquids or solids are burned and converted to gases and liquids, and also for the additional heat capacity and for the transition heat of the reaction products. Of course, chemical reactions other than burning of fuel can be analyzed in this way as well. In a slightly modified version, the adiabatic calorimeter can also serve to analyze, for example, weak bases and weak acids that do not easily respond to other analysis methods. Similarly, the heat of mixing two solutions, of dissolving a solid in a solvent, of diluting a mixture, and even the heat of wetting can be determined in this so-called *solution calorimeter.*

17.5 Thermal Analysis of Materials and Their Behavior with Temperature

Apart from directly measuring chemical reactions, one can also analyze materials by exposing them to a temperature sweep, either in an inert atmosphere (nitrogen or helium) or in an oxidizing atmosphere.

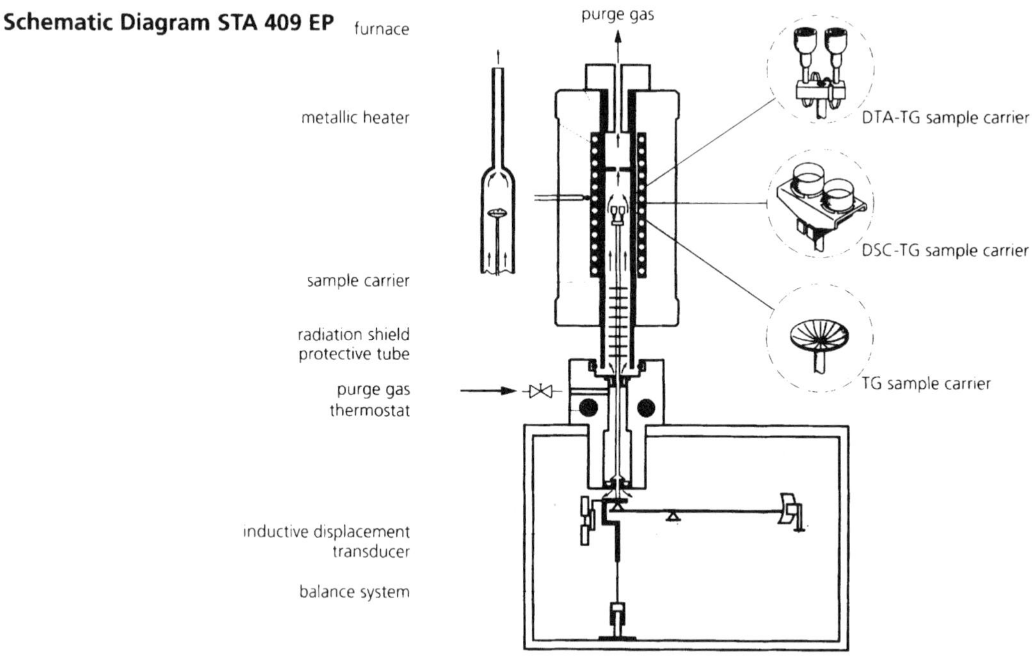

FIGURE 17.3 In a DSC-TG combined thermal analysis instrument, heat is transferred to the sample from the oven by the enveloping purge gas, with radiation shields to diminish heat loss by infrared radiation. The entire sample carrier is placed on a balance to enable thermogravimetric analysis. (Drawing courtesy of Netzsch GmbH.)

Then, all effects — such as glass transition, crystallization, melting, evaporation, decomposition, and even oxidation — can be detected. For this, a DSC is again utilized. In the more expensive analysis systems, DSC instruments often offer the possibility of incorporating other techniques. Mass changes due to oxidation and evaporation are detected by thermogravimetry (TG). This is accomplished by placing the entire experimental chamber on a balance, that continuously measures the sample mass. These data are available as a function of temperature as well, parallel to the heat data. The DSC inner gas atmosphere is usually refreshed continuously using purge gas. This makes it possible to collect the gases coming from the calorimeter furnace and subject them to further analysis, such as mass spectrometry (MS), gas chromatography (GC), Fourier transform infrared analysis (FTIR), and other analysis methods. With sensitive DSC, it is thus possible to detect almost any structural change in matter as a function of temperature. Figure 17.3 shows a drawing of a combined DSC and TG instrument, capable of analyses up to 2000°C. The DSC-TG sample carrier is accurate up to 1500°C, using heat transfer from oven to samples carrier by the surrounding purge gas. Radiation shields are required to reduce heat losses by infrared radiation, which is significant at high temperatures.

Biological Analysis from Cells to Entire Human Beings

Calorimetry is not just the measurement of thermal effects in 1-mL samples. Calorimeters receiving entire human beings are available as well. In practice, there is a wide range of application of calorimeters to biology. This is not surprising, since all forms of live produce heat in activity and often also in rest. The first calorimeters were already used to measure the heat produced by animals. But also, the study of cells (with their heat production of about 1 pW [10]) and the efficiency of enzymatic conversions can be studied using calorimetry. Microcalorimetry is used for very small effects, sometimes using microtechnology to fabricate very small and very sensitive calorimeters [11, 12]; see [10] for an overview and

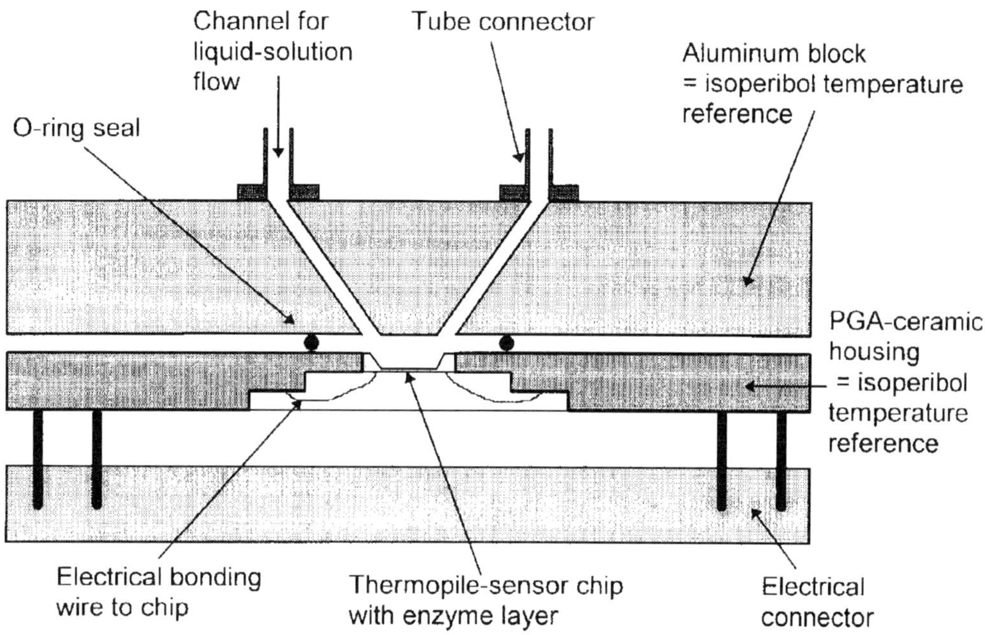

FIGURE 17.4 In microcalorimetry for (bio)chemical analyses, flow-through systems allow continuous, on-line measurement of (bio)chemical quantities in microliter reaction chambers, using 5 × 5-mm silicon microchips coated with enzymes or living cells.

many references. Some microcalorimeters are already being commercialized, using integrated-circuits technology to make very sensitive sensors. Figure 17.4 shows such microcalorimeters encapsulated in a ceramic housing. So far, the use of these sensors is restricted to isoperibol operation around room temperature [11], using aluminum heat sinks to provide the reference temperature. Applications of calorimetry are also found in the food industry, for routine analyses, ecology, plants, etc. See [13, 14] for collections of papers on these subjects.

Summary

The applications of thermal analysis are overwhelming in number. In many cases, when wanting to learn more about materials, thermal analysis can add insight. Some applications were mentioned above. In Table 17.2 some of the often-encountered applications for various disciplines are listed.

17.6 Choosing the Proper Calorimeter for an Application

When faced with a practical analysis problem, one can choose between many instruments. The three main categories are the DSC, the calorimetric bomb or solution calorimeter, and the large reaction or fermentation calorimeter.

What Do You Want To Measure?

If one wants to measure specific heats of reaction, of oxidation, solution, etc., at room temperature, a calorimetric bomb or a solution calorimeter might be the first choice. They are accurate and economical. If one wants to optimize a chemical or biological process, reaction calorimeters or fermentation calorimeters might be the choice. In almost all other cases, a DSC will be the most effective instrument. A

TABLE 17.2 Applications of Calorimeters

Area of interest	Parameter that can be measured with calorimeters
Material characterization and all other areas listed below	Specific heat
	Melting and boiling temperature and behavior
	Fusion and reaction heat and kinetics
	Heat of solution, dilution, mixing, wetting
	Thermal safety
	Glass transition
	Rate and degree of cure
	Crystallization time, temperature, and percentage
	Purity and solid–liquid ratios
	Thermal and oxidative stability
	Identification of multicomponent systems
	Dehydration
Polymers	Effect of the thermomechanical history
	Compatibility tests
	Effects of additives
Pharmaceuticals/Cosmetics	Purity and compatibility of active ingredients
	Polymorphism
	Effects of storage and hydrolysis
	Tablet-compression characteristics
	Influence of emulsifiers
	Concentration of medicines in polymers
	Melting and crystallization behavior of waxes
Foods	Melting, solidification behavior, and specific heat of fats and oils
	Polymophism
	Denaturation of proteins
	Gelatinization of starch
	Freezing-thawing behavior
Biology	Metabolism of cells, organs, animals, and human beings
	Influence of nutrition, toxins, and others on organisms
	Enzymatic efficiency and selectivity
	Concentration of solutions using enzymes or organisms
	Environmental monitoring

DSC is especially useful when one wants to obtain the thermal behavior of materials as a function of temperature. In turn, the temperature-dependent behavior of materials can tell a lot about their structure, their properties, and even their thermomechanical history. Important is the scanning range and rate of the DSC. In some cases, the measurement problem cannot be solved by a standard available calorimeter, and one is forced to either use an experimental calorimeter or develop a special-purpose calorimeter.

Budget, and the Need of Pre- and After-Sales Service

A cost-effective DSC will cost about $25,000, the most expensive models can cost up to $150,000 and include thermogravimetric measurement. Calorimetric bombs are somewhat cheaper, starting at $18,000. The large reaction calorimeters are more costly, at about $120,000 and above. Finally, one can buy experimental instruments or make one's own system. An adiabatic system consisting of a Dewar in a polystyrene housing, using a commercially available stirrer and platinum resistance (measured with any 5½ digit DMM) will put you in business, although at reduced accuracy. However, the budget for this will not need to exceed much more than about $1000 (excluding the DMM). It is less advisable to make one's own DSC (for cost reasons) or calorimetric bomb (for cost and safety reasons).

17.7 Can the Instrument of Choice Measure the Signals Desired?

Here, various parameters must be considered. Accuracy, the degree to which an instrument reading approaches the true value, lies around a few percent for DSCs and parts-of-percent for adiabatic calorimeters. Repeatability of the major calorimeter measurement results — such as heat peak area and peak starting temperature — depends, among other things, on such matters as baseline noise/drift, influence of sample preparation, and positioning in the DSC sample site. Resolution should be considered for two parameters: heat and temperature. First, there is the sensitivity of the instrument for generated heat or power, related to its noise. For DSCs, resolution for power is usually around 0.1 μW to 10 μW; for bomb calorimeters, resolution for heat is around 0.1 J to 10 J. But there is also the resolution of a DSC for separating two heat pulses at two nearby temperatures. This can be designated as the temperature resolution, and depends on time constant and heating rate. Finally, there is the point of linearity and time constant. In case the time constant of the instrument is much larger than that of the process, good linearity is required if one wants to "desmear" the measured curve and obtain the actually produced heat as a function of time. But, a current DSC is a very fast instrument, and desmearing is not really necessary anymore. So, linearity is less important for fast DSCs.

Instrument Control and Data Management

Software is becoming more and more important. Since it will control the measurement, its user-friendliness determines how easy and how error-free one can operate the instrument. It also can take care of analysis of the measurement results, and graphical and numerical presentation of the results. The software can also take care of quality-control aspects of analysis, such as writing to file all the measurement details (not only what was measured, but also how it was measured). Presently, top models of all three types described above will perform these functions. With DSC instruments, software can often be used to control all kinds of thermal analysis instruments apart from the DSC, such as TG, DMA, etc., and merge results obtained with DSCs, TG, DMA, and other analysis techniques. This facilitates interpretation of measurements.

17.8 Commercially Available Calorimeters

Tables 17.3 and 17.4 describe the different instruments and vendors. Table 17.3 gives an overview of instruments and some characteristics, while Table 17.4 gives vendor information. Tables 17.3 and 17.4 are (necessarily) incomplete since only the major vendors have been listed; thus, if looking for a calorimeter, please complete the list with local (and up-to-date) information.

17.9 Advanced Topic: Modulated or Dynamic DSC Operation

Modulating the Temperature Scan Improves Insight into the Measurement

One of the recent developments in DSC is the use of a nonuniform temperature scan. On the standard temperature increase (for example, 1 K min^{-1}), an alternating fast temperature change is superimposed, which can be a sinusoidal signal. Alternatively, a stepped temperature profile can be used. Here, the temperature is increased during, for example, 0.5 min with 2°C, and then stabilized for 0.5 min, resulting in an overall scan rate of 2°C min^{-1}. In fact, the scan is now modulated with a block wave of amplitude 2°C min^{-1}, see Figure 17.5(a). For an instrument with small time constant, such as the DSC-7 of Perkin

TABLE 17.3 Some Commercially Available Calorimeters and Their Specifications

Type	Instrument	Vendor	Cost[a] ($1000)	Range[b] (°C)	Scan rate[c] (K min^{-1})	Resolution[d]
PC DSC[e]	Pyris	Perkin Elmer	55	–170 to +725	500	0.2 μW
CHF DSC[f]	DSC 12E	Mettler	20	–40 to 400	20	10 μW
	DSC 821	Mettler	30	–150 to 700	100	0.7 μW
	DSC 200	Netzsch	45	–170 to 530	40	4 μW
	DSC 6	Perkin Elmer	25	–120 to 450	50	
	Exstar 6000	Seiko	40	–150 to 1500	100	0.2 μW
	DSC 141	Setaram	40	–150 to 600	100	10 μW
	DSC 50	Shimadzu	30	20 to 725	100	10 μW
	DSC 2920	TA Instr	45	–180 to 725	200	0.2 μW
	DSC 2010	TA Instr	30	–180 to 725	200	1 μW
FFHF DSC[g]	DSC 404	Netzsch	60	–120 to 1500	50	8 μW
FFHC DSC + TG	STA 409	Netzsch	70	–160 to 2000	100	8 μW
	Labsys	Setaram	45	20 to 1600	100	10 μW
C DSC[h]	DSC VII	Setaram	55	–45 to 120	1.2	1 μW
	DSC 111	Setaram	80	–120 to 830	30	5 μW
Bomb	1425	Parr	18	Ambient	—	4 J
	1271	Parr	45	Ambient	—	2 J
	C 5000	IKA	33	Ambient	—	6 J
	C 7000	IKA	35	Ambient	—	0.5 J
Solution	1455	Parr	18	0 to 70	—	0.4 J
Calvet	MS 80D	Setaram	95	20 to 200	—	0.1 μW
	HT 1000	Setaram	220	20 to 1000	1	10 μW
Process	RC1	Mettler	120	–50 to 300	30	
	BFK	Berghof	150	20 to 60		40 mW
Micro	LCM-2526	Xensor	—	Ambient	—	0.1 μW

[a] Cost: simplest complete system.
[b] Range: addition of the widest ranges available.
[c] Scan Rate: the highest controlled scan rate (usually for heating).
[d] Resolution: vendor specification or 2x rms noise, for the most accurate system.
[e] PC DSC: Power-Compensated DSC.
[f] CHF DSC: Circular-Disk Heat-Flux DSC.
[g] FFHF DSC: Floating-Foil Heat-Flux DSC.
[h] C DSC: Calvet DSC.

Elmer at about 1 to 2 s, this means that the heat flow for increasing the experiment temperature in the stabilization time will be completely stopped. Any heat flows remaining result from processes in the sample itself; see Figure 17.5(a). Proper interpretation of the measurement results obtained by this method is still under discussion by Reading [15], Schawe [16], and many others.

PET Is Often Used as Example

Figure 17.5(b) gives the analysis of PET (**p**oly**e**thylene **t**erephthalate), a polymer often used as a reference material because of its convenient and exemplary behavior in polymer analysis [17]. The curve is that of shock-cooled PET, i.e., the PET (from an ordinary beverage bottle) is heated to 300°C in a nitrogen atmosphere, and then quench-cooled in 1 min to maintain an amorphous structure. Six regions can be seen. From 50°C to about 75°C, the heat capacity of the PET sample requires power to achieve the temperature increase. At about 75°C, the so-called "glass transition" takes place, where reordering of the PET molecules takes place, increasing the specific heat of PET (an endothermic process), as can be seen from the lifting of the baseline (even at zero temperature increase, heat has to be supplied to the PET). Then follows a region with increased specific heat, caused by the higher freedom of movement of the

TABLE 17.4 Companies That Sell Calorimeters

Berghof GmbH Harretstrasse 1 D-72800 Eningen/Reutlingen, Germany Tel: + 49-7121-8940, Fax: + 49-7121-894100	Perkin-Elmer Corp. 761 Main Ave. Norwalk, Connecticut 06859-0012 Tel: + 1-203-762-1000, Fax: + 1-203-762-6000
IKA Analysentechnik GmbH P.O. Box 1240 D-79420 Heitersheim, Germany Tel: + 49-7633-8310, Fax: + 49-7633-83198	Seiko Instruments 1-8, Nakase, Mihami-Ku, Chiba-shi Chiba 261, Japan Tel: + 81-43-211-1340, Fax: + 81-43-211-8067
Linseis GmbH P.O. Box 1404 D-95088 Selb, Germany Tel: + 49-9287-8800, Fax: + 49-9287-70488	Setaram 7, rue de l'Oratoire, BP 34 F-69641 Caluire Cedez, France Tel: + 33-72 10 2525, Fax: + 33-78 28 6355
Mettler Toledo AG Analytical Sonnenbergstrasse 74 CH-8603 Schwerzenbach, Switzerland Tel: + 41-1-806-7711, Fax: + 41-1-806-7350	Shimadzu Corp. 3. Kanda-Nishikicho 1-chome, Chiyoda-ku Tokyo 101, Japan Tel: + 81-3-3219-5641, Fax: + 81-3-3219-5710
Netzsch Gerätebau GmbH P.O. Box 1460 D-95088 Selb/Bavaria, Germany Tel: + 49-9287-8810, Fax: + 49-9287-88144	TA Instruments Inc New Castle, Delaware 19720 Tel: + 1-302-427-4000, Fax: + 1-302-427-4001
Parr Instrument Company 211 Fifty-Third Street Moline, Illinois 61265 Tel: + 1-309-762-7716, Fax: + 1-309-762-9453	Xensor Integration P.O. Box 3233 2601 DE Delft, the Netherlands Tel. + 31-15-2578040, Fax: + 31-15-2578050

PET molecules compared to the structure below glass transition. Around 140°C, cold crystallization of the material occurs: again a reordering of the material. This is a strongly exothermic process, which again displaces the baseline. The power to be supplied to the sample then steadily increases, and also, with rising temperature, the baseline starts to fall again until a maximum just below melting at 250°C. This is the result of further crystallization, which is facilitated by the higher energy in the sample, and the much better mobility of the molecules just before melting. In the modulated DSC, recrystallization and melting are two concurring phenomena that can be distinguished. Similarly, the hope is that glass transition and cold crystallization can be better distinguished in materials where they overlap (in PET, they are clearly distinct).

Acknowledgments

The author wishes to thank Dr. P. J. van Ekeren of Utrecht University and Dr. G. W. H. Höhne of Universität Ulm for their suggestions in improving this chapter.

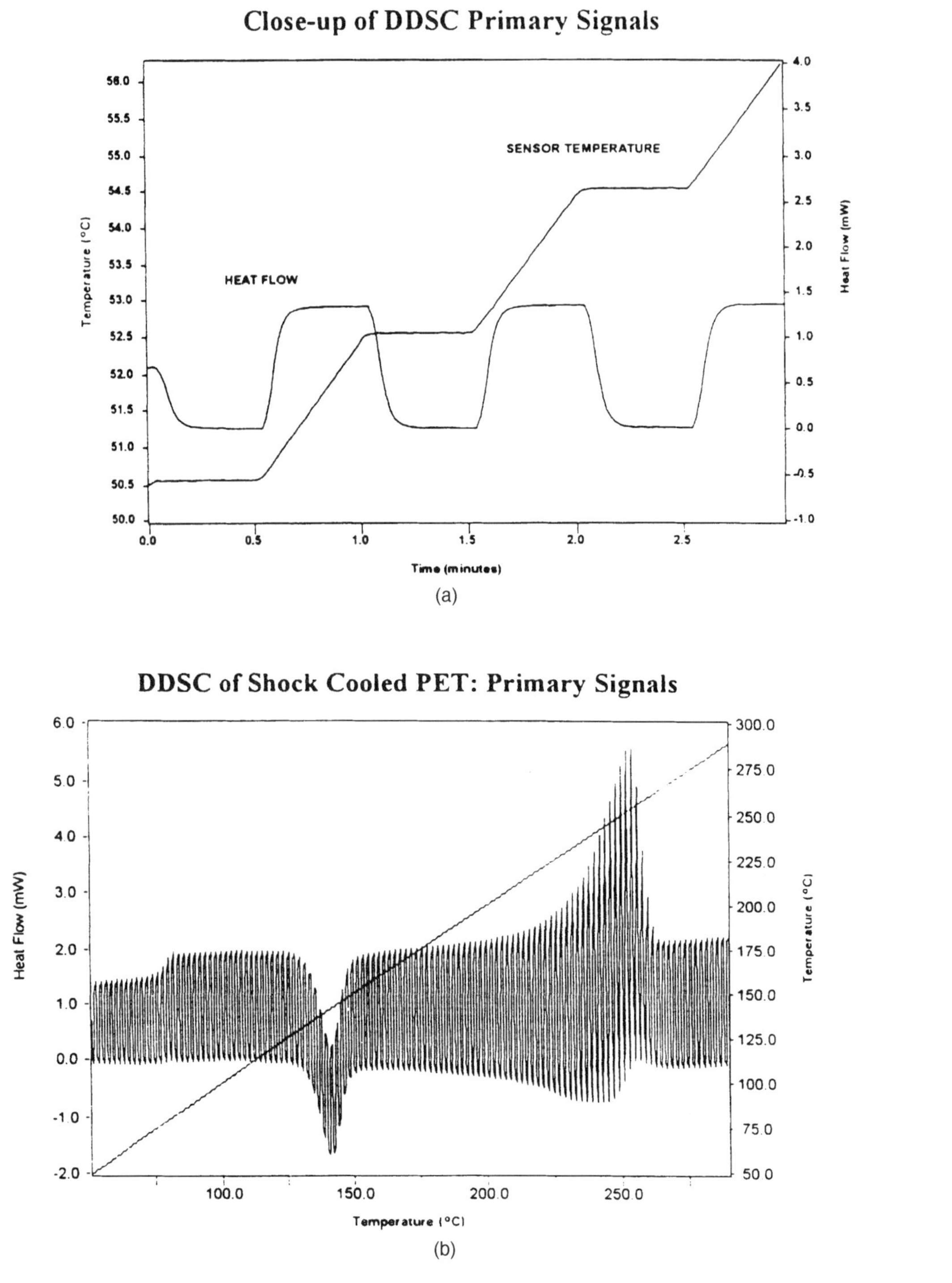

FIGURE 17.5 In modulated or dynamic DSCs, the temperature is not increased at a given rate, but modulated with a sinusoidal or stepwise deviation (a) to separate heat transfer due to temperature increase (specific heat), and heat transfer due to phase-changes in the material (glass transition, crystallization, melting). The heat flow in dynamic a DSC for shock-cooled PET (b) clearly shows the absence of heat flow in regions without material change (only temperature increase), and also the residual heat flow when not increasing temperature due to glass transition (≈75°C), cold crystallization (≈140°C), and melting (≈250°C). (Drawings courtesy of Perkin Elmer.)

References

1. S. Middelhoek and S.A. Audet, *Silicon Sensors*, London, Academic Press, 1989.
2. A.W. van Herwaarden, Physical principles of thermal sensors, *Sensors and Materials*, 8, 373-387, 1996.
3. H.B. Callen, *Thermodynamics and an Introduction to Thermostatistics*, 2nd ed., New York, John Wiley & Sons, 1985.
4. W. Hemminger and G.W.H. Höhne, *Calorimetry — Fundamentals and Practice*, Weinheim, Verlag Chemie, 1984.
5. W. Hemminger, Calorimetric methods, in V.B.F. Mathot (ed.), *Calorimetry and Thermal Analysis of Polymers*, Munich, Hanser Publishers, 1994.
6. K.N. Marsh (ed.), *Recommended Reference Materials for Realization of Physicochemical Properties*, Oxford, Blackwell Scientific, 1987.
7. G.W.H.Höhne, Fundamentals of differential scanning calorimetry and differential thermal analysis, in V.B.F. Mathot (ed.), *Calorimetry and Thermal Analysis of Polymers*, Munich, Hanser Publishers, 1994.
8. E. Gmelin and St.M. Sarge, Calibration of differential scanning calorimeters, *Pure Appl. Chem.*, 67, 1789-1800, 1995.
9. T.M.V.R. de Barros, R.C. Santos, A.C. Fernandes, and M.E. Minas da Piedade, Accuracy and precision of heat capacity measurements using a heat flux differential scanning calorimeter, *Thermochim. Acta*, 269/272, 51-60, 1995.
10. P. Bataillard, Calorimetric sensing in bioanalytical chemistry: principles, applications and trends, *Trends in Anal. Chem.*, 12, 387-394, 1993.
11. A.W. van Herwaarden, P.M. Sarro, J.W. Gardner, and P. Bataillard, Liquid and gas micro-calorimeters for (bio)chemical measurements, *Sensors and Actuators*, A43, 24-30, 1994.
12. G.W.H. Höhne, A.E. Bader, and St. Höhnle, Physical properties of a vacuum-deposited thermopile for heat measurements, *Thermochim. Acta*, 251, 307-317, 1995.
13. J. Lamprecht, W. Hemminger, and G.W.H. Höhne (eds.), Calorimetry in the biological sciences, *Thermochim. Acta*, 193, 1991.
14. R.B. Kemp and B. Schaarschmidt (eds.), Calorimetric and thermodynamic studies in biology, *Thermochim. Acta*, 251, 1995.
15. M. Reading, A. Luget, and R. Wilson, Modulated differential scanning calorimetry, *Thermochim. Acta*, 238/239, 295-307, 1994.
16. J.E.K. Schawe, Principles for the interpretation of modulated temperature DSC measurement. Part 1. Glass transition, *Thermochim. Acta*, 261, 183-194, 1995.
17. Characterization of Amorphous Polyethylene Terephtalate by Dynamic Differential Scanning Calorimetry, *Perkin Elmer Thermal Analysis Newsletter* 69, Perkin-Elmer Corp., Norwalk, CT.

Further Information

Much is being published on Calorimetry each year. Some very good books to obtain a basic understanding of calorimetry are:

W. Hemminger and G.W.H. Höhne, *Calorimetry — Fundamentals and Practice*, Weinheim, Verlag Chemie, 1984.

V.B.F. Mathot (ed.), *Calorimetry and Thermal Analysis of Polymers*, Munich, Hanser Publishers, 1994.

B. Wunderlich, *Thermal Analysis*, San Diego, CA, Academic Press, 1990.

M. Brown, *Introduction to Thermal Analysis*, London, Chapman and Hall, 1988.

On DSC, an up-to-date book is:

G.W.H. Höhne, W. Hemminger, and H.-J. Flammersheim, *Differential Scanning Calorimetry: An Introduction for Practitioners*, Berlin, Springer-Verlag, 1996.

Apart from these, many relevant papers appear in:

Thermochimica Acta, Amsterdam, Elsevier Science Publishers.
Journal of Thermal Analysis, Chichester, U.K., John Wiley & Sons.

Leafing through the latest volumes will bring one up to date on calorimetry research. Regular conferences on Calorimetry and Thermal Analysis are being held, organized by local and global institutes, such as the ICTAC (International Confederation on Thermal Analysis and Calorimetry). Proceedings of these conferences are often published in *Thermochimica Acta* or the *Journal of Thermal Analysis*.

Very interesting is the publication of ICTA from 1991, which contains significant information on nomenclature, literature, suppliers of instrumentation, etc.:

J.O. Hill (ed.), *For Better Thermal Analysis and Calorimetry, Edition III*, ICTA, 1991.

Index

D

E

F

G

H

I

Q

R

S

T

U

V

W

X

Y